Corrosion and Degradation of Materials

Corrosion and Degradation of Materials

Editors

Jian Chen
Yanxin Qiao
Fanjiang Meng
Yuxin Wang

MDPI • Basel • Beijing • Wuhan • Barcelona • Belgrade • Manchester • Tokyo • Cluj • Tianjin

Editors
Jian Chen
Chinese Academy of Sciences
China

Yanxin Qiao
Jiangsu University of Science and Technology
China

Fanjiang Meng
Shanghai Nuclear Engineering Research and Design Institute
China

Yuxin Wang
Jiangsu University of Science and Technology
China

Editorial Office
MDPI
St. Alban-Anlage 66
4052 Basel, Switzerland

This is a reprint of articles from the Special Issue published online in the open access journal *Coatings* (ISSN 2079-6412) (available at: https://www.mdpi.com/journal/coatings/special_issues/Corros_Degrad).

For citation purposes, cite each article independently as indicated on the article page online and as indicated below:

LastName, A.A.; LastName, B.B.; LastName, C.C. Article Title. *Journal Name* **Year**, *Volume Number*, Page Range.

ISBN 978-3-0365-4895-1 (Hbk)
ISBN 978-3-0365-4896-8 (PDF)

Contents

coatings

MDPI

Editorial

Corrosion and Degradation of Materials

Jian Chen [1,*], Yanxin Qiao [2], Fanjiang Meng [3] and Yuxin Wang [2]

1 CAS Key Laboratory of Nuclear Materials and Safety Assessment, Institute of Metal Research, Chinese Academy of Sciences, Shenyang 110016, China
2 School of Materials Science and Engineering, Jiangsu University of Science and Technology, Zhenjiang 212003, China; yxqiao@just.edu.cn (Y.Q.); ywan943@163.com (Y.W.)
3 Shanghai Nuclear Engineering Research and Design Institute, Shanghai 200233, China; mengfanjiang@snerdi.com.cn
* Correspondence: jchen@imr.ac.cn

The corrosion and degradation of materials, i.e., deteriorating materials via chemical/electro-chemical reactions with their surrounding environments, is extremely common and costly. The global cost of corrosion is estimated to be 3%–5% of the global gross domestic product (GDP), not including individual safety or environmental consequences. To reduce the corrosion cost and to avoid catastrophic corrosion consequences, abundant corrosion studies [1–23] have been performed and effective corrosion control practices [18–22] have been used to understand the corrosion fundamentals, to inspect the corrosion performance of structural components or materials, to evaluate the safety issues, to analyze and report the corrosion failure, and to develop advanced corrosion control techniques. This benefits the selection and design of the corrosion-resistant materials used in modern society, and their safety assessment. It can not only facilitate the development of new corrosion mechanisms of the materials in service, but can also improve the integrity and longevity of the materials and eventually decrease corrosion loss.

This Special Issue of *Coatings*, entitled "Corrosion and Degradation of Materials", was intended to consolidate recent findings related to the corrosion and degradation of materials and to provide an opportunity for researchers to publish the latest results, reviews, methodology, and failure case reports to understand or solve material corrosion issues.

With the great efforts of the Guest Editor team, the enthusiastic support of the Editorial Board and the valuable contribution of the participates, this Special Issue has reached a milestone, successfully publishing 31 peer-reviewed papers. It covers a series of research areas ranging from the microstructure characterization of the studied materials which has a great impact on their corrosion performance [11,12], the corrosion behaviors and mechanisms of the structural or novel-designed materials in potential service environments [13–17], the development of new coatings acting as corrosion barriers to protect the materials [18–22], the corrosion model built to predict the corrosion progress of the material investigated [23], and other related areas [24–27].

Author Contributions: Writing—original draft preparation, J.C.; writing—review and editing, Y.Q., F.M. and Y.W. All authors have read and agreed to the published version of the manuscript.

Funding: This research received no external funding.

Conflicts of Interest: The authors declare no conflict of interest.

Citation: Chen, J.; Qiao, Y.; Meng, F.; Wang, Y. Corrosion and Degradation of Materials. *Coatings* **2022**, *12*, 969. https://doi.org/10.3390/coatings12070969

Received: 30 June 2022
Accepted: 6 July 2022
Published: 8 July 2022

References

1. Chen, J.; Wang, J.; Han, E.; Dong, J.; Ke, W. AC impedance spectroscopy study of the corrosion behavior of an AZ91 magnesium alloy in 0.1 M sodium sulfate solution. *Electrochim. Acta* **2007**, *52*, 3299–3309. [CrossRef]
2. Chen, J.; Wang, J.; Han, E.; Dong, J.; Ke, W. States and transport of hydrogen in the corrosion process of AZ91 Alloy in aqueous solution. *Corros. Sci.* **2008**, *50*, 1292–1305. [CrossRef]
3. Chen, J.; Asmussen, R.M.; Zagidulin, D.; Noel, J.J.; Shoesmith, D.W. Electrochemical and corrosion behavior of a 304 stainless-steel-based metal alloy wasteform in dilute aqueous environments. *Corros. Sci.* **2013**, *66*, 142–152. [CrossRef]
4. Chen, J.; Qin, Z.; Shoesmith, D.W. Long-term corrosion of copper in a dilute anaerobic sulfide solution. *Electrochim. Acta* **2011**, *56*, 7854–7861. [CrossRef]
5. Chen, J.; Qin, Z.; Wu, L.; Noël, J.J.; Shoesmith, D.W. The influence of sulphide transport on the growth and properties of copper sulphide films on copper. *Corros. Sci.* **2014**, *87*, 233–238. [CrossRef]
6. Chen, J.; Qin, Z.; Martino, T.; Shoesmith, D.W. Non-uniform film growth and micro/macro-galvanic corrosion of copper in aqueous sulphide solutions containing chloride. *Corros. Sci.* **2017**, *114*, 72–78. [CrossRef]
7. Chen, J.; Qin, Z.; Martino, T.; Guo, M.; Shoesmith, D.W. Copper transport and sulphide sequestration during copper corrosion in anaerobic aqueous sulphide solutions. *Corros. Sci.* **2018**, *131*, 245–251. [CrossRef]
8. Guo, M.; Chen, J.; Lilja, C.; Dehnavi, V.; Behazin, M.; Noël, J.J.; Shoesmith, D.W. The anodic formation of sulfide and oxide films on copper in borate-buffered aqueous chloride solutions containing sulfide. *Electrochim. Acta* **2020**, *362*, 137087. [CrossRef]
9. Yang, Y.; Yin, B.; Shang, Y.; Wang, X.; Chen, J. Preparation and characterization of novel TiB_2-12 (Fe-Co-Cr-Ni) cermets and their corrosion resistance in molten aluminum. *Corros. Sci.* **2021**, *190*, 109643. [CrossRef]
10. Qiao, Y.; Wang, X.; Yang, L.; Wang, X.; Chen, J.; Wang, Z.; Zhou, H.; Zou, J.; Wang, F. Effect of aging treatment on microstructure and corrosion behavior of a Fe-18Cr-15Mn-0.66N stainless steel. *J. Mater. Sci. Technol.* **2022**, *107*, 197–206. [CrossRef]
11. Wang, J.; Ming, H.; Zhang, Z.; Chen, J.; Wang, J. Microstructure and micro-hardness of dissimilar metal cladding from a pipe–nozzle mockup for PWR. *Coatings* **2022**, *12*, 525. [CrossRef]
12. Ma, Y.; Yuan, B.; Liu, Y.; Wang, J.; Su, X. Effect of annealing and oxidation on the microstructure evolution of hot-dipped aluminide Q345 steel with silicon sddition. *Coatings* **2022**, *12*, 503. [CrossRef]
13. Wang, B.; Hou, J.; Luan, J.; Xu, D.; Sun, H.; Sun, J. The corrosion behaviors of an as-rolled Mg-8Li (in wt.%) alloy in two differently concentrated NaCl solutions. *Coatings* **2022**, *12*, 406. [CrossRef]
14. Xie, X.; Yin, B.; Yin, F.; Ouyang, X. Corrosion behavior of FeB-30 wt.% $Al_{0.25}$FeNiCoCr cermet coating in liquid zinc. *Coatings* **2021**, *11*, 622. [CrossRef]
15. Fang, K.; Luo, K.; Wang, L.; Li, C.; Wang, L.; Qiao, Y. Environmental fatigue behavior of a Z3CN20.09M stainless steel in high temperature water. *Coatings* **2022**, *12*, 317. [CrossRef]
16. Sheng, S.; Zhou, H.; Wang, X.; Qiao, Y.; Yuan, H.; Chen, J.; Yang, L.; Wang, D.; Liu, Z.; Zou, J.; et al. Friction and wear behaviors of Fe-19Cr-15Mn-0.66N steel at high temperature. *Coatings* **2021**, *11*, 1285. [CrossRef]
17. Hu, Q.; Wang, X.; Shen, X.; Fu, F.; Tan, Z. Wear and corrosion resistance of CoCrFeNiSiMoW medium-entropy alloy coatings on Q235 steel. *Coatings* **2021**, *11*, 1053. [CrossRef]
18. Huang, J.; Yang, M.; Zhu, W.; Tang, K.; Zhang, H.; Chen, J.; Noël, J.J.; Barker, I.; Zhang, H.; Zhu, J. Extrusion-free fabrication of zinc-rich powder coatings: Press bonding. *Chem. Eng. J.* **2022**, *442*, 135925. [CrossRef]
19. Yang, S.; Huang, J.; Chen, J.; Noël, J.J.; Barker, I.; Henderson, J.D.; He, P.; Zhang, H.; Zhang, H.; Zhu, J. A comparative study on the anti-corrosive performance of zinc phosphate in powder coatings. *Coatings* **2022**, *12*, 217. [CrossRef]
20. Cheng, B.; Chen, H.; Asempah, I.; Wang, J.; Zhu, Y.; Wan, J.; Jiang, F.; Wang, Z.; Shui, Y.; Wang, L.; et al. Self-formed diffusion layer in Cu(Re) alloy film for barrierless copper metallization. *Coatings* **2022**, *12*, 613. [CrossRef]
21. Liu, S.; Chen, J.; Zhang, D.; Wang, Y.; He, Z.; Guo, P. Properties of micro-arc oxidation coatings on 5052 Al alloy sealed by SiO_2 nanoparticles. *Coatings* **2022**, *12*, 373. [CrossRef]
22. Zhang, S.; Yu, C.; Tan, J.; Wang, Y.; He, Z. Ti/SnO_2-Sb_2O_x-TiO_2 electrodeposited from methanesulfonate electrolytes: Preparation, properties, and performance. *Coatings* **2022**, *12*, 366. [CrossRef]
23. Ding, J.; He, W.; Liu, Y.; Zhang, C.; Wang, H.; Han, E.-H. Numerical simulation of crevice corrosion of stainless steel–titanium in NaCl solution. *Coatings* **2022**, *12*, 592. [CrossRef]
24. Ni, Z.; Cao, X.; Wang, X.; Zhou, S.; Zhang, C.; Xu, B.; Ni, Y. Facile synthesis of copper(I) oxide nanochains and the photo-thermal conversion performance of its nanofluids. *Coatings* **2021**, *11*, 749. [CrossRef]
25. Liu, S.; Zhang, X.; Zhou, Y.; Gao, F. Optimization study of fluffy materials flocking drainage pipes to resist blockage based on MD binding energy. *Coatings* **2021**, *11*, 853. [CrossRef]
26. Li, Z.; Chen, S.; Meng, L.; Gao, Y.; Yang, Z.; Shi, M.; Chen, X.; Zhang, H.; Zhang, Y. On the effects of high and ultra-high rotational speeds on the strength, corrosion resistance, and microstructure during friction stir welding of Al 6061-T6 and 316L SS alloys. *Coatings* **2021**, *11*, 1550. [CrossRef]
27. Baghbaderani, M.Z.; Abazari, S.; Bakhsheshi-Rad, H.R.; Ismail, A.F.; Sharif, S.; Najafinezhad, A.; Ramakrishna, S.; Daroonparvar, M.; Berto, F. Dual synergistic effects of MgO-GO fillers on degradation behavior, biocompatibility and antibacterial activities of chitosan coated Mg alloy. *Coatings* **2022**, *12*, 63. [CrossRef]

coatings

Article

An Experimental Study on Dynamic Characteristics of Coarse-Grained Soil under Step Cyclic Loading

Peisen Wang [1], Wenjun Hu [2,*], Pingyun Liu [2], Zhenqiang Yan [3], Xianghui Kong [2,*], Quanman Zhao [2] and Wenhao Yin [2]

1 School of Civil Engineering, Shandong Jianzhu University, Jinan 250101, China; wangpeisen@sdjzu.edu.cn
2 School of Transportation Engineering, Shandong Jianzhu University, Jinan 250101, China; lpy1802@163.com (P.L.); zhaoquanman@sdjzu.edu.cn (Q.Z.); ywh456321@163.com (W.Y.)
3 Jinan Jinqu Road Survey Design Research Co., Ltd., Jinan 250101, China; fangzhang_july@163.com
* Correspondence: huwenjun@sdjzu.edu.cn or huwenjun_82@163.com (W.H.); kongxh@sdjzu.edu.cn (X.K.)

Abstract: The accumulated plastic deformation induced by a cyclic traffic load will lead to destruction of the subgrade. Coarse-grained soil is a widely used subgrade filler. The GDS dynamic triaxial test was carried out on typical coarse-grained soil fillers to investigate the influence of different confining pressures, consolidation ratios and numbers of cyclic actions on the hysteresis curve under step cyclic loading. The results show that surrounding pressure can significantly reduce the energy lost from a soil sample under cyclic loading. Under the same stress level, increasing the consolidation ratio can effectively reduce the area enclosed by the hysteresis curve. When the stress is increased above a certain value, the strain of the response clearly changes with an increase in the number of cyclic loading. The research findings can theoretically guide the design of coarse-grained soil roadbeds in practical engineering.

Keywords: subgrade engineering; coarse-grained soil; step cyclic loading; dynamic triaxial test; hysteresis curve

Citation: Wang, P.; Hu, W.; Liu, P.; Yan, Z.; Kong, X.; Zhao, Q.; Yin, W. An Experimental Study on Dynamic Characteristics of Coarse-Grained Soil under Step Cyclic Loading. *Coatings* **2022**, *12*, 640. https://doi.org/10.3390/coatings12050640

Academic Editor: Joaquim Carneiro

Received: 30 March 2022
Accepted: 2 May 2022
Published: 7 May 2022

1. Introduction

Research on the dynamic properties of coarse-grained soils mainly focuses on the dynamic deformation problems [1–3] caused by road traffic loads, which have resulted in a large number of disease phenomena. Normally, traffic loading on roads is found to be transient, uncertain and repetitive [4]. Thus, it is difficult to formulate traffic loads. The dynamic stress–strain relationship provides an intuitive understanding of the strain development pattern of coarse-grained soils under dynamic loading. Long Y. et al. [5] introduced the effects of the surrounding pressure and cyclic stress ratio on the accumulated plastic strain of coarse-grained soil into the hyperbolic prediction model showing proper prediction results, which consider the influence of cyclic stress, static deviational stress and consolidation pressure. Chen Y.P. et al. [6] proposed an improved exponential model to describe the development law of cumulative plastic strain before the soil was destroyed. Tang et al. [7] introduced the concept of equivalent vibration times and proposed a model for predicting the cumulative plastic strain index in soils under multi-stage dynamic stress loading conditions. Bian [8] investigated the effects of the circumferential pressure and dynamic amplitude on the properties of ballast, such as axial strain and volumetric strain. Trinh et al. [9] investigated the effects of the water content and saturation on the mechanical properties of coarse-grained soil specimens in contaminated ballast layers under old railroad structures and developed a principal structure model, which considers the stress level, number of cycles and water content of the soil. Leng W.M et al. [10] studied the effects of the dynamic stress amplitude, water content and surrounding pressure on the cumulative deformation pattern of coarse-grained soils and proposed the stability limits and discrimination criteria for dynamic deformation of coarse-grained soils. In summary, investigations on the dynamic properties of coarse-grained soils under step cyclic loading through experimental methods are limited. In this paper, the GDS dynamic triaxial test was carried out to investigate the effect of different confining pressures, consolidation ratios on the hysteresis curve of typical coarse-grained soil roadbed under step cyclic loading.

In conclusion, it was necessary to study the dynamic characteristics of coarse-grained soil under step cyclic loading by experimental methods. The GDS dynamic triaxial system is an indoor geotechnical triaxial test instrument controlled by a motor and produced by GDS instrument and Equipment Co., Ltd. (Hook Hampshire, UK) The dynamic triaxial test belonged to the dynamic test range of soil. It is a commonly used method to study the dynamic characteristics of soil in a laboratory. The dynamic triaxial apparatus is a commonly used indoor geotechnical test device. In order to understand the dynamic characteristics of coarse-grained soil, this indoor dynamic triaxial test of coarse-grained soil was carried out, which considers the key influencing factors of a soil dynamic triaxial test: the confining pressure, consolidation ratio and so on. The influence of different confining pressures and consolidation ratios on the hysteretic curve of the soil samples under step cyclic loading was explored.

2. Test Preparation

2.1. Raw Materials

The soil sample used in the test was taken from coarse-grained soil along a new highway, and the granulometry test was performed according to the Test Methods of Soils for Highway Engineering (JTG 3430-2020) [11], which is shown in Table 1. The particle size accumulation curve is shown in Figure 1.

Table 1. Particle analysis results.

Mesh Size (mm)	10	5	2	1	0.5	0.25	0.1	0.075
Pass quality percentage (%)	100	95.3	74.8	62.1	50	33.8	19.5	8.3

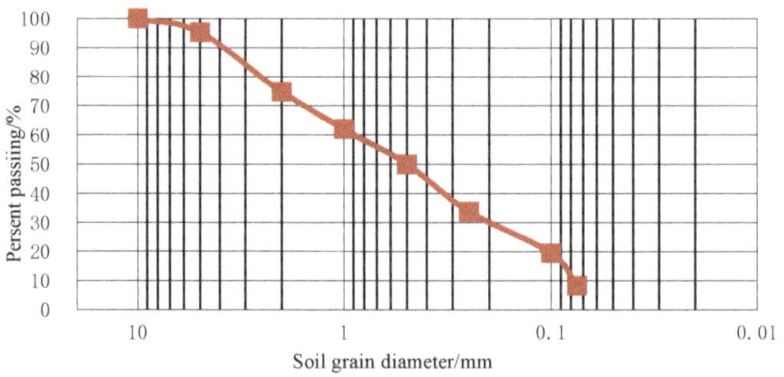

Figure 1. Cumulative particle size curve of soil samples.

The mass percentage of soil particles with a diameter above 0.075 mm ($d \geq 0.075$ mm) in the coarse-grained group of soil samples is 91.7%, which was greater than 50% of the total mass. The sample soil is classified as a coarse-grained soil, according to Chinese standard JTG 3430-2020 [11], whose basic physical and mechanical property indexes are shown in Table 2. It can be seen that the particle distribution of this coarse-grained soil was good.

Table 2. Basic performance indicators of soil samples.

Natural Water Content (%)	Maximum Dry Density /(g/cm^3)	Optimum Moisture Content (%)	CBR (%)	Nonuniformity Coefficient C_U (>5)	Curvature Factor C_C (1~3)	Gradation
2.1	2.13	8.03	51	5.6	1.1	well

The nonuniformity coefficient C_U is calculated with the following equation:

$$C_U = \frac{d_{60}}{d_{10}} \tag{1}$$

The curvature factor C_C is calculated with the following equation:

$$C_C = d_{30}^2/(d_{60} \times d_{10}) \tag{2}$$

where d_{60}, d_{30} and d_{10} = the cumulative percentage of soil weighing less than a certain particle size: 60%, 30% and 10%, respectively.

2.2. Test Principle

The dynamic triaxial test was used to analyze the dynamic response of the specimen under dynamic loading by applying a periodic axial dynamic principal stress to the specimen on the basis of the applied axial static stress. Based on the relative relationship of dynamic indicators, such as stress, strain and pore pressure, the dynamic properties [12–19] of the soil and the properties of the soil specimen under dynamic stress are derived. Stress is usually expressed in terms of axial major principal stress σ_1 and circumferential pressure σ_3, which refer to the stress state of the soil under static and dynamic conditions. The dynamic conditions are various parameters of the simulated dynamic load, mainly including its vibration direction, waveform, frequency, amplitude, etc. The test uses the GDS vibration triaxial instrument (GDS Instruments, Hook Hampshire, UK), which consists of a triaxial pressure chamber, an axial and lateral pressurization system, a counterpressure pressure volume controller, a data collector and a computer, which can accurately complete the small strain triaxial test under dynamic stress.

Referring to the experience of previous soil dynamic triaxial research, the failure standard of the specimen was 5% of the axial dynamic strain. Considering that some specimens were not damaged, we took 25,000 vibrations as the second test termination standard. The dynamic loading frequency range of the GDS dynamic triaxial apparatus used in the test was 0–2 Hz; the dynamic axial pressure could be loaded $\pm$ 10 kN, and the accuracy was 0.1% of the full scale. The displacement range was 100 mm, the displacement resolution was 0.208 mm, and the axial displacement accuracy was 0.07% of the full range. The above ranges met the test requirements. The GDS dynamic triaxial apparatus is a fully automatic dynamic triaxial apparatus. It can realize the saturation, detection, consolidation and dynamic loading of a sample. Except for installation of a sample, the other operations were completely controlled by the computer, which accurately completed each step of operation, effectively reduced human errors and ensured smooth progress of the test. The complete device is shown in Figure 2.

Figure 2. The complete device.

2.3. Test Program

In order to simulate the stress state of the soil sample in the actual situation more accurately, we selected four aspects of the sample parameters: soil conditions, drainage conditions, stress conditions and dynamic conditions. We also chose the main influencing factors, including the confining pressure, dynamic stress amplitude, loading frequency [20] and number of cycles. The specific test scheme is shown in Table 3.

(1) Specimen specification: φ 50 mm × h 100 mm.
(2) Moisture content of the specimen: We adopted the optimum moisture content, and the standard compaction test result was 8%.
(3) Specimen compaction: We adopted compaction of 100%, divided into three layers and tamped manually.
(4) Loading method: We adopted the loading method step-by-step. The dynamic stress amplitude started from 20 kPa, and then increased by 20 kPa at each stage of loading. Loading was stopped when the dynamic stress amplitude reached 200 kPa or the cumulative deformation reached 5% of the initial height. The number of cycles during each stage of loading was 20.
(5) Drainage conditions: without drainage.

Table 3. Dynamic triaxial test scheme.

Surrounding Pressure (KPa)	Dynamic Stress Amplitude (KPa)	Solidification Ratio	Control Mode	Frequency (Hz)
50	20 40 — — — — —	1/1.5	Stress control	1
100	20 40 60 80 100 — —			
150	20 40 60 80 100 120 140			

The entire research process is shown in Figure 3.

2.4. Test Steps

(1) Triaxial sample preparation: We prepared the crushed coarse-grained soil sample after drying at the optimum moisture content, mixed them evenly after preparation, and sealed and stored them for 24 h, to make the soil sample fully wet.
(2) Sample saturation: We adopted vacuum saturation, as shown in Figure 4. After completing vacuum saturation, we tested the B value of the sample. After the B value reached 0.95 or above, the sample was saturated.
(3) Sample installation: We adjusted the base to proper position, by exhausting the base to ensure that the center axis of the sample coincides with the center of the upper

fixed axis. Then, we installed the pressure chamber and, finally, filled the pressure chamber with water.

(4) Sample consolidation: After saturation, we set the consolidation pressure and consolidated the sample. When the volume rate of the back-pressure drainage was lower than 5 $mm^3/5$ min, the consolidation of the sample was considered completed.

(5) Applying dynamic load: According to the requirements of the test scheme, we gradually applied the load to the sample.

(6) Unloading and removing sample: After completing loading, we reduced the confining pressure and back pressure to 0, drained the water, removed the samples and cleaned the instrument.

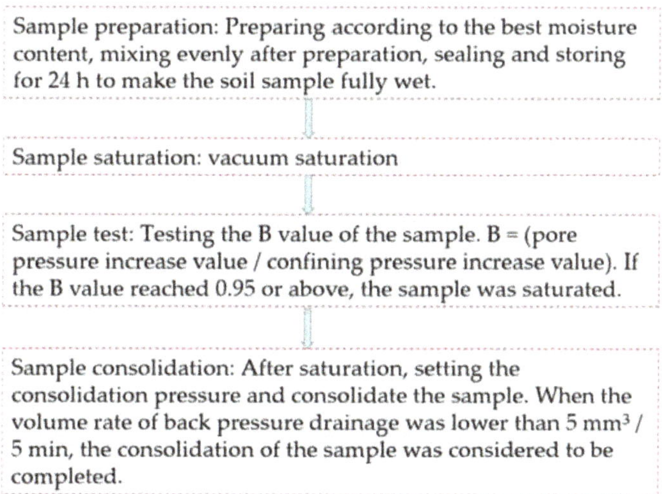

Figure 3. The entire research process.

(**a**) Vacuum saturated containers (**b**) During vacuum saturation (**c**) End of vacuum saturation

Figure 4. Vacuum saturation.

The adopted breakdown criteria were:

(1) When the sample was subjected to a cyclic dynamic load, the increase in pore water pressure reached the initial consolidation confining pressure for the first time. At this time, the effective stress was zero or minimum, and the test was terminated.

(2) When the specimen was subjected to a cyclic dynamic load and the double amplitude axial strain reached 5%, the test was terminated.

In our experiments, (1) was selected as the failure standard of the soil liquefaction resistance strength, and (2) was selected as the failure standard of the soil dynamic strength. The breakdown samples are shown in Figure 5.

Figure 5. Breakdown samples.

3. Analysis of Test Results for Hysteretic Curve of Coarse-Grained Soil

When cyclic stress was applied to the soil, the corresponding response strain also performed cyclically. Through the characteristics of the hysteretic curve, the deformation characteristics and energy consumption characteristics of soil under cyclic stress can be learned.

3.1. Analysis of Hysteretic Curve Characteristics under Different Confining Pressures

Figure 6 shows the influence of different confining pressures on hysteretic curves under isobaric consolidation. It can be intuitively found that the confining pressure of different sizes has a great influence on the hysteretic curve. When the dynamic stress amplitude was 40 kPa, and the confining pressure was 50, 100 and 150 kPa, the maximum deformation of the sample was 0.275%, 0.061% and 0.057%, respectively. The increase in confining pressure induces a denser internal spatial structure of the soil. Furthermore, when the stress is constant, the increase in confining pressure reduces the strain of the specimen, which shows that increasing the confining pressure can effectively inhibit the development of soil deformation.

In Figure 7, it can be seen that the confining pressure has a great impact on the surrounding area of the hysteretic curve. With the increase in confining pressure, the surrounding area of the hysteretic curve showed the characteristics of a significant decrease. The confining pressure increased from 50 to 150 kPa under a dynamic stress amplitude of 40 kPa. In the isobaric consolidation, the area enclosed by the hysteretic curve decreases by 86% and 93%; In the bias consolidation, the surrounding area of hysteretic curve decreased by 94%. The confining pressure has a great influence on the dissipated energy of coarse-grained soil specimens. Increasing the confining pressure can significantly reduce the energy consumption under cyclic stress. The stiffness and overall stability of the coarse-grained soil specimens improved, which demonstrates improved strength characteristics.

Figure 8 shows the second-degree parabola fitting relationship of the surrounding area of the hysteretic curve under different dynamic stress amplitudes. The correlation coefficients were 0.99 and above, indicating that the second-degree parabola fitting has a high correlation. In isobaric consolidation, under the same dynamic stress amplitude, the slope of the parabola–tangent fitting decreased with an increase in the confining pressure, and the bias consolidation also had the same trend. It showed that under the same stress level, increasing the confining pressure can effectively reduce the area surrounded by the hysteretic curve. An increase in the confining pressure can effectively inhibit the rate of increase in its area with the increase in stress, thus controlling its energy consumption at a high stress level and showing better overall stability.

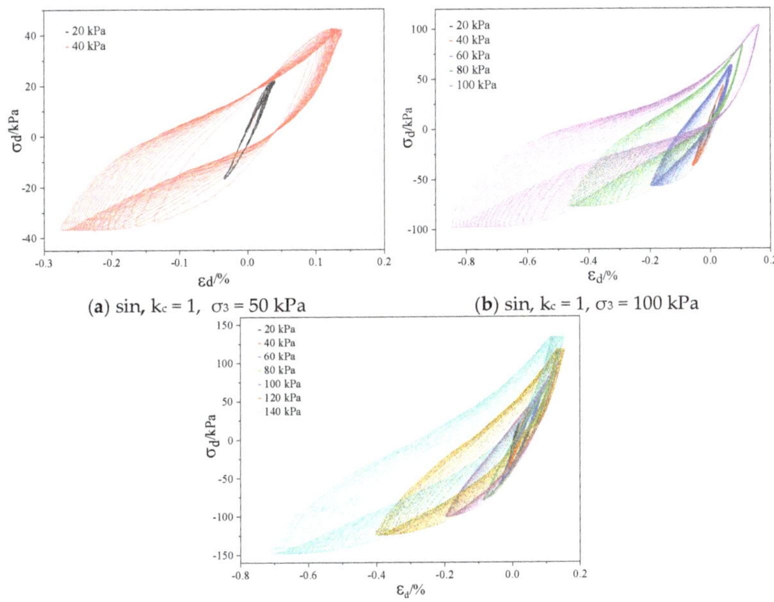

(**a**) sin, k_c = 1, σ_3 = 50 kPa (**b**) sin, k_c = 1, σ_3 = 100 kPa

(**c**) sin, k_c = 1, σ_3 = 150 kPa

Figure 6. Analysis of hysteretic curve characteristics under different confining pressures.

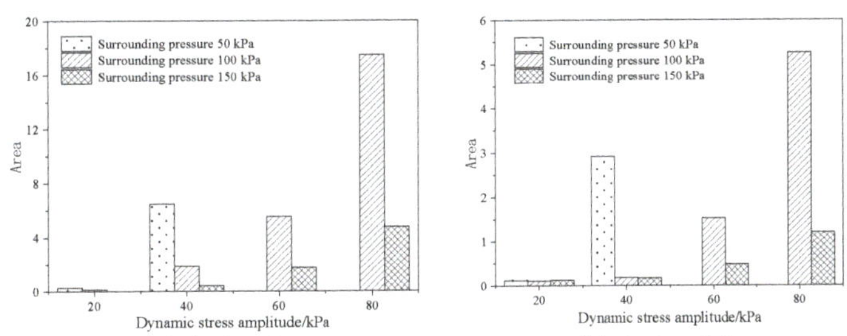

(**a**) sin, k_c = 1 (**b**) sin, k_c = 1.5

Figure 7. Effect of confining pressure on the surrounding area of hysteretic curve.

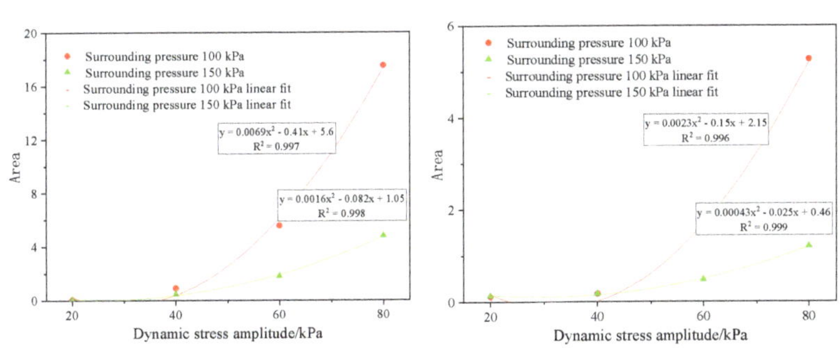

(**a**) sin, k_c = 1 (**b**) sin, k_c = 1.5

Figure 8. The second-degree parabola fitting of enclosed area for hysteretic curve under different confining pressures.

3.2. Analysis of Hysteretic Curve Characteristics under Different Consolidation Ratios

Figure 9a–f shows the influence law of different consolidation ratios on the hysteretic curve of the coarse-grained soil samples. The hysteretic curve generally had similar laws. When the dynamic stress amplitude was small, the hysteretic curve was compact, the slope was large and its shape was closer to an ellipse. With the gradual increase in the dynamic stress amplitude, the corresponding dynamic strain also increased gradually, the curve was elongated and the slope decreased. Under the same amplitude, the hysteretic curve became increasingly loose with the increase in cycle times.

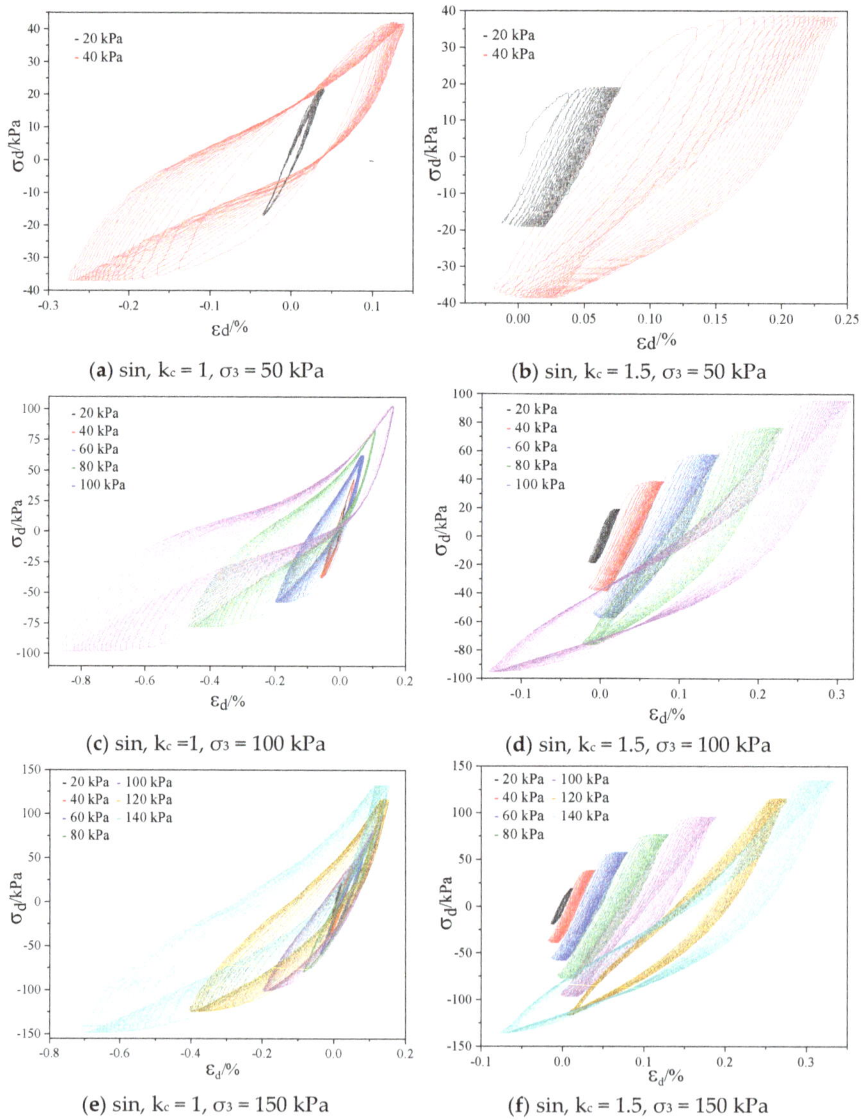

(**a**) sin, $k_c = 1$, $\sigma_3 = 50$ kPa; (**b**) sin, $k_c = 1.5$, $\sigma_3 = 50$ kPa; (**c**) sin, $k_c = 1$, $\sigma_3 = 100$ kPa; (**d**) sin, $k_c = 1.5$, $\sigma_3 = 100$ kPa; (**e**) sin, $k_c = 1$, $\sigma_3 = 150$ kPa; (**f**) sin, $k_c = 1.5$, $\sigma_3 = 150$ kPa

Figure 9. Effect of consolidation ratio on hysteretic curve.

Under the same dynamic stress level, the strain of the specimen response under bias consolidation was small. To be specific, in the hysteretic curve under the confining pressure of 100 kPa, when the dynamic stress amplitude was 60 kPa, the strain amplitude of the isobaric consolidation was 0.175%, and the strain amplitude of the bias consolidation was 0.138%. When the dynamic stress amplitude was 80 kPa, the strain amplitude of the isobaric consolidation was 0.47%, and that of the eccentric consolidation was 0.231%. When the dynamic stress amplitude was 100 kPa, the strain amplitude of the isobaric consolidation was 0.858%, and that of the anisotropic consolidation was 0.316%, showing that the method of bias consolidation inhibited strain development of the coarse-grained soil specimens relatively well with the increasing stress levels.

We selected the hysteresis curve of the tenth cycle under each stage of loading and calculated the area surrounded, as shown in Figure 10. This analyzes the area variation characteristics enclosed by the hysteretic curve of the coarse-grained soil specimens under different consolidation ratios. Compared with isobaric consolidation, the area surrounded by the hysteretic circle was reduced by way of the bias consolidation. At a confining pressure of 100 kPa, its area decreased by 22%, 91%, 73% and 70%, respectively; at the confining pressure of 150 kPa, its area decreased by 90%, 61%, 73% and 74%, respectively. Increasing the consolidation ratio of the specimen was also conducive to reducing energy loss during the process of cyclic loading.

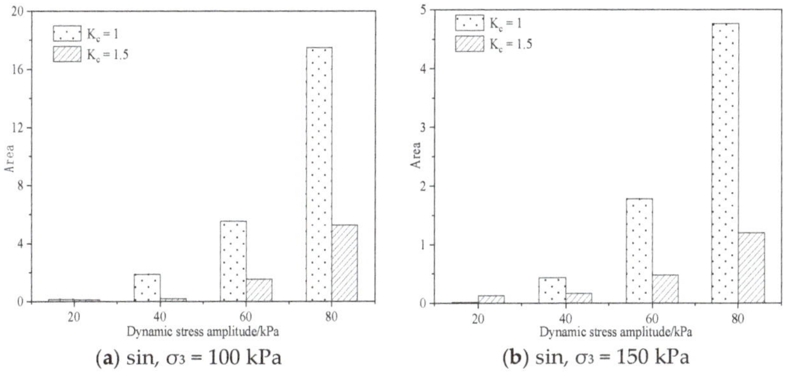

Figure 10. Effect of consolidation ratio on surrounding area of hysteretic curve.

The area surrounded by the hysteretic curve of the specimen under different consolidation ratios was the fitted second-degree parabola, as shown in Figure 11. The correlation coefficients of the linear fitting were all 0.99 and above. Under the same dynamic stress amplitude, the slope of the parabola tangent fitting for the eccentric consolidation was lower than that for the isobaric consolidation. It showed that under the same stress level, increasing the consolidation ratio can effectively reduce the area surrounded by the hysteretic curve. Increasing the consolidation ratio can effectively inhibit the increasing rate of its area with an increase in stress. This could control the energy consumption under a high stress level, showing better overall stability.

3.3. Analysis of Hysteretic Curve Characteristics under Different Cycle Times

Figure 12a–f shows the hysteretic curve under the test conditions of a confining pressure of 150 kPa and isobaric consolidation. The numbers chosen for the cyclic loading are: 1, 5, 10 and 15.

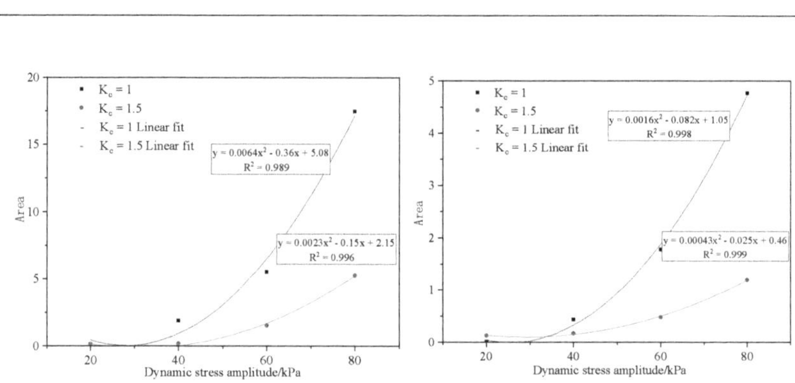

(**a**) sin, σ_3 = 100 kPa (**b**) sin, σ_3 = 150 kPa

Figure 11. The second-degree parabola fitting of surrounding area of hysteretic curve under different consolidation ratio.

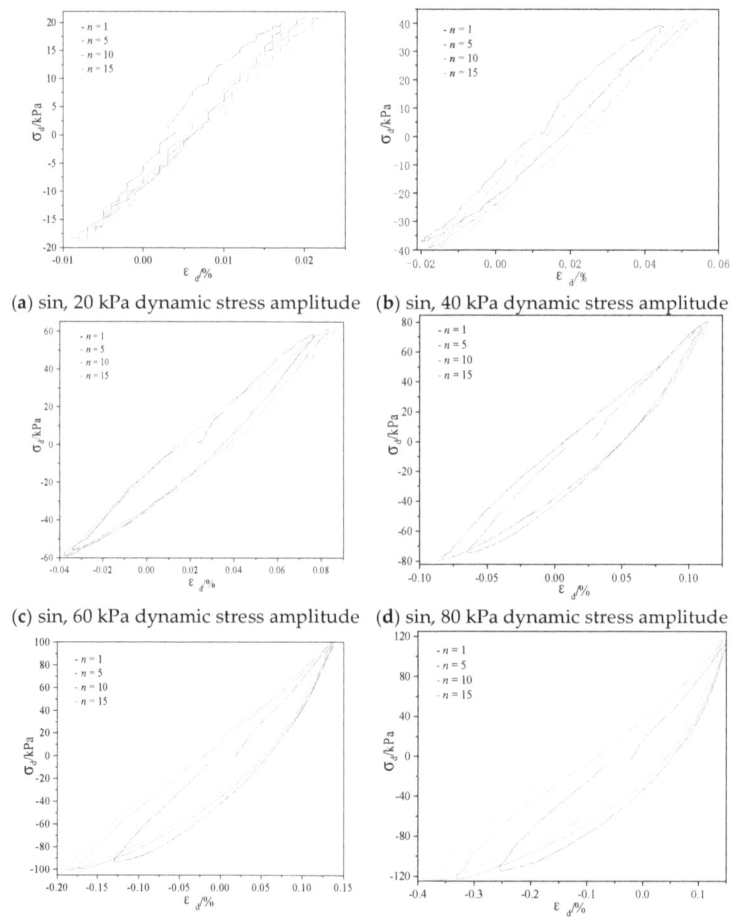

(**a**) sin, 20 kPa dynamic stress amplitude (**b**) sin, 40 kPa dynamic stress amplitude

(**c**) sin, 60 kPa dynamic stress amplitude (**d**) sin, 80 kPa dynamic stress amplitude

(**e**) sin, 100 kPa dynamic stress amplitude (**f**) sin, 120 kPa dynamic stress amplitude

Figure 12. Characteristic analysis of hysteretic curve under different cycle times.

It can be seen that the number of cycles has little effect on the hysteretic curve at a low dynamic stress amplitude. The hysteresis curve is narrow. Under different cycle times, each hysteretic curve basically coincided. The amplitude of the strain, the area surrounded by the hysteresis curve and the slope performed little changes. When the dynamic stress amplitude reached 120 kPa, the hysteretic curve began to deviate greatly with the increase in cyclic loading times. In the case of a small dynamic stress amplitude, within a certain number of cycles, the hysteretic curve of the coarse-grained soil specimen was less affected by the number of cycles. In the case of a large dynamic stress amplitude, the hysteretic curve of the coarse-grained soil specimen was greatly affected by the number of cycles. Therefore, it can be concluded that when the stress reaches a certain level, the response strain changes greatly with the increase in cyclic loading times.

4. Conclusions

In this paper, the GDS dynamic triaxial apparatus, imported from Britain, was used to study the effects of the confining pressure, consolidation ratio and cycle action times on the hysteretic curve of coarse-grained soil; some achievements are summarized as follows:

(1) Increasing the confining pressure can densify the internal spatial structure of soil. Under the same stress level, an increase in the confining pressure reduced the strain of the specimen, which shows that increasing the confining pressure can effectively inhibit the development of soil deformation. Increasing the confining pressure can significantly reduce the energy consumption of the coarse-grained soil specimen under cyclic load. The decrease in energy consumption shows that the stiffness of the coarse-grained soil specimen improved; the overall stability of the specimen improved, and it shows better strength characteristics.

(2) Compared with isobaric consolidation, biased consolidation can inhibit the strain development of coarse-grained soil specimens better with an increasing stress level. Under the same stress level, increasing the consolidation ratio can effectively reduce the area surrounded by the hysteretic curve. Increasing the consolidation ratio can effectively inhibit the increase rate of its area with the increase in stress, thus controlling energy consumption under a high stress level, so as to show better overall stability.

(3) In the case of a small dynamic stress amplitude, the hysteretic curve of the coarse-grained soil specimen was less affected by the number of cycles within a certain number of cycles. In the case of a large dynamic stress amplitude, the hysteretic curve of the coarse-grained soil specimen was greatly affected by the number of cycles. When the stress reaches a certain level, the response strain changes greatly with the increase of cyclic loading times.

The purpose of this paper is to solve the problem of dynamic deformation caused by road traffic load and intuitively explore the strain development law of coarse-grained soil under a dynamic load. Later, through numerical simulation analysis and field stress-strain monitoring, we compare the test data to obtain better conclusions and methods for solving practical problems. In the future, we will analyze the backbone curve, dynamic modulus and damping ratio of coarse-grained soil. Further, we will try to use IRI performance models [21] and Fuzzy logic [22] to predict subgrade dynamic performance.

Author Contributions: Data curation, W.H. and Z.Y.; Investigation, P.L. and W.Y.; Methodology, P.W. and Q.Z.; Supervision, X.K.; Writing—original draft, P.W. All authors have read and agreed to the published version of the manuscript.

Funding: This research was funded by Shandong Provincial Natural Science Foundation (Grant No. ZR2021ME238), Science and Technology Project of Universities in Shandong Province (Grant No. J18KA216), and Science and Technology Project of Universities in Shandong Province (Grant No. J18KA192).

Institutional Review Board Statement: Not applicable.

Informed Consent Statement: Not applicable.

Data Availability Statement: The authors confirm that the data supporting the findings of this study are available within the article.

Conflicts of Interest: The authors declare no conflict of interest.

References

1. Hui, P.E.; Wei, M.A.; Yan-hu, M.U.; Long, J.; Dong-peng, Z. Analysis of disease investigation and long-term deformation characteristics of common fill embankment of the Qinghai-Tibet Highway. *Rock Soil Mech.* **2015**, *36*, 2049–2056.
2. Yu, W.; She, S. On the particular settlement features of the soft soil subgrade in the area of Yangtze River Delta. *J. Saf. Environ.* **2016**, *16*, 108–114.
3. Shi, Y.J.; Li, M.G.; Chen, J.J.; Wang, J.H. Long—term settlement behavior of a highway in land subsidence area. *J. Perform. Constr. Facil.* **2018**, *32*, 04018063. [CrossRef]
4. Luo, J. Determination of traffic load parameters of indoor test Highway. *Low Temp. Build. Technol.* **2015**, *37*, 121–123.
5. Long, Y.; Zhang, J.S.; Ding, J.Y.; Wang, W.B. The cyclic load experiments and an accumulated deformation model for coarse—grained soil filling. *J. Vib. Shock* **2017**, *36*, 128–133.
6. Chen, Y.Y.; Huang, B.; Chen, Y.M. Raliability analysis of high level backfill based on chaotic optimization. *Chin. J. Geotech. Eng.* **2008**, *30*, 764–768.
7. Tang, L.; Yan, M.H.; Ling, X.Z.; Tian, S. Dynamic behaviours of railway base course materials subjected to long-term low-level cyclic loading: Experimental study and empirical model. *Geo-Tech.* **2017**, *67*, 537–545.
8. Bian, X.; Jiang, J.; Jin, W.; Sun, D.; Li, W.; Li, X. Cyclic and post cyclic triaxial testing of ballast and subballast. *J. Mater. Civ. Eng. (ASCE)* **2016**, *28*, 1–11.
9. Trinh, V.N.; Tang, A.M.; Cui, Y.J.; Dupla, J.C.; Canou, J.; Calon, N.; Lambert, L.; Robinet, A.; Schoen, O. Mechanical characterisation of the fouled ballast in ancient railway track substructure by large-scale triaxial tests. *Soils Found.* **2012**, *52*, 511–523. [CrossRef]
10. Leng, W.M.; Liu, W.J.; Zhou, W.Q. Experimental research on dynamic failure rules of compacted coarse-grained soil filling in heavy haul railway subgrade. *Rock Soil Mech.* **2015**, *36*, 640–646.
11. *JTG 3430-2020*; Test Methods of Soils for Highway Engineering (JTG 3430-2020). People's Communications Press: Beijing, China, 2020.
12. Ling, H.I.; Yang, S. Unified sand model based on the critical state and generalized plasticity. *J. Eng. Mech. ASCE* **2006**, *132*, 1380–1391. [CrossRef]
13. Liu, M.C.; Gao, Y.F.; Liu, H.L. Large scale triaxial test study on dilatancy characteristics of rockfill. *J. Geotech. Eng.* **2008**, *29*, 205–211.
14. Chen, J.H.; Zhang, J.S.; Li, X.P. Rock blasting damage model considering rock mass integrity and its application. *J. Geotech. Eng.* **2016**, *38*, 857–866.
15. Li, L.; Du, X.L.; Zhao, C.G. Study on elastoplastic dynamic constitutive model of saturated sand. *J. Rock Mech. Eng.* **2005**, *24*, 3380–3385.
16. Zhang, G.D.; Wei, C.F. Constitutive model of sandy soil under cyclic loading. *J. Solid Mech.* **1998**, *19*, 19–24.
17. Li, L.; Zhao, C.G. Research Progress on dynamic constitutive model of saturated soil. *World Earthq. Eng.* **2004**, *20*, 138–148.
18. Yuan, X.Q.; Liu, H.Y.; Liu, J.P. Three-dimensional composite damage constitutive model of non-penetrating fractured rock mass. *J. Geotech. Eng.* **2016**, *38*, 91–99.
19. Xie, D.Y. *Soil Dynamics*; Higher Education Press: Beijing, China, 2011.
20. Zhang, X.X.; Zhang, J.M.; Wen, Y.F. Practical method to predict settlement of subgrade induced by long-term traffic loads. *Chin. J. Geotech. Eng.* **2015**, *37*, 2067–2072.
21. Pérez-Acebo, H.; Linares-Unamunzaga, A.; Rojí, E.; Gonzalo-Orden, H. IRI performance models for flexible pavements in two-lane roads until first maintenance and/or rehabilitation work. *Coatings* **2020**, *10*, 97. [CrossRef]
22. Olowosulu, A.T.; Kaura, J.M.; Murana, A.A.; Adeke, P.T. Development of framework for performance prediction of flexible road pavement in Nigeria using Fuzzy logic theory. *Int. J. Pavement Eng.* **2021**, 1–10. [CrossRef]

coatings

MDPI

Article

Self-Formed Diffusion Layer in Cu(Re) Alloy Film for Barrierless Copper Metallization

Bin Cheng [1], Hongmei Chen [1], Isaac Asempah [2], Jiheng Wang [1], Yilin Zhu [1], Jun Wan [1], Feiran Jiang [1], Ze Wang [1], Yang Shui [1], Lei Wang [1,2,*], Lei Jin [3,*] and Yanxin Qiao [1,*]

1 National Demonstration Center for Experimental Materials Science and Engineering Education, Jiangsu University of Science and Technology, Zhenjiang 212003, China; cheng_bin1996@163.com (B.C.); hmchen@just.edu.cn (H.C.); wangjiheng27@163.com (J.W.); 18362882656@163.com (Y.Z.); w13862099702@163.com (J.W.); jfrjfr1105@163.com (F.J.); wangze0660@163.com (Z.W.); 15751770131@163.com (Y.S.)

2 Department of Mechanical Engineering, Accra Technical University, Accra P.O. Box 561, Ghana; nanasempah@yahoo.com

3 Stomatology Department, Nanjing General Hospital, Medical School, Nanjing University, Nanjing 210002, China

* Correspondence: wangl_ray@just.edu.cn (L.W.); ljin@nju.edu.cn (L.J.); yxqiao@just.edu.cn (Y.Q.)

Abstract: The barrier properties and diffusion behavior of Cu(Re) alloy films were studied. The films were deposited onto barrierless SiO_2/Si by magnetron sputtering. X-ray diffraction patterns and electric resistivity results proved that the Cu(Re) alloy films without a barrier layer were thermally stable up to 550 °C. Transmission electron microscopy images and energy-dispersive spectrometry employing scanning transmission electron microscopy provided evidence for a self-formed Re-enriched diffusion layer between the Cu(Re) alloy and SiO_2/Si substrate. Furthermore, the chemical states of Re atoms at the Cu(Re)/SiO_2 interface were analyzed by X-ray photoemission spectroscopy. The self-formed diffusion layer was found to be composed of Re metal, ReO, ReO_2 and ReO_3. At 650 °C, the Cu(Re) layer was completely destroyed due to atom diffusion. The low electrical resistivity in combination with the high thermal stability suggests that the Cu(Re) alloy could be the ultimate Cu interconnect diffusion barrier.

Keywords: diffusion barrier; Cu interconnect; self-formed; Cu(Re) alloy

Citation: Cheng, B.; Chen, H.; Asempah, I.; Wang, J.; Zhu, Y.; Wan, J.; Jiang, F.; Wang, Z.; Shui, Y.; Wang, L.; et al. Self-Formed Diffusion Layer in Cu(Re) Alloy Film for Barrierless Copper Metallization. *Coatings* **2022**, *12*, 613. https://doi.org/10.3390/coatings12050613

Academic Editor: Adrian David

Received: 28 March 2022
Accepted: 27 April 2022
Published: 29 April 2022

1. Introduction

To avoid rapid copper diffusion in integrated circuits (ICs), Cu interconnects require an effective barrier to prevent interdiffusion or reaction between the Cu and adjoining materials [1–5]. Various refractory metals and their nitride diffusion barriers, such as Ti, Ta, Ru and Mn, which have a low electrical resistivity, high stability and good interface adhesion, are in high demand [6–9]. To avoid the undesired formation of high-resistivity copper silicide, an ultra-thin Cu diffusion barrier with excellent barrier performance and low interconnect resistivity is needed to enhance IC performance [10,11]. The continuous scaling down of ICs to a few-nanometer regime has led to the issue of fabricating an ultra-thin barrier layer with step coverage on the sidewalls and bottom corners of trenches and vias [12]. Moreover, barrier layers occupy an increasing fraction at the cross-sectional area of conductors, leading to an increase in the electrical resistivity of the Cu interconnects [4]. Therefore, it is urgent to explore alternative methods of removing barrier layers which will also reduce the manufacturing cost using simplified technology.

Due to the difficulty in achieving uniform depositions of ultra-thin barrier layers, more attention has been paid to self-formed diffusion barriers (barrierless metallization) in recent years [13–15]. The self-formed barrier layer offers low electrical resistivity, resistance to Cu diffusion, resistance to electromigration and compatibility with conformal deposition techniques [16–21]. A self-formed barrier scheme is achieved by doping with diffusion barrier elements as well as their nitrides and carbides, such as Ti, Zr, Mn and WN [22–28]. All these doped elements have low-concentration solutes in Cu, which are supposed to segregate at the interface as a diffusion layer during annealing [16,29]. The thin self-formed layer at the interface acts as an adhesion and diffusion barrier layer. Due to the high melting point, strong bonding to resist interdiffusion and low (almost zero) solubility of Cu for an easy phase separation, Re might be a potential alloying element for barrierless metallization [6,30]. Chang et al. reported that the electroless-plated Cu(Re) alloy can serve as a barrierless metallization material to prevent Cu diffusion [6]. The mechanism of influences that Re might have on the diffusion-hindering behavior should be further studied. In the present study, rather than focusing on the influence of preparation methods on the barrier property, attempts were mainly made to provide more detailed information to clarify the growth mechanism of Re self-formed barrier layers, which will be used in the composition design of Cu–Re alloy film compositions.

2. Experimental

(111)-oriented Si wafers with a ~20 nm-thick oxidized SiO_2 layer were used as substrates (Lijing, Wenzhou, Wenzhou, China). The substrates were cleaned successively by ultrasonication in acetone and ethanol for 20 min. After drying with ultra-high purity N_2 gas, the substrates were immediately loaded into the sputtering system. Before deposition, the target and Re chips were pre-sputtered for 10 min to remove the impurities and oxide layers. Cu(Re) thin films were deposited on the SiO_2/Si substrates using a direct current (DC) magnetron sputtering system (Sky Technology development, Shenyang, China). The Cu films with ~2.3 at.% Re were obtained by co-sputtering the Cu target and Re chips (Ketai, Nanchang, China). Pure Re chips (99.99%, 5 mm × 5 mm) were pasted onto a pure Cu target (99.999%, diameter of 75 mm × thickness of 5 mm). The deposition was carried out at a background pressure of 5×10^{-4} Pa and an Ar gas deposition pressure of 1 Pa, with a target power of 100 W. The substrates were not heated during deposition, and the thickness of the Cu(Re) thin film after the deposition was about 150 ± 10 nm.

Isothermal annealing was performed in vacuum at a pressure of 6×10^{-5} Pa for 30 min, at temperatures ranging from 350 to 650 °C. Characterization of the phase composition and crystalline structure of the samples was determined by X-ray diffraction (XRD; D8 Venture, Bruker, Billerica, MA, USA) in the 2θ = 40°–70° range at a step size of 0.04°. The four-point probe (FPP; ST2258C, JingGe, Suzhou, China) method was used to measure the sheet resistance of the films before and after annealing. The interfacial behavior of the films was investigated by cross-section transmission electron microscopy at 200 kV (TEM; JEM2100F, JEOL, Tokyo, Japan). The cross-section TEM samples were ground with SiC paper until the total thickness obtained was about 50–100 μm and then thinned down to electron transparency by an Ion Polishing System (Gatan, Pleasanton, CA, USA). The component distribution of the Cu(Re)/SiO_2/Si films was analyzed by EDX-STEM (energy-dispersive spectrometry employing scanning transmission electron microscopy). The EDX-STEM instrument used was a JEM2100F operated at 200 kV and equipped with a solid-state X-ray detector (HAADF—high-angle annular dark field). EDX spectrum images were acquired with Oxford AZtec software (version 3.1) using the X-Max^N detector system with a dwell time of 30 μs and a typical image size of 512 × 512 pixels. The X-ray photoelectron spectroscopy measurement (XPS, PHI 5000, VersaProbeIII, ULVAC-PHI, Chigasaki, Japan) was carried out on the Ar-ion-sputtered film surface at the beam energy of 1.5 keV. During the test, the photoelectron take-off angle was 45°, monochromated Al Kα (1486.6 eV) was used as the X-ray source and the X-ray beam surveyed 100 μm. In the binding energy (BE) analysis, a correction was made by aligning the peak positions with reference to the standard adventitious carbon peak (C1s) at 284.8 eV.

3. Results and Discussion

Figure 1a presents the XRD analysis which was performed to investigate the phase evolutions of Cu(Re)/SiO_2/Si and Cu/SiO_2/Si samples before and after annealing at various temperatures. The 2θ diffraction peaks were found at 43.11° and 50.26°, corresponding to Cu (111) and Cu (200) for the as-deposited sample, as indexed in Figure 1a. No peaks corresponding to copper silicide were observed for the samples annealed at 350–550 °C. However, when the annealing temperature was increased to 650 °C, the existence of copper silicide was confirmed by the diffraction peaks related to Cu_3Si (012) and Cu_3Si (300). The presence of copper silicide indicates that Cu had diffused and reacted with the SiO_2/Si substrate. However, the formation of Cu_3Si was detected at 350 °C for the pure Cu on SiO_2/Si, implying a total failure of the copper metallization layer. The evolution of the sheet resistance as a function of annealing temperatures is plotted in Figure 1b. The resistivity of the as-deposited Cu(Re) film was found to be 6.41 μΩ·cm, higher than that of the pure Cu films (5.17 μΩ·cm) [12]. The resistivity of Cu(Re)/SiO_2/Si stacks slightly decreased to 5.89 μΩ·cm after annealing at 350 °C. This decrease was attributed to defect annihilation and stress relief in the film during annealing [31]. Then, it increased slowly at a temperature range of 350–550 °C. According to the XRD results, highly resistant copper silicide was not detected. Thus, the slight increase in the resistivity could be due to the Re segregation at the surface and interface that prevents the diffusion of the Cu atoms into the substrate [32]. A dramatic increase in resistivity was observed at 650 °C, as shown in Figure 1b. The increase in the sheet resistance indicated the rapid interdiffusion of copper and silicon toward the interface, finally leading to the formation of copper silicide, as observed in the XRD results. The above results prove that the Cu(Re)/SiO_2/Si interconnect structure can improve thermal stability against Cu diffusion even at 550 °C, which suggests that Re is a good candidate for use as a barrier alloy.

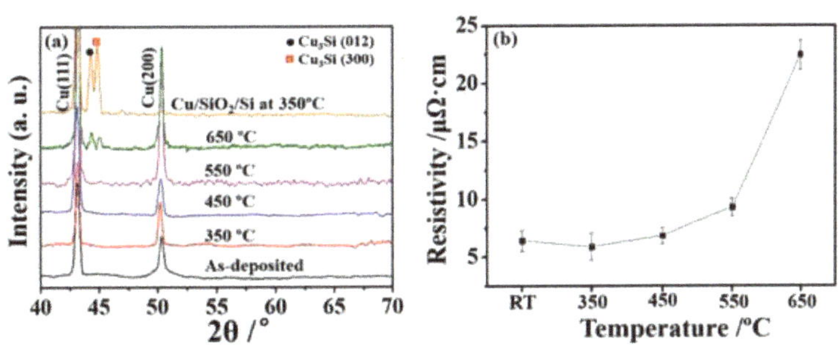

Figure 1. (**a**) The phase evolutions for Cu(Re)/SiO_2/Si annealed at various temperatures and Cu/SiO_2/Si annealed at 350 °C. (**b**) Electrical results for Cu(Re)/SiO_2/Si samples before and after annealing at various temperatures.

To further observe the interface diffusion behavior, typical TEM cross-sectional images of the Cu(Re)/SiO_2/Si interconnect structure are displayed in Figure 2. Figure 2a reveals a uniform thickness of ~150 nm for the Cu(Re) layer with a columnar grain and a ~20 nm amorphous SiO_2 layer for the as-deposited sample. Flat and free of intermixing, Cu/SiO_2 and SiO_2/Si interfaces were clearly revealed. The interconnect structure of the samples annealed at 350 and 450 °C remained unchanged compared with that of the as-deposited sample. The columnar structure of the Cu(Re) layer was maintained, and no considerable grain growth was observed during annealing due to the presence of Re. No obvious formation of copper silicide was observed in Figure 2b,c. With a further increase in the annealing temperature to 550 °C, atom diffusion eliminated the column grain structure of the Cu(Re) layer, as depicted in Figure 2d. Moreover, the Cu(Re) layer no longer maintained its columnar structure. As shown in Figure 2e, the Cu layer did not retain its lateral integrity in the sample annealed at 650 °C. As seen in Figure 2e, the ~55 nm Cu(Re) layer was destroyed by atom diffusion. It is worth noting that a self-formed layer existed at the Cu(Re)/SiO_2 interface, as indicated in Figure 2e, which was revealed in the following EDX-STEM and XPS analysis.

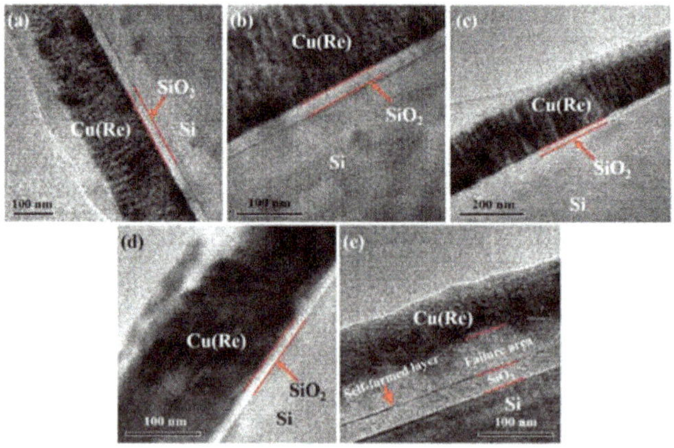

Figure 2. TEM cross-sectional images of the Cu(Re)/SiO_2/Si interconnect structure at various temperatures: (**a**) as-deposited, (**b**) 350 °C, (**c**) 450 °C, (**d**) 550 °C and (**e**) 650 °C.

The high-resolution cross-sectional TEM (HRTEM) images of the interfaces for the Cu(Re)/SiO_2/Si structure are depicted in Figure 3. Clear interfaces were observed in the as-deposited Cu(Re)/SiO_2/Si interconnect structure, as shown in Figure 3a. The Cu layer and the SiO_2 layer were polycrystalline and amorphous, respectively. For the samples annealed at 350 and 450 °C, the interfaces remained unchanged compared with those of the as-deposited sample (Figure 3b,c). With the aid of HRTEM, a clear interfacial reaction product, approximately 2 nm in thickness, was found at the Cu(Re)/SiO_2 interface after annealing at 550 °C, as indicated in Figure 3d. Moreover, the white line at the Cu(Re)/SiO_2 interface implied a slight atomic diffusion. We consider that the white line might be the copper silicide, even though the XRD results reveal the absence of copper silicide. For the sample annealed at 650 °C, the ~55 nm Cu(Re) layer was completely destroyed due to atom diffusion, as depicted in Figure 3e. The discontinuous black line still existed at the Cu(Re)/SiO_2 interface. This indicates that the Cu(Re) alloy began to fail as a barrier layer.

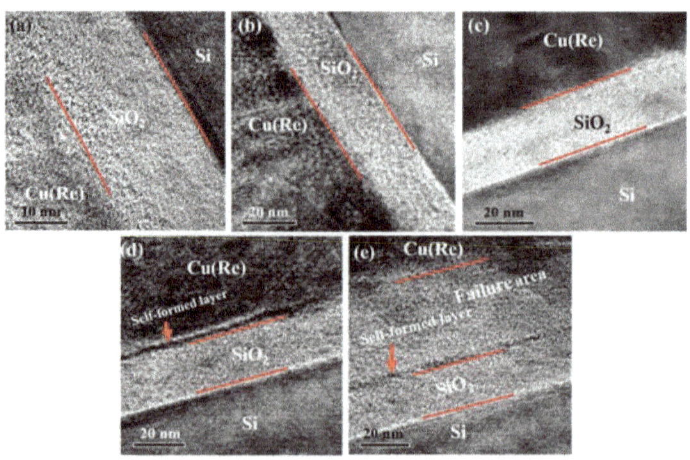

Figure 3. High-resolution image of the Cu(Re)/SiO_2/Si interconnect structure at various temperatures: (**a**) as-deposited, (**b**) 350 °C, (**c**) 450 °C, (**d**) 550 °C and (**e**) 650 °C.

Figure 4a,b present the fourteen points, which were perpendicular to the interfaces, selected for the EDX-STEM experiment. The component distribution of the as-deposited sample and Cu(Re)/SiO_2/Si sample annealed at 650 °C was assessed using EDX-STEM experiments, as shown in Figure 4c,d. The approximate interfaces of Cu(Re)/SiO_2 and SiO_2/Si are indicated by black dashed lines in Figure 4c. The component distribution of the O element cannot be measured correctly by the means of the EDX method. Consequently, it is not shown in Figure 4c,d. According to the EDX-STEM depth profiles, as shown in Figure 4c, the as-deposited Cu(Re) film contained about 2.1 at.% Re. The component distribution of the as-deposited sample showed a sharp variation, indicating a distinct interface structure. After annealing, the slopes of the Cu and Si signals in Figure 4d were flatter than those in Figure 4c. This implies that Cu atoms diffused further into the Si substrate in the Cu(Re)/SiO_2/Si sample annealed at 650 °C. Figure 4e displays the profile composition of Re for the as-deposited sample and Cu(Re)/SiO_2/Si sample annealed at 650 °C. The distribution of Re in the Cu layer in the as-deposited sample maintained a steady value of ~2.1 at.%, and then the value decreased to 0 at.% as it approached the Si substrate. On the contrary, the Re signal was higher in the SiO_2 layer in the stack annealed at 650 °C. The above experimental results prove that an obvious enrichment of Re can be detected at the Cu(Re)/SiO_2 interface. Thus, the black line as observed in Figure 3d,e was interpreted as the self-formed rhenium-enriched layer. It can be concluded that a minor amount of Re doping plays an indispensable role in the improvement of the barrier properties of pure Cu layers.

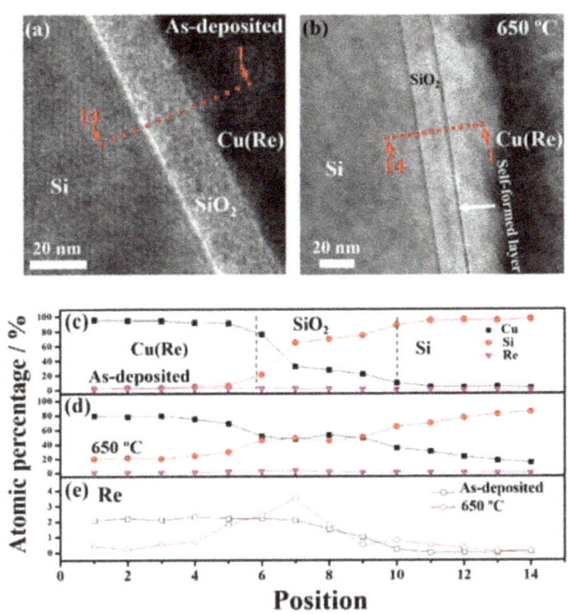

Figure 4. The positions of STEM-EDX testing of the (**a**) as-deposited sample and (**b**) sample annealed at 650 °C. The component distribution of the Cu(Re) film in the (**c**) as-deposited and (**d**) 650 °C annealing conditions. (**e**) The profile composition of Re for the as-deposited sample and Cu(Re)/SiO_2/Si sample annealed at 650 °C.

The chemical states of Re at the Cu(Re)/SiO_2 interface were analyzed using the XPS method [33,34]. Figure 5 displays the XPS spectra of Re 4f measured at the Cu(Re)/SiO_2 interface of the as-deposited film and Cu(Re) film annealed at 650 °C and their fits. For the as-deposited sample in Figure 5a, the peaks at 42.5 and 40.1 eV are characteristic of Re metal, and those at 43.5 and 41.1 eV are characteristic of ReO [35]. This indicates that Re exists at the interface in the chemical form of Re metal and ReO. The fits of the Re 4f photoemission spectra for the sample annealed at 650 °C prove the existence of two additional rhenium oxides in Figure 5b. The doublet at 44.5/42.1 eV was assigned to the Re (IV) oxidation state (ReO_2 oxide), and that at higher binding energies 47.7/45.3 eV was assigned to Re (VI) oxidation states (ReO_3 oxide) [36]. Specifically, the self-formed layer consisted of Re, ReO_2 and ReO_3. Generally, ReO_3 is known to be very unstable at high temperature and disproportionates to Re 7+ (Re_2O_7) and Re 4+ (ReO_2) upon heating above 400 °C [37,38]. The thermal decomposition is accompanied by the disproportionation reaction [38]

$$3ReO_3 \rightarrow ReO_2 + Re_2O_7 \quad (1)$$

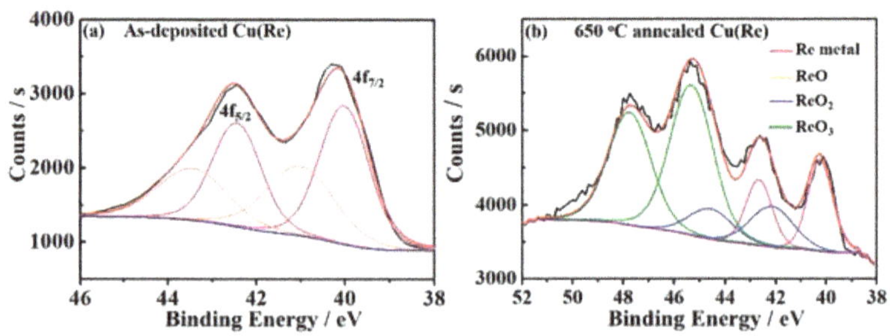

Figure 5. XPS spectra of Re 4f at the Cu(Re)/SiO_2 interface of the (**a**) as-deposited film and (**b**) Cu(Re) film annealed at 650 °C.

ReO_2 is stable against disproportionation up to 850 °C, while Re_2O_7 sublimes at a relatively low temperature (reported to be between 225 and 315 °C) [38]. After annealing, only Re metal, ReO_2 and ReO_3 were observed for the sample annealed at 650 °C. ReO_3 should be formed during cooling. Moreover, the Re content of the sample annealed at 650 °C was ~4.3 at.%, which was higher than that of the as-deposited sample (~2.3 at.%). Thus, we consider that the Re alloy diffused to the Cu(Re)/SiO_2 interface and reacted with SiO_2 to self-form a rhenium-enriched barrier layer during annealing, as observed in the TEM results.

It is well known that the Cu-Re system is essentially immiscible. The solid solubility of Re in Cu is $<10^{-4}$ at.% [39]. Thus, Re atoms may be driven out from the Cu grains during magnetron sputtering, which would cause Re segregation at the grain boundaries. After annealing, the alloyed Re diffused toward the SiO_2/Si interface, which was confirmed by EDX-STEM and XPS testing. Additionally, the formation energy of ReO_3 at 400–537 °C was −214.6 KJ/mol, which was less than that of Cu_2O at 650 °C (−100.977 KJ/mol) [38,40]. Therefore, Re was more likely to diffuse into the interface and react with SiO_2. A self-formed barrier layer was produced which effectively prevented atom interdiffusion.

4. Conclusions

The barrier properties and diffusion behavior of post-annealed Cu(Re) thin films were studied. In summary, a ~2 nm self-formed barrier layer was confirmed by TEM, EDX-STEM and XPS analysis. After doping, Re diffused into the interface of Cu(Re)/SiO_2 and reacted with SiO_2, which can serve as a barrier layer. The self-formed diffusion layer was found to be composed of Re metal, ReO, ReO_2 and ReO_3. It can be claimed that excellent thermal stability was observed for the Cu(Re) alloy without a barrier layer after annealing up to 550 °C. The high thermal stability makes the Cu(Re) alloy a promising metallization material for Cu interconnect applications.

Author Contributions: Conceptualization, B.C, Y.Q. and L.W.; methodology, B.C., J.W. (Jun Wang) and Y.Z.; software, H.C.; validation, Y.Z., J.W. (Jun Wang) and F.J.; formal analysis, Z.W. and Y.S.; investigation, B.C., Y.Z. and J.W. (Jun Wang); data curation, Y.Z. and J.W. (Jiheng Wang); writing—

original draft preparation, B.C.; writing—review and editing, I.A., Y.Q. and L.W.; supervision, L.J.; fund-ing acquisition, L.W. and L.J. All authors have read and agreed to the published version of the manuscript.

Funding: This work was supported by the Science and Technology Department of Jiangsu Province (BK20130468 and BK 20180984), the National Natural Science Foundation of China (51501074, 82071094, 51905110 and 52005228), the China Postdoctoral Science Foundation (2016M602983), the Jiangsu Overseas Visiting Scholar Program for University Prominent Young & Middle-aged Teachers and Presidents, and the Postdoctoral Science Foundation of Jiangsu (1601047A), Jiangsu Province Medical Key Talents Project (No: ZDRCA2016095).

Institutional Review Board Statement: Not applicable.

Informed Consent Statement: Not applicable.

Data Availability Statement: The data used to support the findings of this study are available from the corresponding author upon request.

Acknowledgments: The authors thank I. Asempah for his valuable help in improving the language of the manuscript.

Conflicts of Interest: The authors declare no conflict of interest.

References

1. Wang, F.; Li, D.; Zhang, Z.; Wu, M.; Yan, C. Improvement on interfacial structure and properties of Sn–58Bi/Cu joint using Sn–3.0Ag–0.5Cu solder as barrier. *J. Mater. Sci. Mater. Electron.* **2017**, *28*, 19051–19060. [CrossRef]
2. Wang, L.; Cao, Z.H.; Xu, J.H.; Yu, L.H.; Huang, T.; Meng, X.K. Influence of electric field annealing on atom diffusion in Cu/Ta/Si stacks. *Appl. Phys. A-Mater.* **2014**, *114*, 1091–1095. [CrossRef]
3. Wang, L.; Cao, Z.H.; Syed, J.A.; Hu, K.; She, Q.W.; Meng, X.K. Electric field accelerating interface diffusion in Cu/Ru/TaN/Si stacks during annealing. *Electrochem. Solid-State Lett.* **2012**, *15*, H188. [CrossRef]
4. Wang, L.; Cao, Z.H.; Hu, K.; She, Q.W.; Meng, X.K. Improved diffusion barrier performance of Ru/TaN bilayer by N effusion in TaN underlayer. *Mater. Chem. Phys.* **2015**, *135*, 806–809. [CrossRef]
5. Meng, Y.; Song, Z.X.; Qian, D.; Dai, W.J.; Wang, J.F.; Ma, F.; Li, Y.H.; Xu, K.W. Thermal stability of RuZr alloy thin films as the diffusion barrier in Cu metallization. *J. Alloys Compd.* **2014**, *588*, 461–464. [CrossRef]
6. Chang, S.Y.; Liang, L.P.; Kao, L.C.; Lin, C.F. Electroless- and electroplating of Cu(Re) alloy films for self-forming ultrathin re diffusion barrier. *J. Electrochem. Soc.* **2015**, *162*, D96–D101. [CrossRef]
7. Wang, L.; Asempah, I.; Dong, S.-T.; Yin, P.-P.; Jin, L. Quantitative studies of electric field intensity on atom diffusion of Cu/Ta/Si stacks during annealing. *Appl. Surf. Sci.* **2017**, *399*, 215–219. [CrossRef]
8. Wang, L.; Jin, L.; Yu, L.H.; Dong, S.T.; Chen, J.; Xu, J.H. Evaluation of electric field intensity on atom diffusion of Cu/Ta/Si stacks during annealing. *Appl. Phys. A-Mater.* **2016**, *122*, 3. [CrossRef]
9. Chang, S.-Y.; Li, C.-E.; Chiang, S.-C.; Huang, Y.-C. 4-nm thick multilayer structure of multi-component (AlCrRuTaTiZr)N(x) as robust diffusion barrier for Cu interconnects. *J. Alloys Compd.* **2012**, *515*, 4–7. [CrossRef]
10. Tsai, D.C.; Huang, Y.L.; Lin, S.R.; Jung, D.R.; Chang, S.Y.; Chang, Z.C.; Deng, M.J.; Shieu, F.S. Characteristics of a 10 nm-thick (TiVCr)N multi-component diffusion barrier layer with high diffusion resistance for Cu interconnects. *Surf. Coat. Technol.* **2011**, *205*, 5064–5067. [CrossRef]
11. Liu, B.; Lin, L.W.; Ren, D.; Zhang, Y.P.; Jiao, G.H.; Xu, K.W. Cu(Ge) alloy films with zirconium addition on barrierless Si for excellent property improvement. *J. Phys. D Appl. Phys.* **2013**, *46*, 155305. [CrossRef]
12. Wang, Y.; Tang, B.-H.; Li, F.-Y. The properties of self-formed diffusion barrier layer in Cu(Cr) alloy. *Vacuum* **2016**, *126*, 51–54. [CrossRef]
13. Hsu, K.C.; Perng, D.C.; Wang, Y.C. Robust ultra-thin RuMo alloy film as a seedless Cu diffusion barrier. *J. Alloys Compd.* **2012**, *516*, 102–106. [CrossRef]
14. Wang, Y.; Zhang, M.L.; Cao, F.; Liu, Y.T.; Shao, L. Interficial stability of Cu/Cu(Ru)/Si contact system for barrier-free copper metallization. *J. Alloys Compd.* **2011**, *509*, L180–L182. [CrossRef]
15. Wang, Y.; Yang, X.D.; Song, Z.X.; Liu, Y.T. Property improvement of copper films with zirconium additive for ULSI interconnects. *J. Alloys Compd.* **2009**, *486*, 418–422. [CrossRef]
16. Li, X.N.; Wang, M.; Zhao, L.R.; Bao, C.M.; Chu, J.P.; Dong, C. Thermal stability of barrierless Cu–Ni–Sn films. *Appl. Surf. Sci.* **2014**, *297*, 89–94. [CrossRef]
17. Fang, J.S.; Chen, Y.T. Passivation of copper–hafnium thin films using self-forming hafnium oxide. *Surf. Coat. Technol.* **2013**, *231*, 166–170. [CrossRef]
18. Zhang, X.Y.; Li, X.N.; Nie, L.F.; Chu, J.P.; Wang, Q.; Lin, C.H.; Dong, C. Highly stable carbon-doped Cu films on barrierless Si. *Appl. Surf. Sci.* **2011**, *257*, 3636–3640. [CrossRef]

19. Lozano, J.G.; Lozano-Perez, S.; Bogan, J.; Wang, Y.C.; Brennan, B.; Nellist, P.D.; Hughes, G. Interdiffusion and barrier layer formation in thermally evaporated Mn/Cu heterostructures on SiO_2 substrates. *Appl. Phys. Lett.* **2011**, *98*, 123112. [CrossRef]
20. Liu, B.; Song, Z.X.; Li, Y.H.; Xu, K.W. An ultrathin Zr(Ge) alloy film as an exhaustion interlayer combined with Cu(Zr) seed layer for the Cu/porous SiOC:H dielectric integration. *Appl. Phys. Lett.* **2008**, *93*, 174108. [CrossRef]
21. Zheng, Y.H.; Li, X.N.; Cheng, X.T.; Sun, W.; Liu, M.; Liu, Y.B.; Wang, M.; Dong, C. Ni-V(or Cr) Co-addition Cu alloy films with high stability and low resistivity. *Mater. Chem. Phys.* **2018**, *205*, 253–260. [CrossRef]
22. Clearfield, R.; Railsback, J.G.; Pearce, R.C.; Hensley, D.K.; Fowlkes, J.D.; Fuentes-Cabrera, M.; Simpson, M.L.; Rack, P.D.; Melechko, A.V. Reactive solid-state dewetting of Cu–Ni films on silicon. *Appl. Phys. Lett.* **2010**, *97*, 253101. [CrossRef]
23. Chu, J.P.; Lin, C.H. Formation of a reacted layer at the barrierless Cu(WN)/Si interface. *Appl. Phys. Lett.* **2005**, *87*, 211902. [CrossRef]
24. Wang, Y.; Cao, F.; Zhang, M.L.; Liu, Y.T. Comparative study of Cu-Zr and Cu-Ru alloy films for barrier-free Cu metallization. *Thin Solid Films* **2011**, *519*, 3169–3172. [CrossRef]
25. Lin, P.C.; Chen, H.; Hsieh, H.-C.; Tseng, T.-H.; Lee, H.Y.; Wu, A.T. Co-sputtered Cu(Ti) thin alloy film for formation of Cu diffusion and chip-level bonding. *Mater. Chem. Phys.* **2018**, *211*, 17–22. [CrossRef]
26. Cao, F.; Wang, Y.; Li, F.-Y.; Tang, B.-H. Evaluation of Cu(Ti) and Cu(Zr) alloys in barrier-less Cu metallization. *Mater. Chem. Phys.* **2018**, *217*, 412–420. [CrossRef]
27. Chu, J.; Lin, C.; John, V. Barrier-free Cu metallization with a novel copper seed layer containing various insoluble substances. *Vacuum* **2008**, *83*, 668–671. [CrossRef]
28. Li, X.; Cheng, B.; Asempah, I.; Shi, Q.; Long, A.Q.; Zhu, Y.L.; Wang, Q.; Li, Y.L.; Wang, L.; Jin, L. Effect of Different Ni Contents on Thermal Stability of Cu(Ni) Alloy Film. *J. Electron. Mater.* **2020**, *49*, 5674–5680. [CrossRef]
29. Wang, Y.; Cao, F.; Zhang, M.-L.; Liu, Y.-T. Effects of thermal annealing on Zr-N doped magnetron sputtered copper. *Thin Solid Films* **2011**, *519*, 3407–3410. [CrossRef]
30. Leverant, G.R. Diffusion Barrier for Protective Coatings. U.S. Patent 5,556,713, 17 September 1996.
31. Nie, L.F.; Li, X.N.; Chu, J.P.; Wang, Q.; Lin, C.H.; Dong, C. High thermal stability and low electrical resistivity carbon-containing Cu film on barrierless Si. *Appl. Phys. Lett.* **2010**, *96*, 182105. [CrossRef]
32. Fei, C.; Gao-hui, W.; Long-tao, J.; Guo-qin, C. Feasibility study on the use of Cu(Co) alloy for barrierless copper metallization. *Thin Solid Films* **2016**, *599*, 31–36. [CrossRef]
33. Qiao, Y.; Wang, X.; Yang, L.; Wang, X.; Chen, J.; Wang, Z.; Zhou, H.; Zou, J.; Wang, F. Effect of aging treatment on microstructure and corrosion behavior of a Fe-18Cr-15Mn-0.66N stainless steel. *J. Mater. Sci. Technol.* **2022**, *107*, 197–206. [CrossRef]
34. Yang, L.; Zhou, Z.; Yang, R.; Wang, J.; Chen, M.; Qiao, Y.; Zhu, S.; Wang, F. Effect of Al and Cr on the oxidation behavior of nanocrystalline coatings at 1050 °C. *Corros. Sci.* **2022**, *200*, 110191. [CrossRef]
35. Morant, C.; Galán, L.; Sanz, J.M. X-ray photoelectron spectroscopic study of the oxidation of polycrystalline rhenium by exposure to O2 and low energy O2+ ions. *Anal. Chim. Acta.* **1994**, *297*, 179–186. [CrossRef]
36. Tysoe, W.T.; Zaera, F.; Somorjai, G.A. An XPS study of the oxidation and reduction of the rhenium-platinum system under atmospheric conditions. *Surf. Sci.* **1988**, *200*, 1–14. [CrossRef]
37. Greiner, M.T.; Rocha, T.C.R.; Johnson, B.; Klyushin, A.; Knop-Gericke, A.; Schlögl, R. The Oxidation of Rhenium and Identification of Rhenium Oxides During Catalytic Partial Oxidation of Ethylene: An In-Situ XPS Study. *Z. Für Phys. Chem.* **2014**, *228*, 521–541. [CrossRef]
38. Shcheglov, P.A.; Drobot, D.V. Heterogeneous equilibria in the rhenium-oxygen system. *Russ. J. Phys. Chem.* **2006**, *80*, 1819–1825. [CrossRef]
39. Shabalin, I.L. Rhenium. In *Ultra-High Temperature Materials I*; Springer: Manchester, UK, 2014; pp. 317–357.
40. Holmes, R.D.; Kersting, A.B.; Arculus, R.J. Standard molar Gibbs free energy of formation for Cu_2O: High-resolution electrochemical measurements from 900 to 1300 K. *J. Chem. Thermodyn.* **1989**, *21*, 351–361. [CrossRef]

coatings

MDPI

Article

Study of Mask Electrochemical Machining for Ring Narrow Groove under the Action of Multiple Physical Fields

Ruochen Zhao [1,*], Liang Huang [2], Haiyue Zhao [2], Yan Cao [3], Weijun Tian [4] and Ning Wang [5]

1 Department of Science and Technology, Xi'an Technological University, Xi'an 710021, China
2 School of Mechatronic Engineering, Xi'an Technological University, Xi'an 710021, China; huangliang@xatu.edu.cn (L.H.); weilai543135137@163.com (H.Z.)
3 School of Computer Science and Engineering, Xi'an Technological University, Xi'an 710021, China; caoyan@xatu.edu.cn
4 School of Mechatronic Engineering, Northwestern Polytechnical University, Xi'an 710124, China; tianwj@nwpu.edu.cn
5 Ministry of Technology, Xi'an KunLun Industry (Group) Company with Limited Liability, Xi'an 710021, China; ningtop@163.com
* Correspondence: zhaorc2022@126.com

Abstract: Deep and narrow groove structures are widely used in aviation, aerospace, weapons, and other industries, and play a very important role. In order to solve the problems of machining tool deformation, machining flying edge, burr in traditional Computerized Numerical Control (CNC) milling for deep and narrow grooves, and the problems of serious motor loss and low machining efficiency in non-contact electrical discharge machining (EDM), electrochemical mask machining through the mask treatment of the non-processed part mask processing, and with no loss of the processing cathode tool, was suggested as an efficient way to solve these problems. Considering that the corrosion removal of the anodic workpiece is mainly subject to the multi-physical field coupling action between the electric field, the flow field, and the temperature field, it is necessary to construct a multi-physical field coupling model of electrochemical mask machining and combine this with the numerical simulation analysis to realize the distribution state of the multi-physical field, so as to realize the optimization guidance of the overall processing process.

Keywords: ring narrow groove; mask electrochemical machining; numerical simulation of the multiple physical fields

Citation: Zhao, R.; Huang, L.; Zhao, H.; Cao, Y.; Tian, W.; Wang, N. Study of Mask Electrochemical Machining for Ring Narrow Groove under the Action of Multiple Physical Fields. *Coatings* **2022**, *12*, 605. https://doi.org/10.3390/coatings12050605

Academic Editor: Claudio Mele

Received: 16 February 2022
Accepted: 30 March 2022
Published: 29 April 2022

1. Introduction

Deep and narrow groove structures are widely used in aviation, aerospace, weapons, and other industries, and play a very important role. These deep and narrow groove structures require not only smaller width sizes, but also a large and deep aspect ratio [1]. To achieve the production of such complex small-size structures, on the one hand, KH Oh. took the lead in combining laser processing and wet etching technology to achieve the efficient processing of the deep narrow groove structure (the groove is 50 μm wide, 150 μm deep, and 50 mm long) of an aero-engine micro-cooling tube group [2,3]. On the other hand, in order to improve the processing accuracy of the cooling groove of a rocket engine nozzle, Lei H et al. [4] used digital milling technology to realize the 784 cooling groove processing on a circular pipe with a depth of 2.6 mm in the group (the outer diameter of the rocket nozzle is 1500 mm and its axial height is 940 mm). Therefore, the current processing methods for the above deep narrow grooves mainly include traditional mechanical machining and non-traditional machining [5].

Traditional mechanical machining methods mainly include milling, drilling, cutting, tapping, grinding, etc. Because the machining operation is simple and adaptable, it is a common machining method, and for the deep and narrow groove structure, more machining methods are used for milling processing. In the milling process, the tool is an important factor affecting processing efficiency and processing precision. The heat in the machining process can cause tool deformation, causing machining error; secondly, milling machining is contact machining, therefore, the force applied can also cause deformation of the tool, which can also cause machining error. The tool surface roughness, material characteristics, and tool wear degree can all cause certain machining errors [6,7]. Many scholars have put forward the multi-milling method [8] and using a five-axis precision CNC milling machine [9,10] to realize high efficiency and machining. However, the milling and processing of deep narrow groove structures face an inevitable problem, which is that the edge and bottom of the groove will have residues, such as flying edge, burr, and so on, that are difficult to remove, especially when processing nickel-based high-temperature alloys and titanium alloys where the phenomenon of residue is more obvious. Therefore, the application of milling technology in titanium alloys is restricted [11–13]. To solve the above narrow groove processing problem for hard cutting alloys, non-traditional machining is introduced. While in the process of deep narrow groove electro-spark forming, with an increase in the processing depth, the concentration of electro-corrosion products and the working liquid temperature in the processing gap gradually increase. If the working liquid environment can not be well improved, the stability and processing efficiency of the electro-spark processing process will be seriously affected [14]. Then, the problems of low processing efficiency, serious electrode loss, and high processing costs make this technology unable to be used in a field with high precision requirements and less material removal [15]. With the development of the times, based on the principle of the above electro-spark forming, Soviet scientists proposed electric-spark line cutting processing technology. Electric-spark line cutting technology, with its good flexibility, a wide range of applicable materials, high processing efficiency, and very good processing precision, has been widely used in national defense, medical treatment, and other industrial fields, and is especially suitable for the processing of narrow grooves and narrow joints of parts. However, its application to large shape sizes, blind cavities, and blind grooves has certain limitations [16]. Compared with electro-spark forming processing technology, electrochemical machining technology is also copied by the forming cathode in this study. In contrast, electrochemical machining is used to realize the dissolution and removal of workpiece materials through the redox reaction between the cathode and anticathode, so as to achieve the purpose of forming and processing. This processing method is especially applicable to high-strength and high hardness materials that are difficult to process in the traditional mechanical machining methods. On the basis of theory, electrochemical mask machining is a special method of contactless electrochemical machining, through which the anode workpiece does not process part of the insulation mask. Using electrochemical machining between the electrode liquid in the electrode field of the anode, not the mask field, electrolyte flow field and temperature field coupling realizes the anode material and cathode hydrogen with low cost, efficient processing (as shown in Figure 1). Considering that the corrosion removal of the anodic workpiece is mainly subject to the multi-physical field coupling action between the electric field, the flow field, and the temperature field, it is necessary to construct a multi-physical field coupling model of electrochemical mask machining and combine this with numerical simulation analysis to realize the distribution state of the multi-physical field, so as to realize the optimization guidance of the overall processing process. If a very good grasp of the influence of the process can be achieved, the processed parts can achieve high size precision and surface quality, and also inspire scholars to explore the process and the scientific research results of the production in practice [17,18]. In order to solve the problem of short circuits caused by the flow field instability in the electrochemical machining of the array of deep narrow grooves, Liu et al. studied the influence law of the cathode superimposed low-frequency vibration on the flow field through numerical simulation and analysis of the flow field, and the one-time processing of 30 deep narrow groove structures was realized [19]. Silva et al. studied the influence law of the inter-relationship between electrolyte conductivity and process parameters on the processing precision of narrow groove and narrow seam, and established and analyzed a mathematical model of processing surface-forming law, and a circular array narrow groove structure that meets the requirements was processed through a process test [20]. Meanwhile, Zhao et al. found that for a razor net cover with an outer circle of 90 and an inner circle of 45 deep, narrow groove structure processing proved to be a difficult problem. However, using a forming tool cathode, through electric field stimulation, flow field simulation, cathode feed mode, processing the groove width of 0.27 ± 0.02 mm and a wall thickness is 0.4–0.6 mm, gave a better consistency of deep narrow groove structure, and this was successfully used in enterprising mass production [21]. Although narrow-groove electrochemical machining technology has good application prospects, most of the research is still in the early exploration stage and less applied in production practice. This is mainly because there are many above-discussed factors affecting the electrochemical machining process, and mastering the influence law of these factors on the processing process is not comprehensive enough.

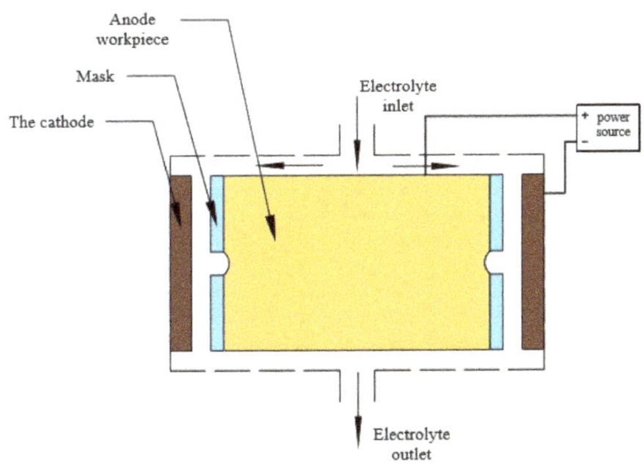

Figure 1. Schematic diagram of mask electrochemical machining of annular narrow groove.

In order to ensure processing efficiency, the electrochemical processing stability of the slot workpiece was studied [22–25]. In this paper, a mathematical model of the electric field, flow field, and temperature field coupling is established by means of mask electrochemical machining, and the above model is numerically simulated based on COMSOL Multiphysics. The variation law of each parameter in the machining process is also analyzed. According to the simulation result, the feasibility of the mask electrolytic machining of the annular narrow groove is verified, the ring groove process parameters are obtained, and the slotting precision and the surface quality are improved, therefore, a new method and a theoretical foundation are provided for the in-depth study of the electrolytic processing of the complex special-shaped cavity.

2. Theoretical Analysis of Mask Electrochemical Machining

2.1. The Principle of Mask Electrolysis

The principle of the mask electrolytic machining ring cell is shown in Figure 1. Mask insulation, anodic workpiece exposed area, and electrolyte contact follow Faraday's law and continue to dissolve until etched to meet the requirements of the shape [26,27].

2.2. Forming Law of Mask Electrochemical Machining

In the process of mask electrolytic machining, the surface shape and size of the cathode will not change when the cathode is fixed. As shown in Figure 2, the initial gap between the anode workpiece and the cathode is Δ_0. After t time, the depth of the machining slot is H.

After calculation, the depth of the ring groove H after the mask electrochemical machining time t is given in Equation (2) [28]:

$$H = \sqrt{2\eta\omega\kappa(U - \delta E)t + C} - \Delta_0 \tag{1}$$

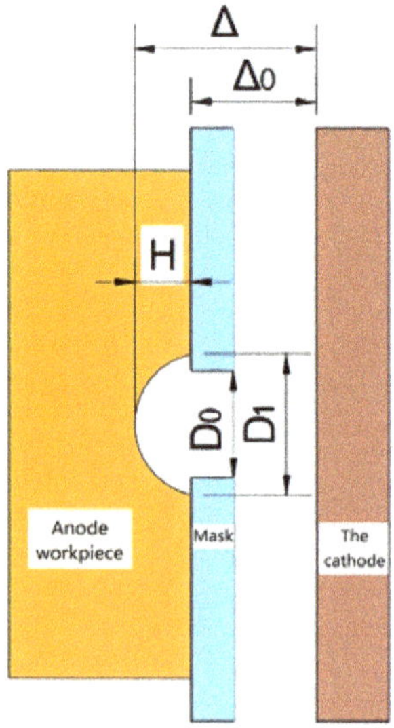

Figure 2. Schematic diagram of the position of the mask electrochemical machining.

In the process of an electrochemical reaction, the distribution of the electric field, flow field, and temperature field in the gap between electrodes is very complex, and its parameters interact with each other and change at all times, which makes it difficult to predict the anodic dissolution characteristics accurately. Therefore, it is necessary to simulate the multi-physical field coupling numerical simulation of the mask electrolytic machining process.

3. Methodology

3.1. Establishment of Geometric Model

The object of processing in this paper is the circular groove in the winding wheel, the overall structure of which is shown in Figure 3a, and the narrow groove structure and enlarged figure are shown in Figure 3b.

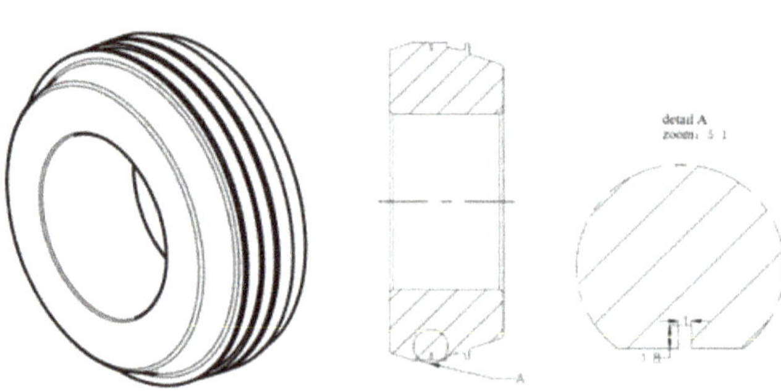

(**a**) (**b**)

Figure 3. Annular narrow slot geometric model. (**a**) Integral structure of winding wheel; (**b**) Narrow slot structure and magnification diagram.

The three-dimensional section view of the electrochemical machining model of the ring groove mask is shown in Figure 4a. Since the whole model is annular, it is assumed that the coupling field in each section is the same. In order to facilitate analysis and calculation, the model is simplified into two dimensions. The red border of the interpolar gap in Figure 4a is taken for analysis. The two-dimensional mathematical model is shown in Figure 4b, where L_2, L_3 is the free boundary of the machining gap, L_1 is the cathode boundary, L_6 is the anode workpiece boundary, L_4, L_5, L_7, L_8 is the insulation boundary.

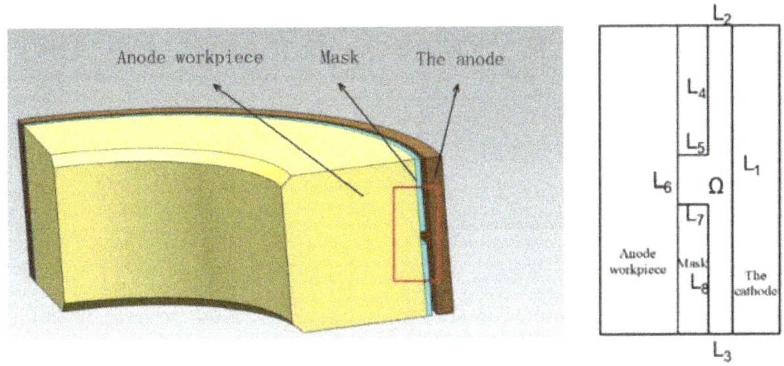

(**a**) (**b**)

Figure 4. Machining area model of ring groove. (**a**) Three-dimensional model of ring groove machining; (**b**) Two-dimensional mathematical model.

3.2. Multi-Physical Field Coupling Analysis

The electrochemical machining process of a ring cell and its complexity involve flow field, electric field, temperature field, and structural field, which vary and affect each other at any time [29].

The flow field and electric field affect the conductivity and current efficiency, and the conductivity increases with the increase in temperature, which leads to the change in workpiece erosion rate and then affects the distribution of the electric field and flow field [30,31]. The electric field model (Equation (2)) and the flow field model (Equation (3)) are substituted into the current density Equation (4) to obtain the coupling Equation (5) of the flow field and the electric field [32–34]:

$$\nabla^2\phi = \frac{\partial^2\phi}{\partial x^2} + \frac{\partial^2\phi}{\partial y^2} = 0 \tag{2}$$

$$\begin{cases} \rho\frac{\partial v}{\partial t} + \rho(v\cdot\nabla)v = -\nabla p + (v+v_T)\left[\Delta v + (\Delta v)^T\right] \\ \rho\nabla\cdot v = 0 \end{cases} \tag{3}$$

$$i = -\sigma\cdot\nabla\varphi \tag{4}$$

$$\rho C_p\left(\frac{\partial T}{\partial t} + v\cdot\nabla T\right) - \nabla\cdot(\sigma_t\nabla T) = -\sigma\cdot\nabla^2\varphi^2 \tag{5}$$

4. Numerical Simulation Analysis

The parameters of the processing area are a groove width of 1mm, the depth of the groove is 1.8 mm, the thickness of the mask is 0.1 mm, the selection quality of the electrolyte is 10% $NaNO_3$ solution [35].

Then, add current density, turbulence, fluid heat transfer, moving grid physical field to COMSOL Multiphysics. In the current density field, the grid is divided (in order to ensure the simulation accuracy, the tetrahedral mesh is needed and the refinement mesh size is 0.4 and the corresponding number of cells is 183,581). Meanwhile, the boundary conditions are adjusted (the inlet pressure is 0.19–0.25 Mpa, the processing gap is 0.1–0.5 mm, and the processing voltage is 8–30 V), and the single-field and the coupling field are simulated by changing the different processing parameters.

4.1. The Influence Law of Process Parameters at the Initial Moment

Under different electrolyte inlet pressures, the flow velocity distribution cloud diagram at the machining gap is shown in Figure 5. With the increase in electrolyte flow rate, the streamline distribution is similar, but the flow rate increases gradually. The flow rate at the slot can reach 11 m/s, the streamline distribution is stable without liquid shortage, and the interpolar products can be taken away in time to ensure stable processing. The temperature distribution in the machining gap under different inlet pressures is shown in Figure 6. The heat generated by the current heats up the electrolyte, the inlet pressure increases, and the interelectrode velocity increases. In order to ensure the uniformity of the flow field, a large electrolyte inlet pressure is used in the machining, but too much pressure makes the flow rate too large, resulting in a tremor affecting the machining accuracy. When the inlet pressure of the electrolyte is 0.23 Mpa, the flow rate between the electrodes is 10.27 m/s, which can meet the processing requirements.

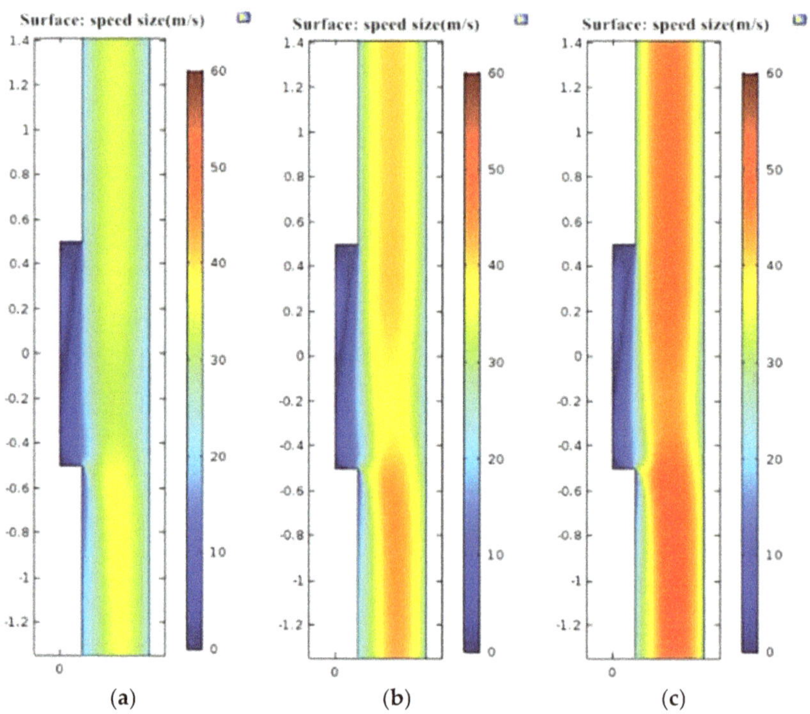

Figure 5. Velocity between poles under different inlet pressures. (**a**) 0.20 Mpa; (**b**) 0.23 Mpa; (**c**) 0.25 Mpa.

As shown in Figure 7, the initial current density around the machining area increases with the decrease in the machining gap, while the current density of the slotted part is basically the same distribution of unit current density under different gaps. Therefore, different machining gaps mainly affect the initial current density distribution at the gap between the anode and cathode, with little difference in the influence on the slotted area. However, too much machining gap can affect the current distribution, weaken the interelectrode electric field, give small clearance for machining electrode current density distribution of the uniform, and can promote ring groove shape precision on the basis of the machining efficiency but if the gap is too small, it is easy to cause short-circuit burns and creates cathode electrolytic product discharge difficulties, so this article chooses an initial machining gap of 0.2 mm.

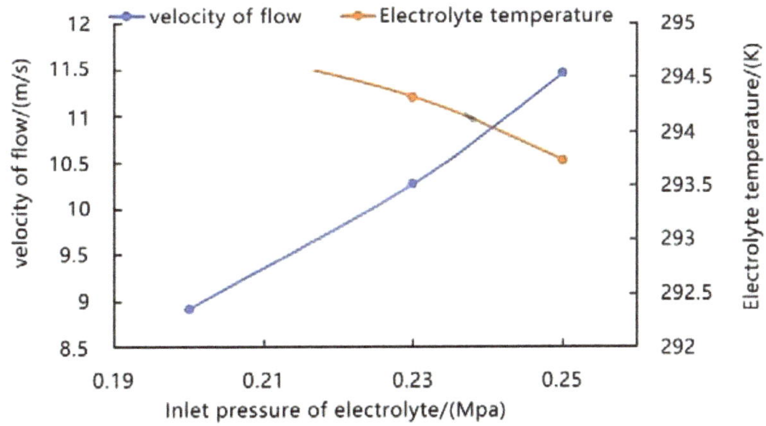

Figure 6. Variation trend of interpolar flow rate and electrolyte temperature.

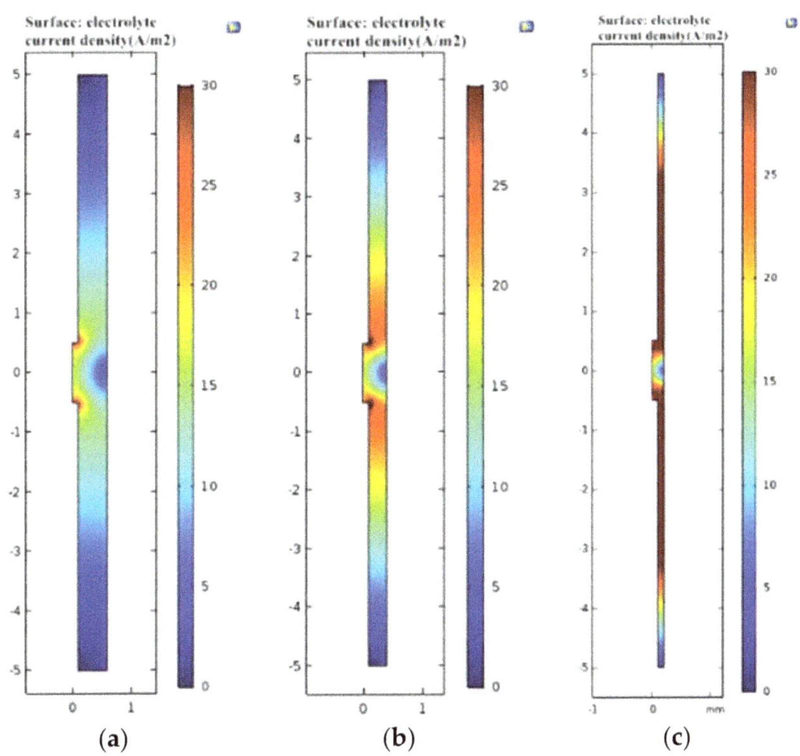

Figure 7. Current density cloud map corresponding to different processing gaps. (**a**) Machining gap 0.5 mm; (**b**) Machining gap 0.2 mm; (**c**) Machining gap 0.1 mm.

4.2. Simulation Analysis of Ring Groove Forming Process

(1). Effect of machining time and voltage on groove depth: On the basis of the above electric field and the coupling simulation of the flow field, the physical field of the moving mesh is added, and the change of the depth of the groove under different machining voltage and processing times is solved by the transient solver, and the variation rule, shown in Figure 8, is obtained by numerical fitting.

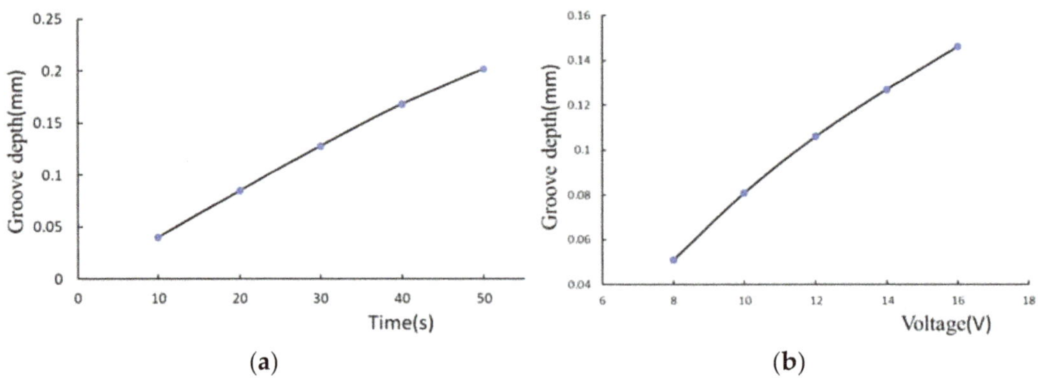

Figure 8. Slot depth variation diagram. (**a**) The relationship between the depth of the slot and the processing time (15 V); (**b**) The relationship between the depth of the slot and the processing time (30 V).

As shown in Figure 8a, as the processing time increases, the depth of the groove deepens, but the growth rate decreases. This is mainly due to the anodic material removal, the interelectrode machining gap becomes larger, and the interelectrode current density decreases under the condition that the processing voltage remains unchanged, leading to a decrease in processing efficiency. As shown in Figure 8b, due to the increase in voltage and conductivity, the depth of the slot increases but the increase in the current density of the machining gap is limited, which leads to the slow growth rate of the slot depth. It is found from the trend diagram that the growth rate of the slot depth is mainly related to the processing time and the inter-pole gap, and the voltage has little effect on it. A processing voltage of 12 V can provide a large interpolar current density to complete the machining of the ring slot and the stray corrosion of slot width can ensure the machining accuracy. Therefore, the processing voltage of 12 V is selected in this paper.

(2). Study on parameter distribution between poles: In that electrolytic process, the constant variation of the profile of the workpiece results in a change in the electric field structure and the electrical conductivity of the electrolyte. Through the multi-physical field coupling simulation, the current density distribution in the machining gap after processing for 120 s is obtained, as shown in Figure 9a, and the current density of the surface of the workpiece is gradually reduced as the processing is carried out, meaning the dissolution rate of the anode is reduced, and the processing erosion amount in the same time is reduced. The erosion rate in the deep direction of the groove becomes slow. Figure 9b shows the electrolyte potential distribution. The anode potential was higher and showed a uniform trend of decline. With the progress of the processing, the potential at the ring groove was higher and evenly distributed.

In order to ensure the smooth progress of the mask electrolytic machining and the timely discharge of electrolytic products and heat, large electrolyte pressure and flow rate are needed at the machining gap. As shown in Figure 9c,d, the flow rate of the small electrolyte between the two poles is faster at the initial time of machining. With the removal of the anode workpiece, the gap between the large electrodes becomes larger with the removal of the anode workpiece because the constant inlet flow rate leads to a decrease in pressure and flow rate. Due to the continuous etching process, the gap between the anode and the cathode will become larger, the flow of the electrolyte in the deep groove will be blocked, the speed of the electrolyte will be slow, and the processing stability will be affected.

In the above, the COMSOL simulation software is used to simulate and study the variation rules of each physical field in the process of circular narrow groove machining under the conditions of multi-physical field coupling of electrochemical machining. According to the simulation results, the machining accuracy can be predicted and the actual electrochemical machining process of a circular groove can be guided to provide a guarantee for mass production of the workpiece.

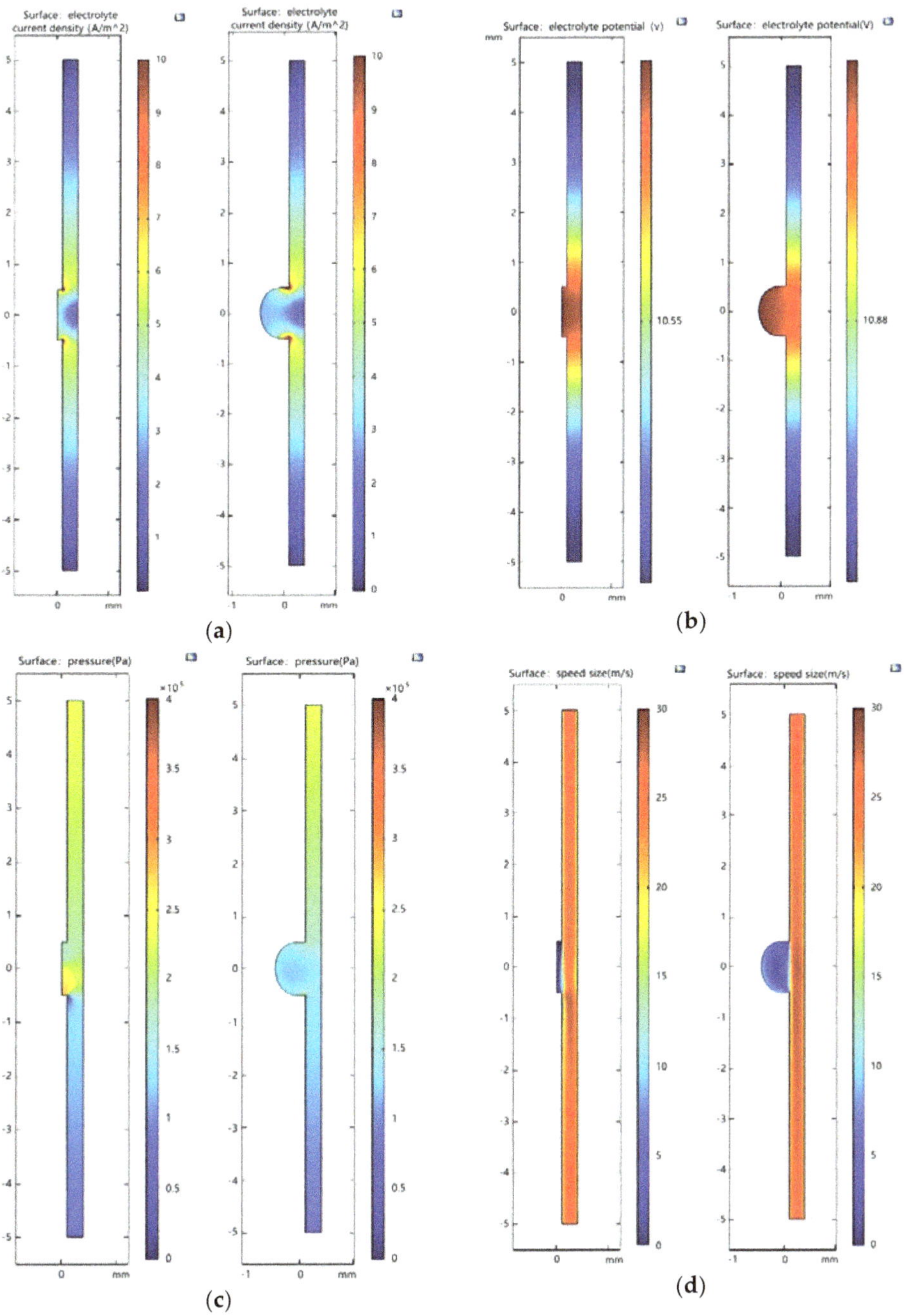

Figure 9. Simulation comparison of different processing times. (**a**) Electrolyte current density; (**b**) electrolyte potential; (**c**) electrolyte pressure change; (**d**) electrolyte flow rate change.

5. Conclusions

With the advantage of the narrow groove of electrolytic mask machining, taking the electrolytic machining of an annular narrow groove as an example, by constructing the coupling action model between the electric field, flow field, and temperature field and using COMSOL numerical simulation analysis software, the mass erosion and process efficiency guidance under the coupling of multiple physical field parameters are realized, as follows:

(1.) It is known that the distribution of current density can be uniform with a small initial machining gap, and the product between electrodes can be taken away in time with large inlet pressure, which can improve the machining efficiency and ensure the accuracy of the ring slot at the same time;

(2.) In the process of the electrolytic machining of ring cells, the current density between electrodes is always evenly distributed. However, with the decrease in the depth current density of the ring slot, the erosion rate decreases, and the machining efficiency decreases;

(3.) The high electrolyte pressure and the flow rate can discharge the electrolytic product and heat in time, and as the inter-electrode gap of the etching is increased, the pressure and the speed of the electrolyte are reduced, and the processing stability can be affected.

By comparing the above findings with references [19–21] in the Introduction, it demonstrates that conclusions (1.) and (3.) above are similar to reference [20] (when using a narrow interelectrode gap (50 μm), size accuracy of $\pm$ 2 μm, surface finish of 0.01 μmRa, (50 μm) for small (100 mm^2) parts), and [19] (using ordinary multi-slot ECM processing where most of the electrolyte is discharged from the channels between the cathode tools), respectively. In conclusion (2.), when the feed rate and the ablation rate change with the processing time, the electric field strength gradually decreases when the processing gap gradually increases, and finally leads to the ablation rate gradually decreasing with the processing time, which is not mentioned in previous studies (that is, the coupling of multiple physical fields and the interpolar structural fields). Through the multi-physical field coupling simulation analysis of annular narrow cell mask electrolytic machining, the process parameters are obtained as follows: initial machining gap 0.1 mm, electrolyte inlet pressure 0.23 Mpa, machining voltage 12 V. The results show that the multi-physical field coupling simulation is carried out on the mask electrolysis process of the annular narrow groove, and the variation rule of the interelectrode parameters can be obtained, and the machining process can be effectively predicted and the test period can be shortened.

Author Contributions: Conceptualization, R.Z. and L.H.; methodology, L.H. and Y.C.; validation, R.Z., L.H. and W.T.; formal analysis, R.Z. and L.H.; investigation, N.W. and H.Z.; resources, Y.C.; writing—original draft preparation, L.H. and R.Z.; writing—review and editing, L.H.; project administration, Y.C.; funding acquisition, Y.C. All authors have read and agreed to the published version of the manuscript.

Funding: This paper is supported by the project of the 2019 Yulin Science and Technology Program under Grant No. K20190176.

Institutional Review Board Statement: Not applicable.

Informed Consent Statement: Not applicable.

Data Availability Statement: Not applicable.

Conflicts of Interest: The authors declare that there are no conflicts of interest regarding the publication of this article.

References

1. Bai, Y.Y.; Lu, M.; Li, W.B. Overview and expectation of research on narrow—deep—groove machining technology. *Manuf. Technol. Mach. Tool* **2014**, *3*, 45–49.
2. Oh, K.H.; Lee, M.K.; Jeong, S.H. Laser micromachining of high-aspect-ratio metallic grooves for application to microthermal devices. *J. Micromechanics Microengineering* **2006**, *16*, 1958. [CrossRef]
3. Oh, K.H.; Lee, M.K.; Kwon, S.J.; Jeong, S.H. Fabrication of high-aspect-ratio microgrooves with laser-assisted wet etching for micro heat pipe. In Proceedings of the Conference on Lasers and Electro-Optics and 2006 Quantum Electronics and Laser Science Conference, San Jose, CA, USA, 21–26 May 2006; pp. 1–2.
4. Han, L.; He, W.D. Digitization process technology of cooling grooves on nozzle of liquid-propellant rocket engine. *J. Rocket. Propuls.* **2014**, *40*, 58–61.
5. Zhang, C.; Guo, J.; Zhang, C.; Chen, X.; Liu, J.; Zhang, Y. Research on flow field of electrochemical milling on deep-narrow groove with tube electrode. *Electromach. Mould.* **2020**, *31*, 31–34.
6. Su, C.; Hou, M.M.; Zhu, L.D.; Wang, W.S. SPH-based numerical simulation for big deformation during metal cutting. *J. Northeast. Univ. Nat. Sci.* **2009**, *30*, 419–421.
7. Bao, J.; Li, L.; He, N.; Huang, L. A brief review of micronmilling technology. *Mech. Sci. Technol. Aerosp. Eng.* **2009**, *28*, 1019–1022.
8. Pan, M.Q.; Li, J.H.; Tang, Y. Development of high-aspect-ratio microchannel heat exchanger based on multi-tool milling process. *J. Cent. South Univ. Technol.* **2008**, *15*, 228–234. [CrossRef]
9. Bang, Y.B.; Lee, K.M.; Oh, S. 5-axis micro milling machine for machining micro parts. *Int. J. Adv. Manuf. Technol.* **2005**, *25*, 888–894. [CrossRef]
10. Shamoto, E.; Saito, A. A novel deep groove machining method utilizing variable-pitch end mill with feed-directional thin support. *Precis. Eng.* **2016**, *43*, 277–284. [CrossRef]
11. Niknam, S.A.; Songmene, V. Analysis of friction and burr formation in slot milling. *Cirp Conf. Manuf. Syst.* **2014**, *21*, 755–759. [CrossRef]
12. Niknam, S.A.; Songmene, V. Factors governing burr formation during high-speed slot milling of wrought aluminum alloys. *Proc. Inst. Mech. Eng. Part B J. Eng. Manuf.* **2013**, *227*, 1165–1179. [CrossRef]
13. Li, A.; Zhao, J.; Zhou, Y.; Chen, X.; Wang, D. Experimental investigation on chip morphologies in high-speed dry milling of titanium alloy Ti-6Al-4V. *Int. J. Adv. Manuf. Technol.* **2012**, *62*, 933–942. [CrossRef]
14. Klocke, F.; Holsten, M.; Welling, D.; Klink, A.; Perez, R. Influence of threshold based process control on sinking EDM of a high aspect ratio geometry in a gamma titanium aluminide. *Procedia Cirp* **2015**, *35*, 73–78. [CrossRef]
15. Abbas, N.M.; Solomon, D.G.; Bahari, M.F. A review on current research trends in electrical discharge machining (EDM). *Int. J. Mach. Tools Manuf.* **2007**, *47*, 1214–1228. [CrossRef]
16. Sarkar, S.; Mitra, S.; Bhattacharyya, B. Parametric optimisation of wire electrical discharge machining of γ titanium aluminide alloy through an artificial neural network model. *Int. J. Adv. Manuf. Technol.* **2006**, *27*, 501–508. [CrossRef]
17. Zhu, Y.W.; Xu, J.W. The basic application study on electrochemical machining integral impeller with big-twisted blades. *Int. Technol. Innov. Conf.* **2006**, *7*, 587–593.
18. Mahdavinejad, R.; Hatami, M. On the application of electrochemical machining for inner surface polishing of gun barrel chamber. *J. Mater. Processing Technol.* **2008**, *202*, 307–315. [CrossRef]
19. Liu, J.; Jiang, X.C.; Zhui, D. Electrochemical machining of multiple slots with low-frequency tool vibrations. *Procedia Cirp* **2016**, *42*, 799–803.
20. Silva, A.K.M.D.; Altena, H.S.J.; Mcgeough, J.A. Influence of electrolyte concentration on copying accuracy of precision-ECM. *CIRP Ann.-Manuf. Technol.* **2003**, *52*, 165–168. [CrossRef]
21. Zhao, J.S.; Wang, F.; Xiao, X.; Li, L. Experiment research on electrochemical machining of meso scale arc-shaped multi-grooves. *J. Mech. Eng.* **2014**, *50*, 187–192. [CrossRef]
22. Wu, L.F.; Li, X.S.; Wang, L. Experimental study on electrolytic machining of micro-pit array based on dry film mask. *Mech. Eng.* **2013**, *11*, 5–7.
23. Wang, Y.N. *Study on Electrochemical Wire Cutting System of Micro Ring Groove and Its Experiment*; Guangdong University of Technology: Guangdong, China, 2016.
24. Yuan, K.; Zhang, C.F.; Ai, H.H.; Cheng, P.Y. Study on micro-electrochemical milling technology of 304 stainless steel micro-ring groove. *Intern. Combust. Engine Parts* **2018**, *44*, 108–109.
25. Lv, Y.M.; Zhao, J.S.; Fan, Y.T.; Liu, D.; Yang, Z. Cathodic structure design of electrochemical machining with large length-width ratio and deep narrow slot. *Aeronaut. Manuf. Technol.* **2018**, *61*, 46–53.
26. Qu, N.S.; Chen, X.L.; Li, H.S.; Zhu, D. Fabrication of PDMS micro through-holes for electrochemical micromachining. *Int. J. Adv. Manuf. Technol.* **2014**, *72*, 487–494. [CrossRef]
27. Song, M. *Study on Electrochemical Machining of Micropits with Active Mask*; Nanjing University of Aeronautics and Astronautics: Nanjing, China, 2010.
28. Cai, W.W. *Study on Electrochemical Machining of Micropits with Active Mask*; Nanjing University of Aeronautics and Astronautics: Nanjing, China, 2015.
29. Wang, F.; Zhao, J.S.; Lv, Y.M.; Yang, Z.; He, Y.; Tian, Z. Experimental research on improving accuracy of electrochemical machining of deep narrow grooves. *Int. J. Adv. Manuf. Technol.* **2018**, *96*, 3217–3225.

30. Chen, S.G.; Yu, Z.Q.; Liu, J.W.; Guo,, Z.N. Simulation analysis and experimental study of hole in mask electrochemical machining. *Mach. Des. Manuf.* **2018**, *15*, 155–158.
31. Li, X.L.; Liu, S.H.; Xiao, H.P. Multi-field coupling simulation analysis for machining process of turbine drill blade. *Oil Field Equip.* **2019**, *48*, 1–6.
32. Wang, Y.; Fu, X.Q.; Wang, Q.Q.; Zhen, Z. Experimental study on electrochemical machining of pit array mask. *Mech. Sci. Technol. Aerosp. Eng.* **2018**, *37*, 896–902.
33. Zhou, X.C.; Cao, C.Y. Simulation of electrochemical machining temperature field based on COMSOL. *J. Qiqihar Univ. Sci. Ed.* **2018**, *34*, 41–44.
34. Liu, G.Q.; Liu, L.; He, C.K.; Han, X. Simulation and experimental study on the coupling of multiple physical fields in electrochemical machining. *J. Mach. Des.* **2018**, *35*, 29–35.
35. Wang, D.Y.; Zhu, Z.W.; Wang, N.F. Investigation of the electrochemical dissolution behavior of Inconel 718 and 304 stainless steel at low current density in $NaNO_3$ solution. *Electrochim. Acta* **2015**, *2*, 301–307. [CrossRef]

coatings

Article

Numerical Simulation of Crevice Corrosion of Stainless Steel–Titanium in NaCl Solution

Jiawei Ding [1,2], Weiping He [3], Yuanhai Liu [3], Chenyu Zhang [3], Haitao Wang [1,4,*] and En-Hou Han [1,4,*]

1 CAS Key Laboratory of Nuclear Materials and Safety Assessment, Institute of Metal Research, Chinese Academy of Sciences, Shenyang 110016, China; jwding17@mail.ustc.edu.cn
2 School of Materials Science and Engineering, University of Science and Technology of China, Hefei 230026, China
3 Aviation Key Laboratory of Science and Technology on Structure Corrosion Prevention and Control, China Special Vehicle Research Institute, Jingmen 448035, China; hwp605@163.com (W.H.); liuyuanhai100@163.com (Y.L.); zhangchenyu@163.com (C.Z.)
4 Institute of Corrosion Science and Technology, Guangzhou 510530, China
* Correspondence: htwang@imr.ac.cn (H.W.); ehhan@imr.ac.cn (E.-H.H.); Tel.: +86-24-2391-5897 (H.W.); +86-24-2389-3841 (E.-H.H.); Fax: +86-24-23894149 (H.W. & E.-H.H.)

Abstract: A multiphysics model based on the finite element method was adopted, emphasizing a deeper insight into the rarely studied crevice corrosion behavior of stainless steel and titanium overlapping. The model takes into account damage due to corrosion inside the crevice, different species transportation, local electrochemical reactions, homogeneous reactions in the electrolyte, and formation of a corrosion product and its influence on electrochemical reaction. The simulation results show that the location of the greatest attack for stainless steel is at the crevice opening; this finding is consistent with the IR drop theory. The potential increases gradually from the tip to the opening of the crevice, and the current changes smoothly following a sharp rise at the opening. The minimum and maximum values of pH and Cl^- concentration are both in the middle and opening of the crevice. The influence of the crevice size on corrosion is also discussed in detail.

Keywords: stainless steel; titanium; crevice corrosion; numerical simulation; multiphysics

Citation: Ding, J.; He, W.; Liu, Y.; Zhang, C.; Wang, H.; Han, E.-H. Numerical Simulation of Crevice Corrosion of Stainless Steel–Titanium in NaCl Solution. *Coatings* **2022**, *12*, 592. https://doi.org/10.3390/coatings12050592

Academic Editor: Stefanos M. Skolianos

Received: 18 March 2022
Accepted: 23 April 2022
Published: 26 April 2022

1. Introduction

With the rapid development of the social economy, the requirements of industrial production on material performance are getting higher and higher, and the types of applied materials are becoming more and more extensive, especially under the condition of meeting performance, trying to reduce costs; thus, the joint use of various materials has become a new trend. For components used in combinations of multiple metals, crevices are easily formed between the metals. When the electrolyte solution enters the crevice, crevice corrosion will occur. For example, thermal spraying prepares a layer of titanium metal coating on the stainless-steel surface to improve the stainless-steel corrosion resistance in the marine environment, but the oxides in the coating will increase the brittleness of the coating. Therefore, once the titanium coating breaks, the coating forms a crevice with the stainless-steel substrate, and the stainless steel will be subjected to severe crevice corrosion, thus causing severe damage.

1.1. Crevice Corrosion Mechanism

Crevice corrosion can be regarded as a combination result of change in crevice solution composition and IR drop [1]. In the initial stage, the metal in contact with the electrolyte dissolves uniformly at the same rate both inside and outside the crevice, but as time goes on (also related to the size of the crevice itself), the oxygen inside the crevice will be gradually consumed, and the amount of dissolved metal ions gradually increases; thus, the pH inside the crevice gradually decreases, and the stability of the passivation film is destroyed. When the oxygen consumption rate is greater than the diffusion replenishment rate, the oxygen within the crevice will be exhausted, and the cathode reaction will be suppressed, but the cathode reaction on the external surface will maintain the anode dissolution reaction in the crevice to continue. At this time, the current flowing through the solution with a resistance value of R is I, and the resulting voltage drop is called the IR drop. A potential value, $E(y)$, inside the crevice wall was produced at position y, and due to this, the IR drop is lower than the potential $E(s)$, which is situated at the crevice opening ($y = 0$). The equation of IR drop can be written as follows [2]:

$$IR_l = E(s) - E(y)$$
$$\nabla\Phi^* = E(s) - E_{pass}$$

where $R_l = \sigma y / A$, σ is the conductivity in the solution, $\nabla\Phi^*$ is critical potential for crevice corrosion, E_{pass} is the transition electrode potential of the active/passive on the crevice wall, and Y_{pass} is the length from the outer surface to E_{pass}. For systems containing metal and electrolyte, when $IR > \nabla\Phi^*$, crevice corrosion occurs.

1.2. Research Progress of Crevice Corrosion Simulation

The detailed information on the reaction kinetics and water chemistry during the crevice corrosion process is extremely important for understanding mechanism and predicting occurrence of crevice corrosion. In this regard, in recent decades, researchers in many fields, including academia and industry, have conducted a large number of scientific experiments by different methods, such as in situ measurement of electrochemical and water chemistry parameters [3], frozen micro-zone solution [4], and acoustic emission technology [5]. However, due to the particularity of the geometry of the crevice, when studying the crevice corrosion through experimental methods, the external small interference signals may have a great influence on the measurement results of the precision test instruments used during the operation process. Therefore, characterizing the crevice corrosion process by experimental means alone cannot obtain very satisfactory information. The computational simulation method can simulate complex physical–chemical processes, visually analyze the results of specific physical or engineering problems, provide us with extremely rich information, and make up for the shortcomings of experiments, so it has attracted more and more attention from researchers [6,7].

For crevice corrosion simulation of stainless steel, Sharland [8] established a model of local corrosion of stainless steel, taking into account the mass transport and homogeneous reaction of ions in solution. The simulation found that the pH value in the crevice decreased as the anode dissolution current increased, and the pH value in the crevice declined as the Cr content rose in the metal. White et al. [9] established a model to study the chemical and electrochemical changes of the solution in the crevice during stainless-steel crevice corrosion. The model assumed that the stainless-steel crevice was in a passivated and activated state, respectively, and compared the simulation results with the experiment. It was found that, when the crevice was in the activated state, the results were consistent with the experimental results. This indicated that, in solutions that are nearly neutral, the typical passivation current density of stainless steel was too low to induce local chemical or potential changes in the crevice. Stroe et al. [10] established a crevice corrosion model of stainless steel in chloride media to investigate the effects of parameters such as pH, Cl^- concentration in bulk solutions, and crevice size on crevice corrosion. The results showed that the pH and crevice size would have a great impact on stainless-steel crevice corrosion, but the Cl^- concentration in the bulk solution had little effect on corrosion. Sun et al. [11] simulated the crevice corrosion of 304 stainless steel. The model assumed that the crevice of 304 stainless steel was activated and considered the effect of precipitation on the corrosion. The simulation results obtained the distribution of the pH value inside the crevice and the effect of the diverse reaction constant of the corrosion product deposition and different current density on the crevice corrosion.

For a crevice corrosion simulation of titanium, Heppner et al. studied the crevice corrosion behavior of titanium immersed in NaCl solution. The model assumed that the crevice was in a passive state, and the fraction of the cathode current provided by the hydrogen ion reduction reaction changed from 0 to 0.8 in increments of 0.2. The pH distribution, conductivity, and IR drop distribution curve in the crevice were obtained [12]. Subsequently, Heppner et al. predicted the effect of the crevice opening size on the IR drop and solution in the titanium crevice. The results showed that reducing the size of the crevice opening would increase the potential inside the crevice, increase the conductivity of the solution, and increase the concentration of the solution metal composite [13].

The crevice corrosion simulations of single stainless steel and single titanium have been performed by many researchers; however, to the best of our knowledge, there has not yet been any report on the crevice corrosion simulation between the dissimilar metals of stainless steel and titanium. The main aim of this research was to study the crevice corrosion growth kinetics and solution water chemistry of 304 stainless steel and titanium overlapping based on a multiphysics model, thereby providing a lot of useful information that is difficult to obtain experimentally, then further supplementing the study of crevice corrosion mechanism, and making corresponding preventive and control measures to reduce crevice corrosion.

2. Crevice Model

2.1. Description of the Problem and Assumption

Assuming that the longitudinal section of the crevice formed by stainless steel and titanium is rectangular, whose opening size (w) is 90 μm, and the height (r) is 8 cm. The schematic diagram of the crevice geometry can be found in Reference [14]. The metal in left side is 304 stainless steel, the right is Ti, and the solution is 0.6 mol/L NaCl with pH = 7.

The present model has the following main assumptions:

(1) For 304 stainless steel, the simulation proceeds directly from the activated state because of the low pH value inside the crevice [15].

(2) For titanium, the model assumes that Ti is always in the passivated state during the simulation. It is reported that, even under acidic conditions, when the potential value is above −0.4 V, titanium can be in the passivated state [16].

(3) It is supposed that no oxygen reduction reaction occurs inside the crevice, and it only takes place on the external surface of crevice because the stainless steel is in an activated state, which can quickly deplete the oxygen within the crevice [17].

(4) Because of the large aspect ratio of the crevice, convection is neglected in this model.

In the model, we considered electrochemical reactions, the chemical species transmission, the deformation of metal corrosion, and the precipitation that occurs during the corrosion process. All the equations needed were solved numerically by COMSOL software.

2.2. Transport Equations

Nernst–Planck equations, which include diffusion and electromigration, are the governing equations for different species in the solution [18]. The mass conservation and the concentration change formula containing Nernst–Planck equations for each species, *i*, in the solution can be depicted as follows:

$$\frac{\partial c_i}{\partial t} + \nabla \cdot \left(-D_i \nabla c_i - \frac{z_i D_i F c_i \nabla \varphi_l}{RT} \right) = R_{i,tot} \quad (1)$$

where c_i is the concentration of species i (mol/m^3), t is time (s), D_i is the diffusion coefficient (m^2/s), z is the number of charges, F is the Faraday constant (C/mol), φ_l is the electrolyte potential (V), R is the gas constant (J/(mol·K)), T is the temperature (K), and $R_{i,tot}$ is the source (mol/(m^3·s)). The conservation equation of mass gives the expression of the concentration of each substance, but there is another unknown quantity, the electrolyte potential, φ_l; thus, it is necessary to solve the unknown quantity in conjunction with the electrical neutral equation:

$$\sum z_i c_i = 0 \tag{2}$$

The electronic conduction in the metal and the directional migration of ions in the solution form a complete current circuit, where electrolyte current density in the solution involved the Faraday constant, and the charge number can be described as follows:

$$i_l = -F\sum D_i z_i \nabla c_i - \frac{F^2}{RT}\nabla\varphi_l \sum z_i^2 D_i c_i \tag{3}$$

2.3. Electrical and Chemical Reactions

The electrochemical reactions contained in the model include the following:

$$\mathrm{Fe} \rightleftharpoons \mathrm{Fe}^{2+} + 2\mathrm{e} \tag{4}$$

$$\mathrm{Cr} \rightleftharpoons \mathrm{Cr}^{3+} + 3\mathrm{e} \tag{5}$$

$$\mathrm{Ni} \rightleftharpoons \mathrm{Ni}^{2+} + 2\mathrm{e} \tag{6}$$

$$\mathrm{Ti} \rightleftharpoons \mathrm{Ti}^{4+} + 4\mathrm{e} \tag{7}$$

Because stainless steel is composed of three different metal elements, the molar mass M_j (kg/mol) and number of charges transferred, z_j, of different metals are diverse; thus, when electrochemical reactions occur, it is necessary to normalize the metal composition, and determine the contribution of the composition coefficient to anodic current. Therefore, f_i is introduced [19]:

$$f_i = \frac{z_i M_i}{\sum_j z_j M_j} \tag{8}$$

Thus, the total flux of metal anodic dissolution can be expressed by following equation:

$$N_i = \sum_m f_i \cdot \frac{v_{i,\,m} i_m}{n_m F} \tag{9}$$

where m is the electrode reactions, n_m represents the electrons transferred number, $\nu_{i,m}$ is reaction coefficient, and i_m represents the localized current density of the electrochemical reaction.

The electrochemical parameters, including the equilibrium potential; exchange current density; and Tafel slope of Fe, Cr, Ni, and O_2 redox reactions used in the simulation can be found in Reference [20]. The homogeneous reactions with equilibrium constants, including Fe ion and Fe oxide, Cr ion and Cr oxide, Ni ion and Ni oxide, and Ti ion and Ti oxide, can be obtained in References [18,21].

2.4. Moving Mesh Method

The moving mesh technology refers to the finite element method in which the divided mesh moves according to the result of the previous calculation or an artificially specified speed during the calculation. The moving mesh method can adjust the density and shape of the mesh according to the change of the solution of the physical field. This method can not only accurately distinguish the material interface but also maintain the geometric characteristics of the mesh, so it has significant advantages in solving the deformation problem of the mesh [22]. The stainless-steel geometry changes when it is actively dissolved, and this change is simulated and solved by the moving mesh technology. The mesh displacement is described by the formula below:

$$\frac{\partial^2}{\partial X^2}\frac{\partial x}{\partial t} + \frac{\partial^2}{\partial Y^2}\frac{\partial y}{\partial t} = 0 \tag{10}$$

$$\frac{\partial^2}{\partial X^2}\frac{\partial y}{\partial t}+\frac{\partial^2}{\partial Y^2}\frac{\partial x}{\partial t}=0 \tag{11}$$

where X, Y, x, and y are the coordinates in material framework and space framework, respectively. Assuming that the metal is dissolved in the normal direction of the metal crevice wall, the normal movement speed of the crevice wall can be expressed as follows:

$$\frac{\partial x}{\partial t}\cdot n=v_{tot} \tag{12}$$

where v_{tot} represents the total movement rate, and it can be written as the sum of the anodic reactions rates of all metals according to Equation (12):

$$v_{tot}=\sum_i\frac{M_i}{\rho_i}\sum_m f_i\cdot\frac{v_{i,m}i_{loc,m}}{n_mF} \tag{13}$$

where ρ represents the metals' density (kg/m^3).

2.5. Level Set Method

The level set method is a method that is suitable for dealing with the discontinuous changes in the physical and chemical properties between the new and original phases when new phases are generated. In the process of stainless-steel crevice corrosion, due to the small crevice size and the high concentration of substances inside the crevice, it is easy to produce precipitation. The resulting precipitate will affect the electrode dynamics during crevice corrosion, and the diffusion of ions in the solution and the precipitate will also change discontinuously. Therefore, the effect of simulated precipitation on crevice corrosion is realized by the level set method. The level set equation can be written as follows:

$$\frac{\partial\varphi}{\partial t}+u\cdot\nabla\varphi=\gamma\nabla\cdot\left(\varepsilon\nabla\varphi-\varphi(1-\varphi)\frac{\nabla\varphi}{|\nabla\varphi|}\right) \tag{14}$$

The terms on the left-hand side give the correct motion of the interface, while those on the right-hand side are necessary for numerical stability. Where the interface moves with the velocity u (m/s), φ stands for the different phases in the system, the parameter γ (m/s) represents the amount of reinitialization or stabilization of the level set function, and ε (m) is the thickness of the transition zone. Because there are no reports on the precipitation kinetic parameters of stainless steel in the literature during corrosion in experiment, we assume the influence of the formation of precipitation as shown below:

$$\frac{1}{\varepsilon_d}\times\frac{\partial\varepsilon_d}{\partial t}=-c_{Fe^{2+}}\cdot c_{Cl^-}\cdot c_{O_2}\cdot k \tag{15}$$

where parameter ε_d represents precipitation porosity and is assumed to be 1, which indicates that no precipitation occurs; and $k = 10^{-11}$ m^{-2}.

Due to the production of precipitation, it will also affect the mass flux [23]; the influence can be expressed by the following equation:

$$n\cdot\left(N_{total}-N_{eff}\right)_i=n\cdot N_{i,red}=\left(D_i-D_{eff}\right)\left(-\nabla c_i-\frac{z_iF}{RT}c_i\nabla\varphi_l\right) \tag{16}$$

where N_{total} is the flux without precipitation, N_{eff} describes the effective flux, $N_{i,red}$ represents the reduced flux, and D_{eff} is the effective diffusion coefficient after precipitation emerging.

3. Results and Discussion

3.1. Analysis of Crevice Corrosion Kinetics

Figure 1 exhibits the potential and current distributions predicted vs. distance from the crevice tip inside the crevice in this model at t = 90 h. The crevice potential gradually becomes more negative from −0.21 V in the opening toward the direction of the crevice tip, −0.34 V, for the reason that the oxygen concentration decreases from the crevice mouth into the crevice. In DeForce's experiment, the researchers investigated crevice corrosion of stainless steel in a NaCl solution; they found potential distribution along the crevice within −0.34 to −0.15 V, though their crevice length was 1.2 cm, with opening size of 125 μm, and the experiment time was 1 h [2]. A similar potential range of −0.4 to −0.2 V was given by Onishi's stainless-steel model, with a length of 2 cm and opening size of 25 μm [24]. According to the simulation results of the 304–304 stainless-steel model, compared with Alavi's experiment [15] in the early stage, under acidic conditions, when the stainless-steel potential is above −0.35 V, the stainless steel is already in an activated state. This potential value is consistent with this model, so it can be considered that the stainless steel is in an activated state. Because stainless steel is in direct contact with Ti, the potential distribution on the Ti surface is the same as that of stainless steel, with a minimum potential of −0.34 V, which is above −0.4 V during the simulation, so Ti can remain in a passivated state, which is consistent with the assumption. It is worth noting that, for all spontaneously passive or active metals in which crevice corrosion does not occur immediately, the commencement of crevice corrosion through the IR mechanism can only emerge after an induction period, in which enough corrosion environment is formed inside the crevice to gradually destroy the stability of the passivation film, and the active peak in polarization curve will gradually form. As the active peak increases to a critical size at the end of the induction period, due to the solution continuing to acidify inside the crevice, the IR drop will exceed the critical potential, $\nabla\Phi^*$; thus, crevice corrosion starts [25].

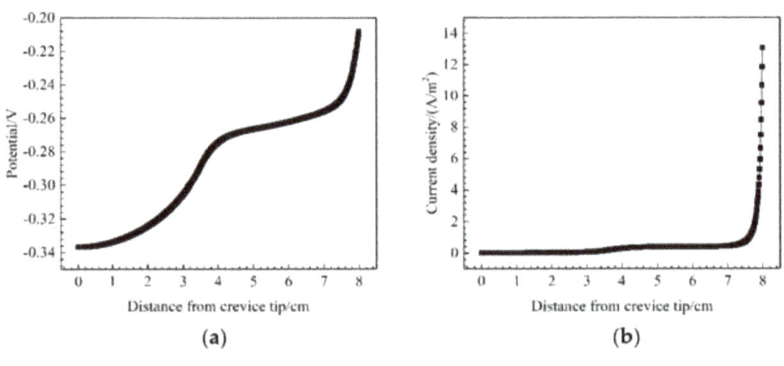

Figure 1. Predicted (**a**) potential and (**b**) current distributions vs. distance from crevice tip for 304 stainless steel at t = 90 h.

The current inside the crevice changes smoothly, but it increases rapidly at the crevice opening, and the maximum value is 13 A/m^2. The distribution and magnitude for current are consistent with the results of Wang's simulation [20]. Under the activation condition, the current density inside the crevice is affected by both the potential and the precipitation porosity. Figure 2 exhibits the precipitation porosity distribution vs. distance from crevice tip at t = 90 h. A porosity value of 1 indicates that the precipitate is loose and porous and will not affect the corrosion process. A porosity value of 0 indicates that there is no pore in the precipitate, and this will hinder the corrosion process. As shown in Figure 2, from the crevice tip to the crevice opening, the precipitation porosity first decreases and then increases, and there is an inflection point near the crevice opening with the minimum value 0.44. This change in porosity is related to the type of precipitated crystals that contain low-porosity crystalline precipitates that are easy to form when the precipitation growth rate is higher, and high-porosity amorphous precipitates emerge when the precipitation formation rate is higher [26]. The ions concentration near the crevice opening is high, and the precipitation formation rate is quick, leading to the formation of amorphous precipitates with high porosity near the crevice opening; thus, the porosity increases correspondingly.

Only at the crevice opening do the potential and porosity values increase at the same time, so the current at the opening rises rapidly.

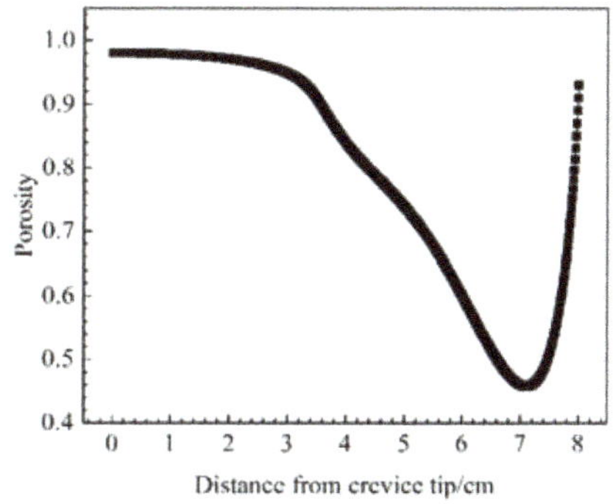

Figure 2. Predicted precipitation porosity distribution vs. distance from crevice tip at t = 90 h.

Figure 3 shows the propagation of crevice wall with time traced by moving mesh at t = 0 and t = 90 h. The legend on the right in the figure is the situation of mesh deformation. Green means less deformation, and red represents more deformation. It can be observed that 304 stainless steel undergoes activated dissolution to cause corrosion deformation. The corrosion depth inside the crevice increases with time, and the corrosion depth at the crevice mouth is the largest; the closer to the tip of the crevice, the lower the depth of corrosion is. This change in corrosion morphology corresponds to the current distribution in Figure 1. Titanium is always in a passive state, so that no corrosion deformation occurs. The crevice corrosion morphology in experiments has been reported in many works in the literature. Al-Khamis et al. [1] studied the crevice corrosion of stainless steel in an acidic Cl^- containing environment and found that the corrosion morphology is similar to the shape of the polarization curve with passivation–activation transition. The area near the crevice opening is the most severely corroded, and the closer to the crevice tip, the less severe the corrosion is. In addition, for passivable metals such as nickel, similar morphologies are also found when measuring the crevice-corrosion dimension changes in the experiment [27]. This corrosion morphology is a typical crevice corrosion morphology for passivable metals. The corroded crevice shape and the peak position of the corrosion depth predicted by this model qualitatively agree with the published experimental studies mentioned above; meanwhile, this morphology is rarely seen in other crevice corrosion simulations.

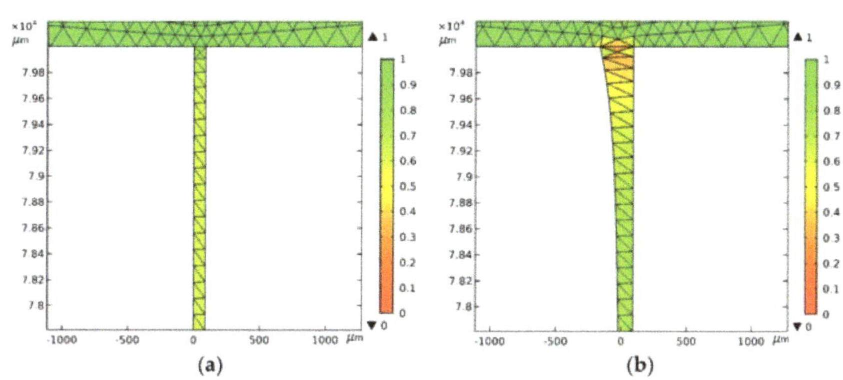

Figure 3. Propagation of crevice wall with time traced by moving mesh at (**a**) t = 0 and (**b**) t = 90 h for 304 stainless steel–Ti. The left side is 304 stainless steel, and the right is Ti. The scale on the right is the state of mesh deformation.

3.2. Crevice Solution Chemistry Evaluation and Ions Release

Figure 4 is the curve of pH change, with time at different positions, together with the pH distribution within the crevice at t = 90 h. For the crevice opening, the pH descends rapidly during the initial 5 h and then slowly reduces to 2.3. For the crevice middle, the pH basically stabilizes to approximately 1.93 after the initial 7 h; nevertheless, it is smoothly reduced to 2.4 after 30 h at the crevice tip. The reasons for the change of pH distribution shape are that, on the one hand, H^+ comes from the hydrolysis of Cr^{3+}, Fe^{2+}, and Ni^{2+} produced by the activation dissolution of 304 stainless steel; on the one hand, it comes from the hydrolysis of Ti^{4+} dissolved under the effect of a passivation current. However, because Ti is uniformly corroded, the pH distribution is mainly affected by the dissolution of stainless steel. For 304 stainless steel, it corrodes quickly at the crevice opening, leading to more H^+ produced; at the same time, OH^- generated due to the oxygen reduction reaction will make the pH of the crevice opening increase. Ultimately, the joint action of the both makes the pH of the crevice opening higher than it is for other parts. Regardless, the amount of metal ions generated at the tip of crevice and the middle of crevice is approximately the same (it can be known from the fact that the current value is similar inside the crevice in Figure 1). However, the highest concentration of metal ions emerges in the middle of the crevice in the process of simulation, and a comparable situation was reported in Reference [28], which manifests that the metal ions produced at the crevice opening will spread into the crevice middle to decrease the pH as a result of the concentration gradients.

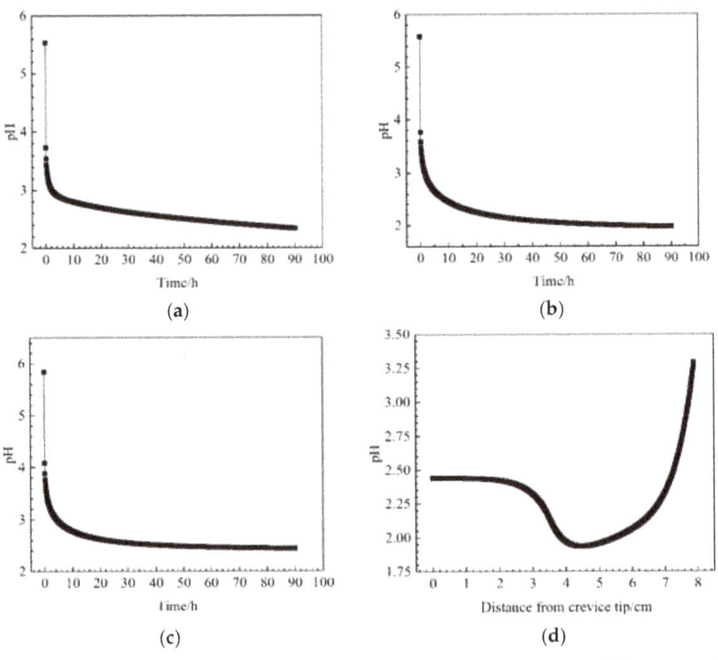

Figure 4. Predicted pH profile change with time at (**a**) opening, (**b**) middle, (**c**) tip of the crevice, and (**d**) pH distribution vs. distance from crevice tip at t = 90 h.

The Cl^- time-varying concentration and distribution within the crevice at t = 90 h are shown in Figure 5. As time goes on, the concentration of Cl^- first increases to 1.08 mol/L and then decreases to 0.82 mol/L at the opening of the crevice; at the middle of the crevice, Cl^- first descends, then changes smoothly, and finally reaches 0.13 mol/L; and at the crevice tip, Cl^- rises initially and reaches 0.8 mol/L at last. The distribution trend of Cl^- shows that it decreases from the tip to the middle of the crevice, and then it increases toward the crevice opening. Sharland simulated the change of Cl^- concentration in the crevice with time of the stainless-steel crevice under the current density of 10 A/m^2. The results showed that, when the time is t = 90 h, the Cl^- concentration in the crevice was about 1.3 mol/L [8]. The current density in the crevice opening is 13 A/m^2, and the corresponding Cl^- concentration is 1.6 mol/L in our model; these results are very consistent with the experiment results. Moreover, the Cl^- concentration obtained in this model is similar with Zuo [29] and Heppner's [18] results in the range of 1.4–3 mol/L. The Cl^- ultimate concentration distribution is determined by the difference between the increasing concentration, due to maintaining the electrical neutrality inside the crevice, and the consumption, because of the precipitation formation. The crevice opening is close to the bulk solution, and the metal dissolution rate is the largest, so the Cl^- concentration at the opening is the highest. In the middle of the crevice, the concentration valley of Cl^- (the lowest concentration) appears. This phenomenon is related to the diffusion of metal ions at the opening into the crevice to form a precipitate, leading to heavy Cl^- consumption.

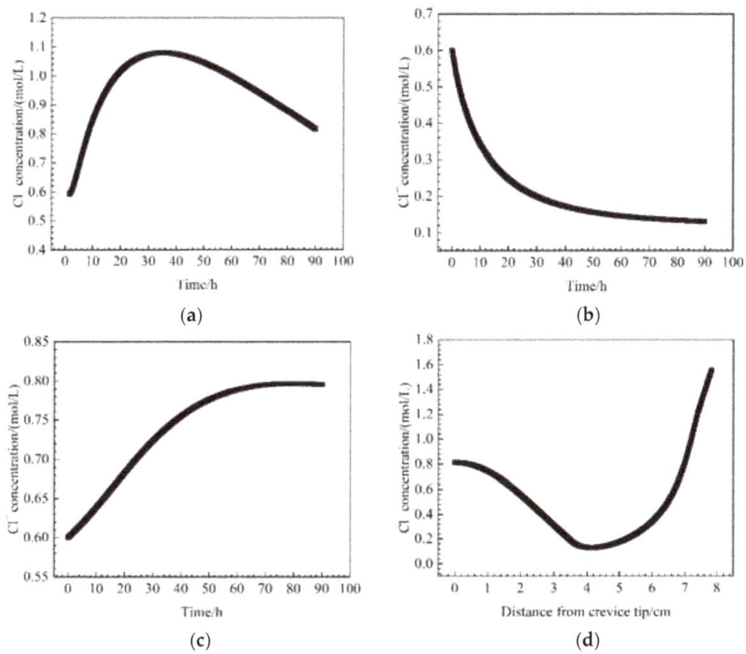

Figure 5. Predicted Cl^- concentration profile change with time at (**a**) opening, (**b**) middle, (**c**) tip of the crevice, and (**d**) Cl^- concentration distribution vs. distance from crevice tip at t = 90 h.

3.3. The Influence of Crevice Geometry

One of the crucial objectives in investigating the electrochemical status inside the crevice is to forecast the metal dissolution rate, which is importantly influenced by the crevice size. In order to investigate the effect of the crevice size on corrosion, parametric crevice geometry factor $\log_{10}(r^2/w)$ is introduced, where r represents the crevice depth, and w describes the crevice opening size. For a narrower crevice, the value of $\log_{10}(r^2/w)$ becomes large, while for a wider crevice, the $\log_{10}(r^2/w)$ becomes a small value. The results aforementioned match $\log_{10}(r^2/w)$ = 8. We add $\log_{10}(r^2/w)$ = 5, 6, 7 to compare simulation results mutually in this part. Figure 6 exhibits the potential and current distributions along the crevice with different $\log_{10}(r^2/w)$ values at t = 90 h. When the value of $\log_{10}(r^2/w)$ decreases from 8 to 5, the potential at the crevice tip rises from −0.34 to −0.26 V, and the potential at the opening is between −0.23 and −0.21 V; as the crevice geometry factor decreases, the potential distribution becomes gentler. However, for the current density, when the $\log_{10}(r^2/w)$ reduces from 8 to 5, the maximum current density decreases quickly from 13 to 5.6 A/m^2 at the crevice opening. This change trend is due to the reduction of the $\log_{10}(r^2/w)$, making the diffusion process more and more dominant within the crevice, and diffusion can eliminate the unevenness of the material inside the crevice in time.

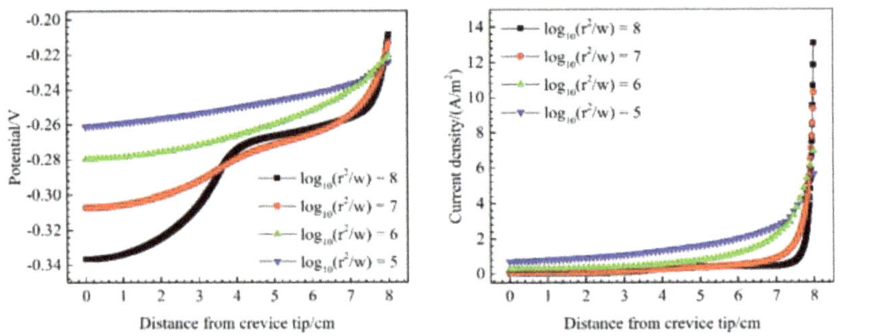

(a) (b)

Figure 6. Predicted (**a**) potential and (**b**) current distributions vs. distance from crevice tip under different $\log_{10}(r^2/w)$ at t = 90 h.

Figure 7 exhibits the pH and Cl^- distributions vs. distance from crevice tip inside the crevice in this model with different $\log_{10}(r^2/w)$ values at t = 90 h. As the $\log_{10}(r^2/w)$ decreases, the pH value increases from 3.3 to 3.7 at the crevice opening. When the $\log_{10}(r^2/w)$ = 8 and 7, the minimum pH value appears near the crevice middle, 1.93 and 2.9, respectively. Moreover, when the $\log_{10}(r^2/w)$ = 6 and 5, the pH minimum appears at the tip of the crevice, as 3.4 and 3.6. The pH becomes gentle as the geometry factor decreases. The pH rises as the $\log_{10}(r^2/w)$ increases, and this behavior is consistent with Stroe's simulation results [10]. The reason why the pH becomes flat is related to the enhanced diffusion caused by the expansion of the crevice size. Nevertheless, the Cl^- concentration distribution inside the crevice becomes convoluted as a result of the rivalry emerging between precipitation and diffusion for consumption and supplement. When the $\log_{10}(r^2/w)$ drops from 8 to 7, the current density at the crevice opening decreases, the metal dissolution rate declines, and the amount of metal ions generated is reduced. Therefore, metal ions diffused from the opening to the inside of the crevice consume Cl^- to form precipitation that drops off, so the Cl^- concentration rises for $\log_{10}(r^2/w)$ = 7. Nonetheless, when the crevice size is further increased and the $\log_{10}(r^2/w)$ continues to decrease, the diffusion effect is further enhanced, and Cl^- is not easy to accumulate in the crevice, so the Cl^- concentration will become closer to the bulk solution; this is why the Cl^- concentration of $\log_{10}(r^2/w)$ = 6 is higher than $\log_{10}(r^2/w)$ = 5.

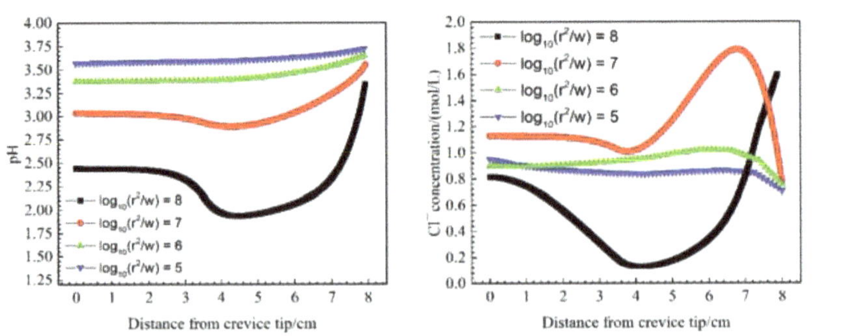

(a) (b)

Figure 7. Predicted (**a**) pH and (**b**) Cl^- concentration distributions vs. distance from crevice tip under different $\log_{10}(r^2/w)$ at t = 90 h.

4. Conclusions

The present study was designed to determine the circumstances of crevice corrosion of dissimilar metal stainless steel–titanium overlapping. The model considers the effect of multi-field coupling: electric field, mass transfer, moving mesh simulating corrosion deformation, and level set for precipitation generation. From the predicted results, the following is concluded:

(1) The results show that, for stainless steel in the process of crevice corrosion, it corrodes most severely at the opening of the crevice; the closer to the tip of the crevice, the less corrosion deformation. However, titanium did not corrode, because it is in a passive state.
(2) The potential increases gradually from the tip to the opening of the crevice and distributes between −0.34 and −0.21 V. The current inside the crevice changes smoothly, but it rises rapidly to 13 A/m^2 at the crevice opening.
(3) The pH value in the crevice decreases from 2.4 at the crevice tip, reaches a minimum value of 1.93 near the middle, and then rises to a maximum value of 3.3 at the crevice opening. The Cl^- concentration is 0.8 mol/L at the tip of the crevice and drops to a minimum value of 0.13 mol/L in the middle, reaching the maximum value of 1.6 mol/L at the crevice opening.
(4) The effect of the crevice size on corrosion is as follows: when the value of $\log_{10}(r^2/w)$ is reduced from 8 to 5, the potential at the tip of the crevice rises from −0.34 to −0.26 V, and the potential at the crevice opening is in between −0.23 to −0.21 V. As the $\log_{10}(r^2/w)$ decreases, the pH value increases from 3.3 to 3.7 at the crevice opening. When the $\log_{10}(r^2/w)$ = 8 and 7, the minimum pH value appears near the middle, which is 1.93 and 2.9, and when the $\log_{10}(r^2/w)$ = 6 and 5, the pH minimum appears at the crevice tip, which is 3.4 and 3.6. As the $\log_{10}(r^2/w)$ decreases, the overall Cl^- concentration appears to increase first and then decrease.

Author Contributions: Conceptualization, H.W. and E.-H.H.; methodology, J.D. and W.H.; validation, J.D.; investigation, J.D., Y.L. and C.Z.; writing—original draft preparation, J.D.; writing—review and editing, H.W.; supervision, H.W. and E.-H.H.; project administration, W.H., Y.L. and C.Z. All authors have read and agreed to the published version of the manuscript.

Funding: This work was supported by the National Key Research and Development Program of China (Grant No. 2021YFB3702200), the Aviation Science Foundation of China (No. 20200038092002), and the LingChuang Research Project of China National Nuclear Corporation.

Institutional Review Board Statement: Not applicable.

Informed Consent Statement: Not applicable.

Data Availability Statement: Data sharing is not applicable to this article.

Conflicts of Interest: The authors declare no conflict of interest.

References

1. Al-Khamis, J.N.; Pickering, H.W. IR Mechanism of Crevice Corrosion for Alloy T-2205 Duplex Stainless Steel in Acidic-Chloride Media. *J. Electrochem. Soc.* **2001**, *148*, B314–B321. [CrossRef]
2. DeForce, B.S. Revisiting the Crevice Corrosion of Stainless Steel and Aluminum in Chloride Solutions—The Role of Electrode Potential. Ph.D. Thesis, The Pennsylvania State University, State College, PA, USA, 2010.
3. Lee, Y.H.; Takehara, Z.; Yoshizawa, S. The enrichment of hydrogen and chloride ions in the crevice corrosion of steels. *Corros. Sci.* **1981**, *21*, 391–397. [CrossRef]
4. Brown, B.F.; Fujii, C.T.; Dahlberg, E.P. Methods for Studying the Solution Chemistry Within Stress Corrosion Cracks. *J. Electrochem. Soc.* **1969**, *116*, 218–219. [CrossRef]
5. Klassen, R.D.; Roberge, P.R.; Hyatt, C.V. A novel approach to characterizing localized corrosion within a crevice. *Electrochim. Acta* **2001**, *46*, 3705–3713. [CrossRef]
6. Radovanović, M.B.; Tasić, Ž.Z.; Mihajlović, M.B.P.; Simonović, A.T.; Antonijević, M.M. Electrochemical and DFT studies of brass corrosion inhibition in 3% NaCl in the presence of environmentally friendly compounds. *Sci. Rep.* **2019**, *9*, 16081. [CrossRef] [PubMed]

7. Ilawe, N.V.; Zimmerman, J.A.; Wong, B.M. Breaking Badly: DFT-D2 Gives Sizeable Errors for Tensile Strengths in Palladi-um-Hydride Solids. *J. Chem. Theory Comput.* **2015**, *11*, 5426–5435. [CrossRef]
8. Sharland, S.M. A mathematical model of the initiation of crevice corrosion in metals. *Corros. Sci.* **1992**, *33*, 183–201. [CrossRef]
9. White, S.P.; Weir, G.J.; Laycock, N. Calculating chemical concentrations during the initiation of crevice corrosion. *Corros. Sci.* **2000**, *42*, 605–629. [CrossRef]
10. Stroe, M.; Oltra, R.; Vuillemin, B. Modeling of crevice corrosion of Fe-Cr alloys in chloride media. *J. Electrochem. Soc.* **2008**, *11*, 12. [CrossRef]
11. Sun, W.; Wang, L.D.; Wu, T.T.; Liu, G.C. An arbitrary Lagrangian—Eulerian model for modelling the time-dependent evolution of crevice corrosion. *Corros. Sci.* **2014**, *78*, 233–243. [CrossRef]
12. Heppner, K.L.; Evitts, R.W. Modelling of the effect of hydrogen ion reduction on the crevice corrosion of titanium. *Environ.-Induc. Crack. Mater.* **2008**, *1*, 95–104.
13. Heppner, K.L.; Evitts, R.W.; Postlethwaite, J. Effect of the Crevice Gap on the Initiation of Crevice Corrosion in Passive Metals. *Corrosion* **2004**, *60*, 718–728. [CrossRef]
14. Ding, J.W.; Wang, H.T.; Han, E.-H. A multiphysics model for studying transient crevice corrosion of stainless steel. *J. Mater. Sci. Technol.* **2021**, *60*, 186–196. [CrossRef]
15. Alavi, A.; Cottis, R. The determination of pH, potential and chloride concentration in corroding crevices on 304 stainless steel and 7475 aluminium alloy. *Corros. Sci.* **1987**, *27*, 443–451. [CrossRef]
16. Thomas, N.T.; Nobe, K. The Electrochemical Behavior of Titanium. *J. Electrochem. Soc.* **1969**, *116*, 1748. [CrossRef]
17. Chen, D.X.; Han, E.-H.; Wu, X.Q. Effects of crevice geometry on corrosion behavior of 304 stainless steel during crevice corrosion in high temperature pure water. *Corros. Sci.* **2016**, *111*, 518–530. [CrossRef]
18. Heppner, K.L.; Evitts, R.W.; Postlethwaite, J. Prediction of the Crevice Corrosion Incubation Period of Passive Metals at Elevated Temperatures: Part II-Model Verification and Simulation. *Can. J. Chem. Eng.* **2002**, *80*, 857–864. [CrossRef]
19. Oldfield, J.W.; Sutton, W.H. Crevice corrosion of stainless steels: I. A mathematical model. *Br. Corros. J.* **2013**, *13*, 13–22. [CrossRef]
20. Wang, W.; Sun, H.Y.; Sun, L.J.; Song, Z.W.; Zang, B.N. Numerical simulation for crevice corrosion of 304 stainless steel in sodium chloride solution. *Chem. Res. Chin. Univ.* **2010**, *26*, 822–828.
21. Yaya, K.; Khelfaoui, Y.; Malki, B.; Kerkar, M. Numerical simulations study of the localized corrosion resistance of AISI 316L stainless steel and pure titanium in a simulated body fluid environment. *Corros. Sci.* **2011**, *53*, 3309–3314. [CrossRef]
22. Deshpande, K.B. Validated numerical modelling of galvanic corrosion for couples: Magnesium alloy (AE44)–mild steel and AE44–aluminium alloy (AA6063) in brine solution. *Corros. Sci.* **2010**, *52*, 3514–3522. [CrossRef]
23. Höche, D. Simulation of Corrosion Product Deposit Layer Growth on Bare Magnesium Galvanically Coupled to Aluminum. *J. Electrochem. Soc.* **2014**, *162*, C1. [CrossRef]
24. Onishi, Y.; Takiyasu, J.; Amaya, K.; Yakuwa, H.; Hayabusa, K. Numerical method for time-dependent localized corrosion analysis with moving boundaries by combining the finite volume method and voxel method. *Corros. Sci.* **2012**, *63*, 210–224. [CrossRef]
25. Abdullah, A.M. The Use of Microelectrodes in the Study of Localized Corrosion of Aluminum 6111-like Alloys. Ph.D. Thesis, The Pennsylvania State University, State College, PA, USA, 2002.
26. Rossmann, M.; Braeuer, A.; Dowy, S.; Gallinger, T.G.; Leipertz, A.; Schluecker, E. Solute solubility as criterion for the appearance of amorphous particle precipitation or crystallization in the supercritical antisolvent (SAS) process. *J. Supercrit. Fluids* **2012**, *66*, 350–358. [CrossRef]
27. Abdulsalam, M.I.; Pickering, H.W. Effect of the applied potential on the potential and current distributions within crevices in pure nickel. *Corros. Sci.* **1998**, *41*, 351–372. [CrossRef]
28. Walton, J.C.; Cragnolino, G.; Kalandros, S.K. A numerical model of crevice corrosion for passive and active metals. *Corros. Sci.* **1996**, *38*, 1–18. [CrossRef]
29. Zuo, J.Y.; Jin, Z.Q. Investigation of chemical and eletrochemical changes within corrosion cracks :I. Interrelations between currents, Cl-migration and pH. *Chin. J. Chem. Eng.* **1984**, *3*, 123–133.

coatings

MDPI

Article

Effect of Edible Coating on the Quality and Antioxidant Enzymatic Activity of Postharvest Sweet Cherry (*Prunus avium* L.) during Storage

Wenzhong Hu [1,*], Sarengaowa [1] and Ke Feng [2,3]

1 School of Pharmacy and Food Science, Zhuhai College of Science and Technology, Zhuhai 519041, China; kuailexiaosa@sina.com
2 LiveRNA Therapeutics Inc., Zhuhai 519041, China; fengkesky@163.com
3 College of Life Science and Technology, Huazhong Agricultural University, Wuhan 430070, China
* Correspondence: wenzhonghu666@sina.com; Tel.: +86-135-9116-2665

Abstract: The effects of edible chitosan coating (0.1%, 0.3%, 0.5% and 0.75% *w/v*) on the changes in the quality, respiration rate, total phenolic content and anthocyanin of postharvest sweet cherry (*Prunus avium* L.) at 10 °C were investigated. The activities of superoxide dismutase (SOD), catalase (CAT) and ascorbate peroxidase (APX) were also determined. The result showed that the treatments of chitosan edible coating were effective at delaying the evolution of the parameters related to postharvest ripening, such as color and firmness, and respiration rate. The edible coatings also showed that the lower total phenolics and total antioxidant activity were maintained compared to that in the control associated with the overripening. It was suggested that the optimal quality and enhanced antioxidant enzymatic activities of postharvest cherry fruits were obtained by an edible coating of chitosan 0.5% up to 24 days at 10 °C. The chitosan edible coating could be favorable for extending shelf-life, maintaining the quality of sweet cherries.

Keywords: sweet cherry; edible coating; quality; antioxidant enzymatic activities

Citation: Hu, W.; S.; Feng, K. Effect of Edible Coating on the Quality and Antioxidant Enzymatic Activity of Postharvest Sweet Cherry (*Prunus avium* L.) during Storage. *Coatings* **2022**, *12*, 581. https://doi.org/10.3390/coatings12050581

Academic Editor: JunSang Ham

Received: 15 March 2022
Accepted: 19 April 2022
Published: 24 April 2022

1. Introduction

Sweet cherries (*Prunus avium* L.) are more and more popular fruits due to their color, nutritional value and taste. Sweet cherries have been such an attractive fruit crop for consumption due to their precocity and excellent quality. Recent attention on the health benefits of cherries has further helped boost their consumption. Sweet cherries, in particular, have been found to offer a good source of phenolic compounds that boost the fruit antioxidant activity including fiber, carotenoids, vitamin C and anthocyanins [1].

Fruit size, bright red color, firmness and flavor are major postharvest sweet cherry quality attributes [2]. The quality of cherries deteriorates rapidly after harvest. The main cause is their relatively high metabolic activity, which leads to loss of weight, changes of chroma, surface pitting, stem browning, softening and loss of acidity [3,4]. The food industry is constantly searching for effective and safe means to control these problems. Of the various preservation techniques, sweet cherry stored at low temperatures is the most preferred by the consumers. The preservation method reduces water loss rate and inhibits pathogen growth and maintains the freshness of sweet cherry. However, biochemical reactions of the quality deterioration of sweet cherry is not inhibited completely at low temperatures. Moreover, temperature-sensitive fruits such as sweet cherries need to be stored at an appropriate temperature to prevent cold damage [4]. The application of fruit edible coatings is another new trend in sweet cherry preservation. Edible coatings including various polysaccharides, proteins, lipids composite coatings have certain advantages in maintaining quality attributes. The edible coating has been widely used in recent years to maintain quality of fruit [5]. It leads to reductions in respiration rate and transpiration and to slowing the ripening process by decreasing permeability to O_2, CO_2, and water vapor [6–8].

Chitosan has received a great deal of attention from the food industries. This derivative of chitin is a copolymer of N-acetylglucosamine and glucosamine residues linked by β-1,4-glycosidic bonds and is insoluble in dilute acids. Chitosan has been proven one of the best edible and biologically safe preservative coatings for different types of fruits and vegetables because of its film forming properties, antimicrobial actions, nontoxic, biodegradability and biocompatible properties [9]. Chitosan edible coatings can modify the internal atmosphere, decrease transpiration loss, and delay the ripening of fruits and vegetables due to its ability to form a semipermeable film that is durable and flexible. All these properties provide advantages to chitosan compared to other edible coatings [10]. Several studies have indicated that chitosan has the beneficial effects for food preservation in recent years [8–11]. A study showed chitosan treatment significantly inhibited the bacterial growth on the surface of Indian oil sardines and reduced the formation of volatile bases and oxidation products [12]. Another study demonstrated that the combination of lemon essential oil and chitosan induces a beneficial preservation of strawberries with fungal decay [13].

It is worth noting that when chitosan comes into contact with plant tissue, it is associated with the antioxidant capability of the fruit and the control of oxidative stress in fresh fruit [14]. In general, enzymatic and non-enzymatic systems are two types of antioxidant defense systems. Superoxide dismutases (SOD), ascorbate per-oxidases (APX) and catalases (CAT) are major enzymatic antioxidants. These enzymes are effective in scavenging reactive oxygen species (ROS). Although some reports showed the impact of the application of several edible coatings on the quality of food, no information was shown on the effect of chitosan coating on sweet cherry in terms of antioxidant capacity. Thus, the aim of this study was to analyze the effect of chitosan, applied as an edible coating at four concentrations (0.1%, 0.3%, 0.5% and 0.75% *w*/*v*), on the quality, antioxidant activity and bioactive compounds of sweet cherry cultivar during storage time.

2. Materials and Methods

2.1. Fruit Materials and Treatments

The sweet cherry (*Prunus avium* L. cv. Summit) used in this study was obtained from Yingyuan farm near Jinzhou, Dalian, Liaoning, China. Sweet cherries with the same size and color were selected and randomized for the experiments. The quality of sweet cherry involving chroma, firmness, total phenols and total anthocyanins were analyzed. Chitosan film-forming solution was prepared according to the procedure of Petriccione et al., (2015) [15]. Chitosan (Solarbio, Beijing, China) was dissolved in an aqueous solution including acetic acid (0.5% *v*/*v*). The solution was heated and stirred at 45 °C to dissolve the chitosan. The pH was adjusted to 5.6 with 1.0 mol/L NaOH, and Tween-80 (0.05 g/100 mL) for all solutions. Sweet cherries were soaked and coated with different concentrations of 0.1%, 0.3%, 0.5% and 0.75% *w*/*v* chitosan in this study. The treatments consisted of immersing the sweet cherry sample for 5 min at 20 °C in: (a) 0.1% *w*/*v* chitosan; (b) 0.3% *w*/*v* chitosan; (c) 0.5% *w*/*v* chitosan; and (d) 0.75% *w*/*v* chitosan. The uncoated sweet cherry was the control. The treated and control samples were air-dried for 2 h at 20 °C, then packaged into plastic polyethylene bags at 10 °C.

2.2. Extraction

Sweet cherry (100 g) was soaked in liquid nitrogen and ground using a blender. Samples were mixed with an extraction solution of 10mL including acetic acid 2%, water 28%, and acetone 70%. Cherry tissues were centrifuged for 15 min at 10,000× *g* in a Multifuge 3S-R (Kendro, Hanau, Germany). The extract of cherry was made up to 25 mL with acidified water (0.01% HCl [*v*/*v*] in deionized, distilled water) and stored at −80 °C until subsequent analyses.

2.3. Determination of Fruit Colour

Surface color (L^*, a^* and b^* values) was detected using a colorimeter (Minolta, Model CR-400, Osaka, Japan). L^* represents color lightness, 0 is black and 100 is white. The color of each sample was detected at 0, 6, 12, 18, 24 days. Chroma was calculated as $C^* = (a^{*2} + b^{*2})^{1/2}$ for color change measurements in the fruit [16]. The measurements were performed at the same marked sample zone of each sweet cherry. A high chroma value represents a highly saturated and intense color while a low value stands for dull colors.

2.4. Determination of Fruit Firmness

The firmness of the sweet cherries was determined using the texture analyzer (model TA.XT2; Stable Micro Systems Texture Technologies, Scarsdale, NY, USA). Firmness was measured using a 3 mm-diameter flat-plate probe. The maximum force (N) generated during probe travel was used for data analysis. Results were expressed as the slope of the curve in N cm^{-2}.

2.5. Determination of Respiration Rate

The respiration rate was measured with a CO_2 Gas Detector (Alnor Compu-flow, Model 8650, Alnor USA, Los Angeles, CA, USA). For 20 min, 200 g of sweet cherries were enclosed in tightly-sealed hermetic flasks. The sensor was used to measure CO_2 and calculate the respiration rate. Results were expressed as milligrams CO_2 per kilogram per hour.

2.6. Determination of Antioxidant Contents

2.6.1. Total Phenol (TP) Content

Chitosan coating samples (100 g) were mixed with 400 mL of 80% ethanol and homogenized until reaching a uniform consistency using a digital homogenizer (T25, Guangzhou Guangpeng, Guangzhou, China) to obtain 0.25 g mL^{-1} ethanol extract. The obtained mixtures were preserved in covered centrifuge tubes for ultrasound extraction for 40 min in darkness at 40 °C and were subsequently centrifuged at 12,000× *g* for 20 min and the TP Content of the supernatant was determined using a modification of the Folin–Ciocalteu method [17]. One milliliter of 0.25 g mL^{-1} supernatant was added to 1 mL of Folin–Ciocalteu reagent. After that, 10 mL of 7.5% (*w*/*v*) Na_2CO_3 solution and 13 mL of distilled water was added, and the mixtures were incubated at 25 °C for 90 min before measuring at 765 nm. TP content was expressed as mg kg^{-1} of gallic acid equivalent on a fresh weight tissue basis, based on a standard curve (Y = 0.0051 X + 0.0178, R^2 = 0.99) prepared with a standard gallic acid solution.

2.6.2. Total Anthocyanin Content

Anthocyanin quantification of the chitosan film samples (100 g) was performed on a UV-visible spectrophotometer (Shimadzu UV-1601, Norcross, GA, USA) by the pH-differential method [18]. Calculation of the anthocyanins concentration was based on a cyanidin-3-glucoside molar extinction coefficient of 26,900 and a molecular mass of 449.2 g/mol. Measurements were replicated three times with means being reported. Results were expressed as milligrams of cyanidin-3-glucoside equivalent (CGE) per 100 g of fresh weight.

2.7. Determination of Antioxidant Enzyme Activity

Freeze-dried samples (5 g) were homogenized with a potassium phosphate buffer (pH 7.0) of 50 mM containing EDTA of 3 mM and polyvinyl pyrrolidone (PVP) of 0.1 g. The mixture was centrifuged for 20 min at 15,000× *g* at 4 °C. The activity of catalase (CAT), superoxide dismutase (SOD) and ascorbate peroxidases (APX) was determined immediately.

SOD activity was assayed by the method of Duan et al., (2011) [19]. One unit (U) of SOD activity was defined as the amount of enzyme required to cause a 50% inhibition of the reduction of nitro blue tetrazolium (NBT), as monitored at 560 nm. APX activity was determined spectrophotometrically at 290 nm according to the method of Karagiannis et al., (2018) [20]. CAT activity was measured by monitoring the disappearance of H_2O_2 by recording the decrease in absorbance at 240 nm of a reaction mixture containing 50 mmol/L sodium phosphate buffer (pH 7.0), 12.5 mmol/L H_2O_2, and 20 μL of enzyme extract. One unit of CAT activity is defined as the amount of enzyme that decomposes 1 μmol of H_2O_2 per minute per milligram of protein under the conditions of the assay, according to the method of Acero et al., (2019) [17]. The specific activity of the enzyme was expressed as unit mg protein$^{-1.}$ The experiment was repeated three times.

2.8. Statistical Analyses

All experiments were conducted in triplicate, as independent experiments. Data were measured using the Statistical Package for the Social Sciences (SPSS, Version 14.0, IBM Corp., Armonk, NY, USA). The significance of the differences between variables was tested using a one-way ANOVA (between groups) and repeated measures of ANOVA (within group). The means were compared using Duncan's multiple range test. Statistical significance was determined at $p < 0.05$.

3. Results and Discussion

3.1. Colour and Firmness

Appearance, color and firmness are the most important quality attributes because consumers usually judge the quality of fruits on the basis of appearance [21]. Thus, in this study, we first investigated the effect of chitosan with different concentrations on the appearance of sweet cherries in order to search for an effective and safe means to control physiological problems such as browning and softening.

As shown in Figure 1A, the chroma index value at harvest was 27.46 and did not significantly change ($p > 0.05$) during the first 12 days of chitosan treatment. The surface color saturation rapidly decreased with increasing storage time. However, this change was slowed down for sweet cherry treated with 0.5% w/v chitosan. After 18 days of storage, the fruit coated with 0.5% w/v chitosan maintained significantly higher chroma values (19.2 ± 0.72) than those with other chitosan concentrations and the control (10.59 ± 6.36). The differences were even much higher between 0.5% w/v chitosan and other treatments ($p < 0.05$) when observed on the 24th day. Instead, no significant difference ($p > 0.05$) was observed between control and chitosan concentration-coated fruit (0.1% and 0.3% w/v). Interestingly, the addition of 0.75% chitosan led to a greater reduction in chroma index as compared to cherries treated with 0.5% w/v chitosan, and no significant change was observed between cherries treated with the highest chitosan concentration and those treated with the low concentrations (0.1% and 0.3%). The reason is probably that the chitosan with a remarkably high concentration leads to higher intermolecular and interparticle forces, and thereby decreases the liquidity of the films, increases their stiffness and enhances their elongation at break, thus leading to the crack of films and disabling the anti-oxidant protection for the sweet cherries. Therefore, we propose that chitosan with 0.5% concentration is most sufficient in controlling browning and maintaining the commercial value of fresh products.

In addition, firmness is the parameter of greatest concern in sweet cherry storage and marketing, because flesh softening is associated with senescence and fruit injuries. the firmness of control and chitosan-treated (0.1% w/v) sweet cherries exhibited similar trends of softening, and decreased during the first 6 days of storage, showing that such low concentrations of chitosan could not protect sweet cherries from oxidant effects (Figure 1B). On the other hand, chitosan with high concentrations (0.5% and 0.75% w/v) efficiently delay softening trends. However, the firmness values of fruit treated at the two concentrations diverged markedly after 12 days. The final values were 25.63 ± 0.57 N/cm^2 for 0.5% w/v and 11.83 ± 0.76 N/cm^2 for 0.75% w/v at 24 days. This result indicates that the highest concentration is not the most efficient concentration, probably because it induces the crisp of the film. Since firmness at 25 N/cm^2 or higher is an indication of fruit not softening fully [22], it is reasonable to indicate that 0.5% chitosan markedly slows down fruit softening, and thus is suitable for maintaining the firmness of sweet cherries during storage.

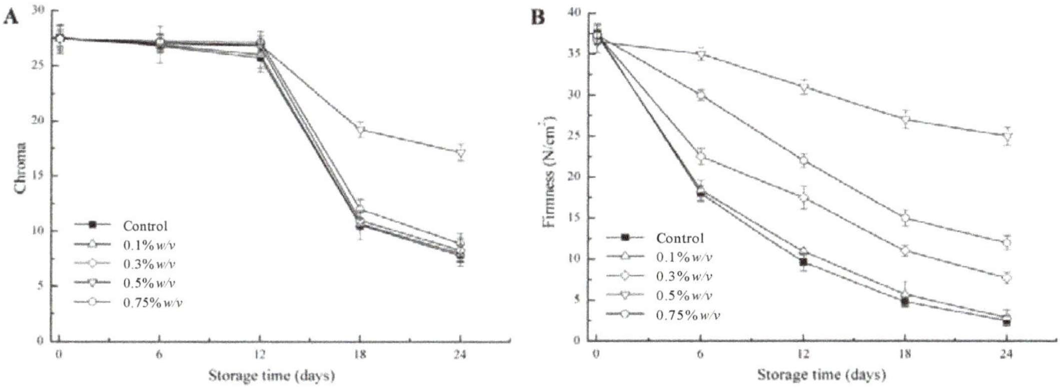

Figure 1. Change of chroma (**A**) and firmness (**B**) in sweet cherry stored at 10 °C during storage time. The data are the means ± standard-deviation (SD).

3.2. Respiration Rate

Respiration rate is a critical index for evaluating the storability of fruit, and it increases with tissue damage and aging [23]. Changes of respiration and storability (delay respiration and consume nutrition) happen during the metamorphic process [24]. In this study, the respiration rate of sweet cherry increases during storage and chitosan treatment can delay the increase to some extent (Figure 2). The respiration rate of high concentrations (0.5% and 0.75% w/v) of chitosan-treated sweet cherries had significant decreases compared with the control and low concentrations (0.1% and 0.3% w/v) of chitosan-treated fruit during the storage ($p < 0.05$), which is consistent with the results of Dong et al. [25]. From the 12th day of storage, the respiration rate of sweet cherry in the 0.5% chitosan treatment group increased more slowly than that in the 0.75% chitosan group. The results clearly indicate that 0.5% w/v chitosan exhibits the maximal response in the inhibition of respiratory intensity in sweet cherries.

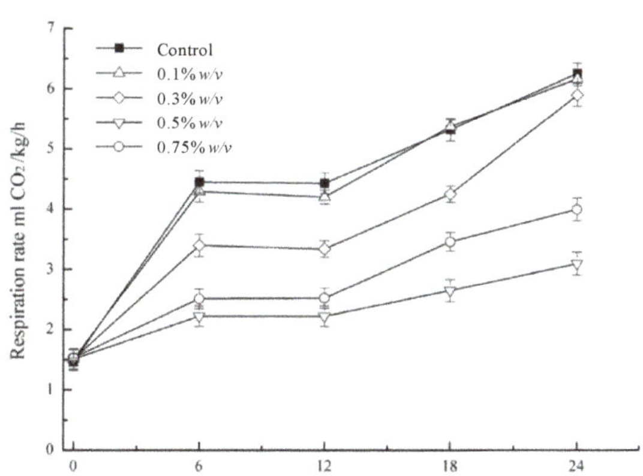

Figure 2. Change of respiration rate in sweet cherry stored at 10 °C during storage time. The data are the means ± standard-deviation (SD).

3.3. *Effect on Phenolic and Anthocyanin Compounds*

Phenolic compounds can improve fruit quality and nutritional value by improving fruit color, taste, aroma and flavor. Anthocyanins are responsible for the red, blue and orange colors of fruits [26,27]. Anthocyanins and phenolics compounds have attracted much attention as food ingredients [28,29]. Hence it is necessary to screen the phenolic properties of sweet cherry to further evaluate the coating efficiency of chitosan [30,31]. As for phenolic compounds, the content of phenolic compounds was established at harvest (70.2 mg100 g^{-1}) in uncoated fruits until day 12 of storage, then decreased significantly until the end of storage (52.00 mg 100 g^{-1}) (Figure 3A). However, total phenolics increased continuously in sweet cherries treated with the chitosan coating. Different concentrations have different effects on the total phenols of sweet cherry. The sweet cherry in a 0.3% chitosan coating showed the highest phenolic compounds with 66.50 $\pm$ 2.37 mg 100 g^{-1} at 24 days. Hence, the quality of sweet cherries coated with chitosan was better than that of uncoated sweet cherries [28]. The main phenolic compounds of sweet cherry are hydroxylcinnamic acids caffeoyl tartaric acid and 3-p-coumaroyl-quinic acid, which contribute to the flavor quality of sweet cherry [32]. Chitosan can form a transparent coating on the surface of the fruit; this coating allows the passage of O_2, not CO_2 and H_2O. Thereby, this high CO_2 low O_2 environment can inhibit the respiration of strawberries and delay the degradation of respiratory substrate, thus delaying the accumulation of secondary metabolites, such as anthocyanins and phenolic substances, to achieve the purpose of keeping fresh.

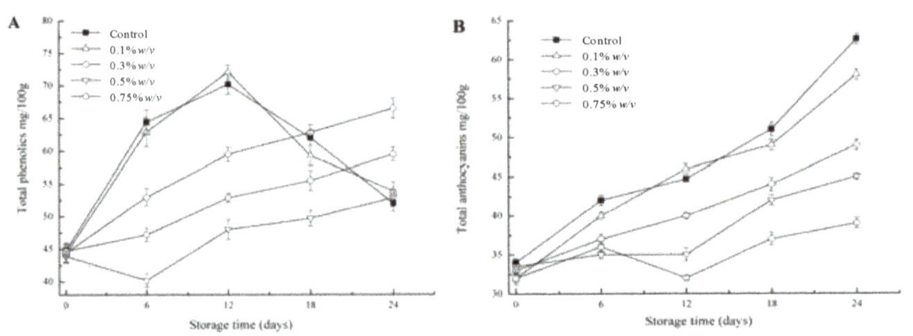

Figure 3. Change of total phenolic contents (**A**), total anthocyanin contents (**B**) in sweet cherry stored at 10 °C during storage time. The data are the means $\pm$ standard-deviation (SD).

Anthocyanin content is easily affected and degraded by light and temperature in fruit. Genotype characteristics, irrigation, plant density and fertilization also affect anthocyanin content in fruits. In this study, the anthocyanin amount of sweet cherry treated with chitosan edible coating is lower than that of the control. The concentration increased during the storage period (Figure 3B), and this increase was delayed in sweet cherry coated with chitosan in a concentration-dependent manner. In the fruits treated with chitosan at 0.75%, anthocyanin concentration was 39.00 mg 100 g^{-1} after 24 days, while these values were significant different with 58.00 mg 100 g^{-1}, 49.00 mg 100 g^{-1} and 45.00 mg 100 g^{-1} for fruits coated with chitosan at 0.1%, 0.3% and 0.5% *w*/*v* at day 24, respectively ($p < 0.05$). Some studies have also demonstrated that the chitosan edible coating decelerated anthocyanin synthesis in treated strawberries and the effects increased with higher chitosan concentration [33,34]. Anthocyanin concentration can increase after harvest during cold storage in pomegranate, sweet cherry, strawberry and raspberry [35–39]. In addition, Kim et al. have reported that anthocyanin contents ranged from 1–432 mg 100g^{-1} in different sweet cherry genotypes [1], showing the reliability of our data.

3.4. *Antioxidant Enzymes*

In plants, antioxidant enzymes, mainly including SOD, CAT and APX, catalyze reactions to neutralize free radicals and ROS, thereby forming the body's endogenous defense mechanisms to help protect against free radical-induced cell damage [40]. In this study, SOD, CAT and APX were detected in sweet cherries to further estimate the effect of chitosan film on antioxidant enzymatic activities.

SOD as a major antioxidant enzyme in cells catalyzes the breakdown of such an anion into oxygen and hydrogen peroxide [41,42]. As shown in Figure 4A, 0.3%, 0.5% and 0.75% chitosan treatment have an obvious enhancement of SOD activity in sweet cherry after 12 days of storage ($p < 0.05$). However, the effect of chitosan on SOD activity after 12 days of storage was significantly, obviously higher than before 12 days of storage. This result indicated that the effect of chitosan film on SOD activity was significantly influenced by storage time, probably because the superoxide anion content at a later storage was higher and then stimulated the SOD activity [19]. In addition, the result showed that the SOD activity of sweet cherry treated with chitosan edible coating (0.5%) is higher than those in chitosan edible coating (0.75%). This might be because the higher concentration chitosan edible coating (0.75%) prevents more oxygen transmittance and inhibits the production of reactive oxygen species (ROS) on the surface microenvironment of sweet cherry. Some studies have demonstrated that lower ROS do not effectively regulate SOD enzyme activities by activating the signal molecules of a plant [43,44]. Therefore, the SOD and APX activities of sweet cherry treated with chitosan coating (0.75%) were lower than those in the 0.5% treatment group. The chitosan coating treatment could be an effective method for enhancing SOD activity in postharvest fruits and vegetables and the influence of chitosan on SOD activity is comparable to other treatments such as methyl-jasmonate and 1-MCP [45,46]. The increase of SOD may reduce free radical accumulation on fruit peel after chitosan treatment, thus leading to increased dismutation of radicals to hydrogen peroxide [47].

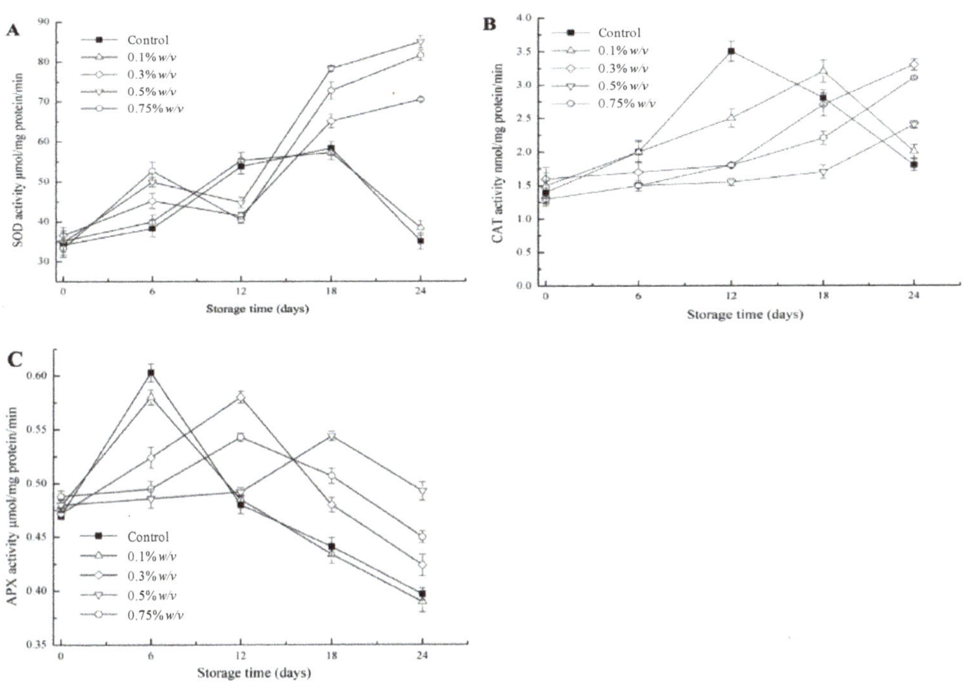

Figure 4. Change of SOD activity (**A**), CAT activity (**B**) and APX activity (**C**) in sweet cherry stored at 10 °C during storage time. The data are the means ± standard-deviation (SD).

CAT is a common enzyme found in nearly all living organisms that are exposed to oxygen, where it functions to catalyze the decomposition of hydrogen peroxide to water and oxygen in order to maintain H_2O_2 homeostasis in cells. As shown in Figure 4B, CAT activity sharply decreased after the 12th day in the control; however, 0.3%, 0.5% and 0.75% *w*/*v* chitosan coating retarded this reduction. Chitosan coating treatment has enhanced CAT activity, which only occurred after 18 days storage. Probably because the enhancement of SOD activity leads to H_2O_2 accumulation and then reduces CAT activity. The results show that 0.5% chitosan is most efficient in maintaining high CAT activity in sweet cherry ($p < 0.05$), which may be important in the defense mechanisms against oxidative stress. Except for CAT, APX is another efficient method for reducing power for H_2O_2 detoxification [48].

For APX, this type of enzyme uses ascorbate as a specific electron donor to catalyze the conversion of H_2O_2 into H_2O, and thereby detoxifies ROS in the ascorbate—glutathione cycle in plant cells [49]. In this study, APX activity in the control and 0.1% chitosan sweet cherry increased rapidly and then decreased significantly from the 6th to the 24th day of storage (from 0.59 to 0.40 μmol/mg/min). Exposure to 0.3%, 0.5% and 0.75% chitosan retard the reduction of APX activity, with the latest peak (0.54 ± 0.04 μmol/mg/min) found in the 0.5% *w*/*v* treatment at the 18th day. As shown in Figure 4C, the APX activity in 0.3%, 0.5% and 0.75% chitosan treatment groups increased significantly compared with the control; this result was similar to the change trend of SOD. An enhancement in APX activity observed in chitosan coating could be due to the availability of ascorbic acid in sweet cherry during storage [15]. These results suggested that chitosan coating has an obvious improvement on antioxidant enzymatic activity in postharvest sweet cherry during 10 °C storage.

4. Conclusions

The edible chitosan coating treatment is an effective method for delaying the postharvest ripening process of sweet cherry, which is manifested by reduced color changes, firmness losses and respiration rate, as well as a positive effect in maintaining higher concentrations of total phenolics and anthocyanins, and in delaying the increases of CAT, SOD and APX activities. Overall, it is suggested that the 0.5% chitosan coating is the most effective factor in storing the sweet cherries for 24 days at 10 °C.

Author Contributions: Conceptualization, W.H.; Methodology, W.H.; Formal analysis, K.F.; Investigation, S.; Resources, K.F. and W.H.; Data curation, K.F. and S.; Software, K.F.; Writing (original draft preparation) W.H. and K.F.; Writing (review and editing), S.; Supervision, K.F.; Project administration, W.H. and S.; Funding acquisition, W.H. All authors have read and agreed to the published version of the manuscript.

Funding: This research was funded by the "13th Five-Year Plan" for the National Key Research and Development Program (Grant No. 2016YFD0400903). Zhuhai College of Science and Technology "Three Levels" Talent Construction Project.

Institutional Review Board Statement: Not applicable.

Informed Consent Statement: Not applicable.

Data Availability Statement: The data used to support the findings of this study are available from the corresponding author upon request.

Conflicts of Interest: The authors declare no conflict of interest.

References

1. Kim, D.O.; Heo, H.J.; Kim, Y.J.; Yang, H.S.; Lee, C.Y. Sweet and sour cherry phenolics and their protective effects on neuronal cells. *J. Agric. Food Chem.* **2005**, *53*, 9921–9927. [CrossRef]
2. Michailidis, M.; Karagiannis, E.; Polychroniadou, C.; Tanou, G.; Karamanoli, K.; Molassiotis, A. Metabolic features underlying the response of sweet cherry fruit to postharvest UV-C irradiation. *Plant Physiol. Biochem.* **2019**, *144*, 49–57. [CrossRef]

3. Bernalte, M.J.; Sabio, E.; Herna´ndez, M.T.; Gervasini, C. Influence of storage delay on quality of sweet cherry. *Postharvest Biol. Technol.* **2003**, *28*, 303–312. [CrossRef]
4. Mirto, A.; Iannuzzi, F.; Carillo, P.; Ciarmiello, L.F.; Woodrow, P.; Fuggi, A. Metabolic characterization and antioxidant activity in sweet cherry (*Prunus avium* L.): Campania accessions Metabolic characterization of sweet cherry accessions. *Food Chem.* **2018**, *240*, 559–566. [CrossRef]
5. Kokalj, D.; Zlatić, E.; Cigić, B.; Vidrih, R. Postharvest light-emitting diode irradiation of sweet cherries (*Prunus avium* L.) promotes accumulation of anthocyanins. *Postharvest Biol. Technol.* **2019**, *148*, 192–199. [CrossRef]
6. Zapata, P.J.; Guillén, F.; Martínez-Romero, D.; Castillo, S.; Valero, D.; Serrano, M. Use of alginate or zein as edible coatings to delay postharvest ripening process and to maintain tomato (*Solanum lycopersicon* Mill) quality. *J. Sci. Food Agric.* **2008**, *88*, 1287–1293. [CrossRef]
7. Zhao, H.; Wang, B.; Cui, K.; Cao, J.; Jiang, W. Improving postharvest quality and antioxidant capacity of sweet cherry fruit by storage at near-freezing temperature. *Sci. Hortic.* **2019**, *246*, 68–78. [CrossRef]
8. Navarro-Tarazaga, M.L.; Sothornvit, R.; Pérez-Gago, M.B. Effect of plasticizer type and amount on hydroxypropyl methylcellulose-beeswax edible film properties and postharvest quality of coated plums (cv. Angeleno). *J. Agric. Food Chem.* **2008**, *56*, 9502–9509. [CrossRef]
9. Tharanathan, R.N.; Kittur, F.S. Chitin—The undisputed biomolecule of great potential. *Crit. Rev. Food Sci. Nutr.* **2003**, *43*, 61–87. [CrossRef]
10. Tokatlı, K.; Demirdöven, A. Effects of chitosan edible film coatings on the physicochemical and microbiological qualities of sweet cherry (*Prunus avium* L.). *Sci. Hortic.* **2020**, *259*, 108656. [CrossRef]
11. Jeon, Y.I.; Kamil, J.Y.V.A.; Shahidi, F. Chitosan as an edible invisible film for quality preservation of herring and Atlantic cod. *J. Agric. Food Chem.* **2002**, *20*, 5167–5178. [CrossRef] [PubMed]
12. Mohana, C.O.; Ravishankara, C.N.; Lalithab, K.V.; Srinivasa Gopalc, T.K. Effect of chitosan edible coating on the quality of double filleted Indian oil sardine (*Sardinella longiceps*) during chilled storage. *Food Hydrocoll.* **2012**, *26*, 167–174. [CrossRef]
13. Perdonesa, A.; Sánchez-Gonzáleza, L.; Chiralta, A.; Vargas, M. Effect of chitosan–lemon essential oil coatings on storage-keeping quality of strawberry. *Postharvest Biol. Technol.* **2012**, *70*, 32–41. [CrossRef]
14. Li, J.; Yan, J.; Wang, J.; Zhao, Y.; Cao, J.; Jiang, W. Effects of chitosan coating on oxidative stress in bruised Yali pears (*Pyrus bretschneideri* Rehd.). *Intern. J. Food Sci. Technol.* **2010**, *45*, 2149–2154. [CrossRef]
15. Petriccione, M.; De Sanctis, F.; Pasquariello, M.S.; Mastrobuoni, F.; Rega, P.; Scortichini, P.; Mencarelli, F. The effect of chitosan coating on the qualityand nutraceutical traits of sweet cherry during postharvest life. *Food Bioprocess Technol.* **2015**, *8*, 394–408. [CrossRef]
16. Al-Said, F.A.; Opara, U.L.; Al-Yahyai, R.A. Physico-chemical and textural quality attributes of pomegranate cultivars (*Punicagranatum* L.) grown in the Sultanate of Oman. *J. Food Eng.* **2009**, *90*, 129–134. [CrossRef]
17. Acero, N.; Gradillas, A.; Beltran, M.; García, A.; Mingarroc, D.M. Comparison of phenolic compounds profile and antioxidant properties of different sweet cherry (*Prunus avium* L.) varieties. *Food Chem.* **2019**, *279*, 260–271. [CrossRef]
18. Giusti, M.M.; Wrolstad, R.E. Characterization and Measurement of Anthocyanins by UV-Visible Spectroscopy. In *Current Protocols in Food Analytical Chemistry*; Wrolstad, R.E., Schwartz, S.J., Eds.; John Wiley & Sons: New York, NY, USA, 2001; pp. F1.2.1–F1.2.13.
19. Duan, X.; Liu, T.; Zhang, D.; Su, X.; Lin, H.; Jiang, Y. Effect of pure oxygen atmosphere on antioxidant enzyme and antioxidant activity of harvested litchi fruit during storage. *Food Res. Intern.* **2011**, *44*, 1905–1911. [CrossRef]
20. Karagiannis, E.; Michailidis, M.; Karamanoli, K.; Lazaridou, A.; Minas, I.S.; Molassiotis, A. Postharvest responses of sweet cherry fruit and stem tissues revealed by metabolomic profiling. *Plant Physiol. Biochem.* **2018**, *127*, 478–484. [CrossRef]
21. Crisosto, C.H.; Crisosto, G.M.; Metheney, P. Consumer acceptance of 'Brooks' and 'Bing' cherries is mainly dependent on fruit SSC and visual skin color. *Postharvest Biol. Technol.* **2003**, *28*, 159–167. [CrossRef]
22. Chen, C.; Hu, W.; He, Y.; Jiang, A.; Zhang, R. Effect of citric acid combined with UV-C on the quality of fresh-cut apples. *Postharvest Biol. Technol.* **2016**, *111*, 126–131. [CrossRef]
23. Giménez, M.J.; Valverde, J.M.; Valero, D.; Zapata, P.J.; Castillo, S.; Serrano, M. Postharvest methyl salicylate treatments delay ripening and maintain quality attributes and antioxidant compounds of 'Early Lory' sweet cherry. *Postharvest Biol. Technol.* **2016**, *117*, 102–109. [CrossRef]
24. Ji, L.; Pang, J.; Li, S.; Xiong, B.; Cai, L.G. Application of new physical storage technology in fruit and vegetable industry. *Afr. J. Biotechnol.* **2012**, *11*, 6718–6722.
25. Dong, F.; Wang, X. Guar gum and ginseng extract coatings maintain the quality of sweet cherry. *LWT—Food Sci. Technol.* **2018**, *89*, 117–122. [CrossRef]
26. Mozetic, B.; Simcic, M.; Trebse, P. Anthocyanins and hydroxycinnamic acids of Lambert Compact cherries (*Prunus avium* L.) after cold storage and 1-methylcyclopropene treatment. *Food Chem.* **2006**, *97*, 302–309. [CrossRef]
27. Pasquariello, M.S.; Di Patre, D.; Mastrobuoni, F.; Zampella, L.; Scortichini, M.; Petriccione, M. Influence of postharvest chitosan treatment on enzymatic browning and antioxidant enzyme activity in sweet cherry fruit. *Postharvest Biol. Technol.* **2015**, *109*, 45–56. [CrossRef]
28. Tomás-Barberán, F.A.; Espín, J.C. Phenolic compounds and related enzymes as determinants of quality in fruits and vegetables. *J. Sci. Food Agric.* **2001**, *81*, 853–876. [CrossRef]

29. Kiprovski, B.; Borković, B.; Malenčić, Đ.; Veberič, R.; Štampar, F.; Mikulič-Petkovšek, M. Postharvest changes in primary and secondary metabolites of sweet cherry cultivars induced by *Monilinia laxa*. *Postharvest Biol. Technol.* **2018**, *144*, 46–54. [CrossRef]
30. Liu, X.; Ji, D.; Cui, X.; Zhang, Z.; Li, B.; Xu, Y.; Chen, T.; Tian, S. *p*-Coumaric acid induces antioxidant capacity and defense responses of sweet cherry fruit to fungal pathogens. *Postharvest Biol. Technol.* **2020**, *169*, 111297. [CrossRef]
31. Zhao, H.; Liu, B.; Zhang, W.; Cao, J.; Jiang, W. Enhancement of quality and antioxidant metabolism of sweet cherry fruit by near-freezing temperature storage. *Postharvest Biol. Technol.* **2019**, *147*, 113–122. [CrossRef]
32. Serrano, M.; Díaz-Mula, H.M.; Zapata, P.J.; Castillo, S.; Guillén, F.; Martínez-Romero, D.; Valverde, J.M.; Valero, D. Maturity stage at harvest determines the fruit quality and antioxidant potential after storage of sweet cherry cultivars. *J. Agric. Food Chem.* **2009**, *57*, 3240–3246. [CrossRef] [PubMed]
33. Varasteh, F.; Arzani, K.; Barzegar, M.; Zamani, Z. Changes in anthocyanins in arils of chitosan-coated pomegranate (*Punica granatum* L. cv. Rabbab-e-Neyriz) fruit during cold storage. *Food Chem.* **2012**, *130*, 267–272. [CrossRef]
34. El Ghaouth, A.; Ponnampalam, R.; Boulet, M. Chitosan coating effect on storability and quality of fresh strawberries. *J. Food Sci.* **1991**, *56*, 1618–1621. [CrossRef]
35. Goncalves, B.; Silva, A.P.; Moutinho-Pereira, J.; Bacelar, E.; Rosa, E.; Meyer, A.S. Effect of ripeness and postharvest storage on the evolution of colour and anthocyanins in cherries (*Prunus avium* L.). *Food Chem.* **2007**, *103*, 976–984. [CrossRef]
36. Artes, F.; Tudela, J.A.; Villaescusa, R. Thermal postharvest treatments for improving pomegranate quality and shelf life. *Postharvest Biol. Technol.* **2000**, *18*, 245–251. [CrossRef]
37. Holcroft, D.M.; Gil, M.I.; Kader, A.A. Effect of carbon dioxide on anthocyanins, phenylalanine ammonia lyase and glucosyltransferase in the arils of stored pomegranates. *J. Am. Soc. Hortic. Sci.* **1998**, *123*, 136–140. [CrossRef]
38. Miguel, G.; Fontes, C.; Antunes, D.; Neves, A.; Martins, D. Anthocyanin concentration of 'Assaria' pomegranate fruits during different cold storage conditions. *J. Biomed. Biotechnol.* **2004**, *5*, 338–342. [CrossRef]
39. Han, C.; Zhao, Y.; Leonard, S.W.; Traber, M.G. Edible coating to improve storability and enhance nutritional value of fresh and frozen strawberries (*Fragaria ananassa*) and raspberries (*Rubus ideaus*). *Postharvest Biol. Technol.* **2004**, *33*, 67–78. [CrossRef]
40. Azarabadi, S.; Abdollahi, H.; Torabi, M.; Salehi, Z.; Jaber, N. ROS generation, oxidative burst and dynamic expression profiles of ROS-scavenging enzymes of superoxide dismutase (*SOD*), catalase (*CAT*) and ascorbate peroxidase (*APX*) in response to Erwinia amylovora in pear (*Pyrus communis* L). *Eur. J. Plant Pathol.* **2017**, *147*, 279–294. [CrossRef]
41. Apel, K.; Hirt, H. Reactive oxygen species: Metabolism oxidative stress, and signal transduction. *Annu. Rev. Plant Biol.* **2004**, *55*, 373–399. [CrossRef]
42. Mittler, R. Oxidative stress antioxidants and stress tolerance. *Trends Plant Sci.* **2002**, *7*, 405–410. [CrossRef]
43. Jacobo-Velázquez, D.A.; González-Agüero, M.; Cisneros-Zevallos, L. Cross-talk between signaling pathways: The link between plant secondary metabolite production and wounding stress response. *Sci. Rep.* **2015**, *5*, 8608. [CrossRef] [PubMed]
44. Cisneros-Zevallos, L.; Jacobo-Velaźquez, D.A. Controlled abiotic stresses revisited: From homeostasis through hormesis to extreme stresses and the impact on nutraceuticals and quality during pre- and postharvest applications in horticultural crops. *J. Agric. Food Chem.* **2020**, *68*, 11877–11879. [CrossRef]
45. Sharma, M.; Jacob, J.K.; Subramanian, J.; Paliyath, G. Hexanal and 1-MCP treatments for enhancing the shelf life and quality of sweet cherry (*Prunus avium* L.). *Sci. Hortic.* **2010**, *125*, 239–247. [CrossRef]
46. Jin, P.; Shang, H.; Chen, J.; Hong, Z.; Zhao, Y.; Zheng, Y. Effect of 1-methylcyclopropene on chilling injury and quality of peach fruit during coldstorage. *J. Food Sci.* **2011**, *76*, 485–491. [CrossRef]
47. Srednicka-Tober, D.; Ponder, A.; Hallmann, E.; Głowacka, A.; Rozpara, E.l.Z. The profile and content of polyphenols and carotenoids in local and commercial sweet cherry fruits (*Prunus avium* L.) and their antioxidant activity in vitro. *Antioxidants* **2019**, *8*, 534. [CrossRef] [PubMed]
48. Racchi, M.L. Antioxidant defences in plants with attention to Prunus and Citrus spp. *Antioxidants* **2013**, *2*, 340–369. [CrossRef]
49. Cao, J.; Tang, D.; Wang, Y.; Li, X.; Li, H.; Sun, C. Characteristics and immune-enhancing activity of pectic polysaccharides from sweet cherry (*Prunus avium* L). *Food Chem.* **2018**, *25*, 447–454. [CrossRef]

coatings

Article

Microstructure and Micro-Hardness of Dissimilar Metal Cladding from a Pipe–Nozzle Mockup for PWR

Jiazhen Wang [1], Hongliang Ming [2,*], Zhiming Zhang [3], Jian Chen [2,*] and Jianqiu Wang [2,3]

[1] Nuclear Power Institute of China, Chengdu 610213, China; jzwangnpic@163.com
[2] CAS Key Laboratory of Nuclear Materials and Safety Assessment, Institute of Metal Research, Chinese Academy of Sciences, Shenyang 110016, China; wangjianqiu@imr.ac.cn
[3] Institute of Corrosion Science and Technology, Guangzhou 510530, China; zmzhang@icost.ac.cn
* Correspondence: hlming12s@imr.ac.cn (H.M.); jchen@imr.ac.cn (J.C.)

Abstract: In this study, the dissimilar metal cladding from a pressure vessel pipe–nozzle mockup for PWR was studied using an optical microscope, scanning electron microscopy, energy-dispersive X-ray spectrometry, electron back-scattering diffraction, and micro-hardness measurement. The microstructure of the SA508 side is non-uniform along the fusion boundary, especially at the concave and convex areas. Martensitic layer (type I and type II) boundaries are found at the fusion boundary area. The chemical composition, residual strain, and microstructure across the SA508–309L fusion boundary are very complex and hence result in a complicated micro-hardness distribution.

Keywords: dissimilar metal cladding; pressure vessel pipe–nozzle; microstructure; micro-hardness

Citation: Wang, J.; Ming, H.; Zhang, Z.; Chen, J.; Wang, J. Microstructure and Micro-Hardness of Dissimilar Metal Cladding from a Pipe–Nozzle Mockup for PWR. *Coatings* **2022**, *12*, 525. https://doi.org/10.3390/coatings12040525

Academic Editor: Alina Vladescu

Received: 22 March 2022
Accepted: 11 April 2022
Published: 13 April 2022

1. Introduction

SA508, one kind of low-alloy steel, is usually utilized for the fabrication of pressure vessels and pipe–nozzles in a pressurized water reactor (PWR) nuclear power plant due to its relatively low cost and high strength [1,2]. However, its corrosion resistance is very low when it is directly exposed to the high-temperature pressurized primary water in a PWR. As a result, a stainless steel cladding layer with a high corrosion resistance is necessary to protect the SA508 low-alloy steel, acting as a barrier layer [3]. As the chemical composition, mechanical properties, and microstructure of low-alloy steel are significantly different from stainless steel (filler metal), the microstructure of the dissimilar metal cladding, especially at the interfacial area, should be very complex. Hence, many problems arise during the cladding process, including sharp changes in element distribution, mechanical property, and microstructure across the fusion boundary; carbon migration; weld residual stress–strain; and the formation of special boundaries at the fusion boundary area [4–7].

In light of the above problems, the dissimilar metal cladding area usually becomes the weak point of the pipe–nozzle. In fact, many stress corrosion cracking (SCC) failures have occurred at this kind of dissimilar metal welding area [8,9]. As a result, lots of research on the microstructure and service behavior of this area has been published. Our group has worked on dissimilar metal weld joints (DMWJs) in PWR nuclear power plants for more than ten years and has investigated different kinds of DMWJs with different filler metals, base metals, welding methods, and welding parameters [10–15]. Even though the microstructures of an SA508–309L/308L–316L domestic dissimilar metal weld safe-end joint and the low-alloy SA508–309/308L stainless steel dissimilar metal weld for pressure vessel lining have been studied previously [1,16], the dissimilar metal cladding of the inner pipe–nozzle has not been investigated in detail before.

During the fabrication of the nuclear power plant, the pipe–nozzle is welded to the primary loop, forming the safe-end DMWJ. In one previous failure case, the SCC crack from the surface of the cladding, propagated to the fusion boundary of the dissimilar metal cladding, then grew just along the fusion boundary of the safe-end DMWJ, leading to a primary water leakage [8]. In fact, SCC is the main failure mode of DMWJ in nuclear power plants [5,17]. SCC behavior is related to the material, stress, and environment. When the stress and environment (which refers to the water chemistry of the primary water in a PWR) are well controlled, the material becomes the key factor affecting SCC. Consequently, there is a requirement to understand and control the microstructure of the DMWJ. Since the microstructure at the weld interface area and the heat-affected zone (HAZ) is highly dependent on the welding parameters, it is necessary to characterize the microstructure and the basic mechanical properties to guide the optimization of the welding parameters. This is equally applicable to dissimilar metal cladding.

2. Materials and Experiment

2.1. Materials

The dissimilar metal cladding samples observed in this study were cut from the inner surface of a pipe–nozzle formed by SA 508 (Gr.3) low-alloy steel. The cladding materials were 309L and 308L stainless steel. A schematic diagram of the pipe–nozzle, the safe-end pipe, the DMWJ, and the exact position of the dissimilar metal cladding investigated in this study is presented in Figure 1 [18]. Comparable dissimilar metal welds have been studied before; however, the welding parameters are different [1,16]. The chemical compositions of SA508 and the filler metals are listed in Table 1. Submerged arc welding (SAW) was used for the cladding, in which 309L was prepared as the first layer and 308L as the second and third layers. The welding parameters are presented in Table 2. The current range was 760–800 A, the voltage was well controlled at 27.5 V, and the welding speed was 150 mm/min. The SA508 base metal was preheated to 150–180 °C before welding, and the inter-pass temperature was controlled between 161 and 197 °C. The cladding surface was machined to achieve the final size requirement.

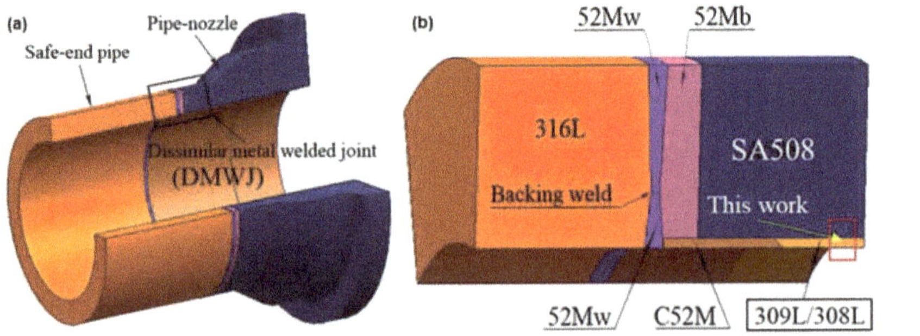

Figure 1. Schematic diagram of the pipe–nozzle, safe-end pipe, and DMWJ (**a**); and the position of the investigated dissimilar metal cladding (**b**) [18].

Table 1. Chemical composition of materials (wt.%).

Materials	C	Si	Mn	P	S	Cr	Ni	Mo	Cu	V	Al	Nb	Ti	B	Co	Fe
SA508	0.17	0.16	1.49	0.005	0.002	0.14	0.76	0.51	0.02	0.007	0.03	/	/	0.0002	0.005	Bal.
309L	0.011	0.33	1.69	0.011	<0.0005	23.49	13.33	0.07	0.039	0.047	/	0.05	<0.005	/	0.021	Bal.
308L	0.012	0.36	1.71	0.012	0.010	20.23	10.25	0.08	0.043	0.065	/	0.01	<0.005	/	0.026	Bal.

Table 2. Welding parameters of the cladding process.

Preheat Temperature (°C)	Inter-Pass Temperature (°C)	Current Range (A)	Voltage Range (V)	Welding Speed (mm/min)
150–180	161–197	760–800	27.5	150

2.2. Microstructure Characterization

Firstly, the samples were ground using silicon carbide abrasive paper with grit sizes ranging from 120# to 2000#, and then mechanically polished with 2.5 μm polishing pastes at a speed of 800 rpm. When no scratches were visible on the polished surfaces, the samples were etched for optical microscope (OM, Axio Observer Z1, Zeiss, Oberkochen, Germany) observation. For the SA508 base metal and its HAZ, a 4% nital solution was used for a duration of 5–10 s, while the weld metal was electro-etched in a 10% oxalic acid+ 90% H_2O solution.

A scanning electron microscopy with X-ray microanalysis (SEM/EDS, FEI XL30, Hillsboro, OR, USA) was used to investigate the chemical composition distribution across the fusion boundary, at an accelerating voltage of 20 kV. Before the EDS line scanning, the sample was slightly etched using the 4% nital solution for a duration of 3–5 s to obtain the position of the fusion boundary.

Electron back-scattering diffraction (EBSD, Oxford, UK) was used to study the grain size and residual strain distribution across the fusion boundary. The EBSD scanning voltage was 25 kV at a scanning step size of 0.5 μm. The raw data were then analyzed using the orientation imaging microscopy (OIM) software. The process to prepare the sample for an EBSD observation has been described in detail previously [1].

2.3. Micro-Hardness Test

The micro-hardness test was performed using an MHVD-1000AP micro-hardness tester (Shanghai, China) at a load of 0.98 N and a holding time of 15 s. Before testing, the sample was also slightly etched to show the fusion boundary.

3. Results and Discussion

3.1. Microstructure by OM

3.1.1. OM Microstructure of Base Metal and Weld Metals

The OM image at a low magnification detailing both the base metal and the weld metal is shown in Figure 2a. The base metal and its HAZ, the first cladding layer (309L), and the 308L cladding layers (the second and third layers) are significantly distinct due to their different microstructures. In addition, the fusion boundary (or the interface) is wavelike.

The microstructure in the SA508 base metal is not uniform; it contains many "black strips" along the axial direction of the pipe–nozzle (Figure 2a). The microstructure of the normal area is shown in Figure 2b, suggesting the presence of a bainite microstructure which was composed of ferrite laths and inter-lath carbides. Figure 2c shows the microstructure of the black strip area. More carbides were found in these areas and they were more uniformly distributed than in the normal area, not only at the boundaries of the ferrite laths but also within the ferrite laths. As these strips are far from the fusion boundary, they may result from macro-segregation during the manufacturing process rather than the welding process.

Figure 2d shows the microstructure of the 309L weld metal. An almost fully austenite structure was found, which was different from the normal microstructure that consists of coarse grains of dendrite-like austenitic structure and ferrites between the dendrites, as demonstrated in Figure 3a of Reference [1]. As described in Reference [19], the solidification mode of the weld metal was determined by the chemical compositions, Cr_{eq} and Ni_{eq}, where the Cr_{eq} = Cr% + Mo% + 1.5 Si% + 0.5 Nb% and the Ni_{eq} = Ni% + 30 C% + 0.5 Mn%. The chemical composition of the 309L layer should deviate from its original composition due to the dilution effect of the melted SA508. On the other hand, the dilution phenomenon had a greater effect on Ni_{eq} than Cr_{eq} as the coefficient factor for C was 30. As a result, the diffusion of carbon from the SA508 to the 309L layer led to a significant increase in the value of Ni_{eq}. When the ratio of Ni_{eq} to Cr_{eq} is high enough, the microstructure of 309L should be fully austenite. The microstructure of the 308L was dendrite-like austenite with dot and rod ferrites between these dendrites (Figure 2e).

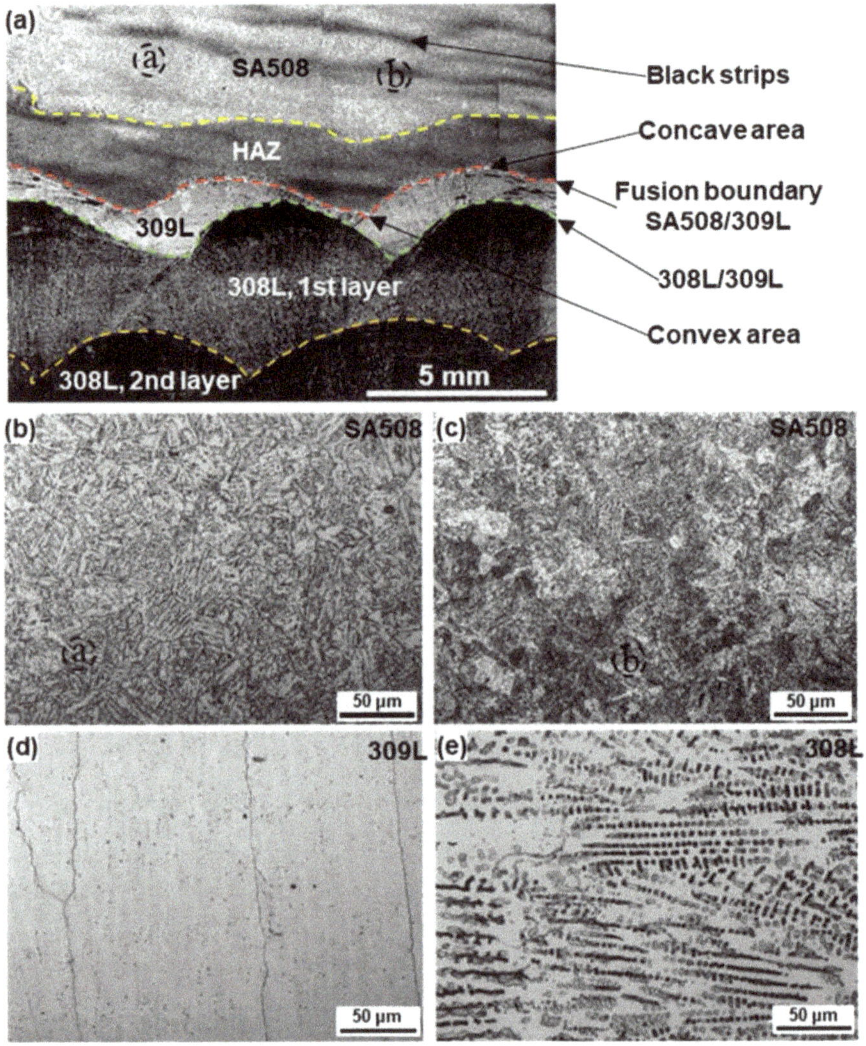

Figure 2. OM microstructure of dissimilar metal cladding: (**a**) image at a low magnification showing both the base metal and the weld metal; (**b**) microstructure of the SA508 base metal; (**c**) microstructure of the black strip area in SA508 base metal; (**d**) microstructure of the 309L weld metal; and (**e**) microstructure of the 308L weld metal.

3.1.2. OM Microstructure across Fusion Boundary

Since the etching methods for SA508 and the weld metal were different, the detailed microstructures across the fusion boundary were observed separately. The microstructures of the 309L side and the SA508 side are shown in Figures 3 and 4, respectively.

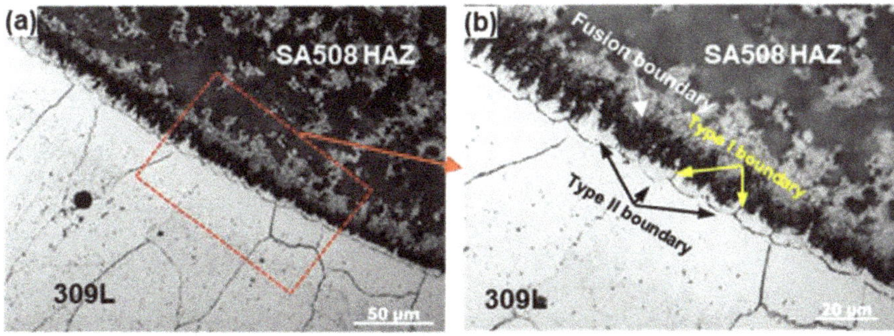

Figure 3. OM microstructure of the 309L side adjacent to the fusion boundary: (**a**) image with low magnification and (**b**) local image with higher magnification of (**a**).

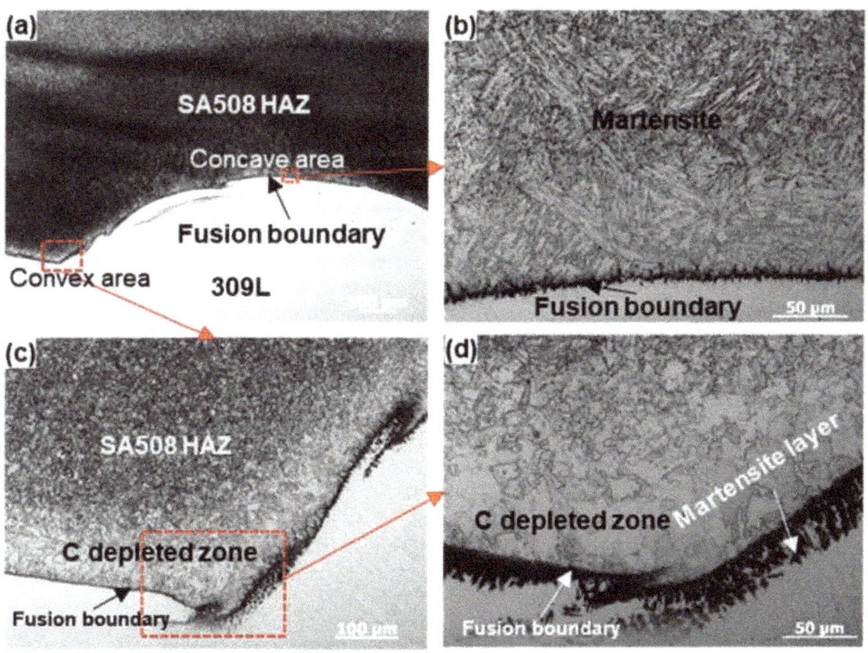

Figure 4. OM microstructure of the SA508 side adjacent to the fusion boundary: (**a**) image at a low magnification; (**b**) microstructure of the concave area in (**a**); (**c**) microstructure of the convex area in (**a**); (**d**) local image at a higher magnification of (**c**).

It is obvious that the microstructure of the SA508 side was not uniform along the fusion boundary, as the heat flux was not homogeneous for every bead or for multi-bead welding (Figure 4a). Overall, a light layer of carbon depletion in the HAZ just adjacent to the fusion boundary was found, although its width was not uniform, which was caused by carbon diffusion from SA508 to 309L. The microstructure of the concave area near the fusion boundary was coarse martensite (Figure 4b), and the average width of the carbon-depleted zone in this area was about 80 μm. A thin dark layer was observed along the fusion boundary, which was shown to be martensite in Reference [4]. In addition, carbides were found at this area, formed by the reaction of the migrated carbon from SA508 with the chromium in 309L. The microstructure of the convex area is shown in

Figure 4c,d, where ferrite was the main structure in the carbon-depleted zone. Furthermore, the carbon-depleted zone in this area was larger, at an average width of about 120 μm, and more obvious than the zone in the concave area. This kind of phenomenon had also been observed in our previous study [1]. Carbon migration is related to the temperature and the high-temperature duration of the welding process [11]. The convex area was double-heated during the multi-bead welding process, and its interface for carbon migration was longer, which promoted carbon migration during welding. A martensite layer was also found in this area, and its thickness was also greater than the layer in the concave area, as its formation was linked to the more pronounced carbon migration in the convex area.

On the 309L side, the most obvious feature was the formation of the type I and type II boundaries (Figure 3). A type II boundary refers to the grain boundary that is parallel to the fusion boundary, while a type I boundary connects the fusion boundary with the type II boundary [20,21]. Although these boundaries are usually found in dissimilar metal (metals with different microstructure) weld joints [22], some researchers have demonstrated that they can also be found at the 316L-182 fusion boundary area since they have the same structure (austenite) [23]. The formation mechanism of these boundaries was described in detail in our previous publications [10,14]. In addition, these boundaries were usually high-angle boundaries and were easy to mechanically crack during the three-points bending test [11]. As a result, they were hazardous to the SCC resistance of the fusion boundary area.

Figure 5 shows the microstructure at the 309L–308L interface. Again, more ferrites were found in 308L than in 309L. In addition, epitaxial growth was observed at the interface area as 309L and 308L have similar crystal structure and chemical composition. This phenomenon was also observed at the 308L and 316L fusion boundary area and was verified using EBSD [1].

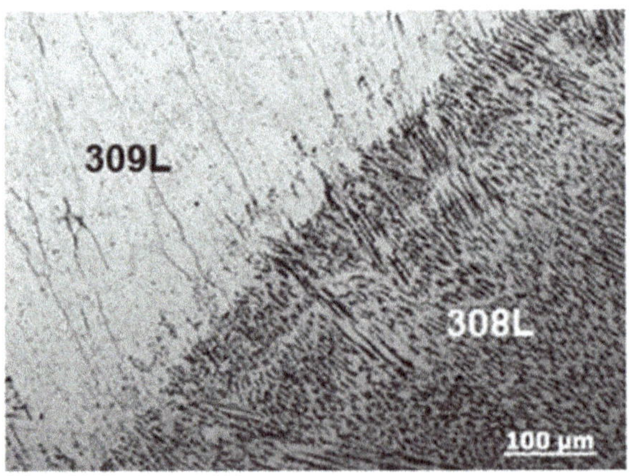

Figure 5. OM microstructure of the 309L–308L interface.

3.1.3. OM Microstructure of SA508 HAZ

As the heat flux during the welding process changed from the fusion boundary to the SA508 base metal, HAZs with widths ranging from 2 mm to 3 mm and a microstructure transition formed, as shown in Figure 6. Microstructure transitions from the fusion boundary at both the concave area (Figure 6a) and at the convex area (Figure 6b) are given in Figure 6. Figure 6a1–a4 demonstrate higher magnifications of the typical areas in Figure 6a. From the fusion boundary, the microstructure transition was coarse martensite + carbide (Figure 6a1) → small amounts of granular bainite + fine martensite (Figure 6a2) → bainite + small amounts of fine martensite (Figure 6a3) → bainite (base metal, Figure 6a4), which is consistent with the findings in Reference [1]. However, the microstructure transition at the convex area was somewhat different due to the presence of the "black strips" (Figure 2) in the original base metal. In fact, microstructure change also occurred in the "black strips" area under the thermal cycles during cladding. From the fusion boundary to the base metal in Figure 6b, the microstructure transition was ferrite (Figure 6b1) → small amounts of granular bainite + fine martensite (Figure 6b2) → bainite + small amounts of fine martensite (Figure 6b3) → bainite (base metal, Figure 6b4). Although the microstructure was similar in Figure 6a2,b2, the grain size was different. The formation of these "black strips" resulted in a smaller grain size in Figure 6b2.

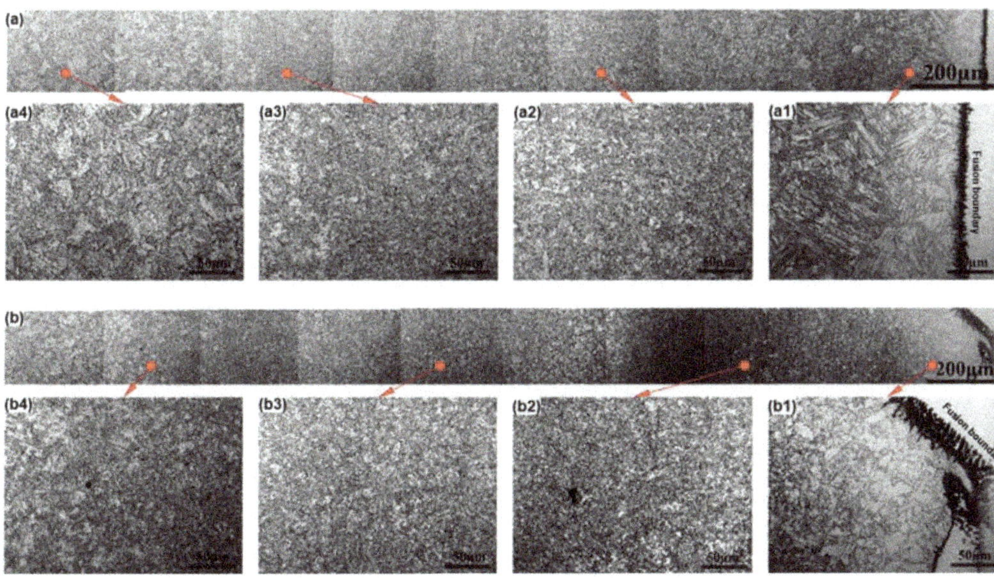

Figure 6. OM microstructure transition in the SA508 HAZ: (**a**) image at a low magnification of the HAZ at the concave area; (**a1–a4**) local images with higher magnification of (**a**); (**b**) image at a low magnification of the HAZ at the convex area; (**b1–b4**) local images at higher magnification of (**b**).

3.2. Chemical Compositions across the Fusion Boundary

The chemical composition distributions across the SA508–309L fusion boundary were characterized using EDS line scanning, and the results are shown in Figure 7. Since the microstructure of the concave area was different from the convex area, the chemical composition distributions were measured separately. Figure 7a,b display the results at the concave area, showing that the SA508 had a higher content of Fe but lower contents of Cr and Ni. In addition, though a dramatic change was found across the fusion boundary, a thin diluted area with a width of about 30 μm was observed.

However, the chemical composition change at the convex area was much more complex than at the concave area (Figure 7c,d). From point B to point C, the chemical composition was similar to the one in Figure 7a, while the chemical transition zone was only about 15 μm in width. The area between points C and D, with the chemical composition between SA508 and 309L, was the martensite area in Figure 4d. An unmixed zone (point C to point D) with a similar chemical composition to the base metal was also found. The reasons for the formation of the unmixed zone have also been described in previous work [1].

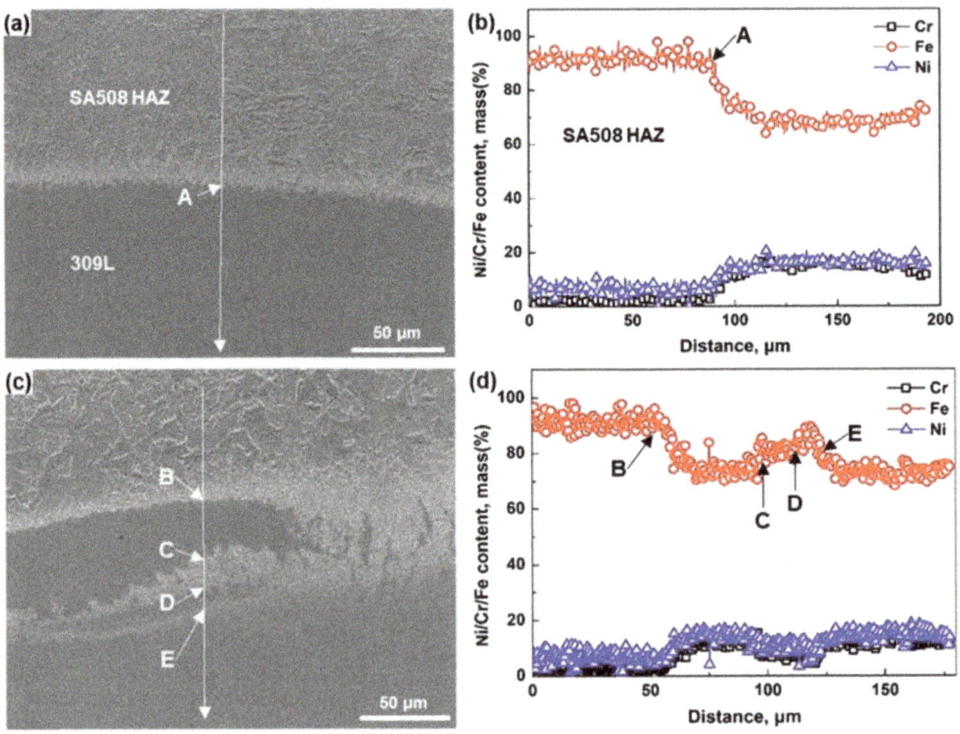

Figure 7. Chemical composition distribution across the SA508–309L fusion boundary: (**a**) the positon of the EDS scanning line at the concave area; (**b**) the corresponding chemical composition distribution along the line in (**a**); (**c**) the positon of the EDS scanning line at the convex area; (**d**) the corresponding chemical composition distribution along the line in (**c**).

3.3. Microstructure by EBSD

The EBSD result of the SA508–309L interface area is shown in Figure 8, including the inverse pole figure (IPF) in Figure 8a, and the kernel average misorientation (KAM) distribution map in Figure 8b,c. From the IPF figure, the grain size from the fusion boundary to the base metal was very similar to the results in Figure 6. However, no type I or type II boundaries were found at this area, indicating that not all of the fusion boundary areas had these kinds of special boundaries. Usually, the KAM is used to estimate the residual strain in materials, that is, the higher the KAM value, the more severe the residual strain [24–27]. From Figure 8b, we can see that the residual strain in SA508 was higher than in 309L cladding. In addition, the residual strains in both materials were nearly homogeneous, with very low fluctuation.

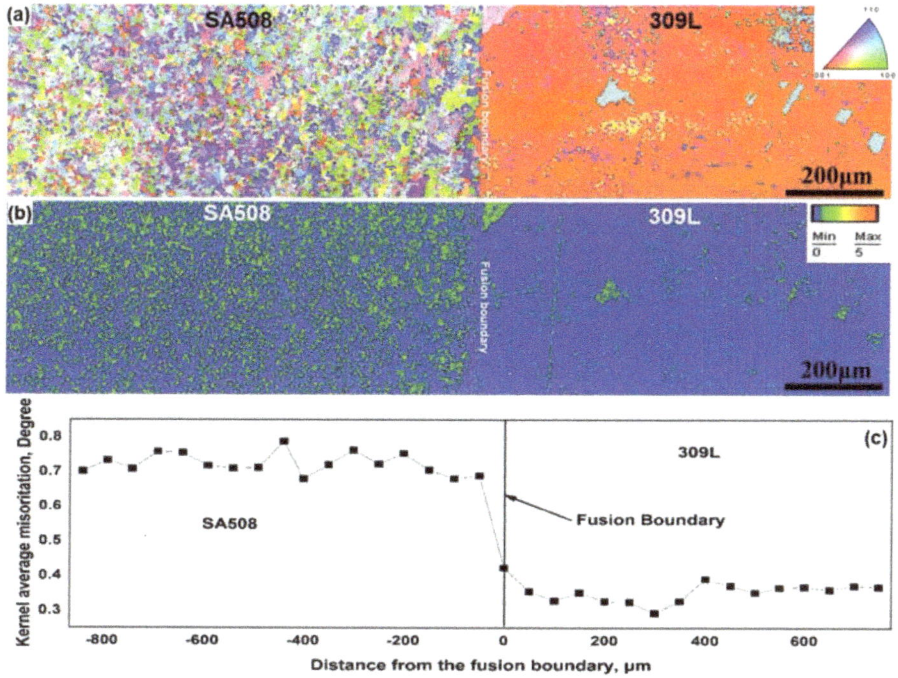

Figure 8. EBSD results of SA508–309L fusion boundary: (**a**) inverse pole figure (IPF); (**b**) kernel average misorientation (KAM) distribution map; and (**c**) the KAM value apart from the fusion boundary.

3.4. Micro-Hardness Distribution

The micro-hardness distribution from the SA508 base metal to the 309L cladding layer and then to the 308L layers is shown in Figure 9a, and the images of the indentations at the SA508–309L interface area are shown in Figure 9b,c, as there was a dramatic change in micro-hardness at the interface area.

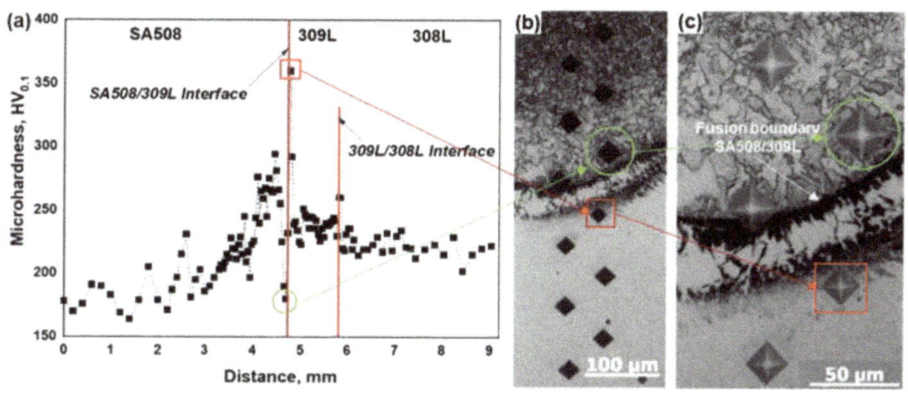

Figure 9. Micro-hardness distribution of the dissimilar metal cladding: (**a**) the hardness values across the dissimilar metal cladding, from SA508 to 309L and finally to 308L; (**b**) the OM image of the indentations across the SA508–309L fusion boundary; and (**c**) the OM image at a higher magnification of (**b**).

In Figure 9a, the micro-hardness of the SA508 base metal was measured to be about 175 HV, which then increased as it moved to the SA508–309L interface. In addition, two obvious fluctuations in hardness were found at the positions of 1.8 mm and 2.5 mm (abscissa axis). This may have resulted from the harder "black strips" areas in the base metal, as these areas had more carbides and their widths were similar to the widths of the fluctuations in the curve. The greater hardness of the HAZ was caused by the formation of martensite and the grain size being finer than in the base metal, as noted by both the OM and EBSD observations. However, a dramatic change was found adjacent to the fusion boundary due to the carbon migration which occurred during the cladding process. The hardness of the carbon-depleted zone was only ~170 HV, while it was ~360 HV for the martensitic layer. Furthermore, the hardness of the 309L just adjacent to the interface was higher (~290 HV) than the normal 309L (~240 HV), and this may have also been caused by the carbon migration. The overall hardness of the 309L was about ~15 HV higher than that of the 308L. The hardness of the 309L–308L interface area was greater than both the weld materials as this area usually had greater residual strain.

4. Conclusions

The microstructure and micro-hardness of some dissimilar metal cladding from a pressure vessel pipe–nozzle for PWR nuclear power plants were studied in detail using the characterization of OM, SEM, EDS, and EBSD, and the measurement of micro-hardness, and the following conclusions can be drawn:

(1) The SA508 base metal has a bainite microstructure and black strip area with more carbides. The 309L cladding has an almost fully austenite structure, while the microstructure of the 308L layers is dendrite-like austenite with dot and rod ferrites between the dendrites.
(2) The microstructure on the SA508 side is not uniform along the fusion boundary. At the concave area, coarse martensite just adjacent to the fusion boundary and a thinner carbon-depleted zone are found. At the convex area, ferrite is the main structure in the carbon-depleted zone and the carbon-depleted area is wider and more obvious than that of the concave area.
(3) A martensitic layer, with type I and type II boundaries, is found at the SA508–309L fusion boundary area. In addition, a microstructure transition with a width of 2–3 mm is observed in the SA508 HAZ.
(4) The chemical composition, residual strain, and micro-hardness distributions across the SA508–309L fusion boundary are very complex. The local hardness is highly dependent on the microstructure and element distribution.

Author Contributions: Conceptualization, J.W. (Jiazhen Wang) and H.M.; methodology, J.W. (Jiazhen Wang); validation, J.C., H.M. and Z.Z.; investigation, J.W. (Jiazhen Wang); resources, J.W. (Jiazhen Wang); data curation, J.W. (Jiazhen Wang); writing—original draft preparation, J.W. (Jiazhen Wang); writing—review and editing, H.M.; visualization, Z.Z.; supervision, J.W. (Jianqiu Wang); project administration, H.M.; funding acquisition, J.W. (Jianqiu Wang). All authors have read and agreed to the published version of the manuscript.

Funding: This work is jointly supported by Youth Innovation Promotion Association CAS (2022187), Hundred Talent Project (E155F207), and Opening project fund of Materials Service Safety Assessment Facilities (MSAF-2021-003).

Institutional Review Board Statement: Not applicable.

Informed Consent Statement: Not applicable.

Data Availability Statement: Data sharing is not applicable to this article.

Conflicts of Interest: The authors declare no conflict of interest.

References

1. Ming, H.; Zhang, Z.; Wang, J.; Han, E.-H.; Ke, W. Microstructural characterization of an SA508–309L/308L–316L domestic dissimilar metal welded safe-end joint. *Mater. Charact.* **2014**, *97*, 101–115. [CrossRef]
2. Wang, H.T.; Wang, G.Z.; Xuan, F.Z.; Liu, C.J.; Tu, S.T. Local mechanical properties of a dissimilar metal welded joint in nuclear power systems. *Mat. Sci. Eng. A Struct.* **2013**, *568*, 108–117. [CrossRef]
3. Okonkwo, B.O.; Ming, H.; Wang, J.; Han, E.-H.; Rahimi, E.; Davoodi, A.; Hosseinpour, S. A new method to determine the synergistic effects of area ratio and microstructure on the galvanic corrosion of LAS A508/309 L/308 L SS dissimilar metals weld. *J. Mater. Sci. Technol.* **2021**, *78*, 38–50. [CrossRef]
4. Wang, S.; Ding, J.; Ming, H.; Zhang, Z.; Wang, J. Characterization of low alloy ferritic steel–Ni base alloy dissimilar metal weld interface by SPM techniques, SEM/EDS, TEM/EDS and SVET. *Mater. Charact.* **2015**, *100*, 50–60. [CrossRef]
5. Chung, W.C.; Huang, J.Y.; Tsay, L.W.; Chen, C. Microstructure and Stress Corrosion Cracking Behavior of the Weld Metal in Alloy 52-A508 Dissimilar Welds. *Mater. Trans.* **2011**, *52*, 12–19. [CrossRef]
6. Hou, J.; Peng, Q.; Takeda, Y.; Kuniya, J.; Shoji, T. Microstructure and stress corrosion cracking of the fusion boundary region in an alloy 182-A533B low alloy steel dissimilar weld joint. *Corros. Sci.* **2010**, *52*, 3949–3954. [CrossRef]
7. Choi, K.-J.; Shin, S.-H.; Kim, J.-J.; Jung, J.-A.; Kim, J.-H. Nano-structural and nano-chemical analysis of ni-base alloy/low alloy steel dissimilar metal weld interfaces. *Nucl. Eng. Technol.* **2012**, *44*, 491–500. [CrossRef]
8. Li, G.F.; Yang, W. Cracking of dissimilar metal welds in nuclear power plants and methods to evaluate its susceptibility to stress corrosion cracking. *Nucl. Saf.* **2003**, *2*, 37–40.
9. Li, G.F.; Congleton, J. Stress corrosion cracking of a low alloy steel to stainless steel transition weld in PWR primary waters at 292 °C. *Corros. Sci.* **2000**, *42*, 1005–1021. [CrossRef]
10. Ming, H.; Zhang, Z.; Wang, J.; Han, E.-H.; Su, M. Microstructure and local properties of a domestic safe-end dissimilar metal weld joint by using hot-wire GTAW. *Acta Metall. Sin.* **2016**, *53*, 57–69.
11. Ming, H.; Wang, J.; Han, E.-H. Comparative study of microstructure and properties of low-alloy-steel/nickel-based-alloy interfaces in dissimilar metal weld joints prepared by different GTAW methods. *Mater. Charact.* **2018**, *139*, 186–196. [CrossRef]
12. Ming, H.; Zhang, Z.; Wang, J.; Han, E.-H. Microstructure of a domestically fabricated dissimilar metal weld joint (SA508-52M-309L-CF8A) in nuclear power plant. *Mater. Charact.* **2019**, *148*, 100–115. [CrossRef]
13. Ming, H.; Zhang, Z.; Wang, J.; Han, E.-H.; Wang, P.; Sun, Z. Microstructure of a safe-end dissimilar metal weld joint (SA508-52-316L) prepared by narrow-gap GTAW. *Mater. Charact.* **2017**, *123*, 233–243. [CrossRef]
14. Ming, H.; Zhu, R.; Zhang, Z.; Wang, J.; Han, E.-H.; Ke, W.; Su, M. Microstructure, local mechanical properties and stress corrosion cracking susceptibility of an SA508-52M-316LN safe-end dissimilar metal weld joint by GTAW. *Mater. Sci. Eng. A* **2016**, *669*, 279–290. [CrossRef]
15. Tang, Y.; Shen, X.; Liu, Z.; Qiao, Y.; Yang, L.; Lu, D.; Zou, J.; Xu, J. Corrosion behaviors of selective laser melted Inconel 718 alloy in NaOH solution. *Acta Metall. Sin.* **2022**, *58*, 324–333.
16. Okonkwo, B.O.; Ming, H.; Wang, J.; Meng, F.; Xu, X.; Han, E.-H. Microstructural characterization of low alloy steel A508–309/308L stainless steel dissimilar weld metals. *Int. J. Pres. Ves. Pip.* **2021**, *190*, 104297. [CrossRef]
17. Seifert, H.P.; Ritter, S.; Shoji, T.; Peng, Q.J.; Takeda, Y.; Lu, Z.P. Environmentally-assisted cracking behaviour in the transition region of an Alloy182/SA 508 Cl.2 dissimilar metal weld joint in simulated boiling water reactor normal water chemistry environment. *J. Nucl. Mater.* **2008**, *378*, 197–210. [CrossRef]
18. Ding, J.; Zhang, Z.-M.; Wang, J.-Z.; Wang, J.-Q.; Han, E.-H. Corrosion behavior of different parts of the weld of 316L/52M/A508 dissimilar metal welded joint in simulated pressurized water reactor primary water. *Mater. Corros.* **2015**, *66*, 1435–1444. [CrossRef]
19. Schaeffler, A.L. Data Sheet 680-B. *Met. Prog.* **1949**, *56*, 680–688.
20. Nelson, T.W.; Lippold, J.C.; Mills, M.J. Nature and evolution of the fusion boundary in ferritic–austenitic dissimilar weld metals, Part 1—Nucleation and growth. *Weld. J.* **1999**, *78*, 329s–337s.
21. Nelson, T.W. Nature and evolution of the fusion boundary in ferritic–austenitic dissimilar metal welds—Part 2: On-cooling transformations. *Weld. J.* **2000**, *79*, 267s–277s.
22. Yoo, S.C.; Choi, K.J.; Bahn, C.B.; Kim, S.H.; Kim, J.Y.; Kim, J.H. Effects of thermal aging on the microstructure of Type-II boundaries in dissimilar metal weld joints. *J. Nucl. Mater.* **2015**, *459*, 5–12. [CrossRef]
23. Wang, W.; Lu, Y.; Ding, X.; Shoji, T. Microstructures and microhardness at fusion boundary of 316 stainless steel/Inconel 182 dissimilar welding. *Mater. Charact.* **2015**, *107*, 255–261. [CrossRef]
24. Lu, Z.; Shoji, T.; Yamazaki, S.; Ogawa, K. Characterization of microstructure, local deformation and microchemistry in Alloy 600 heat-affected zone and stress corrosion cracking in high temperature water. *Corros. Sci.* **2012**, *58*, 211–228. [CrossRef]
25. Yang, M.S.; Huang, J.; Chen, J.; Noël, J.J.; Barker, I.; Henderson, J.D.; He, P.; Zhang, H.; Zhang, H.; Zhu, J. A comparative study on the anti-corrosive performance of zinc phosphate in powder coatings. *Coatings* **2022**, *12*, 217. [CrossRef]
26. Wu, B.; Ming, H.; Zhang, Z.; Meng, F.; He, G.; Wang, J.; Han, E.-H. Microstructure and stress corrosion cracking behavior of Alloy 690TT steam generator tubes with internal bulge defect. *J. Nucl. Mater.* **2022**, *563*, 153629. [CrossRef]
27. Wu, B.; Ming, H.; Zhang, Z.; Meng, F.; Li, Y.; Wang, J.; Han, E.-H. Effect of surface scratch depth on microstructure change and stress corrosion cracking behavior of alloy 690TT steam generator tube. *Corros. Sci.* **2021**, *192*, 109792. [CrossRef]

coatings

MDPI

Article

Effect of Annealing and Oxidation on the Microstructure Evolution of Hot-Dipped Aluminide Q345 Steel with Silicon Addition

Yujian Ma [1,2], Binbin Yuan [1], Ya Liu [1,2,*], Jianhua Wang [1,2] and Xuping Su [1,2,*]

1 School of Materials Science and Engineering, Changzhou University, Changzhou 213164, China; m18351221537@163.com (Y.M.); y18261165372@163.com (B.Y.); wangjh@cczu.edu.cn (J.W.)
2 Jiangsu Key Laboratory of Material Surface Science and Technology, Changzhou University, Changzhou 213164, China
* Correspondence: yliu@cczu.edu.cn (Y.L.); clsj@cczu.edu.cn (X.S.)

Abstract: Q345 steel was coated by hot dipping into molten pure aluminum and Al-Si baths. The coatings were annealed at 800 and 900 °C for 1–3 h and subsequently oxidized at 900 °C for 15 h in air. The results revealed that the thickness of the intermetallic layer increased with increasing hot-dipping time in the range of 700–750 °C, while it decreased when the hot-dipping aluminizing temperature was 800 °C. As the silicon content in the aluminum bath increased, the thickness of the intermetallic layer decreased, and the intermetallic layer/steel-substrate interface transformed from an irregular morphology into a flat morphology. The hot-dipped Al-2.5Si samples were subjected to annealing; the higher the annealing temperature and longer the annealing time, the faster the transformation of the intermediate phase in the coating. The Fe_2Al_5 phase was fully transformed into the ductile FeAl phase after the hot-dipped samples annealed at 900 °C for 3 h. When the outermost layer of Q345 steel was the FeAl phase, oxidation resistance of the oxide was the best.

Keywords: Q345 steel; hot-dipping aluminum-silicon; annealing; FeAl phase; Al_2O_3

Citation: Ma, Y.; Yuan, B.; Liu, Y.; Wang, J.; Su, X. Effect of Annealing and Oxidation on the Microstructure Evolution of Hot-Dipped Aluminide Q345 Steel with Silicon Addition. *Coatings* **2022**, *12*, 503. https://doi.org/10.3390/coatings12040503

Academic Editors: Pier Luigi Bonora and Giorgos Skordaris

Received: 14 February 2022
Accepted: 30 March 2022
Published: 8 April 2022

1. Introduction

In order to reduce room-temperature gas emissions and oil consumption, the use of bioethanol in the automotive industry is gradually increasing; bioethanol and the concentration of ethanol used as a bioadditive in gasoline has increased to about 20% in EU countries [1,2]. It is well-known that after burning bioethanol in engines, the flue gas consists of large amounts of H_2O and CO_2, which can significantly increase the reaction rate of steel in motorcycle- and automotive-exhaust-system components, resulting in the components of the exhaust system being subjected to massive oxidation and severe corrosion of steel and its alloys [3–6]. Low-alloy high-strength steels (Q345) are increasingly used in the automotive industry because of their high strength, good welding properties, and better corrosion resistance, which can save metal materials and reduce the weight of equipment. However, after long-term use of low-alloy high-strength steel in a high-temperature environment with water vapor, the steel surface can very easily form iron oxide, which seriously reduces the surface quality of the steel and limits its use in high-temperature conditions.

Hot-dipping aluminizing is an effective and inexpensive method to form a protective coating on the steel surface [7,8]. By inhibiting the rapid diffusion of aluminum atoms in Fe_2Al_5 and reducing the thickness of the alloy layer, the addition of a certain amount of silicon to the aluminum bath is beneficial to improve the quality of the coating [9–13]. The mechanism of the reduction of the intermetallic layer by the addition of silicon to the hot-dip-coated aluminum solution has been studied widely [9,10,13,14]. Heumann and Dittrich [9] stated that the reduction in the Fe_2Al_5 layer is due to the fact that silicon occupies vacant sites in the c-axis of the Fe_2Al_5 crystal structure, thus slowing down the growth of the Fe_2Al_5 layer. In contrast, Lainer and Kurakin [13] proposed that the growth of the intermetallic compound layer can be slowed down due to the nucleation rate of the Fe-Al-Si intermetallic compound layer. Studies have shown that the high-temperature oxidation resistance of aluminized steel can be improved by using a suitable diffusion-annealing treatment to produce a phase-microstructure transformation in the coating [15–21]. The mechanism to improve the oxidation resistance of aluminized steel is the formation of a dense and stable Al_2O_3 protective layer on the surface of the aluminized layer during the oxidation process, accompanied by a phase transition from the Al-rich phase to the Fe-rich intermetallic compound in the aluminized layer. The improvement of high-temperature oxidation of low-carbon steel exposed to the ethanol combustion product by hot-dip-aluminized coating has been investigated widely [21–23]; however, there are few studies on hot-dip-aluminizing silicon of Q345 steel and its high-temperature oxidation properties. Therefore, it is important to study how to improve the high-temperature resistance of low-alloy high-strength steels in order to reduce the effect of alternative fuels on the high-temperature corrosion of low-alloy high-strength steels.

Previous studies indicated that the mechanical properties of the Fe-Al intermetallic compounds were different [24–26], in which the Fe_3Al and $FeAl_2$ phases show plastic characteristics, and FeAl, Fe_2Al_5, $FeAl_3$, and Fe_4Al_{13} show brittle characteristics. The coatings evolved from the plastic phase had better adhesion. Therefore, the purpose of this paper is to study the effects of hot-dipping aluminizing and diffusion annealing on the evolution of the Fe-Al phase, and to explore ways of improving the high-temperature oxidation resistance of Q345 steel.

2. Experimental Materials and Methods

2.1. Experimental Materials

The substrate used in this experiment was Q345 steel plate; rectangular specimens were cut to the dimensions of 10 mm × 10 mm × 3 mm by a water-cooled cutting machine. The composition of Q345 high-strength steel is shown in Table 1. The hot-dipping Al-Si solution was prepared by pure aluminum and Al-15Si master alloy.

Table 1. Comparison of chemical composition (wt.%) of Q345 steel in this study with standard values.

Sources	Fe	Mn	S	Si	C	P
This study	Bal.	1.3	0.015	0.3	0.16	0.015
Standard	Bal.	≤1.70	≤0.035	≤0.50	≤0.20	≤0.035

2.2. Hot-Dipping Aluminizing

The samples were soaked in 15 wt.% sodium hydroxide aqueous solution at 70 °C for 3 min, and then washed with water and soaked in 15 wt.% hydrochloric acid solution for pickling for 3 min. After washing with water, the samples were treated in 200~300 g/L KCl-K_2ZrF_6-NaF solution at 90 °C for 3 min with a mass ration of $KCl:K_2ZrF_6:NaF$ of 100:100:1. Then, the samples were removed and dried at 100 °C for 8 min. The hot-dipping aluminizing was carried out in a bath that was placed in a KGPS-30-8 medium-frequency induction furnace (Shenguang electric furnace, Ningbo, China) at 700, 750, and 800 °C, respectively. The working thermocouple was periodically tested against a calibrated standard thermocouple. The overall temperature accuracy was estimated to be ±5 °C. After the samples were immersed in the hot-dipping bath for 30 s and 3 min, respectively, they were quickly removed from the bath and cooled in room-temperature water.

2.3. Hot-Dipping Al-Si

The Al-Si molten bath was prepared from industrial pure Al and Al-15Si master alloy. The experiments were divided into six groups, with nominal composition of each group was Al-0.5Si, Al-1Si, Al-2.5Si, Al-5Si, Al-7.5Si, and Al-10Si, weighing 200 g in each. The hot-dip temperature was 750 °C, and the hot-dip time was 30 s.

2.4. Annealing and Oxidation

After removing the surface oxides, the hot-dipped samples were sealed in evacuated quartz tubes and placed in SK2-4-12 tubular resistance furnace (Shanghai Y-feng Electrical Furnace Co. Ltd., Shenyang, China) for diffusion annealing. The annealing temperatures were 800 °C and 900 °C, and the annealing times were 1 h, 2 h, and 3 h, respectively. The oxidation experiments were carried out in three groups: the first group of Q345 steel was hot-dipped in Al-2.5Si solution at 750 °C for 30 s and then directly oxidized in air at 900 °C for 15 h; the second group of Q345 steel was hot-dipped in Al-2.5Si solution at 750 °C for 30 s and then annealed in vacuum at 800 °C for 1 h; followed by oxidation in air at 900 °C for 15 h, and the third group of Q345 steel was hot-dipped in Al-2.5Si solution at 750 °C for 30 s and then annealed in vacuum at 900 °C for 3 h, followed by oxidation in air at 900 °C for 15 h. The three groups of samples were noted as D1, D2, and D3, respectively.

2.5. Characterization

Surface morphology, cross-sectional microstructure, and chemical composition of the coating were analyzed by a JSM-6510 Scanning Electron Microscope (SEM) (JEOL, Tokyo, Japan) equipped with Energy Dispersive X-ray Spectrometer (EDS) (Oxford Instrument, Cambridge, UK). In the observation of SEM, a focused electron beam with energy of 20 keV and working distance of 15 mm interacted with the sample in a teardrop-shape (around 1 μm deep into the sample).

3. Results and Discussion

3.1. Cross-Sectional Microstructure Analysis of Hot-Dipping Pure Aluminum

Figure 1 shows the cross-sectional microstructure of Q345 steel hot-dipped in aluminum solution at 700, 750, and 800 °C for 30 s and 3 min, respectively. It can be seen that the cross-sectional structure of the coating was divided into three different areas: external aluminum layer, intermetallic compound layer, and matrix steel. Composition of the intermetallic layer determined by EDS are listed in Table 2. The representative positions of the components of each phase determined in Figure 1 are marked with symbols a, b, c, respectively. In fact, the composition of each phase was the average of the components at five similar positions. According to the EDS results and combined with the Fe-Al phase diagram, the intermetallic compound layer consisted of two sublayers: the inner part at the metal-coating interface was the Fe_2Al_5 phase, and the composition determined by EDS point analyses was 70.9–71.9 at.% Al, 27.9–28.8 at.% Fe, and $\leq$0.6 at.% Mn. The outer part was the $FeAl_3$ phase, and the composition found by EDS point analyses was 75.6–75.9 at.% Al, 23.8–24.1 at.% Fe, and 0.3 at.% Mn. The Fe_2Al_5 phase was formed by reactive diffusion of iron atoms and aluminum atoms at the interface during hot dipping [4,6]. There was an irregular interface between the alloy layer and the steel matrix, which was in a tongue shape [9,10]. However, the $FeAl_3$-phase layer was not so obvious until the hot-dipping temperature increased to 800 °C. In addition, it could be found that there was also a small amount of discrete-phase precipitation in the external aluminum layer. According to EDS analysis, this discrete phase was the $FeAl_3$ phase, which was precipitated through eutectic reaction in the solidification structure as the cooling speed was not quick enough while taking out the sample from the hot-dipping bath. Because the precipitated $FeAl_3$ phase was relatively small, the average Al content measured by EDS was relative higher than that in the $FeAl_3$ sublayer.

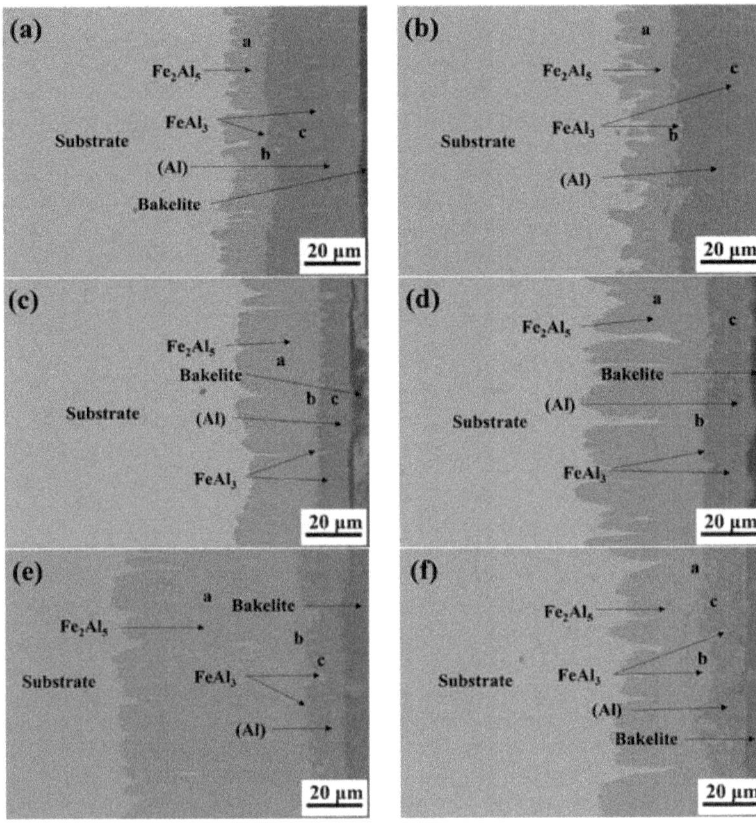

Figure 1. Cross-sectional microstructure of coating of Q345 steel hot-dipped in pure aluminum solution at 700, 750, and 800 °C for 30 s and 3 min, respectively. (**a**) at 700 °C for 30 s, (**b**) at 700 °C for 3 min, (**c**) at 750 °C for 30 s, (**d**) at 750 °C for 3 min, (**e**) at 800 °C for 30 s, (**f**) at 800 °C for 3 min.

Table 2. Phase Analysis by EDS in Figure 1 (at.%).

No.	Typical Positions	Fe	Al	Mn	Phase
(a)	a	28.2	71.2	0.6	Fe_2Al_5
	b	24.2	75.3	0.5	$FeAl_3$
	c	23.3	76.4	0.3	$FeAl_3$
(b)	a	28.8	70.9	0.3	Fe_2Al_5
	b	24.1	75.3	0.6	$FeAl_3$
	c	23.4	76.1	0.5	$FeAl_3$
(c)	a	28.5	71.4	0.1	Fe_2Al_5
	b	24.3	75.3	0.4	$FeAl_3$
	c	23.3	76.2	0.5	$FeAl_3$
(d)	a	28.1	71.6	0.3	Fe_2Al_5
	b	24.5	75.1	0.4	$FeAl_3$
	c	23.2	76.2	0.6	$FeAl_3$
(e)	a	27.9	71.9	0.2	Fe_2Al_5
	b	23.8	75.9	0.3	$FeAl_3$
	c	23.1	76.3	0.6	$FeAl_3$
(f)	a	28.1	71.5	0.4	Fe_2Al_5
	b	24.1	75.6	0.3	$FeAl_3$
	c	23.1	76.5	0.4	$FeAl_3$

The intermetallic layer thickness of Q345 steel hot-dipped in pure aluminum solution at 700, 750, and 800 °C for 30 s and 3 min is shown in Figure 2. Comparing Figure 1a,b, it can be found that when the hot-dip temperature was 700 °C, the external pure aluminum layer was relatively thick, while the Fe_2Al_5 phase area was relatively thin. The reason for this obvious difference was that at this temperature, the thermal diffusion was relatively slow and the fluidity of the aluminum solution was poor, resulting in a considerable part of thick pure aluminum layer still adhering to the sample when the sample was taken out after hot dipping. At the same time, it can be found that when the hot-dip temperature was 700 °C, the type and thickness of the intermetallic layer did not change significantly with the extension of hot-dip time. When the dipping time was extended from 30 s to 3 min, the thickness of the intermetallic layer increased from 17 to 20 μm.

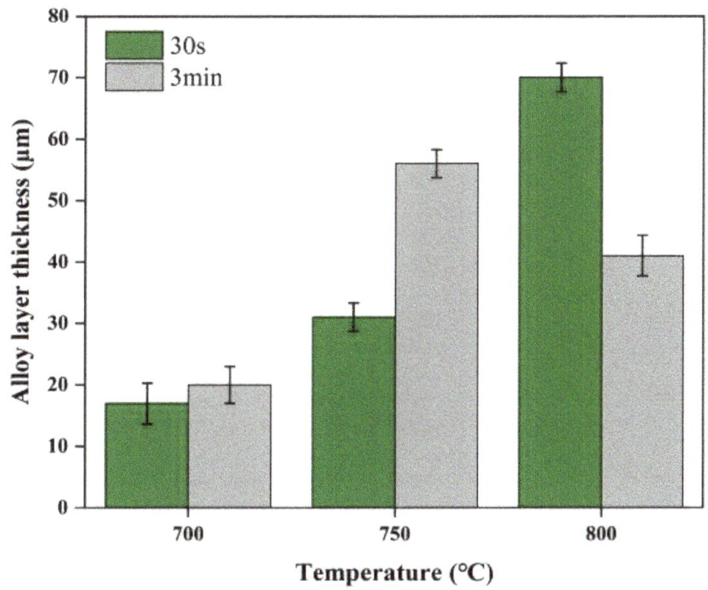

Figure 2. Thickness of intermetallic layer in the coating of Q345 steel after hot dipping at different temperature for 30 s and 3 min.

By observing Figure 1c,d, it can be found that when the hot-dip temperature was increased to 750 °C, the fluidity of aluminum solution was improved, so the external pure aluminum layer of the sample was significantly reduced compared with Figure 1a,b. At the same time, due to the increase of temperature, the diffusion of aluminum and iron was accelerated, and the intermetallic layer was obviously thickened. At 750 °C, the thickness of the intermediate alloy layer increased greatly with the extension of hot-dip time. When the dipping time was 30 s, the thickness was about 31 μm. When the dipping time was 3 min, the thickness was about 56 μm. The thickness increased by about 25 μm.

When the hot-dip temperature was raised to 800 °C, as shown in Figure 1e,f, the thickness and type of intermetallic layer changed significantly. When the dipping time was 30 s, the thickness of the intermetallic layer was about 70 μm. Compared with 750 °C, the thickness of the intermetallic layer had increased further. However, with the extension of the hot-dipping time to 3 min, the thickness of the intermetallic layer decreased. According to the results reported by Chen et al. [27], the average thickness of Fe_2Al_5 also decreased with increasing reaction time when studying the interfacial reaction between liquid aluminum and ST12 cold-roll steel sheets, though the growth of Fe_2Al_5 almost followed the parabolic law at 700, 800 and 900 °C. They had attributed the decrease in thickness into the austenitizing of the substrate (austenitizing temperature of ST12 was 935 °C). The crystal structure of steel transformed from the body-centered cubic phase (α-Fe) into the face-centered cubic phase (γ-Fe) when the temperature of steel exceeded the austenitizing temperature. The diffusion coefficient of Al atoms in the γ-Fe phase was significantly lower than that in the α-Fe phase [28], which means that the migration of the Fe_2Al_5/γ-Fe interface was suppressed relative to that of the Fe_2Al_5/α-Fe interface. Thus, the growth of Fe_2Al_5 was hindered. Yousaf et al. had also found that the intermetallic layer decreased radically or even ceased when the relative amount of α-Fe decreased substantially or when the single γ-Fe phase came in contact with the growing front of the intermetallic layer, in the temperature range 775–950 °C [29]. In the present study, as the critical temperatures for Q345 steel undergoing phase transition were Ac_1 = 756 °C and Ac_3 = 900 °C [30], the Q345 substrate was undergoing austenitizing while hot-dipping at 800 °C; thus, the migration of the Fe_2Al_5/γ-Fe interface was suppressed. With the extension of hot-dipping time, Al atoms reacted with the Fe_2Al_5 phase to form the $FeAl_3$ phase, which consumed the Fe_2Al_5 phase, leading to further thinning of Fe_2Al_5 phase layer.

3.2. *Effect of Si Addition in Molten Bath on Microstructure of Coating*

Since Q345 steel had a moderate, flat, and uniform diffusion layer after hot dipping in pure aluminum at 750 °C for 30 s, so the temperature of hot dipping aluminum-silicon was also set to be 750 °C and the dipping time was 30 s. Figure 3 shows the cross-sectional microstructure of Q345 steel hot-dipped in different Al-Si molten pools. Phase analysis by EDS in Figure 3 is listed in Table 3. Similarly, the typical positions (a, b, c …) used to determine the composition of each phase were also added. It should be noted that the compositions of two ternary phases ($Al_2Fe_3Si_3$ and $Al_4Fe_{1.7}Si$) in Figure 3 are not listed in Table 3. This is mainly because their size was too small to be analyzed by EDS. These two phases were inferred by combining the phase diagram and literature, which will be discussed in detail later. Figure 4 shows the relationship between the thickness of the intermetallic layer and the silicon content. When the Si content was 0.5 at.%, the thickness of the alloy layer was about 19 μm. When the Si content was increased to 10 at.%, the thickness of the alloy layer was reduced to about 7 μm. It is also found that when the Si content was 0.5 at.%, the interface between the alloy layer and the matrix was still in a classic tongue shape, but when the Si content was increased to 2.5 at.% or above, the interface between the alloy layer and the matrix gradually changed from tongue-like to flat. These observations indicated that the Si addition in pure Al reduced the growth rate of the intermetallic layer and changed its morphology.

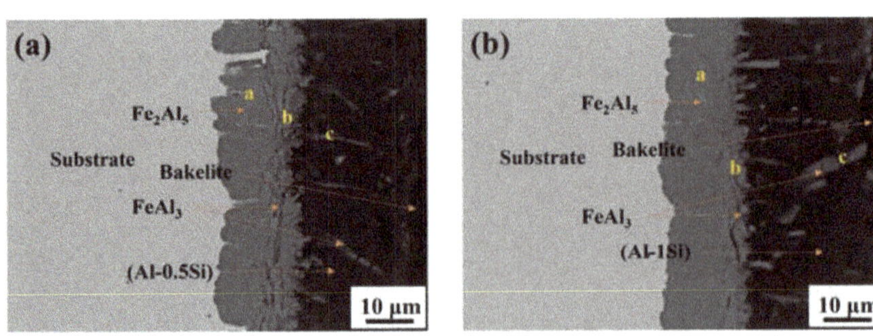

Figure 3. *Cont.*

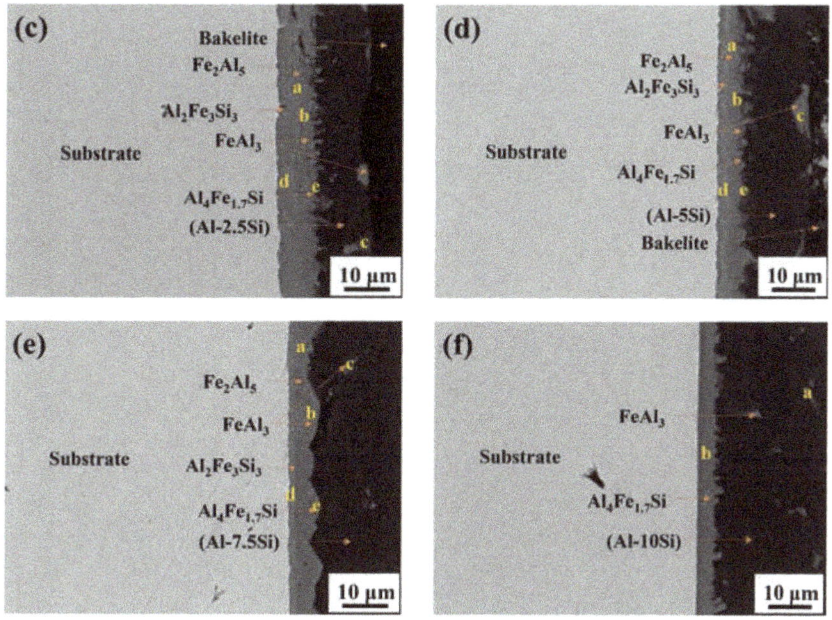

Figure 3. Cross-sectional micrographs of coatings of Q345 steel hot-dipped in baths with different Si addition at 750 °C for 30 s. (**a**) 0.5Si, (**b**) 1Si, (**c**) 2.5Si, (**d**) 5Si, (**e**) 7.5Si, and (**f**) 10Si.

Table 3. Phase analysis by EDS in Figure 3 (at.%).

No.	Typical Positions	Fe	Al	Si	Phase
(a)	a	28.1	70.5	1.4	Fe_2Al_5
	b	24.5	74.3	1.2	$FeAl_3$
	c	23.6	74.9	1.5	$FeAl_3$
(b)	a	27.3	70.6	2.1	Fe_2Al_5
	b	23.3	74.9	1.8	$FeAl_3$
	c	22.9	75.2	1.9	$FeAl_3$
(c)	a	26.5	70.4	3.1	Fe_2Al_5
	b	23.3	74.3	2.4	$FeAl_3$
	c	22.3	75.2	2.5	$FeAl_3$
(d)	a	27.7	69.1	3.2	Fe_2Al_5
	b	23.4	73.5	3.1	$FeAl_3$
	c	23.4	74.1	3.5	$FeAl_3$
(e)	a	27.6	68.2	4.2	Fe_2Al_5
	b	24.2	69.7	6.1	$FeAl_3$
	c	23.0	70.4	6.6	$FeAl_3$
(f)	a	24.1	68.6	7.3	$FeAl_3$

The intermetallic layer of hot-dip pure aluminum at 750 °C consisted mainly of the Fe_2Al_5 phase, while the intermetallic layer of hot-dipped Al-0.5Si consisted of the outer $FeAl_3$ phase layer and the inner Fe_2Al_5 phase layer. According to the Al-Si binary-phase diagram [31], the addition of Si elements can lower the melting point of the Al-Si alloy and makes the microstructure formation temperature of Al-0.5Si hot-dip coating lower than that of pure aluminum hot-dip coating. At the same time, the lower temperature during the hot dip of the Al-Si alloy slowed down the interdiffusion between the steel substrate and the bath, resulting in a thinner alloy-layer thickness. In addition, when the Si content was added to 2.5 at.%, white granular intermetallic compounds appeared in the inner layer of the coating. The size of the white granular phase was so small that it was difficult to analyze its exact composition by EDS. According to the Al-Fe-Si ternary-phase diagram [32], at 700–800 °C, $Al_2Fe_3Si_3$ is in equilibrium with Fe_2Al_5 and $FeAl_3$ phases; therefore, these white intermetallic-compound phases were probably the $Al_2Fe_3Si_3$ phase. Meanwhile, a new phase layer appeared in the outer layer of coating, and this new phase was distributed between the (Al) phase and the $FeAl_3$ phase. By a similar analysis of the Al-Fe-Si ternary-phase diagram, it can be estimated that this intermetallic-compound phase was the $Al_4Fe_{1.7}Si$ phase. It can be seen that when the Si content was added above 2.5 at.%, the coating consisted of the $Al_4Fe_{1.7}Si$ phase, the $FeAl_3$ phase, the Fe_2Al_5 phase, and the $Al_2Fe_3Si_3$ phase. When the Si content was increased to 10 at.%, the coating consisted mainly of the $FeAl_3$ phase and the $Al_4Fe_{1.7}Si$ phase. Based on the above analysis, a plausible reason for reducing the growth rate of the intermetallic layer can be attributed to the formation of new phases due to the Si addition in molten aluminum. Meanwhile, a small amount of Si was dissolved into the Fe_2Al_5 phase, as listed in Table 3, so the reduction of the Fe_2Al_5 layer might also be due to the fact that silicon occupies vacant sites in the c-axis of the Fe_2Al_5 crystal structure, thus hindering the diffusion of Al to the interface Fe_2Al_5/substrate.

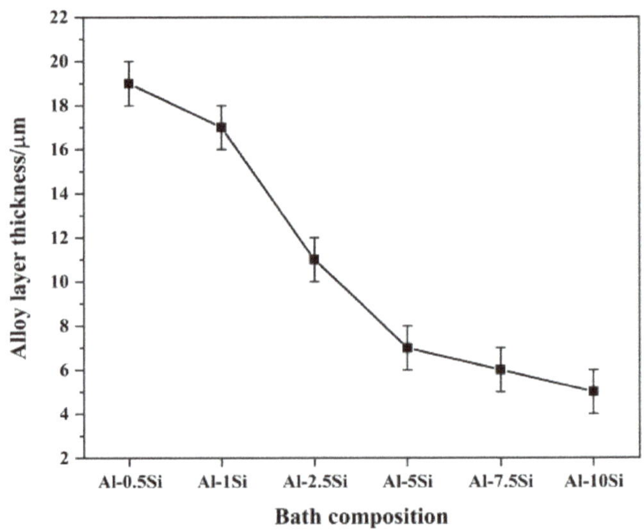

Figure 4. Relationship between intermetallic thickness of Q345 steel after hot-dip aluminum with different silicon content.

3.3. *Effect of Annealing Temperature and Time on the Phase Evolution of the Coating*

Since the thickness of the diffusion layer of Q345 steel was moderate, flat, and uniform after hot dipping in Al-2.5Si solution at 750 °C for 30 s, this group of samples was selected for the subsequent diffusion-annealing treatment to study the effects of annealing temperature and annealing time on the evolution of microstructure in the coating. Figure 5 shows the cross-sectional microstructures of hot-dipped samples annealed at 800 °C and 900 °C for 1, 2, and 3 h, respectively. Phase analysis by EDS in Figure 5 is listed in Table 4. The typical positions used to determine the composition of each phase were also added in Figure 5. As the composition contrast between the α-Fe phase layer and the Fe-Al phases shown in BSE was not obvious in Figure 5, EDS line profiles were supplemented to distinguish these phase boundaries. In addition, compositions of each phase were determined by point analysis of EDS, and the typical positions marked as a, b, c, and d were also added in the figure.

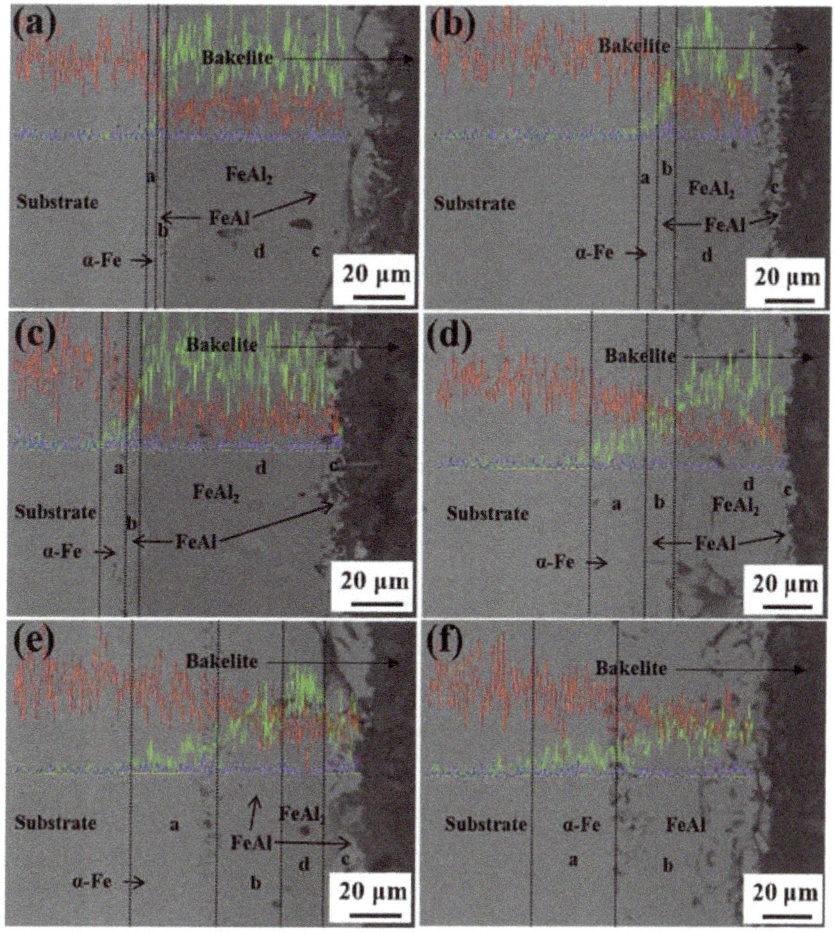

Figure 5. Cross-sectional microstructure of Q345 after hot dipping in Al-2.5Si solution and annealed at 800 °C or 900 °C for different times, and its corresponding EDS line profiles of elements Fe (red line), Al (green line), and Si (blue line) distributions across the coating layer on the steel substrate. (**a**) at 800 °C for 1 h, (**b**) at 800 °C for 22 h, (**c**) at 800 °C for 3 h, (**d**) at 900 °C for 1 h, (**e**) at 900 °C for 2 h, (**f**) at 900 °C for 3 h.

Figure 5a shows the cross-sectional microstructure of Q345 steel after being hot-dipped in Al-2.5Si solution and annealed at 800 °C for 1 h. According to the EDS results and the analysis of the Fe-Al binary-phase diagram, it is found that the gray phase in the outermost layer of the coating was the $FeAl_2$ phase. There was a small amount of light-gray circle areas, which were composited of the $FeAl_2$ phase according to the EDS analysis. This was due to the inward diffusion of Al elements and the outward diffusion of Fe elements from the matrix during annealing at 800 °C, resulting in the transformation of the Fe_2Al_5 phase in the original coating into the $FeAl_2$ phase. Moreover, the Fe_2Al_5 phase had a rhombohedral crystal structure with an atomic saturation concentration of only 70% [15], which allowed the inner Fe atoms to diffuse through the phase and further accelerates transformation of the external Fe_2Al_5 phase. Between the inner $FeAl_2$ phase layer and the substrate in Figure 5a, there were another two layers, though the color contrast was not obvious. According to the EDS analysis, these two layers consisted of the FeAl phase and the α-Fe phase, respectively.

Figure 5b shows the cross-sectional microstructure of Q345 steel after being hot-dipped in Al-2.5Si solution and annealed at 800 °C for 2 h. The outermost light-gray region and the middle gray-circle region were composed of the FeAl phase. Compared with Figure 5a, it can be found that the thickness of the FeAl phase presented between the α-Fe phase and the $FeAl_2$ phase increased and the outermost region gradually changed from the $FeAl_2$ phase to the FeAl phase. This indicates that the rate of diffusion of Fe atoms into the outer $FeAl_2$ phase layer increased with time. In Figure 5c, the thickness of the FeAl phase layer between the α-Fe phase and the $FeAl_2$ phase further increased.

Figure 5d shows the cross-sectional microstructure of Q345 steel after being hot-dipped in Al-2.5Si solution and annealed at 900 °C for 1 h. According to the EDS analysis and its corresponding line profiles of elements Fe, Al, and Si distributions across the coating layer on the steel substrate, the outermost layer was the $FeAl_2$ phase, inwardly followed by the FeAl phase, the α-Fe phase, and the matrix. In Figure 5e, the outermost layer was the FeAl phase, inwardly followed by the $FeAl_2$ phase, the FeAl phase, the α-Fe phase, and the matrix. In Figure 5f, the $FeAl_2$ phase disappeared and all of them were transformed into the FeAl phase. By analyzing the experimental results in Figure 5, it is known that in order to transform the brittle Fe_2Al_5 phase in the coating into the FeAl phase with better toughness, it is better to anneal the coating at 900 °C for 3 h.

Table 4. Phase Analysis by EDS in Figure 5 (at.%).

No.	Typical Positions	Fe	Al	Si	Phase
(a)	a	74.6	22.3	3.1	α-Fe
	b	45.2	41.9	12.9	FeAl
	c	52.1	36.7	11.2	FeAl
	d	31.7	67.2	1.1	$FeAl_2$
(b)	a	78.2	19.1	2.7	α-Fe
	b	46.6	45.5	7.9	FeAl
	c	46.4	47.5	6.1	FeAl
	d	30.2	67.5	2.3	$FeAl_2$
(c)	a	75.2	22.1	2.7	α-Fe
	b	48.5	44.1	7.4	FeAl
	c	44.8	7.1	48.1	FeAl
	d	31.1	67.2	1.7	$FeAl_2$
(d)	a	78.8	18.7	2.5	α-Fe
	b	56.9	40.4	2.7	FeAl
	c	62.3	35.6	2.1	FeAl
	d	33.8	65.5	0.7	$FeAl_2$
(e)	a	78.8	18.8	2.4	α-Fe
	b	50.4	46.5	3.1	FeAl
	c	49.4	44.9	5.7	FeAl
	d	34.1	64.8	1.1	$FeAl_2$
(f)	a	78.3	21.3	0.4	α-Fe
	b	52.5	46.4	1.1	FeAl

3.4. *Effect of Oxidation Process on the Formation of Al_2O_3 Film*

Figure 6 shows the surface morphology and cross-sectional microstructure of these three groups of samples after oxidation and the corresponding EDS line profiles of elements Fe (red line), Al (green line), O (blue line), and Si (purple line) distributions across the coating layer on the steel substrate. Compositions of each phase are listed in Table 5. In Figure 5a–c were typical positions used to determine the composition of each phase.

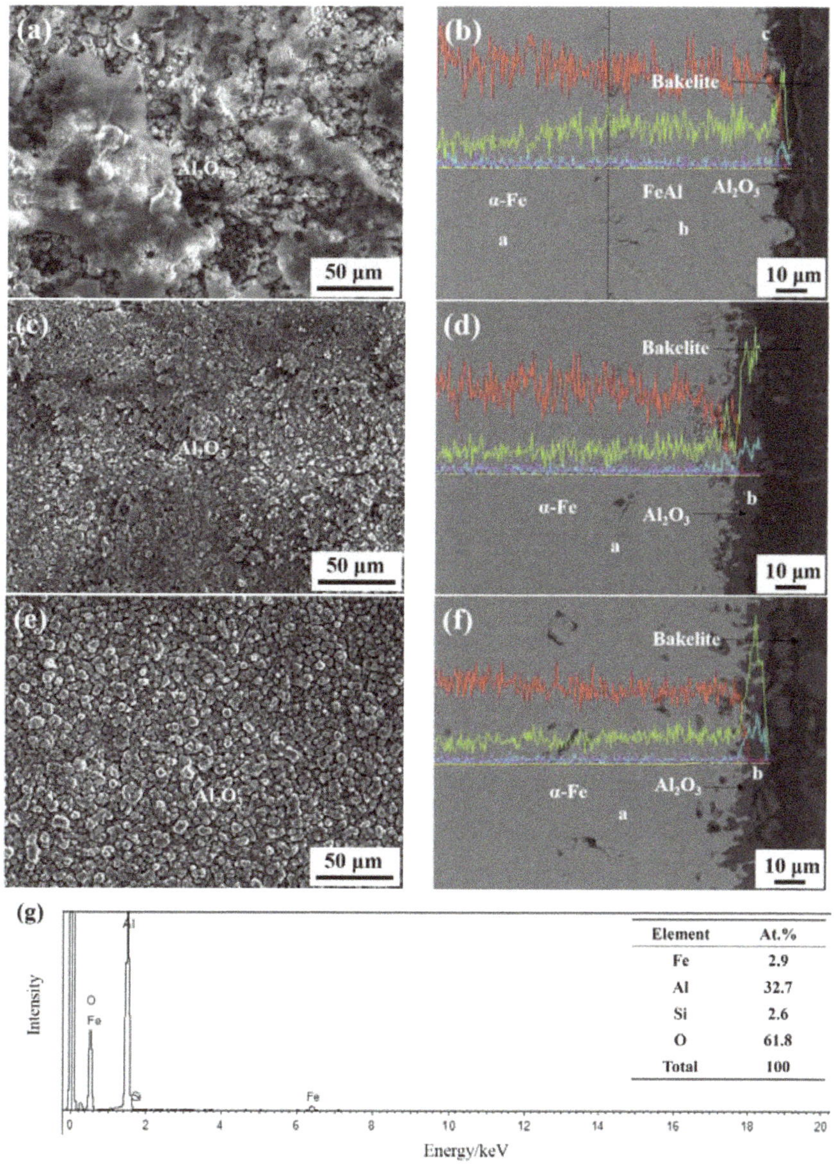

Figure 6. Surface morphology and cross-sectional microstructure of Q345 steel and its corresponding EDS line profiles of elements Fe (red line), Al (green line), O (blue line), and Si (purple line) distributions across the coating layer on the steel substrate after oxidation at 900 °C for 15 h with different preoxidation heat treatment. (**a**,**b**) Oxidation directly after hot dip in Al-2.5Si; (**c**,**d**) oxidation after hot dip in Al-2.5Si and annealing at 800 °C for 1 h; (**e**,**f**) oxidation after hot dip in Al-2.5Si and annealing at 900 °C for 3 h; and (**g**) the typical EDS pattern of the Al_2O_3 phase.

Table 5. Phase Analysis by EDS in Figure 6 (at.%).

No.	Typical Positions	Fe	Al	Si	O	Phase
(b)	a	76.6	22.2	1.2	-	α-Fe
	b	63.9	35.3	0.8	-	FeAl
	c	2.9	32.7	2.6	61.8	Al_2O_3
(d)	a	78.2	19.1	2.7	-	α-Fe
	b	2.6	29.2	2.8	65.4	Al_2O_3
(f)	a	72.9	24.9	2.2	-	α-Fe
	b	2.4	29.8	2.6	65.2	Al_2O_3

Figure 6a,b show the surface morphology and cross-sectional microstructure of the D1 sample after oxidation at 900 °C for 15 h in air. The phase near the matrix in the cross section contained about 76 at.% Fe, 22 at.% Al, and 1 at.% Si, which was determined to be the α-Fe phase, and the phase near the outer part in the cross section contained about 64 at.% Fe, 35 at.% Al and 1 at.% Si, which was determined to be the FeAl phase according to the Fe-Al-Si phase diagram [32]. According to the EDS pattern shown in Figure 6g, the dark-gray area was the Al_2O_3 film that formed in the surface, which had poor continuity and was not dense, with a thickness of about 4–6 μm.

Figure 6c,d show the surface morphology and cross-sectional microstructure of the D2 sample after oxidation at 900 °C in air for 15 h. From the figures, it can be found that the Al_2O_3 film on the surface of this group of samples was denser after oxidation. In the surface morphology shown in Figure 6c, Al_2O_3 was covered with particles in the outermost layer, and in the cross-sectional microstructure shown in Figure 6d, the light-gray phase contained about 74 at.% Fe, 25 at.% Al and 1 at.% Si, so this phase was the α-Fe phase. The dark-gray area was the Al_2O_3 film formed in the surface, which had good continuity and denseness and was about 6–8 μm thick.

Figure 6e,f show the surface morphology and cross-sectional organization of the D3 sample after oxidation at 900 °C for 15 h in air. It can be found that Al_2O_3 was covered in the outermost layer in the form of strips, and the light-gray phase in the cross-section contained about 77 at.% Fe, 22 at.% Al, and 1 at.% Si, which was the α-Fe phase, and the dark gray area was the Al_2O_3 film formed in the surface layer, which had better continuity and denseness than the other two groups, with a thickness of about 5~7 μm.

3.5. High-Temperature Oxidation Resistance of Al_2O_3 Film

In order to test the high-temperature oxidation resistance of the oxidized samples, Q345 steel and oxidized D1, D2, and D3 samples were placed in corundum crucibles and oxidized at 900 °C in a chamber furnace. The weight increased was recorded at regular intervals, and each sample was oxidized for a total of 63 h. Figure 7 reveals the variation of the weight gain with time. It can be seen from the Figure 7 that the oxidation weight gain of the original Q345 steel was the most obvious as the high-temperature oxidation continued, while slopes of the oxidation weight-gain curves of D1, D2, and D3 samples were relatively small, indicating that the Al_2O_3 film on the surface of the samples could play a good role in inhibiting further high-temperature oxidation and protecting the Q345 steel substrate. Among the three oxidized samples, sample D3, i.e., the sample with Al_2O_3 film prepared after annealing at 900 °C for 3 h, showed the smallest weight gain in the subsequent high-temperature oxidation. This indicates that the Al_2O_3 film formed from the FeAl phase in the outermost layer of the coating has the best resistance to high-temperature oxidation.

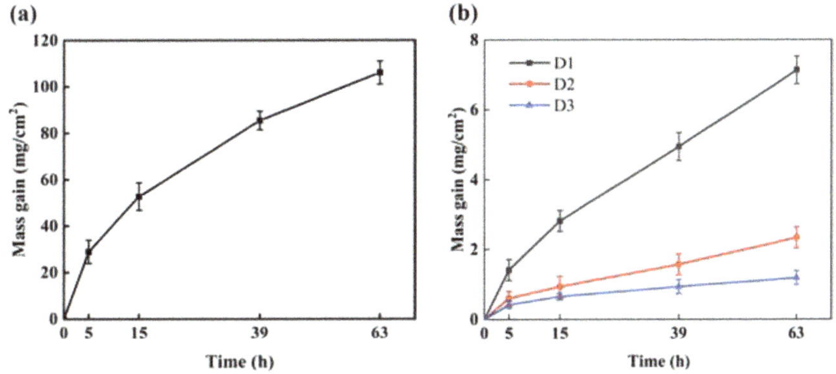

Figure 7. Relationship of weight gain of Q345 steel with oxidation time: (**a**) pure Q345 steel; (**b**) D1, D2, and D3 samples prepared by hot dipping and annealing.

4. Conclusions

The addition of silicon in aluminum baths changes the microstructure and phase constitution of the coating. The main findings of the study are as follows:

(1) When Q345 steel was hot-dipped in pure aluminum, the intermediate phase layer was mainly composed of the Fe_2Al_5 phase. When the hot-dipping temperature was 750 °C and hot-dipping time was 30 s, the interface between the intermetallic phase layer and the substrate was relatively flat.

(2) As the silicon content in the aluminum bath increased, the thickness of the intermetallic layer decreased, and the intermetallic layer/steel-substrate interface transformed from an irregular morphology into a flat morphology.

(3) The annealing time affected the transformation of the intermediate phase in the coating. In addition, the longer the time, the more favorable the transformation of the brittle Fe_2Al_5 phase into the FeAl phase with good toughness. When the aluminized sample was annealed at 900 °C for 3 h, the intermediate phase in the diffusion layer was completely transformed into the FeAl phase.

In conclusion, the addition of Si into the aluminum bath made the thickness of the aluminizing layer decrease, and the outermost aluminized layer could be transformed into FeAl phase by annealing. This was found to be beneficial to improve the oxidation resistance of Q345 steel. Although annealing at 900 °C was beneficial to promote the transformation from the Fe_2Al_5 phase to the FeAl phase, the effect of high-temperature annealing on the mechanical properties of the steel matrix should be further studied.

Author Contributions: Y.M.: investigation, writing—original draft; B.Y.: investigation; Y.L.: conceptualization, methodology, investigation, writing—review and editing; J.W.: investigation; X.S.: supervision, validation. All authors have read and agreed to the published version of the manuscript.

Funding: Financial supports from the National Nature Science Foundation of China (Grant Nos. 51871030 and 52171003) and a project funded by the Priority Academic Program Development of Jiangsu higher education institutions are greatly acknowledged.

Institutional Review Board Statement: Not applicable.

Informed Consent Statement: Not applicable.

Data Availability Statement: Not applicable.

Conflicts of Interest: The authors declare no conflict of interest.

References

1. Shafiee, S.; Topal, E. When will fossil fuel reserves be diminished? *Energy Policy* **2009**, *37*, 181–189. [CrossRef]
2. The Council of the European Union. Directive 2003/30/EC of the European Parliament and of the Council of 8 May 2003 on the promotion of the use of biofuels or other renewable fuels for transport. *Off. J. Eur. Union* **2003**, *123*, 43–46.
3. Grzesik, Z.; Smoła, G.; Adamaszek, K.; Jurasz, Z.; Mrowec, S. Thermal shock corrosion of valve steels utilized in automobile industry. *Oxid. Met.* **2013**, *80*, 147–159. [CrossRef]
4. Sazzad, B.S.; Fazal, M.A.; Haseeb, A.; Masjuki, H.H. Retardation of oxidation and material degradation in biodiesel: A review. *RSC Adv.* **2016**, *6*, 60244–60263. [CrossRef]
5. Drożdż, M.; Kyzioł, K.; Grzesik, Z. Chromium-based oxidation-resistant coatings for the protection of engine valves in automotive vehicles. *Mater. Tehnol.* **2017**, *51*, 603–607. [CrossRef]
6. Badaruddin, M.; Kurniawan, R.D.; Pambudi, A. Low cycle fatigue properties of aluminizing coating on cold-drawn AISI 1018 steel. *IOP Conf. Ser. Mater. Sci. Eng.* **2020**, *807*, 012019. [CrossRef]
7. Ryabov, V.R. *Aluminizing of Steel (Translated from the Russian Work: Alitirovanie Stali)*; Amerind Publishing Co. Pvt. Ltd.: New Delhi, India, 1985.
8. Bahadur, A.; Mohanti, O.N. Structural studies of hot dip aluminized coatings on mild steel. *Mater. Trans. JIM* **1991**, *32*, 1053–1061. [CrossRef]
9. Heumann, T.; Dittrich, N.A. Structure character of the Fe_2Al_5 intermetallics compound in hot dip aluminizing process. *Z Met.* **1959**, *50*, 617–623.
10. Eggeler, G.; Auer, W.; Kaesche, H. On the influence of silicon on the growth of the alloy layer during hot dip aluminizing. *J. Mater. Sci.* **1986**, *21*, 3348–3350. [CrossRef]
11. Dey, P.P.; Modak, P.; Chakrabarti, D.; Banerjee, P.S.; Ghosh, M. A Study on the phase formation and physical characteristics of hot-dip aluminized coating at 750 °C. *Metallogr. Microstruct. Anal.* **2021**, *10*, 823–838. [CrossRef]
12. Richards, R.W.; Jones, R.D.; Clements, P.D.; Clarke, H. Metallurgy of continuous hot dip aluminizing. *Int. Mater. Rev.* **1994**, *39*, 191–212. [CrossRef]
13. Lainer, D.I.; Kurakin, A.K. Mechanism of the effect of silicon in aluminum on the process of reactive diffusion of iron into aluminum. *Fiz. Metal. Metalloved.* **1964**, *18*, 145.
14. Zarei, F.; Nuranian, H.; Shirvani, K. Effect of Si addition on the microstructure and oxidation behaviour of formed aluminide coating on HH309 steel by cast-aluminizing. *Surf. Coat. Technol.* **2020**, *394*, 125901. [CrossRef]
15. Wang, C.J.; Chen, S.M. The high-temperature oxidation behavior of hot-dipping Al–Si coating on low carbon steel. *Surf. Coat. Technol.* **2006**, *200*, 6601–6605. [CrossRef]
16. Huang, X.; Xiao, K.; Fang, X.; Xiong, Z.; Wei, L.; Zhu, P.; Li, X. Oxidation behavior of 316L austenitic stainless steel in high temperature air with long-term exposure. *Mater. Res. Express* **2020**, *7*, 066517. [CrossRef]
17. Abro, M.A.; Hahn, J.; Lee, D.B. High temperature oxidation of hot-dip aluminized T92 steels. *Met. Mater. Int.* **2018**, *24*, 507–515. [CrossRef]
18. Zhang, G.; Yang, F.; Lu, G.; Xiang, X.; Tang, T.; Wang, X. Fabrication of Al_2O_3/FeAl coating as tritium permeation barrier on tritium operating component on quasi-CFETR scale. *J. Fusion Energy* **2018**, *37*, 317–324. [CrossRef]
19. Chen, J.; Li, X.; Hua, P.; Wang, C.; Chen, K.; Wu, Y.; Zhou, W. Growth of inter-metallic compound layers on CLAM steel by HDA and preparation of permeation barrier by oxidation. *Fusion Eng. Des.* **2017**, *125*, 57–63. [CrossRef]
20. Cheng, W.J.; Wang, C.J. Study of microstructure and phase evolution of hot-dipped aluminide mild steel during high-temperature diffusion using electron backscatter diffraction. *Appl. Surf. Sci.* **2011**, *257*, 4663–4668. [CrossRef]
21. Badaruddin, M. Improvement of high temperature oxidation of low carbon steel exposed to ethanol combustion product at 700 C by hot-dip aluminizing coating. *Makara J. Technol.* **2012**, *15*, 137–141. [CrossRef]
22. Badaruddin, M.; Suharno, S.; Wijaya, H.A. Isothermal oxidation behavior of aluminized AISI 1020 steel at the temperature of 700 °C. *J. Tek. Mesin* **2014**, *15*, 15–19.
23. Hakam, M.; Wang, C.J.; Wardono, H.; Asmi, D. High temperature oxidation of low carbon steel with and without an Al coating in an atmosphere containing burning ethanol. *AIP Conf. Proc.* **2018**, *1983*, 050002.
24. Xue, Z.; Hao, X.; Huang, Y.; Gu, L.; Ren, Y.; Zheng, R. Wear resistance and wear mechanism of a hot dip aluminized steel in sliding wear test. *Surf. Rev. Lett.* **2016**, *23*, 1550098. [CrossRef]
25. Li, Y.; Liu, Y.; Yang, J. First principle calculations and mechanical properties of the intermetallic compounds in a laser welded steel/aluminum joint. *Opt. Laser Technol.* **2020**, *122*, 105875. [CrossRef]
26. Xu, G.; Wang, K.; Dong, X.; Yang, L.; Ebrahimi, M.; Jiang, H.; Wang, Q.; Ding, W. Review on corrosion resistance of mild steels in liquid aluminum. *J. Mater. Sci. Technol.* **2021**, *71*, 12–22. [CrossRef]
27. Chen, S.; Yang, D.; Zhang, M.; Huang, J.H.; Zhao, X.K. Interaction between the growth and dissolution of intermetallic compounds in the interfacial reaction between solid iron and liquid aluminum. *Metall. Mater. Trans. A* **2016**, *47*, 5088–5100. [CrossRef]
28. Zhang, Q.; Chen, J.H.; Li, Y.J. Microstructure and oxidationresistance property of aluminizing layer of two kinds of high-chromium steel. *Heat Treat. Met.* **2000**, *10*, 7–10.
29. Yousaf, M.; Iqbal, J.; Ajmal, M. Variables affecting growth and morphology of the intermetallic layer (Fe_2Al_5). *Mater. Charact.* **2011**, *62*, 517–525. [CrossRef]

30. Lin, W.; Zhang, X.; Zhao, T.K.; Li, H.Y. Continuous cooling transformation curve of under cooling austenite about Q345 steel. *Mater. Sci. Technol.* **2009**, *17*, 247–250.
31. Massalski, T.B. *Binary Alloy Phase Diagrams*; ASM International: Phoenix, AZ, USA, 1986.
32. Marker, M.; Skolyszewska-Kühberger, B.; Effenberger, H.S.; Schmetterer, C.; Richter, K.W. Phase equilibria and structural investigations in the system Al–Fe–Si. *Intermetallics* **2011**, *19*, 1919–1929. [CrossRef]

coatings

Article

Obtaining Uniform High-Strength and Ductility of 2A12 Aluminum Alloy Cabin Components via Predeformation and Annular Channel Angular Extrusion

Kai Chen [1], Xi Zhao [1,2,*], Deng-Kui Wang [1], La-Feng Guo [1,*] and Zhi-Min Zhang [2]

1 School of Mechanical and Electrical Engineering, North University of China, Taiyuan 030051, China; 13643511918@163.com (K.C.); 15735658590@163.com (D.-K.W.)
2 National Defense Industry Innovation Center for Complex Component Extrusion Technology, Taiyuan 030051, China; nucforge@126.com
* Correspondence: zhaoxi_1111@163.com (X.Z.); 20040027@nuc.edu.cn (L.-F.G.); Tel.: +86-13934204597 (X.Z.)

Abstract: A 2A12 aluminum alloy component with uniform high-strength and ductility was developed via predeformation (one-pass repetitive upsetting extrusion) and annular channel angular extrusion (ACAE). Moreover, the microstructure evolution and age-hardening behavior were investigated. The results show that the upsetting-extrusion predeformation improved the cumulative strain of the component and refined the grain size, and that the second Al–Cu–Mg phases were obviously broken and refined, and that, especially, the distribution of the second phases along the extrusion direction was weakened. Thus, compared with directly ACAE-formed components, after the T6 heat treatment, the axial ultimate tensile strength (UTS) of the cabin increased from 476 to 484 MPa, and the elongation (EL) increased from 12.9% to 17.5%. The circumferential UTS increased from 426 to 482 MPa, and the EL increased from 9.24% to 16.8%. A large number of dislocations were introduced into the upsetting extrusion (UE) + ACAE method, which resulted in strain hardening and higher precipitation strengthening in the late artificial aging process. The finer and denser grains and s precipitates significantly enhanced the strength and ensured the good ductility of the alloy. It is suggested that the combination of predeformation and annular channel angular extrusion is an effective method for forming aluminum alloy cabin components with higher and more uniform mechanical properties.

Keywords: annular channel angular extrusion; predeformation; 2A12 aluminum alloy; heat treatment; microstructure; uniform mechanical properties

Citation: Chen, K.; Zhao, X.; Wang, D.-K.; Guo, L.-F.; Zhang, Z.-M. Obtaining Uniform High-Strength and Ductility of 2A12 Aluminum Alloy Cabin Components via Predeformation and Annular Channel Angular Extrusion. *Coatings* **2022**, *12*, 477. https://doi.org/10.3390/coatings12040477

Academic Editor: Hideyuki Murakami

Received: 4 March 2022
Accepted: 30 March 2022
Published: 31 March 2022

1. Introduction

Lightweight and high-strength aluminum alloy cabin components can meet the demands of aerospace and weapons equipment for the integration and lightweight of the major load-bearing members [1–3]. The annular channel angular extrusion (ACAE) process is a kind of large plastic deformation process for preparing cabin shells in a short process with high efficiency, which enjoys a bright prospect in the application of industrial production. Evolving from the traditional backward extrusion (BE) process, the ACAE process can form the small-diameter billet into a large-diameter cabin and shell at one time, without complicated processes such as free upsetting and punching and reaming, with a large height-to-diameter ratio. Shatermashhadi et al. [4] proposed this new method and applied it to the preparation experiment of high-purity lead components. The results show that, compared with BE, the ACAE method reduced the extrusion force in the forming process, and the prepared components had a larger equivalent strain and a more uniform distribution along the height direction, so that the final components had a uniform microstructure and performance. This new back extrusion method was expected to be applied to the preparation of ultrafine crystal samples of aluminum, copper, and magnesium alloys. On the basis of this, S.H. Hosseini et al. [5] improved the geometric shape of the ACAE process deformation zone to obtain larger plastic strain and a more uniform distribution of strain. The improved method was applied to the preparation of industrial pure aluminum cup-shaped parts. This work focused on the microstructure evolution, the equivalent strain distribution, and the microhardness of the materials during deformation. The results show that: (1) Two shear zones provided high shear strain on the material, which led to significant grain refinement; and (2) The extrusion component had good uniformity of equivalent strain and hardness along the axial direction while obtaining a high equivalent strain and hardness. Zhao et al. [6,7] conducted a comparative study on the microstructure and properties of same-sized AZ80 magnesium alloy cabins (height: 210 mm; outer diameter: 200 mm: wall thickness: 16 mm), prepared by the ACAE method and the BE method, respectively, which showed that, compared with BE, the introduction of two strong shear deformations in the ACAE process helped refine the microstructure and significantly improved the deformation uniformity of the extruded components. This gave rise to more uniform mechanical properties along the height and thickness directions, but still exhibited anisotropy along the axial and circumferential directions. Gao et al. [8] applied the improved ACAE process for the preparation of a commercial 2A12 aluminum alloy cabin. After the T6 heat treatment, the axial UTS of the extruded cabin was increased to 476.1 Mpa, and the EL was 12.9%.

In the above works, a casting billet or extruded billet was directly used for the ACAE forming. Although the components obtained good uniformity of the strain distribution and higher mechanical properties along the axial direction, the consistency of the mechanical properties along the axial direction and the circumferential direction was ignored [9]. The cabin components not only have to withstand the load along the axial (longitudinal) direction during service but must also withstand the overload with the circumferential (transverse) direction. Highly mechanical properties with only a single direction cannot guarantee that the entire component will withstand enough outer load and keep the profile. Therefore, it is a difficult problem to prepare cabin components with the same axial and circumferential performance. The irregular second phase is usually the source of cracks and corrosion [10,11]. In this paper, the size and distribution of the second phase of an 2A12 aluminum alloy billet was controlled by upsetting-extrusion (UE) deformation. Then, the cabin was prepared by the ACAE forming, combined with a reasonable heat treatment process, and the effects of the deformation process and the heat treatment on the microstructure and mechanical properties of the 2A12 aluminum alloy were studied. The purpose was to improve the mechanical properties of the cabin and to improve the difference between the axial and circumferential performances, and this is expected to inhibit the occurrence of corrosion and the crack sources.

2. Materials and Methods

In this study, a commercial 2A12 aluminum alloy extrusion billet was employed, the chemical composition is shown in Table 1. A cylindrical billet with a diameter of 90 mm and a height of 360 mm was used for the experiments. The prepared billet was heated to 460 °C for 3 h, and the UE deformation of φ 90 mm-φ 140 mm-φ 90 mm was carried out to achieve the purpose of preforming the billet before the ACAE process. Then, the prepared billet was heated to 430 °C and was held for 3 h before the ACAE process.

Table 1. Chemical composition.

Element	Weight %
Al	Others
Mg	1.5
Cu	4.5
Zn	≤0.30
Mn	0.4
Si	0.1
Ti	≤0.15
Ni	≤0.10

The whole process is shown in Figure 1. In the process, MoS_2 was used as lubricant, the extrusions were conducted at 4-THP-630, and the extrusion speed was 1 mm/s. Finally, a cabin, with a height of 160 mm, an outer diameter of 200 mm, and a wall thickness of 16 mm, was extruded. The anatomy of the cabin is shown in Figure 2a.

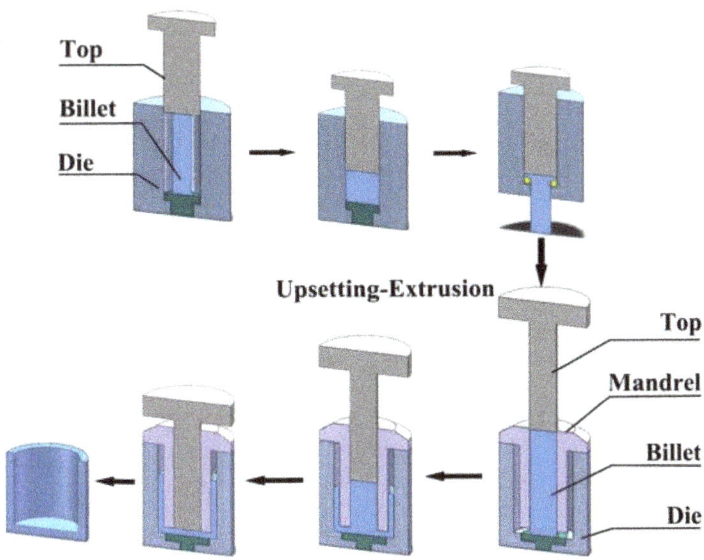

Figure 1. Schematic diagram of forming test.

In order to study the microstructure evolution of the material during the whole deformation process, the extrusion piece was refined into five zones, as shown in Figure 2a: (I) the billet zone; (II) Shear Zone 1; (III) the bottom zone; (IV) Shear Zone 2; and (V) the wall zone. The heat-treated samples were all taken from the wall zone, and the observation section of the samples was in an axial–circumferential direction. The sampling diagram and sample size are shown in Figure 2b. Optical microscopy (OM, Carl Zeiss, ZEIS-Image, Jena, Germany), scanning electron microscopy (SEM, Hitachi, SU-5000, Tokyo, Japan), and the electron backscattering diffraction (EBSD, AMETEK, Pegasus System, Berwyn, PA, USA) apparatus were used to observe the microstructure of the extrusion. In order to obtain more intuitive results, we compared and studied the microstructure and the properties of the extrusion billet (unupset-extruded deformation) directly applied to the ACAE-forming cabin in our team's previous research work.

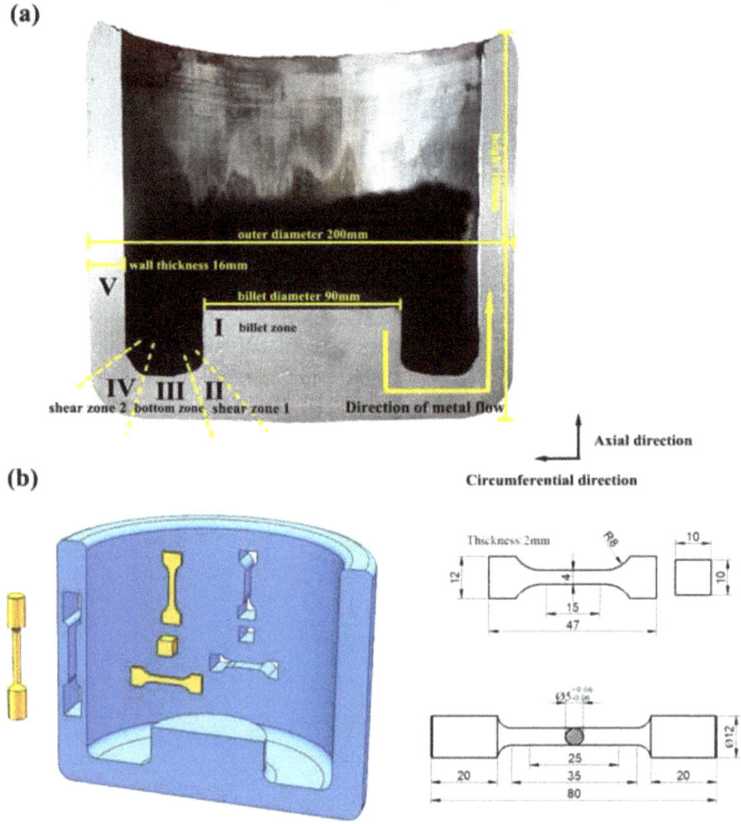

Figure 2. (**a**) anatomical diagram of extrusion cabin; (**b**) schematic diagram of microstructure observation and mechanical properties sampling.

3. Results and Discussion

3.1. Finite Element Simulation

In this study, Deform-3d software was used to simulate the direct ACAE- and UE + ACAE-forming processes. The mesh flow tracking method was used to represent the deformation characteristics of the metal during the forming process. In addition, the effective strain distribution of the two forming processes was shown. According to the flow grid distribution of Figure 3a, it can be seen, in Zone I, that the mesh had almost no deformation, and that the mesh shape was still the same as the square in the initial billet. After the billet passed through Shear Zone 1 (Zone II), the metal was subjected to shear force, and the mesh shape changed from a square to a parallelogram that was elongated along the diagonal. As it flowed through Zone III, the mesh continued to maintain its shape, which indicated that the metal was subjected to small strain. In Shear Zone 2 (Zone IV), the mesh was elongated along another vertical diagonal direction, which proved that the two corner zones provided shear forces that were perpendicular to each other. Finally, as shown in Figure 3b, a large strain level was accumulated in the wall area of the cabin in Zone V, and the average effective strain of the wall became 3.68.

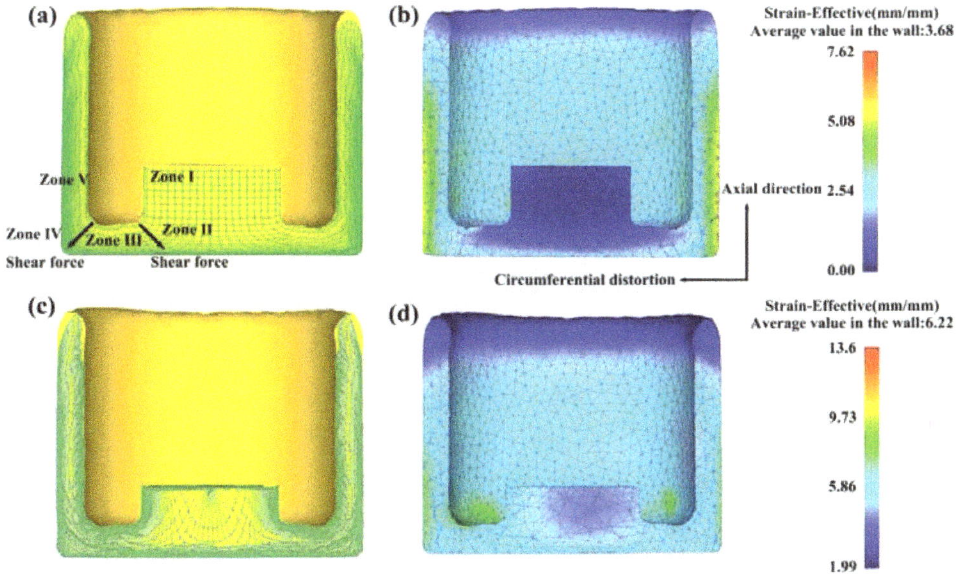

Figure 3. Mesh flow simulation of: (**a**) ACAE; and (**c**) UE + ACAE. Effective strain distribution simulation of: (**b**) ACAE; and (**d**) UE + ACAE.

By comparing and observing the FE simulation results of the UE + ACAE-deformed member in Figure 3c, the mesh in Zone I was found to be in an elongated shape, which was due to the advanced upsetting extrusion. In the process of deformation, the cumulative strain of the material increased, and the mesh deformation became more and more obvious. The final expression was 5.04 for the effective strain value, which was caused by the upsetting extrusion in the undeformed zone (Zone I) in Figure 3d, and it was 6.22 for the average effective strain value in the wall area (Zone V). The UE provided the initial strain and the larger cumulative equivalent strain for the subsequent ACAE process.

3.2. *Microstructure Evolution*

Figures 4 and A1 shows the BSE image of the material during the whole deformation process. The energy dispersive spectrometer (EDS) shows that the bright white parts are mainly the Al–Cu–Mg phase (Figure 5) [12,13]. Figure 4a shows, in the coarse second phase, which is often agglomerated, that the microstructure of the 2A12 extrusion bar billet is unevenly distributed, with sizes in these phases that were mostly 15–20 µm.

However, large, irregular, and inhomogeneous component phases (especially insoluble component phases containing impurities such as Fe and Si) often become the source of crack initiation and corrosion, and may promote crack propagation, thereby affecting the performance of the components [12]. In addition, a large number of rod-shaped T phases ($Al_{20}Cu_2Mn_3$) [14] with sizes below 1 µm can be clearly observed from the higher multiples BSE diagram. The T-phase particles in the 2A12 aluminum alloy precipitate in the process of the homogenization heat treatment, and gradually grow up with the extension of the homogenization time. These particles are very stable in the subsequent solid solution and aging heat treatment process, and their main effect is to prevent grain boundary slippage, and to play the role of high-temperature strengthening during the high-temperature heat treatment or deformation heat treatment. The large-sized second-phase particles usually belong to the brittle phase, and its deformation ability for the coordination of the Al matrix is poor. When the particles have irregular shapes, large sizes, and an inhomogeneous distribution, especially, they are often broken during the deformation.

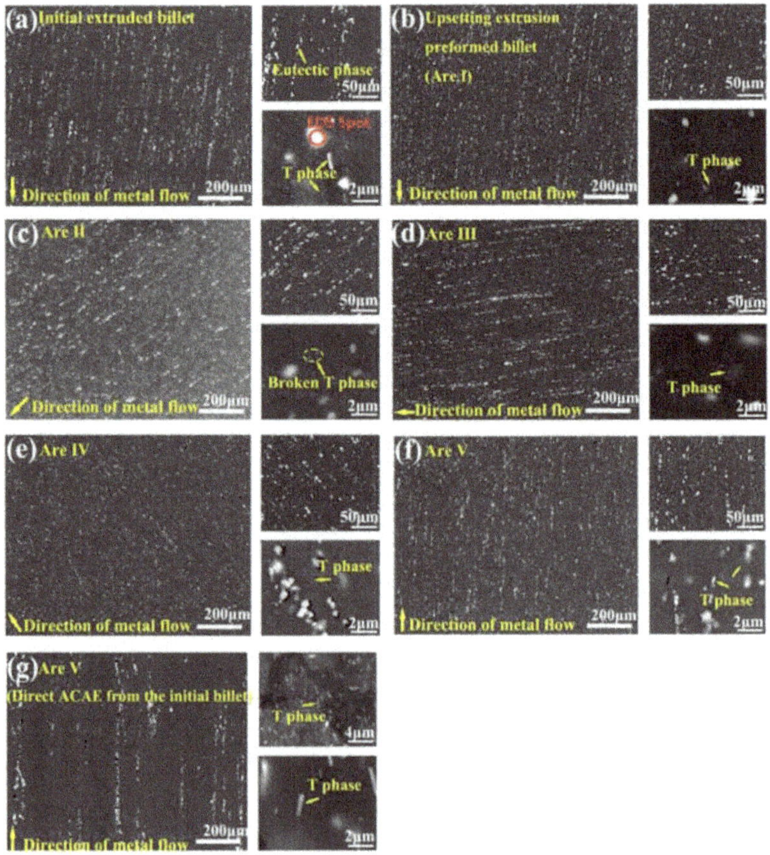

Figure 4. BSE image of the second-phase evolution during deformation: (**a**) initial billet; (**b**) billet deformed by UE; (**c**) UE + ACAE Zone II; (**d**) UE + ACAE Zone III; (**e**) UE + ACAE Zone IV; (**f**) UE + ACAE Zone V; and (**g**) ACAE Zone V.

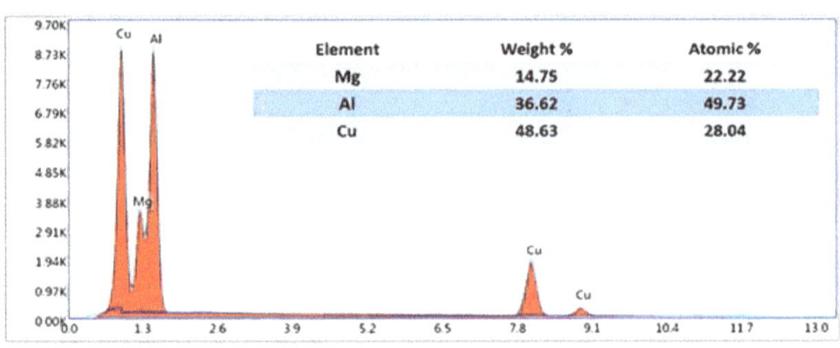

Figure 5. EDS results for the massive phases.

As is shown in Figure 4b, the size and distribution of the second phase in the billet were regulated by the UE. During the hot extrusion process, the component phases were broken down to less than 10 μm by the UE and were then dispersed again. Compared with the initial extrusion bar billet, the distribution of the component phases along the extrusion direction was weakened to a large degree, and the distribution of the rod-shaped T phase was more dispersed. After the predeformation, the size of the second phase in the blank was refined, the shape tended to be regular, and the distribution was dispersive, and it provided a large cumulative strain on the subsequent ACAE deformation.

Figure 4c,e show two shear zones, Zone II and Zone III, in the process of ACAE forming. The metal was subjected to two strong shear stresses in these two zones. Although a small amount of aggregation in the second phase occurred when it passed through Zone III (Figure 4d), the direction of the strong shear stress provided by Zone IV was mutually perpendicular to that provided by Zone II, and so the degree of dispersion of the second phase increased greatly in the deformation process. In this process, the component phases were broken and refined to a diameter of less than 10 μm, with a high degree of dispersion. The broken T phases, most of which are 0.2–0.3 μm in size, were observed in Zone II. As shown in Figure 4f, the microstructure with a small size and a high degree of dispersion of the second phase was finally obtained in the wall area of Zone V. The second phase of the crushing dispersion increased the deformation capacity of the coordinated Al matrix in all directions.

Figure 4g shows the wall area of the cabin that was obtained by the direct ACAE process of the same initial billet. Compared with Figure 4f, the second phase showed obvious intergranular agglomeration distribution, and the T-phase particles with the size of 2 μm were not fully broken.

The UE can effectively break the second phase and weaken the second-phase agglomeration along the extrusion direction. The fine dispersive distribution of the second phase can be continued in the subsequent Severe Plastic Deformation (SPD) process. Compared with direct ACAE-forming components, UE + ACAE-forming components represent a higher degree of second-phase refinement and dispersion.

In order to further analyze the microstructure changes of the sample after the UE deformation and ACAE forming, an EBSD analysis was used in the microstructure of the billet and in the Zone V microstructure of the cabin, as shown in Figures 6–8. Figure 6a shows the initial microstructure of the extruded billet; Figure 6c shows the Zone V structure of the direct ACAE-forming cabin; Figure 6b shows the microstructure of the billet after the UE process; and Figure 6d shows the Zone V structure of the UE + ACAE-forming cabin. The low-angle grain boundary (LAGB,2°–15°) and the high-angle grain boundary (HAGB,15°–180°) were identified by misorientation angle quantification and are represented by the white line and black line, respectively. Figures 7 and 8 show the corresponding grain size distribution and dynamic recrystallization grain distribution.

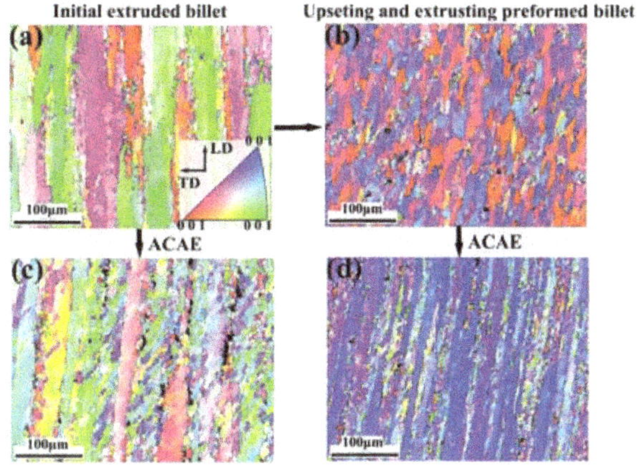

Figure 6. The inverse pole fig (IPF) of: (**a**) initial billet; (**b**) billet deformed by UE; (**c**) Zone V of direct ACAE-forming cabin [7]; (**d**) Zone V of UE + ACAE-forming cabin.

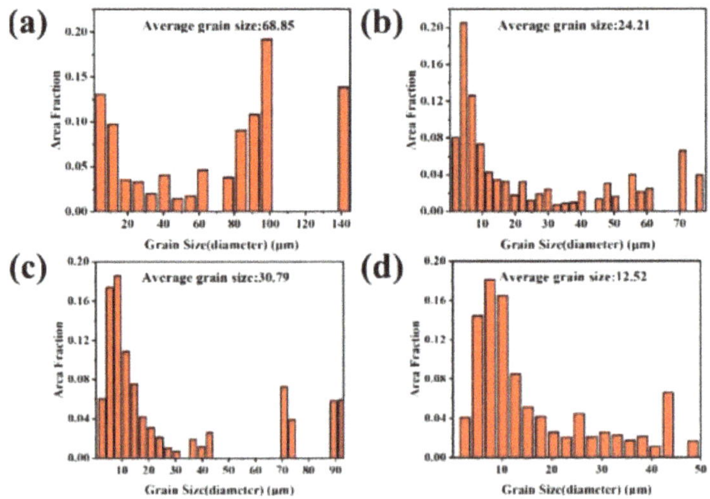

Figure 7. Grain size distribution of: (**a**) initial billet; (**b**) billet deformed by UE; (**c**) Zone V of direct ACAE-forming cabin; (**d**) Zone V of UE + ACAE-forming cabin.

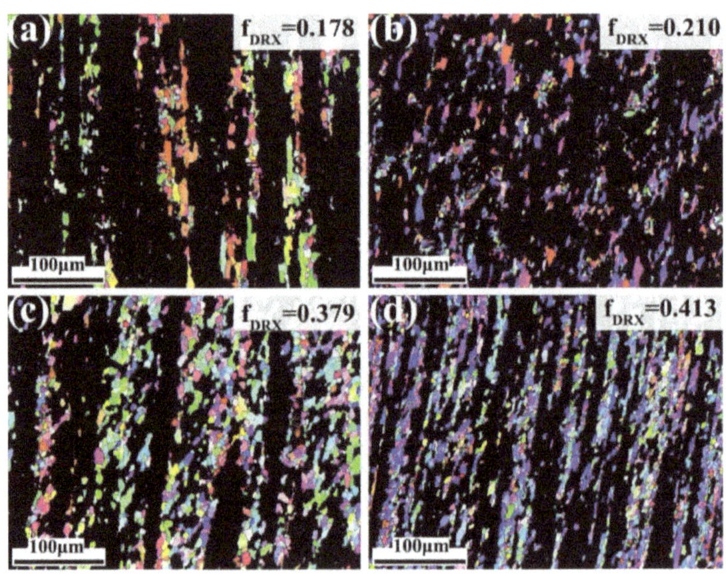

Figure 8. Dynamic recrystallization grain distribution of: (**a**) initial billet; (**b**) billet deformed by UE; (**c**) Zone V of direct ACAE-forming cabin; (**d**) Zone V of UE + ACAE-forming cabin.

As can be seen from Figure 6, compared with the direct ACAE structure, the internal colors of the large grains in the component structure formed by the UE + ACAE changed dramatically. Large grains turned into more elongated strips, which were surrounded by a large number of small dynamically recrystallized (DRXed) grains. This is because the UE process provided a larger cumulative strain for the whole deformation process (the average effective strain increased from 3.68 to 6.22), which made the elongation of the grains and the rotation of the lattice more intense. When the force was constantly applied in the process of thermal deformation, plastic deformation occurred in the metal with the applied force, which resulted in the intracrystalline dislocation accumulation [15].

Along with the continuous strain accumulation and the thermal deformation, the dislocation density increased, which resulted in the recovery process. When the dislocation proliferation rate and disappearance rate reached equilibrium, a stable state was achieved. At this time, the dislocation was mainly concentrated on the cell wall to form subgrains, and then constantly evolved into new dynamic recrystallized grains, which was a typical continuous dynamic recrystallization (cDRX) process [16,17]. In the end, the two deformation structures of the ACAE cabin and the UE + ACAE cabin were both banded structures that were composed of elongated deformed grains and relatively fine DRXed grains.

According to Figure 7, the average grain size of the blank was refined from 68.85 to 30.79 μm by the direct ACAE forming, but a small number of large ones with sizes of 90 μm still existed. After the UE, the average grain size of the billet was refined to 24.21 μm, and the maximum was reduced by half compared with the initial billet. The average grain size was further refined to 12.52 μm, and its distribution was more uniform after the UE + ACAE. This shows that the large grains can be broken significantly by upsetting extrusion, and that the grain size can be further reduced in the subsequent ACAE forming process.

The grains with a Grain Orientation Spread (GOS) less than 2° were defined as DRXed grains and were displayed separately to obtain the dynamic recrystallized grain distribution, as shown in Figure 8. Meanwhile, the corresponding DRX ratio was calculated. As the initial billet was extruded, the dynamic recrystallization ratio was 17.8%. The DRX ratio (37.9%) was obtained by the direct ACAE forming. In contrast, the DRX ratio of the UE + ACAE Zone V structure was as high as 41.3%, and the DRXed grain size was significantly smaller. This may be because the billet, after the upsetting extrusion, had a finer grain size and more grain boundaries, which provided a favorable region for recrystallization nucleation in the subsequent ACAE deformation process. Moreover, at the initial stage of thermal deformation, the second-phase particles could accelerate the DRX process through particle-stimulated nucleation (PSN) [18,19]. For the UE + ACAE-forming component, more dislocation pinning points and nucleation sites could be generated by thinning and dispersing the second-phase particles. At the same time, the large cumulative strain could store more deformation energy in the material, and it generated more dynamic recrystallized grains through grain boundary migration or subgrain rotation [15–17]. Moreover, due to the Zener resistance effect, the broken fine particles impeded the grain boundary movement and inhibited the growth of DRXed grains [18,20]. Therefore, the fine microstructure after large plastic UE deformation enabled the subsequent ACAE-forming members to obtain fine DRXed grains [15].

3.3. Heat Treatment Microstructure

In order to improve the mechanical properties of the cabin, the UE + ACAE-formed component in this paper was conducted with a T6 heat treatment. The specific parameter was as follows: 515 °C × 1 h solid solution + 190 °C × 6 h artificial aging. The microstructure of the heat treatment is shown in Figure 9. After the T6 heat treatment, part of the residual insoluble phases was still arranged along the crystal, most of the eutectic phases were dissolved back, and the stable structure of the spiculate S phases (Al2CuMg) precipitated during the aging treatment of the 2A12 aluminum alloy [21–25]. A high-temperature solution treatment led the alloy to static recovery, the subgrains merged into recrystallized grains, and the number of HAGBs increased, which resulted in the grain boundary strengthening of the alloy. After the high-temperature solution treatment, the average grain size increased slightly to 18.46 μm; however, at the same time, the equiaxed degree of the recrystallized grains increased. This uniform structure can improve the deformation uniformity of the component in different directions after heat treatment. Moreover, several UE + ACAE plastic deformations resulted in the fine dispersion of the Al–Cu–Mg-soluble components, and a large number of subgrain boundaries and high-density dislocations provided diffusion channels for the migration of the Cu atoms, which caused the Cu-rich phases to be more fully redissolved during the solution process; at the same time, a large number of dispersed nucleation sites were provided for the S phases, and a large number of dislocations and vacancies served as diffusion channels of the Cu atoms to promote the precipitation of the aging strengthening phase [26,27]. Finally, the difference between the axial and circumferential mechanical properties of the components was reduced after the heat treatment.

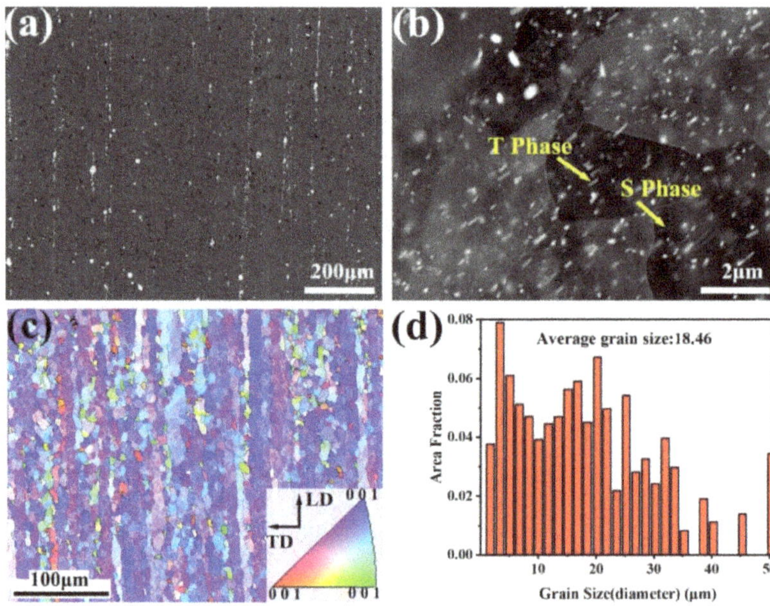

Figure 9. Heat treatment microstructure of UE + ACAE-forming cabin Zone V: (**a**) BSE Figure 1; (**b**) BSE Figure 2; (**c**) IPF figure; (**d**) grain size distribution.

3.4. Mechanical Properties

Figure 10 shows the average mechanical properties of the directly ACAE-forming cabin and the UE + ACAE-forming cabin. Axial-direction and circumferential-direction samples were used to test the mechanical properties of Cabin Zone V, according to the extrusion state and the T6 state, respectively, and the results are shown in Figure 10.

After, the billet was formed by the direct ACAE process, and the extruded mechanical properties are shown in Figure 10a. The extruded axial UTS reaches 340 MPa, the YS is 223.7 MPa, and the EL is 16.2%, while the circumferential properties are relatively low, with the UTS at 266 MPa, the YS at 155 MPa, and the EL at 9.3%. As can be seen from Figure 10b, the performance of the cabin after the T6 heat treatment shows that it maintains this characteristic. The UTS is 476 MPa, the YS increased to 371 MPa, and the EL is 12.9%. Meanwhile, the circumferential UTS is lower than the longitudinal UTS (426 MPa), the YS is 317 MPa, and the EL is 9.24%. In general, the mechanical properties of the component that was obtained by direct ACAE forming are significantly different in the axial and circumferential directions.

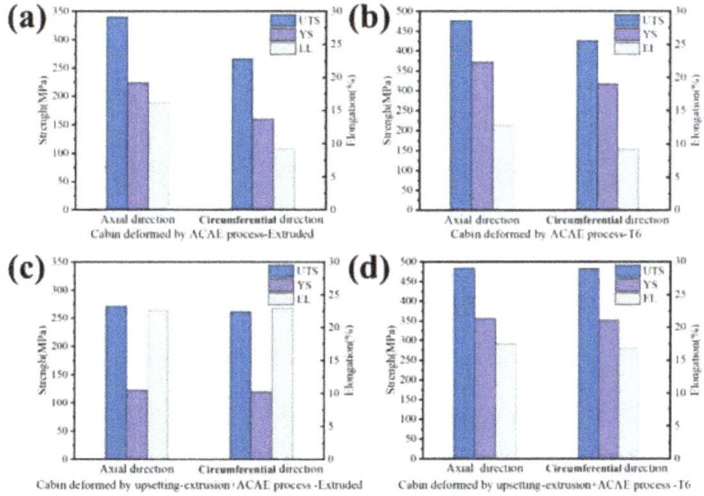

Figure 10. Tensile properties of Cabin Zone V in axial and circumscribed directions: (**a**) direct ACAE-extruded; (**b**) direct ACAE-T6; (**c**) UE + ACAE-extruded; (**d**) UE + ACAE-T6.

After, the billet was formed by the UE + ACAE process, and the extruded mechanical properties are shown in Figure 10c. The axial UTS is 271 MPa, the YS is 123 MPa, and the EL is 22.7%. The circumferential UTS is 262 MPa, the YS is 120 MPa, and the EL is 23%. The mechanical properties of the component after the T6 heat treatment are shown in Figure 10d. The axial UTS reaches 484 MPa, the YS is 356 MPa, and the EL is 17.5%. The circumferential UTS is 482 MPa, the YS is 351 MPa, and the EL is 16.8%.

The axial and circumferential mechanical properties of the cabin formed by the UE + ACAE process are more uniform. The lower UTS and YS of the extruded state are mainly due to the higher DRX ratio, which makes the material soften. The upsetting-extrusion process provides a more uniform grain size and fine dispersive second-phase particles for the component, which greatly enhances the deformation coordination ability of the material and may be beneficial to increase the degree of the solid solution and the nucleation sites that are precipitated by the aging in the subsequent solid solution and the artificial aging heat treatment. The results show that the mechanical properties of the components prepared by the UE + ACAE process are uniform in both the axial and circumferential directions in the extrusion state and the T6 state, and the EL increased at the small sacrifice of the YS.

4. Conclusions

The microstructure evolution and age-hardening behavior of 2A12 aluminum alloy components were studied by comparing predeformation (one-pass repetitive upsetting extrusion) and annular channel angular extrusion. The UE + ACAE process can significantly improve the mechanical uniformity and elongation of the 2A12 aluminum alloy ACAE component along the axial and circumferential mechanical properties. The results show that:

1. Preforming the initial billet of the 2A12 aluminum alloy by the upsetting extrusion of several plastic deformations before the ACAE can effectively obtain the regular, fine, and dispersing second phases, and can increase the cumulative deformation, and obtain more refined grains;
2. The second phases, with a regular shape and a fine and dispersive distribution, represent a higher ability to coordinate the deformation of the Al matrix, which plays a key role in improving the mechanical uniformity and elongation of the 2A12 aluminum alloy ACAE component along the axial and circumferential mechanical

properties. After the UE + ACAE process, the axial UTS and EL of the cabin in the extrusion state are 271 MPa and 22.7%, respectively. At the same time, the circumferential UTS and EL are 262 MPa and 23%, respectively. The UE process can significantly improve the mechanical properties of materials by affecting the second phase;

3. After proper T6 heat treatment, the UE + ACAE-formed component can still maintain the characteristics of the uniform axial and circumferential mechanical properties and high elongation. After a solution treatment of 515 °C × 1 h, and an artificial aging treatment of 190 °C × 6 h, the 2A12 aluminum alloy cabin enjoys good mechanical properties and uniformity. The axial UTS is 484 MPa, and the EL is 17.5%. The circumferential UTS is 482 MPa, and the EL is 16.8%. The mechanical uniformity is better than in the ACAE process. High-performance cabins with a uniform performance can be prepared by combining the upsetting extrusion of several plastic deformations with the annular channel angular extrusion process.

Author Contributions: Conceptualization, methodology, X.Z. and L.-F.G.; software, writing—review and editing, K.C.; validation, resources, writing—original draft preparation, D.-K.W.; validation, Z.-M.Z. All authors have read and agreed to the published version of the manuscript.

Funding: This research was funded by the Key Core Technology and Generic Technology Research and Development Project of Shanxi province, Grant No. 2020XXX015.

Institutional Review Board Statement: Not applicable.

Informed Consent Statement: Not applicable.

Data Availability Statement: All data included in this study are available upon request by contact with the corresponding author.

Conflicts of Interest: The authors declare no conflict of interest.

Appendix A

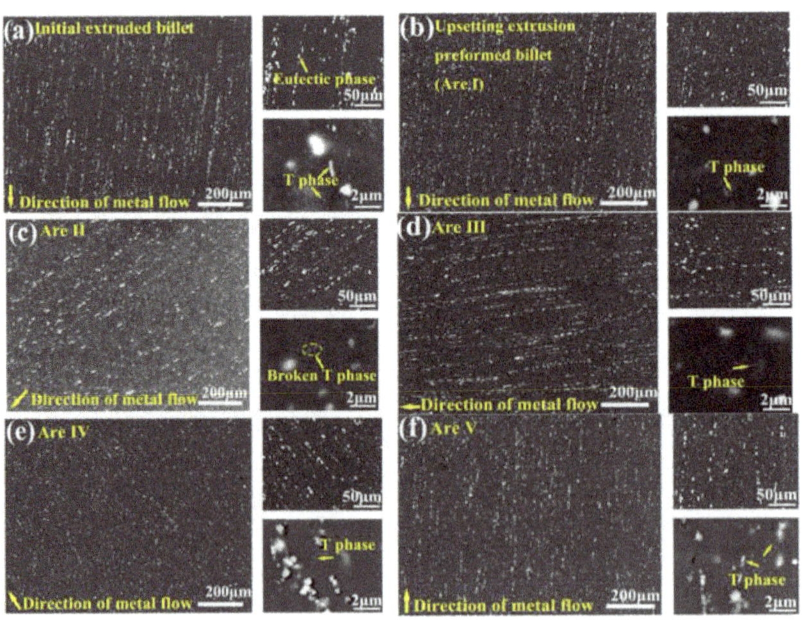

Figure A1. *Cont.*

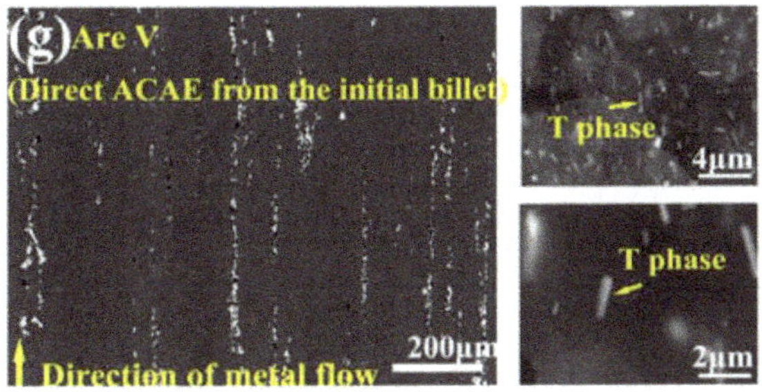

Figure A1. High resolution of Figure 4: (**a**) initial billet; (**b**) billet deformed by UE; (**c**) UE + ACAE Zone II; (**d**) UE + ACAE Zone III; (**e**) UE + ACAE Zone IV; (**f**) UE + ACAE Zone V; (**g**) ACAE Zone V.

References

1. Dursun, T.; Soutis, C. Recent developments in advanced aircraft aluminium alloys. *Mater. Des.* **2014**, *56*, 862–871. [CrossRef]
2. Williams, J.C.; Starke, E.A., Jr. Progress in structural materials for aerospace systems. *Acta Mater.* **2003**, *51*, 5775–5799. [CrossRef]
3. Zhang, X.; Chen, Y.; Hu, J. Recent advances in the development of aerospace materials. *Prog. Aerosp. Sci.* **2018**, *97*, 22–34. [CrossRef]
4. Shatermashhadi, V.; Manafi, B.; Abrinia, K.; Faraji, G.; Sanei, M. Development of a novel method for the backward extrusion. *Mater. Des.* **2014**, *62*, 361–366. [CrossRef]
5. Hosseini, S.H.; Abrinia, K.; Faraji, G. Applicability of a modified backward extrusion process on commercially pure aluminum. *Mater. Des.* **2015**, *65*, 521–528. [CrossRef]
6. Zhao, X.; Li, S.; Zhang, Z.; Gao, P.; Kan, S.; Yan, F. Comparisons of microstructure homogeneity, texture and mechanical properties of AZ80 magnesium alloy fabricated by annular channel angular extrusion and backward extrusion. *J. Magnes. Alloy.* **2020**, *8*, 624–639. [CrossRef]
7. Zhao, X.; Li, S.; Yan, F.; Zhang, Z.; Wu, Y. Microstructure Evolution and Mechanical Properties of AZ80 Mg Alloy during Annular Channel Angular Extrusion Process and Heat Treatment. *Materials* **2019**, *12*, 4223. [CrossRef]
8. Gao, P.; Zhao, X.; Yan, F.; Li, S.; Zhang, Z. Effects of annular channel angular extrusion process and heat treatment on microstructure and mechanical properties of 2A12 aluminum alloy cabin. *Mater. Res. Express* **2019**, *6*, 126578. [CrossRef]
9. Balasundar, I.; Raghu, T. Severe Plastic Deformation (SPD) Using a Combination of Upsetting and Extrusion. *J. Metall. Eng.* **2013**, *2*, 130–139.
10. Xya, B.; Am, A.; Lz, A. Crystallographic anisotropy of corrosion rate and surface faceting of polycrystalline 90Cu-10Ni in acidic NaCl solution. *Mater. Des.* **2022**, *215*, 1–14.
11. Tang, Y.; Shen, X.; Liu, Z. Corrosion Behaviors of Selective Laser Melted Inconel 718 Alloy in NaOH Solution. *Acta Metal. Sin.* **2022**, *58*, 324–333.
12. Haigen, W.; Fuzhong, X.; Mingpu, W. Effect of ingot grain refinement on the tensile properties of 2024 Al alloy sheets. *Mater. Sci. Eng. A* **2017**, *682*, 1–11. [CrossRef]
13. Liu, X.Y.; Pan, Q.L.; Lu, Z.L.; Cao, S.F.; He, Y.B.; Li, W.B. Effects of solution treatment on the microstructure and mechanical properties of Al–Cu–Mg–Ag alloy. *Mater. Des.* **2010**, *31*, 4392–4397. [CrossRef]
14. Chen, Z.; Chen, P.; Li, S. Effect of Ce addition on microstructure of Al20Cu2Mn3 twin phase in an Al–Cu–Mn casting alloy. *Mater. Sci. Eng. A* **2012**, *532*, 606–609. [CrossRef]
15. Sakai, T.; Belyakov, A.; Kaibyshev, R.; Miura, H.; Jonas, J.J. Dynamic and post-dynamic recrystallization under hot, cold and severe plastic deformation conditions. *Prog. Mater. Sci.* **2014**, *60*, 130–207. [CrossRef]
16. Huang, K.; Logé, R. A review of dynamic recrystallization phenomena in metallic materials. *Mater. Des.* **2016**, *111*, 548–574. [CrossRef]
17. Maizza, G.; Pero, R.; Richetta, M.; Montanari, R. Continuous dynamic recrystallization (CDRX) model for aluminum alloys. *J. Mater. Sci.* **2017**, *53*, 4563–4573. [CrossRef]
18. Wang, F.; Zheng, R.; Chen, J.; Lyu, S.; Li, Y.; Xiao, W.; Ma, C. Significant improvement in the strength of Mg-Al-Zn-Ca-Mn extruded alloy by tailoring the initial microstructure. *Vacuum* **2019**, *161*, 429–433. [CrossRef]
19. Chen, X.H.; Liu, L.Z.; Liu, J.; Pan, F.S. Enhanced Electromagnetic Interference Shielding of Mg–Zn–Zr Alloy by Ce Addition. *Acta Metall. Sin.* **2015**, *28*, 492–498. [CrossRef]
20. Robson, J.D.; Henry, D.T.; Davis, B. Particle effects on recrystallization in magnesium–manganese alloys: Particle pinning. *Mater. Sci. Eng. A* **2011**, *528*, 4239–4247. [CrossRef]
21. Styles, M.J.; Hutchinson, C.R.; Chen, Y.; Deschamps, A.; Bastow, T.J. The coexistence of two S (Al2CuMg) phases in Al-Cu-Mg alloys. *Acta Mater.* **2012**, *60*, 6940–6951. [CrossRef]
22. Wang, S.C.; Starink, M.J. Two types of S phase precipitates in Al-Cu-Mg alloy. *Acta Mater.* **2007**, *55*, 933–941. [CrossRef]

23. Hu, Z.Y.; Fan, C.H.; Tong, S.H.; Ling, O.U.; Dai, N.S.; Lu, W.A. Effect of aging treatment on evolution of S′ phase in rapid cold punched Al-Cu-Mg alloy. *Trans. Nonferrous Met. Soc. China* **2021**, *31*, 1930–1938. [CrossRef]
24. Hu, Z.Y.; Fan, C.H.; Zheng, D.S.; Liu, W.L.; Chen, X.H. Microstructure evolution of Al-Cu-Mg alloy during rapid cold punching and recrystallization annealing. *Trans. Nonferrous Met. Soc. China* **2019**, *29*, 1816–1823. [CrossRef]
25. Feng, Z.; Yang, Y.; Huang, B.; Han, M.; Luo, X.; Ru, J. Precipitation process along dislocations in Al–Cu–Mg alloy during artificial aging. *Mater. Sci. Eng. A* **2010**, *528*, 706–714. [CrossRef]
26. Liu, Z.; Bai, S.; Zhou, X.; Gu, Y. On strain-induced dissolution of θ′ and θ particles in Al–Cu binary alloy during equal channel angular pressing. *Mater. Sci. Eng. A* **2011**, *528*, 2217–2222. [CrossRef]
27. Cabibbo, M.; Evangelista, E.; Vedani, M. Influence of severe plastic deformations on secondary phase precipitation in a 6082 Al-Mg-Si alloy. *Met. Mater. Trans. A* **2005**, *36*, 1353–1364. [CrossRef]

coatings

Article

The Corrosion Behaviors of an As-Rolled Mg-8Li (in wt.%) Alloy in Two Differently Concentrated NaCl Solutions

Baojie Wang [1,*], Jiyu Hou [1], Jiyu Luan [1], Daokui Xu [2,*], Haijing Sun [1] and Jie Sun [1]

[1] School of Environmental and Chemical Engineering, Shenyang Ligong University, Shenyang 110159, China; jyhou@126.com (J.H.); jyluan@126.com (J.L.); hjsun@sylu.edu.cn (H.S.); sunjie@sylu.edu.cn (J.S.)

[2] Key Laboratory of Nuclear Materials and Safety Assessment, Institute of Metal Research, Chinese Academy of Sciences, Shenyang 110016, China

* Correspondence: bjwang@alum.imr.ac.cn (B.W.); dkxu@imr.ac.cn (D.X.)

Abstract: By means of an electrochemical workstation, hydrogen evolution device, optical microscope (OM) and scanning electron microscope (SEM), the corrosion behaviors of an as-rolled Mg-8%Li alloy with a dual phase structure in 0.9 wt.% NaCl and 3.5 wt.% NaCl solutions have been investigated and compared. The results show that when the immersion time exceeds 8 h, the hydrogen evolution rate of the alloy in the 0.9 wt.% NaCl is 3 times higher than that in the 3.5 wt.% NaCl solution. Moreover, the corrosion behaviors of the alloy are obviously different in the two differently concentrated NaCl solutions. In the 3.5 wt.% NaCl solution, the localized corrosion is much more severe and can occur simultaneously in the interior of both the α-Mg and β-Li matrix phases. However, the localized corrosion in the 0.9 wt.% NaCl solution is obviously weak and mainly occurs at the α-Mg phase.

Keywords: magnesium–lithium alloy; microstructure; localized corrosion; corrosion product film

Citation: Wang, B.; Hou, J.; Luan, J.; Xu, D.; Sun, H.; Sun, J. The Corrosion Behaviors of an As-Rolled Mg-8Li (in wt.%) Alloy in Two Differently Concentrated NaCl Solutions. *Coatings* **2022**, *12*, 406. https://doi.org/10.3390/coatings12030406

Academic Editor: Stefanos M. Skolianos

Received: 24 February 2022
Accepted: 17 March 2022
Published: 18 March 2022

1. Introduction

Magnesium–lithium (Mg–Li) alloys are the lightest metallic materials known and their density is between 1.35~1.65 g/cm^3, which is only 1/2 of that of aluminum alloys. Compared with traditional magnesium alloys, Mg–Li alloys have great application prospects in the fields of automotive, aerospace and the military due to their excellent deformation capacity, specific strength and stiffness [1–5]. Generally, the microstructure of Mg–Li alloys varies with the added Li content. The dual phase Mg–Li alloys is comprised of the α-Mg phase and β-Li phase when the Li content is less than 10.3 wt.% and higher than 5.5 wt.% [5]. It has been widely reported that the duplex structured Mg–Li alloys exhibit a good combination of mechanical strength and plasticity [6–12], ensuring their wide application prospects compared to other Mg alloys. However, micro galvanic corrosion can easily occur between these two phases because the electrode potential of the β-Li phase is lower than that of the α-Mg phase, resulting in the poor corrosion resistance of dual phase Mg–Li alloys [3,13–17].

Since the chloride ion (Cl^-) is an aggressive ion and has strong activity, the corrosion behavior of magnesium alloys is generally reported to be positively correlated with chloride ion concentration [18–20]. It has been reported that the corrosion severity of uncoated and coated AZ31 samples was significantly increased with the concentration of chloride ions [20]. In the research of the corrosion behavior of AZ31D magnesium alloys in NaCl solutions, Shan et al. reported that due to the increased concentration of chloride ions, the high frequency capacitive loop shrank, and the open circuit potential shifted to a more negative direction [19]. However, it is also reported that the induced pitting corrosion in the simulated acid rain solutions was not sensitive to the chloride ion concentrations, but due to the AlMn phases, it was located at the grain boundary [21]. Accordingly, it can be concluded that the corrosion behavior of Mg alloys in NaCl solutions depends on the Cl^- concentration and the microstructural characteristics. For Mg–Li based alloys, although their corrosion behaviors in NaCl solutions have been widely reported [22–27], no related work about the effect of Cl^- concentrations on the corrosion behavior of Mg–Li alloys has been performed.

Previous work has demonstrated that severely localized corrosion was visible for the duplex structured Mg–Li alloy [28,29]. Based on the mechanism of pitting, the interior of the pits becomes acidic and acts as the filament tip, whilst the back end of the filament is alkaline and covered with corrosion products. The OH^- can respectively react with Mg^{2+} and Li^+ to form $Mg(OH)_2$ and LiOH, and subsequently react with CO_2 to produce $MgCO_3$ and Li_2CO_3 [3,5,29–31]. Xu et al. reported that a thin and dense outer layer of Li_2CO_3 formed on the surface of the β-Li phase could effectively weaken the corrosion attack to the β-Li phase [30]. On the contrary, the porous and loose $MgCO_3$ and $Mg(OH)_2$ formed on the surface of the α-Mg phase can be easily destroyed by the Cl^- and then induce the initiation of pitting corrosion [3,5]. Based on the description mentioned above, it can be predicted that for the dual phase Mg–Li alloys, their corrosion resistance should be quite sensitive to the concentration of Cl^-. However, previous work seldomly focused on whether the concentration of Cl^- can influence the protectiveness of the corrosion product films to the α-Mg and β-Li phase or not. Moreover, it is still unknown whether the concentration of Cl^- can change the corrosion mechanism of the dual-phase Mg–Li alloys.

In this work, through performing the hydrogen evolution test, corrosion morphology characterization and electrochemical experiments of an as-rolled dual phase Mg-8%Li alloy in different concentrated NaCl solutions, the target is to answer the following questions: (1) can the corrosion resistance of the alloy can be significantly influenced by the concentrations of chloride ions? (2) If yes, what is the difference of the corrosion mechanisms in different concentrated NaCl solutions? In addition, the corrosion process and corrosion mechanism of the alloy were deeply disclosed.

2. Materials and Methods

2.1. Sample Preparation

The experimental material used in this work is a typical (α-Mg + β-Li) dual phase Mg-8%Li (in wt.%) alloy. Through the inductively coupled plasma atomic emission spectrum (ICP-AES) apparatus, its chemical composition was determined, as listed in Table 1. The alloy was prepared by melting the high purity Mg and Li bulks in a vacuum reaction furnace under the atmosphere of flux argon to avoid burning. After keeping for 40 min at 710 °C to homogenize it, the molten alloy was cast into plate ingots with a cross-section of 200 mm × 300 mm and a thickness of 50 mm. Then, thin plates with a cross-section of 150 mm × 150 mm and a thickness of 18 mm were cut from the ingot. After being homogenized at 300 °C for 30 min, the plates were rolled at room temperature and their thickness was controlled to be 6 mm. To eliminate the residual stress induced by the rolling process, plates were held at 300 °C for 1 h in an air furnace. Then, block samples with a dimension of 10 mm × 5 mm × 5 mm for electrochemical measurements were cut from the plates and mounted by the epoxy resin AB glue with an exposed working area of 5 mm × 10 mm. Then, samples were grinded gradually with 400#, 800#, 1000# and 2000# water abrasive papers to remove the surface scratches, followed by washing with acetone and drying with cold air. Moreover, samples for the immersion and hydrogen evolution tests were polished with 2.5 μm diamond paste after grinding. To clearly reveal the microstructure of the alloy, the polished surface of one sample was etched by an enchant solution of (4% HNO_3 + 96% C_2H_5OH) and subsequently observed by using an optical microscope (OM). Based on the "ImageJ" software, the occupied area fractions of two matrix phases were measured.

Table 1. Chemical composition of Mg-8%Li alloy.

Chemical Composition	Li	Fe	Mg
Content (wt.%)	7.85	<0.001	balance

2.2. Electrochemical Tests

The CS350 electrochemical workstation (Corrtest Co. Ltd., Wuhan, China) was used for the electrochemical test. The traditional three-electrode set-up was applied, i.e., the exposed areas of the mounted samples were the working electrode, the saturated calomel electrode (SCE) was the reference electrode and the platinum plate was the auxiliary electrode. After being pre-immersed in 0.9 wt.% NaCl and 3.5 wt.% NaCl solutions, respectively, for 0, 4, 8 and 24 h, their electrochemical measurements are performed. Before the measurements of potentiodynamic polarization curves and electrochemical impedance spectrum, a delay of 10 min was required to obtain a stable open circuit potential. The scanning rate of potentiodynamic polarization measurements is 0.166 mv/s. Potentiodynamic polarization curves were fitted by CorrView software. Electrochemical impedance spectra (EIS) were measured in a frequency range from 100 kHz to 10 mHz with a disturbed potential amplitude of 10 mV and subsequently fitted by using the ZSimDemo 3.30 software. To reveal the deviation of the measured data, each electrochemical experiment was carried out at least three times. Finally, the electrochemical curves were drawn by Origin 2019 software.

2.3. Hydrogen Evolution Measurements and Immersion Test

It has been reported that when Mg alloys are exposed to the aqueous solutions, the hydrogen evolution reaction due to the cathodic reaction (i.e., $2H^+ + 2e \rightarrow H_2\uparrow$) could take place [32,33]. Thus, the measured volume of the evolved hydrogen is related to the anodic dissolution of the substrate of Mg alloys [32]. The schematic diagram of the hydrogen evolution device used in the current investigation is shown in Figure 1. The polished samples were respectively immersed into the 0.9 wt.% NaCl and 3.5 wt.% NaCl solutions. The ratio of the corrosion medium volume to the exposed area of samples was set to 200:1. The measurements of hydrogen evolution volume were performed up to 48 h and the data were recorded every 2 h. After being respectively immersed in 0.9 wt.% NaCl and 3.5 wt.% NaCl solution for 4, 8 and 24 h, the surface morphologies and their three-dimensional (3D) profiles were observed by using a stereo optical microscope (VHX-900F, Keyence International, Mechelen, Belgium). The operational principle for obtaining 3D images with precise changes in height includes two steps: (1) taking the image at the same location automatically layer by layer; (2) combining all images into one. Moreover, the cross-sectional morphologies of the immersed samples were observed by using a scanning electron microscope (SEM; FEI Quanta 450, Hillsboro, OR, USA).

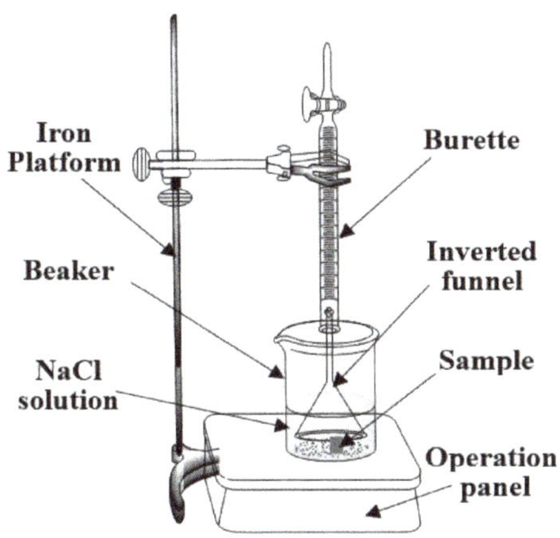

Figure 1. Schematic diagram of hydrogen evolution apparatus.

3. Results

3.1. Microstructure

Figure 2 shows the optical microstructure of the as-rolled Mg-8%Li alloy. It can be seen that the alloy exhibits a typical dual phase structure. Generally, for the Mg–Li binary alloys, they are composed of α-Mg and β-Li phases when the added Li content varies from 5.5 to 10.3 wt.% [3,5,10,14]. Moreover, for the duplex structured Mg–Li alloys, the α-Mg and β-Li matrix phases respectively appear in white and black when observed by using OM [3,5,10,14]. Based on the Mg–Li binary phase diagram proposed by Masing et al. [34], the start melting temperature of β-Li phase is lower than that of the α-Mg phase and decreases from 592 °C to 180.6 °C with the increase of Li content in the β-Li phase. Therefore, the α-Mg phase first nucleates and grows at random sites during solidification process, resulting in their irregular and messy distribution. On the basis of the measurement from Image J software, the area fraction of the α-Mg phase is determined to be 60%.

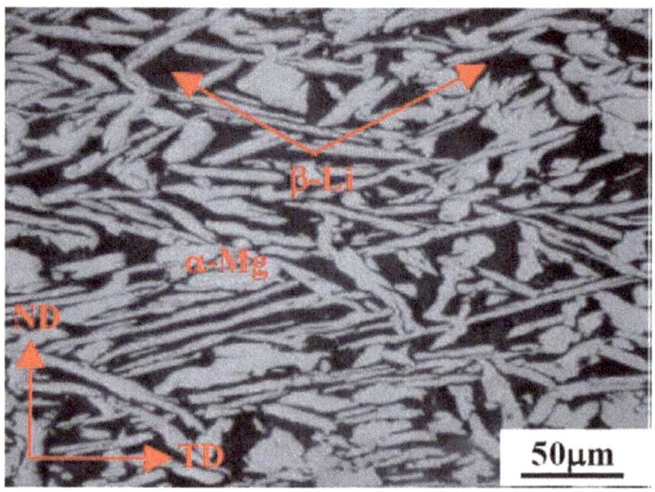

Figure 2. Optical microstructure of the as-rolled Mg-8%Li alloy.

3.2. Electrochemical Responses

Figure 3 shows the potentiodynamic polarization curves of the samples being respectively pre-immersed in the two NaCl solutions for 0, 4, 8 and 24 h. It can be clearly seen that their cathodic branches are basically overlapped, indicating that the Cl^- concentration has almost no impact on the cathodic process. Since the negative difference effect and anodic dissolution occur during anodic polarization and then have a great impact on the sample surface, the cathodic and anodic branches of the polarization curves are asymmetric and the current density in the anodic section increases rapidly. As a result, the cathodic branch is generally used for fitting analysis [28]. The Tafel fitted results of potentiodynamic polarization curves are listed in Table 2. This reveals that the corrosion current densities of the samples being pre-immersed in the 0.9 wt.% NaCl are less than those being pre-immersed in the 3.5 wt.% NaCl. In addition, the corrosion potentials of the samples are much higher in the 0.9 wt.% NaCl than those in the 3.5 wt.% NaCl. After performing pre-immersion for up to 24 h, the corrosion current density of the sample in the 0.9 wt.% NaCl increased from the initial value of 1.9735×10^{-5} to 5.104×10^{-5} A/cm^2. However, the current density of the sample in the 3.5 wt.% NaCl corrosion medium increased from the initial value of 2.1679×10^{-5} to 6.1598×10^{-5} A/cm^2. Thus, the corrosion current density of samples in two concentrated solutions increases continuously with the prolongation of pre-immersion time. Moreover, the increased rate of corrosion current density of the samples due to the pre-immersion in the 3.5 wt.% NaCl is much higher than those in the

0.9 wt.% NaCl. Generally, the lower corrosion current density ensures the alloys have better corrosion resistance [35]. Compared with the corrosion in the 3.5 wt.% NaCl solution, samples immersed in the 0.9 wt.% NaCl have better corrosion resistance.

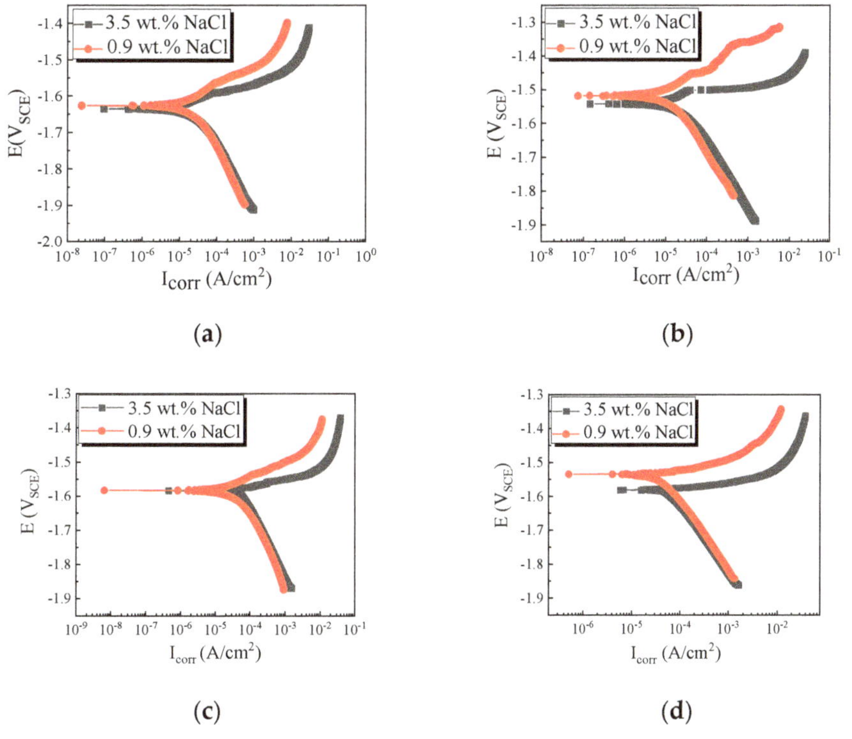

Figure 3. Potentiodynamic polarization curves of samples being pre-immersed in 0.9 wt.% NaCl and 3.5 wt.% NaCl solutions for different time: (**a**) 0 h, (**b**) 4 h, (**c**) 8 h and (**d**) 24 h.

Table 2. Fitted results of polarization curves.

Pre-Immersion Time	In 0.9 wt.% NaCl		In 3.5 wt.% NaCl	
	E_{corr} (V_{SCE})	i_{corr} ($\mu A/cm^2$)	E_{corr} (V_{SCE})	i_{corr} ($\mu A/cm^2$)
0 h	1.625 ± 0.005	20 ± 2	1.643 ± 0.005	22 ± 2
4 h	1.524 ± 0.003	24 ± 2	1.545 ± 0.003	43 ± 3
8 h	1.585 ± 0.003	41 ± 4	1.586 ± 0.003	55 ± 4
24 h	1.538 ± 0.003	51 ± 4	1.572 ± 0.003	62 ± 5

Figure 4 shows the AC impedance spectrum of the sample immersed in the 0.9 wt.% NaCl and 3.5 wt.% NaCl solutions for 0, 4, 8 and 24 h. For the samples measured in two concentrations of NaCl solutions, a high-frequency capacitive loop and a middle to low-frequency capacitive loop can be observed (Figure 4a). Moreover, the radius of the capacitive loop of the sample immersed in 3.5 wt.% NaCl is smaller than that in 0.9 wt.% NaCl. Generally, the larger the radius of the capacitive loop, the better the corrosion resistance of the sample [3,14,27,30]. Thus, the corrosion resistance of the sample immersed in 0.9 wt.% NaCl is much better. After being pre-immersed for 4 h, the impedance spectrum of the sample in 0.9 wt.% NaCl shows a high-frequency capacitive loop and a middle to low-frequency capacitive loop, whilst it contains a high frequency capacitive and a low-frequency inductive loop for the sample in the 3.5 wt.% NaCl solution (Figure 4b). Generally, the appearance of the low-frequency inductive loop indicates that the sample surface is prone to pitting or localized corrosion [21]. Therefore, it indirectly reflects that the occurrence of pitting or localized corrosion may happen at the surface of the sample being immersed in the 3.5 wt.% NaCl solution for 4 h. After being pre-immersed for 8 h, the sample immersed in 0.9 wt.% NaCl contains a high-frequency capacitive loop, a middle to low-frequency capacitive loop and a low-frequency inductive loop, whilst the sample immersed in 3.5 wt.% NaCl contains a high-frequency capacitive and a low-frequency inductive loop (Figure 4c). Moreover, the radius of each of the capacitive loop of the two curves is reduced, demonstrating that the corrosion resistance decreases with increasing immersion time in both concentrated NaCl solutions. After being pre-immersed for 24 h, the AC impedance spectrums of the samples measured in two concentrated NaCl solutions contain a high frequency capacitive and a low-frequency inductive loop (Figure 4d). Moreover, the radius of their capacitive loops is further reduced. For the sample immersed in the 0.9 wt.% NaCl solution, the middle to low-frequency capacitive loop disappears. Based on the changes of the capacitive loop radius, this reflects that the corrosion resistance of the samples in the two NaCl solutions decreases further with the prolongation of immersion time. Additionally, the localized corrosion attack may occur earlier in the 3.5 wt.% NaCl solution, indicating that the corrosion resistance of the sample is much better in the 0.9 wt.% NaCl solution.

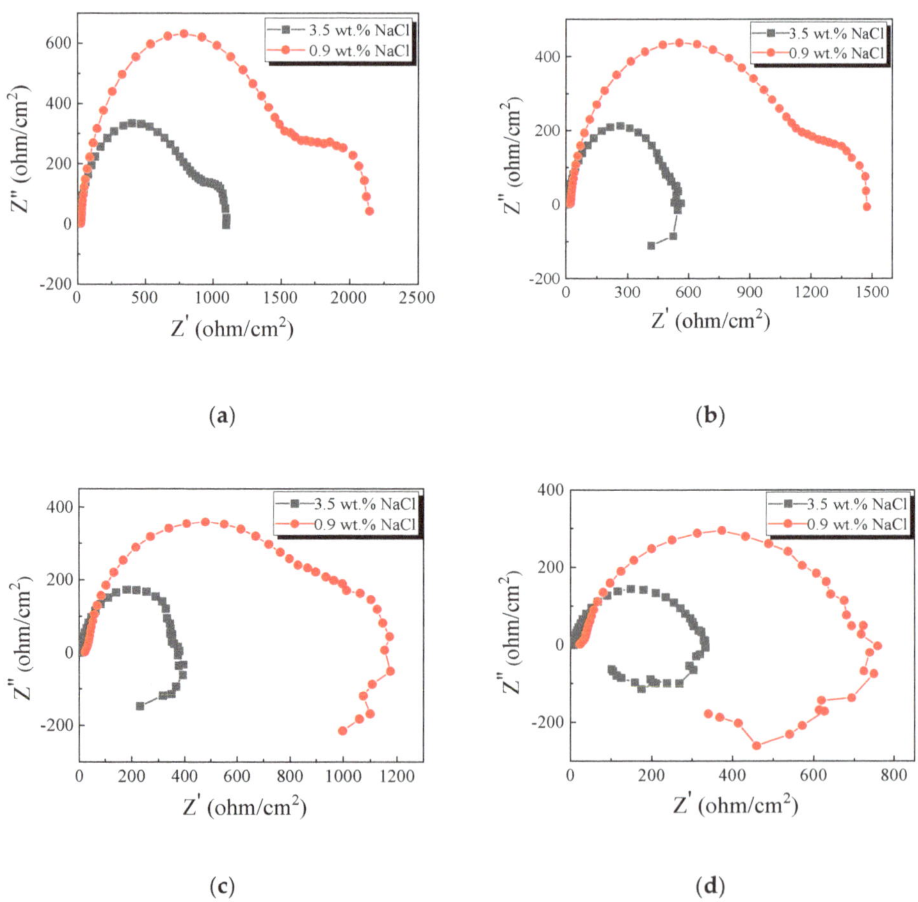

Figure 4. AC impedance spectrum of the Mg-8%Li alloy pre-immersed in 0.9 wt.% NaCl and 3.5 wt.% NaCl solution for different time: (**a**) 0 h, (**b**) 4 h, (**c**) 8 h and (**d**) 24 h.

To reflect the corrosion characteristics of the alloy being pre-immersed in two NaCl solutions for different times, electrochemical equivalent circuits were proposed to fit the EIS data, as shown in Figure 5. In the equivalent circuits, R_s is the solution resistance. R_{ct} and Q_{dl} represent charge transfer resistance and electric double layer, respectively, at the interface between substrate and electrolyte, which are used to describe the capacitance loop at high frequency. Q_{dl} is a constant phase element that replaces an ideal capacitor to account for the non-homogeneity in the system, which is defined by the two values of Y_{dl} and n_{dl}. Among them, n_{dl} is the dispersion coefficient of Q_{dl}. If n_{dl} = 1, Q_{dl} is identical to a capacitor; if n_{dl} = 0, Q_{dl} represents a resistance. R_f and Q_f (defined by Y_f and n_f) represent film resistance and capacity in the medium frequency capacitance loop, respectively. R_L and L stand for resistance and inductance, respectively, which are used to describe the low frequency inductance loop and related to the localized corrosion [36].

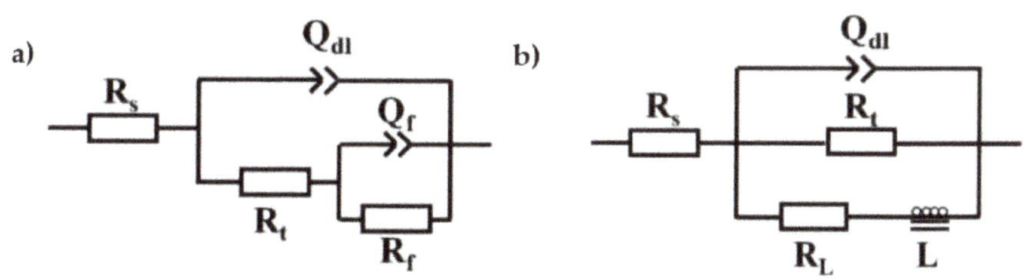

Figure 5. Equivalent circuits for the Mg-8%Li alloy pre-immersed in different NaCl solutions for different time: (**a**) 0 h and 4 h in 0.9 wt.% NaCl solution, 0 h in 3.5 wt.% NaCl solution, (**b**) 8 h and 24 h in 0.9 wt.% NaCl solution, 4 h, 8 h and 24 h in 3.5 wt.% NaCl solution.

The fitted results are listed in Table 3. It reveals that for the samples measured in two differently concentrated NaCl solutions, their R_{ct} values decrease with the prolongation of pre-immersed time. Moreover, the R_{ct} values measured in the 3.5 wt.% NaCl solution are obviously lower than those in the 0.9 wt.% NaCl solution. Thus, the corrosion resistance of the Mg-8%Li alloy in the 0.9 wt.% NaCl solution is relatively higher and decreases with the prolongation of pre-immersed time. In addition, the low frequency inductance loop in the 3.5 wt.% NaCl solution occurs earlier than that in the 0.9 wt.% NaCl solution, indicating that the occurrence of localized corrosion in the 3.5 wt.% NaCl solution is much easier.

Table 3. The fitted EIS data on the basis of the equivalent circuits.

Conditions		R_s (Ω cm^2)	Y_{dl} ($\mu\Omega^{-1}$cm^{-2}s^n)	n_{dl}	R_{ct} (Ω cm^2)	Y_f ($\mu\Omega^{-1}$cm^{-2}s^n)	n_f	R_f (Ω cm^2)	L (H cm^{-2})	R_L (Ω cm^2)
In 0.9 wt.% NaCl	0 h	24.2 ± 0.5	20 ± 2	0.91 ± 0.03	1447 ± 90	1537 ± 150	0.71 ± 0.03	738 ± 75	-	-
	4 h	20.4 ± 0.5	18 ± 2	0.91 ± 0.03	991 ± 80	1245 ± 125	0.61 ± 0.02	521 ± 65	-	-
	8 h	22.3 ± 0.4	88 ± 5	0.91 ± 0.03	831 ± 80	-	-	-	2619 ± 348	175 ± 58
	24 h	24.8 ± 0.4	102 ± 8	0.83 ± 0.02	726 ± 70	-	-	-	6840 ± 465	382 ± 95
In 3.5 wt.% NaCl	0 h	7.0 ± 0.2	27 ± 2	0.89 ± 0.02	764 ± 70	1829 ± 176	0.61 ± 0.02	394 ± 35	-	-
	4 h	7.6 ± 0.2	89 ± 5	0.89 ± 0.02	523 ± 60	-	-	-	2185 ± 315	146 ± 54
	8 h	7.4 ± 0.2	119 ± 8	0.91 ± 0.03	378 ± 40	-	-	-	3695 ± 385	204 ± 69
	24 h	7.3 ± 0.2	96 ± 5	0.91 ± 0.03	316 ± 40	-	-	-	1482 ± 175	111 ± 45

3.3. Hydrogen Evolution

Since the cathodic process of electrochemical corrosion of Mg alloys is due to the hydrogen evolution reaction, their corrosion rate can be calculated by the volume fraction of evolved hydrogen per unit of exposed area. Figure 6 shows the hydrogen evolution curves of the samples immersed in two concentrated NaCl solutions for up to 48 h. It reveals that after immersion for 48 h, the hydrogen evolution volumes of the samples immersed in 0.9 wt.% and 3.5 wt.% NaCl solutions are 1.16 and 2.69 mL/cm^2, respectively, indicating that the corrosion rate of the sample in the 3.5 wt.% NaCl solution is faster than that in the 0.9 wt.% solution. This result is consistent with the electrochemical measurements. Moreover, the variations of hydrogen evolution rates of the samples measured in the two concentrated solutions are different during the immersion process. When the immersion time is less than 8 h, the hydrogen evolution rates of the samples measured in the two concentrated solutions are low and their difference is quite small. When the immersion time exceeds 8 h, the hydrogen evolution rates of the samples immersed in the two concentrated solutions increase remarkably. Based on the slopes of hydrogen evolution curves, the hydrogen evolution rate of the alloy in the 3.5 wt.% NaCl solution is 3 times higher than that in the 0.9 wt.% NaCl solution.

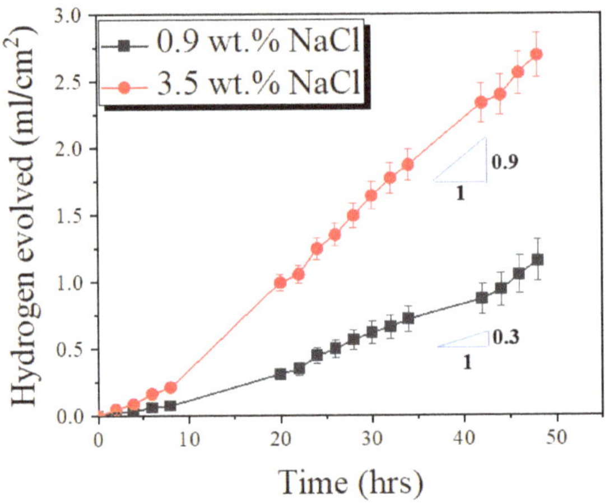

Figure 6. Hydrogen evolution curves of samples immersed in 0.9 wt.% NaCl and 3.5 wt.% NaCl solutions.

3.4. Observations to the Corrosion Morphologies

To reveal the different corrosion mechanisms of samples immersed in the two concentrated NaCl solutions, their surface corrosion morphologies and 3D images were observed and compared, as shown in Figures 7 and 8, respectively. At the early stage of corrosion, obvious filiform-like corrosion occurs on the surfaces of the different samples. Moreover, the corrosion severity of the samples in the two solutions intensifies with the prolongation of immersion time. When the immersion time is 4 h, the corrosion severity of the sample immersed in the 0.9 wt.% NaCl solution is significantly slighter than that in the 3.5 wt.% NaCl solution (Figures 7a and 8a). Based on the measured results from 3D images, the maximum depths of corrosion pits in the 0.9 wt.% and 3.5 wt.% NaCl solutions are 6.65 μm and 14.07 μm, respectively. Since a dense layer of Li_2CO_3 can be formed on the surface of the β-Li phase [3,5,29–31], the subsequent corrosion attack to the substrate can be effectively inhibited. For the sample that is immersed in the 3.5 wt.% NaCl solution, the corrosion attack is much more severe and occurs widely in the α-Mg phase. Moreover, the corrosion severity increases, and the depths of the corrosion pits become deeper when the immersion time increases. It is well known that the head of the filiform tip has the stronger acidity and can accelerate the corrosion, whilst the back of the filiform tip is alkaline and corrosion products are deposited [3]. When the immersion time is up to 24 h, some β-Li phases can still be present for the sample immersed in the 0.9 wt.% NaCl solution, whilst the whole surface of the sample that is immersed in the 3.5 wt.% NaCl solution is almost completely covered by the corrosion products. However, the maximum depths of the corrosion pits in the two solutions after immersion for 24 h are basically same and their values are 20.55 μm and 21.56 μm, respectively. Combined with surface corrosion morphology and 3D images, it can be concluded that the corrosion attack of the Mg-8%Li alloy is much more severe in the 3.5 wt.% NaCl solution, which is consistent with the results of the hydrogen evolution and electrochemical tests.

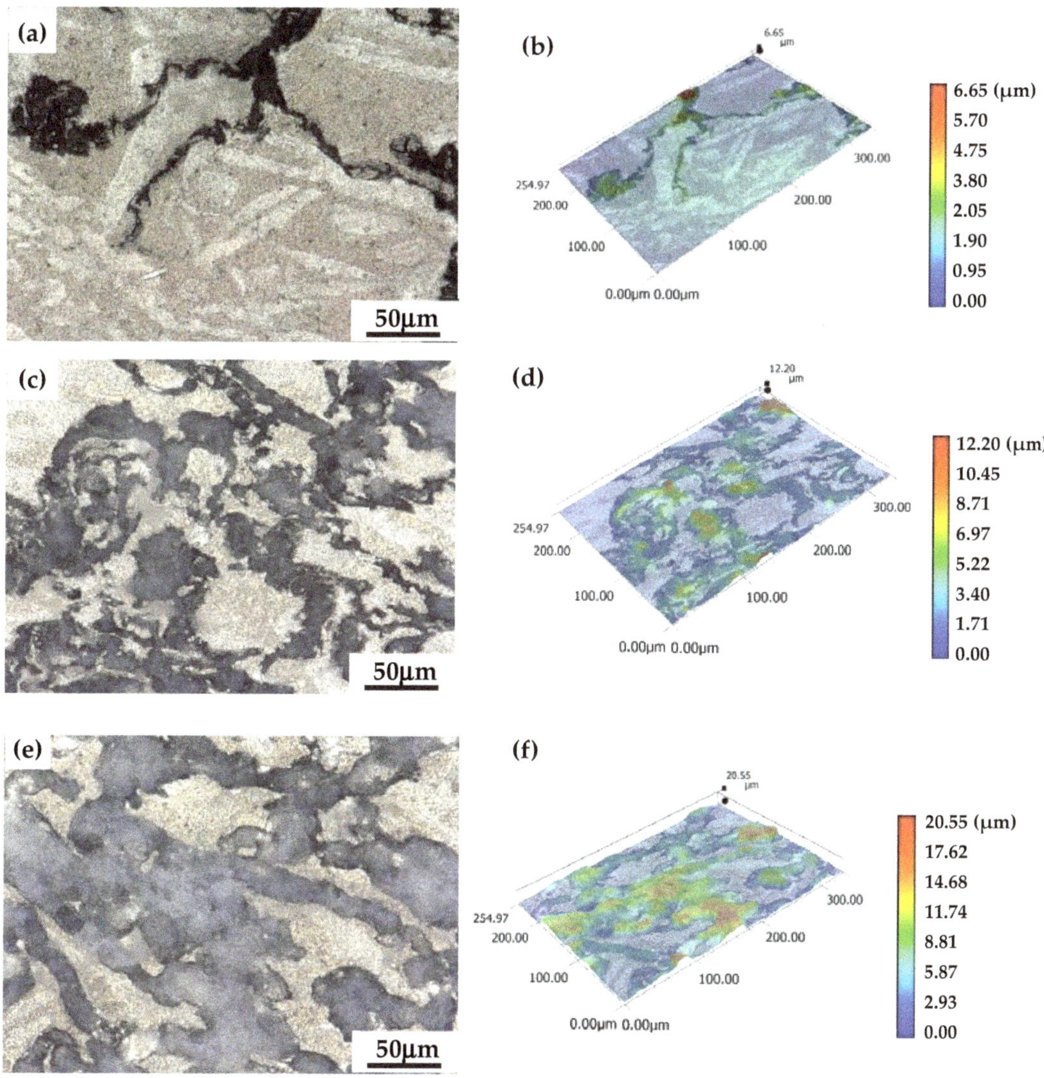

Figure 7. OM observations to the surface corrosion morphologies of the sample immersed in 0.9 wt.% solution for: (**a**) 4, (**c**) 8 and (**e**) 24 h. Images (**b**,**d**,**f**) are 3D profiles of surface corrosion morphologies of samples immersed in 3.5 wt.% solution for 4, 8 and 24 h.

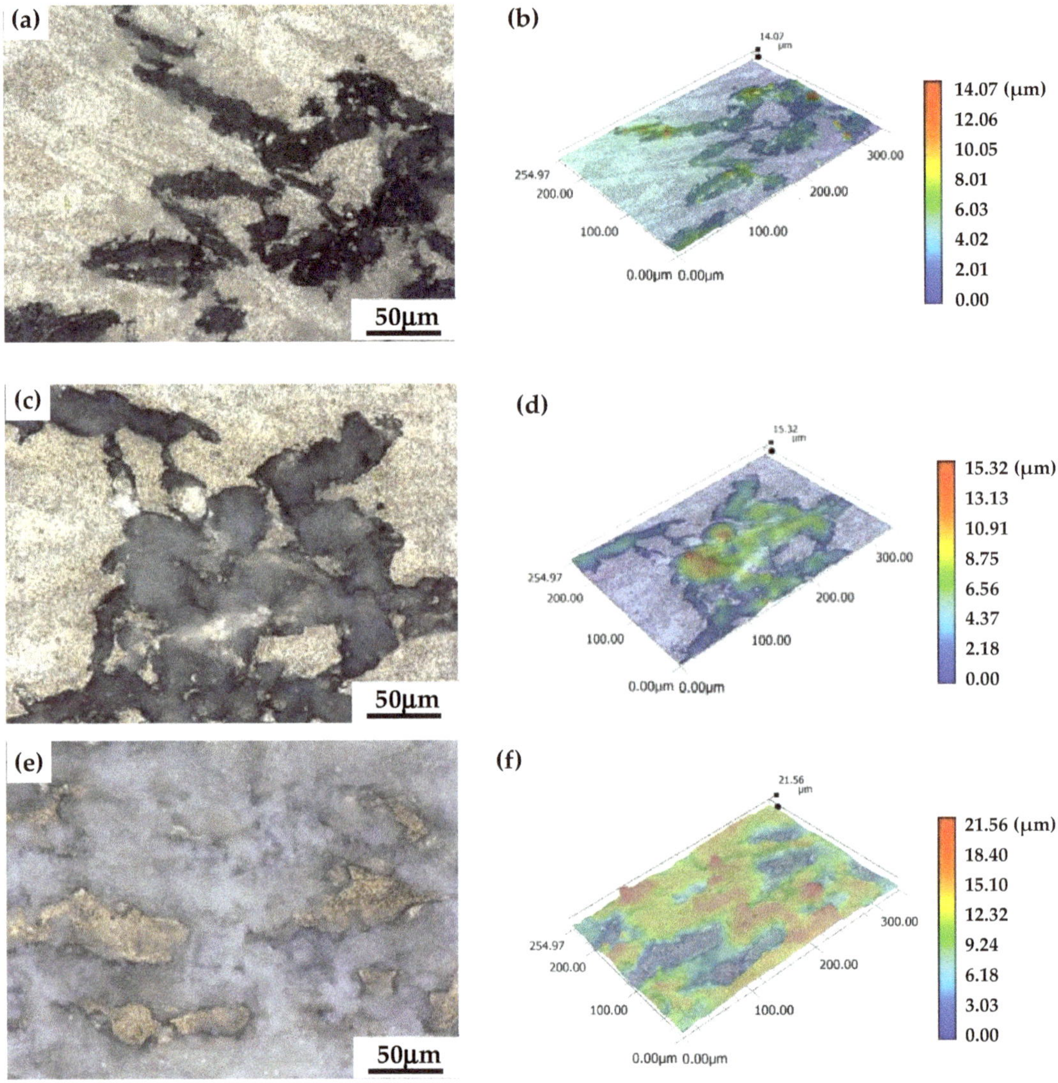

Figure 8. OM observations to the surface corrosion morphologies of the sample immersed in 3.5 wt.% solution for: (**a**) 4, (**c**) 8 and (**e**) 24 h. Images (**b**,**d**,**f**) are 3D profiles of surface corrosion morphologies of samples immersed in 3.5 wt.% solution for 4, 8 and 24 h.

Figure 9 shows the cross-sectional corrosion morphologies of the samples that are immersed in the two concentrated NaCl solutions for 4 h, respectively. It reveals that the localized corrosion attack of the sample immersed in the 3.5 wt.% NaCl is much more severe than that in the 0.9 wt.% NaCl (Figure 9a,b). Based on the high magnification images (Figure 9c), there are almost no visible corrosion pits on the surface. However, for the sample in 3.5 wt.% NaCl, the maximum depth of the corrosion pits can reach 10 μm (Figure 9b,d). Compared with the surface and cross-sectional observations, this indicates that although the filiform corrosion can easily occur on the surface of the Mg-8%Li alloy in differently concentrated NaCl solutions, the localized pitting is quite sensitive to the concentration of Cl^-. High magnification images clearly show that the pitting can only occur in the 3.5 wt.% NaCl solution and preferentially nucleates in the interior of the α-Mg matrix phases. Moreover, with the intensification of pitting corrosion, the electrolyte can reach underneath matrix and cause the dissolution of the β-Li phases because the Li_2CO_3 film is hardly formed due to the limited concentration of CO_2 in the solution, resulting in the corrosion of the whole surface of the sample that is immersed in the 3.5 wt.% NaCl solution for 24 h (Figure 8e). In a previous work, Song et al. reported that when the duplex structured Mg-8 wt.%Li alloy immersed in NaCl solution, the formed surface film was composed of MgO, Li_2O, $Mg(OH)_2$ and LiOH [37]. Moreover, due to the reaction between Li_2O and CO_2, the Li_2CO_3 can be formed on the surface of β-Li matrix phase. Generally, the Pilling–Bedworth ratio (PBR) of the formed film is between 1 and 2 [29]. It can have good protective capability for the underneath substrate. Zeng et al. reported that the PBR values of MgO, Li_2O, $Mg(OH)_2$, LiOH and Li_2CO_3 are 0.88, 0.57, 1.77, 1.26 and 1.35, respectively [29]. In the previous work, Xu et al. reported that the formed Li_2CO_3 film was dense and could protect the corrosion attack of the β-Li matrix phase, whereas the $MgCO_3$ and $Mg(OH)_2$ films formed on the surface of α-Mg phase were porous and loose [30]. Moreover, the Cl^- can easily destroy the $MgCO_3$ and $Mg(OH)_2$ films formed on the surface of the α-Mg phase [3]. Since the Cl^- concentration in the 3.5 wt.% NaCl solution is high, the stronger electrical conductivity of the solution can accelerate the electrochemical reactions and increase the corrosion rate of the alloy. Moreover, the high concentration of Cl^- can cause much more severe attacks to the surface films and reduce their corrosion protectiveness to the substrate. Thus, the difference of corrosion resistance of the alloys in the two concentrated NaCl solutions reaches the maximum when the immersion time is less than 4 h. With the increasing of the immersion time, the quantity of formed corrosion products on the sample surfaces will increase considerably. Then, the corrosion resistance due to the Cl^- attack can be offset by the increased corrosion products on the surfaces. This explains why the difference of corrosion resistance between the samples immersed in the 0.9 wt.% and 3.5 wt.% NaCl solutions decreases with the prolongation of immersion time. When the immersion time reaches 24 h, the corrosion severity on the surfaces of the differently immersed samples in the 0.9 wt.% and 3.5 NaCl solutions are basically the same (Figures 7f and 8f). Moreover, the initially localized corrosion increases the exposed area of the sample surfaces. Thus, when the immersion time exceeds 8 h, the hydrogen evolution rates of the samples immersed in the two concentrated NaCl solutions increase slightly.

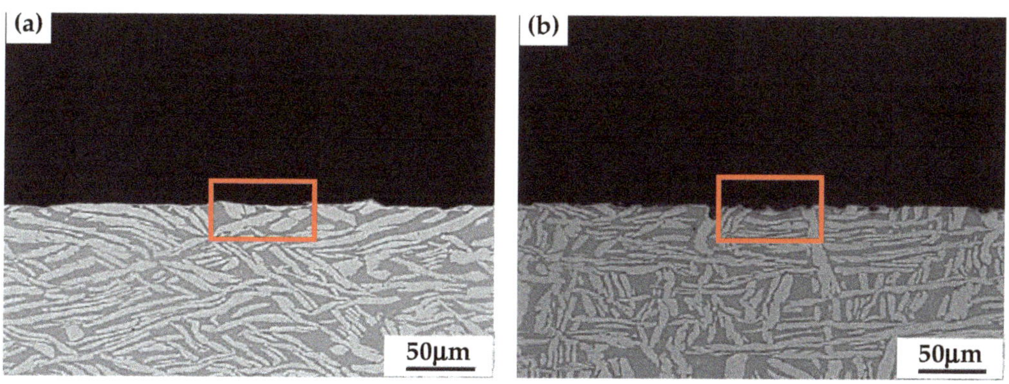

Figure 9. *Cont.*

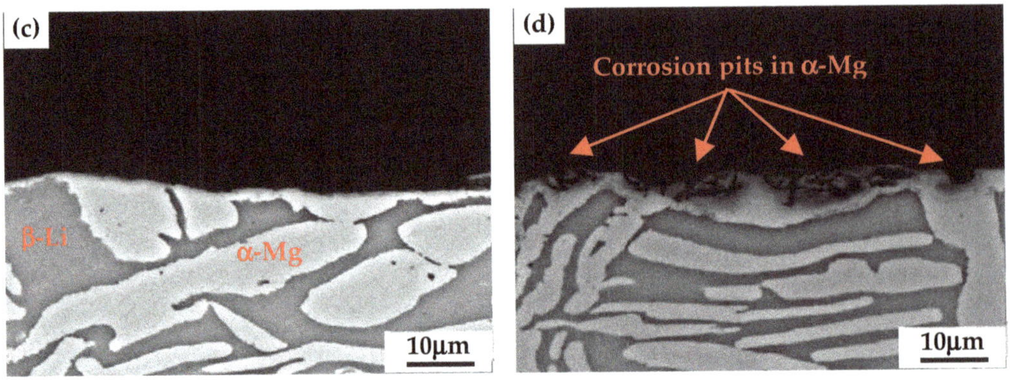

Figure 9. SEM observations to the cross-sectional morphologies of the immersed samples in solutions of: (**a**) 0.9 wt.% and (**b**) 3.5 wt.% NaCl solutions for 4 h, respectively. Images (**c**,**d**) are high magnification observations to the areas squared in images (**a**,**b**), respectively.

4. Conclusions

Through investigating and comparing the corrosion behavior of an as-rolled Mg-8%Li alloy in differently concentrated NaCl solutions, two conclusions can be drawn:

(1) Based on the determined slopes of the measured hydrogen evolution curves, the corrosion rate of the alloy in the 3.5 wt.% NaCl solution is 3 times higher than that in the 0.9 wt.% NaCl solution when the immersion time exceeds 8 h.

(2) The corrosion mechanisms of the as-rolled Mg-8%Li alloy are obviously different in the two concentrated NaCl solutions. In the 0.9 wt.% NaCl solution, the corrosion attack mainly occurs at the α-Mg phase. In the 3.5 wt.% NaCl solution, the corrosion attack initially occurs in the α-Mg phase and then gradually extends to the β-Li phase.

Author Contributions: Conceptualization, D.X. and B.W.; methodology, J.H.; software, J.L.; validation, B.W., H.S. and J.S.; formal analysis, B.W.; investigation, J.H.; resources, B.W.; data curation, H.S.; writing—original draft preparation, J.H.; writing—review and editing, J.S.; visualization, D.X. and B.W.; supervision, D.X.; project administration, D.X. and B.W.; funding acquisition, D.X. All authors have read and agreed to the published version of the manuscript.

Funding: This work was supported by the National Natural Science Foundation of China Projects under Grant [Nos. 52071220, 51871211, U21A2049, 51701129 and 51971054], Liaoning Province's project of "Revitalizing Liaoning Talents" (XLYC1907062), the Doctor Startup Fund of Natural Science Foundation Program of Liaoning Province (No. 2019-BS-200), High level achievement construction project of Shenyang Ligong University (SYLUXM202105), Liaoning BaiQianWan Talents Program, the Domain Foundation of Equipment Advance Research of 13th Five-year Plan (61409220118), National Key Research and Development Program of China under Grant [Nos. 2017YFB0702001 and 2016YFB0301105], the Innovation Fund of Institute of Metal Research (IMR), Chinese Academy of Sciences (CAS), the National Basic Research Program of China (973 Program) project under Grant No. 2013CB632205, and the Fundamental Research Fund for the Central Universities under Grant [No. N2009006].

Institutional Review Board Statement: Not applicable.

Informed Consent Statement: Not applicable.

Data Availability Statement: Not applicable.

Conflicts of Interest: The authors declare no conflict of interest.

References

1. Zhu, W.H.; Zheng, Z.X.; Feng, J.Z.; Xu, H.H.; Wang, Y.T.; Lu, L.L. Dynamic Design Method for Magnesium Alloys Wheel of New Energy Vehicles. *Adv. Mater. Res.* **2010**, *961*, 930–934. [CrossRef]
2. Kannan, M.B.; Raman, P.K.S. In vitro degradation and mechanical integrity of calcium-containing magnesium alloys in modified-simulated body fluid. *Biomaterials* **2008**, *29*, 2306–2314. [CrossRef]
3. Wang, B.J.; Xu, D.K.; Cai, X.; Qiao, Y.X.; Sheng, L.Y. Effect of rolling ratios on the microstructural evolution and corrosion performance of an as-rolled Mg-8 wt.%Li alloy. *J. Magnes. Alloy.* **2021**, *9*, 560–568. [CrossRef]
4. Zhang, Y.; Xu, J.; Ruan, Y.C.; Yu, M.K.; O'Laughlin, M.; Wise, H.; Chen, D.; Tian, L.; Shi, D.; Wang, J.; et al. Implant-derived magnesium induces local neuronal production of CGRP to improve bone-fracture healing in rats. *Nat. Med.* **2016**, *22*, 1160–1169. [CrossRef]
5. Wang, B.J.; Xu, K.; Xu, D.K.; Cai, X.; Qiao, Y.X.; Sheng, L.Y. Anisotropic corrosion behavior of hot-rolled Mg-8 wt.%Li alloy. *J. Mater. Sci. Technol.* **2020**, *53*, 102–111. [CrossRef]
6. Tang, Q.B.; Huang, S.D.; Wu, T.B. Application analysis of magnesium alloys in weapon components. *Ordnance. Mater. Sci. Eng.* **2007**, *30*, 69–72.
7. Lv, Y.; Wang, L.; Li, Y.; Jin, Y.; Feng, J.; Ren, Y.M.; Cao, D.; Wang, G.; Zhang, M. The effect of different concentrations of Na_2SnO_3 on the electrochemical behaviors of the Mg-8Li electrode. *Ionics* **2014**, *20*, 1573–1578. [CrossRef]
8. Chen, Z. *Magnesium Alloy*; Chemical Industry Press: Beijing, China, 2004.
9. Li, M.; Hao, H.; Zhang, A.; Song, Y.; Zhang, X. Effects of Nd on microstructure and mechanical properties of as-cast Mg-8Li-3Al alloy. *J. Rare Earths* **2012**, *30*, 492–496. [CrossRef]
10. Xu, D.K.; Li, C.Q.; Wang, B.J.; Han, E.H. Effect of icosahedral phase on the crystallographic texture and mechanical anisotropy of duplex structured Mg-Li alloys. *Mater. Des.* **2015**, *88*, 88–97. [CrossRef]
11. Xu, D.; Wang, B.; Li, C.; Zu, T.; Han, E. Effect of icosahedral phase on the thermal stability and ageing response of a duplex structured Mg-Li alloy. *Mater. Des.* **2015**, *69*, 124–129. [CrossRef]
12. Li, C.Q.; Xu, D.K.; Wang, B.J.; Sheng, L.Y.; Qiao, Y.X.; Han, E.H. Natural ageing responses of duplex structured Mg-Li based alloys. *Sci. Rep.* **2017**, *7*, 40078. [CrossRef]
13. Lv, Y.Z.; Xu, Y.; Wang, X.X.; Tian, H.L. The Electrochemical Performances of Mg-8Li, Mg-8Li-0.5Ce and Mg-8Li-1Ce Alloys in 0.7 $mol{\cdot}L^{-1}$ NaCl Solution. *Acta. Chim. Sinica* **2011**, *69*, 2248–2252.
14. Xu, D.K.; Han, E.H. Effect of quasicrystalline phase on improving the corrosion resistance of a duplex structured Mg-Li alloy. *Scr. Mater.* **2014**, *71*, 21–24. [CrossRef]
15. Yang, L.H.; Jiang, Q.T.; Zheng, M.; Hou, B.; Li, Y. Corrosion behavior of Mg-8Li-3Zn-Al alloy in neutral 3.5% NaCl solution. *J. Magnes. Alloy* **2016**, *4*, 22–26. [CrossRef]
16. Yuan, X.; Yu, D.; Gao, L.-L.; Gao, H. Effect of Phosphate-Buffered Solution Corrosion on the Ratcheting Fatigue Behavior of a Duplex Mg-Li-Al Alloy. *J. Mater. Eng. Perform* **2016**, *25*, 1802–1810. [CrossRef]
17. Dobkowska, A.; Adamczyk-Cieslak, B.; Mizera, J.; Kubásek, J.; Vojtěch, D. Corrosion Behaviour of Magnesium Lithium Alloys in NaCl Solution. *Solid. State. Phenom.* **2015**, *3763*, 87–90. [CrossRef]
18. Thirumalaikumarasamy, D.; Shanmugam, K.; Balasubramanian, V. Comparison of the corrosion behaviour of AZ31B magnesium alloy under immersion test and potentiodynamic polarization test in NaCl solution. *J. Magnes. Alloy.* **2014**, *2*, 36–49. [CrossRef]
19. Shan, D.Y.; Zhou, W.Q.; Han, E.H.; Ke, W. Corrosion and electrochemical behavior of AZ31D magnesium alloys in sodium chloride. *Trans. Nonferr. Met. Soc. China* **2006**, *16*, 1789–1792.
20. Thirumalaikumarasamy, D.; Shanmugaml, K.; Balasubramanian, V. Influence of chloride ion concentration on immersion corrosion behaviour of plasma sprayed alumina coatings on AZ31B magnesium alloy. *J. Magnes. Alloy.* **2014**, *2*, 325–334. [CrossRef]
21. Liu, F.; Song, Y.W.; Shan, D.Y.; Han, E.H. Corrosion behavior of AZ31 magnesium alloy in simulated acid rain solution. *Trans. Nonferrous Met. Soc. China* **2010**, *20*, 638–642. [CrossRef]
22. Li, M.M.; Qin, Z.; Yang, Y.; Xiong, X.M.; Zhou, G.; Cui, X.F.; Jiang, B.; Peng, X.D.; Pan, F.S. Microstructure and Corrosion Properties of Duplex-Structured Extruded Mg-6Li-4Zn-xMn Alloys. *Acta. Metall. Sin.* **2022**. [CrossRef]
23. Ma, X.C.; Jin, S.Y.; Wu, R.Z.; Wang, J.X.; Wang, G.X.; Krit, B.; Betsofen, S. Corrosion behavior of Mg-Li alloys: A review. *Trans. Nonferrous Met. Soc. China* **2021**, *31*, 3228–3254. [CrossRef]
24. Li, C.Q.; Liu, X.; Dong, L.J.; Shi, B.Q.; Tang, S.; Dong, Y.; Zhang, Z.R. Simultaneously improved mechanical strength and corrosion resistance of Mg-Li-Al alloy by solid solution treatment. *Mater. Lett.* **2021**, *301*, 130305. [CrossRef]
25. Dong, L.J.; Liu, X.; Liang, J.X.; Li, C.Q.; Dong, Y.; Zhang, Z.R. Corrosion behavior of a eutectic Mg-8Li alloy in NaCl solution. *Electrochem. Commun.* **2021**, *129*, 107087. [CrossRef]
26. Dobkowska, A.; Adamczyk-Cieslak, B.; Kubasek, J.; Vojtech, D.; Kuc, D.; Hadasik, E.; Mizera, J. Microstructure and corrosion resistance of a duplex structured Mg-7.5Li-3Al-1Zn. *J. Magnes. Alloy.* **2021**, *9*, 467–477. [CrossRef]
27. Li, C.Q.; He, Y.B.; Huang, H.P. Effect of lithium content on the mechanical and corrosion behaviors of HCP binary Mg-Li alloys. *J. Magnes. Alloy.* **2021**, *9*, 569–580. [CrossRef]
28. Wang, B.J.; Xu, D.K.; Sun, J.; Han, E.H. Effect of grain structure on the stress corrosion cracking (SCC) behavior of an as-extruded Mg-Zn-Zr alloy. *Corros. Sci.* **2019**, *157*, 347–356. [CrossRef]

29. Zeng, R.C.; Sun, L.; Zheng, Y.F.; Cui, H.Z.; Han, E.H. Corrosion and characterization of dual phase Mg-Li-Ca alloy in Hank's solution: The influence of microstructural features. *Corros. Sci.* **2014**, *79*, 69–82. [CrossRef]
30. Xu, W.; Birbilis, N.; Sha, G.; Wang, Y.; Daniels, J.; Xiao, Y.; Ferry, M. A high-specific-strength and corrosion-resistant magnesium alloy. *Nat. Mater.* **2015**, *14*, 1229–1235. [CrossRef]
31. Liu, G.; Xie, W.; Wei, G.; Yang, Y.; Liu, J.; Xu, T.; Xie, W.; Peng, X. Dynamic Recrystallization Behavior and Corrosion Resistance of a Dual-Phase Mg-Li Alloy. *Materials* **2018**, *11*, 408. [CrossRef]
32. Xin, Y.C.; Chu, P.K. Influence of Tris in simulated body fluid on degradation behavior of pure magnesium. *Mater. Chem. Phys.* **2010**, *124*, 33–35. [CrossRef]
33. Wang, B.J.; Xu, D.K.; Zhao, T.Y.; Sheng, L.Y. Effect of $CaCl_2$ and $NaHCO_3$ in physiological saline solution on the corrosion behavior of an as-extruded Mg-Zn-Y-Nd alloy. *Acta Metall. Sin.* **2021**, *34*, 239–247. [CrossRef]
34. Masing, G.; Tamman, G. Behavior of lithium, toward sodium, potassium, yin, cadmium and magnesium. *Z. Fur Anorg. Allg. Chem.* **1960**, *67*, 197–198.
35. Wang, B.J.; Xu, D.K.; Dong, J.H.; Ke, W. Effect of corrosion product films on the in vitro degradation behavior of Mg-3%Al-1%Zn (in wt%) alloy in Hank's solution. *J. Mater. Sci. Technol.* **2018**, *34*, 1756–1764. [CrossRef]
36. Li, C.Q.; Xu, D.K.; Chen, X.B.; Wang, B.J.; Wu, R.Z.; Han, E.H.; Birbilis, N. Composition and microstructure dependent corrosion behaviour of Mg-Li alloys. *Electrochim. Acta* **2018**, *260*, 55–64. [CrossRef]
37. Song, Y.W.; Shan, D.Y.; Chen, R.S.; Han, E.H. Investigation of surface oxide film on magnesium lithium alloy. *J. Alloys Compd.* **2009**, *484*, 585–590. [CrossRef]

coatings

MDPI

Article

Properties of Micro-Arc Oxidation Coatings on 5052 Al Alloy Sealed by SiO_2 Nanoparticles

Siqi Liu, Jiahuan Chen, Dongdong Zhang, Yuxin Wang *, Zhen He * and Pingyi Guo

School of Materials Science and Engineering, Jiangsu University of Science and Technology, Zhenjiang 212003, China; lsq15376988099@163.com (S.L.); 15606107161@163.com (J.C.); zhangdong0820@just.edu.cn (D.Z.); pyguo@just.edu.cn (P.G.)
* Correspondence: ywan943@163.com (Y.W.); hezhen@just.edu.cn (Z.H.)

Abstract: Micro-arc oxidation (MAO) treatment can effectively improve the wear resistance, corrosion resistance, and mechanical strength of aluminum alloy substrates. Improving the porous structure of MAO film and effectively sealing the pores is a significant research issue. In this study, the MAO treatment of 5052 aluminum alloy was carried out in silicate electrolytes. The MAO films were sealed with different concentrations of SiO_2 nanoparticles. The effects of SiO_2 nanoparticle content on the MAO films' microstructure, mechanical properties, and corrosion performance were systematically investigated. When adding SiO_2 nanoparticles to electrolytes, the particles were deposited at the micropores of the film, which could effectively seal the porous MAO film and significantly improve its corrosion and wear resistance. The corrosion resistance and wear resistance properties were optimal with 5.0 g/L SiO_2 addition. Compared to the unsealed film, the corrosion current density and corrosion rate decreased from 1.24×10^{-9} A/cm^2 and 1.47×10^{-5} mm/a to 7.78×10^{-10} A/cm^2 and 9.15×10^{-6} mm/a, respectively. Moreover, the average friction coefficient of the sealed film was 0.606, which was ~19.3% lower than that of the substrate and 3.3% lower than for the unsealed film.

Keywords: micro-arc oxidation (MAO); SiO_2 nanoparticles; corrosion resistance; wear property

Citation: Liu, S.; Chen, J.; Zhang, D.; Wang, Y.; He, Z.; Guo, P. Properties of Micro-Arc Oxidation Coatings on 5052 Al Alloy Sealed by SiO_2 Nanoparticles. *Coatings* **2022**, *12*, 373. https://doi.org/10.3390/coatings12030373

Academic Editor: Emanuele Galvanetto

Received: 8 February 2022
Accepted: 3 March 2022
Published: 11 March 2022

1. Introduction

Aluminum and its alloy materials are widely used in automotive, aerospace, and other industrial applications because of their high specific strength, strong corrosion resistance, superior machinability, good thermal conductivity, and recyclability [1,2]. However, aluminum has some disadvantages as a structural material, such as soft texture, poor wear resistance, low corrosion resistance, and poor heat resistance, which seriously affect the scaled-up applications [3,4].

To improve the performance of aluminum alloy, various surface engineering methods have been developed and applied. Common surface treatment methods include anodic oxidation [5–7], electroplating [8,9], plasma spraying, laser surface modification, and micro-arc oxidation [10,11]. Micro-arc oxidation (MAO) is a novel surface treatment technology that allows in situ deposition of autogenous ceramic films on magnesium, aluminum, and titanium [12–14]. MAO films are characterized by solid adhesion, compact structure, high strength, exceptional corrosion resistance [15,16], and excellent high-temperature impact resistance. However, many microcracks and discharge micropores inevitably appear on MAO films due to the special reaction process [17–19]. The corrosive medium can easily penetrate the membrane and interface layers, leading to the corrosion of substrate material [20].

In addition to optimizing the MAO process, it is necessary to seal the pores of the MAO film and enhance the shielding effect in the corrosive medium [21,22]. It has been found that the corrosion resistance of MAO films can be improved by applying different sealing materials and sealing processes [23,24]. The traditional sealing methods can be divided into hydrated, inorganic, and organic seals [25,26]. Inorganic seals have promising application prospects due to their simple process, low cost, and easy implementation [27–29]. Introducing highly stable oxide nanoparticles, such as TiO_2, ZrO_2, and SiO_2, can effectively enhance corrosion protection [30,31]. The suspension of SiO_2 nanoparticles can be used to prepare coatings with enhanced thermomechanical properties, reduced surface porosity, and improved wear resistance [32–34]. SiO_2 nanoparticles can act as sintering aids to promote the formation of MAO films [35–38]. Meanwhile, SiO_2 nanoparticles with high melting point and small size can reduce surface cracks and increase the density of ceramic layers, thus improving the corrosion and wear resistance of surface layers [39,40].

In the present study, the surface of 5052 Al alloy was treated by MAO process with different addition (0–10 g/L) of SiO_2 nanoparticles (particle size of 20 nm) in silicate electrolytes. The nanoparticles were uniformly suspended in the electrolyte by high-speed stirring and ultrasonic shaking to seal the generated pores. The effects of SiO_2 nanoparticle content on the microstructure, mechanical properties and corrosion resistance of MAO films were systematically investigated.

2. Experimental Method

2.1. Sample Preparation

The substrates used in this research were 5052 Al alloys with dimensions of 20 mm × 20 mm × 6 mm. The substrates were carefully pretreated before being subject to the MAO process. The samples were polished and ultrasonically cleaned in ethanol and deionized water for 5 min to remove the surface oil. The chemical composition of the 5052 Al alloy is detailed in Table 1.

Table 1. Chemical Composition of 5052 Al Alloy.

Element	Si	Fe	Cu	Mn	Mg	Cr	Zn	Al
Content (wt.%)	25	0.4	0.1	0.3	2.5	0.2	0.1	Bal

The MAO treatment was conducted using micro-arc oxidation equipment (WHD-30, Harbin, China). The electrolyte composition and process parameters are shown in Table 2. The electrolyte was prepared with deionized water. All reagents were purchased from the Sigma company with analytical purity. We used the bidirectional pulse current. The positive and negative current density was 6 A/dm^2. The positive and negative duty cycles were 30% & 40%.

Table 2. Electrolyte Composition and MAO Process Parameters.

Reagents and Conditions	Values
$Na_2SiO_3 \cdot 9H_2O$	10 g/L
KOH	1 g/L
$Na_2WO_4 \cdot 2H_2O$	1 g/L
SiO_2 nanoparticle	0–10 g/L
pH	12~14
Current density	6 A/dm^2
Frequency	400 Hz
Duty cycle	30%/40%
Pulse ratio	1/1
Oxidation time	10 mins

2.2. Sample Characterization

The microstructure and chemical distribution of the MAO films was characterized by a scanning electron microscope (SEM, Phenom Pro X, Eindhoven, The Netherlands) equipped with an energy dispersive spectrometer (EDS). The phase structure of the MAO films was studied using an X-ray diffractometer (XRD, Shimadzu XRD-6000, Kyoto, Japan).

The electrochemical impedance spectroscopy and potentiodynamic polarization curve of the MAO films were tested using an electrochemical workstation (PGSTAT302 N, Metrohm Autolab B.V., Utrecht, The Netherlands). For the three-electrode corrosion cell, the counter electrode was made of platinum mesh, the reference electrode was a saturated calomel electrode, and the sample was the working electrode. The corrosion results were collected in 3.5 wt.% NaCl solution at room temperature. The microhardness of the MAO films was tested by a semi-automatic microhardness tester (HXS-1000 TAC, Shanghai, China). The load was 10 N and the loading time was set as 15 s. A reciprocating wear tester (HSR-2 M, Lanzhou, China) examined the friction properties of the MAO films. A $Si_3 N_4$ ceramic ball with a diameter of 4 mm was used for the test and the total sliding distance was set as 20 m for each sample. All the wear tests were conducted at room temperature for 10 mins under 5 N load.

3. Results and Discussion

3.1. Microstructures

The microstructures of the MAO films sealed using different concentrations of SiO_2 nanoparticles were observed, as presented in Figure 1. Figure 1a shows the microstructure of unsealed MAO film prepared by the common electrolyte. It can be clearly seen that there are dense "honeycomb" holes on the film surface. These holes provide channels for the flow of corrosive media. During the MAO process, the surface film layer is constantly broken down by high voltage. The molten $Al_2 O_3$ was ejected and cooled rapidly in the external electrolyte, resulting in the formation of holes. When 2.5 g/L SiO_2 nanoparticles were added into the electrolyte, SiO_2 nanoparticles were deposited outside the discharge micropores and a small amount of aggregation occurred, as shown in Figure 1b. When the addition of SiO_2 increased to 5 g/L, the micropores were more obviously covered by SiO_2, as shown in Figure 1c, thereby blocking the corrosion channel. However, when excess SiO_2 was added, as shown in Figure 1d–e, the nanoparticles gathered on the film surface under discharge, adsorbing a large number of free ions, and accentuating the micro-arc discharge, resulting in high surface energy and ablation, so that the molten oxide could not solidify well on the surface of the film. When the SiO_2 concentration reached 10.0 g/L, film roughness was higher and surface oxide ablation was accentuated.

The elemental content of MAO films sealed with different concentrations of SiO_2 nanoparticles was determined by EDS analysis, as shown in Table 3. The main elements in the MAO films prepared by common electrolyte were O, Al and Si. O was mainly derived from $SiO_3{}^{2-}$ in the electrolyte and occurred as SiO_2 and $Al_2 O_3$, while Al was derived from the matrix, and Si was derived from the residual $SiO_3{}^{2-}$ and SiO_2 generated by the discharge of $SiO_3{}^{2-}$. The atomic percentage of Si in the film increased continuously with addition of SiO_2 nanoparticles and reached 27.22% when the concentration of SiO_2 nanoparticles in the electrolyte was 10 g/L, confirming that the SiO_2 nanoparticles had entered the film layer.

Table 3. EDS Analysis (at%) of Film.

Samples	O	Al	Si	W
Unsealed MAO film	61.02	22.65	15.60	0.72
MAO film sealed with 2.5 g/L SiO_2	63.80	12.13	20.97	3.10
MAO film sealed with 5.0 g/L SiO_2	63.91	9.51	23.89	2.69
MAO film sealed with 7.5 g/L SiO_2	65.86	6.25	25.55	2.34
MAO film sealed with 10.0 g/L SiO_2	66.69	2.68	27.22	3.41

Figure 2 shows the cross-sectional morphology of MAO films prepared with different concentrations of SiO_2 nanoparticles. The results showed that the thickness of the film layer increased with increase in SiO_2 concentration. Figure 2a shows the MAO film layer without SiO_2 nanoparticle addition, having a thickness of ~16 μm. The dense area near the substrate had fewer pores, while the outer area had some pores penetrating the film. Figure 2b shows the MAO film layer with 2.5 g/L SiO_2 addition, having a thickness of ~20 μm. Many cracks and large inner pores appeared in the film. Figure 2c shows the MAO film layer prepared by adding 5.0 g/L SiO_2 nanoparticles, having a thickness of ~33 μm. The internal dense area showed a certain degree of porosity with small pores, while cracks and holes were still evident. There were some particles deposited on the surface, blocking the micropores generated in the MAO process. Figure 2d shows the MAO film layer prepared by adding 7.5 g/L SiO_2. The thickness of the film continued to increase to ~45 μm, yet the number of inner pores increased with larger pore diameter. Figure 2e shows that there was no significant increase in thickness when the nanoparticle concentration was further increased. The larger pore size holes inside them can seriously damage corrosion resistance.

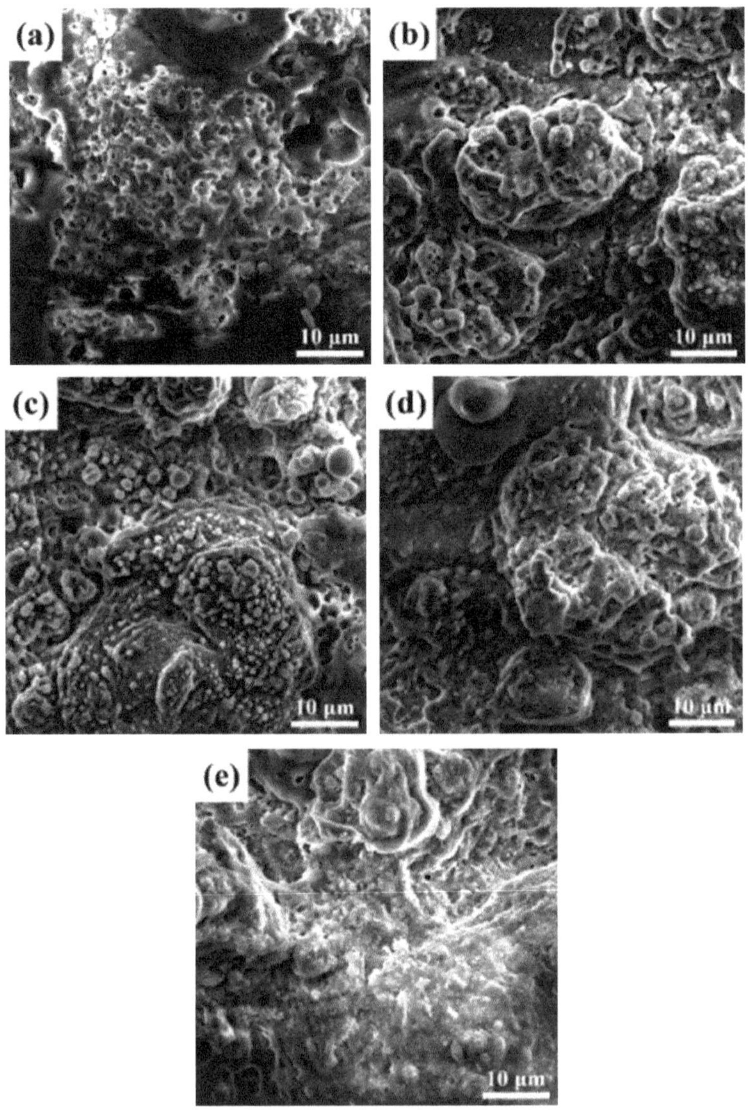

Figure 1. Surface SEM images of (**a**) unsealed MAO film and (**b**) MAO films sealed by different concentrations of SiO_2 nanoparticles: 2.5 g/L, (**c**) 5 g/L (**d**) 7.5 g/L and (**e**) 10 g/L.

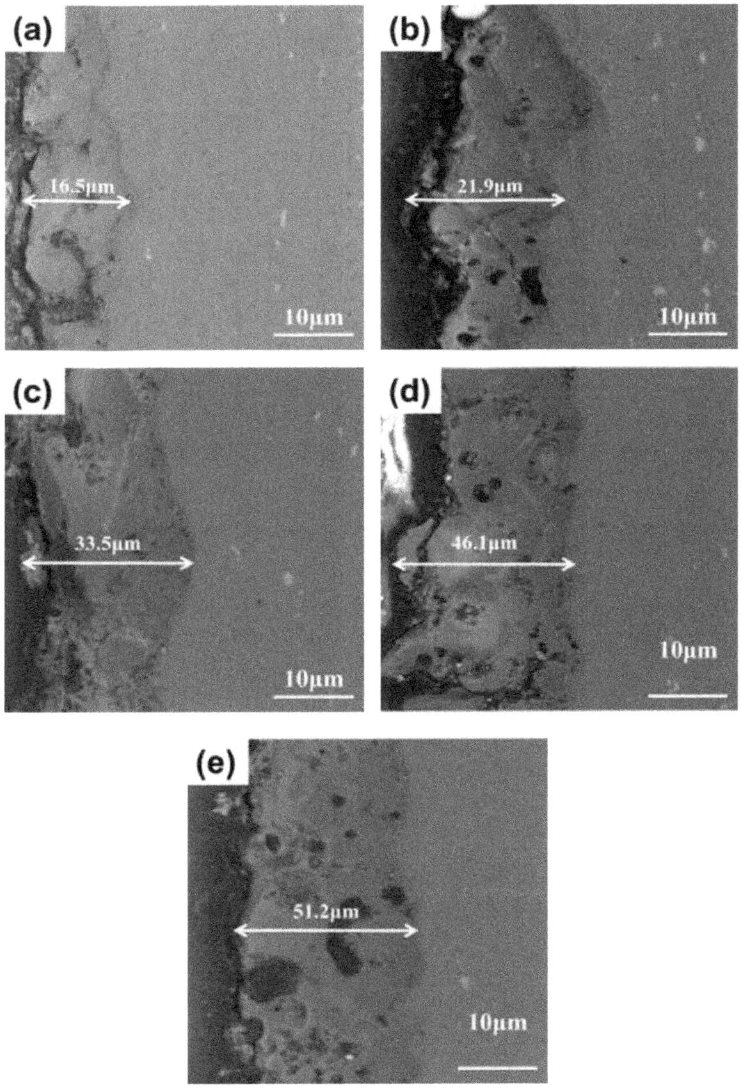

Figure 2. Cross-sectional SEM images of (**a**) unsealed MAO film and, (**b**) MAO films sealed by different concentrations of SiO_2 nanoparticles: 2.5 g/L, (**c**) 5 g/L (**d**) 7.5 g/L and (**e**) 10 g/L.

Figure 3 presents the cross-sectional elemental distribution of unsealed and 5.0 g/L SiO_2 sealed films. The MAO film prepared with the addition of 5 g/L nano-SiO_2 had a higher content and a more uniform distribution of Si compared to the unsealed layer. The Si concentrated in the outer sparse area, indicating that the nano-SiO_2 particles were well-deposited on the surface. The distribution density of Si increased in the pores and cracks. It can be inferred that SiO_2 nanoparticles were continuously deposited in the micropores, blocking the micropores and filling the cracks. Moreover, a certain amount of Si was also detected in the inner areas, implying that the SiO_2 nanoparticles were successfully incorporated into the MAO film.

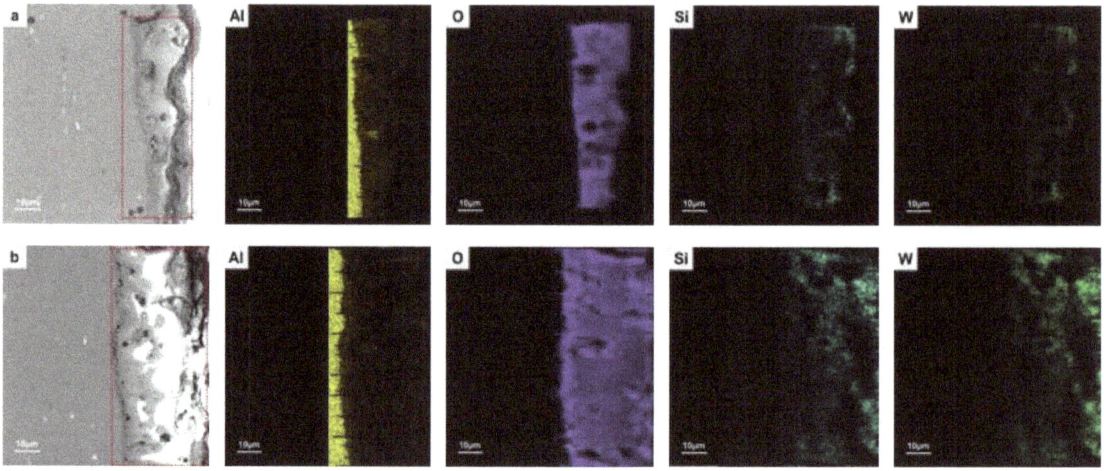

Figure 3. Cross-sectional images and element distribution of (**a**) unsealed MAO film, and (**b**) MAO film sealed with 5 g/L concentration of SiO_2 nanoparticles.

3.2. XRD Analysis

Figure 4 shows the XRD patterns of MAO thin films and MAO thin films sealed by SiO_2 nanoparticles. It can be seen from the figure that the phase structure of MAO thin films was mainly composed of Al, α-$Al_2 O_3$, γ-$Al_2 O_3$ and mullite. The intensity of the diffraction peaks of Al was mainly from the matrix. During the MAO process, $Al_2 O_3$ with different crystal structures was formed in the MAO films as a result of the combined effect of chemical and electrochemical reactions. When aluminum is placed in alkaline electrolyte, a dense oxide film is rapidly formed on the substrate surface. The formation of different crystalline types of $Al_2 O_3$ was related to the cooling rate and cooling temperature. When the cooling rate was relatively fast, the molten $Al_2 O_3$ was converted mostly to γ-$Al_2 O_3$. When the cooling rate was relatively slow, the molten $Al_2 O_3$ was converted mostly to α-$Al_2 O_3$, and some γ-$Al_2 O_3$ was also converted to α-$Al_2 O_3$. At the same time, the presence of mullite peaks indicated the successful incorporation of nano-SiO_2 into the film layer during the MAO process.

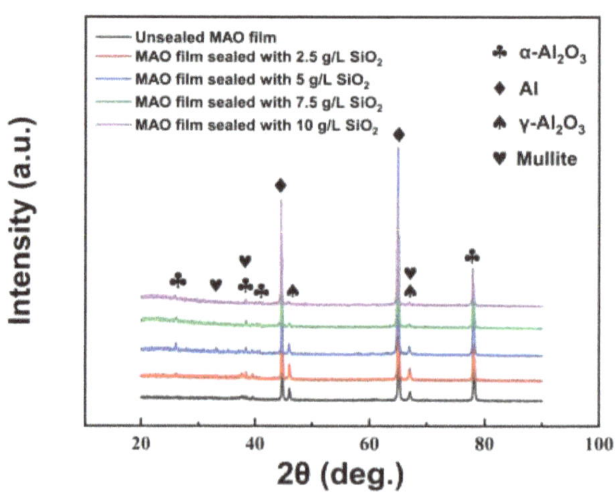

Figure 4. XRD patterns of MAO film and MAO films sealed by SiO_2 nanoparticles.

3.3. Mechanical Properties

Figure 5 shows the microhardness of unsealed MAO film and MAO film sealed by SiO_2 nanoparticles. The microhardness of unsealed MAO film was ~HV 1241. The microhardness of the MAO films increased significantly after sealing. When the concentration of SiO_2 nanoparticles in the electrolyte was 5 g/L, the microhardness of sealed MAO film reached a maximum of ~HV 1628, which was about 31.2% higher than that of the unsealed film. As the concentration of SiO_2 nanoparticles increased, the hardness of the sealed MAO film decreased. However, the hardness of the MAO films after sealing was always higher than that of the film without sealing.

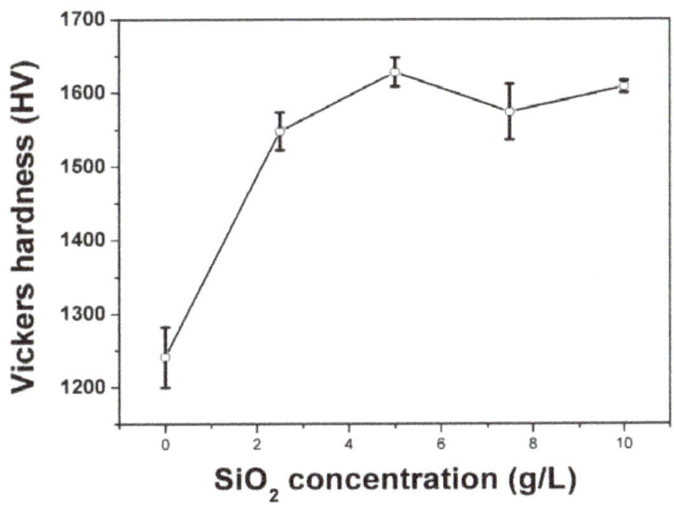

Figure 5. Microhardness of unsealed MAO film and MAO films sealed by SiO_2 nanoparticles.

The above results are related to the microstructure evolution of the MAO films. Appropriate addition of SiO_2 nanoparticles can effectively adsorb free ions and promote a more uniform voltage breakdown on the film surface. Consequently, the hardness of the surface ceramic layer was significantly improved due to the better crystallization of Al_2O_3 ejected from the homogeneously distributed holes.

3.4. Corrosion Resistance

Figure 6 shows the friction coefficient curves of different MAO films. The friction coefficient of the MAO film was more stable and fluctuated less than the matrix during the friction process. Table 4 presents the average friction coefficients of the substrate and the MAO films. The average friction coefficient of the MAO layer first decreased and then increased after sealing with SiO_2 nanoparticles. When the concentration of SiO_2 nanoparticles was 5 g/L, the average friction coefficient of the sealed film was 0.606, which was about 19.3% lower than the value of 0.773 for the matrix and 3.3% lower than the value of 0.627 for the unsealed film. This phenomenon was mainly attributed to the uniform surface structure and the reduced surface roughness after sealing. Further addition of SiO_2 nanoparticles gradually increased the film's average friction coefficient. Aggregation of nanoparticles aggravated the surface discharge of the film, resulting in the ablation of the film and the increase in surface roughness. When the concentration of SiO_2 nanoparticles was 10 g/L, the average friction coefficient of the film increased to 0.91.

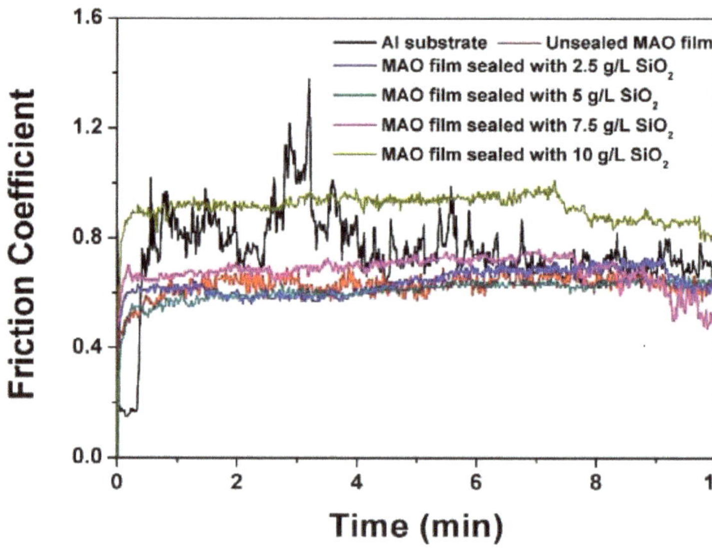

Figure 6. Wear track images of unsealed MAO film and MAO films sealed by SiO_2 nanoparticles.

Table 4. Average friction coefficient of different specimens.

Sample	Average Friction Coefficient
Substrate	0.773
Unsealed MAO film	0.627
MAO film sealed with 2.5 g/L SiO_2	0.637
MAO film sealed with 5 g/L SiO_2	0.606
MAO film sealed with 7.5 g/L SiO_2	0.680
MAO film sealed with 10 g/L SiO_2	0.910

Figure 7 shows a comparison of wear track widths for different samples. Table 4 presents the average friction coefficient of different specimens. The wear track width of unsealed MAO film was ~595 µm. After sealing with SiO_2 nanoparticles, the wear track width of the film was significantly reduced. Consistent with the variation trend of the average friction coefficient, the wear track width of the sealed film first decreased and then increased with increase in SiO_2 nanoparticle concentration. When the addition of SiO_2 nanoparticles was 5 g/L, the edge of the wear track was more uniform, and the wear track width decreased to ~425 µm. Due to the adsorption of nanoparticles on the film surface, further increase in the concentration of TiO_2 nanoparticles led to film ablation, increasing surface roughness and deterioration in wear resistance. When the concentration of SiO_2 nanoparticles was 10 g/L, the wear track width of the film rose to ~530 µm.

To investigate the effect of the sealing process on MAO film corrosion resistance, potentiodynamic polarization tests and electrochemical impedance spectroscopy measurements were performed on the samples. The polarization curves are shown in Figure 8. The fitting results of the polarization curves are shown in Table 5. E_{corr}, i_{corr}, β_a, β_c were derived from the Tafel extrapolation method; Rp was calculated based on the Stern–Geary equation [41]. The corrosion potential of the films sealed with 5 g/L SiO_2 nanoparticles increased to -1.05 V compared to the unsealed films, which showed a significant passivation trend in the anodic region. This was attributed to the formation of a barrier layer after sealing with SiO_2 nanoparticles, which hindered the intrusion of the corrosive medium and thus prevented corrosion attack in the electrolyte. At the same time, the corrosion current density and corrosion rate decreased from 1.24×10^{-9} A/cm^2 and 1.47×10^{-5} mm/a to

$7.78 \times 10–10\ A/cm^2$ and 9.15×10^{-6} mm/a, respectively. The corrosion rate decreased by ~37%, indicating that the corrosion resistance of the sealed film was significantly improved.

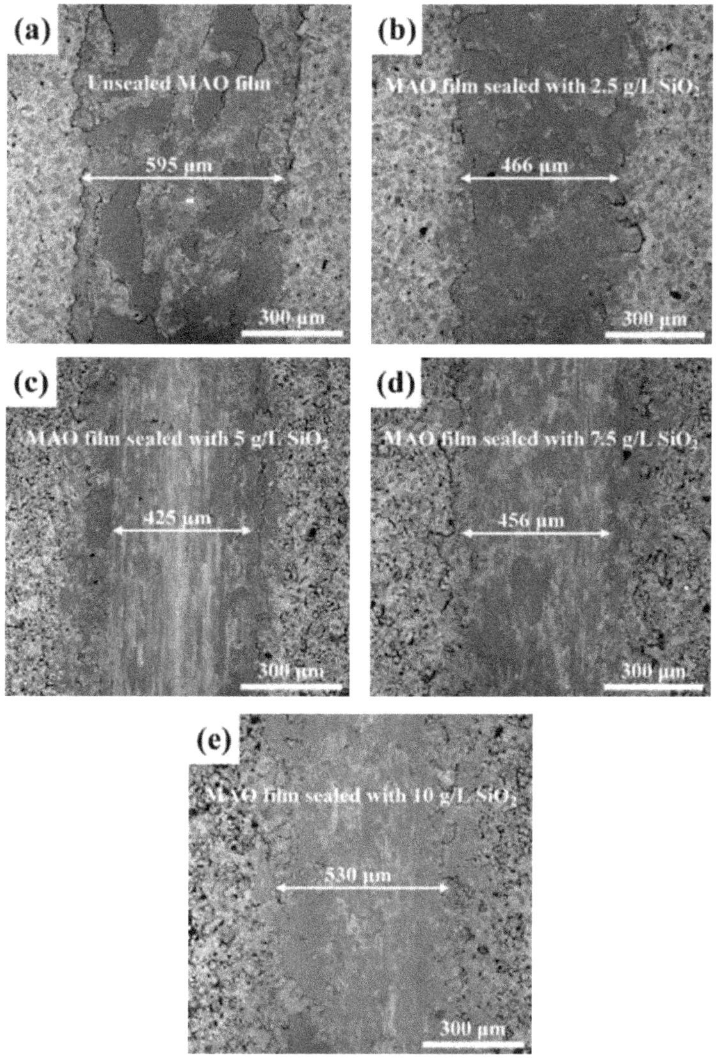

Figure 7. Wear track images of unsealed MAO film (**a**) and MAO films sealed by different concentrations of SiO_2 nanoparticles: (**b**) 2.5 g/L, (**c**) 5 g/L (**d**) 7.5 g/L and (**e**) 10 g/L.

Figure 9 shows the Nyquist plots of MAO film and MAO films sealed by SiO_2 nanoparticles. With increase in SiO_2 nanoparticle concentration in the electrolyte, the capacitive reactance arc radius of the film first increased and then decreased. When the concentration was 5 g/L, the impedance value reached a maximum, showing the best corrosion resistance. The increasing SiO_2 adsorbed the free ions in the solution under the discharge, resulting in excessive high surface energy, intensified discharge process, and finally, ablation. The molten oxide could not solidify uniformly on the film surface, which increased the cracks and reduced the corrosion resistance. Accordingly, the impedance value decreased to a minimum when the concentration of SiO_2 was 10 g/L.

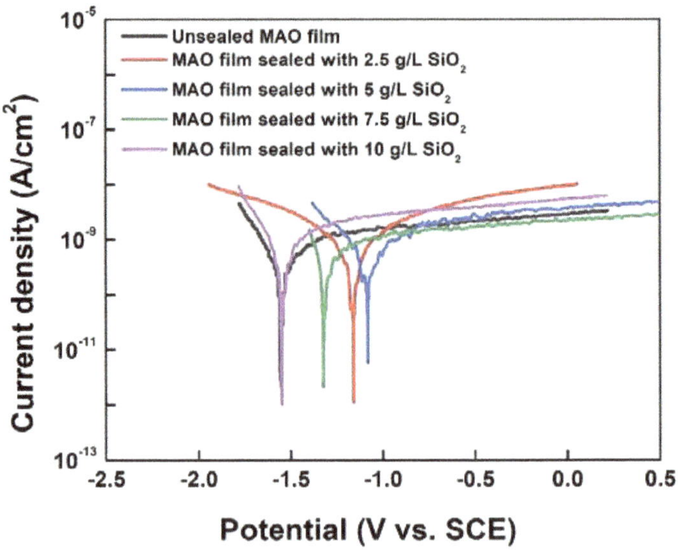

Figure 8. Polarization curves of unsealed MAO film and MAO films sealed by SiO_2 nanoparticles.

Table 5. Fitting results of polarization curves.

Sample	E_{corr} (V)	I_{corr} (A/cm^2)	Corrosion Rate (mm/a)	B_a (mV/Decade)	B_c (mV/Decade)	R_p (Ω·cm^2)
Unsealed MAO film	−1.56	1.24×10^{-9}	1.47×10^{-5}	559.09	−238.59	5.88×10^{6}
MAO film sealed with 2.5 g/L SiO_2	−1.12	8.80×10^{-10}	9.37×10^{-6}	503.18	−489.07	1.22×10^{8}
MAO film sealed with 5 g/L SiO_2	−1.05	7.78×10^{-10}	9.15×10^{-6}	638.93	−368.26	1.31×10^{8}
MAO film sealed with 7.5 g/L SiO_2	−1.32	8.56×10^{-10}	9.87×10^{-6}	482.89	−109.68	4.3×10^{7}
MAO film sealed with 10 g/L SiO_2	−1.55	1.15×10^{-9}	2.06×10^{-5}	504.01	−223.34	5.8×10^{7}

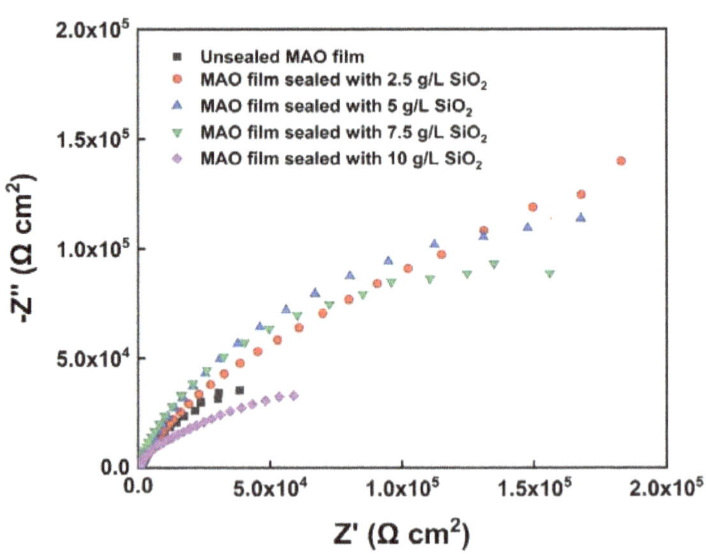

Figure 9. Nyquist plots of unsealed MAO film and MAO films sealed by SiO_2 nanoparticles.

Figure 10 shows the fitted equivalent circuit diagram of the electrochemical impedance spectrum, which consists mainly of two constant phase elements, since the MAO film usually consists of a dense inner layer and a sparse outer layer. R_1 is the solution resistance depending on the corrosive medium, and CPE_1 and CPE_2 are the constant phase elements, corresponding to the outer sparse layer and the dense inner layer, respectively. R_{ct1} and R_{ct2} represent the outer and inner film layer charge transfer resistance values (Table 6 shows the fitted parameters of each component). As can be seen from Table 6, the addition of SiO_2 nanoparticles increased the external Rct value by 1–2 orders of magnitude. When the concentration of SiO_2 nanoparticles was 5 g/L, the value of Rct1 reached its maximum value, which was about 11 times higher than that of the unsealed membrane layer. The Bode diagram reflects the impedance to a certain extent.

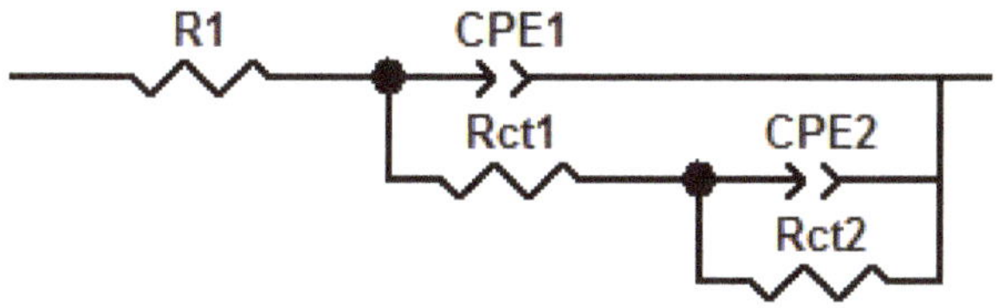

Figure 10. The equivalent circuit diagram of EIS.

Table 6. Equivalent circuit fitting parameters of EIS for different samples.

Sample	R_1 ($\Omega \cdot cm^2$)	CPE1-T (Ω^{-1} cm^2 s^{-P})	CPE1-P	R_{ct1} ($\Omega \cdot cm^2$)	CPE2-T (Ω^{-1} cm^2 s^{-P})	CPE2-P	R_{ct2} ($\Omega \cdot cm^2$)
Unsealed MAO Film	37.30	1.26×10^{-6}	0.62	9.93×10^{3}	4.75×10^{-7}	0.76	1.23×10^{5}
MAO Film Sealed with 2.5 g/L SiO_2	65.40	1.50×10^{-7}	0.81	5.30×10^{4}	3.59×10^{-7}	0.71	4.91×10^{5}
MAO Film Sealed with 5 g/L SiO_2	31.60	5.72×10^{-8}	0.88	1.13×10^{5}	5.80×10^{-7}	0.68	4.76×10^{5}
MAO Film Sealed with 7.5 g/L SiO_2	50.80	1.49×10^{-7}	0.74	3.31×10^{4}	1.52×10^{-7}	0.89	3.31×10^{5}
MAO Film Sealed with 10 g/L SiO_2	63.00	1.12×10^{-7}	0.89	1.12×10^{4}	5.53×10^{-7}	0.69	1.42×10^{5}

Figure 11 presents the Bode diagrams of unsealed MAO film and MAO films sealed by SiO_2 nanoparticles. The polarization resistance of the films after sealing was higher than that of the film without sealing as shown in Figure 11a. There were two time constants in the medium and high-frequency region of the unsealed film, as shown in Figure 11b. The unsealed film comprised a loose outer layer and a dense inner layer. When the unsealed film was in contact with the corrosive solution, the corrosive medium penetrated the porous outer layer and reacted with the compact inner layer. The time constant corresponding to the intermediate frequency region mainly derives from the contribution of the loose outer layer. In contrast, the time constant corresponding to the high frequency region primarily derives from the assistance of the dense inner layer. However, the phase angle diagram of the films after sealing has only one time constant in the high-frequency region. SiO_2 nanoparticles filled the micropores of the films, and the corrosive medium could not quickly penetrate into the inner layer, thus improving the corrosion resistance of the film.

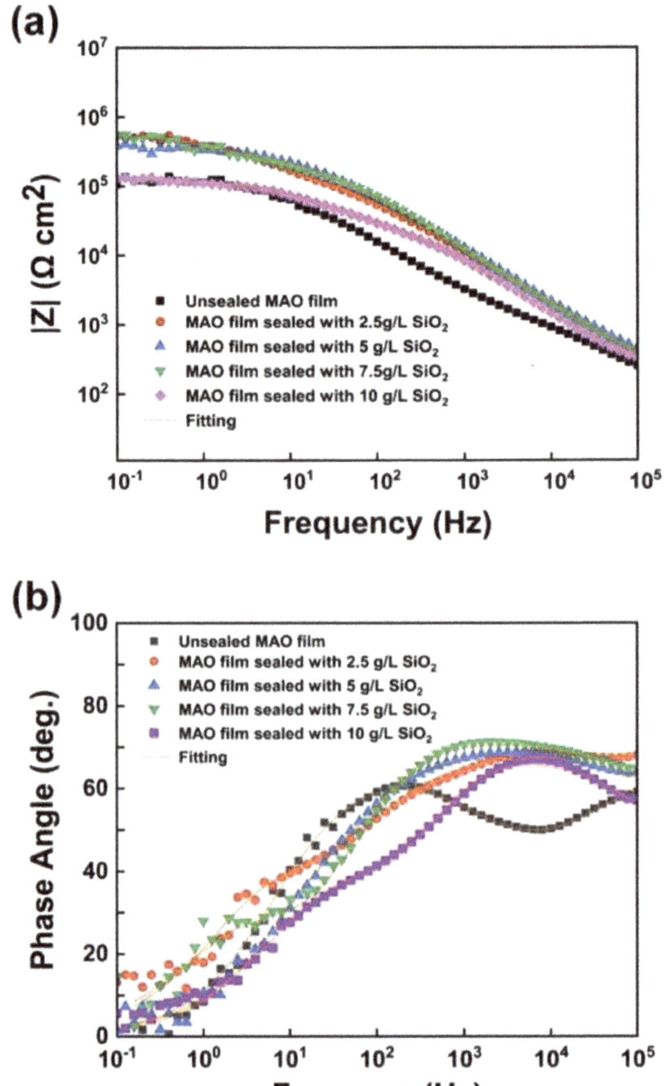

Figure 11. Bode plots of unsealed MAO film and MAO films sealed by SiO_2 nanoparticles: (**a**) relationship between frequency and Z; (**b**) relationship between frequency and phase angle.

4. Conclusions

The discharge micropores on the surface of MAO films can be uniformly filled and effectively sealed by SiO_2 nanoparticles. When the concentration of SiO_2 nanoparticles in the electrolyte was 5 g/L, the MAO films' microhardness, wear resistance, and corrosion resistance were significantly improved and reached an optimal value. Compared with unsealed film, the microhardness of the film surface increased by ~31.2%, while the wear track width decreased from ~595 μm to ~425 μm. In addition, the corrosion rate reduced by ~37%. The current research provides valuable information on the MAO processing of Al alloy and solid grounds for future industrial application.

Author Contributions: Conceptualization, S.L. and Y.W.; methodology, S.L. and Y.W.; formal analysis, S.L. and J.C.; investigation, J.C. and D.Z.; resources: Y.W. and P.G.; writing—original draft preparation: S.L.; writing—review and editing: Y.W. and Z.H.; supervision, Y.W. and Z.H.; project administration, Y.W. and Z.H. All authors have read and agreed to the published version of the manuscript.

Funding: The research was funded by the National Natural Science Foundation of China (51601073), the Research Project of Jiangsu Provincial Natural Science Fund (BK20211344), Jiangsu Provincial Six Talent Peaks (2018 XCL-028) and the Postgraduate Research and Practice Innovation Program of Jiangsu Province (KYCX21_3451).

Institutional Review Board Statement: Not applicable.

Informed Consent Statement: Not applicable.

Data Availability Statement: Not applicable.

Acknowledgments: The authors would like to express their appreciation to Chris Goode (Cirrus Materials Science Ltd., NZ) for his generous assistance.

Conflicts of Interest: The authors declare no conflict of interest.

References

1. Fagiolari, L.; Bini, M.; Costantino, F.; Gatto, G.; Jeremy Kropf, A.; Marmottini, F.; Nocchetti, M.; Wegener, E.C.; Zaccaria, F.; Delferro, M.; et al. Iridium-Doped Nanosized Zn-Al Layered Double Hydroxides as Efficient Water Oxidation Catalysts. *ACS Appl. Mater. Interfaces* **2020**, *12*, 32736–32745. [CrossRef] [PubMed]
2. Tacikowski, M.; Kamiński, J.; Rudnicki, J.; Borowski, T.; Trzaska, M.; Wierzchoń, T. The effect of the diffusive, composite chromium nitride layers produced by a hybrid surface treatment on the corrosion behavior of AZ91D magnesium alloy. *Vacuum* **2011**, *85*, 938–942. [CrossRef]
3. Wang, Y.; Guan, L.; He, Z.; Zhang, S.; Singh, H.; Hayat, M.D.; Yao, C. Influence of pretreatments on physicochemical properties of Ni-P coatings electrodeposited on aluminum alloy. *Mater. Des.* **2021**, *197*, 109233. [CrossRef]
4. Guo, P.Y.; Sun, H.; Shao, Y.; Ding, J.; Li, J.; Huang, M.; Mao, S.; Wang, Y.; Zhang, J.; Long, R.; et al. The evolution of microstructure and electrical performance in doped Mn-Co and Cu-Mn oxide layers with the extended oxidation time. *Corros. Sci.* **2020**, *172*, 108738. [CrossRef]
5. Wang, Y.; Cao, D.; Gao, W.; Qiao, Y.; Jin, Y.; Cheng, G.; Gao, W.; Zhi, Z. Microstructure and properties of sol-enhanced Co-P-TiO_2 nano-composite coatings. *J. Alloys Compd.* **2019**, *792*, 617–625. [CrossRef]
6. Zhang, W.; Cao, D.; Qiao, Y.; He, Z.; Wang, Y.; Li, X.; Gao, W. Microstructure and Properties of Duplex Ni-P-TiO_2/Ni-P Nanocomposite Coatings. *Mater. Res.* **2019**, *22*, 748. [CrossRef]
7. Shi, P.; Ng, W.F.; Wong, M.H.; Cheng, F.T. Improvement of corrosion resistance of pure magnesium in Hanks' solution by microarc oxidation with sol–gel TiO_2 sealing. *J. Alloys Compd.* **2009**, *469*, 286–292. [CrossRef]
8. Yang, S.; Cai, W.; Zeng, H.; Xu, X. Ultra-fine β-SiC quantum dots fabricated by laser ablation in reactive liquid at room temperature and their violet emission. *J. Mater. Chem.* **2009**, *19*, 7119. [CrossRef]
9. Wang, Y.Q.; Deng, Y.Z.; Shao, Y.W.; Wang, F.H. New sealing treatment of microarc oxidation coating. *Surf. Eng.* **2013**, *30*, 31–35. [CrossRef]
10. Wang, Y.; Gao, W.; He, Z.; Jin, Y.; Qiao, Y.; Cheng, G. Cu–Sn–Zn nanocomposite coatings prepared by TiO_2 sol-enhanced electrodeposition. *J. Appl. Electrochem.* **2020**, *50*, 875–885. [CrossRef]
11. Bahramian, A.; Raeissi, K.; Hakimizad, A. An investigation of the characteristics of Al_2O_3/TiO_2 PEO nanocomposite coating. *Appl. Surf. Sci.* **2015**, *351*, 13–26. [CrossRef]
12. Fatimah, S.; Kamil, M.P.; Kwon, J.H.; Kaseem, M.; Ko, Y.G. Dual incorporation of SiO_2 and ZrO_2 nanoparticles into the oxide layer on 6061 Al alloy via plasma electrolytic oxidation: Coating structure and corrosion properties. *J. Alloys Compd.* **2016**, *707*, 358–364. [CrossRef]
13. Kaseem, M.; Ko, Y.G. Electrochemical Response of Al_2O_3-MoO_2-TiO_2 Oxide Films Formed on 6061 Al Alloy by Plasma Electrolytic Oxidation. *J. Electrochem. Soc.* **2016**, *163*, C587. [CrossRef]
14. Arunnellaiappan, T.; Ashfaq, M.; Krishna, L.R.; Rameshbabu, N. Fabrication of corrosion-resistant Al_2O_3–CeO_2 composite coating on AA7075 via plasma electrolytic oxidation coupled with electrophoretic deposition. *Ceram. Int.* **2016**, *42*, 5897–5905. [CrossRef]
15. Raj, V.; Ali, M.M. Formation of ceramic alumina nanocomposite coatings on aluminium for enhanced corrosion resistance. *J. Mater. Processing Tech.* **2009**, *209*, 5341–5352. [CrossRef]
16. Li, T.; Cao, M.; Liang, J.; Xie, X.; Du, G. Mechanism of Base-Catalyzed Resorcinol-Formaldehyde and Phenol-Resorcinol-Formaldehyde Condensation Reactions: A Theoretical Study. *Polymers* **2017**, *9*, 426. [CrossRef]
17. Wang, L.; Zhou, J.; Liang, J.; Chen, J. Microstructure and corrosion behavior of plasma electrolytic oxidation coated magnesium alloy pre-treated by laser surface melting. *Surf. Coat. Technol.* **2012**, *206*, 3109–3115. [CrossRef]
18. Cui, X.-J.; Liu, C.-H.; Yang, R.-S.; Li, M.-T.; Lin, X.-Z. Self-sealing micro-arc oxidation coating on AZ91D Mg alloy and its formation mechanism. *Surf. Coat. Technol.* **2015**, *269*, 228–237. [CrossRef]

19. Xiong, Y.; Hu, X.; Song, R. Characteristics of CeO_2/ZrO_2-HA composite coating on ZK60 magnesium alloy. *J. Mater. Res.* **2017**, *32*, 1073–1082. [CrossRef]
20. Zahedi Asl, V.; Zhao, J.; Anjum, M.J.; Wei, S.; Wang, W.; Zhao, Z. The effect of cerium cation on the microstructure and anti-corrosion performance of LDH conversion coatings on AZ31 magnesium alloy. *J. Alloys Compd.* **2020**, *821*, 153248. [CrossRef]
21. Zhou, M.; Pang, X.; Wei, L.; Gao, K. Insitu grown superhydrophobic Zn–Al layered double hydroxides films on magnesium alloy to improve corrosion properties. *Appl. Surf. Sci.* **2015**, *337*, 172–177. [CrossRef]
22. Gong, C.; Zhou, Z.; Zhou, H.; Liu, R. Vacuum-assisted synthesis of tiny Au nanoparticles entrapped into mesoporous carbon matrix with superior catalytic activity for 4-nitrophenol reduction. *Adv. Powder Technol.* **2019**, *30*, 649–655. [CrossRef]
23. Xi, J.J.; Zhao, J. Influence of Organic Sealed on the Corrosion Behaviors of Micro Arc Oxidation Coated ZM5 Magnesium Alloy. *Adv. Mater. Res.* **2011**, *420*, 844–847. [CrossRef]
24. Kaseem, M.; Lee, Y.H.; Ko, Y.G. Incorporation of MoO_2 and ZrO_2 particles into the oxide film formed on 7075 Al alloy via micro-arc oxidation. *Mater. Lett.* **2016**, *182*, 260–263. [CrossRef]
25. Wang, Z.-H.; Zhang, J.-M.; Li, Y.; Bai, L.; Zhang, G. Enhanced corrosion resistance of micro-arc oxidation coated magnesium alloy by superhydrophobic Mg−Al layered double hydroxide coating. *Trans. Nonferrous Met. Soc. China* **2019**, *29*, 2066–2077. [CrossRef]
26. Li, H.; Song, R.; Ji, Z. Effects of nano-additive TiO_2 on performance of micro-arc oxidation coatings formed on 6063 aluminum alloy. *Trans. Nonferrous Met. Soc. China* **2013**, *23*, 406–411. [CrossRef]
27. Guo, H.F.; An, M.Z.; Huo, H.B.; Xu, S.; Wu, L. Microstructure characteristic of ceramic coatings fabricated on magnesium alloys by micro-arc oxidation in alkaline silicate solutions. *Appl. Surf. Sci.* **2006**, *252*, 7911–7916. [CrossRef]
28. Hung, J.C.; Ku, C.Y.; Fan, Z.W. Fabrication of an Electrode Insulated by Using Hot Dip Aluminizing and Micro-arc Oxidation Method for Electrochemical Microhole Machining. *Procedia CIRP* **2018**, *68*, 438–443. [CrossRef]
29. Zhang, Y.; Bai, K.; Fu, Z.; Zhang, C.; Zhou, H.; Wang, L.; Zhu, S.; Guan, S.; Li, D.; Hu, J. Composite coating prepared by micro-arc oxidation followed by sol–gel process and in vitro degradation properties. *Appl. Surf. Sci.* **2012**, *258*, 2939–2943. [CrossRef]
30. Sun, W.W.; Li, M.Q.; Gao, Y.; Liu, J. Double Sealing of Ultrasonic Micro-Arc Oxidation Coating on Pure Magnesium by Nano-SiO_2 Particles and SiO_2 Sol Sealing Agent. *Adv. Mater. Res.* **2014**, *1030–1032*, 48–51. [CrossRef]
31. Wang, X.; Zhu, L.; He, X.; Sun, F. Effect of cerium additive on aluminum-based chemical conversion coating on AZ91D magnesium alloy. *Appl. Surf. Sci.* **2013**, *280*, 467–473. [CrossRef]
32. Tang, Y.; Shen, X.; Liu, Z.; Qiao, Y.; Yang, L.; Lu, D.; Zou, J.; Xu, J. Corrosion behaviors of laser melted inconel 718 alloy in NaOH solution. *Acta Metallurgica Sinica* **2022**, *58*, 324–333.
33. Arun, S.; Arunnellaiappan, T.; Rameshbabu, N. Fabrication of the nanoparticle incorporated PEO coating on commercially pure zirconium and its corrosion resistance. *Surf. Coat. Technol.* **2016**, *305*, 264–273. [CrossRef]
34. Erfanifar, E.; Aliofkhazraei, M.; Nabavi, H.F.; Sharifi, H.; Rouhaghdam, A.S. Growth kinetics and morphology of plasma electrolytic oxidation coating on aluminum. *Mater. Chem. Phys.* **2017**, *185*, 162–175. [CrossRef]
35. Asgari, M.; Aliofkhazraei, M.; Darband, G.B.; Rouhaghdam, A.S. How nanoparticles and submicron particles adsorb inside coating during plasma electrolytic oxidation of magnesium? *Surf. Coat. Technol.* **2020**, *383*, 152252. [CrossRef]
36. Matykina, E.; Arrabal, R.; Skeldon, P.; Thompson, G.E. Incorporation of zirconia nanoparticles into coatings formed on aluminium by AC plasma electrolytic oxidation. *J. Appl. Electrochem.* **2008**, *38*, 1375–1383. [CrossRef]
37. Pezzato, L.; Rigon, M.; Martucci, A.; Brunelli, K.; Dabalà, M. Plasma Electrolytic Oxidation (PEO) as pre-treatment for sol-gel coating on aluminum and magnesium alloys. *Surf. Coat. Technol.* **2019**, *366*, 114–123. [CrossRef]
38. Lu, X.; Mohedano, M.; Blawert, C.; Matykina, E.; Arrabal, R.; Kainer, K.U.; Zheludkevich, M.L. Plasma electrolytic oxidation coatings with particle additions—A review. *Surf. Coat. Technol.* **2016**, *307*, 1165–1182. [CrossRef]
39. Pezzato, L.; Angelini, V.; Brunelli, K.; Martini, C.; Dabalà, M. Tribological and corrosion behavior of PEO coatings with graphite nanoparticles on AZ91 and AZ80 magnesium alloys. *Trans. Nonferrous Met. Soc. China* **2018**, *28*, 259–272. [CrossRef]
40. Gnedenkov, S.V.; Sinebryukhov, S.L.; Egorkin, V.S.; Vyaliy, I.E. Wettability and electrochemical properties of the highly hydrophobic coatings on PEO-pretreated aluminum alloy. *Surf. Coat. Technol.* **2016**, *307*, 1241–1248. [CrossRef]
41. Stern, M.; Geary, A.L. A Theoretical Analysis of the Shape of Polarization Curves. *J. Electrochem. Soc.* **1957**, *104*, 56–63. [CrossRef]

coatings

MDPI

Article

Ti/SnO_2-Sb_2O_x-TiO_2 Electrodeposited from Methanesulfonate Electrolytes: Preparation, Properties, and Performance

Shengping Zhang [1], Chen Yu [1], Jingyi Tan [2], Yuxin Wang [1] and Zhen He [1,*]

1 School of Materials Science and Engineering, Jiangsu University of Science and Technology, Zhenjiang 212100, China; zsp960920@163.com (S.Z.); yuchen080797@163.com (C.Y.); ywan943@163.com (Y.W.)

2 School of Health, Guangzhou Xinhua University, Guangzhou 510520, China; jytan1204@163.com

* Correspondence: hezhen@just.edu.cn

Abstract: In this study, Ti/SnO_2-Sb_2O_x-TiO_2 electrodes were produced using a sol-enhanced electrodeposition technique from methanesulfonate electrolytes. The surface microstructures of Ti/SnO_2-Sb_2O_x-TiO_2 were observed, and their phase constituents were determined. The surface features were analyzed by X-ray photoelectron spectroscopy. Linear sweep voltammetry and degradation tests were also conducted to determine the degradation performance. The results show that the addition of TiO_2 sol affects the microstructures of Ti/SnO_2-Sb_2O_x-TiO_2 electrodes, while a uniform coating surface can be obtained at a proper sol concentration in electrolytes. Adding TiO_2 sol also causes deep oxidation of Sb and generates more adsorbed oxygen on the electrode surface. The favorable surface features and the well-dispersed TiO_2 in the coatings of 10 mL/L TiO_2 modified Ti/SnO_2-Sb_2O_x-TiO_2 electrodes award them the best electrocatalytic performance, and their uniform coating surface prolongs the electrode service life.

Keywords: SnO_2-Sb_2O_x; sol-enhanced electrodeposition; organic degradation

Citation: Zhang, S.; Yu, C.; Tan, J.; Wang, Y.; He, Z. Ti/SnO_2-Sb_2O_x-TiO_2 Electrodeposited from Methanesulfonate Electrolytes: Preparation, Properties, and Performance. *Coatings* **2022**, *12*, 366. https://doi.org/10.3390/coatings12030366

Academic Editor: László A. Péter

Received: 9 February 2022
Accepted: 7 March 2022
Published: 9 March 2022

1. Introduction

The demand for effective effluents treatment technology has raised increasing research interests under growing public concern in environmental issues. The treatment of poorly biodegradable organics is essential in the effluents treatment industry. Advanced electrochemical oxidation technology (AEOT) is considered to be a promising method to deal with this issue, owing to its advantages in efficiency and flexibility [1–3]. Organic degradation reactions mostly occur on the anode surface, which plays a crucial role in the electrochemical treatment process. The primary anode materials applied in effluent treatment technology are precious metals (Pt, Au, etc.), carbon, boron-doped diamond (BDD), and metal oxide electrodes [4–6].

Titanium-supported antimony-tin oxide (Ti/SnO_2-Sb_2O_x) is a type of metal oxide electrode that can suppress the oxygen evolution reaction (OER) in order to allow for the effective degradation of organics by hydroxyl radicals. Originally, SnO_2-Sb_2O_x was utilized as a catalyst for organic conversion, whereas this mixed oxide was recently employed as an anode coating due to its adequate conductivity and chemical resistance. Moreover, SnO_2-Sb_2O_x can be efficiently and economically coated on a Ti substrate, making it into large areas for industrial applications. The inert Ti/SnO_2-Sb_2O_x electrodes, therefore, show unique advantages in effluent treatment technology due to their low preparation cost compared to precious metal and BDD, high OER potential, and solid electrocatalytic ability.

The conventional fabrication route for the Ti/SnO_2-Sb_2O_x electrode is thermal decomposition, in which, the precursor solution is transformed into SnO_2-Sb_2O_x after calcination [7]. However, the inferior stability and inadequate electrocatalytic ability limit the scale-up application of Ti/SnO_2-Sb_2O_x electrodes made from thermal decomposition. The electrodeposition fabrication route was recently proposed to manufacture Ti/SnO_2-Sb_2O_x-TiO_2 electrodes. However, despite an enhanced stability, the electrodeposited Ti/SnO_2-Sb_2O_x-TiO_2 electrode suffers from a relatively low electrocatalytic performance.

Previous literature has investigated various modification methods for electrode performance enhancement. Venkatesha et al. [8] and Wang et al. [9] constructed a porous SnO_2-Sb_2O_x coating based on anodized TiO_2 nanotubes, boosting the electrode electrocatalytic performance and long-term stability. Chen et al. [10] synthesized the dense spherical Ti/Sb–SnO_2 with a superior electrocatalytic performance using colloidal electrodeposition. Moreover, doping other elements and nanoparticles could effectively modify the electrodeposited Ti/SnO_2-Sb_2O_x [11]. Liu et al. [12] and Qiao et al. [13] fabricated the Ti/SnO_2-Sb_2O_x–Ce electrodes for efficient electrocatalytic oxidation. Wu et al. [14] prepared a high-performance duplex-structured Ti/SnO_2-Sb_2O_x-CNT composite anode using the electrodeposition method. In our recent studies, a sol-enhanced electrodeposition method has been proposed to efficiently dope inert nanoparticles and achieve their well-dispersion into the coating [15–18], which can serve as a promising method for Ti/SnO_2-Sb_2O_x modification.

This work explores the sol-enhanced electrodeposition route for the Ti/SnO_2-Sb_2O_x-TiO_2 electrodes in methanesulfonate electrolytes. Compared to the traditional oxidizing sulfuric sources, the methanesulfonate electrolyte better reveals the advantages of a low toxicity, excellent chemical stability, and outstanding biodegradability [19–21]. The influence of TiO_2 sol addition on the electrode's properties and performance is systematically investigated. We propose a novel modification procedure for Ti/SnO_2-Sb_2O_x, which will benefit the scale-up application for Ti/SnO_2-Sb_2O_x in the effluent treatment industry. We also aim to thoroughly unveil the underlying mechanism and potential of its electrodeposition fabrication route.

2. Materials and Methods

2.1. Preparation of Ti/SnO_2-Sb_2O_x-TiO_2

The analytical grade chemicals were bought from Aladdin Reagent, Shanghai, China. A pure titanium (99.9%) plate was used in our work. The titanium substrate was pretreated as described in the procedures below: (1) grinding and polishing; (2) soaking into 15% wt sodium hydroxide (NaOH) solution for 2 h; and (3) soaking in 12% wt oxalic acid solution for 3 h. The TiO_2 sol was prepared as reported in the previous papers [22,23].

The proposed Ti/SnO_2-Sb_2O_x-TiO_2 electrodes were fabricated from the electrodeposition method. The electrodeposition fabrication route consists of the electrodeposition process and the following heat-treatment process.

In the electrodeposition process, the electrolyte consisted of 60 mL/L tin methanesulfonate ($C_2H_6O_6S_2Sn$, 50% wt in H_2O), 50 mL/L methane sulfonic acid (CH_4O_3S, 99%), antimony trichloride ($SbCl_3$, 99%), 2 g/L gelatin, and 2 g/L hydroquinone. The electrodes were prepared in an electrolytic cell as follows: the tin plate (40 mm × 40 mm) was used as the anode and the pretreated titanium substrate (20 mm × 30 mm × 0.1 mm) was used as the cathode. TiO_2 sol was prepared following the procedures in previous work [24–26]. A certain amount (5, 10, 15 mL/L) of as-prepared TiO_2 sol was slowly dropped into the stirring electrolytes. The electrodeposition was carried out at a current density of 30 mA/cm^2 for 20 min at 30 °C, and the electrolyte was stirred at 300 rpm. Then, in the heat treatment process, the samples were heat-treated in a muffle furnace at temperatures of 600 °C for 10 h to obtain the Ti/SnO_2-Sb_2O_x-TiO_2 electrodes. The thickness of electrodeposited SnO_2-Sb_2O_x coating was ~26 μm, whereas TiO_2 sol addition hardly modified the coating thickness.

The Ti/SnO_2-Sb_2O_x was also made by the traditional dip-coating method for comparison. Firstly, the pretreated titanium substrate was dipped into a mixture of 1.8 g of tin tetrachloride ($SnCl_4$), 2 mL of antimony oxide (Sb_2O_3), and 8 mL of isopropyl alcohol for 15 s, followed by a drying treatment at 100 °C for 10 min and then a heating treatment in a muffle furnace for 10 min at 600 °C. The drying and heating treatment process was repeated six times, and, finally, the titanium substrate was placed in a muffle furnace for ten hours at 600 °C to acquire the Ti/SnO_2-Sb_2O_x.

2.2. Structural and Elemental Characterization

X-ray diffraction (XRD-6000X, Shimazu, Kyoto, Japan) identified the phase constituents at a step size of 0.1°/s. The surface morphology was observed by a scanning electron microscope (SEM, Phenom ProX, Eindhoven, The Netherlands) with an energy dispersive spectroscopy (EDS) detector, and the embedded simulation program completed the 3D surface imaging. The X-ray photoelectron spectroscopy (XPS, ESCALAB 250Xi, Thermo Fisher, Waltham, MA, USA) was utilized to analyze elemental details, during which, Al Kα radiation at 1486.6 eV was used.

2.3. Electrochemical and Degradation Tests

The electrochemical performance of the electrode was tested by linear sweep voltammetry (LSV) using an electrochemical workstation (CH660E, Chenhua Instrument, Shanghai, China). A three-electrode system was used in the tests. Ti/SnO_2-Sb_2O_x-TiO_2 electrode (10 mm × 10 mm) was the working electrode. Platinum plate (20 mm × 20 mm) was the counter electrode, and saturated calomel electrode (SCE) was the reference electrode. A 0.25 mol/L Na_2SO_4 solution was used as the electrolyte solution. The scan range was from 0 to 4 V, at a scanning rate of 20 mV/s.

The benzoic acid degradation experiment was carried out in 100 mg/L benzoic acid solution at a volume of 200 mL. We used the Ti/SnO_2-Sb_2O_x-TiO_2 electrode (20 mm ×30 mm) as the anode and the same-sized titanium sheet as the cathode. A direct current power was employed to control the current density at 40 mA/cm^2, and the 300-rpm magnetic stirring was applied. The target solution sample was taken for detection after a certain degradation period. The degradation process was monitored using high-performance liquid chromatography (HPLC, Nexera, Shimadzu, Kyoto, Japan) equipped with a TC-C18 column. The mobile phase consisted of 90% water (0.1% trifluoroacetic acid) and 10% acetonitrile. The accelerated life was tested in 0.25 mol/L Na_2SO_4 at room temperature at 100 mA/cm^2. The prepared sample was the anode, and a titanium sheet was the cathode. The accelerated lifetime was calculated when the potential steeply raised and exceeded 5.0 V.

3. Results and Discussion

3.1. Structural Characterizations

Figure 1 presents the XRD profiles of Ti/SnO_2-Sb_2O_x and Ti/SnO_2-Sb_2O_x-TiO_2 electrodes. The electrode prepared by the conventional dip-coating method reveals broad SnO_2 peaks, which indicates a large extent of the amorphous phase. In addition, the sharp Ti peaks imply the exposure of underneath titanium substrate. Such an observation is in good agreement with previous literature [7,27]. In contrast, distinct phase constituents are attained for these samples from the electrodeposition fabrication route. The crystallinity substantially increases, and several intense peaks of tetragonal rutile SnO_2 (JCPDS 99-0024) are discovered. Similar phenomena have also been reported in some earlier literature [7].

The incorporation of TiO_2 addition affects the phase composition significantly. When the TiO_2 addition was ≤10 mL/L, the SnO_2 peak intensity sharply decreases, which indicates a refinement for the prepared SnO_2 crystallites. As summarized in Table 1, the grain size of SnO_2 is estimated according to the Scherrer equation through full width at half maximum (FWHM) [28]. The results show that adding 10 mL/L sol leads to a significantly decreased grain size, whereas larger grains form when further increasing the TiO_2 sol concertation to 15 mL/L. During the electrodeposition process, the TiO_2 nanoparticles are in situ generated and then co-deposited to generate SnO_2-Sb_2O_x-TiO_2. These well-dispersed nanoparticles can serve as active sites for crystallite nucleation, thereby providing a driving force for nucleation and forming smaller SnO_2 crystallites [29]. However, the sharp SnO_2 peaks show an opposite phase variation at a 15 mL/L TiO_2 concentration. Under this condition, the generated TiO_2 nanoparticles tend to aggregate and decrease the active sites in the electrodeposition. It is noted that no TiO_2 diffraction peak is detected in the composite coatings due to the limited amount of embedded TiO_2 nanoparticles.

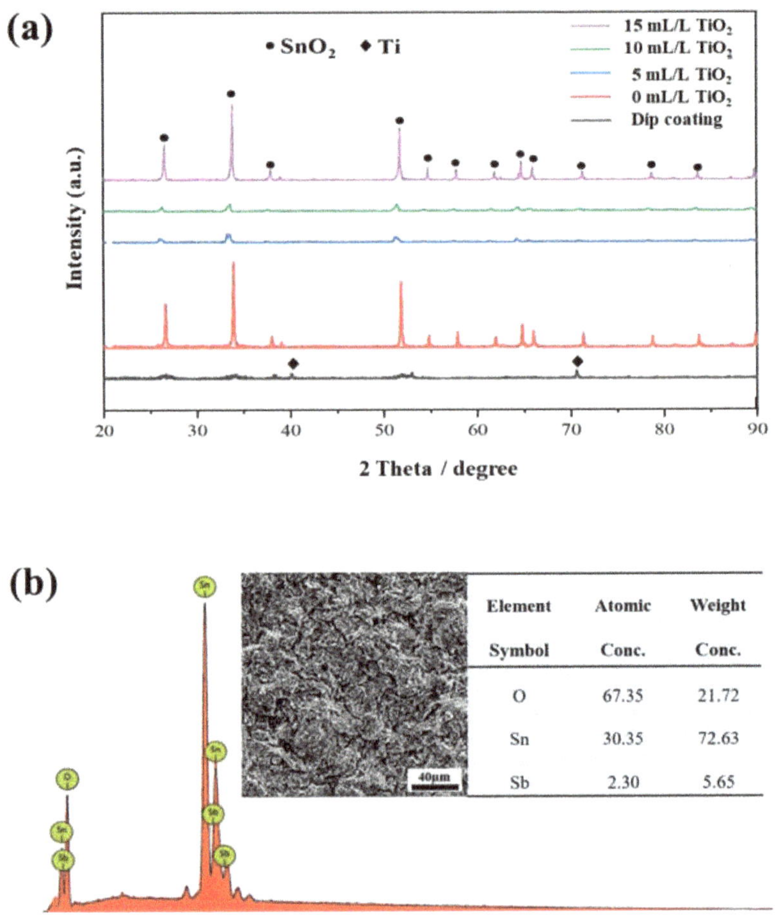

Figure 1. (**a**) XRD profiles recorded on the Ti/SnO_2-Sb_2O_x-TiO_2 with different concentrations of TiO_2 sol, and (**b**) the EDS results recorded on the electrodeposited Ti/SnO_2-Sb_2O_x.

Table 1. Estimated grain size of SnO_2-Sb_2O_x-TiO_2 prepared at different TiO_2 sol additions.

TiO_2 Sol Addition	0 mL/L	10 mL/L	15 mL/L
Grain size (nm)	60	28	45

Figure 1b presents the elemental analysis recorded on the electrodeposited Ti/SnO_2-Sb_2O_x, proving the existence of antimony in the deposited coatings. The ratio of Sn/Sb is ~8 wt.% in SnO_2-Sb_2O_x, which is in good agreement with previous reports [7,30]. Such antimony content in the lattice provides good conductivity for the prepared electrode.

The surface microstructures of prepared Ti/SnO_2-Sb_2O_x-TiO_2 are presented in Figure 2. A typical mud-like coating surface is obtained using the conventional dip-coating method, as shown in Figure 2a. The cracks and holes in such coating surfaces could damage the electrode stability due to the oxide generation in the electrolysis process [7]. In the meantime, the electrodeposited Ti/SnO_2-Sb_2O_x-TiO_2 demonstrates a different surface morphology, showing a relatively compact coating surface with a much-improved uniformity. An increased addition of TiO_2 sol ($\leq$10 mL/L) gives rise to a more compact surface, whereas an addition of 15 mL/L TiO_2 sol causes a relatively non-uniform surface morphology due to the nanoparticle's aggregation.

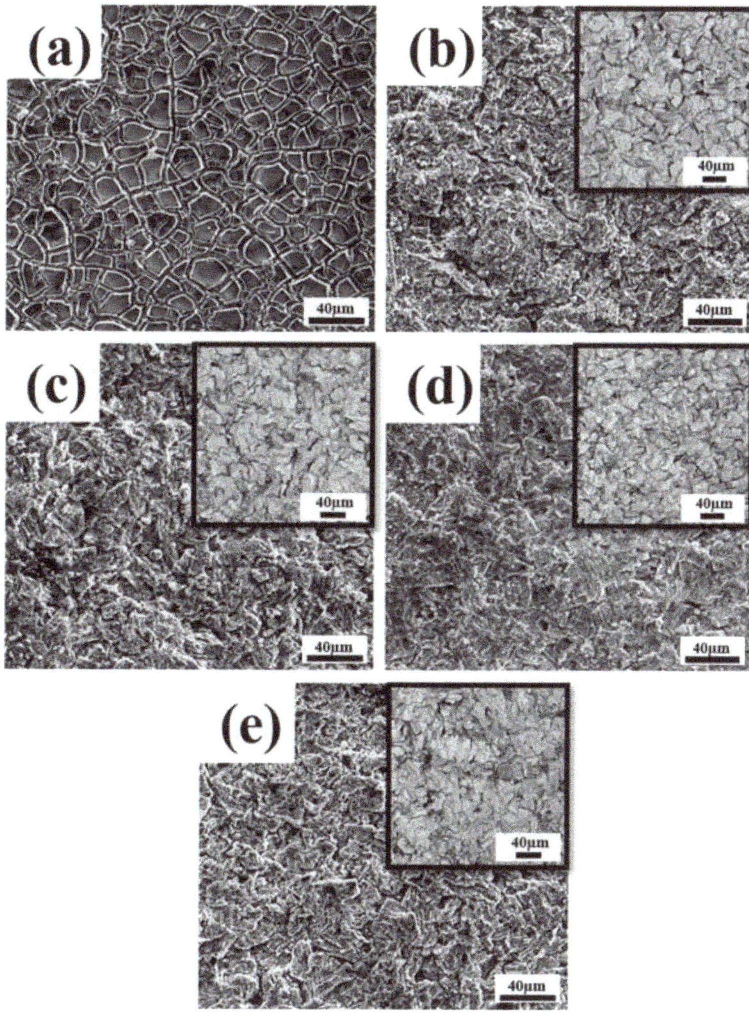

Figure 2. The surface morphologies of (**a**) Ti/SnO_2-Sb_2O_x made by traditional dip-coating, and Ti/SnO_2-Sb_2O_x-TiO_2 electrodeposited with different additions of TiO_2 sol: (**b**) 0 mL/L (**c**) 5 mL/L, (**d**) 10 mL/L, and (**e**) 15 mL/L (insets are the corresponding Sn-Sb-TiO_2 coatings before the heat treatment process).

A similar morphologic variation can be seen for the Sn-Sb-TiO_2 layers before the heat-treatment process, as shown in the inset images. The TiO_2 addition refines the surface morphology of Sn-Sb-TiO_2 at a TiO_2 sol concentration $\leq$ 10 mL/L, whereas an excessive sol concentration caused relatively non-uniform morphology. The surface structures of Ti/SnO_2-Sb_2O_x-TiO_2 electrodes are highly associated with the early electrodeposited Sn-Sb-TiO_2 layers, with some features remaining after the oxidation reactions during heat treatment.

The surface features were further characterized by a 3D imaging technique, as shown in Figure 3. These images show the surface roughness and uniformity for the electrodeposited samples. The proper addition of TiO_2 sol brings a more uniform and compact surface, depicted in Figure 3b,c. The observation corresponds with our earlier XRD and

SEM findings, which proves that the 10 mL/L TiO_2 sol addition refines the surface with a decreased crystallite size and leads to a relatively open surface morphology with an improved uniformity. Nevertheless, the 15 mL/L TiO_2 addition causes a non-uniform coating surface, depicted in Figure 3d.

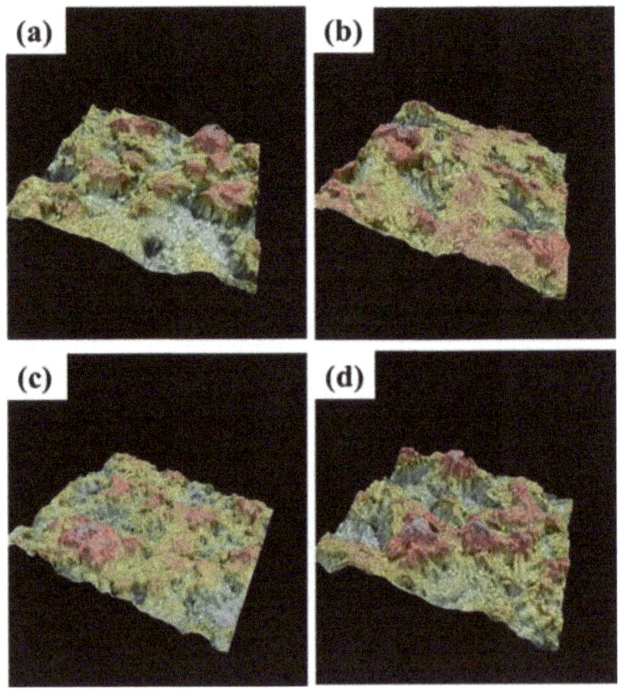

Figure 3. The 3D-imaging surface (depicted with heat map) for the prepared Ti/SnO_2-Sb_2O_x-TiO_2 samples electrodeposited with different additions of TiO_2 sol: (**a**) 0 mL/L, (**b**) 5 mL/L, (**c**) 10 mL/L, (**d**) 15 mL/L.

3.2. Elemental and LSV Analysis

The elemental analysis characterizes the electrodeposited samples, as shown in Figure 4. Figure 4a shows the overall spectra of the electrodeposited Ti/SnO_2-Sb_2O_x and Ti/SnO_2-Sb_2O_x-TiO_2 samples, where prominent peaks of Sb and Sn are detected. The presence of Ti peaks in Figure 4b proves the successful incorporation of TiO_2 into the prepared SnO_2-Sb_2O_x-TiO_2 coatings, whereas the SnO_2-Sb_2O_x sample shows no Ti peak. The weak Ti peaks correlate with the limited content of TiO_2 existing on the electrode surface. During the sol-enhanced electrodeposition, the TiO_2 nanoparticles are in situ generated in electrolytes. The attached organic chains avoid the agglomeration of nanoparticles, which are then co-deposited in the coating with excellent dispersion.

Figure 4c,d compare the spectra of Sb and O on the electrodeposited Ti/SnO_2-Sb_2O_x and Ti/SnO_2-Sb_2O_x-TiO_2. The results elaborate in detail on the state of antinomy oxide and absorbed oxygen with and without TiO_2 sol addition. Both Sb^{3+} and Sb^{5+} exist in the prepared coatings, indicating that there are two oxidation states for antinomy after the heat treatment process. In addition, the detected oxygen can be categorized as O_{lat} (i.e., lattice oxygen species) and O_{ads} (hydroxyl oxygen species), which matches with previous studies [13,31].

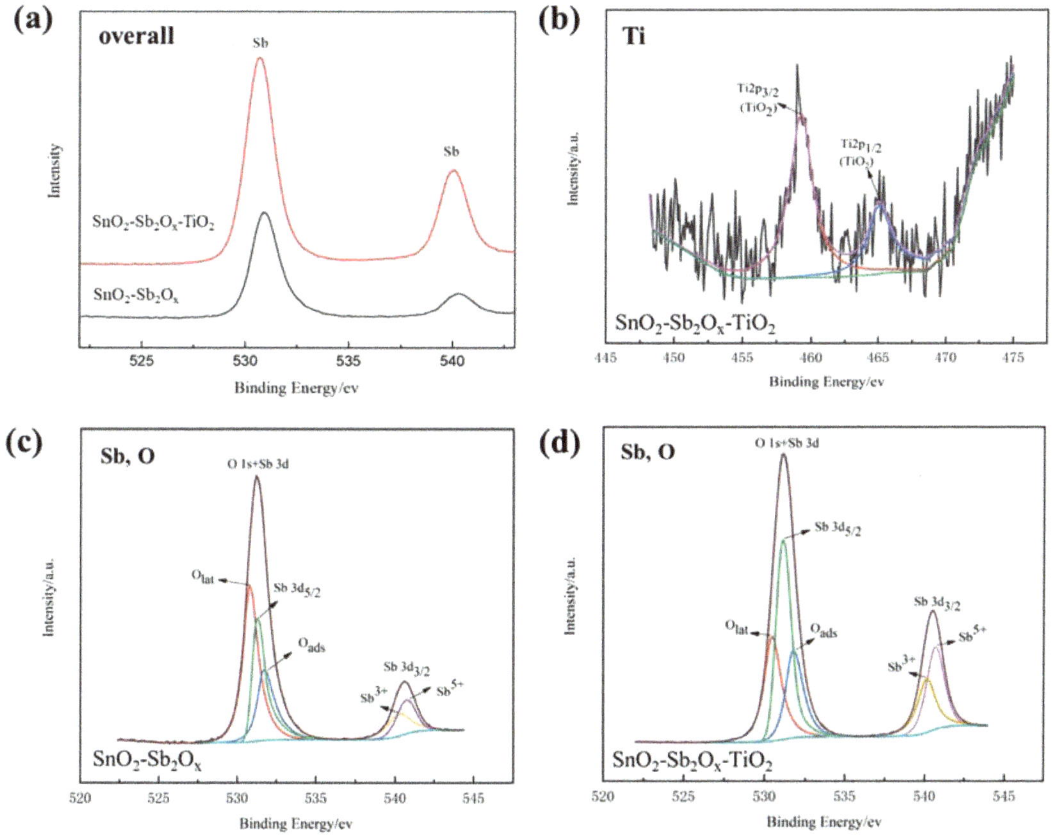

Figure 4. XPS spectra of Ti/SnO_2-Sb_2O_x and Ti/SnO_2-Sb_2O_x-TiO_2 (10 mL/L TiO_2 sol addition): (**a**) the overall spectra, (**b**) Ti spectra for Ti/SnO_2-Sb_2O_x-TiO_2, (**c**) Sb, O spectra for Ti/SnO_2-Sb_2O_x, and (**d**) Sb, O spectra for Ti/SnO_2-Sb_2O_x-TiO_2.

Table 2 summarizes the comparison between the Ti/SnO_2-Sb_2O_x and Ti/SnO_2-Sb_2O_x-TiO_2 in XPS analysis. The listed atom ratios of O_{ads}/O_{lat} and Sb^{5+}/Sb^{3+} provide a qualitative judgment for the element variation. The atom ratio of O_{ads}/O_{lat} is higher in Ti/SnO_2-Sb_2O_x-TiO_2. It is noted that adsorbed oxygen is a potent oxidizing agent. In the meantime, the Sb 3d3/2 is split into two oxidation states, Sb^{5+} and Sb^{3+}, respectively. The higher ratio of Sb^{5+}/Sb^{3+} implies that deeper surface oxidation occurs in the heat treatment process. The higher valence of Sb generally offers more excessive electrons and functions as dominant donors for the SnO_2-Sb semiconductor, improving the conductivity of the prepared electrode [32]. The natural donors of oxygen vacancies are inhibited simultaneously, and the lattice oxygen is reduced. To conclude, the results imply that the addition of TiO_2 sol can help to generate favorable surface features of the chemical composition and oxygen state for the improved electrocatalytic performance.

Table 2. Calculated atom ratios for Ti/SnO_2-Sb_2O_x and Ti/SnO_2-Sb_2O_x-TiO_2.

Samples	Atom Ratio	
	Sb^{5+}:Sb^{3+}	O_{ads}:O_{lat}
0 mL TiO_2 sol sample	1.13	0.49
10 mL TiO_2 sol sample	1.34	0.78

Figure 5 presents the linear sweep voltammetry recorded on the Ti/SnO_2-Sb_2O_x and Ti/SnO_2-Sb_2O_x-TiO_2. The oxygen evolution reaction (OER) potential decreases in the order below: 10 mL/L TiO_2 modified sample (2.34 V) > 5 mL/L TiO_2 modified sample (2.27 V) > 15 mL/L TiO_2 modified sample (2.19 V) > 0 mL/L TiO_2 modified sample (2.15 V). The picture indicates that adding a certain TiO_2 sol increases the electrode's OER potential. When the amount of TiO_2 sol increases to 15 mL/L, the OER potential shows a downward tendency, resulting from the nanoparticle aggregation in the coating [33]. In general, the effective organic degradation process favors a high OER potential, which avoids the OER (i.e., the side-reaction) in the electrolysis process.

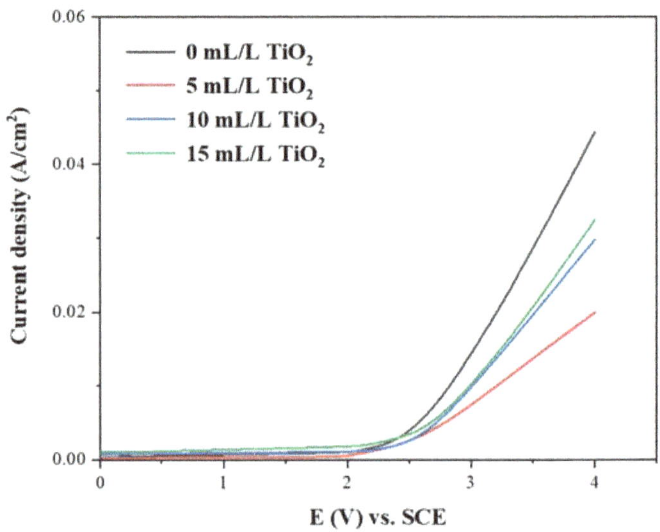

Figure 5. LSV results of electrodeposited Ti/SnO_2-Sb_2O_x and Ti/SnO_2-Sb_2O_x-TiO_2.

3.3. The Degradation Performance of Ti/SnO_2-Sb_2O_x-TiO_2 Electrodes

The electrocatalytic performance of the prepared electrode was studied by degradation tests, as demonstrated in Figure 6. Benzoic acid (BA) is a typical organic pollutant used as a target in degradation tests. The results show that adding TiO_2 sol influences the electrocatalytic ability. The degradation performance decreases following the order below: 10 mL/L TiO_2 modified sample > 5 mL/L TiO_2 modified sample > 15 mL/L TiO_2 modified sample > 0 mL/L TiO_2 modified sample. After the 10 h electrolysis, 98% BA is electrochemically combusted by the 10 mL/L TiO_2 modified SnO_2-Sb electrodes, whereas only 79% BA is degraded on the non-doped sample in the same duration.

Figure 6b shows the kinetic fitting curve of benzoic acid concentration with degradation time. The logarithm of benzoic acid concentration and degradation time showed a good linear relationship during the degradation process, which proved that the degradation process of benzoic acid follows the primary reaction kinetic model:

$$\ln\left(\frac{C_0}{C_t}\right) = kt \tag{1}$$

where C_0 represents the initial concentration of benzoic acid, C_t represents the concentration of benzoic acid at a specific moment, and k is the reaction rate constant. Table 3 lists the calculated k value for the degradation process on different electrodes. The larger k value of the 10 mL/L TiO_2 modified sample represents the electrode's faster degradation rate of benzoic acid.

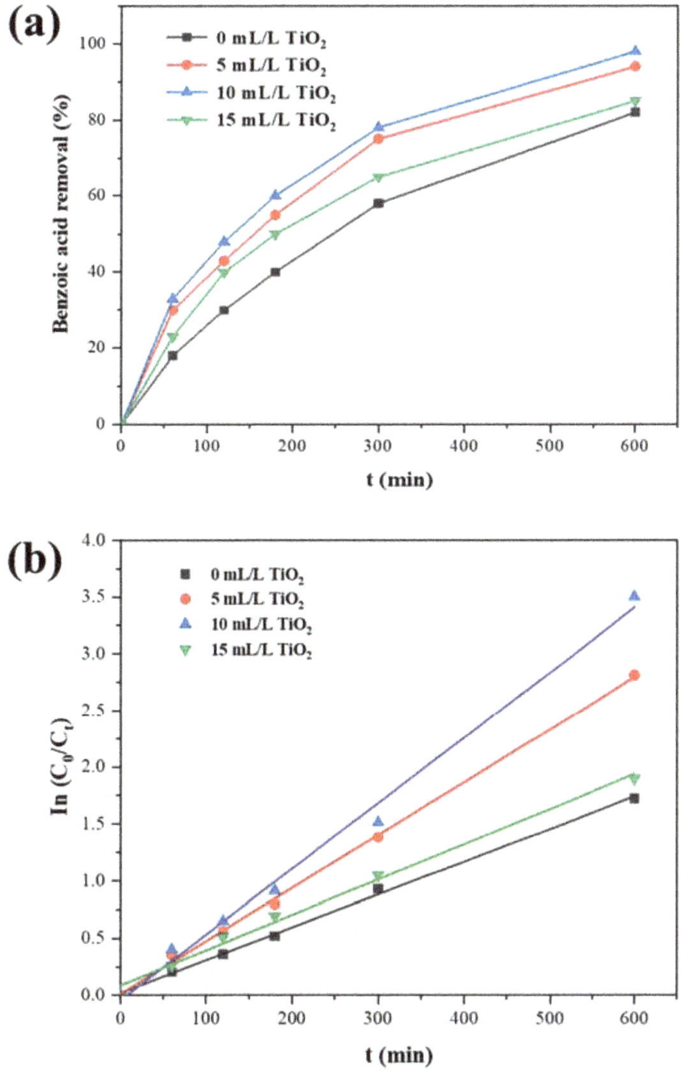

Figure 6. (**a**) The benzoic concentration after a certain duration of degradation on different electrodes, (**b**) the fitted curve for the BA degradation process, where C_t is the BA concentration and C_0 is the initial BA concentration.

Table 3. Degradation ability of Ti/SnO_2-Sb_2O_x-TiO_2 electrodes with different concentrations of TiO_2 sol in electrolytes, where *k* is kinetics coefficients.

	0 mL/L TiO_2	**5 mL/L TiO_2**	**10 mL/L TiO_2**	**15 mL/L TiO_2**
K (min^{-1})	2.87×10^{-3}	4.64×10^{-3}	5.75×10^{-3}	3.09×10^{-3}

The results prove that the proper addition of TiO_2 sol enhances the electrocatalytic performance of Ti/SnO_2-Sb_2O_x-TiO_2 electrodes. In general, the electrochemical combustion of organics is achieved by the hydroxyl radicals—a robust oxidizing agent that breaks the chemical bonds and finally turns the organic into H_2O and CO_2. The organic degradation pathway is proposed as follows in Equations (2) and (3) [13,34]:

$$SnO_2 + H_2O \rightarrow SnO_2\ (OH)_{ads} + H^+ + e \quad (2)$$

$$R + SnO_2\ (OH)_{ads} \rightarrow SnO_2 + H^+ + e + \text{Oxidation products (such as } CO_2) \quad (3)$$

where R stands for the organic and $(OH)_{ads}$ stands for the physic-adsorbed hydroxyl radicals. The higher content of Sb^{5+} increases the electrode conductivity and helps to generate surface adsorbed oxygen species. Besides, the uniform electrode surface with numerous peaks and valleys could provide many active sites for degradation reactions for the 5 and 10 mL/L TiO_2 modified electrodes. The well-dispersed TiO_2 nanoparticles also offer additional reaction sites, further improving the degradation ability for the TiO_2 sol-enhanced electrodes. As for the 10 mL/L TiO_2 modified Ti/SnO_2-Sb_2O_x-TiO_2 electrode, both the superior surface feature and well-dispersed TiO_2 award it the best electrocatalytic performance in the examined samples.

Accelerated life tests were also conducted to investigate the electrode performance, as shown in Figure 7, comprehensively. The dip-coated Ti/SnO_2-Sb_2O_x electrode was tested for comparison, which shows an inferior long-term stability due to its open surface structure being vulnerable to corrosion attack. During the degradation process, the corrosion attack results in the generation of titanium oxide, a non-conductive inert material that inhibits the degradation process and causes coating defoliation.

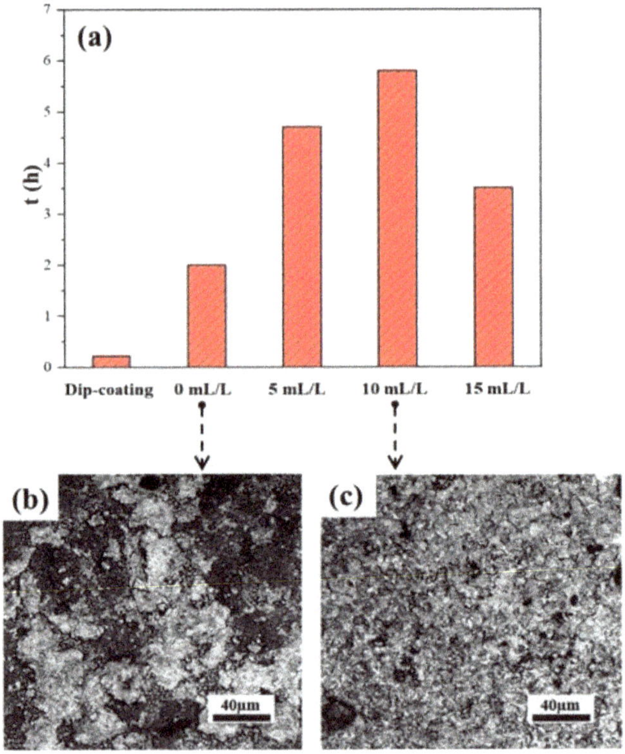

Figure 7. The accelerated service life results, (**a**) the tested accelerated service life for samples preprepared by dip-coating and sol-enhanced electrodeposition with different TiO_2 sol concentrations, (**b**) surface morphology of electrodeposited Ti/SnO_2-Sb_2O_x after 2 h test, and (**c**) surface morphology of Ti/SnO_2-Sb_2O_x-TiO_2 (10 mL/L TiO_2 sol doped) after 2 h test.

The electrodeposited Ti/SnO_2-Sb_2O_x-TiO_2 shows a much-improved stability in the accelerated life tests. A relatively compact coating fully covers underneath the Ti substrate for these samples. An increasing TiO_2 addition gradually prolongs the accelerated life at a concentration less than 10 mL/L. Figure 7b,c demonstrate the surface morphology of the undoped sample and 10 mL/L TiO_2 sol modified samples after 2 h accelerated life tests. The undoped SnO_2-Sb_2O_x coating is almost peeled off from the substrate, whereas the electrolysis process only slightly changes the surface morphology of the Ti/SnO_2-Sb_2O_x-TiO_2 electrode.

The electrode stability is highly correlated to the different morphologic features. The proper addition of TiO_2 sol leads to a uniform surface with fewer cracks and holes, thereby improving the electrode stability. Moreover, well-dispersed nanoparticles in the coating can prevent corrosion penetration and enhance the corrosion resistance for electrodes. In contrast, excessive TiO_2 addition harms the coating uniformity due to the nanoparticle aggregation, negatively affecting the electrode stability.

4. Conclusions

This study develops the Ti/SnO_2-Sb_2O_x-TiO_2 electrodes by applying sol-enhanced electrodeposition from methanesulfonate electrolytes. The results show that adding TiO_2 sol affects the surface morphologies of Ti/SnO_2-Sb_2O_x-TiO_2 electrodes. A uniform coating surface can be obtained at a suitable sol addition. Furthermore, the addition of TiO_2 sol leads to deep oxidation of Sb and helps to generate more adsorbed oxygen. The organic degradation performance decreases in the following order: 10 mL/L TiO_2 sol modified sample > 5 mL/L TiO_2 modified sample > 15 mL/L TiO_2 modified sample > 0 mL/L TiO_2 modified sample. The favored surface feature and well-dispersed TiO_2 in the 10 mL/L TiO_2 modified Ti/SnO_2-Sb_2O_x-TiO_2 electrode results in the best electrocatalytic performance in the examined samples, and its uniform coating surface with fewer cracks and holes prolong the electrode service life.

Author Contributions: Investigation, S.Z.; methodology, C.Y.; writing—review and editing, J.T.; validation, Y.W.; writing—original draft preparation, Z.H. All authors have read and agreed to the published version of the manuscript.

Funding: This work was supported by the Natural Science Foundation of Jiangsu Province under Grant BK20201008, the Key Research Project of Zhenjiang under Grant GJ2020014.

Institutional Review Board Statement: Not applicable.

Informed Consent Statement: Not applicable.

Data Availability Statement: Not applicable.

Conflicts of Interest: The authors declare no conflict of interest.

References

1. Salazar-Banda, G.R.; Santos, G.d.O.S.; Duarte Gonzaga, I.M.; Dória, A.R.; Barrios Eguiluz, K.I. Developments in electrode materials for wastewater treatment. *Curr. Opin. Electrochem.* **2021**, *26*, 100663. [CrossRef]
2. Tan, X.; Zhao, Y.; Sun, W.; Jin, C.; Chen, L.; Wei, H.; Sun, C. Three-dimensional hierarchically porous PbO_2 electrode for electrochemical degradation of m-cresol. *J. Electroanal. Chem.* **2020**, *856*, 113726. [CrossRef]
3. Syam Babu, D.; Nidheesh, P. A review on electrochemical treatment of arsenic from aqueous medium. *Chem. Eng. Commun.* **2020**, *208*, 1–22. [CrossRef]
4. Wu, W.; Huang, Z.-H.; Lim, T.-T. Recent development of mixed metal oxide anodes for electrochemical oxidation of organic pollutants in water. *Appl. Catal. A Gen.* **2014**, *480*, 58–78. [CrossRef]
5. Chaplin, B.P. Critical review of electrochemical advanced oxidation processes for water treatment applications. *Environ. Sci. Process. Impacts* **2014**, *16*, 1182–1203. [CrossRef]
6. Feng, Y.; Yang, L.; Liu, J.; Logan, B.E. Electrochemical technologies for wastewater treatment and resource reclamation. *Environ. Sci. Water Res. Technol.* **2016**, *2*, 800–831. [CrossRef]
7. Ni, Q. Ti/SnO_2-Sb_2O_5 as an Anode Material in Electro-Oxidation of Organic Compounds for Wastewater Treatment. Ph.D. Thesis, University of Toronto, Toronto, ON, Canada, 2016.

8. Rao, A.N.S.; Venkatarangaiah, V.T. Preparation, characterization, and application of Ti/TiO_2-NTs/Sb-SnO_2 electrode in photo-electrochemical treatment of industrial effluents under mild conditions. *Environ. Sci. Pollut. Res.* **2018**, *25*, 11480–11492. [CrossRef]
9. Yu, L.; Chen, Y.; Han, W.; Sun, X.; Li, J.; Wang, L. Preparation of porous TiO_2-NTs/m-SnO_2-Sb electrode for electrochemical degradation of benzoic acid. *RSC Adv.* **2016**, *6*, 19848–19856. [CrossRef]
10. Duan, Y.; Chen, Y.; Wen, Q.; Duan, T. Fabrication of dense spherical and rhombic Ti/Sb-SnO_2 electrodes with enhanced electrochemical activity by colloidal electrodeposition. *J. Electroanal. Chem.* **2016**, *768*, 81–88. [CrossRef]
11. Yang, B.; Wang, J.; Jiang, C.; Li, J.; Yu, G.; Deng, S.; Lu, S.; Zhang, P.; Zhu, C.; Zhuo, Q. Electrochemical mineralization of perfluorooctane sulfonate by novel F and Sb co-doped Ti/SnO_2 electrode containing Sn-Sb interlayer. *Chem. Eng. J.* **2017**, *316*, 296–304. [CrossRef]
12. Yang, K.; Liu, Y.; Liu, J.; Qiao, J. Preparation optimization of multilayer-structured SnO_2–Sb–Ce/Ti electrode for efficient electrocatalytic oxidation of tetracycline in water. *Chin. J. Chem. Eng.* **2018**, *26*, 2622–2627. [CrossRef]
13. Yang, K.; Liu, Y.; Qiao, J. Electrodeposition preparation of Ce-doped Ti/SnO_2-Sb electrodes by using selected addition agents for efficient electrocatalytic oxidation of methylene blue in water. *Sep. Purif. Technol.* **2017**, *189*, 459–466. [CrossRef]
14. Wu, W.; Huang, Z.-H.; Hu, Z.-T.; He, C.; Lim, T.-T. High performance duplex-structured SnO_2-Sb-CNT composite anode for bisphenol A removal. *Sep. Purif. Technol.* **2017**, *179*, 25–35. [CrossRef]
15. Wang, Y.; Gao, W.; He, Z.; Jin, Y.; Qiao, Y.; Cheng, G.; Gao, W.; Zhi, Z. Cu–Sn–Zn nanocomposite coatings prepared by TiO_2 sol-enhanced electrodeposition. *J. Appl. Electrochem.* **2020**, *50*, 875–885. [CrossRef]
16. He, Z.; Cao, D.; Wang, Y.; Yin, L.; Hayat, M.D.; Singh, H. Preparation of Co–P–TiO_2 nanocomposite coatings via a pulsed electrodeposition process. *Surf. Eng.* **2020**, *36*, 975–981. [CrossRef]
17. He, Z.; Cao, D.; Cao, F.; Zhang, S.; Wang, Y. Effects of heat treatment on the properties of Co–P–TiO_2 nanocomposite coatings. *Surf. Eng.* **2020**, *36*, 720–726. [CrossRef]
18. Wang, Y.X.; Cao, D.; Gao, W.D.; Qiao, Y.X.; Jin, Y.X.; Cheng, G.; Gao, W.; Zhi, Z. Microstructure and properties of sol-enhanced Co-P-TiO_2 nano-composite coatings. *J. Alloys Compd.* **2019**, *792*, 617–625. [CrossRef]
19. Liu, Z.; Luo, X.; Ji, D. Effect of phase composition of PbO_2 on cycle stability of soluble lead flow batteries. *J. Energy Storage* **2021**, *38*, 102524. [CrossRef]
20. He, Z.; Hayat, M.D.; Huang, S.; Wang, X.; Cao, P. PbO_2 electrodes prepared by pulse reverse electrodeposition and their application in benzoic acid degradation. *J. Electroanal. Chem.* **2018**, *812*, 74–81. [CrossRef]
21. He, Z.; Hayat, M.D.; Huang, S.; Wang, X.; Cao, P. Physicochemical characterization of PbO_2 coatings electrosynthesized from a methanesulfonate electrolytic solution. *J. Electrochem. Soc.* **2018**, *165*, D670–D675. [CrossRef]
22. Wang, Y.X.; Hu, B.; Tay, S.L.; Hou, F.Y.; Gao, W.; Xiong, C. The microstructure and properties of sol-enhanced Sn-TiO_2 nanocomposite coatings. *Int. J. Mod. Phys. B* **2017**, *31*, 16–19. [CrossRef]
23. Sun, Y.; Cheng, S.; Mao, Z.; Lin, Z.; Ren, X.; Yu, Z. High electrochemical activity of a Ti/SnO_2-Sb electrode electrodeposited using deep eutectic solvent. *Chemosphere* **2020**, *239*, 124715. [CrossRef] [PubMed]
24. He, Z.; Zhang, S.; Yin, L.; Hayat, M.D.; Cao, P. Cu–TiO_2 nanocomposite coatings prepared from sol-enhanced electrodeposition. *Int. J. Mod. Phys. B* **2020**, *34*, 2040038. [CrossRef]
25. Zhang, W.; Cao, D.; Qiao, Y.; He, Z.; Wang, Y.; Li, X.; Gao, W. Microstructure and properties of duplex Ni-P-TiO_2/Ni-P nanocomposite coatings. *Mater. Res.* **2019**, *22*. [CrossRef]
26. Mao, A.; Xiang, H.-Z.; Zhang, Z.-G.; Kuramoto, K.; Yu, H.; Ran, S. Solution combustion synthesis and magnetic property of rock-salt ($Co_{0.2}Cu_{0.2}Mg_{0.2}Ni_{0.2}Zn_{0.2}$) O high-entropy oxide nanocrystalline powder. *J. Magn. Magn. Mater.* **2019**, *484*, 245–252. [CrossRef]
27. Ni, Q.; Kirk, D.W.; Thorpe, S.J. Characterization of the mixed oxide layer structure of the Ti/SnO_2-Sb_2O_5 anode by photoelectron spectroscopy and impedance spectroscopy. *J. Electrochem. Soc.* **2014**, *162*, H40–H46. [CrossRef]
28. Patterson, A.L. The Scherrer formula for X-ray particle size determination. *Phys. Rev.* **1939**, *56*, 978–982. [CrossRef]
29. Allahkaram, S.R.; Nazari, M.H.; Mamaghani, S.; Zarebidaki, A. Characterization and corrosion behavior of electroless Ni-P/nano-SiC coating inside the CO_2 containing media in the presence of acetic acid. *Mater. Des.* **2011**, *32*, 750–755. [CrossRef]
30. Zhang, L.; Xu, L.; He, J.; Zhang, J. Preparation of Ti/SnO_2-Sb electrodes modified by carbon nanotube for anodic oxidation of dye wastewater and combination with nanofiltration. *Electrochim. Acta* **2014**, *117*, 192–201. [CrossRef]
31. Wu, T.; Zhao, G.; Lei, Y.; Li, P. Distinctive tin dioxide anode fabricated by pulse electrodeposition: High oxygen evolution potential and efficient electrochemical degradation of fluorobenzene. *J. Phys. Chem. C* **2011**, *115*, 3888–3898. [CrossRef]
32. Cui, Y.-H.; Feng, Y.-J.; Liu, Z.-Q. Influence of rare earths doping on the structure and electro-catalytic performance of Ti/Sb–SnO_2 electrodes. *Electrochim. Acta* **2009**, *54*, 4903–4909. [CrossRef]
33. Duan, Y.; Wen, Q.; Chen, Y.; Duan, T.; Zhou, Y. Preparation and characterization of TiN-doped Ti/SnO_2-Sb electrode by dip coating for orange II decolorization. *Appl. Surf. Sci.* **2014**, *320*, 746–755. [CrossRef]
34. Chen, Y.; Tu, Y.; Zhang, Y.H.; Lu, J.A.; Fang, B.B. Fabrication and enhanced electrocatalytic activity of TiO_2 nanotubes based three-dimensionally macroporous SnO_2 with mesoporous walls. *Chem. Eng. J.* **2017**, *311*, 100–110. [CrossRef]

coatings

MDPI

Article

Environmental Fatigue Behavior of a Z3CN20.09M Stainless Steel in High Temperature Water

Kewei Fang [1,*], Kunjie Luo [1], Li Wang [1], Chengtao Li [1], Lei Wang [2] and Yanxin Qiao [2,*]

1 Life Management Center, Suzhou Nuclear Power Research Institute, Suzhou 215004, China; luokunjie@cgnpc.com.cn (K.L.); lwongsz@foxmail.com (L.W.); lichengtao@cgnpc.com.cn (C.L.)

2 School of Materials Science and Engineering, Jiangsu University of Science and Technology, Zhenjiang 212003, China; wangl_ray@just.edu.cn

* Correspondence: fangkewei@cgnpc.com.cn (K.F.); yxqiao@just.edu.cn (Y.Q.)

Abstract: The low-cycle fatigue behavior of a Z3CN20.09M austenitic stainless steel was investigated and its fatigue life in high temperature water was compared to that in the air at room temperature. It is found that the fatigue life in water at 300 °C was shorter than that in air, and it decreased with the decreasing strain rate from 0.4% to 0.004%/s. The ductile striations having streamed down features were observed at the strain rate of 0.004%/s, indicating that Z3CN20.09M austenitic stainless steel experienced anodic dissolution. The fatigue life obtained in the present experiment was consistent with that using prediction models.

Keywords: Z3CN20.09M; corrosion fatigue; fatigue life; high temperature water

Citation: Fang, K.; Luo, K.; Wang, L.; Li, C.; Wang, L.; Qiao, Y. Environmental Fatigue Behavior of a Z3CN20.09M Stainless Steel in High Temperature Water. *Coatings* **2022**, *12*, 317. https://doi.org/10.3390/coatings12030317

Academic Editor: Christian Mitterer

Received: 27 December 2021
Accepted: 18 February 2022
Published: 28 February 2022

1. Introduction

Corrosion and environmental fatigue damage is one common material degradation process in nuclear power plants [1–6]. The wide use of austenitic stainless steels (ASSs) to fabricate nuclear power plant components raises the importance to investigate their low-cycle fatigue (LCF) properties in a light water reactor (LWR) environment to ensure the integrity and safety of nuclear power plants [3,4,6,7]. Furthermore, the ASME design fatigue curve cannot explicitly address the contribution of the service environment to the service life of these ASS components [3,5,8]. Recently, many studies related to corrosion fatigue of SSs have been performed and tried to address the effect of the corrosive environment at high temperatures (i.e., 300 °C) on their fatigue life and possible fatigue mechanism [2,5,6,9]. A large number of research data on corrosion fatigue of ASSs in simulating the environment in LWR were collected by Keisler et al. [2] in the Argonne National Laboratory (ANL). Based on these data, Chopra et al. [3] established a statistical model and can use it to predict the fatigue life of ASSs in LWR environments successfully. Higuchi et al. [5,10] have also evaluated the fatigue life data of SSs reported by Japanese researchers and have proposed a parameter named the fatigue life correction factor (F_{en}) to estimate the fatigue life of SSs in LWR environments. Although there are many similarities between the ANL's model and Higuchi's model, differences can also be found in them, such as loading parameters, the working environments, material variability, etc. [8,11].

The Z3CN20.09M ASS is commonly used in PWR nuclear power plants in China. The corrosion fatigue properties of Z3CN20.09M ASS in a simulated LWR environment are closely related to the safe operation of nuclear power plants. In the present work, the LCF tests were conducted to explore the fatigue life of the Z3CN20.09M ASS. The obtained fatigue life of this work was compared with the ANL and Higuchi models. Additionally, the observation of the fractured surface of Z3CN20.09M ASS was conducted to further clarify the corrosion fatigue behaviors in a simulated LWR environment.

2. Materials and Experimental Procedures

The test rig used in the present study is composed of an autoclave chamber, a high temperature and pressurized water loop, a data acquisition system, and control equipment, as shown in Figure 1. It simulates the operating conditions in LWR via controlling the content of dissolved oxygen (DO) in water, pH, and water conductivity.

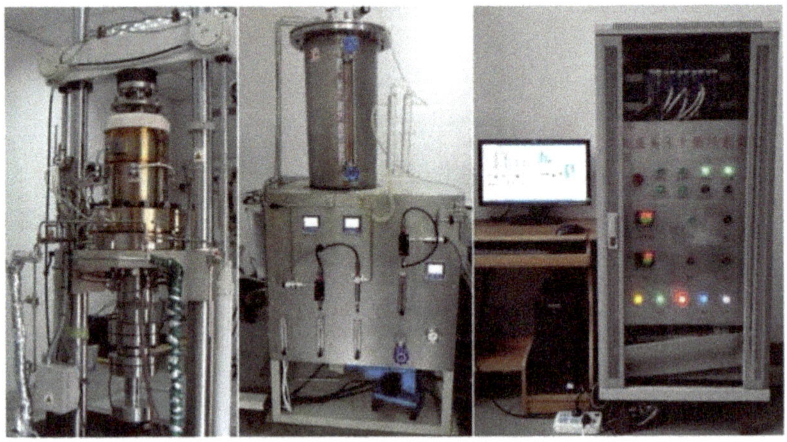

Figure 1. An image showing the instrument setup for a low-cycle fatigue test.

The material studied was the Z3CN20.09M ASS used as the primary piping material in nuclear power plants in China. Its chemical composition (in wt.%) was: 0.015% C, 1.08% Si, 1.09% Mn, 0.018% P, 0.002% S, 19.63% Cr, 9.27% Ni, 0.02% Cu, 0.04% Co, and bal. Fe. The mechanical properties of Z3CN20.09M ASS at 25 °C (room temperature, RT) and 350 °C are given in Table 1. Figure 2 shows the microstructure of the Z3CN20.09M ASS, which has a duplex structure: gray δ-ferrite distributing as a network in a white austenitic matrix, which has been characterized carefully by Xue et al. [12].

The LCF test conditions, including strain rate, strain amplitude, load ratio (R), temperature, pressure, and DO, are shown in Table 2. Fatigue loading was applied using a strain-controlled method. The dimensions of the specimens used in the LCF test are presented in Figure 3. The fatigue life, N_{25}, is the number of cycles at which tensile stress during the fatigue cycle decreased to 75% of peak stress value, was used to evaluate the fatigue behavior of the tested materials. Consequently, the LCF tests were terminated when the tensile stress decreased by 25% of the peak stress value. After tests, the fatigue surface was observed to obtain the morphologies of fatigue crack using a scanning electron microscope (SEM) (SEM, JEOL, JSM-6480, Takeno, Japan).

Table 1. Mechanical properties of the Z3CN20.09M ASS at RT and 350 °C.

Properties	Temperature	Properties	Z3CN20.09M
Tensile Properties	25 °C	$R_{p0.2}$	235 MPa
		R_m	655 MPa
		A% (5d)	56.8 %
	350 °C	$R_{tp0.2}$	155 MPa
		R_m	395 MPa
KV Impact	25 °C	Lowest Average Value	265 J

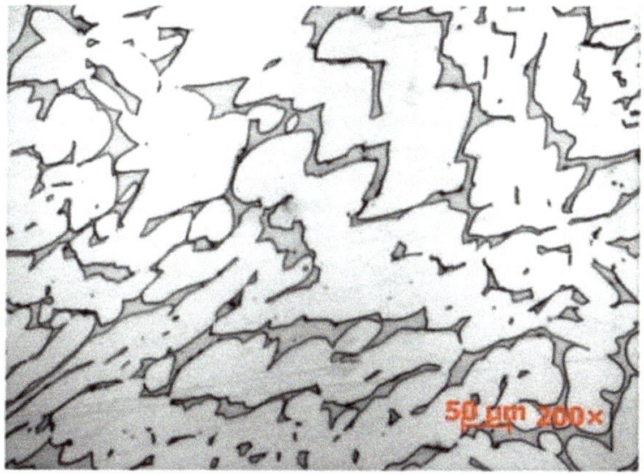

Figure 2. Optical observation of the Z3CN20.09M ASS.

Table 2. LCF test conditions and water chemistry.

Test conditions	Load ratio (R)	−1
	Control mode	Strain
	Wave form	Full reversed triangular
	Strain rate	0.4, 0.04 and 0.004%/s
	Strain amplitude (ε_a)	0.2–1.0%
Water chemistry	Temperature	300 °C
	Pressure	12 MPa
	Conductivity	<0.1 μS/cm
	Dissolved oxygen (DO)	10 ± 1 ppb

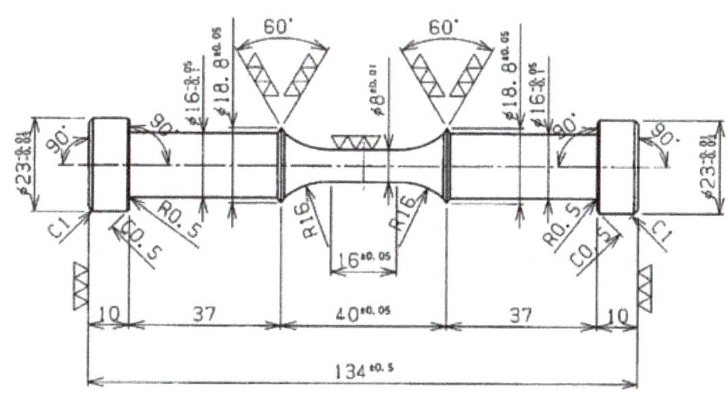

Figure 3. Schematic diagram showing the dimensions of the fatigue specimen.

3. Results and Discussion

3.1. Fatigue Life

The fatigue life, N_{25}, of the Z3CN20.09M austenitic SS in water at 300 °C is plotted against strain and shown in Figure 4. In Figure 4, the ASME design fatigue curve and the fatigue life in air at RT are also presented [4,8]. As is seen, in the same loading condition, the fatigue life of the Z3CN20.09M ASS in water at 300 °C was shorter than that in air at RT [2,3,6,9,13–15]. It indicates that the involvement of corrosive medium or the interactions of corrosive medium with applied load might be responsible for the decrease in the fatigue life of the Z3CN20.09M ASS [16,17]. At the same strain amplitude, the decrease in strain rate yielded a lower fatigue life [15,18–21]. For example, at 0.8% strain amplitude, the decrease in strain rate from 0.04% to 0.004%/s led to 3.5 times lower in the fatigue life.

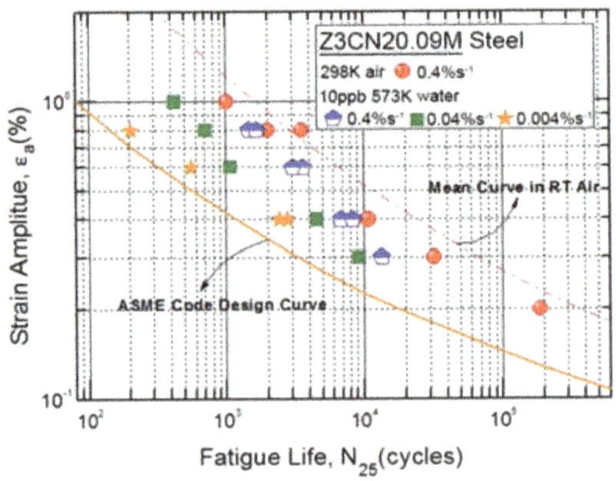

Figure 4. Fatigue life of the Z3CN20.09M ASS in water at 300 °C.

The plot of fatigue life of the Z3CN20.09M ASS in water at 300 °C as a function of strain rate is shown in Figure 5. Generally, the fatigue life of the Z3CN20.09M ASS in water at 300 °C decreased with the decreasing strain rate. Furthermore, the difference in the fatigue life at these three strain rates became more pronounced at higher strain amplitude. It might be caused by the higher corrosion rate of the tested material at higher strain amplitude in the autoclave environments [6,13–27].

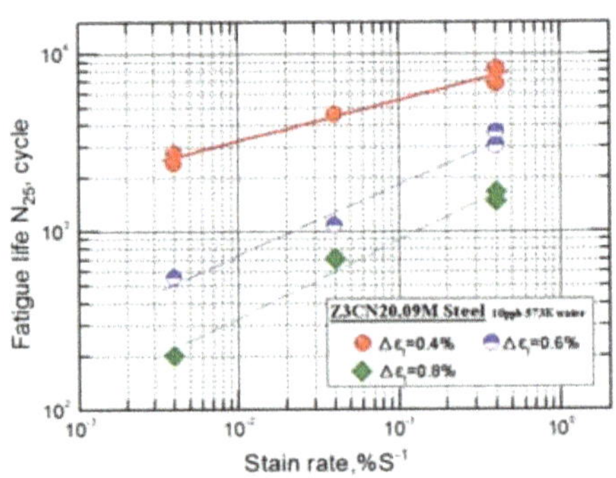

Figure 5. The plot of the fatigue life of the Z3CN20.09M ASS against strain rate in water. The testing corrosive medium was water, and the temperature was 300 °C.

3.2. SEM Observation

To explore the contribution of corrosive medium to the decrease in the fatigue life of the Z3CN20.09M ASS in water at 300 °C, the morphologies of fatigue surfaces at various strain rates were observed, Figure 6. As a comparison, Figure 6 also shows the fractographic surface of the Z3CN20.09M ASS specimen tested in air at RT. It is clear that well-developed ductile striations were present on the fatigue surface of the Z3CN20.09M ASS specimen when exposed to air at RT [12,27,28], Figure 6a. When the Z3CN20.09M ASS specimen was tested in water at 300 °C, its fatigue crack features were different from that in air at RT [4,6,22,23], as shown in Figure 6b,c. Additionally, the characteristics of the fatigue crack morphologies of the specimens tested in high temperature water changed with strain rates [13,22,23]. At a higher strain rate (i.e., 0.4%/s), several flattened regions marked using white dish lines in Figure 6b were observed, where many striations having "streamed down features" were visible. At a lower strain rate [24] (i.e., 0.004%/s), these ductile fatigue striations with streamed down features were dominant at the fracture surface [13], Figure 6c. The presence of these striations with streamed down features is believed to be correlated with the occurrence of metal dissolution during the LCF test [18,25,29,30]. The metal dissolution process was proposed to occur via the following steps: (i) the formation of the oxide film at the crack tip by the anodic metal dissolution, which passivated the crack tip [23], (ii) the rupture of the passive film by applied strain, (iii) the exposure of fresh metal to the corrosive environment, leading to the reactivation, (iv) the dissolution of the freshly exposed metal [24], and (v) the propagation of fatigue crack and the increase in crack growth rate due to the interaction of the metal dissolution and applied load [31,32]. It means that the interaction of metal dissolution and applied load accelerated the crack propagation, leading to the decrease in the fatigue life of the Z3CN20.09M ASS in the simulated LWR water.

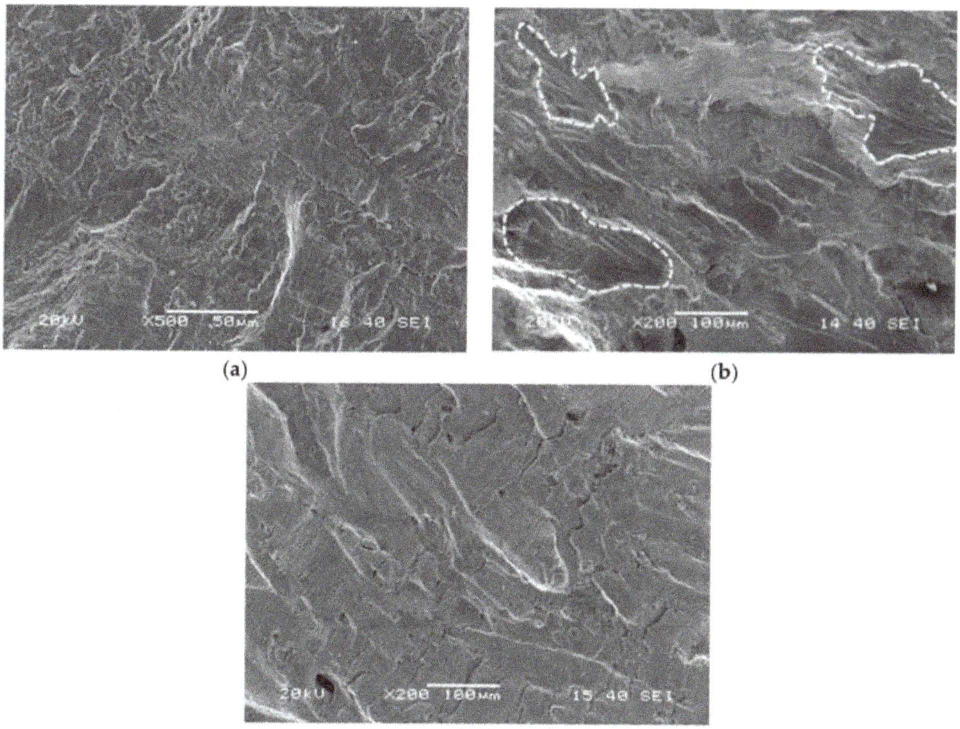

Figure 6. Fractured surfaces of Z3CN20.09M ASS at the strain amplitude of 0.8% and the strain rate of: (**a**) 0.4%/s in air at RT, (**b**) 0.4%/s and (**c**) 0.004%/s in high temperature water.

Figure 7 presents the cracks at the gauge sections of the LCF specimens tested in RT air and in simulated LWR water. At the same strain amplitude and strain rate, the crack propagation path in high temperature water was more tortuous than that in air, Figure 7a,d, suggesting that the corrosion medium was involved in the crack growth. It is consistent with the results of figure life measurements in Figure 5. With the increase of strain rate from 0.004%/s to 0.4%/s, the crack propagation path changed from being perpendicular to the loading direction to being zigzagged, Figure 7b–d.

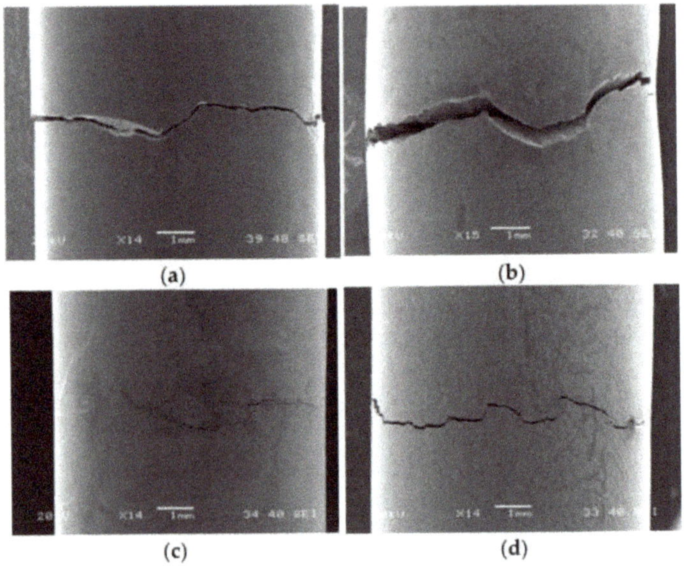

Figure 7. Surface crack morphology of LCF specimens tested at the strain amplitude of 0.8% and the strain rate of: (a) 0.4%/s in air at RT, (b) 0.004%/s, (c) 0.04%/s and (d) 0.4%/s in high temperature water.

The surface morphology at the crack tip of the LCF specimens after testing in high temperature water is shown in Figure 8. It can be seen that the crack tip had the features of interconnecting (Figure 8a), crossing (Figure 8b) and bifurcation [24] (Figure 8c,d). At the crack tip when the stress concentrated, many dislocations initiated were slipping along the slip bands, raising the electrochemical activity of the dislocation slipping regions, and thus accelerating their corrosion processes. It led to the accumulation of corrosion products at the crack tip (Figure 9) and the dislocation slipping regions, which resulted in the blunting of the crack tip. The cyclic application of tensile and compressive load would rupture the corrosion products formed at the crack tip and reactivate the corrosion process by the exposure of the fresh metal surface, leading to the sharpening of the crack tip. These repeating blunting and sharpening processes would prompt the growth of fatigue crack. On the other hand, new slipping bands may generate as the crack propagated, which would favor the initiation of micro-cracks in the front of the main crack because of the applied load and active corrosion interactions [19,29]. Under the action of applied load, the main crack would connect these micro-cracks, possessing the features of coalescing (Figure 8a), crossing (Figure 8b) and bifurcating (Figure 8c). This depended on the orientation of micro-cracks with the main crack.

3.3. Difference between the Fatigue Life Data and the Prediction Models

To ensure the reliability of the experimental data, two prediction models, ANL's model [3] and Higuchi's model [5] were used in the present work. The ANL's statistical model was established based on the Langer equation, in which several terms were added to consider the environmental factors influencing the fatigue life of SS [11,13,33,34]. In LWR environments, this model to predict the fatigue life of austenitic SS can be expressed using Equation (1) [3]:

$$\ln(N_{25}) = 6.157 - 192\ln(\varepsilon_a - 0.112) + T^* \cdot \varepsilon^* \cdot O^* \tag{1}$$

where T^*, ε^*, and O^* is the transformed temperature, strain rate, and DO, respectively. These parameters can be determined by the test conditions in the present work.

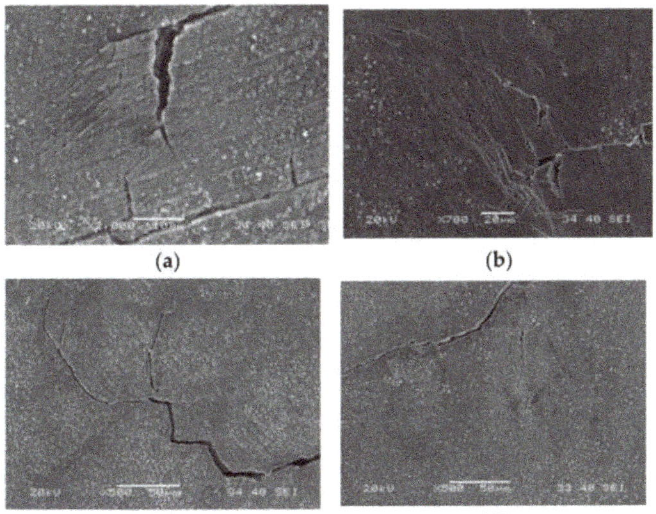

(a) (b) (c) (d)

Figure 8. Crack tip morphology of the LCF specimens at the strain amplitude of 0.8% and the strain rate of 0.4%/s in high temperature water: (**a**) crack coalescing; (**b**) crack crossing; (**c**,**d**) crack bifurcating.

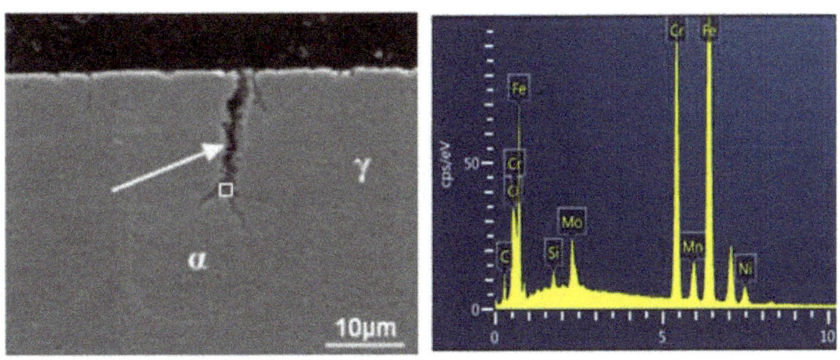

(a) (b)

Figure 9. Cross-section morphology of secondary cracks (**a**) and the EDS analysis of corrosion products accumulated at the crack (**b**).

According to Higuchi et al. [5,10], the fatigue life correction factor, Fen, was proposed and defined using Equation (2):

$$F_{en} = \exp((C - \varepsilon^*)T^*)A^* \tag{2}$$

where, C is a constant determined by the reactor type, and ε^* *is* strain rate, T^* *is* temperature and A^* is strain amplitude. The fatigue life predicted using Higuchi's model was calculated by multiplying the fitted fatigue life of SS in air [9] with F_{en} calculated from our test conditions [7,35,36].

The fatigue life obtained in this work and the predicted ones available using ANL's model and Higuchi's model are plotted in Figure 10. Generally, our experimental results are identical with the predicted fatigue life for these three strain rates [3,35,37]. Although the difference in the experimental data from the predicted one was within a factor of 3, it was considered to be acceptable. Thus, the fatigue life obtained in this work was acceptable and reliable. However, it is necessary to further analyze the variability of data, which is an arduous and important task.

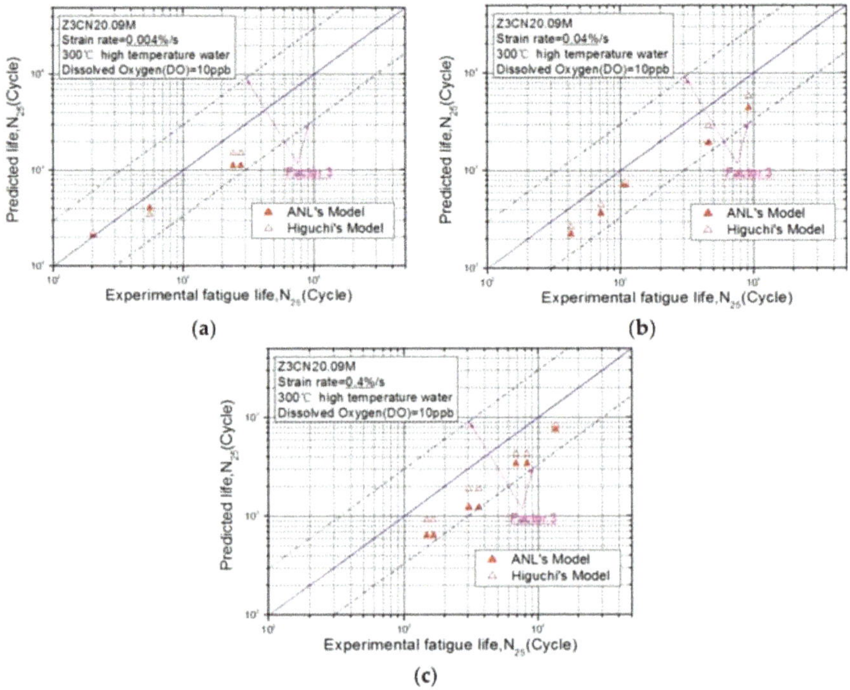

Figure 10. The fatigue life data obtained in the current work to the predicted fatigue life at the strain rate of: (**a**) 0.004%/s, (**b**) 0.04%/s and (**c**) 0.4%/s.

4. Conclusions

The LCF tests of the Z3CN20.09M austenitic SS were performed to investigate the fatigue life in high temperature water. Its fatigue life in 300 °C water was shorter than that in air under present testing conditions depending on strain rate, suggesting the involvement of corrosive medium. When the strain rate decreased from 0.4%/ to 0.004%/s, the reduction in the fatigue life became remarkable. The fractographic observation suggested that the decrease in fatigue life was attributable to the interaction between metal dissolution and applied load. Under the action of applied load, corrosion preferentially occurred at the places where dislocations initiated and accumulated, leading to the deposition of corrosion products. Owing to the expansion stress of corrosion deposit formation and applied stress, stress could easily concentrate at the bottom of corrosion damage, leading to the initiation of crack and the acceleration of the corrosion process. It in turn accelerated the accumulation of corrosion products at the crack tip and the dislocation slipping regions and thus, resulted in the blunting of the crack tip. Additionally, the cyclic application of tensile and compressive load would rupture the corrosion deposits to expose fresh metal surface and reactivate the corrosion process, leading to the sharpening of the crack tip. These repeating blunting and sharpening processes would facilitate the propagation of fatigue crack.

The comparison of the fatigue life obtained experimentally to the one predicted using ANL's model and Higuchi's model confirmed that our experimental data was reliable, and their differences were within the acceptable range.

Author Contributions: Conceptualization, C.L. and Y.Q.; methodology, K.L.; formal analysis, K.F.; investigation, K.F.; data curation, K.F.; writing—original draft preparation, K.F., C.L., and L.W. (Li Wang); writing—review and editing, L.W. (Lei Wang) and Y.Q.; supervision, Y.Q.; funding acquisition, K.F. and C.L. All authors have read and agreed to the published version of the manuscript.

Funding: This research received no external funding.

Institutional Review Board Statement: Not applicable.

Informed Consent Statement: Not applicable.

Data Availability Statement: The data used to support the findings of this study are available from the corresponding author upon request.

Acknowledgments: This work was jointly supported by the National Key Research and Development Program of China under (No. 2016YFB0700404), the Natural Science Foundation of China (No. 51375182) and the Major Natural Science Foundation of China under (No. U1260201).

Conflicts of Interest: The authors declare no conflict of interest.

References

1. Iida, K. A Review of fatigue failures in LWR plants in Japan. *Nucl. Eng. Des.* **1992**, *138*, 297–312. [CrossRef]
2. Keisler, J.M.; Chopra, O.K.; Shack, W.J. Statistical models for estimating fatigue strain-life behavior of pressure boundary materials in light water reactor environments. *Nucl. Eng. Des.* **1996**, *167*, 129–154. [CrossRef]
3. Chopra, O.K. *Effects of LWR Coolant Environments on Fatigue Design Curves of Austenitic Stainless Steels*; NUREG/CR-5704, ANL-98/31; U.S. Nuclear Regulatory Commission: Rockville, MD, USA, 1999.
4. Tang, Y.B.; Shen, X.W.; Liu, Z.H.; Qiao, Y.X.; Yang, L.L.; Lu, D.H.; Zou, J.S.; Xu, J. Corrosion behaviors of selective laser melted inconel 718 alloy in NaOH solution. *Acta Metall. Sin.* **2022**, *58*, 324–333.
5. Higuchi, M.; Tsutsumi, K.; Hirano, A. A proposal of fatigue life correction factor F_{en} for austenitic stainless steels in LWR water environments. *J. Press. Vess. Tech.* **2003**, *125*, 403–410. [CrossRef]
6. Cho, H.; Kim, B.K.; Kim, I.S. Environmental fatigue testing of type 316 stainless steel in 310 °C water. In Proceedings of the ASME 2005 Pressure Vessels and Piping Conference, Denver, CO, USA, 17–21 July 2005.
7. Tice, D.R.; McLennan, A.; Gill, P. *Environmentally Assisted Fatigue (EAF) Knowledge Gap Analysis*; EPRI: Palto Alto, CA, USA, 2018.
8. *ASME Boiler and Pressure Vessel Code III Division 1-Appendices*; The American Society of Mechanical Engineers: New York, NY, USA, 1986.
9. Tsutsumi, K.; Kanasaki, H.; Umakoshi, T. Fatigue life reduction in PWR water environment for stainless steels. In Proceedings of the 2000 ASME PVP Conference, Seattle, WA, USA, 23–27 July 2000.
10. Higuchi, M.; Nakamura, T.; Sugie, Y. Development of an environmental fatigue evaluation method for nuclear power plants in JSME code. *J. Environ. Eng.* **2011**, *6*, 452–468. [CrossRef]
11. Haan, F.H.E.; Wilde, M.H.C.; Blom, F.J. Overview of international implementation of environmental fatigue. In Proceedings of the 2013 ASME PVP Conference, Paris, France, 14–18 July 2013.
12. Xue, F.; Shi, F.; Zhang, C.; Zheng, Q.; Yi, D.; Li, X.; Li, Y. The microstructure and mechanical and corrosion behaviors of thermally aged Z3CN20-09M cast stainless steel for primary coolant pipes of nuclear power plants. *Coatings* **2021**, *11*, 870. [CrossRef]
13. Vincent, L.; Le Roux, J.C.; Taheri, S. On the high cycle fatigue behavior of a type 304L stainless steel at room temperature. *Int. J. Fatigue* **2012**, *38*, 84–91. [CrossRef]
14. Chen, W.; Spätig, P.; Seifert, H.P. Fatigue behavior of 316L austenitic stainless steel in air and LWR environment with and without mean stress. *MATEC Web Conf.* **2018**, *165*, 03012. [CrossRef]
15. Zhang, W.Q.; Fang, K.W.; Wang, X.L. Investigation of stress corrosion cracking initiation in machined 304 austenitic stainless steel in magnesium chloride environment. *J. Mater. Eng. Perform.* **2020**, *29*, 191–204. [CrossRef]
16. Wire, G.L.; Leax, T.R.; Kandra, J.T. Mean stress and environmental effects on fatigue in type 304 stainless steel. In Proceedings of the 1999 ASME PVP Conference, Boston, MA, USA, 1–5 August 1999.
17. Solomon, H.D.; Amzallag, C.; Vallee, A.J.; De Lair, R.E. Influence of mean stress on the fatigue behavior of 304L SS in air and PWR water. In Proceedings of the 2005 ASME PVP Conference, Denver, CO, USA, 17–21 July 2005.
18. Kamaya, M. Influence of strain range on fatigue life reduction of stainless steel in PWR primary water. *Fatigue Fract. Eng. Mater. Struct.* **2017**, *40*, 2194–2203. [CrossRef]
19. Spätig, P.; Heczko, M.; Kruml, T.; Seifert, H.P. Influence of mean stress and light water reactor environment on fatigue life and dislocation microstructures of 316L austenitic steel. *J. Nucl. Mater.* **2018**, *509*, 15–28. [CrossRef]
20. Bradaï, S.; Gourdin, C. Equi-biaxial loading effect on austenitic stainless steel fatigue life. In Proceedings of the 2015 ASME PVP Conference, Boston, MA, USA, 19–23 July 2015.
21. Dong, L.J.; Han, E.H.; Peng, Q.J.; Ke, W.; Wang, L. Environmentally assisted crack growth in 308L stainless steel weld metal in simulated primary water. *Corros. Sci.* **2017**, *117*, 1–10. [CrossRef]

22. Seifert, H.P.; Ritter, S.; Leber, H.J. Corrosion fatigue crack growth behaviour of austenitic stainless steels under light water reactor conditions. *Corros. Sci.* **2012**, *55*, 61–75. [CrossRef]
23. Seifert, H.P.; Ritter, S.; Leber, H.J. Corrosion fatigue initiation and short crack growth behaviour of austenitic stainless steels under light water reactor conditions. *Corros. Sci.* **2012**, *59*, 20–34. [CrossRef]
24. Yuan, X.; Yu, W.; Fu, S.; Yu, D.; Chen, X. Effect of mean stress and ratcheting strain on the low cycle fatigue behavior of a wrought 316LN stainless steel. *Mater. Sci. Eng. A* **2016**, *677*, 193–202. [CrossRef]
25. Miura, N.; Takahashi, Y. High-cycle fatigue behavior of type 316 stainless steel at 288 °C including mean stress effect. *Int. J. Fatigue* **2006**, *28*, 1618–1625. [CrossRef]
26. Chen, W. Experimental Evaluation and Modelling of Fatigue of a 316L Austenitic Stainless Steels in High-Temperature Water and Air Environments. Ph.D. Thesis, EPFL, Lausanne, Switzerland, 2020.
27. Zhang, Z.Y.; Tan, J.B.; Wu, X.Q.; Han, E.H.; Ke, W.; Rao, J.C. Effects of temperature on corrosion fatigue behavior of 316LN stainless steel in high-temperature pressurized water. *Corros. Sci.* **2019**, *146*, 80–89. [CrossRef]
28. Paul, S.K. A critical review of experimental aspects in ratcheting fatigue: Microstructure to specimen to component. *J. Mater. Res. Technol.* **2019**, *8*, 4894–4914. [CrossRef]
29. Ogawa, Y.; Birenis, D.; Matsunaga, H.; Thøgersen, A.; Prytz, Ø.; Takakuwa, O.; Yamabe, J. Multi-scale observation of hydrogen-induced, localized plastic deformation in fatigue-crack propagation in a pure iron. *Scr. Mater.* **2017**, *140*, 13–17. [CrossRef]
30. Yi, X.N.; Ma, A.L.; Zhang, L.M.; Zheng, Y.G. Crystallographic anisotropy of corrosion rate and surface faceting of polycrystalline 90Cu-10Ni in acidic NaCl solution. *Mater. Des.* **2022**, *215*, 110429. [CrossRef]
31. Ford, F.P.; Andresen, P.L. Corrosion Fatigue of Pressure Boundary Materials. In Proceedings of the 7th International Conference On Fracture, Houston, TX, USA, 20–24 March 1989.
32. Kamaya, M. Development of disc bending fatigue test technique for equi-biaxial loading. *Int. J. Fatigue* **2016**, *82*, 561–571. [CrossRef]
33. Nakamura, T.; Miyama, S. Environmental fatigue evaluation in PLM activities of PWR plant. *E-J. Adv. Maint.* **2010**, *2*, 82–100.
34. Spätig, P.; Seifert, H.P. Mean stress effect on fatigue life of 316L austenitic steel in air and simulated boliing water reactor hydrogen water chemistry environment. In Proceedings of the 17th International Conference on Environmental Degradation of Materials in Nuclear Systems—Water Reactors, TMS, Ottawa, ON, Canada, 9–12 August 2015.
35. Procopio, I.; Cicero, S.; Mottershead, K.; Bruchhausen, M.; Cuvilliez, S. Increasing safety in NPPs by covering gaps in environmental fatigue assessment. *Pro. Str. Integrity* **2018**, *13*, 97–103. [CrossRef]
36. Mottershead, K.; Bruchhausen, M.; Métais, T.; Cicero, S.; Tice, D.; Platts, N. Increasing Safety in Nuclear Power Plants by Covering Gaps in Environmental Fatigue Assessment. In Proceedings of the 2016 ASME PVP Conference, Vancouver, BC, Canada, 17–21 July 2016.
37. Kamaya, M.; Kawakubo, M. Mean stress effect on fatigue strength of stainless steel. *Int. J. Fatigue* **2015**, *74*, 20–29. [CrossRef]

coatings

Article

Bridge Structure Damage Identification Based on Dynamic Characteristics

Yunkai Zhang [1,2,*], Xixue Tan [1,2], Guohua Li [1], Jun Dong [1,2], Jingyi Guo [1] and Fanyue Liu [1]

[1] School of Civil and Transportation Engineering, Beijing University of Civil Engineering and Architecture, Beijing 102616, China; 201802020414@stu.bucea.edu.cn (X.T.); liguohua@bucea.edu.cn (G.L.); dongjun@bucea.edu.cn (J.D.); 201802030120@stu.bucea.edu.cn (J.G.); 201802020125@stu.bucea.edu.cn (F.L.)
[2] Yunchuang Intelligent Testing Technology Institute, Beijing University of Civil Engineering and Architecture, Beijing 102616, China
* Correspondence: 201802040124@stu.bucea.edu.cn

Abstract: With the increasing traffic volume and years of usage during the operation process, a bridge structure will experience aging and damage to different degrees, leading to the decline in bridge reliability and seriously affecting its operation safety. In this study, the bridge was abstracted into a beam structure for damage identification. Next, the influence of damage on the bridge structure was explored from the angles of its inherent frequency and displacement mode, respectively. Our results showed that whether the structure was damaged could be accurately judged by its inherent frequency, but the specific damage could not be further judged. Through the structural displacement curve, the rough range of structural damage could be judged; however, the damage could not be accurately positioned. The damage position could be accurately identified to some extent by taking the derivatives from the difference value of the structural displacement curve. The above conclusions were verified based on a double-span beam. We found that the above conclusions still held true for the double-span beam, thus proving their universality.

Keywords: bridge structure; mode; frequency; displacement curve

Citation: Zhang, Y.; Tan, X.; Li, G.; Dong, J.; Guo, J.; Liu, F. Bridge Structure Damage Identification Based on Dynamic Characteristics. *Coatings* **2022**, *12*, 313. https://doi.org/10.3390/coatings12030313

Academic Editors: Jian Chen, Yanxin Qiao, Fanjiang Meng and Yuxin Wang

Received: 31 December 2021
Accepted: 16 February 2022
Published: 25 February 2022

1. Introduction

Presently, about one-fourth of bridges and buildings in China are affected by hidden structural dangers to a certain degree. It is critical to study the effective means of structural flaw detection to repair these damages. The methods resorting to dynamic characteristics are the current research hotspot. Frequency is the number of times of completing periodical changes within unit time, and the inherent frequency will be reduced owing to the increasing structural damages [1]. Besides, the damages can also be identified by paying attention to the frequency change.

Pandey et al. [2] found that the natural frequency of a bridge is related to the structural mode and shape. Given the structural damage, its stiffness will decline while the structural damping ratio increases.

Hearn and Testa found through a study that after the maximum structural frequency change is normalized, the ratio of changes at any two stages is a damage position function, verified through a model experiment. They further found that a certain law exists between the structural damage and dynamic structural characteristics, but the frequency change is uncorrelated with the evolution of structural damage [3].

Ramanjaneyulu et al. [4] studied the damage identification method based on the structural frequency and vibration mode as well as the effects of changes in different parameters such as frequency, vibration mode, mode curvature and strain energy on the damage identification, positioning and quantification.

Weng built the models of 20–40 m T-beam bridges and box girder bridges via finite element software Midas Civil and explored the relationship between the structural natural vibration frequency and the number of spans of a beam bridge [5].

Hu studied the change in the inherent frequency of a simply supported beam bridge using the fingerprint identification method; established three structural damage fingerprint maps based on the frequency change rate, its quadratic sum and regularized frequency change rate, respectively; conducted a model test; and finally validated the sensitivity of the three fingerprint maps [6].

Wu derived the influences of changes in the structural stiffness and mass on the dynamic characteristics, established a damage identification model based on the mass change according to the variable quantity of inherent structural frequency and verified the feasibility of two methods: additional mass method and a variable quantity of inherent frequency [7–13].

Su conducted the quantitative derivation of structural damage degree based on the displacement curvature and stiffness and proposed different treatment methods for various damages, such as total damage at measuring point and crack damage [14].

Jia constructed a corresponding damage detection system according to the change laws of five parameters related to the damage degree, such as curvature mode value, curvature mode difference and curvature mode ratio [15].

Wei put forward a comprehensive damage identification method based on the inherent frequency and curvature mode and conducted a related study [16].

Lian took the mode curvature as the identification signal, analyzed it using the wavelet transform theory and discovered that the modal curvature difference could overcome the problem of the modal curvature method that fails to identify mild damages [17].

Taking a multiple simply supported beam bridge as the study object, Zhao set the working conditions with one and multiple damages and verified the damage identification accuracy of modal curvature [18–21].

Consequently, the structural damage was simulated in this study mainly by creating directed damages in the structure, mainly studying the change laws of inherent frequency, displacement mode and curvature mode before and after beam structure damage. Furthermore, the nondestructive detection method for structural damage was explored through the change laws of dynamic structural characteristics.

2. Theoretical Derivation

The structural vibration equation could be expressed as a differential equation in the form of an n-order matrix as in Equation (1):

$$M\ddot{u} + C\dot{u} + Ku = f(x) \tag{1}$$

where M is an $n \times n$-order mass stiffness matrix, $\ddot{u}$ denotes an $n \times 1$-order acceleration array, $\dot{u}$ is an $n \times 1$-order velocity array, u stands for an $n \times 1$-order displacement array, K is a stiffness matrix with the $n \times n$-order structure and C is an $n \times n$-order damping matrix.

The damping, which had a minor influence on the structural vibration, could be neglected in terms of the structural natural vibration frequency, so the structure was under a free vibration state. However, $f(x)$ was zero when the structure was not subjected to any external force. Hence, the equation could be rewritten as Equation (2):

$$M\ddot{u} + Ku = 0 \tag{2}$$

Suppose that $u = \varphi e^{j\omega l}$, where φ is the amplitude array of free vibration, and thus the above frequency domain Equation (2) could be written as Equation (3):

$$K\varphi = \omega^2 M\varphi \tag{3}$$

Therefore, the solving of vibration mode and frequency was transformed into the solving of eigenvalue. ω [2] was set as the eigenvalue λ_i and φ_i as the i(th)-order displacement modal vector, so Equation (3) could be expressed alternatively as Equation (4):

$$K\varphi_i = \lambda_i M\varphi_i \tag{4}$$

In case of any damage, the structural stiffness will be changed, and so will its stiffness matrix and the corresponding λ_i and φ_i, as in Equation (5):

$$\delta K_{\varphi i} - \delta\lambda_i M_{\varphi i} = -(K - \lambda_i M)\delta_{\varphi i} \tag{5}$$

where δK, $\delta\lambda_i$ and $\delta_{\varphi i}$ represent the variable quantity of structural stiffness matrix, eigenvalue and eigenvector, respectively. The two sides of Equation (5) are simultaneously multiplied by ω_i^T, and Equation (6) can be acquired by combining Equation (4):

$$\delta\lambda_i = \omega_i^T \delta K\omega_i \tag{6}$$

$\delta K \leq 0$ in the case of any structural damage, so $\delta\lambda_i \leq 0$. $\delta\omega_i \leq 0$ because $\lambda = \omega^2$. Therefore, it could be seen that the structural natural vibration frequency was changed owing to the structural damage. Meanwhile, we found from Equation (6) that when the damage occurred at the same position, the variable quantity of structural natural vibration frequency was enlarged with the increase in the damage degree.

The mode at the corresponding position was changed under the structural damage, while no noticeable change occurred at the position without any damage.

The change of displacement mode before and after the structural damage could be expressed as follows:

$$S = |s - s'| \tag{7}$$

where s is the displacement mode under the undamaged circumstance and s' represents the displacement mode after the structural damage.

When the structural damage took place, S in Equation (7) increased with the damage degree.

3. Experimental Verification

3.1. Experimental Scheme

This study chose a simply supported beam for the experimental simulation. Polymethyl methacrylate (PMMA) was selected as the structural material at a density of about 1.15–1.19 g/cm [3], conveniently processed and transformed in the laboratory environment. In this experiment, the models selected were simply supported beam (1500 mm × 200 mm × 12 mm) and double-span beam (3200 mm × 200 mm × 12 mm), as shown in Figure 1.

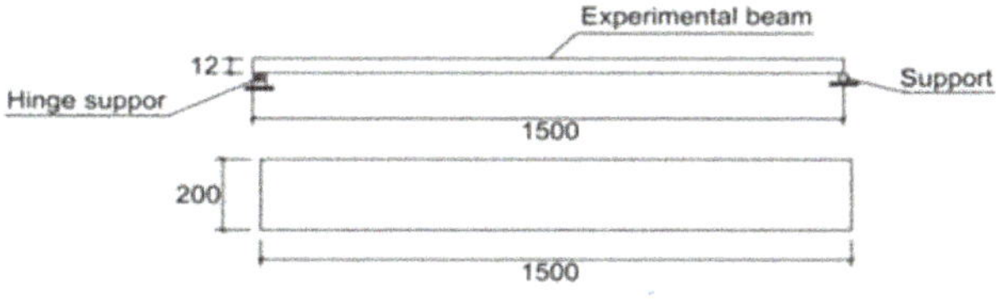

Figure 1. Schematic diagram of experimental beam models.

The working conditions were designed in this experiment using the directed damage assignment method. By reference to many studies, the following three damage conditions were designed (Table 1).

Table 1. Damage conditions.

Condition No.	Span Number	Damage Degree	Damage Position	Number of Damages	Whether Stiffened
Condition 1	Single span	No damage	NO	0	NO
Condition 2	Single span	Single-point 25% damage	1/2 L position	1	NO
Condition 3	Single span	Single-point 25% damage	1/4 L position	1	NO
Condition 4	Single span	Single-point 25% damage	1/2 and 1/4 positions	2	NO

The working conditions are displayed in Figure 2.

Condition 2

(**a**)
Condition 3

(**b**)
Condition 4 (double-damage)

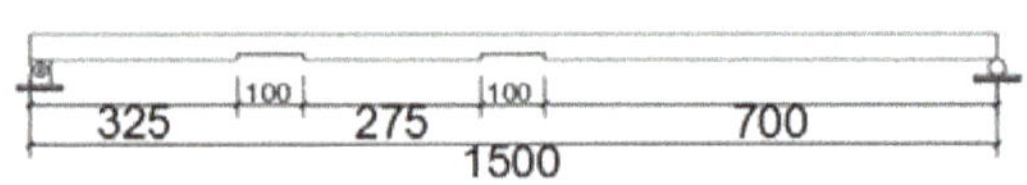

(**c**)

Figure 2. (**a**) Schematic diagram of midspan damage condition; (**b**) schematic diagram of damage condition at 1/4 position; (**c**) schematic diagram of one-span double-damage condition.

To simulate the practical situation in engineering, the models were stiffened. Here, 1400 × 4 × 12 rectangular ribs of the same material were used, and the stiffening forms are as shown in Figure 3.

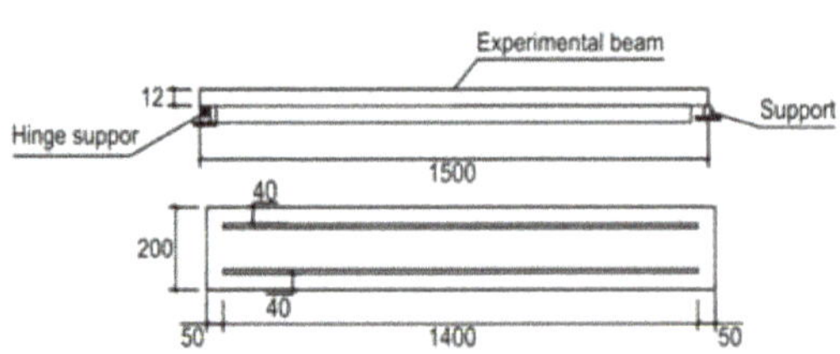

Figure 3. Schematic diagram of models under stiffening conditions.

Therefore, four stiffening conditions were obtained, as listed in Table 2.

Table 2. Summary of stiffened condition.

Condition No.	Span Number	Damage Degree	Damage Position	Number of Damages	Whether Stiffened
Condition 5	Single span	No damage	NO	0	YES
Condition 6	Single span	Single-point 25% damage	1/2 L position	1	YES
Condition 7	Single span	Single-point 25% damage	1/4 L position	1	YES
Condition 8	Single span	Single-point 25% damage	1/2 and 1/4 positions	2	YES

The working conditions are shown in Figure 4.

Condition 5

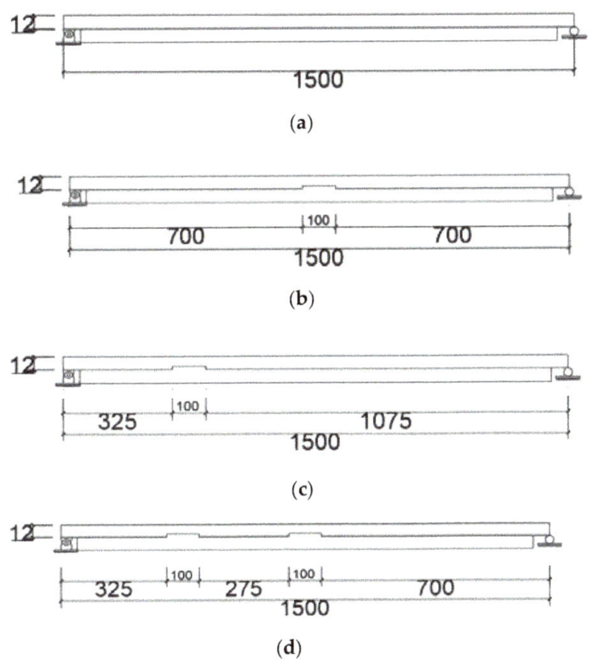

Figure 4. (**a**) Schematic diagram of stiffened and nondamaged condition; (**b**) schematic diagram of stiffened and midspan damage condition; (**c**) schematic diagram of stiffened and damage condition (1/4 position); (**d**) schematic diagram of stiffened and one-span double-damage condition.

The sensor system used in the experiment was composed of INV3018CT 24-bit high-precision data acquisition instrument (Figure 5), computer, force hammer, sensor, double-sided adhesive tape, test piece and two-way crystal interface cable.

The layout of experimental measuring points and excitation points is displayed below. The experimental beam was divided into 15 equal parts, the sensor was fixed at points 5 and 7 (Figure 6), the excitation points were numbered 1–16, the excitation was conducted at each excitation point three times and the average of three results was taken.

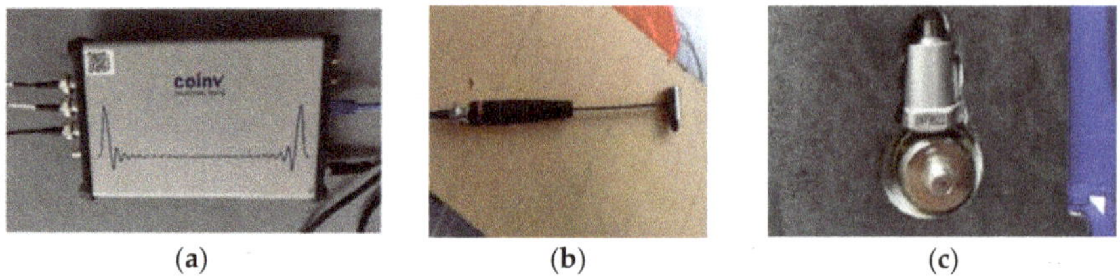

Figure 5. INV3018CT 24-bit high-precision data acquisition instrument, force hammer and sensor. (**a**) Instrument host; (**b**) Stress hammer; (**c**) Sensor.

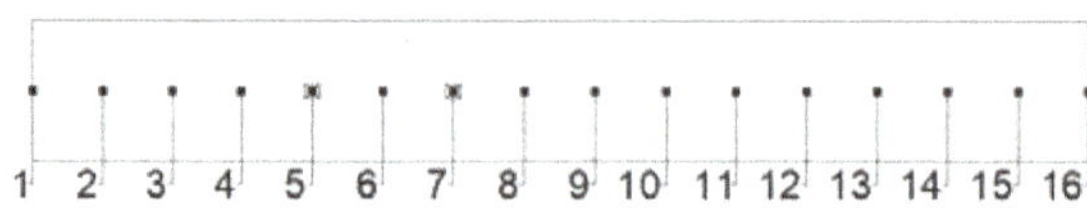

Figure 6. Layout plan of measuring points.

Coinv Dasp V11 software matching with the experimental instrument was used for the data processing. The model graph and animated graph of vibration mode could be automatically generated in the software, as shown in Figures 7 and 8.

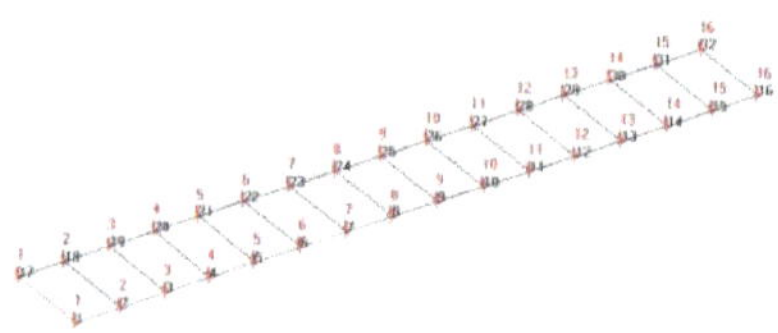

Figure 7. Automatically generated model graph.

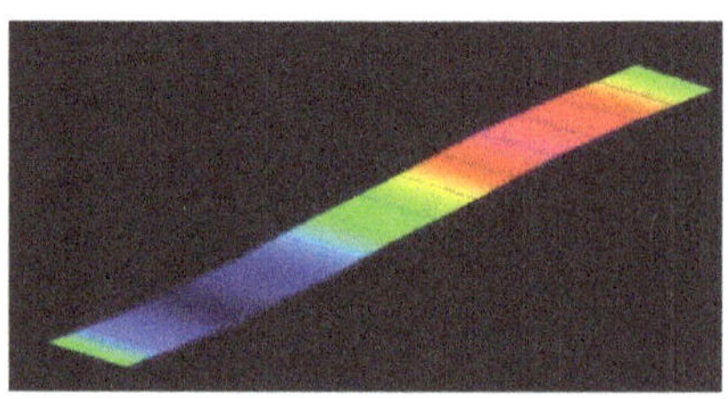

Figure 8. Animated graph of vibration mode.

3.2. *Experimental Data and Analysis*

3.2.1. Damage Identification Analysis Based on the Inherent Frequency

As one of the dynamic structural characteristics, the inherent frequency can reflect the structural damage condition to some extent. Therefore, the inherent frequencies under different damage conditions obtained through the experiment were compared with those under undamaged conditions. Next, the loss rate of inherent frequency under each vibration mode was calculated by taking the nondamaged and stiffened condition and nondamaged and unstiffened condition (Condition 5 and Condition 1) as the criteria. The comparison results are listed in Tables 3 and 4.

Table 3. Frequency loss rates under different damage conditions.

Order of Vibration Mode	Condition 1	Condition 2		Condition 3		Condition 4	
	Frequency (Hz)	Frequency (Hz)	Frequency Loss Rate %	Frequency (Hz)	Frequency Loss Rate %	Frequency (Hz)	Frequency Loss Rate %
1	6.4435	6.376	1.05	6.325	1.84	6.212	3.59
2	19.043	18.176	4.55	18.837	1.08	18.031	5.31
3	40.746	39.826	2.26	39.328	3.48	39.548	2.94
4	69.659	68.301	1.95	68.841	1.17	68.965	1.00
5	111.789	110.365	1.27	111.425	0.33	110.348	1.29

Table 4. Frequency loss rates under different stiffening conditions.

Order of Vibration Mode	Condition 5	Condition 6		Condition 7		Condition 8	
	Frequency (Hz)	Frequency (Hz)	Frequency Loss Rate %	Frequency (Hz)	Frequency Loss Rate %	Frequency (Hz)	Frequency Loss Rate %
1	13.948	13.917	0.22	13.833	0.82	13.803	1.04
2	48.057	46.684	2.86	47.751	0.64	48.363	3.52
3	103.68	102.13	1.49	103.30	0.37	103.29	4.23
4	170.74	167.56	1.86	167.44	1.93	164.45	3.68
5	216.65	215.76	0.41	216.36	0.13	215.78	1.79

The above tables show that the inherent structural frequency was reduced due to the damage. By comparing the frequency loss rates under Conditions 2 and 3 with those under Conditions 6 and 7, it could be known that the structural frequency was significantly more changed by the midspan damage than by the damage at the 1/4 position. Meanwhile, the frequencies of the second- and third-order vibration modes were changed significantly, while the gap of the first order with the fourth and fifth orders became smaller and smaller. Based on the above results, it could be obtained that the frequency loss brought by the current damage condition was smaller and smaller with the order change of vibration mode. By comparing the structural frequency loss rates under Conditions 1–4 with those under Conditions 5–6, it could be seen that the stiffness of stiffened structure was enhanced, so the frequency loss caused by the structural damage was reduced relative to that before the stiffening.

3.2.2. Damage Identification Based on Displacement Mode

According to the previous section, we found that the inherent frequency only reflected whether the structure was damaged but failed to determine the damage position, so the structural displacement mode was exported in this section for the study. Like the frequency study part, two groups were formed: stiffened and nonstiffened groups, and the undamaged conditions within the two groups were compared. The displacement mode under Condition 1 (single-span nondamaged) was compared with that under Condition 2 (midspan damage), as shown in Figure 9.

The comparison graphs of displacement modes under Condition 1 (single-span non-damaged) and Condition 3 (damage at the 1/4 position), as shown in Figure 10.

The comparison results of displacement modes at different orders under Condition 1 (single-span nondamaged) and Condition 4 (single-span double-damage) are shown in Figure 11.

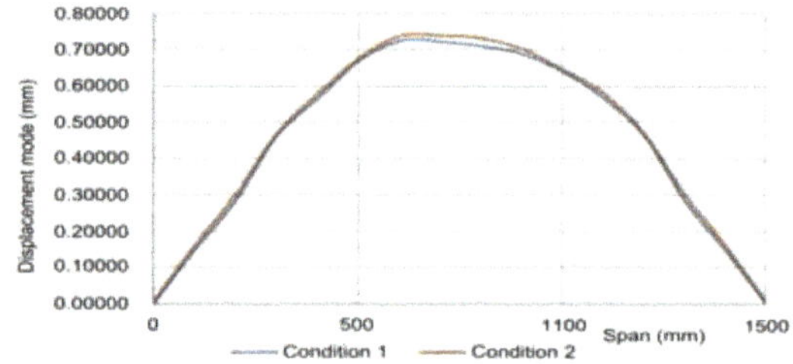

(**a**) Comparison of the first-order vibration modes under Conditions 1 and 2.

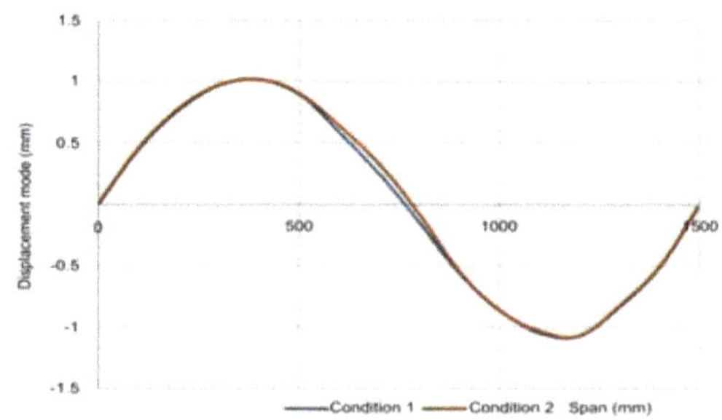

(**b**) Comparison of the second-order vibration modes under Conditions 1 and 2.

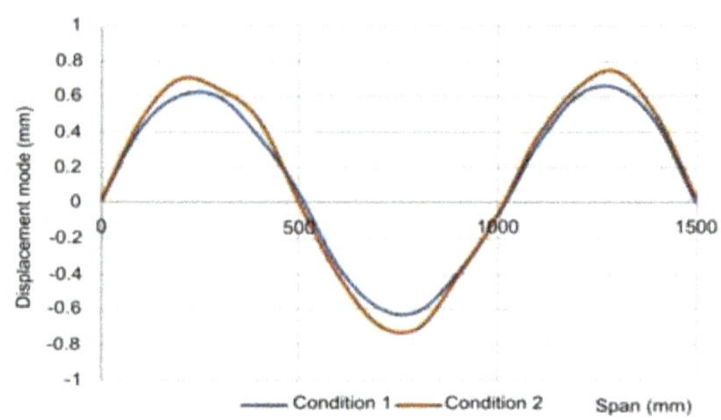

(**c**) Comparison of the third-order vibration modes under Conditions 1 and 2.

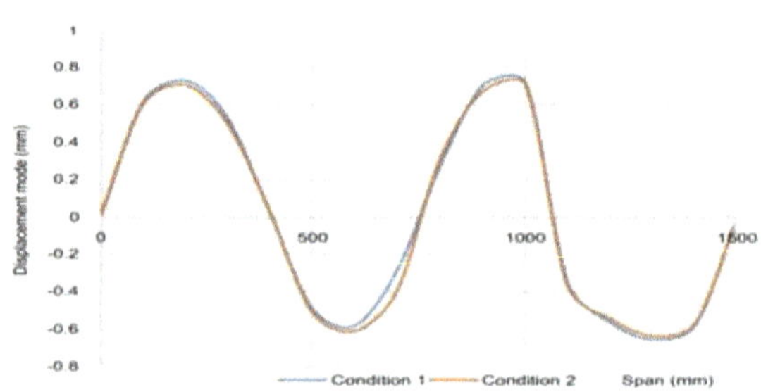

(**d**) Comparison of the fourth-order vibration modes under Conditions 1 and 2.

Figure 9. *Cont.*

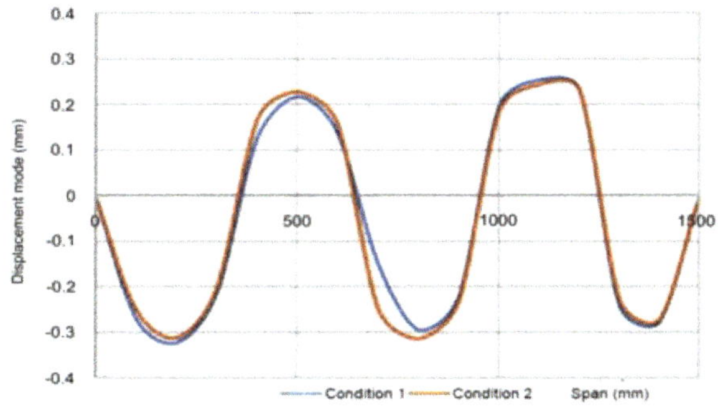

(**e**) Comparison of the fifth-order vibration modes under Conditions 1 and 2.

Figure 9. Comparison of first five orders of vibration modes under Conditions 1 and 2.

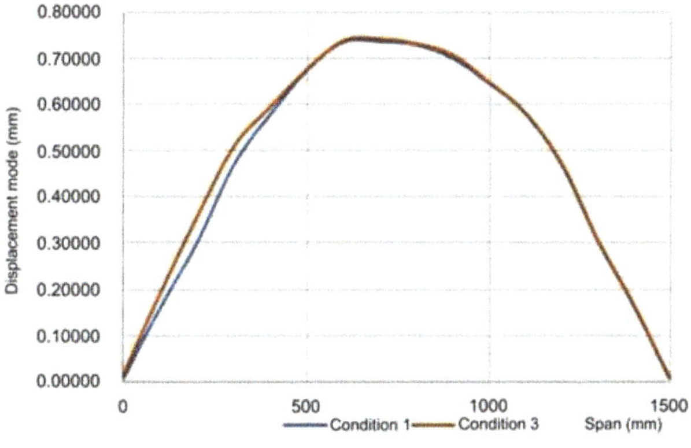

(**a**) Comparison of the first-order vibration modes under Conditions 1 and 3.

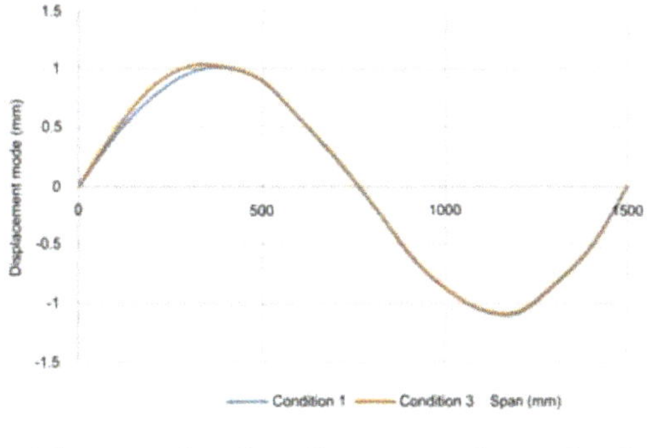

(**b**) Comparison of the second-order vibration modes under Conditions 1 and 3.

Figure 10. *Cont.*

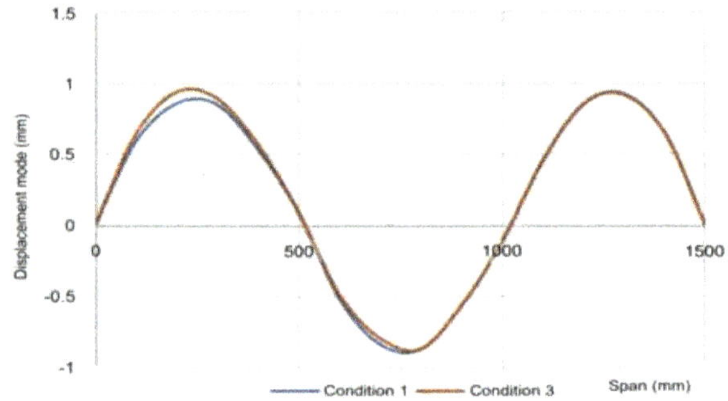

(**c**) Comparison of the third-order vibration modes under Conditions 1 and 3.

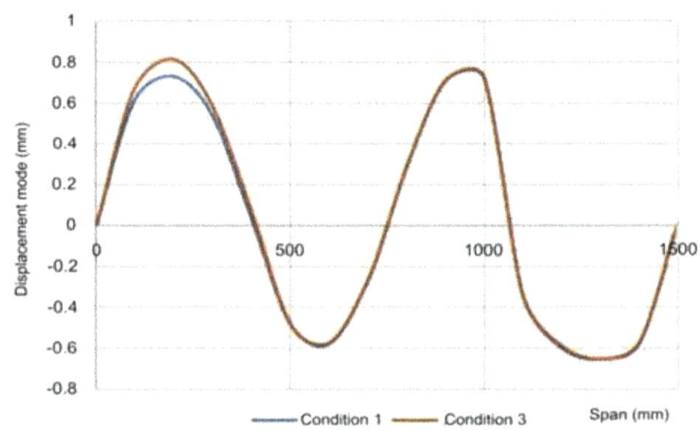

(**d**) Comparison of the fourth-order vibration modes under Conditions 1 and 3.

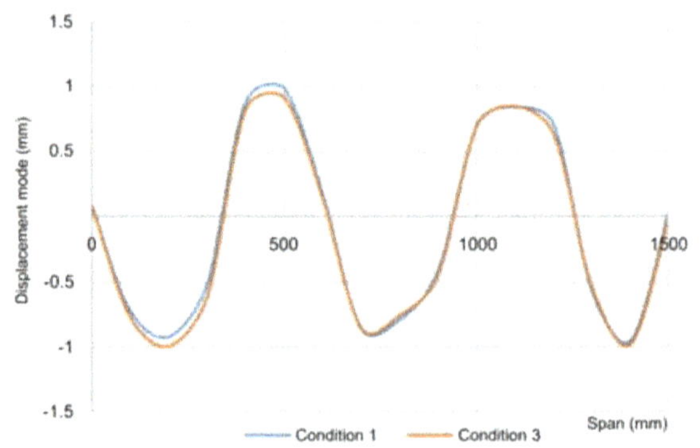

(**e**) Comparison of the fifth-order vibration modes under Conditions 1 and 3.

Figure 10. Comparison of the first five orders of vibration modes under Conditions 1 and 3.

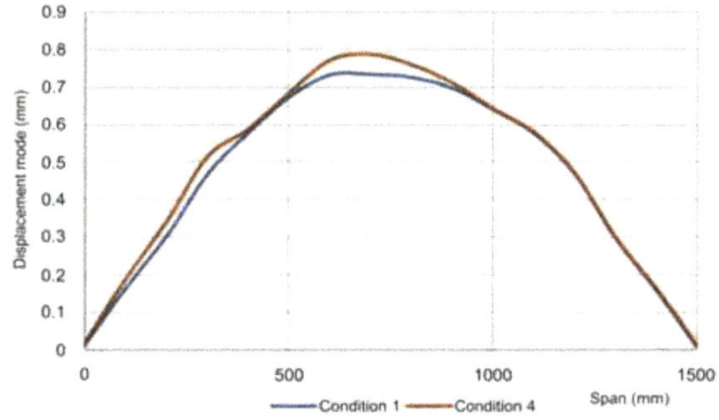

(**a**) Comparison of the first-order vibration modes under Conditions 1 and 4.

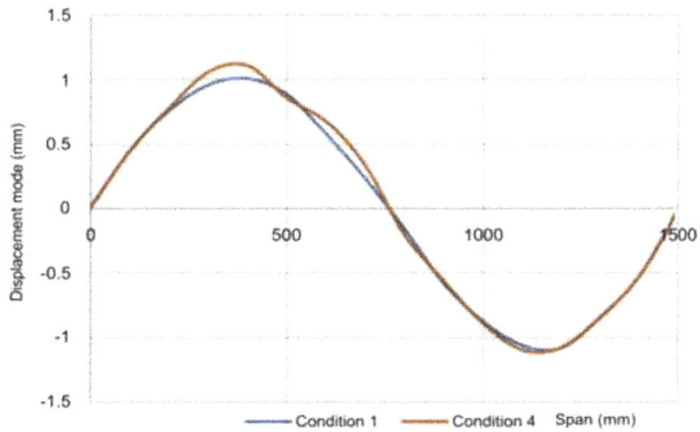

(**b**) Comparison of the second-order vibration modes under Conditions 1 and 4.

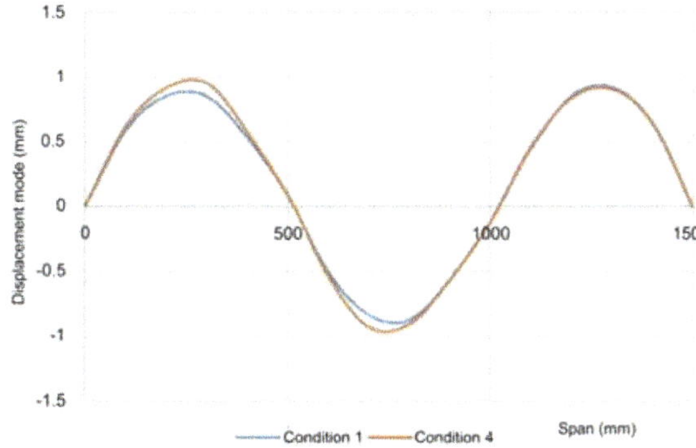

(**c**) Comparison of the third-order vibration modes under Conditions 1 and 4.

Figure 11. *Cont.*

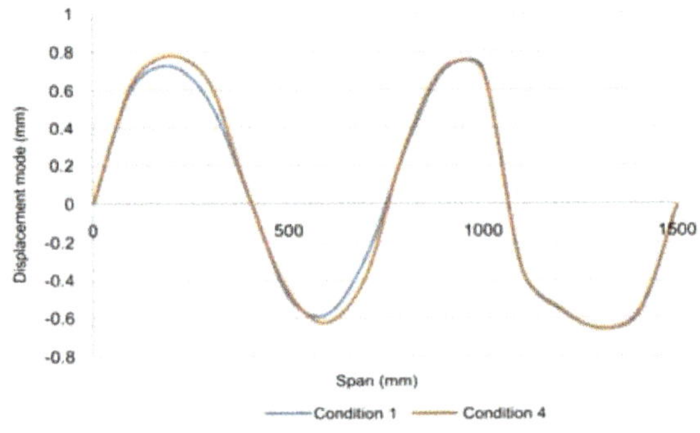

(**d**) Comparison of the fourth-order vibration modes under Conditions 1 and 4.

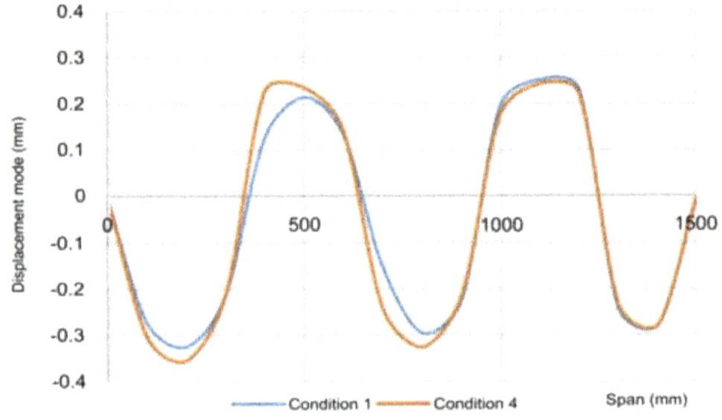

(**e**) Comparison of the fifth-order vibration modes under Conditions 1 and 4.

Figure 11. Comparison of the first five orders of vibration modes under Conditions 1 and 4.

The comparison results of displacement modes under Condition 5 (stiffened and nondamaged) and Condition 6 (stiffened and midspan damage) are shown in Figure 12.

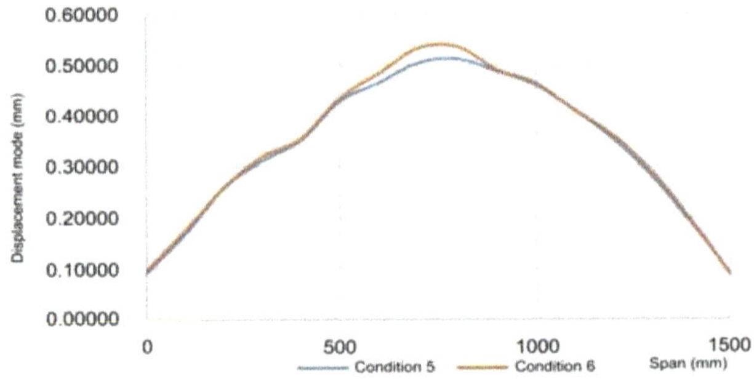

(**a**) Comparison of the first-order vibration modes under Conditions 5 and 6.

Figure 12. *Cont.*

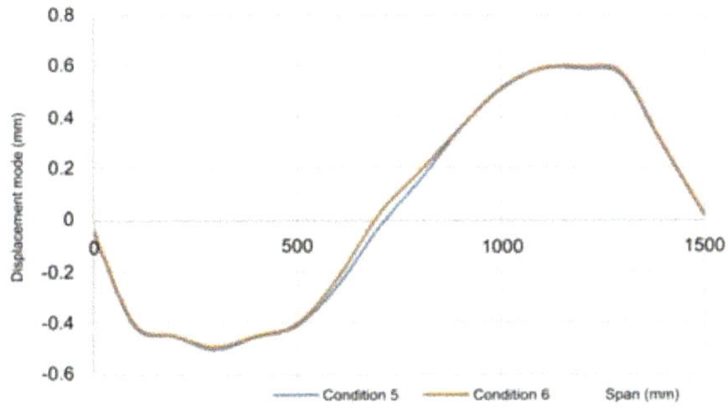

(**b**) Comparison of the second-order vibration modes under Conditions 5 and 6.

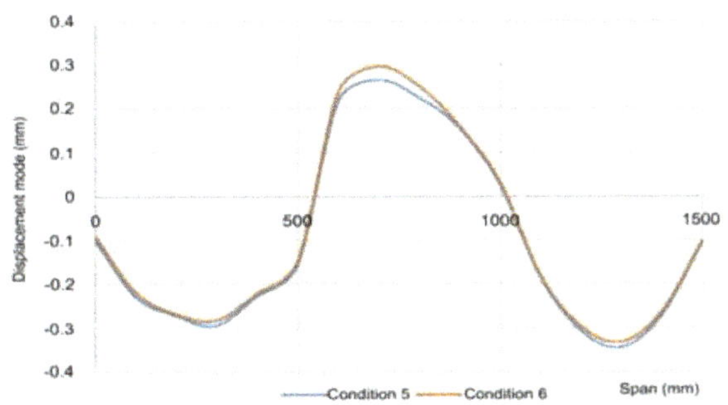

(**c**) Comparison of the third-order vibration modes under Conditions 5 and 6.

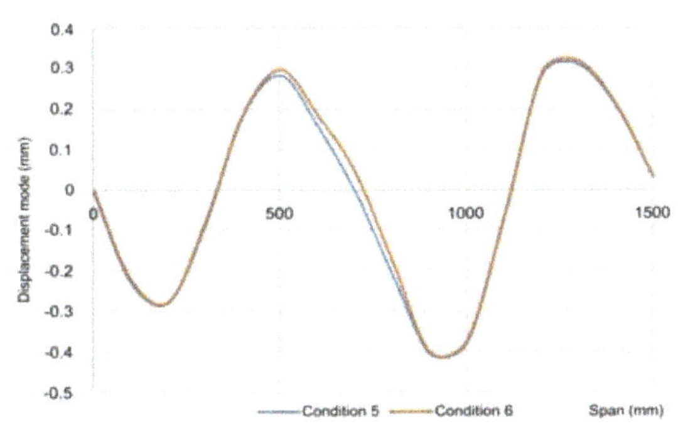

(**d**) Comparison of the fourth-order vibration modes under Conditions 5 and 6.

Figure 12. *Cont.*

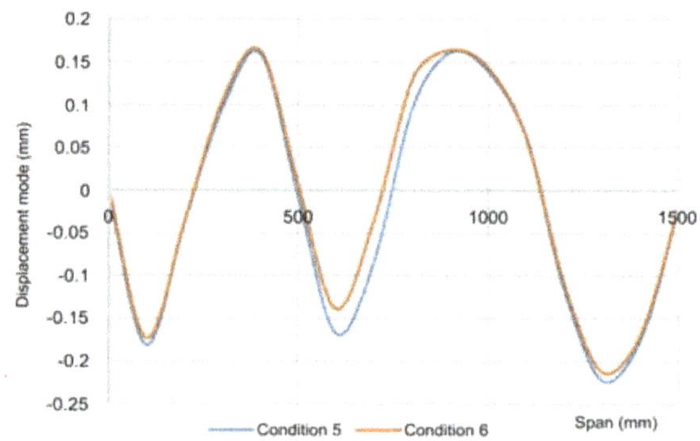

(**e**) Comparison of the fifth-order vibration modes under Conditions 5 and 6.

Figure 12. Comparison of the first five orders of vibration modes under Conditions 5 and 6.

The displacement modes under Condition 5 (stiffened and nondamaged) and Condition 7 (stiffened and damaged at the 1/4 position) were compared, as shown in Figure 13.

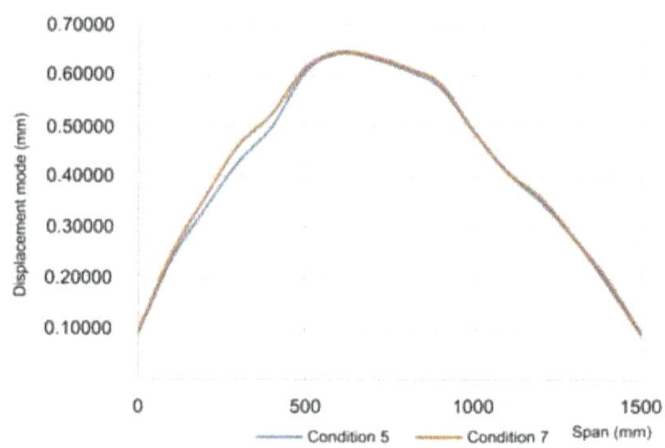

(**a**) Comparison of the first-order vibration modes under Conditions 5 and 7.

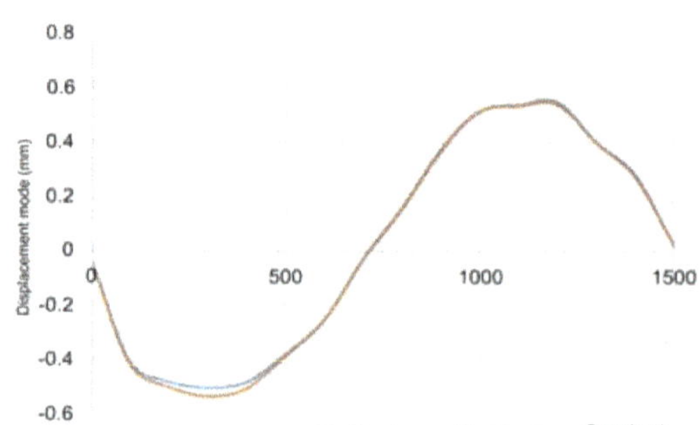

(**b**) Comparison of the second-order vibration modes under Conditions 5 and 7.

Figure 13. *Cont.*

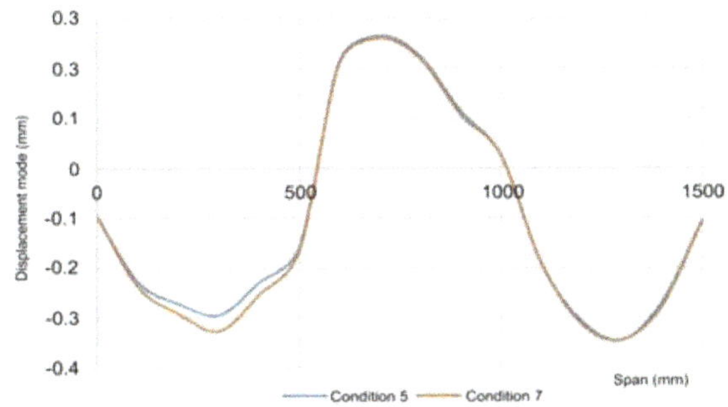

(**c**) Comparison of the third-order vibration modes under Conditions 5 and 7.

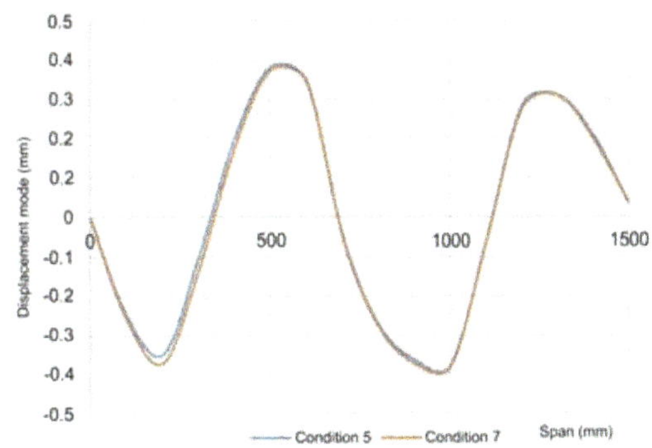

(**d**) Comparison of the fourth-order vibration modes under Conditions 5 and 7.

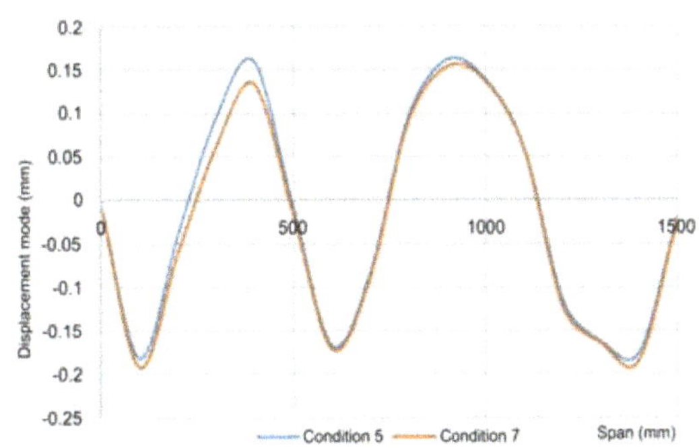

(**e**) Comparison of the fifth-order vibration modes under Conditions 5 and 7.

Figure 13. Comparison of the first five orders of vibration modes under Conditions 5 and 7.

The comparison results of displacement modes under Condition 5 (stiffened and nondamaged) and Condition 8 (stiffened and double-damage) are shown in Figure 14.

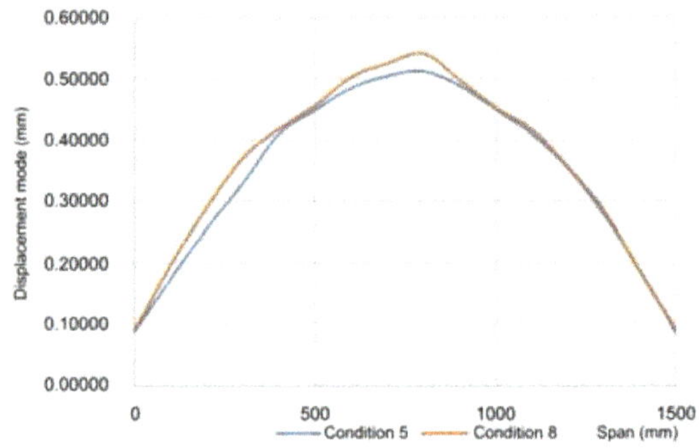

(**a**) Comparison of the first-order vibration modes under Conditions 5 and 8.

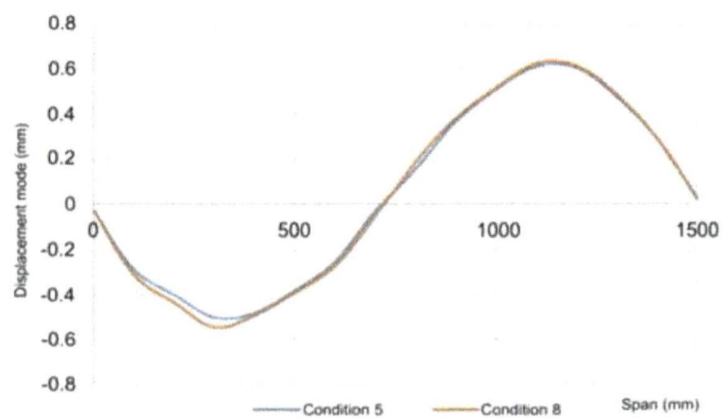

(**b**) Comparison of the second-order vibration modes under Conditions 5 and 8.

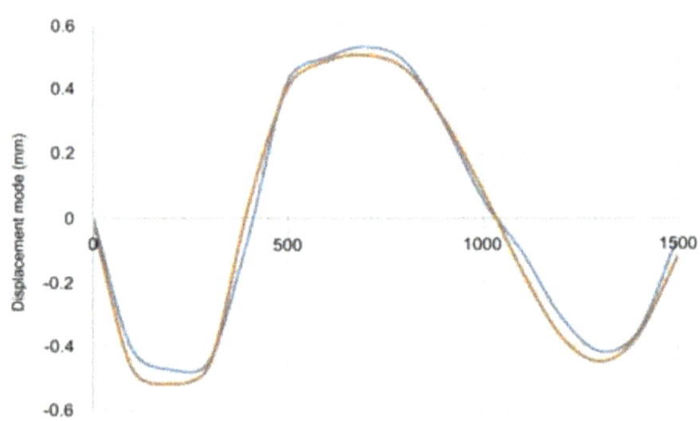

(**c**) Comparison of the third-order vibration modes under Conditions 5 and 8.

Figure 14. *Cont.*

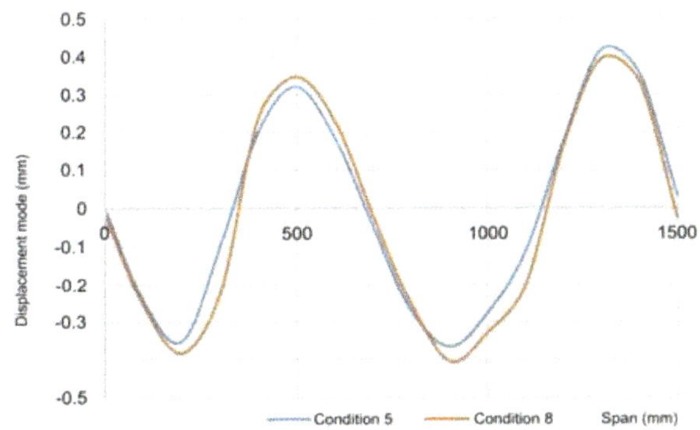

(**d**) Comparison of the fourth-order vibration modes under Conditions 5 and 8.

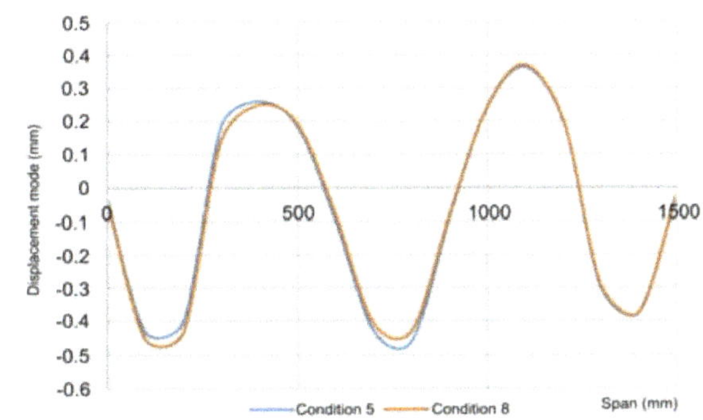

(**e**) Comparison of the fifth-order vibration modes under Conditions 5 and 8.

Figure 14. Comparison of the first five orders of vibration modes under Conditions 5 and 8.

The displacement modes obtained through the experiment under the damaged and undamaged conditions were compared in the same coordinate system. It appeared that the displacement amplitude at the damage position under the damage condition was different from that at the same position under the nondamaged condition, and the displacement amplitude difference under the stiffening condition was smaller than that under the nonstiffened condition. To conveniently view the specific damage position, the displacement mode curve was amplified by taking the derivatives of difference value, which could obtain the first-order and second-order derivatives of displacement mode difference, as shown from Figures 9–14.

By comparing the graphs in Figure 15, the first-order and second-order derivatives of the first-order vibration mode difference fluctuated significantly at two middle points, which conformed to the directed additional working condition in the experiment. However, the difference derivatives of the second- to fifth-order vibration modes did not show any significant fluctuation point, and the position of given damage could not be observed through the curve. Hence, it was concluded that the first-order vibration mode had better sensitivity to the damage during the damage identification. Therefore, the derivatives were continuously taken from the difference between Conditions 1 and 3 and that between Conditions 1 and 4, and the first-order and second-order derivatives of the first-order vibration mode were further obtained, as shown in Figures 16 and 17, respectively.

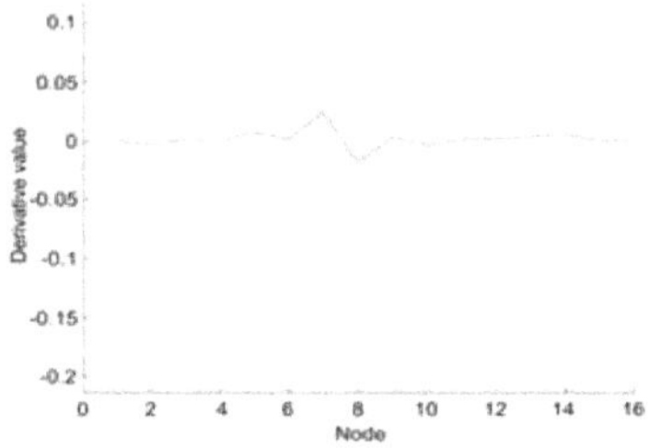

(**a**) First-order derivative of first-order mode difference.

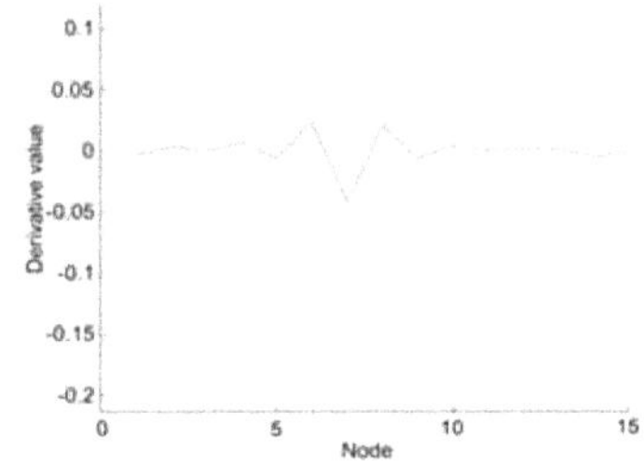

(**b**) Second-order derivative of first-order mode difference.

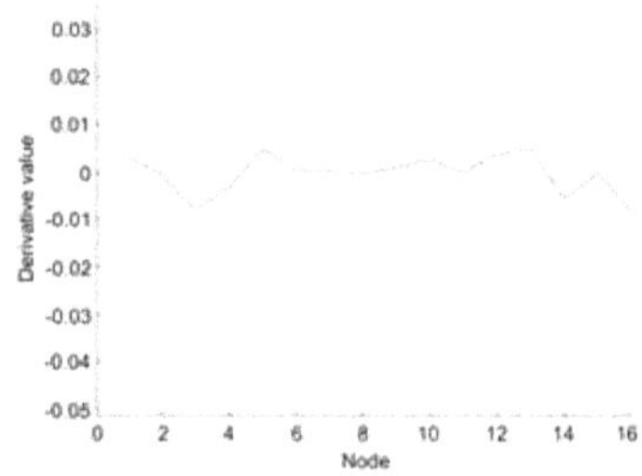

(**c**) First-order derivative of second-order mode difference.

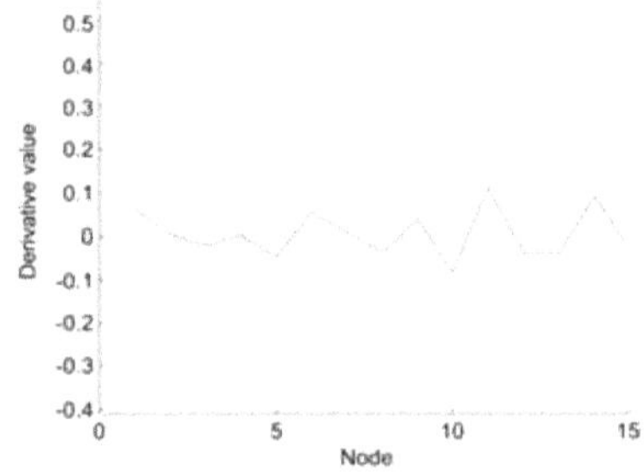

(**d**) Second-order derivative of second-order mode difference.

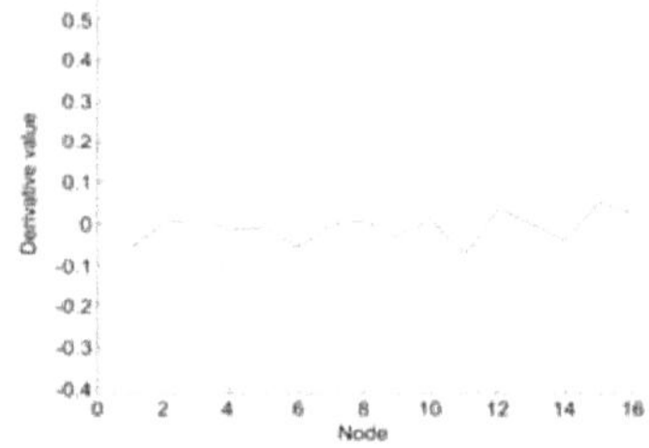

(**e**) First-order derivative of third-order mode difference.

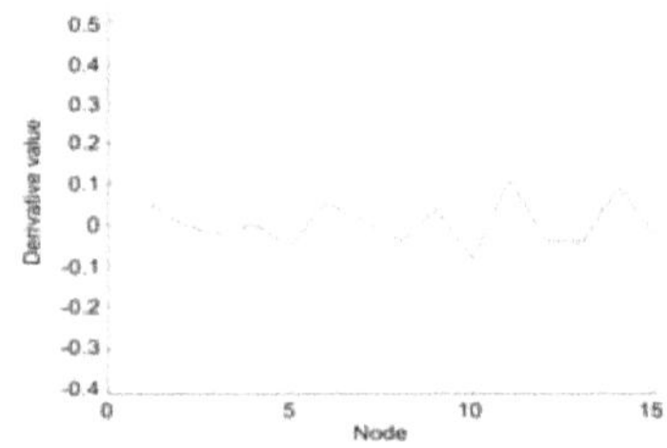

(**f**) Second-order derivative of third-order mode difference.

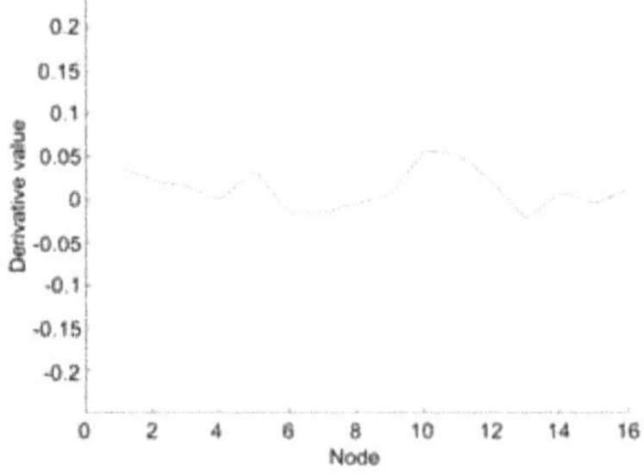

(**g**) First-order derivative of fourth-order mode difference.

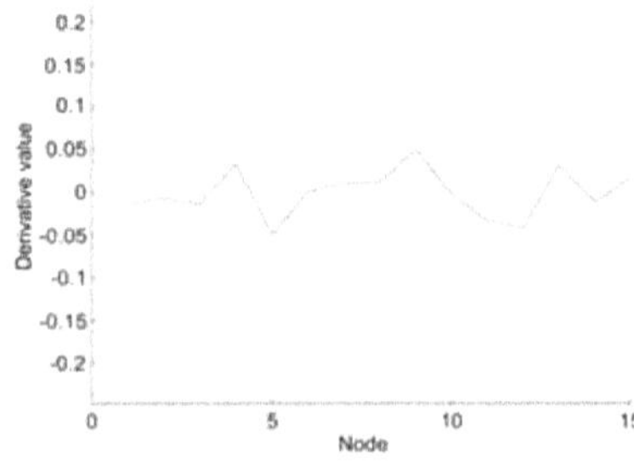

(**h**) Second-order derivative of fourth-order mode difference.

Figure 15. *Cont.*

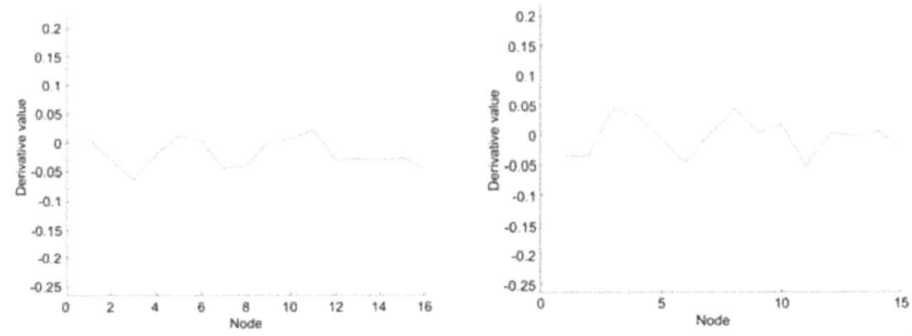

(**i**) First-order derivative of fifth-order mode difference.

(**j**) Second-order derivative of fifth-order mode difference.

Figure 15. First-order and second-order derivative curves of the differences of the first five orders of modes under Conditions 1 and 2.

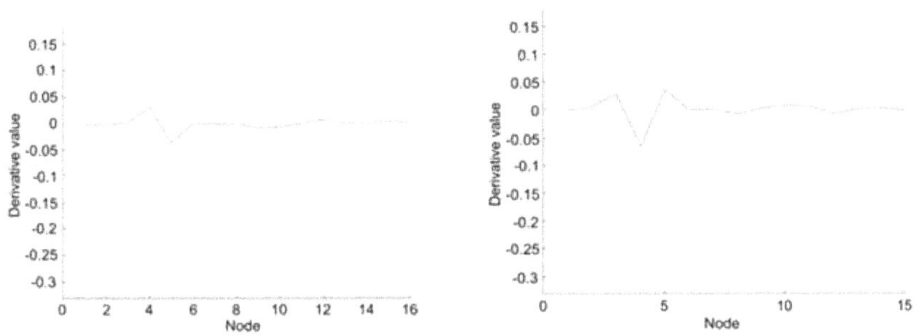

(**a**) First-order derivative of first-order mode difference.

(**b**) Second-order derivative of first-order mode difference.

Figure 16. Difference derivatives under Conditions 1 (single-span and nondamaged) and 3 (damage at the 1/4 position).

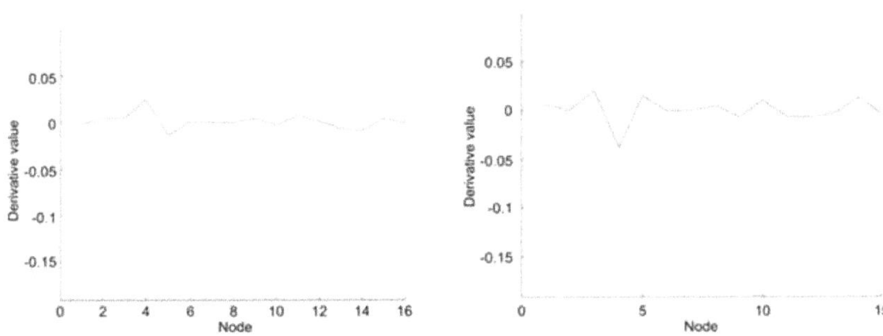

(**a**) First-order derivative of first-order mode difference.

(**b**) Second-order derivative of first-order mode difference.

Figure 17. Difference derivatives under the Conditions 1 (single-span nondamaged) and 4 (single-span double-damage).

Under the single-span double-damage condition, we found that the first-order derivative of the first-order mode difference fluctuated at the damage position. The fluctuations in the other graphs were noticeable but uniform, and the damage position could not be observed. By continuously solving the first-order derivatives of first-order mode difference under the stiffening condition, the above conclusion was still verified to be true, and the first-order derivative curves of first-order mode difference under the stiffening condition are displayed in Figures 18–20.

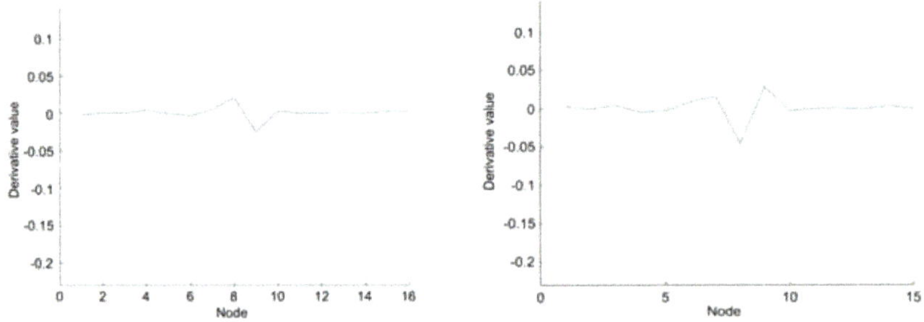

(**a**) First-order derivative of first-order mode difference. (**b**) Second-order derivative of first-order mode difference.

Figure 18. First-order and second-order derivative curves of the differences of the first five orders of modes under Conditions 5 and 6.

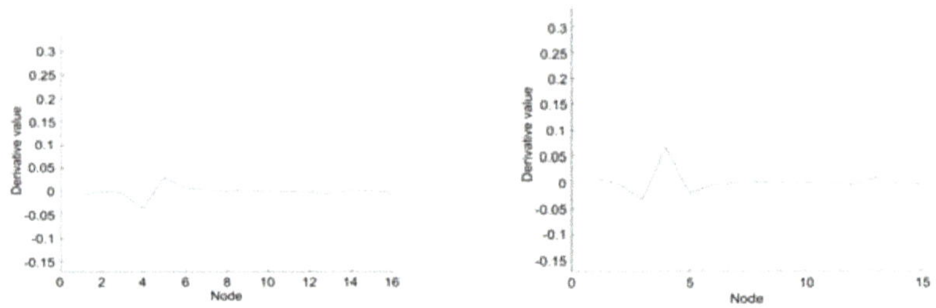

(**a**) First-order derivative of first-order mode difference. (**b**) Second-order derivative of first-order mode difference.

Figure 19. First-order and second-order derivative curves of the differences of the first five orders of modes under Conditions 5 and 7.

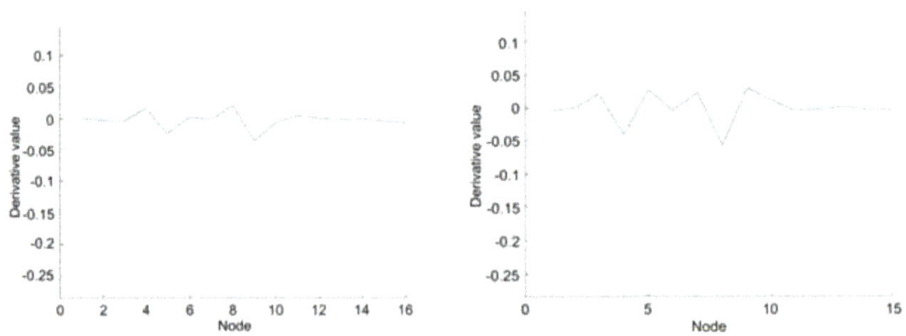

(**a**) First-order derivative of first-order mode difference. (**b**) Second-order derivative of first-order mode difference.

Figure 20. First-order and second-order derivatives of the differences of the first five orders of modes under Conditions 5 and 8.

Similarly, the conclusion consistent with the previous section could be obtained by comparing the first-order and second-order derivative curves of vibration mode differences at different orders: the first-order derivative of first-order vibration mode difference experienced a major fluctuation in the case of structural damage, but the other curves showed no evident trends.

3.3. Finite Element Simulation

3.3.1. Simulation Effect Analysis of Inherent Frequency

The finite element method is a numerical method used to solve the numerical solution to the set of differential equation sets or integral equation sets. In this experiment, the modeling was conducted using ABAQUS, and the modeling effect graphs are shown in Figures 21 and 22.

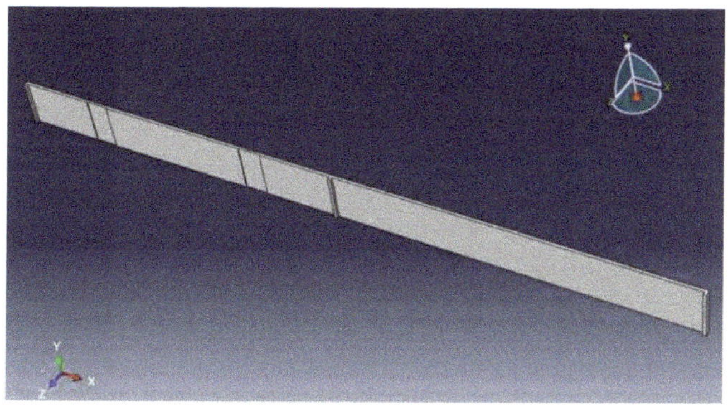

Figure 21. Model graph after the creation of damage and support.

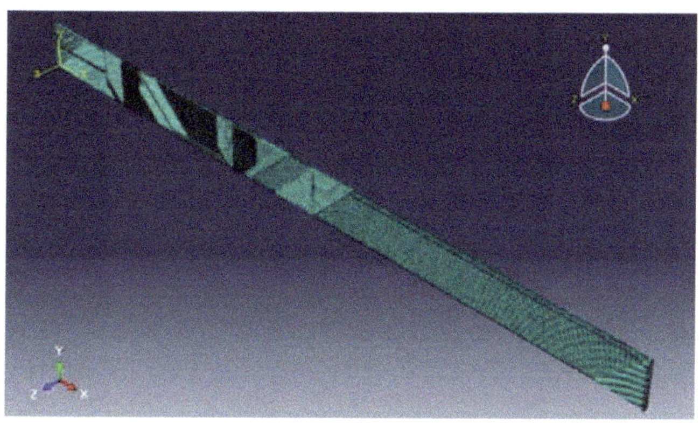

Figure 22. Model graph after the mesh generation.

Under ideal conditions, the inherent structural frequency could be obtained by establishing a finite element model. The theoretical value and experimentally measured value of inherent frequency were compared, as listed in Table 5.

Table 5. Inherent frequency errors.

Order of Vibration Mode	Condition 1			Condition 2		
	Measured Frequency (Hz)	Simulated Frequency (Hz)	Percentage Error %	Measured Frequency (Hz)	Simulated Frequency (Hz)	Percentage Error %
1	6.4435	7.3394	−12.21	6.376	7.2711	−12.31
2	19.043	20.597	−7.54	18.176	20.555	−11.57
3	40.746	41.022	−0.67	39.826	39.912	−0.22
4	69.659	68.751	1.32	68.301	68.302	0.00
5	111.789	103.84	7.66	110.365	102.434	7.74
Order of Vibration Mode	**Condition 3**			**Condition 4**		
	Measured Frequency (Hz)	**Simulated Frequency (Hz)**	**Percentage Error %**	**Measured Frequency (Hz)**	**Simulated Frequency (Hz)**	**Percentage Error %**
1	6.325	7.3327	−13.74	6.212	7.1989	−13.71
2	18.837	20.43	−7.80	18.031	20.134	−10.45
3	39.328	39.714	−0.97	39.548	38.858	1.78
4	68.841	68.242	0.88	68.965	68.366	0.88
5	111.425	101.8	9.45	110.348	101.34	8.89
Order of Vibration Mode	**Condition 5**			**Condition 6**		
	Measured Frequency (Hz)	**Simulated Frequency (Hz)**	**Percentage Error %**	**Measured Frequency (Hz)**	**Simulated Frequency (Hz)**	**Percentage Error %**
1	6.325	7.3327	−13.74	6.212	7.1989	−13.71
2	18.837	20.43	−7.80	18.031	20.134	−10.45
3	39.328	39.714	−0.97	39.548	38.858	1.78
4	68.841	68.242	0.88	68.965	68.366	0.88
5	111.425	101.8	9.45	110.348	101.34	8.89
Order of Vibration Mode	**Condition 7**			**Condition 8**		
	Measured Frequency (Hz)	**Simulated Frequency (Hz)**	**Percentage Error %**	**Measured Frequency (Hz)**	**Simulated Frequency (Hz)**	**Percentage Error %**
1	6.325	7.3327	−13.74	6.212	7.1989	−13.71
2	18.837	20.43	−7.80	18.031	20.134	−10.45
3	39.328	39.714	−0.97	39.548	38.858	1.78
4	68.841	68.242	0.88	68.965	68.366	0.88
5	111.425	101.8	9.45	110.348	101.34	8.89

We found that the experimentally measured result was not much different from the simulated result in the numerical value; the two presented the consistent change laws, they deviated a lot from each other only at the fifth order and both percentage errors did not exceed 15%. This revealed that the construction method of the finite element model was correct and accorded with reality. The inherent frequency loss rates of the vibration modes at different orders under the simulated conditions are listed in Table 6.

As seen in Table 6, the frequency loss rates of the second- and third-order vibration modes were the maximum. It could be observed that the inherent frequency of each order under the stiffening conditions (Conditions 5 to 8) was always greater than that under the nonstiffened conditions (Conditions 1 to 4). Meanwhile, the frequency loss rate brought by the midspan damage was higher than triggered by the damage at the 1/4 position. Both abovementioned laws indicated the variation trends of experimental data, manifesting that the experimental data are approximate to the simulated data and the frequencies of both the second-order and third-order vibration modes can be taken as the criteria for the damage identification.

Table 6. Inherent frequency loss rates at different orders.

Order of Vibration Mode	Condition 1	Condition 2		Condition 3		Condition 4	
	Frequency (Hz)	Frequency (Hz)	Frequency Loss Rate %	Frequency (Hz)	Frequency Loss Rate %	Frequency (Hz)	Frequency Loss Rate %
1	7.3394	7.2711	0.93	7.3327	0.09	7.1989	1.91
2	20.597	20.555	0.20	20.43	0.81	20.134	2.25
3	41.022	39.912	2.71	39.714	3.19	38.858	5.28
4	68.751	68.302	0.65	68.242	0.74	68.366	0.56
5	103.84	102.434	1.35	101.8	1.96	101.34	2.41
Order of Vibration Mode	**Condition 5**	**Condition 6**		**Condition 7**		**Condition 8**	
	Frequency (Hz)	**Frequency (Hz)**	**Frequency Loss Rate %**	**Frequency (Hz)**	**Frequency Loss Rate %**	**Frequency (Hz)**	**Frequency Loss Rate %**
1	13.948	13.717	1.66	13.833	0.82	13.903	0.32
2	48.057	48.684	−1.30	47.751	0.64	48.363	−0.64
3	103.68	102.13	1.49	103.30	0.37	103.29	0.38
4	170.74	167.56	1.86	167.44	1.93	164.45	3.68
5	216.65	215.76	0.41	216.36	0.13	215.78	0.40

3.3.2. Simulation Curve Analysis of Displacement Mode

The mode curves at different orders under Conditions 1 to 4 and those under Conditions 5 to 8 were plotted into the same graphs, as shown in Figures 23 and 24, respectively.

It can be observed from Figure 6-1 that although each curve deviated from the nondamaged condition due to the damage, this was consistent with the experimentally measured result, indicating that it is feasible to take the derivatives from the displacement mode difference. To eliminate the disturbance triggered by the experimental errors and further verify the sensitivity of the first-order derivative of structural first-order mode difference to the damage, the derivatives were also taken from the simulated displacement mode difference under the vibration mode at each order, and whether the variation trends were identical with the experimental results was checked. The derivative curves of vibration mode differences at different orders under the simulated Condition 1 (single-span nondamaged) and Condition 2 (midspan damage) are shown in Figure 25.

By comparing the above curves that after the number of points was expanded, the obvious numerical fluctuations could be manifested by the first-order vibration mode; moreover, the damage scope could be primarily defined by the first-order or third-order derivatives of vibration mode at each order, too. However, the first-order derivative of the first-order vibration mode was still relatively more apparent. The derivative curves of vibration mode differences at different orders under the simulated Conditions 1 (single-span nondamaged) and 3 (damage at the 1/4 position), as shown in Figure 26.

Figure 26 shows that the damage could be positioned by all the first-order derivative graphs of difference values. Besides, the damage position could also be reflected by some second-order derivatives. In the curves, the fluctuation amplitude of first-order derivatives for the first-order vibration mode differences was the minimum. Although the curve of the nondamaged part presented continuous oscillation, it could be regarded as an approximately straight line, so the first-order derivative curves of first-order vibration modes still had the superiority compared to the other curves. The derivative curves of displacement mode differences at different orders under Condition 5 (single-span nondamaged) and Condition 8 (one-span double-damage) are shown in Figure 27.

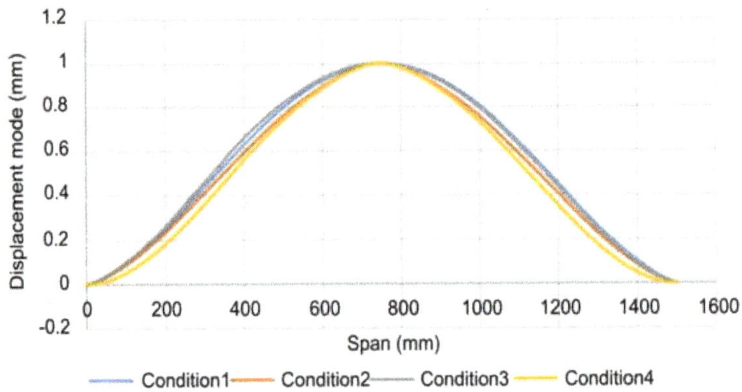

(**a**) First order vibration mode.

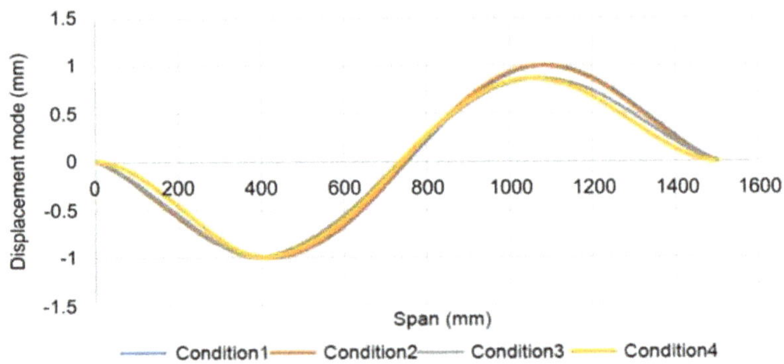

(**b**) Second order vibration mode.

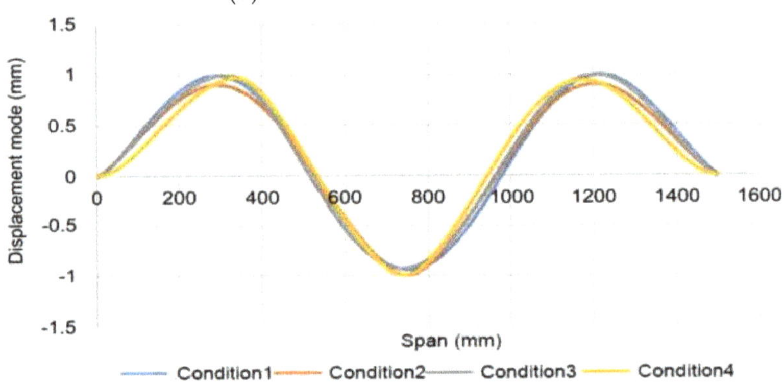

(**c**) Third order vibration mode.

Figure 23. *Cont.*

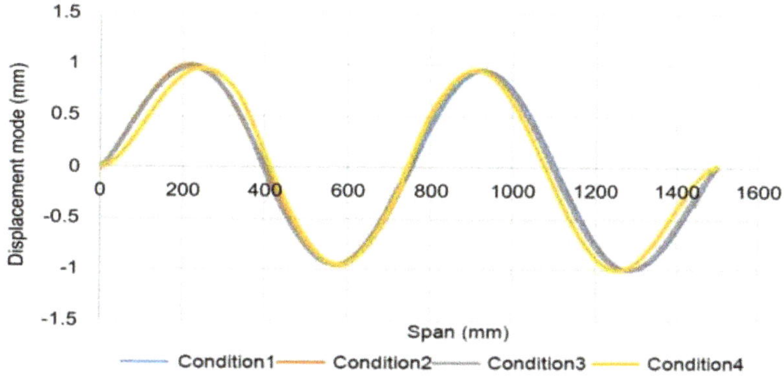

(**d**) Fouth order vibration mode.

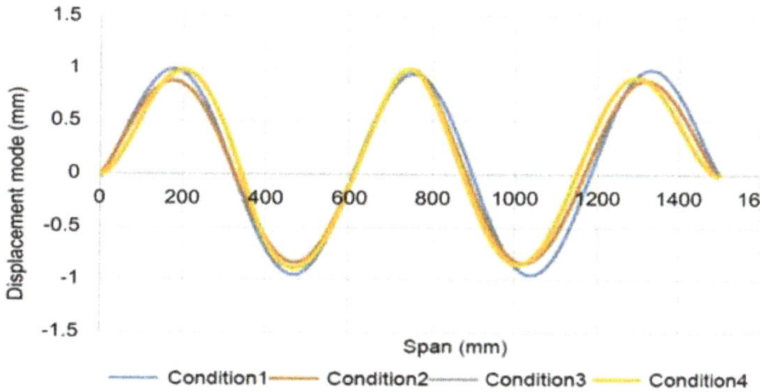

(**e**) 5th order vibration mode.

Figure 23. Comparison of mode curves at different orders under Conditions 1 to 4.

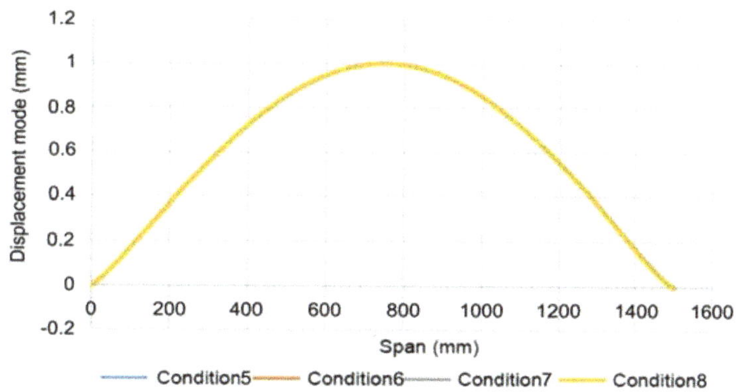

(**a**) First order vibration mode.

Figure 24. *Cont.*

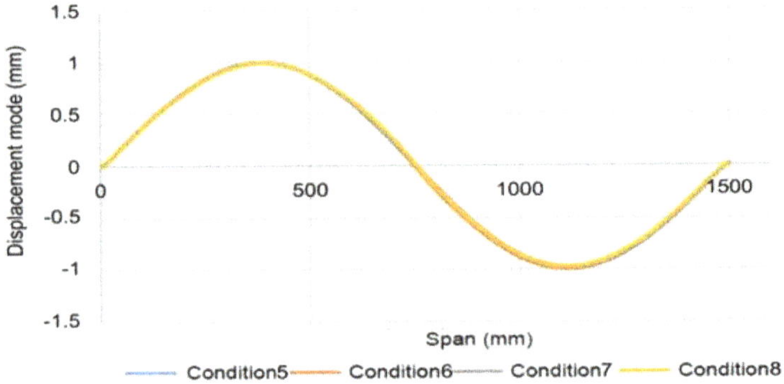

(**b**) Second order vibration mode.

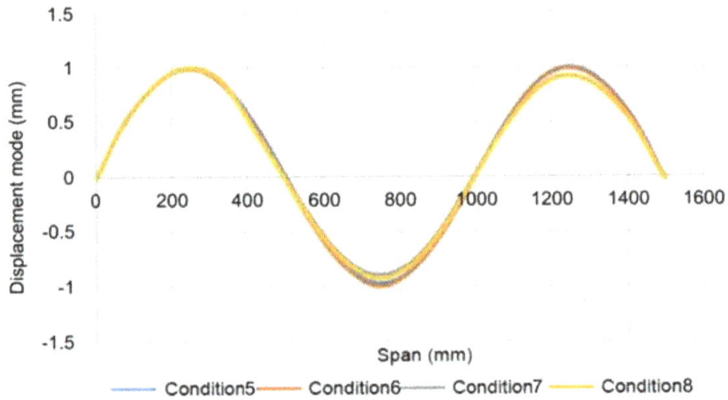

(**c**) Third order vibration mode.

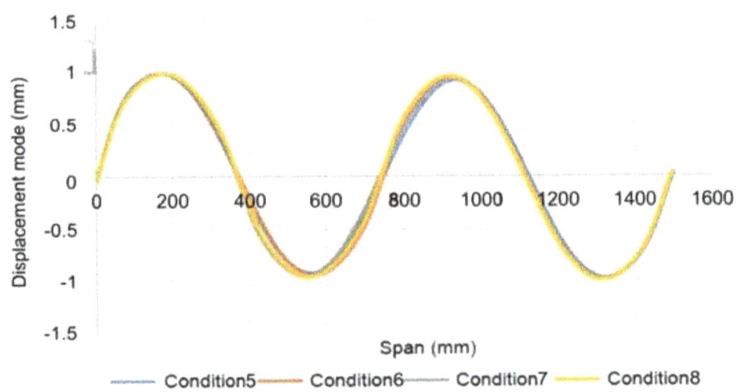

(**d**) Fouth order vibration mode.

Figure 24. *Cont.*

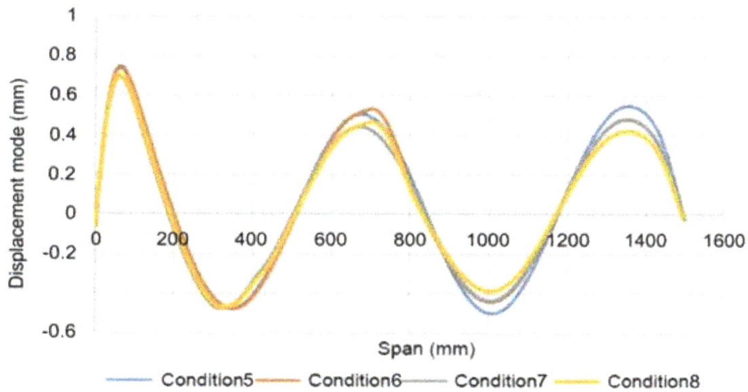

(**e**) 5th order vibration mode.

Figure 24. Displacement mode curves at different orders under Conditions 5 to 8.

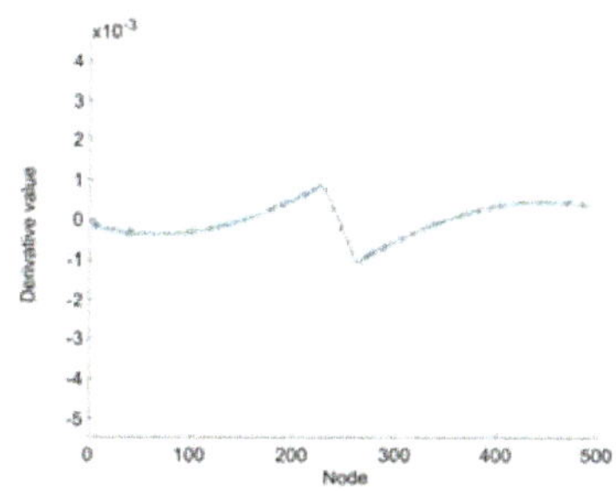

(**a**) First-order derivative of first-order mode difference.

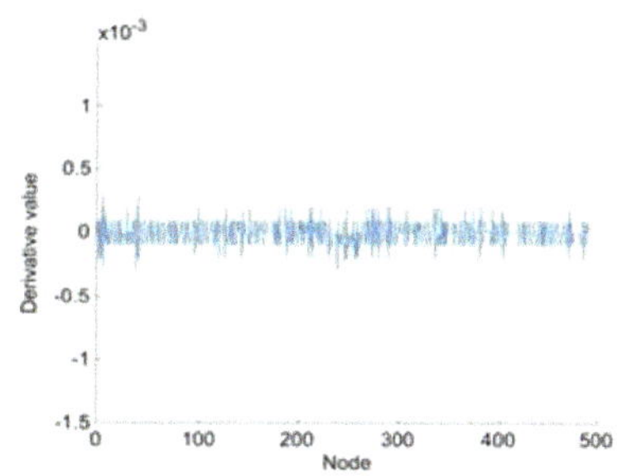

(**b**) Second-order derivative of second-order mode difference.

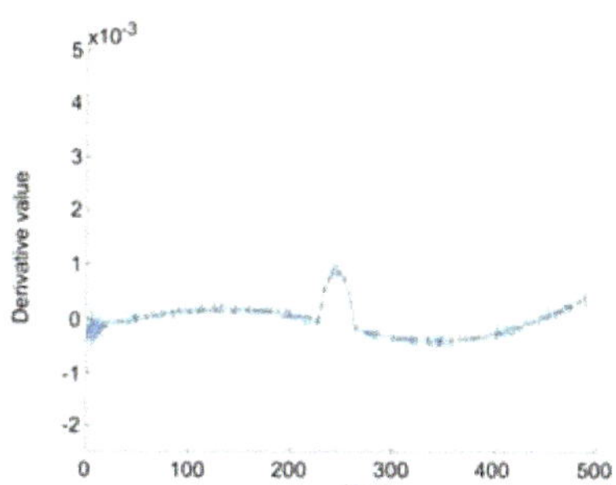

(**c**) First-order derivative of second-order mode difference.

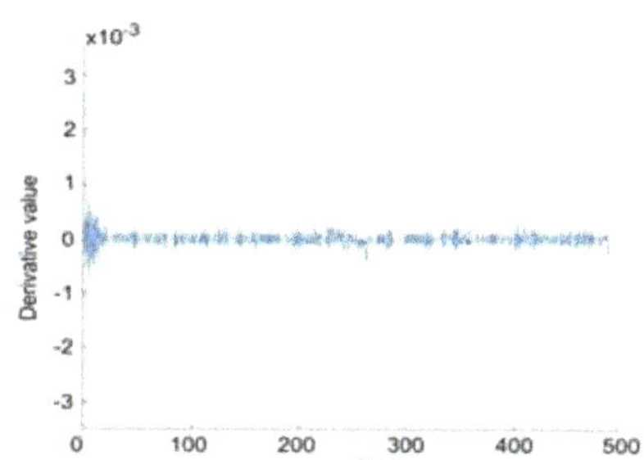

(**d**) Second-order derivative of second-order mode difference.

Figure 25. *Cont.*

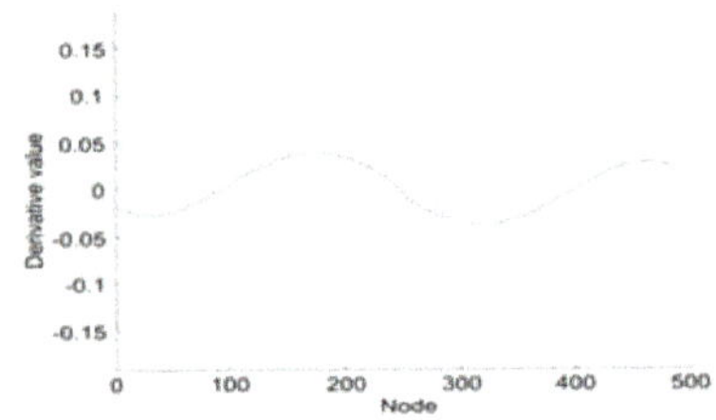

(**e**) First-order derivative of third-order mode difference.

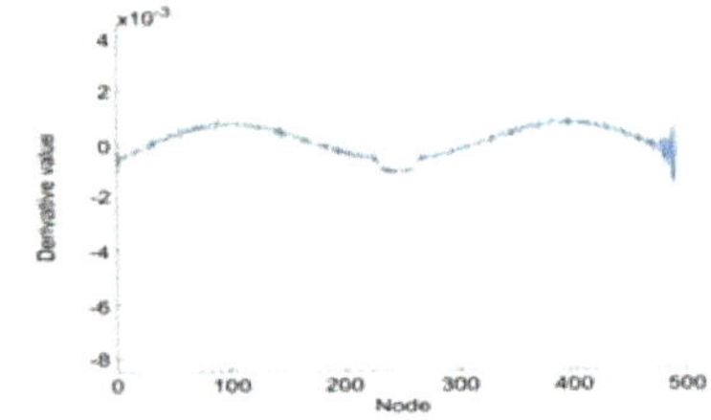

(**f**) Second-order derivative of third-order mode difference.

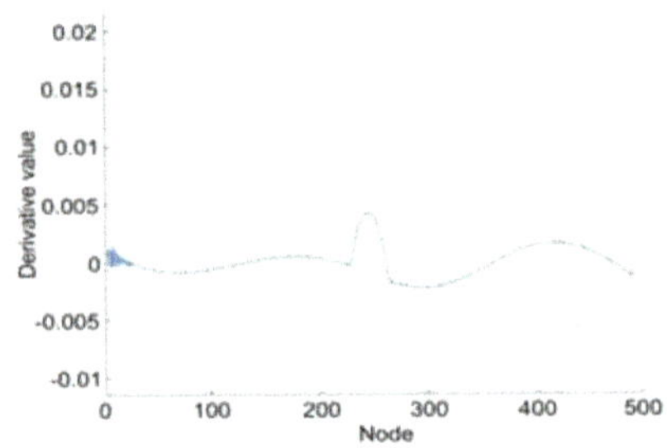

(**g**) First-order derivative of fourth-order mode difference.

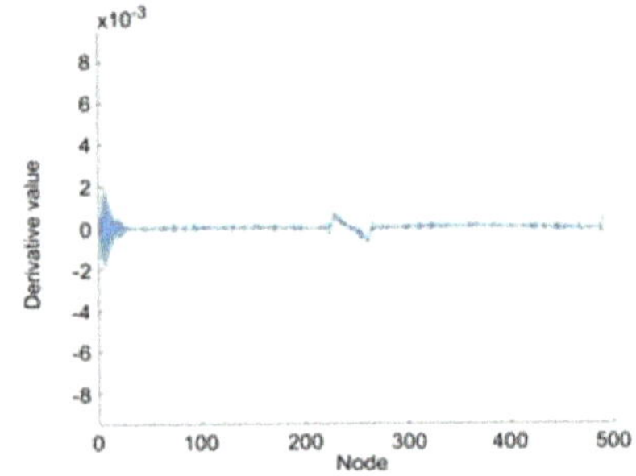

(**h**) Second-order derivative of fourth-order mode difference.

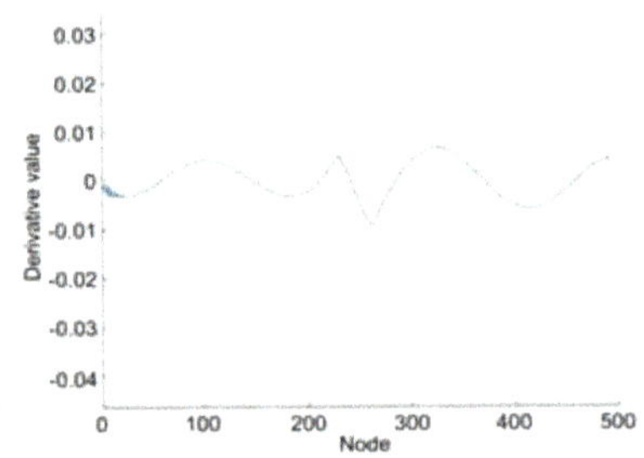

(**i**) First-order derivative of fifth-order mode difference.

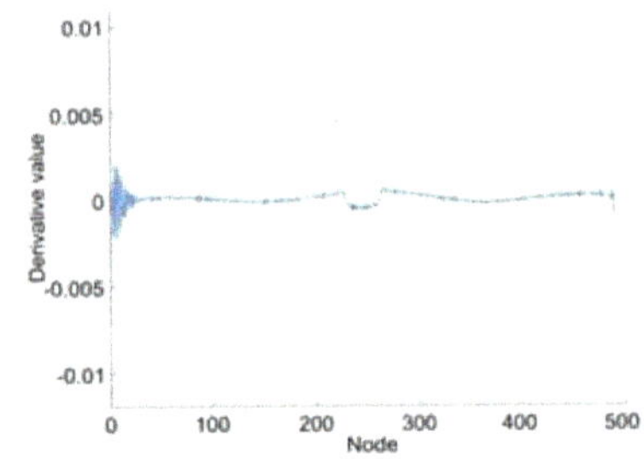

(**j**) Second-order derivative of fifth-order mode difference.

Figure 25. First-order and second-order derivatives of the differences of the first five orders of modes under Conditions 1 and 4.

Restricted by the length of the paper, the other conditions are not described in detail. The damage position could be reflected through the data simulation by both the first-order derivative curves of each vibration mode difference and most second-order derivative curves. Similarly, the first-order derivative of first-order vibration mode difference owned a smoother curve in the undamaged part, which embodied the superiority of first-order derivative curves of first-order vibration mode difference over the other first-order derivative curves.

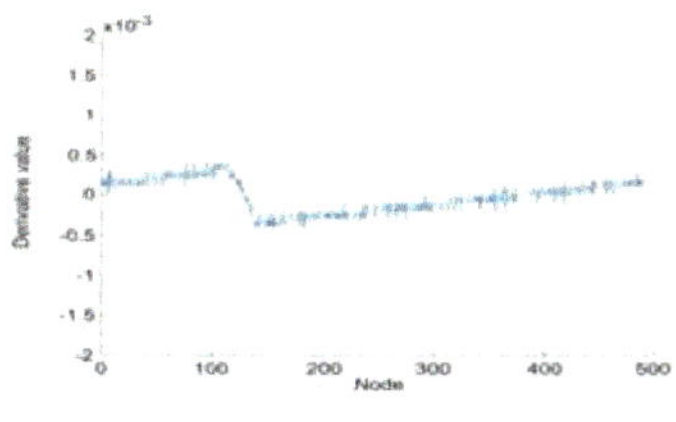

(**a**) First-order derivative of first-order mode difference.

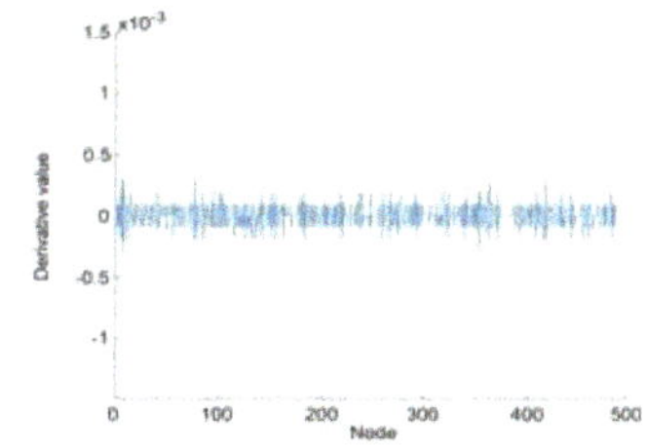

(**b**) Second-order derivative of first-order mode difference.

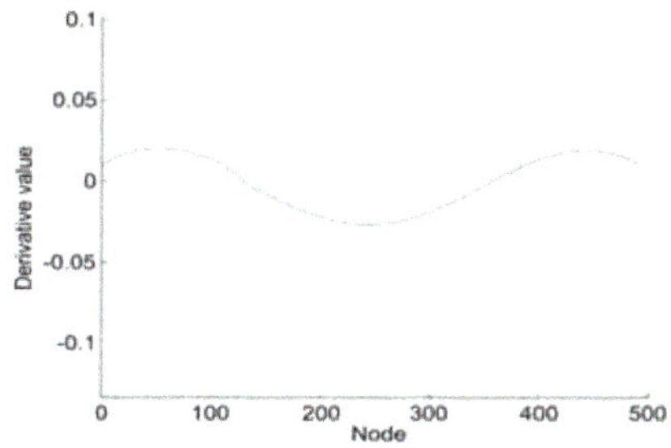

(**c**) First-order derivative of second-order mode difference.

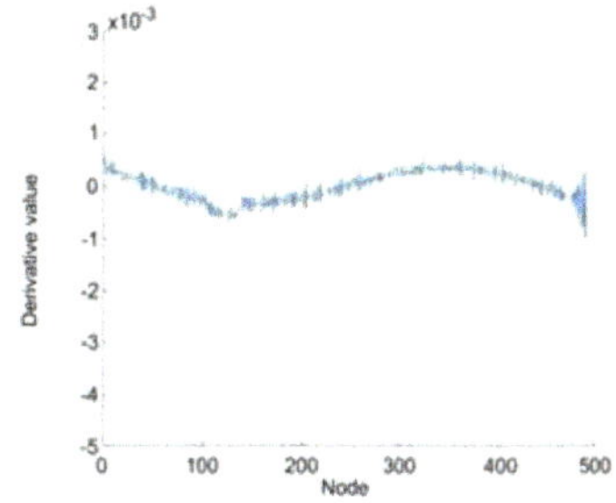

(**d**) Second-order derivative of second-order mode difference.

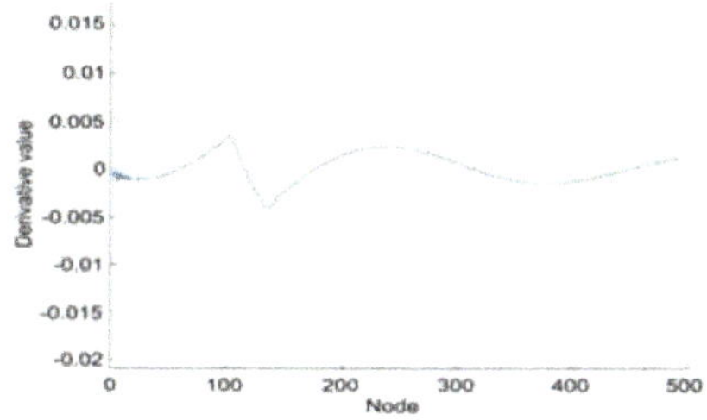

(**e**) First-order derivative of third-order mode difference

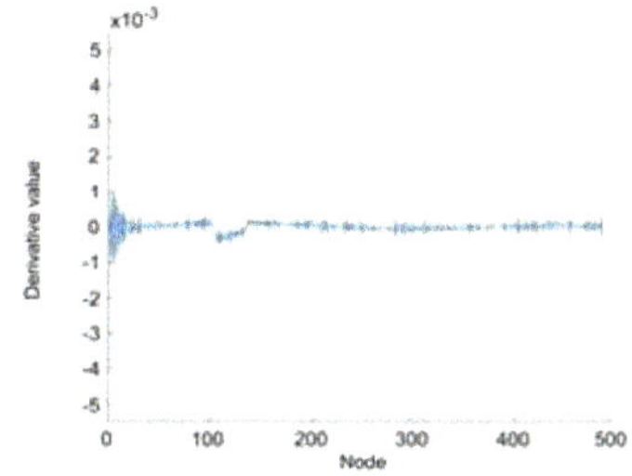

(**f**) Second-order derivative of third-order mode difference.

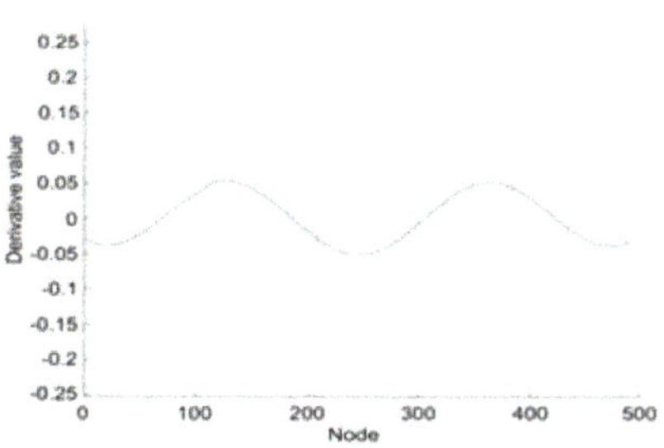

(**g**) First-order derivative of fourth-order mode difference

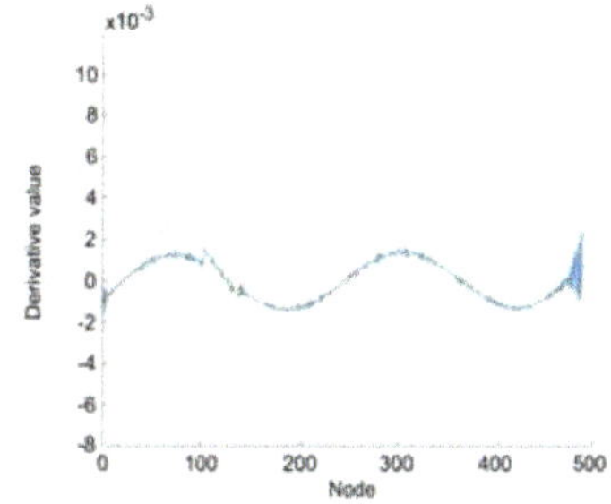

(**h**) Second-order derivative of fourth-order mode difference.

Figure 26. *Cont.*

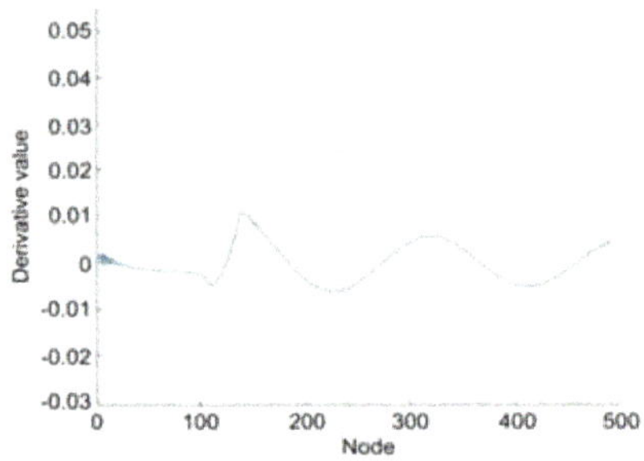

(**i**) First-order derivative of fifth-order mode difference

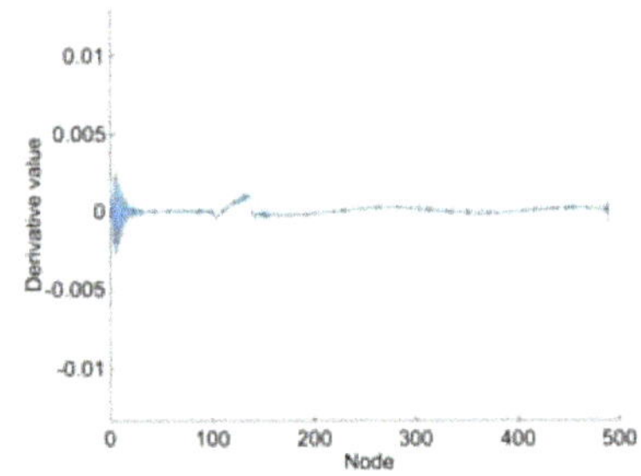

(**j**) Second-order derivative of fifth-order mode difference.

Figure 26. First-order and second-order derivatives of the differences of the first five orders of modes under Conditions 1 and 3.

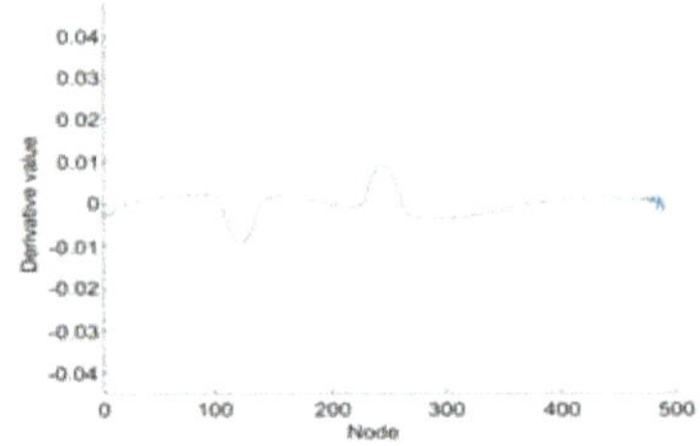

(**a**) First-order derivative of first-order mode difference.

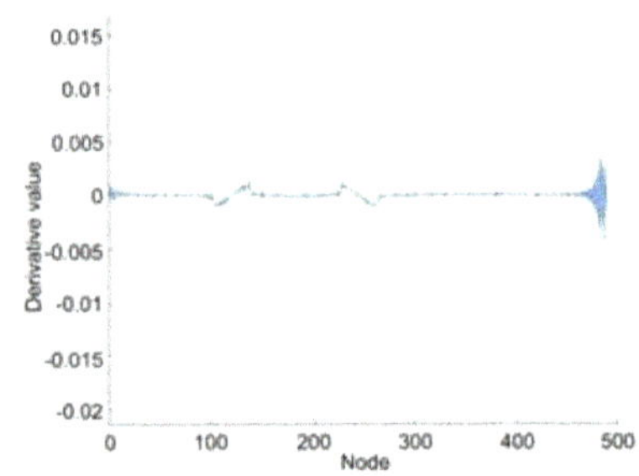

(**b**) Second-order derivative of first-order mode difference.

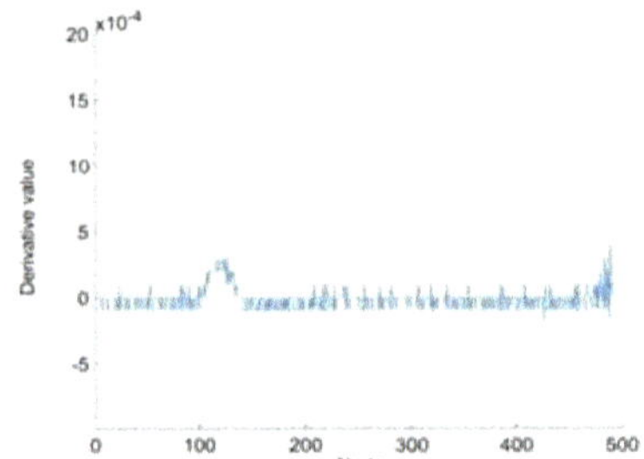

(**c**) First-order derivative of second-order mode difference.

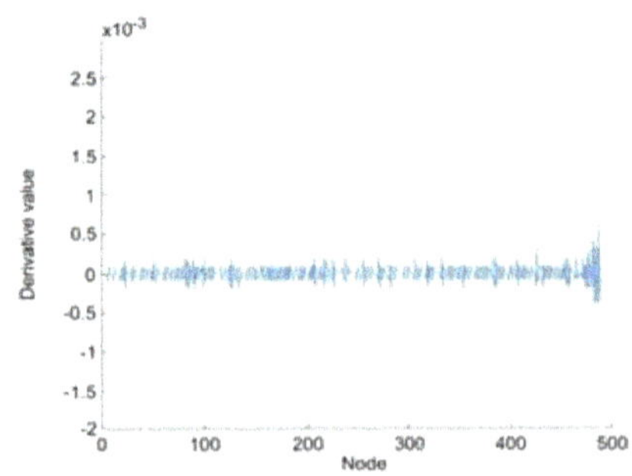

(**d**) Second-order derivative of second-order mode difference.

Figure 27. *Cont.*

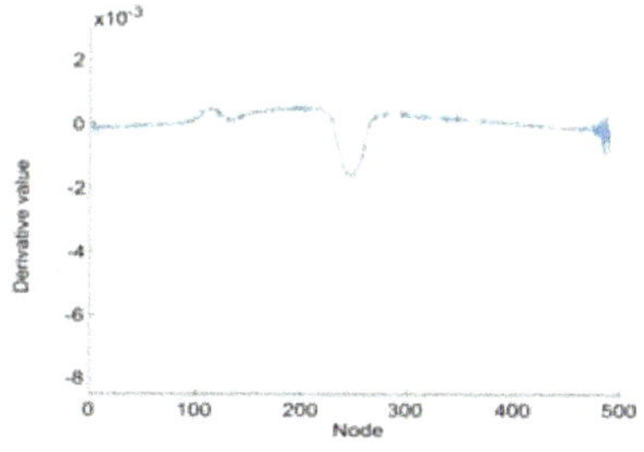

(**e**) First-order derivative of third-order mode difference.

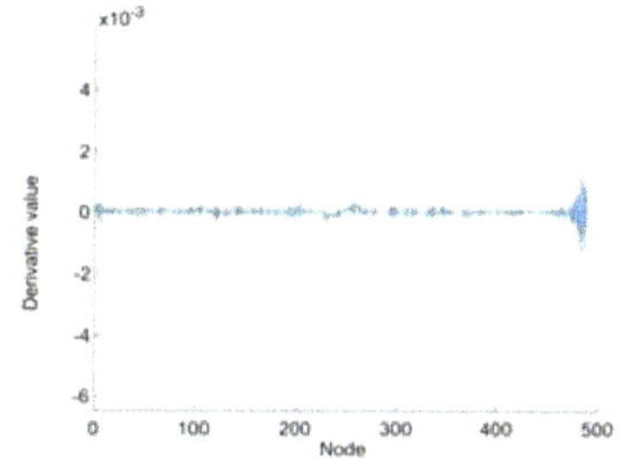

(**f**) Second-order derivative of third-order mode difference.

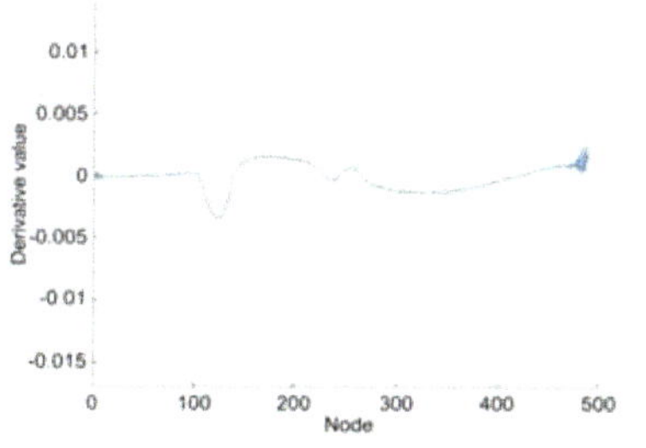

(**g**) First-order derivative of fourth-order mode difference.

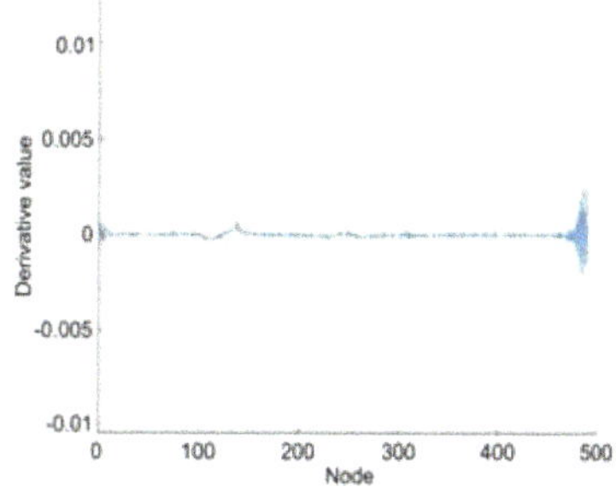

(**h**) Second-order derivative of fourth-order mode difference.

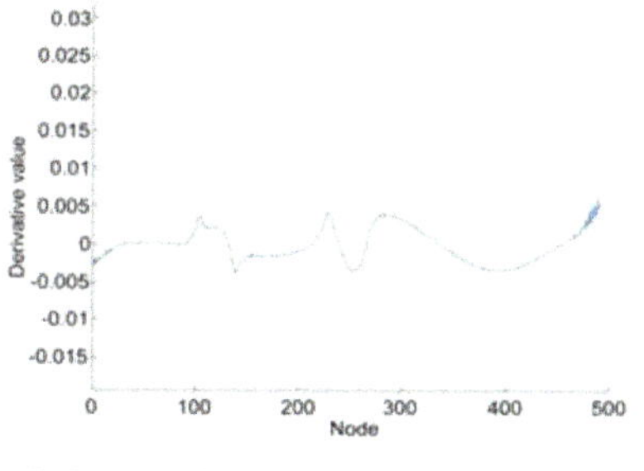

(**i**) First-order derivative of fifth-order mode difference.

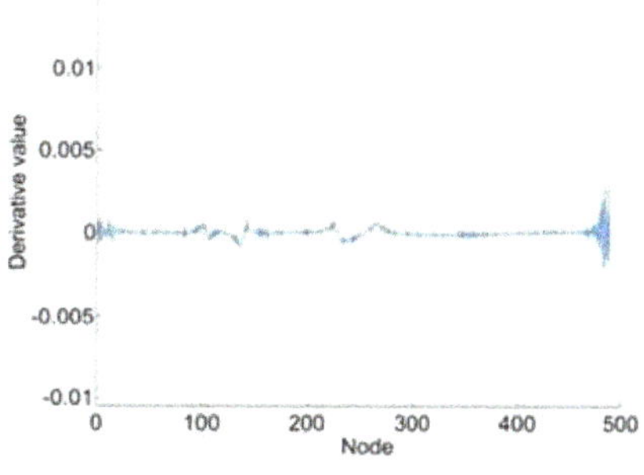

(**j**) Second-order derivative of fifth-order mode difference.

Figure 27. First-order and second-order derivatives of the differences of the first five orders of modes under Conditions 5 and 8.

3.4. Validation by Case Simulation

A double-span continuous beam was designed and different damage conditions were set to validate whether the conclusions above applied to the other structural forms. The case model is displayed in Figure 28, and the working conditions of this case are listed in Table 7.

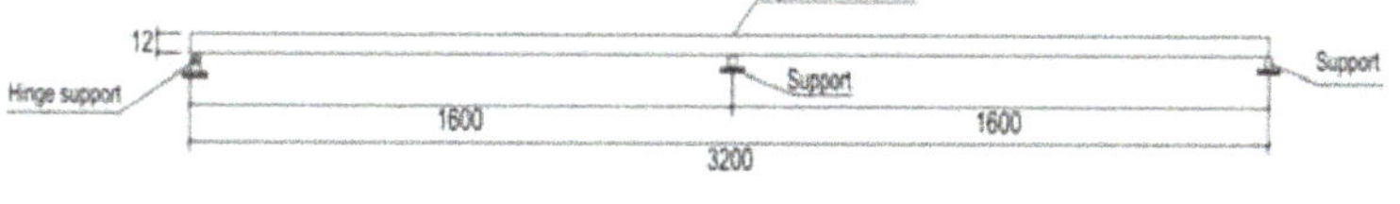

Figure 28. Schematic diagram of case model.

Table 7. Working conditions of the case.

Condition No.	Span Number	Damage Degree	Damage Position	Number of Damages
Condition 1	Two-span	No damage	NO	0
Condition 2	Two-span	Two-point 25% damage	One-span 1/4 position and two-span 3/4 position	2
Condition 3	Two-span	Two-point 25% damage	One-span position 50 mm away from the hinge support, two-span 3/4 position	2
Condition 4	Two-span	Two-point 25% damage	One-span 1/2 position and two-span 3/4 position	2
Condition 5	Two-span	Two-point 25% damage	Midspan position (damage width: 200 mm) and two-span 3/4 position	2
Condition 6	Two-span	Two-point 25% damage	Midspan position (damage width: 300 mm) and two-span 3/4 position	2
Condition 7	Two-span	Two-point 25% damage	Midspan position (damage width: 400 mm) and two-span 3/4 position	2

The inherent frequencies of the test beam at different orders under different conditions were measured through the abovementioned testing method, as listed in Table 8.

Table 8. Inherent frequency loss rates at different orders.

Order of Vibration Mode	Condition 1	Condition 2		Condition 3		Condition 4	
	Frequency (Hz)	Frequency (Hz)	Frequency Loss Rate %	Frequency (Hz)	Frequency Loss Rate %	Frequency (Hz)	Frequency Loss Rate %
1	7.2525	6.8123	6.07	7.0053	3.41	6.839	5.70
2	7.5042	7.0088	6.60	7.2162	3.84	7.00782	6.61
3	20.009	16.929	15.39	17.104	14.52	17.143	14.32
4	20.677	18.669	9.71	18.984	8.19	18.93	8.45
5	39.304	37.345	4.98	34.546	12.11	34.015	13.46
Order of Vibration Mode	**Condition 1**	**Condition 5**		**Condition 6**		**Condition 7**	
	Frequency (Hz)	**Frequency (Hz)**	**Frequency Loss Rate %**	**Frequency (Hz)**	**Frequency Loss Rate %**	**Frequency (Hz)**	**Frequency Loss Rate %**
1	7.2525	6.7894	6.39	6.7458	6.99	6.7252	7.27
2	7.5042	6.8559	8.64	6.8402	8.85	6.8388	8.87
3	20.009	17.128	14.40	17.01	15.00	16.77	16.18
4	20.677	18.991	8.15	18.775	9.20	18.518	10.44
5	39.304	34.015	13.46	33.265	15.36	33.29	15.30

Table 8 shows that the frequency loss rate of third-order vibration mode was higher as abovementioned. Meanwhile, by comparing the frequencies under Conditions 1, 4, 5, 6 and 7, we found that with the expansion of damage width, the frequency was not sensitive to the damage change at the same position.

The related displacement mode curves were directly plotted together to compare the displacement mode curves under Conditions 1 to 4, as shown in Figure 29.

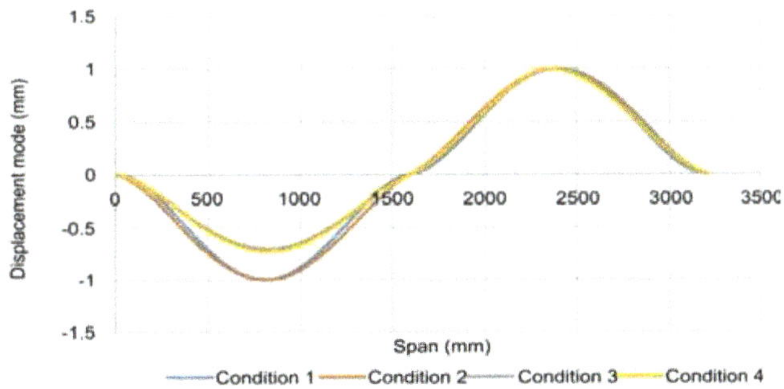

(**a**) Vibration mode curves under Conditions 1 to 4.

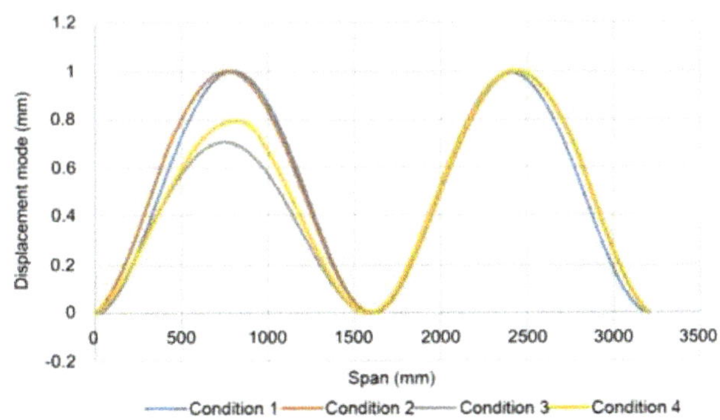

(**b**) Second-order vibration mode curves under Conditions 1 to 4.

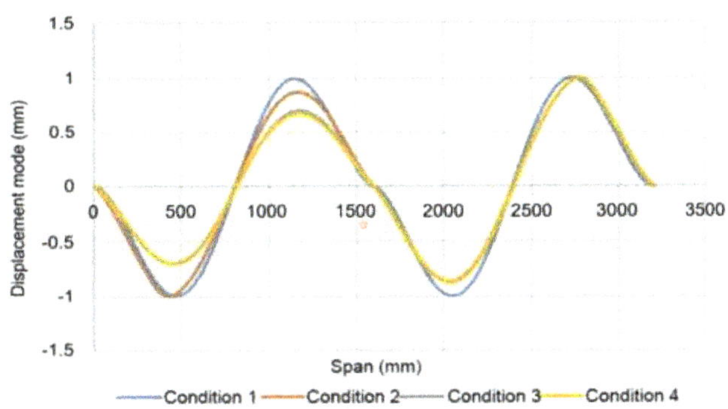

(**c**) Third-order vibration mode curves under Conditions 1 to 4.

Figure 29. *Cont.*

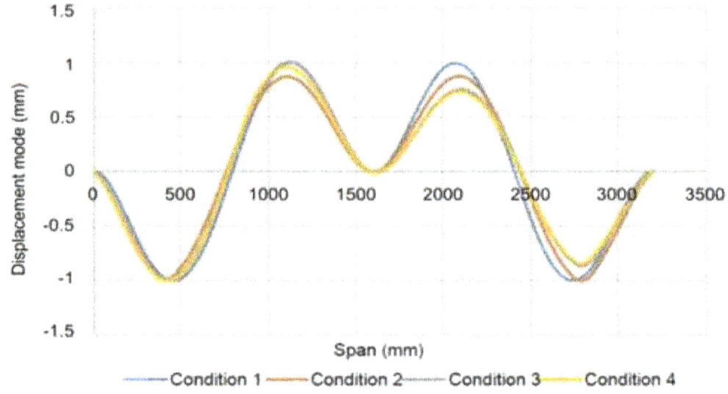

(**d**) Fourth-order vibration mode curves under Conditions 1 to 4.

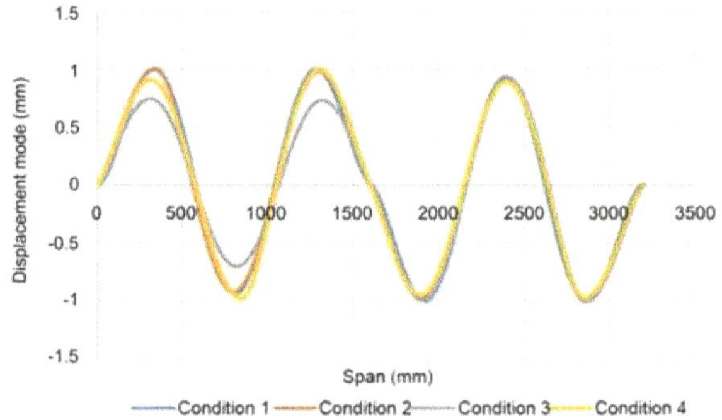

(**e**) Fifth-order vibration mode curves under Conditions 1 to 4.

Figure 29. First five orders of vibration mode curves under Conditions 1 to 4.

As with the working conditions of a single-span beam, the difference between displacement modes was not great, and the displacement difference only appeared at the damage position more or less. The images were approximately overlapped even if damages existed at the second span, which served as the control variable and the comparison span. Given that the difference value of displacement mode curves for the double-span continuous beam was approximately similar to that for the single-span beam, only the first-order derivative of the first-order mode was verified. To facilitate the comparison, the difference values of first-order vibration modes under different working conditions were plotted onto the same graph (Figure 30).

It can be observed from the above curves that the first-order derivative curve of first-order vibration mode difference was located at the midspan damage of the first span, and the fluctuation width of singularity increased with the increase in the damage width, indicating that the first-order derivative curve of first-order vibration mode difference is of particular reference value for identifying the damage scope.

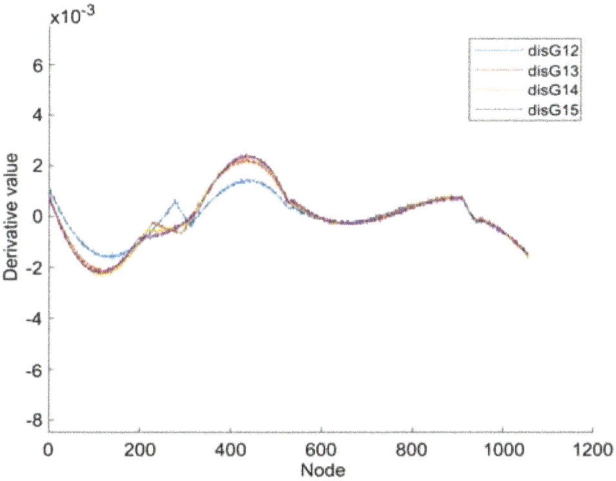

Figure 30. First-order derivatives of first-order mode differences under Conditions 1, 4, 5, 6 and 7.

4. Conclusions

In this study, a beam structure damage identification method was explored through finite element simulation and experiment from structural inherent frequency and displacement mode angles. Directed damages were created for the experimental beam and finite element model. It was discovered that the inherent structural frequency would be reduced in case of any damage, and the frequency loss rate of the third-order vibration mode was the maximum among the first five orders of vibration modes. Through these findings, whether the structure is damaged can be tentatively judged, but the damage position cannot be determined. The damage will affect the structural displacement mode, and the structural displacement will be aggravated at the damaged part. Therefore, the damage scope can be inferred, but the structural displacement mode difference, which is too small, can hardly be observed. The displacement mode difference was amplified by solving the derivatives of mode difference to solve this problem. Through comparing the first-order and second-order derivatives of the first five orders of displacement mode differences under different working conditions, the first-order derivative of the first-order vibration mode difference was considered the reference basis for the damage identification. This parameter is capable of sensitively identifying the rough damage position. However, the fluctuation trend will gradually be smooth when damage with great width is identified, and the damage scope cannot be accurately identified in the case of small data points. Besides, the identification effect on the damage close to the support is unsatisfactory. Defects still exist in this study. Only directed damages were assigned to the experimental beam, but the damage conditions are incredibly complicated in practical engineering. Moreover, the identification effect of the proposed method under unknown damage conditions remains unclear and should be explored in the future.

Author Contributions: Conceptualization, Y.Z., X.T., G.L. and F.L.; methodology, Y.Z., X.T., G.L., J.D., J.G. and F.L.; validation, Y.Z., X.T., G.L., J.D., J.G. and F.L.; formal analysis, Y.Z., X.T., G.L., J.D., J.G. and F.L.; investigation, Y.Z., X.T., G.L., J.D., J.G. and F.L.; resources, Y.Z., X.T., G.L., J.D., J.G. and F.L.; data curation, Y.Z., X.T., G.L., J.D., J.G. and F.L.; writing—original draft preparation, Y.Z., X.T., G.L., J.D., J.G. and F.L.; writing—review and editing, Y.Z., X.T., G.L., J.D., J.G. and F.L.; visualization, Y.Z., X.T., G.L., J.D., J.G. and F.L.; supervision, Y.Z., X.T., G.L., J.D., J.G. and F.L.; project administration, Y.Z., X.T., G.L., J.D., J.G. and F.L.; funding acquisition, Y.Z., X.T., G.L., J.D., J.G. and F.L. All authors have read and agreed to the published version of the manuscript.

Funding: This research was supported by the Natural Science Foundation of Beijing, China (Grant No. 82020120), and the National College Students' Innovation and Entrepreneurship Training Project (Grant No. 202010016018 and Grant No. 202110016019).

Institutional Review Board Statement: Not applicable.

Informed Consent Statement: Not applicable.

Data Availability Statement: All the data can be requested from the corresponding author upon reasonable request.

Conflicts of Interest: The authors declare no conflict of interest.

References

1. Wang, Y. Bridge Damage Identification Based on the Change of Inherent Frequency. Ph.D. Thesis, East China Jiaotong University, Jiangxi, China, 2012.
2. Pandey, A.K.; Biswas, M. Experimental verification of flexibility difference method for locating damage in structures. *J. Sound Vib.* **1955**, *184*, 311–328. [CrossRef]
3. Hearn, G.; Testa, R.B. Modal analysis for damage detection in structures. *J. Struct. Eng.* **1991**, *171*, 3042–3063. [CrossRef]
4. Ramanjaneyulu, K.; Srinivas, V.; Sasmal, S.; Jeyasehar, C.A. Damage identification based on vibration data. *Int. J. Forensic Eng.* **2013**, *1*, 209–226. [CrossRef]
5. Weng, J.C. Study on Damage Identification of Simple Supported-to-Continuous Bridge Based on Curvature Mode Theory. Master's Thesis, Southwest Jiaotong University, Chengdu, China, 2013.
6. Hu, Y.C. Study of Beam Bridge Damage Identification Based on Structural Dynamic Characteristics. Ph.D. Thesis, Wuhan University of Technology, Wuhan, China, 2014.
7. Wu, Q.Y. Study of Beam Structure Damage Identification Based on Dynamic Characteristics. Ph.D. Thesis, Wuhan University of Technology, Wuhan, China, 2019.
8. Liu, C.; Wu, D.; Li, Y.; Du, Y. Large-scale pavement roughness measurements with vehicle crowdsourced data using semi-supervised learning. *Transp. Res. Part C Emerg. Technol.* **2021**, *125*, 103048. [CrossRef]
9. Fang, Q.; Wang, G.; Yu, F.; Du, J. Analytical algorithm for longitudinal deformation profile of a deep tunnel. *J. Rock Mech. Geotech. Eng.* **2021**, *13*, 845–854. [CrossRef]
10. Lu, N.; Wang, H.; Wang, K.; Liu, Y. Maximum Probabilistic and Dynamic Traffic Load Effects on Short-to-Medium Span Bridges. *Comput. Modeling Eng. Sci.* **2021**, *127*, 345–360. [CrossRef]
11. Luo, Y.; Zheng, H.; Zhang, H.; Liu, Y. Fatigue reliability evaluation of aging prestressed concrete bridge accounting for stochastic traffic loading and resistance degradation. *Adv. Struct. Eng.* **2021**, *24*, 3021–3029. [CrossRef]
12. Xiao, X.; Bu, G.; Ou, Z.; Li, Z. Nonlinear in-plane instability of the confined FGP arches with nanocomposites reinforcement under radially-directed uniform pressure. *Eng. Struct.* **2022**, *252*, 113670. [CrossRef]
13. Zhang, H.; Liu, Y.; Deng, Y. Temperature gradient modeling of a steel box-girder suspension bridge using Copulas probabilistic method and field monitoring. *Adv. Struct. Eng.* **2021**, *24*, 947–961. [CrossRef]
14. Su, B.J. Study of Beam Structure Damage Identification Based on Dynamic Characteristics. Ph.D. Thesis, Xiangtan University, Xiangtan, China, 2018.
15. Jia, C. Study of Structural Damage Identification Method Based on Dynamic Characteristics. Ph.D. Thesis, Jilin Jianzhu University, Changchun, China, 2017.
16. Wei, T. Study on Damage Identification of Steel Truss Bridge Based on Inherent Frequency and Curvature Mode. Ph.D. Thesis, Xihua University, Chengdu, China, 2018.
17. Lian, X. Study of Damage Identification Based on Dynamic Characteristics of Bridge. Ph.D. Thesis, North China University of Water Resources and Electric Power, Zhengzhou, China, 2018.
18. Zhao, Y.P. Study on Damage Identification of Concrete Simple-Supported Beam Bridge Based on Dynamic and Acoustic Characteristics. Ph.D. Thesis, Northeast Forestry University, Harbin, China, 2019.
19. Mousavi, A.A.; Zhang, C.; Masri, S.F.; Gholipour, G. Damage detection and localization of a steel truss bridge model subjected to impact and white noise excitations using empirical wavelet transform neural network approach. *Measurement* **2021**, *185*, 110060. [CrossRef]
20. Jahangir, H.; Hasani, H.; Esfahani, M.R. Wavelet-based damage localization and severity estimation of experimental RC beams subjected to gradual static bending tests. *Structures* **2021**, *34*, 3055–3069. [CrossRef]
21. Kordestani, H.; Zhang, C.; Shadabfar, M. Beam damage detection under a moving load using random decrement technique and Savitzky–Golay filter. *Sensors* **2020**, *20*, 243. [CrossRef] [PubMed]

coatings

Article

Impact Analysis of Initial Cracks' Angle on Fatigue Failure of Flange Shafts

Zhiqiang Xu [1,2], Yunxian Cui [1], Baoliang Li [1], Ketong Liu [3], Feiting Shi [4] and Peng Cao [5,*]

1 School of Mechanical Engineering, Dalian Jiaotong University, Dalian 116028, China; xzq3171057@126.com (Z.X.); dlcyx007@126.com (Y.C.); libl@djtu.edu.cn (B.L.)
2 School of Chemical and Mechanical Engineering, Eastern Liaoning University, Dandong 118001, China
3 College of Architecture and Civil Engineering, Xi'an University of Science and Technology, Xi'an 710054, China; ketong-1982@163.com
4 School of Civil Engineering, Yancheng Institute of Technology, Yancheng 224051, China; shifeiting@ycit.cn
5 College of Architecture and Civil Engineering, Beijing University of Technology, Beijing 100124, China
* Correspondence: caopeng518888@126.com

Abstract: A fatigue test on the failure mode of flange shafts was conducted. The propagation characteristics of the initial crack at the junction between the shaft and the flange as well as its angle effect were studied. This study developed an analysis program of fatigue crack propagation, based on the APDL (ANSYS Parametric Design Language). It obtained the effective angle interval within which the initial crack is able to propagate. The fitting calculation formula was derived and the results showed that: (1) The initial crack at the junction between the shaft and the flange would propagate in the radial and axial directions; the unstable crack propagation would cause an abrupt fracture of the cross-section, failing connection; and the angle of initial crack was uncertain. (2) The crack followed the I-II-III mixed mode, which was dominated by mode I. An initial crack with a larger angle showed more noticeable II-III characteristics; K_{II} and K_{III} affected the crack's propagation angle in the radial and axial directions and they also affected the structure's surface direction. (3) The deepest point A of the crack was located at the junction between the shaft and the flange. Its crack propagation can be divided into three stages: rapid growth (stage 1), steady decline (stage 2, buffer stage), and instability (stage 3). The initial crack angle not only affected the propagation rate at stage 1 but also influenced the fatigue life distribution of the structure during propagation. The larger the initial crack angle was, the smaller the proportion of buffer stage in the total fatigue life would be. Moreover, the propagation of crack with a larger initial angle reached instability faster after stage 1, which would cause an abrupt fracture of the cross-section. This was unfavorable for deciding the crack detection time or carrying out maintenance and reinforcement. (4) The crack propagation at the junction between the shaft and the flange was determined by the size relation between ΔK_I and ΔK_{th}, instead of the effective stress intensity factor. The effective stress intensity factor can partly reflect the law of crack propagation, but cannot serve as the only criterion for crack propagation; it must be combined with the effective angle interval, which was negatively correlated with the crack's shape ratio, to determine whether the crack would propagate.

Keywords: flange shaft; fatigue failure; fracture mechanics; stress intensity factor; crack propagation

Citation: Xu, Z.; Cui, Y.; Li, B.; Liu, K.; Shi, F.; Cao, P. Impact Analysis of Initial Cracks' Angle on Fatigue Failure of Flange Shafts. *Coatings* **2022**, *12*, 276. https://doi.org/10.3390/coatings12020276

Academic Editor: Jian Chen

Received: 19 January 2022
Accepted: 14 February 2022
Published: 18 February 2022

1. Introduction

As the support of the transmission parts, a flange shaft can transfer the torque and bear the load. It is widely used in machines and equipment, such as machine tools, aircraft engines, fans, and pumps. However, when it is under cyclic stress when working, it will be subject to fatigue cracks. Studies have shown that most flanges have fatigue cracks after 5–10 years of service [1]. Fatigue cracks are most common at the junction between the shaft and the flange. This is due to the abrupt change of the cross-section. Once the crack occurs, it will propagate in the radial and axial directions, seriously affecting the safety of machinery. Therefore, it is important to investigate the fatigue failure of flange shafts.

In the engineering field, the most common assessment method of fatigue life is the S-N curve nominal stress method. This method greatly simplifies the fatigue analysis and is easy to use. However, some obstacles remain: first, it needs to be checked whether the S-N curve of various weld details is suitable for new steel; secondly, making S-N curve requires previous load information to predict the fatigue life of the structure details, but the historic information is difficult to acquire, let alone determine the fatigue life; third, this method cannot assess the remaining life with the flange's test data, such as the distribution and size of the crack. In recent years, the method that combines fracture mechanics with finite elements has developed rapidly and become one of the main tools in the research of fatigue crack [2–7]. Atanasovska et al. carried out finite element analysis of corrosion fatigue crack and turbine shaft damage based on actual turbine shaft failure cases and proposed a new method to calculate the stress concentration coefficient [8]. Li et al. studied the effects of pre-strain on fatigue crack growth rate, strain energy, crack tip plastic zone size, and fatigue fracture mechanism of commercial titanium alloy (CP-Ti) under a wide range of load amplitude and load ratio, and analyzed how pre-strain affects fatigue crack growth and thus the entire fatigue life [9]. Sola et al. combined the critical plane fatigue theory and URM standard to preliminarily estimate the damage and fatigue life of a crankshaft with increasing power [10]. Pandey performed the analysis of the fracture on the crankshaft and predicted the crack propagation life based on the Pairs formula [11]. Lei used crack beam elements to simulate the failure diagnosis of crankshaft crack [12]. Zhou conducted a fatigue test of a flange shaft under bending stress and obtained the curve of the crack propagation rate and the residual strength model [13].

Currently, global studies on the cracks of mixed-mode mainly focus on cylindrical samples. Yang et al. conducted a series of fatigue tests on C45 steel specimens with local surface cracks and found that carburizing treatment can effectively delay crack propagation and improve the fatigue life of specimens. Surface depth can only improve the fatigue life of materials in a certain range, too much depth may reduce the fatigue strength of materials [14]. Ye et al. obtained the S-N curve of 7A85-T7452 aluminum alloy in air and neutral 3.5% NaCl solution through axial fatigue test and found that corrosion had no obvious influence on fatigue life in high-stress cycling regions, but significant influence in low-stress regions [15]. Fonte designed and built the test machine of bending coupling fatigue and performed fatigue tests on the cylindrical samples [16–18]. The test showed that the crack propagation rate of the cylindrical samples decreased after the superposition of the steady-state torsion load. This was due to the crack closure. Yang carried out a crack propagation test on cylindrical samples under multiaxial load and showed that axial tension greatly affected the crack propagation rate, while axial pressure had no impact on the crack propagation rate [19]. Zangeneh simulated the failure of an agitator shaft in a large vessel with ABAQUS and found that the failure was due to the stress concentration at the root of the shaft [20]. Furthermore, the shape of crack morphology was compared with numerical and experimental results. Azevedo [21] claimed that rounding radius in the shaft base was the cause of failure employing finite element simulation of a sugarcane loader rear shaft. Aliakbari analyzed the failure of a ductile iron crankshaft in a compact pickup truck diesel engine and contributed its failure to the low cycle fatigue [22]. Khameneh focused on high-cycle bending fatigue in the crankshaft and revealed the failure mechanism from a micro perspective [23]. Alvarez et al. compared Tungsten Inert Gas Welded Alloy 718 and Laser Beam Welded Alloy 718 on hot cracking susceptibility by varestraint testing [24]. Coro et al. had proposed a new methodology to evaluate the reliability impact of the replacement of welded components using the first-order reliability method and fracture mechanics [25]. Rodriguez et al. studied the effect of roller burnishing on improving the fatigue performance of austempered iron casting cylindrical parts [26]. A new methodology to evaluate the reliability impact of the replacement of welded components using the first-order reliability method and fracture mechanics was proposed.

Many scholars have fully studied the propagation characteristics of fatigue cracks on welding components of shafts based on the numerical simulation of fracture mechanics [27]. However, the influence of initial fracture direction and size on the fracture initiation and propagation in flank shafts is rarely investigated, which can provide valuable insights into inspection, repair, and scrapping of flank shafts.

Due to the complexity and randomness during the casting, the initial crack on the flange shaft is unavoidable. There is a strong geometric uncertainty about the initial crack at the junction between the shaft and the flange, but it is hardly studied. Therefore, it becomes urgent to analyze the propagation characteristics of the fatigue crack on the flange shaft. Based on the theory of 3D fracture mechanics, the author studied the propagation characteristics of fatigue crack at the junction between shaft and flange and the focus of the study was on the impact of the initial crack angle. The study aims to help researchers understand more about the propagation characteristics of the fatigue crack on the flange shaft and to provide a reference for related research.

2. Calculation Principle of Crack Propagation

Flange shaft fatigue is typical high cycle fatigue, with the size of the plastic zone on the crack tip much smaller than the crack length and a linear elastic state. Therefore, the theory and method of linear elastic fracture mechanics can still solve the problem of the small-scale yield. This paper studied the 3D propagation of the fatigue crack based on linear elastic fracture mechanics.

2.1. Criterion of the Crack Propagation Direction

Based on different forms of load, a crack can have three basic modes: the opening mode (mode I), the sliding mode (mode II), and the tearing mode (mode III). As the flange shaft is subject to the common interaction of the bending moment, the axial force, and the shear force, most of its fatigue cracks are of a mixed-mode, deriving from the combination of two or three cracking modes. The mixed-mode cracks and opening mode cracks are different at the macroscopic level. One significant difference is that the mixed-mode cracks do not necessarily propagate along the plane determined by the initial crack. Instead, the crack propagation will be at a certain angle of the original crack surface. The spatial form is determined jointly by the propagation step length and the propagation angle. Based on the maximum circumferential stress criterion [28], this paper assumed that the crack propagated along with the position of the maximum circumferential stress. The maximum circumferential stress was at the circular failure nuclear perimeter of the crack tip. The propagation angle of the crack surface is θ [29]:

$$\theta = 2\arctan\left(\frac{1-\sqrt{1+8K_{\mathrm{II}}^2/K_{\mathrm{I}}^2}}{4K_{\mathrm{II}}/K_{\mathrm{I}}}\right) \tag{1}$$

where: K_{I} and K_{II} were the stress intensity factors of modes I and II cracks, respectively.

2.2. Effective Stress Intensity Factor

In the mechanics of the linear elastic fracture, the stress intensity factor is an important physical quantity to measure the local stress–strain state of the crack tip. Accurate stress intensity factor at the crack tip is the premise to analyze the fatigue crack propagation and fatigue life. The fatigue crack starts at the junction between the shaft and the flange. As the crack is a mixed-mode (I-II-III), when calculating the fatigue using the stress intensity factor, the impact of various cracking modes on crack propagation should be considered. BS7910 uses Formula (2) to calculate the equivalent stress intensity factor amplitude of the mixed-mode crack [29]. The far-field force that drives crack propagation may be negative. Therefore, considering the possible crack closure effect, this paper used the effective stress intensity factor in Formula (3) instead of ΔK_{I} in the Paris formula for the analysis of fatigue crack propagation and fatigue life.

$$\Delta K_{eq} = \sqrt{\Delta K_{\mathrm{I}}^2 + \Delta K_{\mathrm{II}}^2 + \alpha \Delta K_{\mathrm{III}}^2/(1-v)} \tag{2}$$

$$\Delta K_{eff} = \begin{cases} 0, & K_{eq,\max} < 0 \\ \Delta K_{eq}, & K_{eq,\max} \geq 0 \end{cases} \tag{3}$$

where: ΔK_{eq} and ΔK_{eff} were the amplitudes of equivalent stress intensity factor and effective stress intensity factor, respectively; ΔK_{I}, ΔK_{II} and ΔK_{III} were the amplitudes of stress intensity factor of modes I, II, and III cracks; $\Delta K_{eq,\max}$ was the maximum value of the equivalent stress intensity factor, v was the Poisson's ratio and here we adopted 0.3, and α was the weight coefficient, for which we took 1.0 for safety [30].

2.3. Crack Propagation Analysis Model

Based on the analysis of the fatigue crack propagation of metal structures, Paris et al. proposed a theoretical model to describe the relation between crack length and fatigue life in subcritical propagation areas [31]. For the propagation of mixed-mode fatigue crack, the Paris formula was modified by the amplitude of effective stress intensity factor to analyze the propagation of mixed-mode crack tip, as shown in Formula (4):

$$\frac{da}{dN} = C_{eq}\left(\Delta K_{eff}\right)^m \tag{4}$$

where: da/dN was the crack propagation rate, a was the crack size, N was the number of stress cycles, C_{eq} and m were the fatigue crack propagation parameters. The test showed that among the test specimens with the same geometry and material, the mixed loading modes had little effect on m, but the great effect on C_{eq}, as shown in Formula (5) [32]:

$$C_{eq} = C\left[1 + \beta(M^e - 1)^2\right] \tag{5}$$

where: $M^e = \frac{2}{\pi}\tan^{-1}\left|\frac{K_{\mathrm{I}}}{K_{\mathrm{II}}}\right|$, C and M were the Paris coefficients of the mode I crack; β was the tested coefficient, in which we took $\beta = 3$ for aluminum alloy and steel, as suggested by the test [30].

According to Formula (4), the N (fatigue life) can be obtained by the integral Formula (6) with the given a_0 (initial crack depth), a_f (final crack depth), and ΔK_{eff}.

$$N = \int_{a_0}^{a_f} \frac{da}{C_{eq}\left(\Delta K_{eff}\right)^m} \tag{6}$$

However, the stress distribution of complex structures is a complicated 3D problem. During the crack propagation, the stress is constantly redistributed, so it is difficult to express ΔK_{eff}. The Paris formula is an ordinary differential equation with an initial value, so integral Formula (6) can be converted to summation Formula (7). This can be done based on Euler's formula. Through iterative calculation, the final fatigue life is obtained:

$$N_{n+1} = N_n + \frac{da}{C_{eq}\left(\Delta K_{eff}\right)^m} \tag{7}$$

where: N_{n+1} and N_n were the fatigue life of the structure at No. n+1 and No. n iterative steps, respectively.

Similarly, the expression for the iterative calculation of crack propagation was:

$$a_{n+1} = a_n + \Delta N\left[C_{eq}\left(\Delta K_{eff}\right)^m\right] \tag{8}$$

where: a_{n+1} and a_n were the crack sizes at No. n+1 and No. n iterative steps, respectively; ΔN was the cyclic stress increment applied from a_n to a_{n+1}.

Formulas (7) and (8) reflected the numerical simulation of crack propagation and calculation of the structure's fatigue life.

2.4. Calculation Procedures of the Crack Propagation Process

For example, if a_0 (initial crack depth) and c_0 (initial crack length) were given, the propagation of an elliptical or semi-elliptical crack to limit a_c or c_c can be simulated in the following steps:

(1) Select an appropriate crack propagation step length as Δa_i. Generally, the smaller Δa_i is, the higher the calculation accuracy and the longer the calculation time will become.

(2) The figures to be calculated include: crack depth a_i, crack length c_i, the effective stress intensity factor $\Delta K^{(a)}_{eff,i}$ of the deepest point of crack, crack depth a_{i-1}, crack length c_{i-1}, effective stress intensity factor $\Delta K^{(a)}_{eff,i}$ of the deepest point of crack, and the effective stress intensity factor $\Delta K^{(c)}_{eff,i-1}$ of long axis endpoint of the crack. Among them, i represents the number of steps for current propagation calculation, and n refers to the total number of propagation steps calculation.

(3) With a given tolerance ε, if $\left(\Delta K^{(a)}_{eff,i} - \Delta K^{(a)}_{eff,i-1}\right)/\Delta K^{(a)}_{eff,i-1} < \varepsilon$ reaches accuracy, then we proceed to the following steps, otherwise, let $\Delta a_i = \Delta a_i/2$ and return to step (2).

(4) Suppose that the da/dN remained unchanged when the crack propagated from a_{i-1} to a_i, and

$$\mathrm{d}a/\mathrm{dN} = C_{eq}\left(\Delta K^{(a)}_{eff,i-1}\right)^m \tag{9}$$

Calculate the number of load cycles corresponding to Δa_i:

$$\Delta N_i = \Delta a_i/(\mathrm{d}a/\mathrm{dN})_i \tag{10}$$

Then the cumulative number of load cycles corresponding to the crack depth a_i is:

$$N_i = \sum \Delta N_i \tag{11}$$

(5) Calculate the propagation rate and propagation step length of the long axis endpoint of the crack in No. i propagation:

$$\mathrm{d}c/\mathrm{dN} = C_{eq}\left(\Delta K^{(c)}_{eff,i-1}\right)^m \tag{12}$$

$$\Delta c_i = \Delta N_i/(\mathrm{d}c/\mathrm{dN})_i \tag{13}$$

(6) Repeat step (1)–step (5), until $a_i = a_0 + \sum \Delta a_i \geq a_c$ or $c_i = c_0 + \sum \Delta c_i \geq c_c$.

The above steps can be implemented according to the programming in the program block diagram shown in Figure 1.

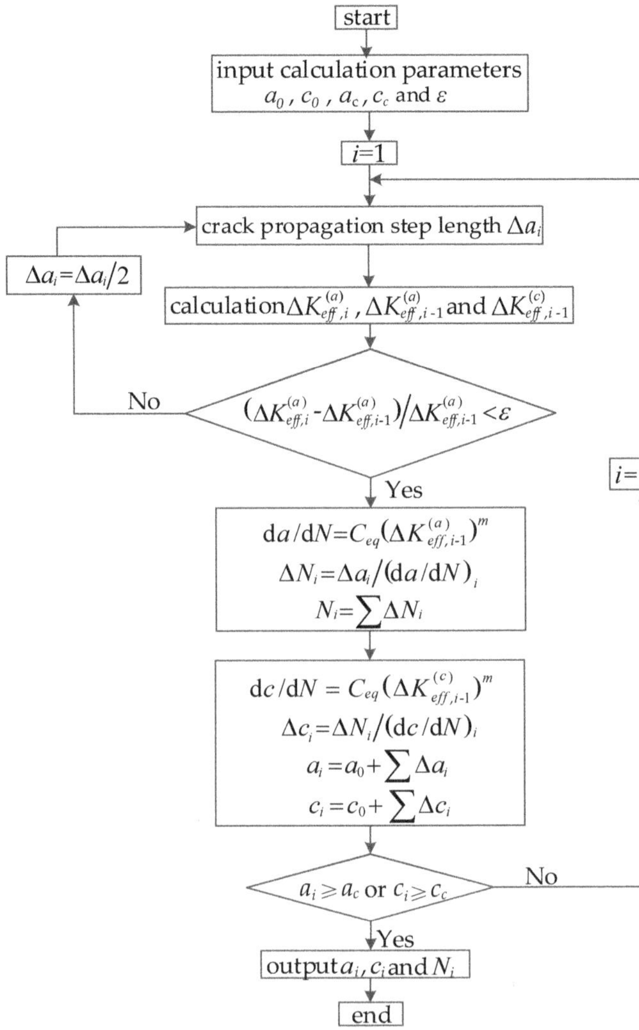

Figure 1. Calculation procedures of the crack propagation process.

3. Verification of Fatigue Test and Numerical Simulation Method for Flange Shaft

3.1. Test Overview

Considering the uncertainty of test results, three test specimens of the same size (T1, T2, and T3) were made. The specimens were of the same material, 45# steel. The detailed size of test specimens, restraint loading device, welding form, and test site layout are shown in Figure 2. To investigate the fatigue cracking performance of the flange shaft, the load of the actuator passed through the clamp to the shaft. Near the upper edge of the junction, which was between the shaft and flange, was subjected to tensile and bending stress by acting the load imposed by the actuator. The test specimen was fixed on the platform by bolts, the flange shaft was connected with the actuator by a fixture in the course of experimental testing, then the actuator imposed a cyclic load on the flange shaft for fatigue test loading. Based on the static load test results, 11 kN was taken as the maximum fatigue load, the stress ratio was 0.05, and the frequency was 2 Hz.

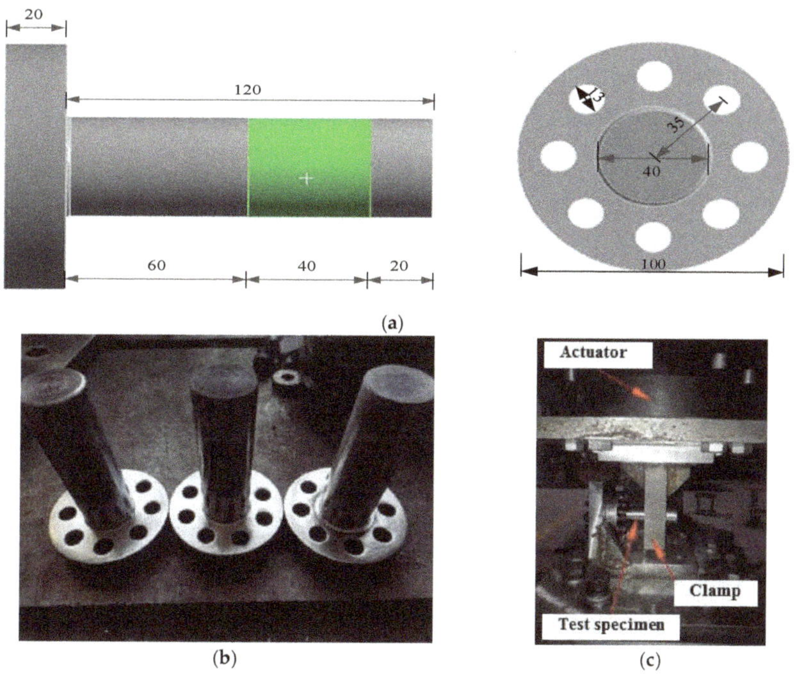

Figure 2. Test specimens and test site layout. (**a**) Test specimen size (unit: mm), (**b**) Test specimens, (**c**) Test site layout.

3.2. Fatigue Failure

Crack locations and propagation are shown in Figure 3. T1, T2, and T3 cracked at the junction between the shaft and the flange. They propagated in radial and axial directions. This caused connection failure between the shaft and the flange. It was about 10° between the failure surface of the three specimens and the flange surface, but the angle between the initial crack surface and the flange surface is uncertain. The initial crack was approximately semi-elliptical or semi-circular. The chamfered surface between the vertical flange and the shaft was set to 0°, as shown in Figure 4. When biasing towards the shaft, the value was positive and when biasing towards the flange, the value was negative. According to measurement and calculation, the maximum normal angle between the initial crack area and the flange surface was about 18° and the minimum normal angle was about −8°.

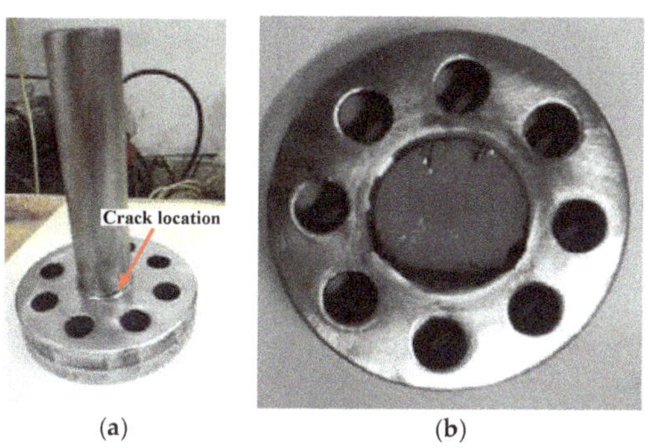

Figure 3. Crack location and propagation. (**a**) Crack location, (**b**) Crack propagation.

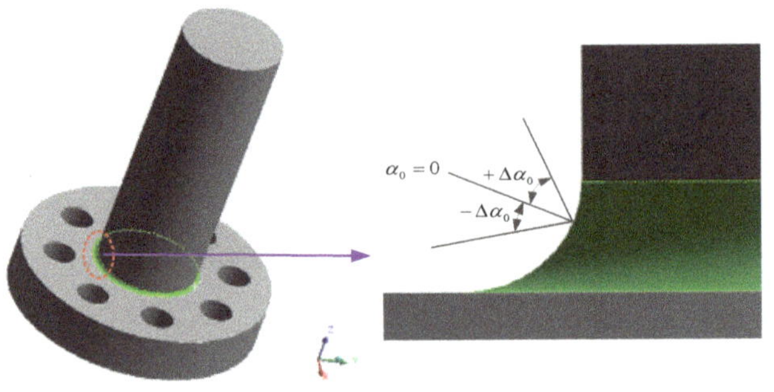

Figure 4. The definition for the initial crack angle.

3.3. *Verification of Numerical Simulation Method for Fatigue Crack*

3.3.1. Verification of Standard Test

The stress intensity factor is a key parameter to calculate crack propagation. Compact tension and shear (CTS) samples selected in the literature [33] were used as validation examples to verify the effectiveness of the numerical algorithm in this paper. The specimen dimension and experimental devices are shown in Figure 5. The specimen has a dimension of 70 mm × 90 mm and a thickness of 14 mm. A total of 6 holes having a diameter of 9.5 mm were drilled for tension. The initial fracture length was 35 mm. As shown in Figure 5c, the 3rd, 4th, 5th holes were used in the test, yielding three different angles between tension direction and fracture direction of 63°, 50°, 36°, respectively. The tension loadings are shown in Table 1.

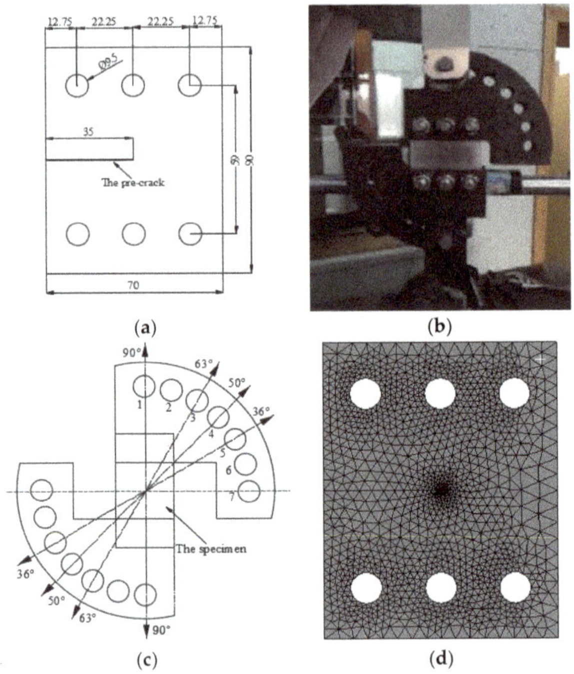

Figure 5. Compact tension and shear test. (**a**) Test specimen, (**b**) Experimental devices [33], (**c**) Test loading direction, (**d**) Finite element model.

Table 1. The tension loadings.

Loading Angle (°)	63	50	36
Load F (kN)	8	9.6	11

For CTS specimens without crack propagation, Richard deducted the approximate analytic formula of the stress intensity factor when the specimens' a/w was between 0.45 and 0.7, as can be shown [34]:

$$K_{\mathrm{I}} = \frac{F}{WB}\sqrt{\pi a}\frac{\cos\alpha}{(1-\frac{a}{W})}\sqrt{\frac{0.26+2.65(\frac{a}{W-a})}{1+0.55(\frac{a}{W-a})+0.08(\frac{a}{W-a})^2}} \tag{14}$$

$$K_{\mathrm{II}} = \frac{F}{WB}\sqrt{\pi a}\frac{\sin\alpha}{(1-\frac{a}{W})}\sqrt{\frac{-0.23+1.40(\frac{a}{W-a})}{1-0.67(\frac{a}{W-a})+2.08(\frac{a}{W-a})^2}} \tag{15}$$

where F denoted the loading force, W represented the width of CTS, B was the depth of CTS, a meant crack length, and α was the angle between the loading line and normal direction of the crack surface.

The analytical solution and finite element numerical simulation results of stress intensity factors under different loading angles were shown in Table 2. As can be seen from the comparison results, the error between the stress intensity factor calculated by the numerical calculation method in this paper and Richard's analytical solution was within ±6%. It can be concluded that the numerical method presented in this paper can accurately calculate the stress intensity factor of cracks [35].

Table 2. The comparison between finite element numerical solution and analytical solution of stress intensity factor.

Loading Angle (°)	The Calculated Value (MPa·m$^{0.5}$)		Richard Analytical Solution (MPa·m$^{0.5}$)		Error (%)	
	K_{I}	K_{II}	K_{II}	K_{II}	K_{II}	K_{II}
36	6.10	4.01	5.85	4.20	4.3	−4.5
50	6.49	2.75	6.64	2.91	−2.3	−5.5
63	6.29	1.61	6.45	1.71	−2.5	−5.9

3.3.2. Verification of Test in This Paper

The numerical analysis was performed on the ANSYS platform. The mesh size and number significantly affected the simulation results. Based on [36,37], we performed a series of simulations to determine the optimal mesh size and number. The SOLID187 element was used to simulate the irregular shape of flange shafts. A total of five cases by varying mesh sizes were simulated as shown in Table 3. Cases 1 and 2 used globally densified mesh with mesh sizes of 3 mm and 2 mm, respectively. In cases 3, 4, and 5, mesh sizes at the connection of flange face and axle were set as 1.75 mm, 1.4 mm, and 1 mm, respectively, and the mesh sizes at other parts were set as 5 mm. The simulation results of five cases are compared in Table 3. The locally densified mesh with a size of 1 mm (case 5) yielded acceptable simulation accuracy and speed, which is used in this study.

Table 3. Finite element grid verification results.

Simulation Response	Integral Uniform Grid		Local Grid Refinement		
	Case 1 (3 mm)	Case 2 (2 mm)	Case 3 (1.75 mm)	Case 4 (1.4 mm)	Case 5 (1 mm)
Configuration					
Elements	100,041	316,514	59,476	106,390	212,872
Simulation time (h)	5.6	27.5	3.7	6.1	15.3
Computed result (cycle number)	9.76×10^5	4.72×10^6	5.16×10^6	6.51×10^6	6.56×10^6

Figures 6 and 7 show the evolutions of Von Mises stress and maximum principal stress. It is evident that the Von Mises stress and maximum principal stress at the tips of fractures increases.

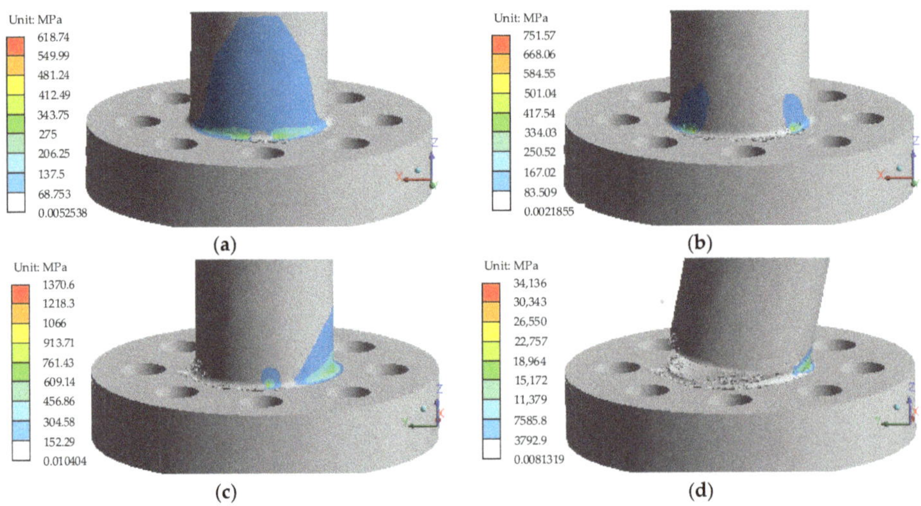

Figure 6. Evolutions of Von Mises stress during crack propagation (Unit: MPa). (**a**) Crack depth $a = 1$ mm, (**b**) Crack depth $a = 10$ mm, (**c**) Crack depth $a = 25$ mm, (**d**) Crack depth $a = 38$ mm.

The propagation path and shape changes of the cracks were compared and shown in Figure 8. In the beginning, the fractures grew in the predefined direction. Then, the fracture growing path became parallel with the flange face after the fracture depth reached 3 mm. The fracture growing path deviated from the flange face by a small angle after the fracture depth reached 15 mm and caused the breakage of the flange shaft eventually. As shown in Figure 8, the numerical results matched experimental results, indicating the effectiveness of the method developed by this paper for predicting the fracture initiation and propagation of the flange shaft.

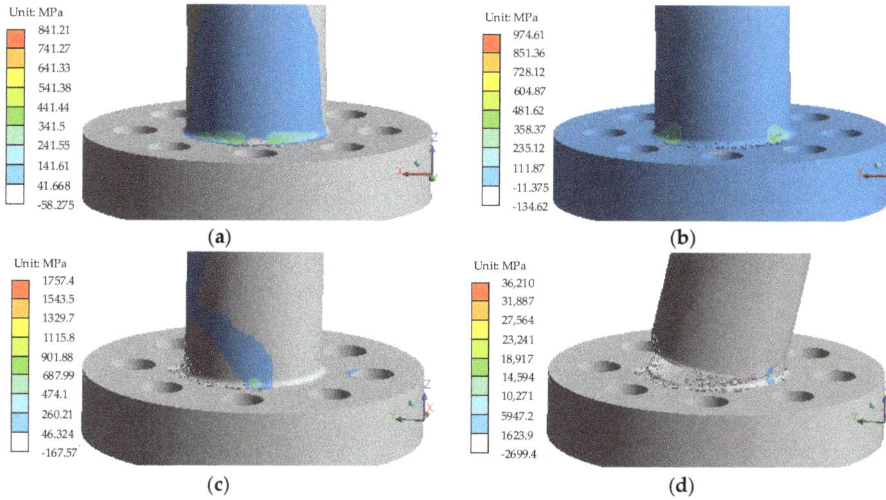

Figure 7. Evolutions of maximum principal stress during crack propagation (Unit: MPa). (**a**) Crack depth a = 1 mm, (**b**) Crack depth a = 10 mm, (**c**) Crack depth a = 25 mm, (**d**) Crack depth a = 38 mm.

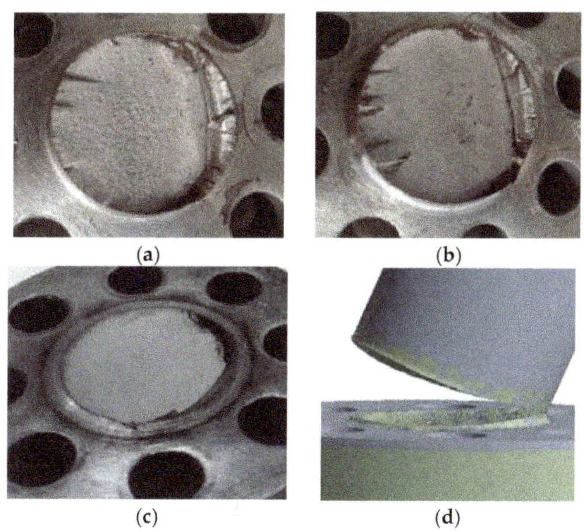

Figure 8. Comparison between numerical and experimental results. (**a**) Fracture surface of T1, (**b**) Fracture surface of T2, (**c**) Fracture surface of T3, (**d**) Numerical simulation results.

As can be seen from Figure 8, the numerical simulation method in this paper can reasonably simulate the crack propagation path and morphology change characteristics. Figure 9 shows the a-N curve of the flange shaft's crack propagation calculated by the numerical simulation method, where a was the projection depth of the deepest point of the crack front on the cross-section, the fatigue life of the flange shaft was 6.56×10^6. Table 4 shows the comparison between the simulation values and the test results of the fatigue life of the flange shaft. The average value of the test results of the three groups was 7.16×10^6, and the numerical solution in this paper was 6.56×10^6. The error between the numerical solution and the test value was 8.4%, which meets the requirements of engineering applications. Therefore, the numerical simulation method proposed in this paper can be used to analyze the crack propagation characteristics of the flange shaft.

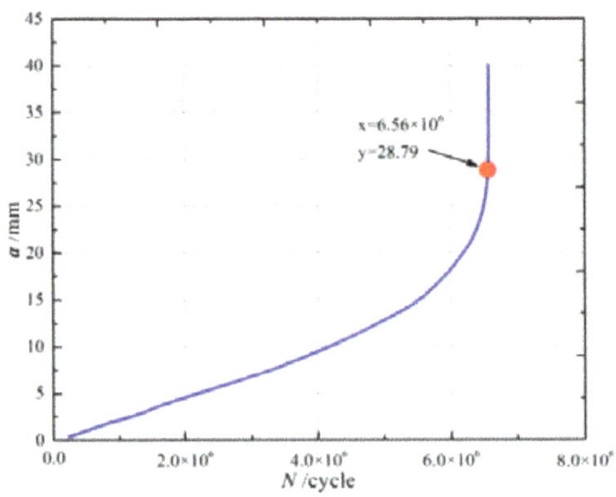

Figure 9. The *a*-*N* curve of the flange shaft's crack propagation.

Table 4. The comparison between the simulation values and the test results of the fatigue life of the flange shaft.

Experiment Results				Numerical Result	Error (%)
T1	T2	T3	Average Value		
6.49×10^6	7.36×10^6	7.64×10^6	7.16×10^6	6.56×10^6	8.4

4. The Numerical Simulation of Crack Propagation on the Flange Shaft

4.1. Numerical Model

ANSYS was utilized for modeling based on the actual size of test specimens, with the element type of Solid187. Figure 10 showed the global finite element model and sub-models of the flange shaft. The initial crack was inserted into the chamfer surface between the shaft and the flange, and the element ring grids at the crack front were divided into three-layer symmetric element rings to ensure accuracy. The location of the inserted initial crack and the division of element ring grids at the crack front are shown in Figure 10c. The 15-node wedge element was used in the innermost circle of the element ring; the 20-node hexahedral element was adopted in the middle layer; the 13-node pyramid element was applied in the transition region of the crack tip and global finite element model, and the 10-node tetrahedral element was taken for the global model outside the element ring. There were approximately 18,519 elements in the initial state. Boundary conditions applied on the simulation are almost the same as those in the experiment, with bolt holes being fixed. A concentration force is performed in the cantilever terminal as shown in Figure 10a. To obtain the bottom limited solution of the plasticity deformation, a trial force is applied for probing the plasticity deformation commence in the structure. When the loading value was 11 KN, the maximal stress arose up in the neck of the cantilever welding zone, being 382.2 Mpa and approaching the plasticity yield strength that is 386 Mpa. The most unfavorable load amplitude was 0.55–11 kN. It was selected for loading in the elastic stage, with the stress ratio $R = 0.05$.

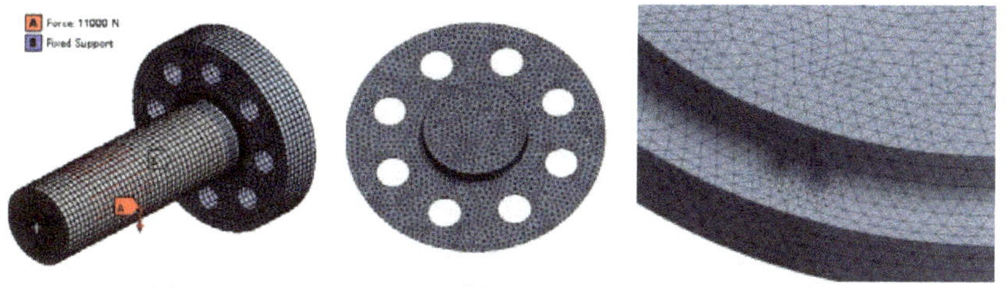

(a) (b) (c)

Figure 10. Analysis technique for the sub-models of flange shaft. (**a**) Global finite element model with initial crack, (**b**) Sub-model, (**c**) Sub-model inserted with initial crack.

4.2. Geometric Parameters of Initial Crack

According to the test results and the literature [1,28], the depth of semi-elliptical initial crack was generally not more than 0.5 mm, which can be used as the initial crack depth to reflect the most unfavorable condition. Figure 11 shows a typical semi-elliptical crack on the surface of the cylinder, point A, F, and G located the leading edge of the semi-elliptical crack front, c represented half of the long axis of the ellipse, a represented half of the short axis, and a represented the crack depth in Figure 11. The geometric parameters of the initial crack were shown in Table 5, where $\omega = a_0/(2c_0)$ was defined as the shape ratio of the initial crack, the shape of the elliptical initial crack can be changed by the value of ω. To investigate the effect of initial crack shape ratio on the crack propagation of flange shaft, five groups of initial cracks as shown in Table 5 were selected to simulate the crack propagation of flange shaft specimens, and the initial crack angle effect of fatigue crack propagation of flange shaft was mainly investigated. The definition of the initial crack angle can be obtained from Figure 4. The simulation results using five types of geometric parameters of initial crack are shown in Figure 12.

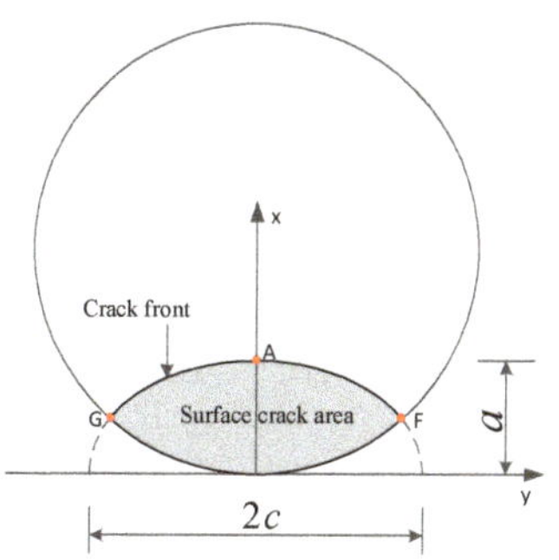

Figure 11. Semi-elliptical crack on the surface of the cylinder.

Table 5. Geometric parameters of initial crack.

Numerical Test Number	Geometry Parameters		
	a_0 (mm)	c_0 (mm)	ω
C1	0.5	2.5	0.1
C2	0.5	1.0	0.25
C3	0.5	0.5	0.5
C4	0.5	0.335	0.75
C5	0.5	0.25	1.0

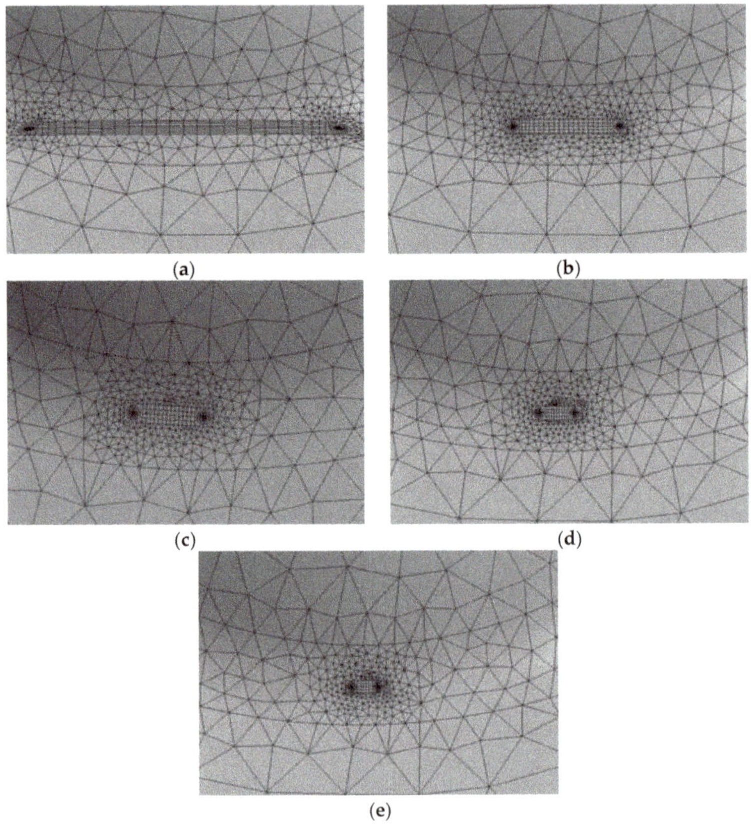

Figure 12. Five types of calculation grids of initial crack. (a) C1, (b) C2, (c) C3, (d) C4, (e) C5.

5. Crack Propagation Characteristics

5.1. Description of Calculation Points at the Crack Front

The calculation points of the crack front were shown in Figure 13, points A, B, C, D, E, F, and G were taken as the calculation points. If angle AOF was equal to θ, then points B, C, D, and E divided θ equally into five parts. As the point F on the surface and its nearby small regions did not satisfy the plane strain hypothesis, the calculation results may have errors. More accurate results can be extrapolated by the calculated values of the intensity factor of the inner point, as recommended in the stress intensity factors handbook [38].

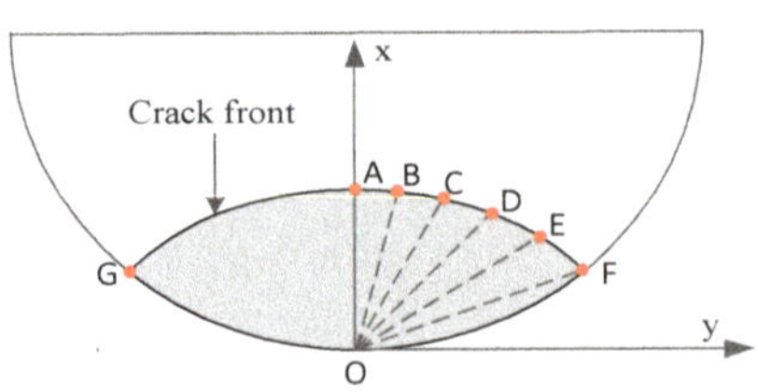

Figure 13. Calculation points at the crack front.

5.2. Impact of Different Angles of the Initial Crack on Fatigue Life

5.2.1. Test Phenomena

When the remaining cross-section of the test specimen cannot withstand the external loads (when the stress of the remaining cross-section reached or exceeded the static strength of the material), the crack instability propagation will appear, resulting in an abrupt fracture of the remaining cross-section. The fracture of the test specimen is shown in Figure 14. As can be seen from Figure 14, there was an obvious abrupt fracture area in the fracture, which was rough and significantly different from the fatigue striation. The average width of the fatigue striation accounted for 25–35% of the total fracture width. As can be seen from Figure 6 above, when crack depth A is 28.79 mm, the flange shaft enters the crack instability propagation. The width of the transient fracture zone is 11.21 mm, about 28% of the total width, which is in good agreement with the size of the transient fracture zone of the specimen.

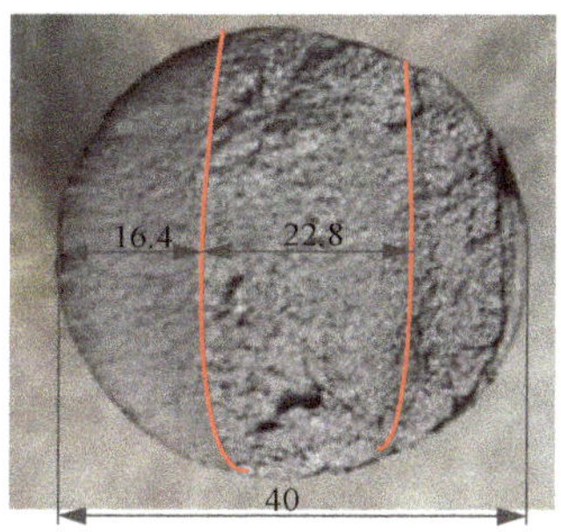

Figure 14. The fracture morphology of the test specimen (unit: mm).

Under different initial angles, the propagation of C3, the semi-circular initial crack, was calculated. The crack propagation was measured by the normalized crack depth $D = a/d$, where a was the projected depth of the deepest point at the crack front on the shaft cross-section, and d was the shaft diameter. To avoid the interference of surface point calculation error in the process of crack propagation, points A and E of the crack front were selected to analyze. The changes in the amplitude of the effective stress intensity factor ΔK_{eff} of C3 with different initial crack angles were shown in Figure 15. C3 had a similar overall propagation trend of effective stress intensity factor under different initial crack angles. When D was between 0.5–0.6, the dispersion of the effective stress intensity factor occurred. This indicated that C3's instability propagation was based on the shape of fatigue striation. When D was between 0.6–0.7, the dispersion of the effective stress intensity factor increased significantly and the propagation simulation ended around $D = 0.7$. The simulation results were consistent with the fracture characteristics of test specimens. The normalized crack depths of C3 with different initial angles were the same when reaching instability, regardless of the initial angle. This showed that the crack instability and the abrupt fracture of the remaining cross-section depended on whether it can withstand the external loads. Based on changes in the amplitude of the effective stress intensity factor, the crack propagation in the length direction was the main cause for the subsequent reduction in the remaining cross-section.

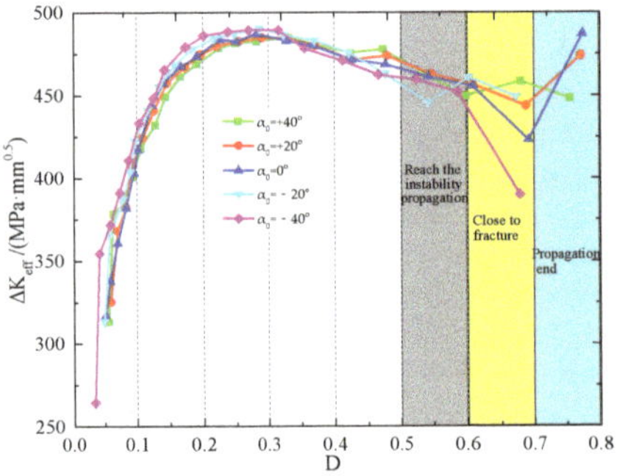

(a)

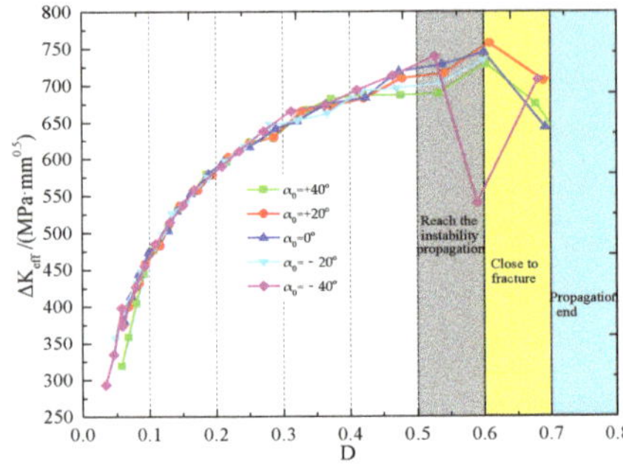

(b)

Figure 15. Changes in the amplitude of effective stress intensity factor. (**a**) Crack front point A, (**b**) Crack front point E.

5.2.2. The Impact of the Initial Crack Angle on Fatigue Life

The propagation rate of the deepest point A in Figure 16 can be divided into three stages: rapid growth (stage 1), steady decline (stage 2, buffer stage), and instability (stage 3). According to the results, $T = 0.35$ was regarded as the normalized crack depth when the crack reached the instability, and the corresponding fatigue life was recorded as the full life N of stage 1. If Nu denoted the life of stage 1, and N_s was the life of stage 2, then $N = N_u + N_s$. Table 6 showed changes in the propagation rate of the deepest point A, as the number of load cycles increased.

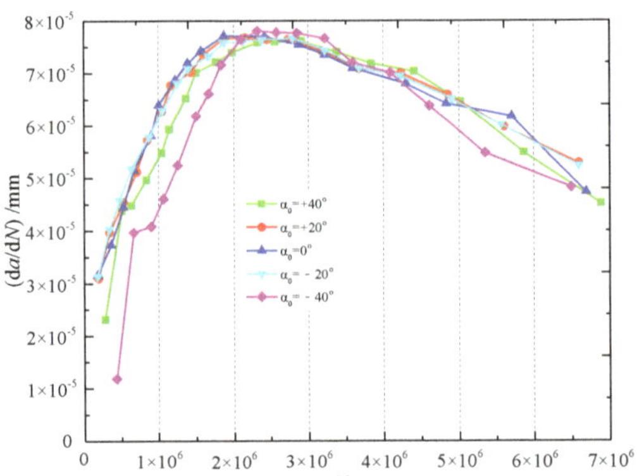

Figure 16. Propagation rate of the deepest point A.

Table 6. Changes of N_u, N_s, and the propagation rate under different α_0.

α_0/(°)	(N_u/N) (%)	Propagation Rate (%)	Average Propagation Rate (%)	N_s/N (%)	Propagation Rate (%)	Average Propagation Rate (%)
−40	60.3	+562.1		39.7	−30.5	
−20	57.9	+144.8		42.1	−31.7	
0	45.7	+144.0	+246	54.3	−38.7	−34.56
20	60.4	+148.2		39.6	−31.2	
40	62.3	+231.0		37.7	−40.7	

Figure 16 and Table 6 showed the propagation rate. The effect α_0 only manifested in the propagation rate of stage 1. Larger $|\alpha_0|$ ($|\alpha_0|$ is taking the absolute value of α_0) would lead to a smaller initial propagation rate. In stage 1, the energy exchange between the crack and the outside world was fast, and the surrounding stress was unstable. α_0 had little impact on the propagation rate of stage 2. As for the fatigue life, larger $|\alpha_0|$ would increase Nu/N and decrease N_s/N. This suggested that a larger initial crack angle $|\alpha_0|$ would reduce the proportion of buffer stage in the total fatigue life. After stage 1, the initial crack with a larger angle would reach the instability propagation faster, which would cause an abrupt fracture of the cross-section. Therefore, the angle of the initial crack at the junction between the shaft and flange mainly affected its fatigue life distribution. The initial crack with a larger angle would adversely influence the flange shaft's detection cycle, maintenance, and reinforcement. The angle of the initial crack should be fully considered in the study and reasonably set before the propagation test or numerical simulation of prefabricated cracks.

5.3. Impact of Initial Crack Angle on Crack Propagation

The initial propagation of crack referred to the change of the initial crack before and after the first load cycle. According to the definition of ΔK_{th}, which was the crack propagation threshold, this state can be calculated and the result can be used to determine whether the crack can propagate at a given load amplitude. Based on the literature [39], when R is 0.05, ΔK_{th} is 212.86 MPa·mm$^{0.5}$ for 45# steel, the stress intensity factor of the initial propagation of initial cracks with different α_0 was calculated at the most unfavorable load amplitude. The results are shown in Figure 17:

(1) With α_{0p} as the benchmark, ΔK_I of the crack front decreased with the increase of $|\alpha_0|$; ΔK_{II} and ΔK_{III} increased first and then decreased as $|\alpha_0|$ increased. Both of them were less than ΔK_{th}. Thus, the propagation of the initial crack at the junction between the shaft and flange resulted from $\Delta K_I > \Delta K_{th}$. ΔK_{II} and ΔK_{III} only affected the propagation angle during the initial propagation. When $|\alpha_0|$ exceeded a certain value, ΔK_I, ΔK_{II} and ΔK_{III} were less than ΔK_{th} and the propagation of the initial crack was not observed.

(2) The changing trend of ΔK_{eff} and ΔK_I was similar. In other words, ΔK_{eff} can reasonably reflect the overall change of stress intensity factors during the crack propagation. However, in cracks with large $|\alpha_0|$ (including welding cracks and other similar surface cracks), $\Delta K_{eff} > \Delta K_{th}$ might still be true. ΔK_I, ΔK_{II} and ΔK_{III} were smaller than ΔK_{th}. Therefore, the amplitude of the effective stress intensity factor cannot be used to assess the crack propagation. In relevant studies, crack propagation should be evaluated by three stress intensity factors and the crack propagation threshold.

We used $|K_{II}/K_I|$ and $|K_{III}/K_I|$ to reflect the size relation among three stress intensity factors and obtained the changing pattern of stress intensity factor on the front of the semi-circular crack C3 with different initial angles. The results are also shown in Figure 17:

(1) The crack was categorized to a mixed I-II-III mode, which was dominated by mode I. At a certain initial angle, it can be considered as the mode I crack. In this example, α_0 was about + 4°, and its angle was expressed as α_{0p}.

(2) The larger α_0, the stronger modes II-III characteristics would be. When α_0 was larger, K_{II} accounted for a large proportion at both the surface and the deepest point of C3. The stress difference between positions of the crack front was large, but the crack still belonged to mode I; K_{III} gradually increased from the deepest point towards the surface. Under the dominant action of K_{II} and K_{III}, the surface and the deepest point of C3 deflected, as shown in Figure 18.

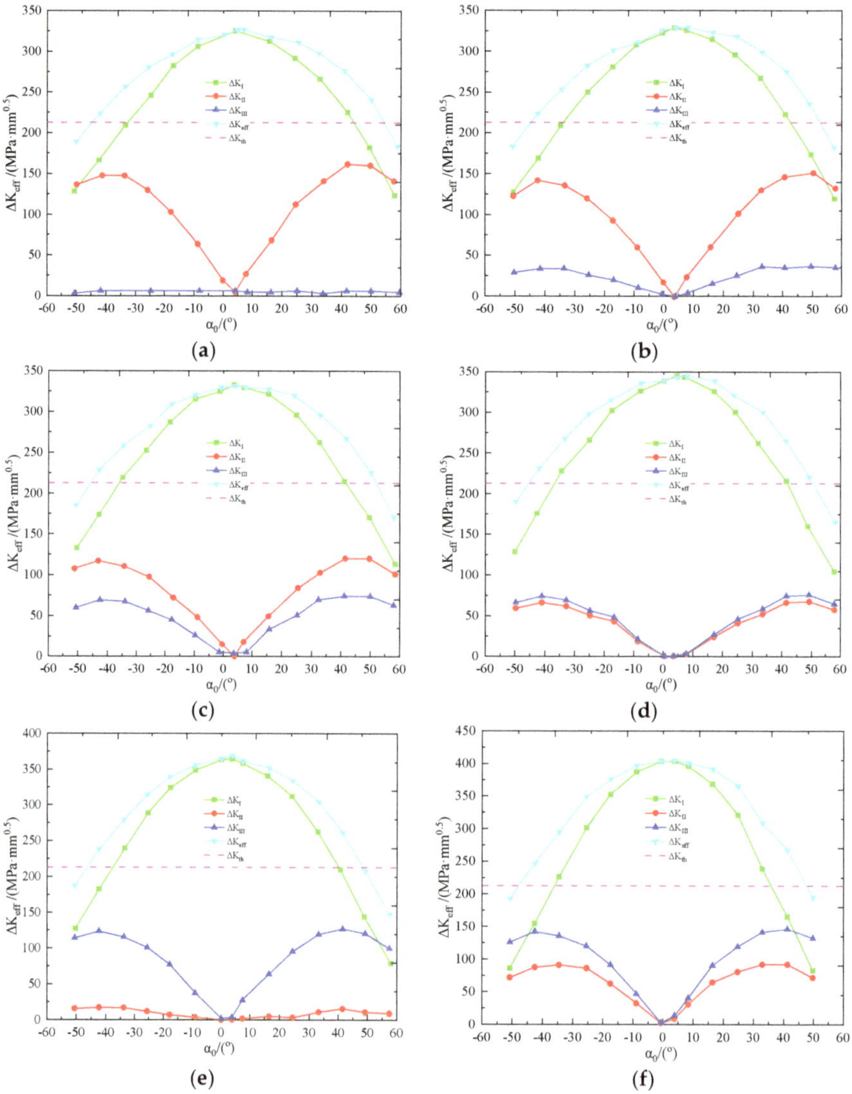

Figure 17. The stress intensity factor amplitude of the initial propagation state of initial cracks. (**a**) Crack front point A, (**b**) Crack front point B, (**c**) Crack front point C, (**d**) Crack front point D, (**e**) Crack front point E, (**f**) Crack front point F.

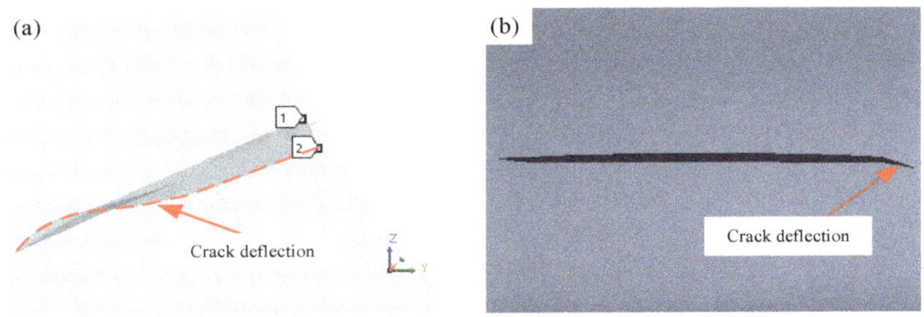

Figure 18. Crack shift during propagation. (**a**) Deflection under KII dominant action; (**b**) Deflection under KIII dominant action.

5.4. Maximum Effective Angle Interval for the Initial Crack Propagation

Previous studies showed that an initial crack would only propagate under an effective angle interval. This may be affected by the geometric parameters, the stress state of the crack, and the steel type. The influence of crack shape ratio on the effective angle interval of crack propagation of flange shaft will be analyzed as follows.

5.4.1. Shape Ratio

To reflect the most unfavorable condition, this paper used 0.5 mm as the maximum initial crack depth in engineering [1,30]. According to previous studies, ΔK_{I} and ΔK_{th} were used to judge the propagation of the initial crack at the junction between the shaft and flange. The effective angle interval for the propagation of C1–C5 initial cracks and its relationship with the shape ratio are shown in Figures 19 and 20 and Table 7, respectively:

(1) $|\Delta\alpha_{0c}| = |\alpha_{0\max} - \alpha_{0\min}|$ was used to define the effective angle interval for the propagation of initial cracks at the junction between the shaft and flange. $\alpha_{0\max}$ and $\alpha_{0\min}$ were the upper and lower limits of the interval's initial angles. Only the initial cracks with an angle within the interval can propagate. The initial cracks with an angle outside the interval did not propagate in the elastic stage. The interval was determined by the size relation between the ΔK_{I} of each point at the crack front and the base metal ΔK_{th}. The α_{0p} of the initial crack did not change with the shape ratio and $|\Delta\alpha_{0c}|$ would decrease as ω grown under the same stress ratio.

(2) The long flat cracks ($\omega < 0.5$) did not propagate in the direction of the surface length at a certain initial angle. They only propagated locally in the depth direction. The crack propagation towards the surface was smaller than that in the depth direction. The narrow and deep cracks ($\omega > 0.5$) experienced the opposite propagation pattern.

Table 7. The effective angle interval for the propagation of initial cracks with different shape ratios.

Numerical Test Number	a_0 (mm)	α_{0p} (°)	ω	$|\Delta\alpha_{0c}|$ (°)
C1	0.5	+4	0.1	108.9
C2	0.5	+4	0.25	100.1
C3	0.5	+4	0.5	79.3
C4	0.5	+4	0.75	72.9
C5	0.5	+4	1.0	68.8

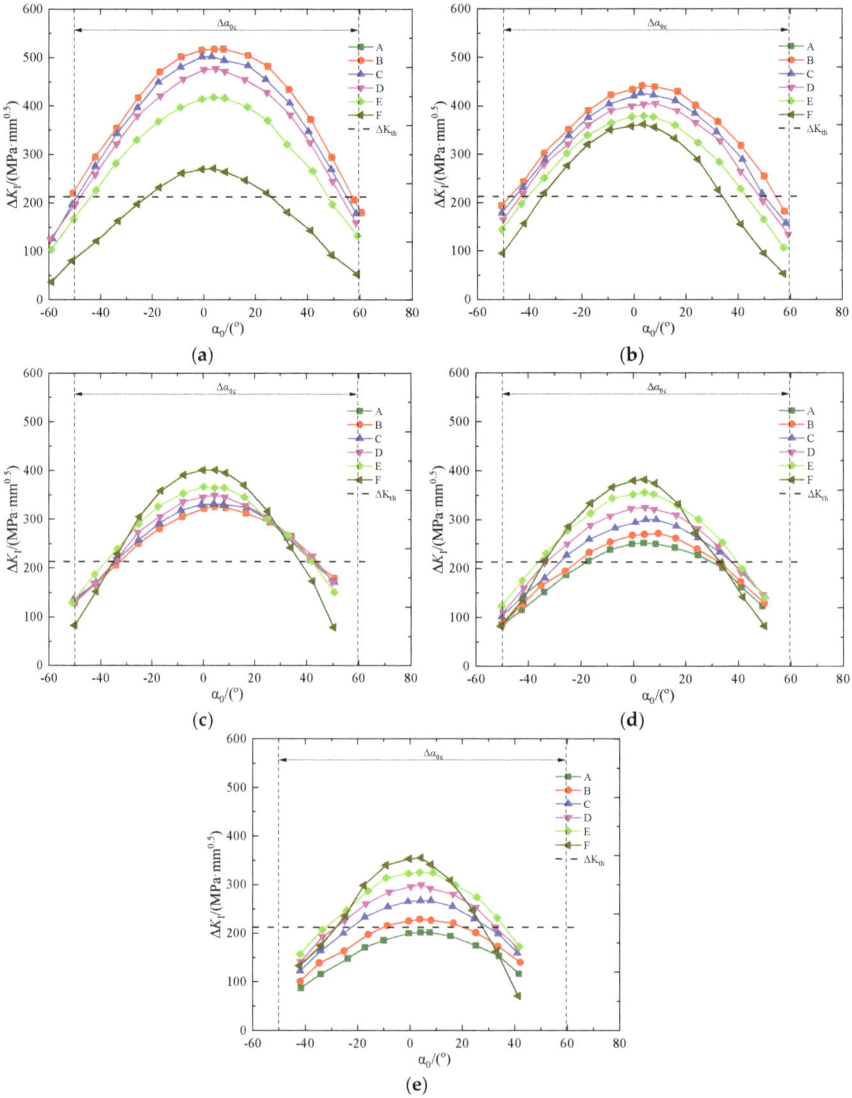

Figure 19. The effective angle interval for the propagation of initial cracks. (**a**) The shape ratio $\omega = 0.1$, (**b**) The shape ratio $\omega = 0.25$, (**c**) The shape ratio $\omega = 0.5$, (**d**) The shape ratio $\omega = 0.75$, (**e**) The shape ratio $\omega = 1.0$.

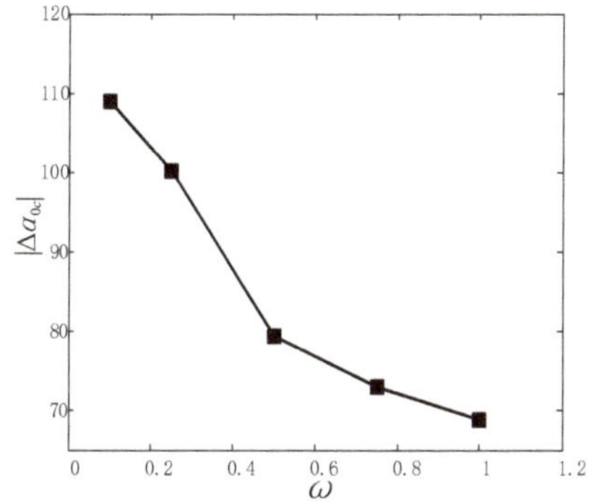

Figure 20. The relationship between the effective angle interval for the propagation of initial cracks and the shape ratio.

5.4.2. Mathematical Model of Effective Angle Interval

Changes in the amplitude of stress intensity factor at points A and E with the shape ratio were given in Figure 21. When $\omega < 0.5$, ΔK_{I} of point A on the initial cracks with different angles, almost changed linearly with ω; when $\omega > 0.5$, ΔK_{I} of point E on the initial cracks with different angles changed linearly with ω. It was assumed that slopes k_{A} and k_{E} were constants, which were less than 0 when the load condition was determined, as can be shown in Figure 21. Therefore, $|\Delta\alpha_{0c}|$ and $\omega = 0.5$ showed a piecewise linear and negative correlation. The general form of fitting formula was obtained by Formula (16):

$$|\Delta\alpha_{0c}| = \begin{cases} k_1\omega + b_1, & 0 < \omega < 0.5 \\ k_2\omega + b_2, & \omega > 0.5 \end{cases} \tag{16}$$

where: k_1, b_1, k_2, and b_2 were the constants related to the initial crack ΔK_{I} and the crack propagation threshold ΔK_{th}. Figure 22 shows the fitting results of the effective angle interval for the propagation of the initial crack of 45# steel flange shaft. Under the most unfavorable stress condition, $k_1 = -74.94$, $b_1 = 117.3$, $k_2 = -21$, $b_2 = 89.42$.

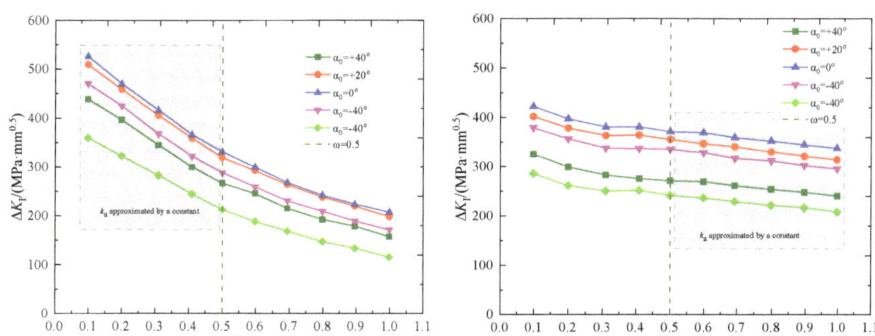

(a) (b)

Figure 21. Changes in the amplitude of stress intensity factor at points A and E with the shape ratio. (**a**) Location of crack front: point A, (**b**) Location of crack front: point E.

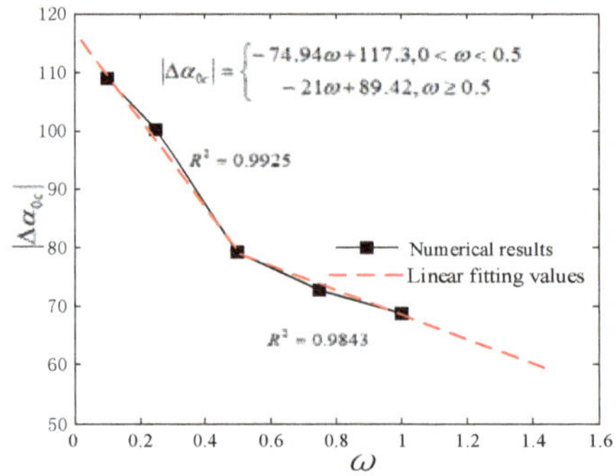

Figure 22. The fitting results of the effective angle interval for the propagation of the initial crack.

Formula (16) was obtained based on the maximum stress amplitude in the elastic stage and the maximum initial crack depth, which was generally accepted in engineering. Under this condition, the results reflected the maximum value of $|\Delta\alpha_{0c}|$. $|\Delta\alpha_{0c}|$ for different materials and stress levels can be calculated in the same way. The fitting calculation formula helps researchers comprehensively consider the initial angle and reasonably set it in the crack propagation calculation. It will deepen researchers' understanding of the propagation characteristics of fatigue crack and provide a reasonable reference for the duration of the crack detection cycle. With this knowledge, engineers can formulate related measures for maintenance and reinforcement.

6. Conclusions

(1) The initial crack appeared at the junction between the shaft and flange and propagated in radial and axial directions; the instability propagation of the crack caused an abrupt fracture of the cross-section, resulting in connection failure; the angle of the initial crack was uncertain.

(2) The crack was the mixed I-II-III mode dominated by mode I. Cracks with larger initial angles would have stronger mode II-III characteristics; KII and KIII affected the propagation angle of the crack in radial and axial directions as well as the direction of the structure surface.

(3) The propagation of the deepest point A of the crack was at the junction between shaft and flange. It can be divided into three stages: rapid growth (stage 1), steady decline (stage 2, buffer stage), and instability (stage 3). First, the initial crack angle only had a significant effect on the propagation rate of stage 1. Second, the initial crack angle affected the distribution of the fatigue life upon the structure during the propagation stage. The crack with a larger initial crack angle had a smaller proportion of buffer stage in the total fatigue life. After stage 1, a larger crack initial angle accelerated the structure to reach the instability propagation, which led to an abrupt fracture of the cross-section. This was more unfavorable for deciding how regular the crack detection should be carried out or pushing forward maintenance and reinforcement.

(4) It was the size relation between ΔK_I and ΔK_{th} that determined the crack propagation at the junction between the shaft and the flange. The effective stress intensity factor can partly reflect the law of crack propagation, but cannot serve as a criterion for crack propagation. Moreover, crack propagation can be determined by the effective angle interval, which was negatively correlated with the shape ratio.

(5) The initial crack at the junction between the shaft and the flange only propagated at a certain initial angle. There was a piecewise linear negative correlation between the effective angle interval and the shape ratio. The fitting calculation formula can provide the basis for the crack propagation research and can also be used to judge the propagation of welding cracks or other similar surface cracks. This paper has studied the 45# steel flange and shaft. It obtained the maximum value of effective

angle interval for the propagation of the initial crack. The formulas for other types of steels and stress levels and the evaluation method based on reliability need to be further explored.

Author Contributions: Z.X.: Investigation, Methodology, Software, Writing—original draft. Y.C.: Data curation, Validation. B.L.: Software, Validation. K.L.: Formal analysis, Writing—review and editing, Funding acquisition. F.S.: Writing—review and editing, Drawing graphs. P.C.: Data curation, Funding acquisition, Project administration, Writing—review and editing. All authors have read and agreed to the published version of the manuscript.

Funding: This research was funded by the National Natural Science Foundation of China (grant number 51769028), and the Natural Science Foundation of Shanxi Province (grant No. 2018JQ5019).

Institutional Review Board Statement: Not applicable.

Informed Consent Statement: Not applicable.

Data Availability Statement: The data used to support the findings of this study are available from the corresponding author and upon request.

Conflicts of Interest: The authors declare no conflict of interest.

References

1. Bangchun, W. *Mechanical Design Manual*; China Machine Press: Beijing, China, 2015.
2. Li, Y.T.; Duan, H.Y.; Wang, R.F. Effect of Notched Parameters on Low Cycle Fatigue Life of Shaft with Annular Notch under Cantilever Bending. *Key Eng. Mater.* **2011**, *462*, 136–141. [CrossRef]
3. Asi, O. Fatigue failure of a rear axle shaft of an automobile. *Eng. Fail. Anal.* **2006**, *13*, 1293–1302. [CrossRef]
4. Fonseca, L.G.A.; Cantisano, A.; Faria, A.R. Numerical modeling of deep rolling influence over crankshaft bending and correlation with fatigue behavior. *Fatigue Fract. Eng. Mater. Struct.* **2020**, *43*, 672–683. [CrossRef]
5. Gardin, C.; Courtin, S.; Bézine, G.; Bertheau, D.; Hamouda, H.B.H. Numerical simulation of fatigue crack propagation in compressive residual stress fields of notched round bars. *Fatigue Fract. Eng. Mater. Struct.* **2007**, *30*, 231–242. [CrossRef]
6. Jamil, M.; Khan, A.M.; Hegab, H.; Sarfraz, S.; Sharma, N.; Mia, M.; Gupta, M.K.; Zhao, G.; Moustabchir, H.; Pruncu, C.I. Internal Cracks and Non-Metallic Inclusions as Root Causes of Casting Failure in Sugar Mill Roller Shafts. *Materials* **2019**, *12*, 2474. [CrossRef]
7. Ktari, A.; Haddar, N.; Rezai-Aria, F.; Ayedi, H.F. On the assessment of train crankshafts fatigue life based on LCF tests and 2D-FE evaluation of J-integral. *Eng. Fail. Anal.* **2016**, *66*, 354–364. [CrossRef]
8. Atanasovska, I.; Jelić, M.; Mitrović, R.; Momčilović, D. *The Influence of Corrosion on Stress Concentration Factor at Shaft to Flange Radius, Power Transmissions*; Springer: Berlin/Heidelberg, Germany, 2013; pp. 657–666.
9. Li, J.; Zhang, P.; Lu, L.; Lv, F.; Miao, X.-T.; Chang, L.; Zhou, B.-B.; He, X.-H.; Zhou, C.-Y. Effect of pre-strain on fatigue crack growth behavior for commercial pure titanium at ambient temperature. *Int. J. Fatigue* **2018**, *117*, 27–38. [CrossRef]
10. Sola, J.F.; Alinejad, F.; Rahimidehgolan, F.; Niazmand, A. Fatigue life assessment of crankshaft with increased horsepower. *Int. J. Struct. Integr.* **2019**, *10*, 13–24. [CrossRef]
11. Pandey, R. Failure of diesel-engine crankshafts. *Eng. Fail. Anal.* **2003**, *10*, 165–175. [CrossRef]
12. Lei, X. Dynamic Modeling of Vibration and Analysis on Crack Fault in I.C.E. Crankshafts. Ph.D. Thesis, Dalian University of Technology, Dalian, China, 2003.
13. Zhou, X. Theoretical and Experimental Study on Fatigue Behavior and Reliability of Engine Crankshaft. Ph.D. Thesis, Zhejiang University, Hangzhou, China, 2006.
14. Yang, F.-P.; Chen, T.; Lu, Y.-C. The Effects of Carburization on the Fatigue Crack Growth Behaviors of Local Surface Cracks in Cylindrical Bars. *J. Mater. Eng. Perform.* **2019**, *28*, 3423–3429. [CrossRef]
15. Ye, Z.-Y.; Liu, D.-X.; Zhang, X.-H.; Zhang, X.-M.; Lei, M.-X.; Yang, Z. Corrosion Fatigue Behavior of 7A85 Aluminum Alloy Thick Plate in NaCl Solution. *Acta Met. Sin. (Engl. Lett.)* **2015**, *28*, 1047–1054. [CrossRef]
16. Fonte, M.; Anes, V.; Duarte, P.; Reis, L.; Freitas, M. Crankshaft failure analysis of a boxer diesel motor. *Eng. Fail. Anal.* **2015**, *56*, 109–115. [CrossRef]
17. Fonte, M.; Freitas, M. Semi-elliptical fatigue crack growth under rotating or reversed bending combined with steady torsion. *Fatigue Fract. Eng. Mater. Struct.* **1997**, *20*, 895–906. [CrossRef]
18. Fonte, M.; Freitas, M.; Reis, L. Failure analysis of a damaged diesel motor crankshaft. *Eng. Fail. Anal.* **2019**, *102*, 1–6. [CrossRef]
19. Yang, F.-P.; Kuang, Z.-B. Fatigue crack growth for a surface crack in a round bar under multi-axial loading condition. *Fatigue Fract. Eng. Mater. Struct.* **2005**, *28*, 963–970. [CrossRef]
20. Zangeneh, S.; Ketabchi, M.; Kalaki, A. Fracture failure analysis of AISI 304L stainless steel shaft. *Eng. Fail. Anal.* **2014**, *36*, 155–165. [CrossRef]

21. Azevedo, T.F.; Sampaio, W.R.V.; Câmara, E.C.B.; Lima, G.D.; dos Santos Silva, W.F.; da Silva Ramos, S. Failure analysis of a sugarcane loader rear shaft. *Eng. Fail. Anal.* **2019**, *109*, 104326. [CrossRef]
22. Aliakbari, K.; Nejad, R.M.; Mamaghani, T.A.; Pouryamout, P.; Asiabaraki, H.R. Failure analysis of ductile iron crankshaft in compact pickup truck diesel engine. *Structures* **2021**, *36*, 482–492. [CrossRef]
23. Khameneh, M.J.; Azadi, M. Evaluation of high-cycle bending fatigue and fracture behaviors in EN-GJS700-2 ductile cast iron of crankshafts. *Eng. Fail. Anal.* **2018**, *85*, 189–200. [CrossRef]
24. Alvarez, P.; Vázquez, L.; Ruiz, N.; Rodríguez, P.; Magaña, A.; Niklas, A.; Santos, F. Comparison of Hot Cracking Susceptibility of TIG and Laser Beam Welded Alloy 718 by Varestraint Testing. *Metals* **2019**, *9*, 2075–4701. [CrossRef]
25. Coro, A.; Macareno, L.M.; Aguirrebeitia, J.; De Lacalle, L.N.L. A Methodology to Evaluate the Reliability Impact of the Replacement of Welded Components by Additive Manufacturing Spare Parts. *Metals* **2019**, *9*, 932. [CrossRef]
26. Rodriguez, A.; De Lacalle, L.N.L.; Pereira, O.; Fernandez, A.; Ayesta, I. Isotropic finishing of austempered iron casting cylindrical parts by roller burnishing. *Int. J. Adv. Manuf. Technol.* **2020**, *110*, 753–761. [CrossRef] [PubMed]
27. Liu, Y.; Lian, Z.; Xia, C.; Qian, L.; Liu, S. Fracture failure analysis and research on drive shaft of positive displacement motor. *Eng. Fail. Anal.* **2019**, *106*, 104145. [CrossRef]
28. Erdogan, F.; Sih, G.C. On the crack extension in plates under plane loading and transverse shear. *J. Basic Eng.* **1963**, *85*, 519–525. [CrossRef]
29. Yangfan, Q.; Te, C.; Yiming, C.; Fengpeng, Y. Experimental Study on the I-II Mixed Mode Fatigue Crack Growth in Polyethylene. *Chin. Q. Mech.* **2021**, *42*, 87–97.
30. British Standards Institution. *Guide on Methods for Assessing the Acceptability of flaws in Metallic Structures*; British Standards Institution: London, UK, 2013.
31. Paris, P.; Erdogan, F. A Critical Analysis of Crack Propagation Laws. *J. Basic Eng.* **1963**, *85*, 528–533. [CrossRef]
32. Ma, S.; Hu, H. The mixed-mode propagation of fatigue crack in CTS specimen. *Chin. J. Theor. Appl. Mech.* **2006**, *38*, 698–704.
33. Lianqing, W.; Jin, K.; Hongying, W. Investigation on mixed mode I-II fatigue crack of 7050 aluminum alloy. *China Meas. Tes* **2021**, *47*, 139–146.
34. Richard, H.A.; Benitz, K. A loading device for the creation of mixed mode in fracture mechanics. *Int. J. Fract.* **1983**, *22*, R55–R58. [CrossRef]
35. Liu, C.; Wu, D.; Li, Y.; Du, Y. Large-scale pavement roughness measurements with vehicle crowdsourced data using semi-supervised learning. *Transp. Res. Part C Emerg. Technol.* **2021**, *125*, 103048. [CrossRef]
36. Alcázar, J.; Abate, G.; Antunez, N.; Simoncelli, A.; Egea, A.J.S.; Krahmer, D.M.; de Lacalle, N.L. Reduction of Die Wear and Structural Defects of Railway Screw Spike Heads Estimated by FEM. *Metals* **2021**, *11*, 1834. [CrossRef]
37. Alshoaibi, A.M.; Fageehi, Y.A. 3D modelling of fatigue crack growth and life predictions using ANSYS. *Ain Shams Eng. J.* **2021**, *13*, 101636. [CrossRef]
38. Establishment, C.A. *Stress Intensity Factor Handbook (Revised Edition)*; Science Press: Beijing, China, 1993.
39. Xinhua, Y.; Chuanyao, C. *Fatigue and Fracture*; Huazhong University of Science and Technology Press: Wuhan, China, 2018.

coatings

Article

Evolution Law and Mechanism of Freeze–Thaw Damage of Cement-Stabilized Weathered Sand

Xianghui Kong [1,2], Shuai Cui [1], Gaoqiang Wang [1], Wenjun Hu [1,*], Yunpeng Liang [1] and Zhibin Zhang [3]

1 School of Transportation Engineering, Shandong Jianzhu University, Jinan 250101, China; kongxh@sdjzu.edu.cn (X.K.); dacuishuai3@163.com (S.C.); wanggq97@163.com (G.W.); liangyunpeng99@163.com (Y.L.)
2 Jinan Rail Transit Group Co., Ltd., Jinan 250100, China
3 School of Municipal & Environmental Engineering, Shandong Jianzhu University, Jinan 250101, China; zhangzhibinzzb@sdjzu.edu.cn
* Correspondence: huwenjun@sdjzu.edu.cn

Abstract: To explore the damage evolution law of cement-stabilized weathered sand under the action of freeze–thaw cycles, relevant experimental research was carried out on stabilized weathered sand with a cement content of 3%, including unconfined compressive strength (UCS), scanning electron microscope (SEM), and mercury intrusion porosimetry (MIP) testing. Using the surface appearance, mass, and UCS of the specimen, combined with the changes of microscopic characteristics, the freeze–thaw damage law and microscopic degradation mechanism of cement-stabilized weathered sand were analyzed. The test results showed that with the increase of the number of freeze–thaw cycles, the surface appearance of the specimen continued to deteriorate, the unconfined compressive strength gradually decreased, and the mass of the specimen first increased and then decreased. After nine freeze–thaw cycles, the specimen was seriously damaged, and the strength loss was as high as 55%. From a microscopic point of view, the freeze–thaw cycles caused the pore water inside the specimen to continuously change between ice crystals and liquid. Frost heave and shrinkage weakened the cement bond between the weathered sand particles and made the pores in the specimen develop and expand continuously. With the decrease of the number of micropores and the increase of the number of medium and large pores, the particle skeleton changed from a dense structure to a porous structure, which eventually led to the deterioration of the macroscopic properties of the cement-stabilized weathered sand.

Keywords: weathered sand; freeze–thaw cycles; damage law; microscopic characteristics; deterioration mechanism

Citation: Kong, X.; Cui, S.; Wang, G.; Hu, W.; Liang, Y.; Zhang, Z. Evolution Law and Mechanism of Freeze–Thaw Damage of Cement-Stabilized Weathered Sand. *Coatings* **2022**, *12*, 272. https://doi.org/10.3390/coatings12020272

Academic Editor: Andrea Nobili

Received: 3 January 2022
Accepted: 16 February 2022
Published: 18 February 2022

1. Introduction

Weathered sand is widely distributed in mountainous and hilly areas of China. It is the product of rock disintegration under natural conditions, and has the general nature of talus materials, such as loose, large pores; easy compression deformation; and low strength. The physical and mechanical properties of weathered sand are poor, and it is easily destroyed under the action of external forces, so it cannot be directly used in engineering construction in most cases [1]. There are only a few studies on the engineering application of weathered sand, and most of them focus on its use as a dam building material [2], road building material [3–5], and for expansive soil improvement [6–8]. With the rapid development of highway construction in China, high-quality filler is increasingly scarce. Making full use of local resources is of great significance for saving engineering investments and protecting the environment. The stability of weathered sand is worse than that of general filler, so it cannot be directly used as a road material and must be stabilized [8]. The chemical stabilization method is mostly used in engineering, and the commonly used stabilizers are cement, lime, and lime-fly ash. An [3] used cement and lime-fly ash respectively to stabilize weathered sand and found that both had good mechanical strength, but the freeze stability and crack resistance of lime-fly ash-stabilized weathered sand were slightly worse. Guo et al. [4] studied the road performance of cement stabilized weathered sand. They found that the strength of cement stabilized weathered sand increased greatly, the water stability and freeze stability were good, and the temperature shrinkage coefficient and dry shrinkage coefficient were small. Wang et al. [5] carried out physical characteristics and field rolling test research on the weathered sand in the Three Gorges Dam area and believed that the compaction of the roadbed filled with weathered sand could meet the engineering requirements. Yang et al. [6] used weathered sand to improve expansive soil and found that weathered sand can better inhibit the expansion of expansive soil. Huang et al. [8] conducted experimental research on the shear strength of expansive soil improved with lime-weathered sand and proposed that the shear strength of the improved expansive soil can meet the soil standard for subgrade filling. Yang et al. [9,10] conducted systematic studies on the weathered sand in the Three Gorges Reservoir area and found that the weathered sand has distinct edges and corners, forming a skeleton by interlocking with each other, and has a large internal friction force. By adding stabilizers such as lime and cement, the cohesion between particles could be increased to improve its overall strength. For lime-ash-stabilized weathered sand, lime content is the primary factor affecting compressive strength, while the lime-ash ratio is the secondary factor [9]. For cement-stabilized weathered sand, considering economic and mechanical properties, the optimal cement content should be controlled at 7% [10].

Frost heaving and melt settling usually occur in inorganic binder-stabilized materials, leading to various road diseases such as mass loss [11], surface appearance defects, and strength attenuation [12,13]. Therefore, it is necessary to study the stability and durability of inorganic binder-stabilized materials under freeze–thaw cycles. Jumassultan et al. [14] studied the mechanical properties of cement-stabilized sand under freeze–thaw conditions and found that the strength and durability of the specimens decreased with the increase of the number of freeze–thaw cycles. Tian et al. [15] analyzed the strength loss law of cement-stabilized crushed rock under freeze–thaw action and suggested that a higher cement content and curing temperature could effectively reduce the strength loss. Hao et al. [16] carried out freeze–thaw cycle tests on red mud-steel slag-modified cement soil and found that the early freeze–thaw action had a greater impact on the strength. With the increase of the number of freeze–thaw cycles, the pores between the modified cement soil particles developed and expanded, the structure was loose, and the surface appearance showed different degrees of damage. Gao et al. [17] studied the static and dynamic characteristics of basalt fiber-reinforced cement soil after freeze–thaw cycles and found that freeze–thaw action has a significant impact on the strength loss, which was serious in cycles 1–3, and the addition of basalt fiber could improve this situation. Li et al. [18] studied the tensile strength of fiber-reinforced soil under freezing-thawing conditions and found that the decline of tensile strength was mainly concentrated in the first five times of freezing-thawing, and the strength remained relatively stable during the sixth to ninth freeze–thaw cycles. Li et al. [19] found in the freeze–thaw cycle that when loads were applied, the freeze–thaw cycle would destroy the structure of expansive soil and form new cracks, resulting in loose pore structure. Lake et al. [20] studied the mechanism of freeze–thaw damage of lightly cement soil before and after freeze–thaw cycle through optical microscope and mercury injection method, and found that the increase of sample hydraulic conductivity, the formation of microcracks and cracks and the damage of matrix after freeze–thaw were obvious, with high permeability coefficient. Eskisar et al. [21] studied the strength characteristics of fat clay and lean clay treated with Portland cement under freeze–thaw cycle. Through unconfined compression test and ultrasonic pulse velocity test, it was found that cement can improve the freeze–thaw durability of clay. Zhang et al. [22] further understood the phase change process of soil water by measuring pore water pressure of silty clay during freezing and thawing.

Macroscopic mechanical properties are the external manifestation of the microstructure, so it is of great significance to study the evolution of microscopic characteristics in a freeze–thaw environment. The microstructure of soil is mainly the relationship between the size, shape, and quantity of pores and particles, among which the change of pores is an important embodiment of structural deformation [23]. Scanning electron microscopy (SEM) is an effective way to study the microscopic pores and particles of soil [24]. The development morphology of pores can be qualitatively analyzed through SEM images, and the development status of porosity abundance can be quantitatively analyzed by image processing software [25]. Liu et al. [26] analyzed the abundance and orientation of pores and structural units of cement soil, and found that it has no obvious orientation. The above methods can also be used to analyze the microscopic degradation mechanism of macroscopic damage under freeze–thaw cycles. Hou et al. [27] analyzed the microstructure changes of cement soil under freeze–thaw action and found that the accumulation of repeated freezing and thawing damage caused the expansion of internal cracks, which eventually led to the overall destruction of the cement soil. Zhang et al. [28] analyzed the deterioration mechanism of cement soil under freeze–thaw cycles from the perspective of energy dissipation. When materials are in a freeze–thaw environment, the water in the pores will undergo a liquid–solid–liquid phase change, and the repeated action of frost heaving force is the main cause of mass loss and strength failure, which is the consensus of many studies [29–31]. Gao et al. [29] studied the changes of the apparent morphology and other properties of mineralized materials in four different freeze–thaw environments, and proposed that the appearance and disappearance of test samples accompanied by frost heave force resulted in the peeling phenomenon of materials in freeze–thaw environments. Ye et al. [30] conducted an experimental study on freeze–thaw cycle of loess ancient soil with different moisture content based on NUCLEAR magnetic resonance, and concluded that the freeze–thaw cycle changed the internal structure of ancient soil, and pore water and fissure water in soil changed the pore structure of soil under the action of frost swelling, resulting in damage and deterioration of soil. Ding et al. [31] studied the effects of fiber, cement, and freeze–thaw cycle on cement treated clay and fiber cement stabilized clay, and they found that the unconfined compressive strength decreased sharply after the first few freeze–thaw cycles and changed little after the last five freeze–thaw cycles. They put forward the empirical model of predicted UCS, and the predicted model is in good agreement with the experimental results.

At present, the utilization rate of weathered sand is low. If the weathered sand is used reasonably as subgrade filler, it can not only protect the ecological environment, but also reduce the project cost and bring good economic benefits. In addition, research on stabilized weathered sand mainly focuses on macroscopic mechanical properties, while the damage evolution of stabilized weathered sand under freeze–thaw action, especially on the microscopic properties, is rarely reported. This paper takes the weathered sand along the Xintai-Taierzhuang expressway project in Shandong Province as the research object, adopts cement to stabilize it, and focuses on the macroscopic deterioration characteristics of the stabilized weathered sand under freeze–thaw cycles. Moreover, by means of SEM, mercury intrusion porosimetry test (MIP), and image processing, the evolution law of the microstructure, and pore and particle characteristics of cement-stabilized weathered sand are analyzed. Based on this, the mechanism of freeze–thaw damage of cement-stabilized sand is discussed from the micro perspective.

2. Materials and Methods

2.1. Materials

The weathered sand used in this study was yellowish brown, the natural water content was 3.1%, the optimal water content was 8.03%, and the maximum dry density was 2.13 g/cm^3. Figure 1 shows the X-ray diffraction (XRD) results of the weathered sand after grinding. The main components of the weathered sand are quartz and albite, accounting for 96% of the total content. The analysis results of particle gradation are shown in Figure 2. The natural weathered sand was mainly composed of sand and gravel, in which the content of the gravel group accounted for 24.5%, and the content of the sand group accounted for

75.5%. The coefficient of nonuniformity *Cu* was 5.6 and the coefficient of curvature *Cc* is 1.1, indicating that the sand was well-graded. The basic physical indexes are shown in Table 1. Ordinary Portland cement 42.5 and tap water were used in the tests.

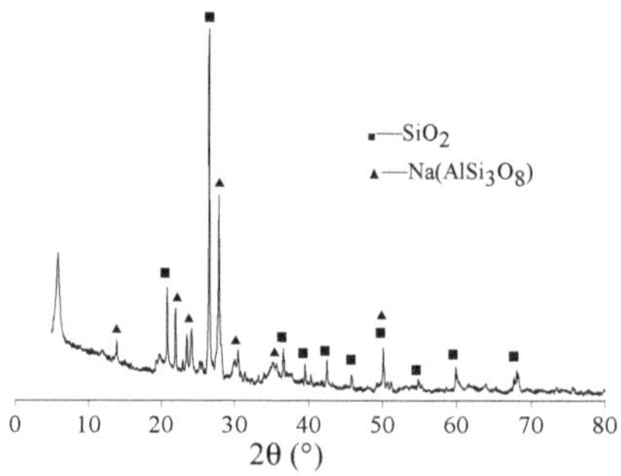

Figure 1. XRD test results of the natural weathered sand.

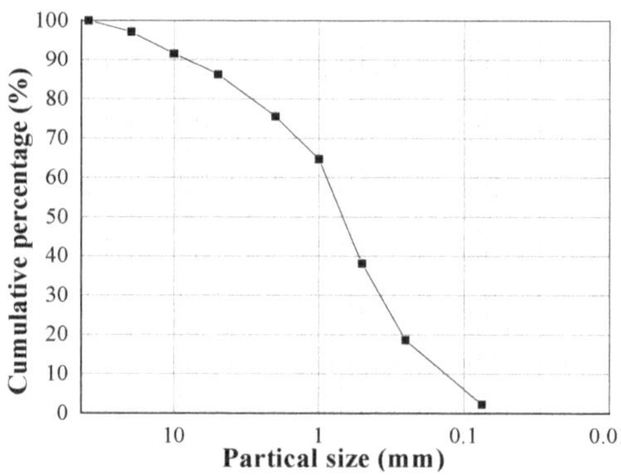

Figure 2. Grain gradation curve of natural weathered sand.

Table 1. Basic physical indexes of the natural weathered sand.

Natural Water Content (%)	Particle Size Distribution (%)			Maximum Dry Density (g/cm^3)	Optimal Water Content (%)	CBR (%)
	<0.5 mm	0.5–5 mm	>5 mm			
3.1	38.1	48.1	13.8	2.13	8.03	51

2.2. Testing Methods

Firstly, the optimum cement content in the stabilized weathered sand was determined. The cement content is the ratio of the mass of cement to dry weathered sand, which was 0, 1%, 2%, 3%, and 4%, respectively in the specimens. The compaction degree of specimens was 98%, and the curing ages were 7, 14, 28, 60, and 90 d. Table 2 shows the unconfined compressive strength (UCS) of the stabilized weathered sand with different cement content levels. Considering the actual engineering demands, the cement content of the stabilized weathered sand was 3% in this study. Secondly, freeze–thaw cycle tests were carried out to analyze the influence of freeze–thaw action on the surface appearance, mass changes, and strength characteristics of the stabilized weathered sand. Finally, based on the macroscopic study, the microstructural properties, and pore and particle characteristics of the cement-stabilized weathered sand subjected to a different number of freeze–thaw cycles were analyzed with the aid of SEM and MIP to explore the degradation mechanism of the cement-stabilized weathered sand in a freeze–thaw environment.

Table 2. Unconfined compressive strength test results.

No.	Cement Content (%)	UCS (MPa)				
		7 d	14 d	28 d	60 d	90 d
C-0	0	0.19	-	-	-	-
C-1	1	0.90	1.00	1.09	1.23	1.42
C-2	2	2.08	2.32	2.36	2.75	2.91
C-3	3	3.15	3.25	3.65	4.44	4.60
C-4	4	3.70	4.28	4.45	4.57	5.23

2.2.1. Sample Preparation

First, the dried weathered sand was mixed with water, and cement was added after being sealed overnight indoors. The water content was controlled to 8.5%, which is the optimal water content of a mixture with a cement content of 3%. Cylindrical specimens with a diameter of 50 mm and a height of 50 mm were made by static pressure and placed in a standard curing box for 28 days [32]. The height-diameter ratio of the sample is 1:1, close to that recommended in ASTM D1633-17, which is 1.15:1 [33]. On the last day of curing, the specimens were immersed in water.

2.2.2. Freeze–Thaw Cycle Test

Following the "Test Methods of Materials Stabilized with Inorganic Binders for Highway Engineering (JTG E51-2009)" [32], before the freeze–thaw cycle test, the surface of the specimen was wiped with a damp cloth until there was no obvious liquid water, and its initial mass was recorded. The specimen was first placed in a low-temperature box and frozen at −18 °C for 16 h, then put into water at 20 °C to thaw for 8 h, and this process was a freeze–thaw cycle. The maximum number of freeze–thaw cycles in this study was 12, and the process of freeze–thaw cycles is shown in Figure 3. During the test, there were three specimens for each freeze–thaw cycle, and they were evenly placed at intervals of 20 mm to facilitate air circulation.

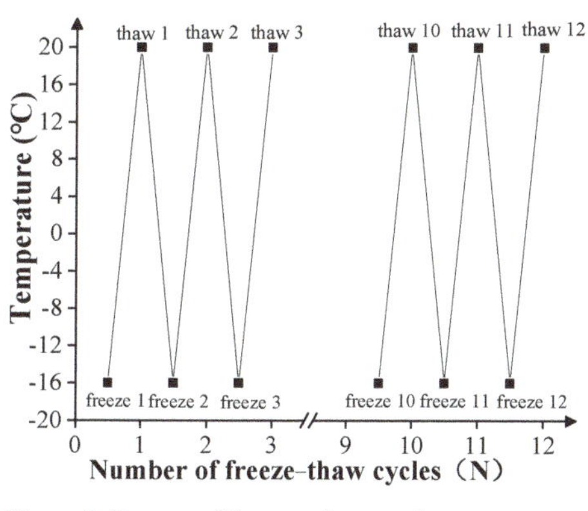

Figure 3. Process of freeze−thaw cycles.

2.2.3. UCS Test

The unconfined compressive strength (UCS) test was carried out with a pavement material strength meter, and the loading rate was maintained at 1 mm/min [32]. The loading rate recommended in ASTM D1633-17 is 1.3 mm/min, which is close to 1 mm/min [33]. The mean average value of the three groups of parallel tests was taken as the test result.

2.2.4. SEM Test

SEM can clearly characterize the internal structure of soil, making it an effective method for qualitatively analyzing the evolution process of the internal structure of soil.

After the freeze–thaw cycles test, the specimens were cut, and the samples with dimensions of less than 10 mm × 5 mm × 5 mm were selected for freeze-drying and gold spraying, and observed and photographed with a JSM-7610F scanning electron microscope.

In this study, the samples that underwent the SEM test and MIP test were all dried by a liquid nitrogen freeze vacuum sublimation drying method. The principle is to directly sublimate the amorphous ice to minimize the volume and microstructure changes caused by water loss. The selected samples were put into liquid nitrogen, and the water in it was frozen into non-expandable amorphous ice. Then the samples were put into a freeze dryer for vacuum treatment, and vacuumed for 8 h at −50 °C.

2.2.5. MIP Test

Mercury intrusion porosimetry test (MIP) is an effective way to quantitatively analyze the internal pore distribution of materials. The test used the AutoPore 9500 mercury intrusion instrument produced by Mike Instruments. A small piece (1 cm^3) was cut from the specimens, then freeze-dried and loaded into the dilatometer for low-pressure and high-pressure tests.

2.2.6. SEM Image Processing

In this stage, SEM images were preprocessed by MATLAB (Matlab 2016b) and PHOTOSHOP software (Photoshop CC 2017). In MATLAB, the disk-shaped structure Disk in the image processing function Imopen was used to estimate the uneven background to eliminate the influence brought by the inconsistency of image light and shade. The function Imadjust was used to adjust the gray level of the image to enhance the contrast of the image. The function Wiener2 was used to control the noise. After the microscopic image was preprocessed, image processing software Image-Pro Plus (IPP) was used to extract and analyze the data.

In the quantitative analysis of the microstructure, the parameters that characterize the microstructure were firstly determined. From the perspective of morphology and geometry, the selected parameters included abundance C and average abundance *Cp* of pores and particles [34].

Abundance C: the ratio of the short axis to the long axis of pores (or particles), which can be calculated as follows: $C = B/L$ where B is the short axis of pores (or particles) and L is long axis of pores (or particles).

Average abundance *Cp*: The average abundance of all pores (or particles). The abundance value is between 0 and 1, and the closer the value is to 0, the more elongated the shape of pores (or particles) is. On the contrary, the closer the value is to 1, the more the shape tends to be round.

3. Results and Discussion

3.1. Deterioration Characteristics of Stabilized Weathered Sand

3.1.1. Degradation of Surface Appearance

Figure 4 shows the appearance characteristics of cement-stabilized weathered sand under different freeze–thaw cycles. It can be seen from Figure 4b,c that the first three freeze–thaw cycles had almost no effect on the integrity of the specimen, and the specimen exhibits no obvious cracks or damage. Compared with when the number of freeze–thaw cycles is 1 ($N = 1$), the surface of the specimen at $N = 3$ is slightly rough, which indicates that under the effect of the early freeze–thaw action, the tiny particles on the surface of the specimen fell off, and this provided conditions for the intrusion of water into the interior. When $N = 6$, partial peeling occurred on the specimen surface (as shown in Figure 4d). When $N = 9$, the peeling area increased and extended to the interior, and large particles became exposed (as shown in Figure 4e). When $N = 12$, the peeling area and depth developed further, and the specimen degraded seriously and lost its integrity (as shown in Figure 4f).

(a) (b) (c) (d) (e) (f)

Figure 4. Effect of freeze–thaw cycles on specimen appearance. (**a**) $N = 0$, (**b**) $N = 1$, (**c**) $N = 3$, (**d**) $N = 6$, (**e**) $N = 9$, (**f**) $N = 12$.

3.1.2. Mass Change

In the process of freezing and thawing, the mass of the specimen changed because of water intrusion and particle shedding. Under repeated freeze–thaw action, with the continuous development of the pore structure inside the specimen, water migrated to the interior. The increased water was involved in the hydration of the cement and remained in the new crack, resulting in the increase of specimen mass. When the freeze–thaw action started to damage the structure of the specimen, the surface particles gradually peeled off, and the mass of the specimen itself was defective. If the defect amount was greater than the amount of water entering, the overall mass of the specimen decreased.

Taking the mass of the specimen before the freeze–thaw cycles test as the reference value, the mass change is defined as the difference between the mass of the specimen after the freeze–thaw cycles and the reference value. Figure 5 shows the effect of freeze–thaw cycles on the mass of the stabilized weathered sand specimens. It can be seen from Figure 5 that the mass change of specimens increased linearly during the first six freeze–thaw cycles. When $N = 7$, the mass of the specimen began to decline, but it was still greater than the reference value, indicating that when $N \leq 7$, the leading factor affecting specimen mass was the amount of water entering, and the number of particles shedding at this stage was small. When $N \geq 8$, specimen mass continued to decline and the mass change was negative. At this time, the hydration reaction and water migration were basically complete, and the macroscopic deterioration of the specimen was aggravated. The controlling factor of mass change was particles shedding. Wang et al. [35] attributed the reasons for the mass changes caused by freeze–thaw action to four factors: the amount of water migration, the amount of chemical reaction, the number of specimen defects, and the number of freeze–thaw cycles.

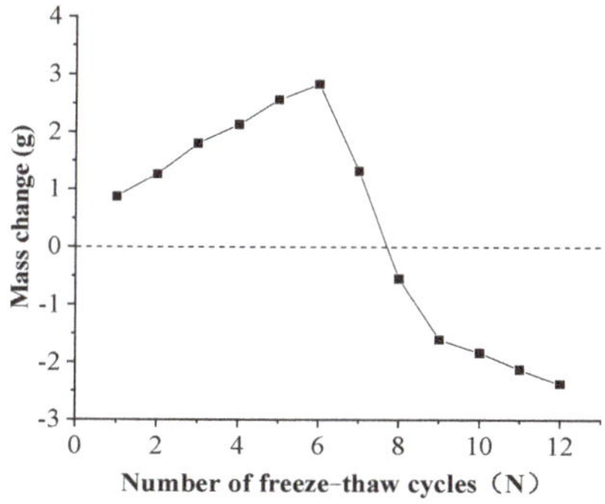

Figure 5. Effect of freeze–thaw cycles on the mass of specimens.

3.1.3. Loss of Strength

The effect of freeze–thaw action on the compressive strength of the stabilized weathered sand is shown in Figure 6. It can be seen from Figure 6 that as the number of freeze–thaw cycles increased, the unconfined compressive strength of the specimen gradually decreased. The strength of the specimen that did not undergo freeze–thaw action ($N = 0$)

was 3.65 MPa, when $N = 1, 3, 6, 9, 12$, the corresponding compressive strengths were 3.05, 2.74, 2.58, 1.64, and 1.61 MPa, and the strength loss rates were 16.44%, 24.93%, 29.32%, 55.07%, 55.89%, respectively. It can be seen that when $N = 9$, the compressive strength of the specimen was greatly reduced and most of the bearing capacity was lost, indicating that the structure of the stabilized weathered sand had been greatly damaged, which is consistent with the law of the surface appearance and mass change of the specimen. After $N = 9$ times, increasing the number of freeze–thaw cycles had little influence on the strength of the specimens. For example, the strength loss rate at $N = 12$ was only 0.82% higher than that at $N = 9$.

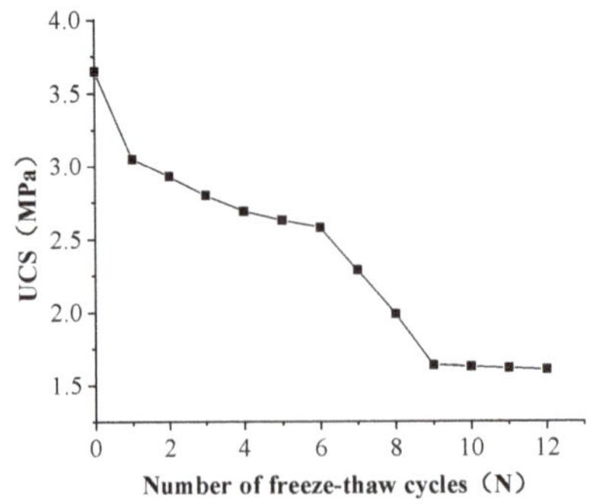

Figure 6. Effect of freeze–thaw cycles on compressive strength.

With the rapid development of machine learning (ML), there have been various studies using machine learning to build predictive models and optimize material responses. For example, artificial neural network (ANN) was used to establish the prediction model of soil–water characteristic curve (SWCC) to study the influence of biochar on soil water retention [36]. ANN was used to predict the fracture behavior of granular polymer composites under different loading rates [37]. The combination of gas sensor array (GSA) and artificial intelligence (AI) algorithm is applied in the fields of environmental detection, food production, and explosive detection [38], and optimizing battery design based on machine learning (ML) models [39]. In the following research, it will be an interesting inspiration to use machine learning (ML) to find the optimal response of weathered sand.

3.2. *Microscopic Characteristics of Stabilized Weathered Sand*

3.2.1. Evolution Characteristics of the Microstructure

Figures 7 and 8 are SEM images of stabilized weathered sand under different freeze–thaw cycles, magnified 200 times and 2000 times, respectively. It can be seen from Figure 7 that the specimen at $N = 0$ has fewer pores and has a compact and dense structure. As the number of freeze–thaw cycles increased, the fine particles wrapped around the coarse particles or filled between the skeletons decreased. When $N = 6$, the continuous 'gullies' and large pores made the skeleton interface of coarse particles clearer, and the bonding effect between particles weakened. According to the figures, When $N = 9$, 12 times, the fine particles are further lost, some particles are displaced, and the cracks are interconnected.

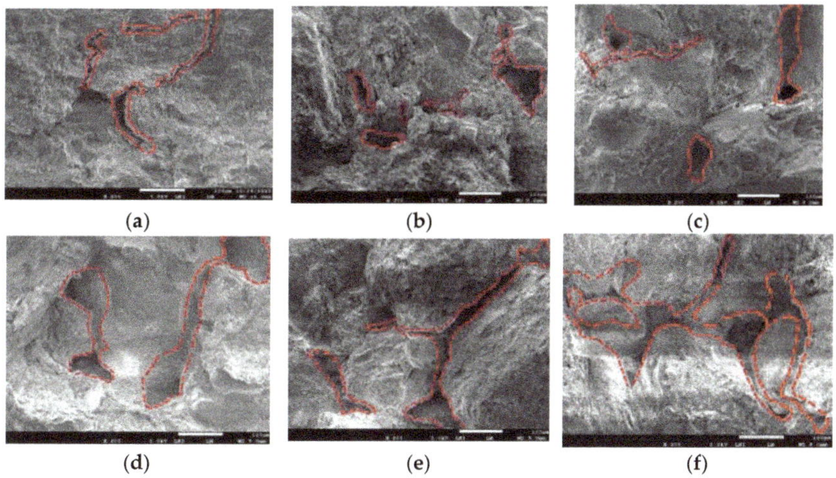

Figure 7. SEM images of specimen at different freeze–thaw cycles (×200 times). (**a**) N = 0, (**b**) N = 1, (**c**) N = 3, (**d**) N = 6, (**e**) N = 9, (**f**) N = 12.

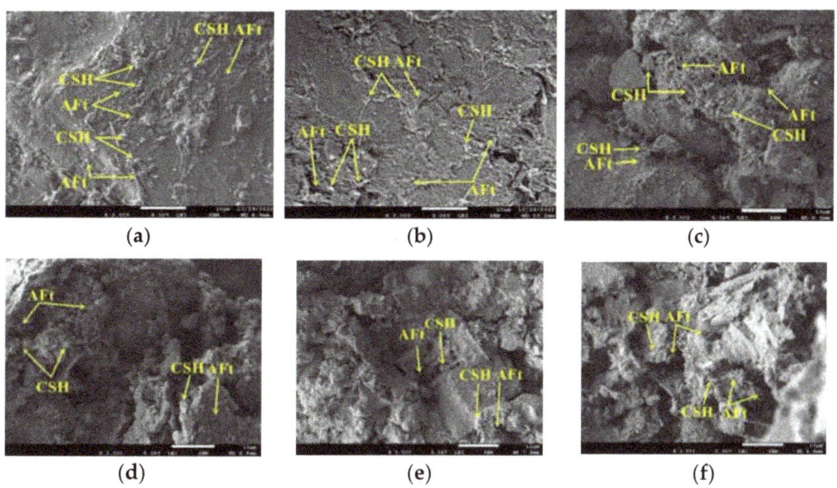

Figure 8. SEM images of specimens at different freeze–thaw cycles (×2000 times). (**a**) N = 0, (**b**) N = 1, (**c**) N = 3, (**d**) N = 6, (**e**) N = 9, (**f**) N = 12.

As can be seen from Figure 8, the specimen at N = 0 contains a large amount of floccule calcium silicate gel (CSH) and acicular ettringite (AFt) produced by cement hydration. CSH gel and AFt crystal interweaved with each other to fill gaps and the contact between particles is extremely tight. With the increase of the number of freeze–thaw cycles, cracks gradually increased, and hydration products were constantly lost. It is difficult for CSH and AFt to fill the pores, and the specimen changed from a stable compact structure to a skeleton structure with multiple cracks and multiple pores.

3.2.2. Evolution Characteristics of Pore Diameter

The results of the mercury intrusion porosimetry test (MIP) are shown in Figure 9. As can be seen from Figure 9, the cumulative pore volume increased with the increase of the number of freeze–thaw cycles. When N = 0, the cumulative mercury intake was 0.171 mL/g, and when N = 12, it reached 0.186 mL/g, increasing by 0.015 mL/g. The increase in cumulative pore volume indicates that under the action of freeze–thaw cycles, the pores inside the specimen gradually expanded and increased, making the structure loose, which is consistent with the SEM analysis results. The shape of the pore distribution curve under different freeze–thaw times is roughly the same. The curve is the steepest in the interval of 0.7–8.0 μm, and the cumulative pore volume changed greatly, indicating that most of the pores are within this interval.

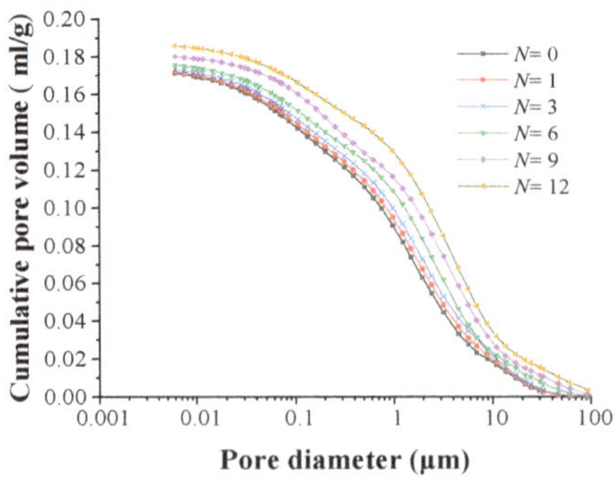

Figure 9. Curve of cumulative pore volume.

To facilitate further analysis, pores were divided into five categories according to the pore diameter d: micro pores ($d \leq 0.04$ μm), small pores (0.04 μm $< d \leq 0.4$ μm), medium pores (0.4 μm $< d \leq 4$ μm), large pores (4 μm $< d \leq 40$ μm), and super large pores ($d > 40$ μm). The distribution of various pore contents (the percentage of each pore volume in the total pore volume) is shown in Figure 10. As can be seen from Figure 10, most pores were medium pores. With the increase of freeze–thaw times, the pore content of $d \leq 4$ μm (including micro pores, small pores, and medium pores) decreased, while the pore content of $d > 4$ μm (large pores and super large pores) increased. When $N \leq 9$, the relationship of all kinds of pore content was medium pore > large pore > small pore > micro pore > super large pore. When $N = 12$, the relationship of pore content was medium pore > large pore > small pore > super large pore > micro pore. When $N = 12$, the micro, small, medium, large and super large pore contents were 4%, 16%, 38%, 35%, and 7%, respectively. Compared with $N = 0$, the pore volume changed by −3%, −6%, −7%, +10%, and +6%, respectively.

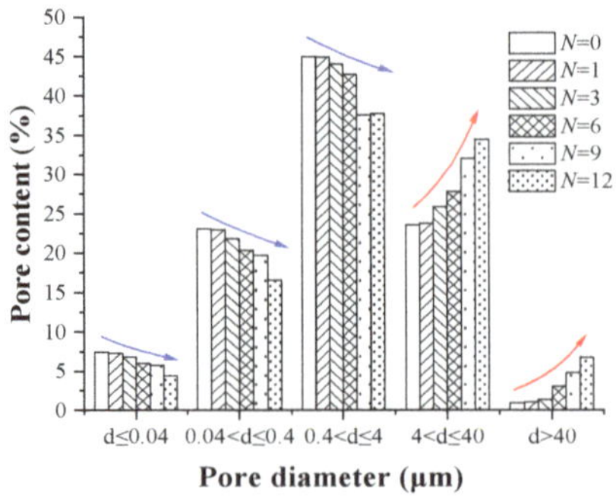

Figure 10. Effect of freeze–thaw cycles on pore diameter.

From the overall trend of change, the effect of freeze–thaw cycles on pore diameter takes $d = 4$ μm as the dividing line. With the increase of the number of freeze–thaw cycles, micro pores, and micro cracks continued to expand and extend, and then penetrate to form large pores. As a result, the pore content of $d \leq 4$ μm continued to decrease, while that of $d > 4$ μm continued to increase.

3.2.3. Evolution Characteristics of Pore Shape

Figure 11 shows the variation of pore abundance and average abundance under different freeze–thaw cycles. It can be seen from Figure 11a that the pore abundance is concentrated in the range of 0.2–0.4, 0.4–0.6, and 0.6–0.8, indicating that the pores were mainly oblate and oval under the freeze–thaw action. With the increase of the number of freeze–thaw cycles, the content of pore abundance in the range of 0–0.2 decreased, while that in the range of 0.4–0.6 increased, indicating that the narrow and long pores gradually changed into ovals during freeze–thaw cycles.

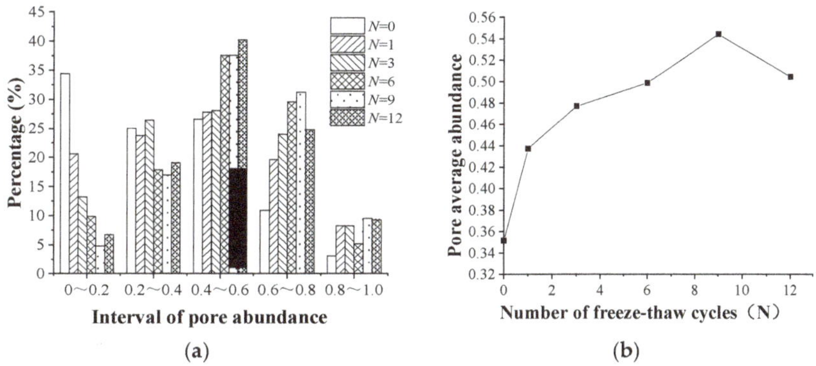

Figure 11. Pore abundance and average abundance under different freeze–thaw cycles. (**a**) Pore abundance. (**b**) Average abundance of pores.

As shown in Figure 11b, with the increase of the number of freeze–thaw cycles, the average abundance of pores first increased and then decreased. The reason is that during the freeze–thaw cycles, water in the pores freezes and expands at low temperature, leading to pores and cracks not developing in a single direction, but gradually connecting with each other. The pore volume increases and the shape of elongated pores changes to oval. After nine cycles, the structure of stabilized weathered sand was destroyed, the oval pores turned to elongated pores, and the average abundance of pores decreased.

3.2.4. Evolution Characteristics of Particle Shape

Figure 12 shows the variation rules of particle abundance and average particle abundance under different freeze–thaw cycles. It can be seen from Figure 12a that the particle abundance is concentrated in the range of 0.2–0.4, 0.4–0.6, and 0.6–0.8, and a small amount appear in the range of 0–0.2, indicating that there were few elongated particles and most of the particles were oblate and oval.

It can be seen from Figure 12b that the average abundance of sample particles first decreased and then increased during the freeze–thaw cycles. The reason is that the hydration products and aggregates attached to the particle surface, gradually falling off and dissolving under the freeze–thaw action, which makes the particle shape become irregular, and the average abundance shows a downward trend.

When the freeze–thaw cycles progressed to a certain extent ($N = 9$), the cements on the surface of particles had basically disappeared, the surface became smooth, and the particles began to develop into a flat shape, so the average abundance of particles increased.

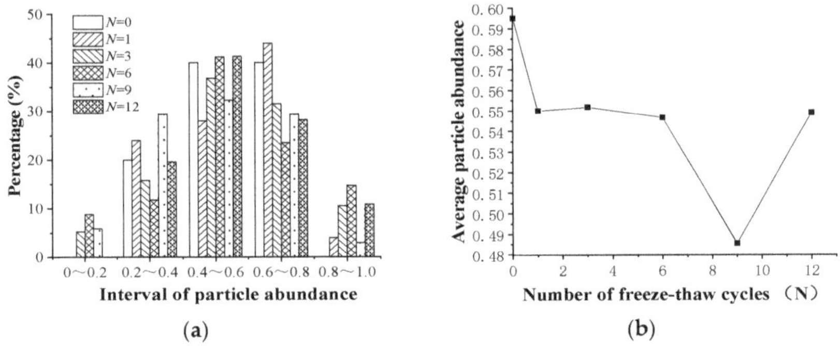

Figure 12. Particle abundance and average abundance under different freeze–thaw cycles. (**a**) Grain abundance. (**b**) Average abundance of particles.

4. Deterioration Mechanism

The main components of weathered sand are quartz and sodium aluminosilicate, which hydrate with cement to form CSH gel and ettringite AFt. In cement-stabilized weathered sand that has not undergone freezing and thawing, CSH gel and AFt crystals fill the internal pores and cracks, sealing some of the pores to a certain extent. Furthermore, the hydration cement adheres to the surface of the particles and strengthens the connection between them. During the freeze–thaw cycles, freezing and thawing occur alternately. The pore water inside the specimen freezes and forms ice crystals at low temperature, and the resulting frost heaving force destroys the hydration cement between particles, making the pore structure inside the specimen develop continuously. As the temperature rises, the solid ice crystals change into liquid, and to a lesser extent, gas, and the melted liquid water enters the new pores created by frost heaving. With the increase of the number of freeze–thaw cycles, the periodic frost heaving force leads to cracks in the hydration cements, which continue to expand until broken, and dissolve in the melting liquid water environment. As a result, the cementation force between particles decreases, the pore structure develops significantly, and the macro level consequences are particle peeling and strength reduction.

The performance deterioration caused by freezing and thawing is a process of micro damage accumulation, mainly due to the generation and expansion of micro cracks and micro pores. The freeze–thaw damage of cement-stabilized weathered sand can be summarized as follows: internal pores → the alternating effects of frost heaving and thawing settlement leads to the development of pores and the migration of water to the interior → under freeze–thaw conditions, the cements are successively dissolved and the loss of cementation force decreases → crack expansion and connection → surface looseness and peeling, and the damage continues to develop to the interior. The freeze–thaw deterioration process is shown in Figure 13.

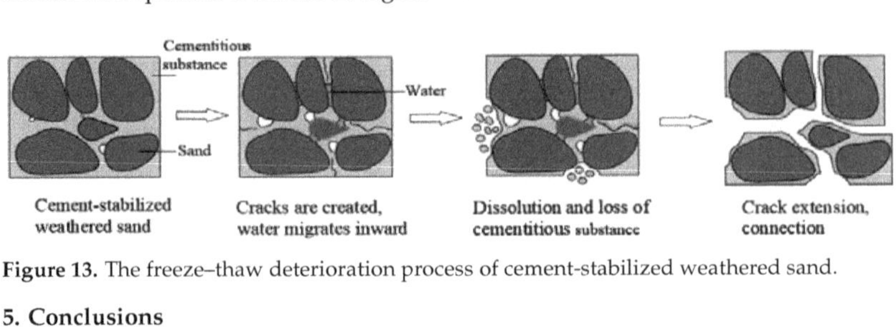

Figure 13. The freeze–thaw deterioration process of cement-stabilized weathered sand.

5. Conclusions

Based on laboratory tests, the macroscopic degradation laws and microscopic characteristics of cement-stabilized weathered sand under the action of freeze–thaw cycles were analyzed, and the following conclusions can be drawn:

(1) Early freeze–thaw action caused tiny particles on the surface of the specimen to peel off. As the number of freeze–thaw cycles increased, the peeling area and depth of the specimen expanded further until the specimen lost its integrity.
(2) When $N \leq 7$, the mass change of the specimen was positive, and the dominant factor affecting specimen mass at this stage was the amount of water entering. When $N \geq 8$, the mass change of specimen was negative, and the controlling factor of the mass change was particles shedding.
(3) As the number of freeze–thaw cycles increased, the unconfined compressive strength of the specimens decreased gradually. When $N = 9$, the strength loss rate of the specimen was 55%. After that, the freeze–thaw action had less influence on the strength of the specimen.
(4) The freeze–thaw action caused the loss of fine particles and hydration products wrapped or filled around the coarse particles, leading to continuous accumulation and expansion of micro cracks and micro pores, and then interpenetrated to form large pores. In this process, the particle skeleton of the stabilized weathered sand changed from dense structure to porous structure.
(5) When $N \leq 9$, the average abundance of pores inside the specimen increased with the increase of number of freeze–thaw cycles, while the average abundance of particles decreased. The main reason is that the freeze–thaw action made the pores expand and connect, and at the same time, the cements attached to the particles' surface continuously dissolves. When $N > 9$, the development trend of pore and particle abundance changed due to the damage of soil structure.
(6) For cement-stabilized weathering sand, the performance deterioration caused by freeze–thaw cycles is a process of continuous accumulation of micro damage. The pore water inside the specimen is constantly transformed between ice crystals and liquid, and the frost heaving and shrinkage caused by this change weaken the connection between particles, which makes the internal pores develop and expand continuously.

In engineering application, cement content varies in different application scenarios. Further study is needed on the deterioration of stabilized weathered sand with different cement content under freeze–thaw conditions. In addition, with the development of new technology, machine learning (ML) methods can be considered to find the best response of weathered sand materials.

Author Contributions: Conceptualization, X.K. and W.H.; Methodology, S.C.; Validation, S.C., G.W. and Y.L.; Formal analysis, S.C.; Investigation, X.K.; Resources, W.H. and Z.Z.; Data curation, G.W. and Y.L.; Writing original draft, X.K.; Writing review, Z.Z.; Supervision, W.H. All authors have read and agreed to the published version of the manuscript.

Funding: This research was funded by Key Research and Development Project of Shandong (grant no. 2020CXGC011404); and Science and Technology Program of Higher Education in Shandong Province (grant no. J18KA216).

Institutional Review Board Statement: Not applicable.

Informed Consent Statement: Not applicable.

Data Availability Statement: The data presented in this study are available on request from the corresponding author.

Conflicts of Interest: The authors declare no conflict of interest.

References

1. Yang, J.; Liu, S.; Zhang, G. Experimental study on CBR value of weathered sand in the Three Gorges Reservoir area stabilized by cement. *J. Yangtze River Sci. Res. Inst.* **2016**, *33*, 98–101.
2. Xiong, G.; Li, Q.; Fan, Y.; Song, J. Long-term stability test of Maopingxi weathered sand in the Three Gorges. *J. Yangtze River Sci. Res. Inst.* **1999**, *16*, 23–27.
3. An, Y. Application of Stabilized Weathered Sand in Expressway. Master's Thesis, Chang'an University, Xi'an, China, 2005.

4. Guo, Y.; An, Y.; Liu, G. Research on the road performance of cement stabilized weathered sand. *China Foreign Highw.* **2005**, *25*, 52–54.
5. Wang, X.; Yang, S.; Chen, C. Application of weathered sand filling at Jiangxia avenue in Three Gorges dam area. *Yangtze River* **2001**, *32*, 16–17.
6. Yang, J.; Yang, Z.; Zhang, G.; Tang, Y. Experimental study on swelling force of expansive soil improved by weathered sand. *J. Chongqing Jiaotong Univ. (Nat. Sci.)* **2014**, *33*, 106–109.
7. Zhuang, X.; Wang, J.; Wang, K.; Li, K.; Hu, Z. Dynamic characteristics of weathered sand for improving expansive soil. *Rock Soil Mech.* **2018**, *39*, 149–156.
8. Huang, M.; Xiong, J.; Tang, J.; Long, C. Experimental study on shear strength of expansive soil improved by lime-weathered sand. *Appl. Mech. Mater.* **2014**, *2799*, 1441–1447. [CrossRef]
9. Yang, J.; Zhu, C.; Zhang, G. Experimental study on unconfined compressive strength of lime-ash stabilized weathered sand. *J. Yangtze River Sci. Res. Inst.* **2015**, *32*, 101–106.
10. Yang, J.; Xiang, F. Cement stabilized weathered sand in the Three Gorges Reservoir area. Resilience modulus test and model evaluation. *J. Chongqing Univ.* **2016**, *39*, 132–139.
11. Zaimoglu, A.S. Freezing-thawing behavior of fine-grained soils reinforced with polypropylene fibers. *Cold Reg. Sci. Technol.* **2010**, *36*, 63–65. [CrossRef]
12. Lu, Y.; Liu, S.; Zhang, Y.; Li, Z.; Xu, L. Freeze-thaw performance of a cement-treated expansive soil. *Cold Reg. Sci. Technol.* **2020**, *170*, 102926. [CrossRef]
13. Huang, M.; Duan, J.; Wang, J. Research on basic mechanical properties and fracture damage of coal gangue concrete subjected to freeze-thaw cycles. *Adv. Mater. Sci. Eng.* **2021**, *2021*, 6701628. [CrossRef]
14. Jumassultan, A.; Sagidullina, N.; Kim, J.; Ku, T.; Moon, S.W. Performance of cement-stabilized sand subjected to freeze-thaw cycles. *Geomech. Eng.* **2021**, *25*, 41–48.
15. Tian, Y.; Ma, B.; Wang, D.; Li, N. Frost resistance of cement stabilized crushed rock under freeze-thaw cycles. *J. Chang'an Univ. (Nat. Sci.)* **2017**, *37*, 84–91.
16. Hao, Y.; Wen, H.; Fan, P.; Dong, X. Experimental study of freeze-thaw cycles on the strength of red mud-steel slag modified cement soil. *J. Taiyuan Univ. Technol.* **2021**, *52*, 117–121.
17. Gao, C.; Du, G.; Guo, Q.; Zhuang, Z. Static and dynamic behaviors of basalt fiber reinforced cement-soil after freeze-thaw cycle. *KSCE J. Civ. Eng.* **2020**, *24*, 3573–3583. [CrossRef]
18. Li, Y.; Ling, X.; Su, L.; An, L.; Li, P.; Zhao, Y. Tensile strength of fiber reinforced soil under freeze-thaw condition. *Cold Reg. Sci. Technol.* **2018**, *146*, 53–59. [CrossRef]
19. Li, T.; Kong, L.; Gao, A. The deformation and microstructure characteristics of expansive soil under freeze-thaw cycles with loads. *Cold Reg. Sci. Technol.* **2021**, *192*, 103393. [CrossRef]
20. Lake, C.; Yousif, M.; Jamshidi, R. Examining freeze/thaw effects on performance and morphology of a lightly cemented soil. *Cold Reg. Sci. Technol.* **2017**, *134*, 33–44. [CrossRef]
21. Eskisar, T.; Altun, S.; Kalipcilar, L. Assessment of strength development and freeze-thaw performance of cement treated clays at different water contents. *Cold Reg. Sci. Technol.* **2015**, *111*, 50–59. [CrossRef]
22. Zhang, K.; Zhang, L. Analysis of influence factors of pore water pressure change in frozen soil. *Adv. Mater. Sci. Eng.* **2021**, *2021*, 3290831. [CrossRef]
23. Tang, C.; Lin, L.; Cheng, Q.; Zhu, C.; Wang, D.; Lin, Z.; Shi, B. Quantification and characterizing of soil microstructure features by image processing technique. *Comput. Geotech.* **2020**, *128*, 103817. [CrossRef]
24. Zhang, X.; Kong, L.; Guo, A.; Tuo, Y. Based on SEM and MIP test structured clay microscopic pores during compression. *Chin. J. Rock Mech. Eng.* **2012**, *31*, 406–412.
25. Xu, Q. Application of Digital Image Analysis Program in Soil Microstructure Research and Data Analysis. Master's Thesis, Taiyuan University of Technology, Taiyuan, China, 2008.
26. Liu, B.; Du, J.; Shen, T.; Wang, Z.; Hu, J.; Xie, P. Quantitative analysis of the microstructure of organic impregnated sand-cement soil. *J. Hainan Univ. (Nat. Sci.)* **2021**, *39*, 68–76.
27. Hou, S.; Chen, L.; Wang, S.; Wan, Y.; Ding, X. Research on damage and degradation characteristics of cement-soil under freeze-thaw cycles. *J. Yangtze River Sci. Res. Inst.* **2016**, *33*, 124–127, 132.
28. Zhang, J.; Duan, X. Damage characteristics and energy dissipation of cement-soil at different ages under freeze-thaw cycles. *Bull. Chin. Ceram. Soc.* **2019**, *38*, 2144–2151.
29. Gao, Y.; Yao, D.; Qin, X.; Li, C. Study on freeze-thaw characteristics of microbial mineralized geotechnical materials under salt erosion environment. *J. Earthq. Prevent. Disaster Mitigat. Eng.* **2018**, *38*, 787–794.
30. Ye, W.; Qiang, Y.; Jing, H.; Zou, Y. Experimental study on freeze-thaw cycles of loess paleosol with different moisture content based on nuclear magnetic resonance. *J. Eng. Geol.* **2021**. [CrossRef]
31. Ding, M.; Zhang, F.; Ling, X.; Lin, B. Effects of freeze-thaw cycles on mechanical properties of polypropylene Fiber and cement stabilized clay. *Cold Reg. Sci. Technol.* **2018**, *154*, 155–165. [CrossRef]
32. In *JTG E51-2009*; Test Methods of Materials Stabilized with Inorganic Binders for Highway Engineering. Beijing People's Communications Press: Beijing, China, 2009; pp. 146–149.

33. In *ASTM D 1633-17*; Standard Test Methods for Compressive Strength of Molded Soil-Cement Cylinders1. ASTM: West Conshohocken, PA, USA, 2017.
34. Kim, J.S.; Chun, S.Y.; Stephan, D.; Han, T.S. Issues on characterization of cement paste microstructures from μ-CT and virtual experiment framework for evaluating mechanical properties. *Constr. Build. Mater.* **2019**, *202*, 82–102. [CrossRef]
35. Wang, L.; Ding, S.; He, H.; Liu, X. Experimental study on freeze-thaw failure of cement mortar with different particle sizes. *J. Exp. Mech.* **2019**, *34*, 138–148.
36. Garg, A.; Wani, I.; Zhu, H.; Kushvaha, V. Exploring efficiency of biochar in enhancing water retention in soils with varying grain size distributions using ANN technique. *Acta Geotech.* **2021**. [CrossRef]
37. Sharma, A.; Kushvaha, V. Predictive modelling of fracture behavior in silica-filled polymer composite subjected to impact with varying loading rates using artificial neural network. *Eng. Fract. Mech.* **2020**, *239*, 107328. [CrossRef]
38. Chen, Z.; Chen, Z.; Song, Z.; Ye, W.; Fan, Z. Smart gas sensor arrays powered by artificial intelligence. *J. Semicond.* **2019**, *40*, 111601. [CrossRef]
39. Turetskyy, A.; Wessel, J.; Herrmann, C.; Thiede, S. Battery production design using multi-output machine learning models. *Energy Storage Mater.* **2021**, *38*, 93–112. [CrossRef]

coatings

MDPI

Article

Reconstruction Method of Old Well Logging Curves Based on BI-LSTM Model—Taking Feixianguan Formation in East Sichuan as an Example

Chao Cheng [1], Yan Gao [1,*], Yan Chen [1], Shixiang Jiao [1], Yuqiang Jiang [1], Juanzi Yi [2] and Liang Zhang [1]

1 School of Geoscience and Technology, Southwest Petroleum University, Chengdu 610500, China; ylksh@163.com (C.C.); carly.chenyan@gmail.com (Y.C.); cwbp.sunshinej@gmail.com (S.J.); xnsyjyq3055@126.com (Y.J.); 15596650069@163.com (L.Z.)

2 Chongqing Gas Mine, PetroChina Southwest Oil and Gas Field Company, Chongqing 400021, China; yadong21@163.com

* Correspondence: ylksh@swpu.edu.cn or gysnzlh@163.com

Abstract: In order to define a favorable oil and gas accumulation area, this study focused on reservoir recognition which is based on logging data of old wells. The Gaofengchang structure in eastern Sichuan is used as a test area to discuss the necessity and feasibility of curve construction by combining new and old wells. Analysis of the reasons for the inaccuracy of the traditional curve reconstruction method is also provided. Given the interdependence of the well log in the depth domain sample sequence, a new intelligent construction method (BI-LSTM) based on the cyclic neural network is proposed. A discussion on the effect of data increments on prediction accuracy is also provided. The following four conclusions were achieved through curve reconstruction experiments: a high-precision CNL pseudo-curve was obtained; an underdetermined equation in optimization logging interpretation method needed to be extended to a positive definite equation; the quantitative processing of the complex lithologic reservoir parameters for the old wells was realized; and the processing result of the lithology physical property were basically consistent with the core data. Therefore, the BI-LSTM proposed in this paper could improve the accuracy of logging curve construction and has a good promotion significance for the old well review.

Keywords: Feixianguan Formation; underdetermined system; curve reconstruction; circulating neural network; old well review

Citation: Cheng, C.; Gao, Y.; Chen, Y.; Jiao, S.; Jiang, Y.; Yi, J.; Zhang, L. Reconstruction Method of Old Well Logging Curves Based on BI-LSTM Model—Taking Feixianguan Formation in East Sichuan as an Example. *Coatings* **2022**, *12*, 113. https://doi.org/10.3390/coatings12020113

Academic Editor: Jian Chen

Received: 1 December 2021
Accepted: 25 December 2021
Published: 20 January 2022

1. Introduction

In 1963, the first oolitic beach gas reservoir of the Feixianguan Formation was discovered by well Ba-3 in the Shiyougou structure, eastern Sichuan, which was 4.26×10^4 m^3/d in gas yield. After that, dozens of independent oolitic beach gas reservoirs of the Feixianguan Formation, Fuchengzhai, Jiannan, Huangcaoxia, Gaofengchang and other structures were discovered, which opened the prelude to the exploration of the oolitic beach gas reservoir in eastern Sichuan. Even so, the exploration of the Feixianguan Formation gas reservoir in the Sichuan Basin has been neglected for a long time. The causes of this phenomenon are the small scale of reserves, which is less than 500 million cubic meters, and the lithology of the gas reservoir which is composed mainly of micritic limestone and oolitic limestone. At that stage, the logging data of the Feixianguan Formation gas reservoir only recorded AC, GR and RT, RXO curves, namely the acoustoelectric combination [1]. In 1995, with further exploration, 10 billion cubic meters of gas reservoir was discovered in the Feixianguan Formation oolitic beach of the Dukouhe structure, which wrote a new chapter of gas reservoir exploration in eastern Sichuan once again. Since then, we have made the logging data more complete mainly with the comprehensive logging series. At present, the oolitic beach resources of the Feixianguan Formation in eastern Sichuan are estimated to be

331.15 × 10^{12} m^3, and the proved reserves are 0.986 × 10^{12} m^3. The low proved rate, 3.36%, warns us to make further exploration. Therefore, it is of great practical significant to study the log optimization processing interpretation of the old well which could help us to find out the oolitic beach reservoir distribution in the Feixianguan Formation.

2. Basic Characteristics and Logging Response of the Feixianguan Formation Reservoir

According to the petrochemical analysis, the mineralogical composition of the Feixianguan Formation in eastern Sichuan is mainly composed of clay, calcite, dolomite and gypsum. Namely: limestone, mudstone, dolomite and gypsum (by lithology) [2,3]. Brief introduction as follows: The limestone, mainly the oolitic limestone and the micrite limestone, is widely developed in this area. The fine-grained mudstone is mainly developed in the bottom of the Tf4 and Tf1. The gypsum is developed in the Tf4, and the rest of the formations appear in the form of the gypsum nodule occasionally. The oolitic dolomite and the oolitic limestone are the main reservoir rocks, and, among them, the oolitic dolomite is the best reservoir rock, but its development frequency is low; the oolitic limestone, in contrast, is high in development frequency, but, to be good reservoir, it has to be reformed by fracture reconstruction. The distribution of the various lithologies in the Feixianguan Formation is shown in Table 1.

Table 1. Statistics of mineralogical composition of the Feixianguan Formation.

Layer	Lithification Analysis Result				Histogram
	Dolomite (%vol)	Calcite (%vol)	Gypsum (%vol)	Clay (%vol)	
Tf4	1–70, 25	5–98, 52	0–55, 4	5–40, 15	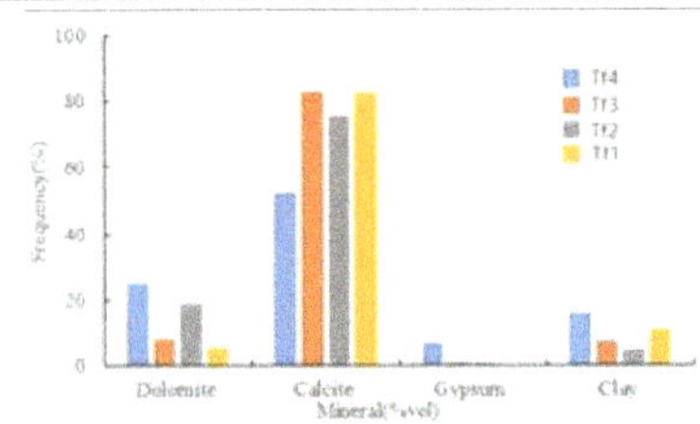
Tf3	0–90, 8	5–100, 83	0–30, 0.5	0–30, 7	
Tf2	0–98, 18	0–100, 75	0.75, 0.5	0–23, 4	
Tf1	0–80, 5	0–100, 82	/	0–63, 10	

According to the core and the thin section calibration, the Feixianguan Formation reservoir is of the fracture-pore type, and the logging response shows low gamma, high acoustic wave, high neutron, low resistivity and positive difference characteristics. Figure 1 and Table 2 show the typical logging response characteristics of various lithologies.

Table 2. Different lithology logging response characteristics in Feixianguan formation.

Lithology		Porosity (%)	Typical Logging Response Characteristics				
			GR (API)	AC (μs/ft)	CNL (%)	DEN (g/cm^3)	RT/RXO
Limestone	Oolitic	1–6	low, 10–20	higher, 48–52	>0	low, 2.6–2.7	lower, plus difference characteristic
	Micritic	0–1.5	higher, 15–25	lower, 47–49	Near the "0" line	lower, 2.69–2.72	higher, approximate coincidence
Dolomite		1.5–12	low, 10–20	high, 48–55	>0	lower, 2.6–2.8	low, obvious plus difference characteristic
Anhydrite		0–1	low, 10–15	low, 43–47	<0	High, >2.9	high, approximate coincidence

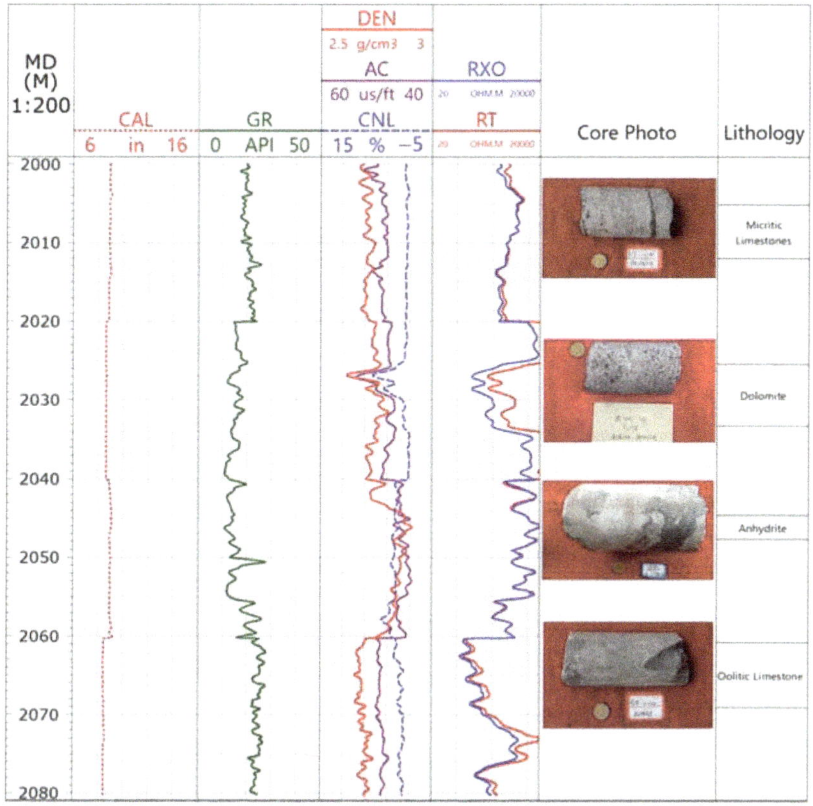

Figure 1. Different lithology logging response characteristics in Feixianguan formation.

3. Pseudo-Curve Reconstruction Method

According to the petrochemical analysis, the reservoir of the Feixianguan Formation is characterized by complex lithology and poor physical property. It can be seen from the logging response analysis that, normally, the complete conventional logging curve could indicate the lithologies and the physical properties of the reservoir better. In the light of the lithologic characteristics, the rock volume physical model can be simplified to five parts, namely, clay, gypsum, calcite, dolomite and porosity [1,4]. In order to solve five unknown quantities, we needed at least five curves to set up a system of simultaneous equations. However, out of a total of twelve wells in this area, only five old wells (X1–X5) had complete conventional logging curves; the remaining seven old wells (X6–X12) just had AC, GR and RT, RXO curves. Obviously, these old wells could only establish one underdetermined equation, including four equations and five unknown quantities, which could not meet the basic requirements of the quantitative solution of parameters of the complex carbonate reservoir. Therefore, it was unrealistic to expect the quantitative processing of complex lithology unless the pseudo-curve could be constructed and the underdetermined equation extended to the positive definite equation.

3.1. Selection of the Reconstruction Curves

In order to reduce the error transfer and the uncertainty of the pseudo-curve, we had to make a choice between two logging curves, CNL and DEN, during the curve construction [5–7]. In this section, we will demonstrate the selection of the reconstruction curves. Firstly, by analyzing the crossplot of the core porosity of the coring section of well

X1 and the 3-porosity curve (Figure 2), we found that the CNL curve was most significantly correlated with the core porosity, with the correlation coefficient of 0.71, less with the AC curve, and the least with the DEN curve, only 0.4. Obviously, the CNL curve could reflect the physical properties of reservoir better.

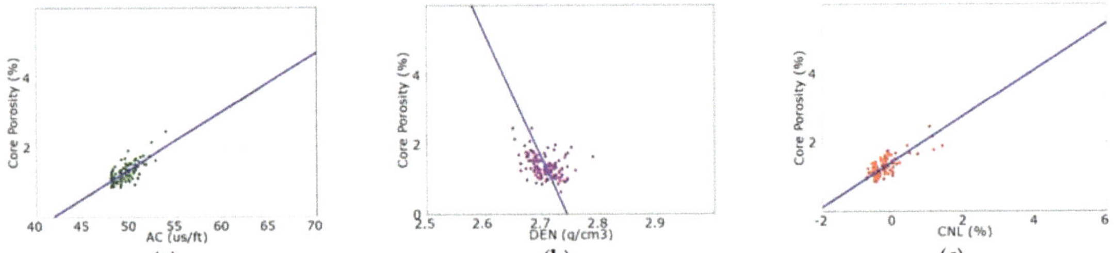

(a) (b) (c)

Figure 2. Cross plot of the core porosity and the 3-porosity curve of well X1. (**a**) the core porosity versus the AC curve; (**b**) the core porosity versus the DEN curve and (**c**) the core porosity versus the CNL curve.

Secondly, we analyzed the relationship between the logging curves. It is known that the GR curve is sensitive to the clay, the AC, DEN and CNL curve can reflect the lithologies and physical properties of the reservoir simultaneously, and the RT curve can reflect the oil/gas-bearing properties [8,9]. We used the CNL curve and the DEN curve of well X1, respectively, to make a correlation analysis with other curves (Figure 3). It turned out that the correlation coefficients of the GR, AC and RT curve with the CNL curve were significantly higher than that with the DEN curve. The reason is that the high-density drilling fluid caused expansion of hole diameter and induced fracture in some sections of wells, which affected the AC curve and the DEN curve [10–12]. Therefore, it was more appropriate to reconstruct the CNL curve than the DEN curve.

3.2. *Selection of Construction Method*

Regression is the essence of the old well curve reconstruction problem. At present, there are a lot of research and reports on the reconstruction methods for curves, including curve fitting, BP neural network, support vector machine, etc. [5,13–15]. These traditional reconstruction methods for curves basically have the same idea: Firstly, a certain number of the parameter eigenvalues are extracted, which form the training sample data set. Secondly, by using the mathematical algorithm, the corresponding curve prediction model is established [16]. The key of this technology is the convergence rate and the error analysis.

It is easy to find that, for the training data set, these traditional methods can obtain a good effect in self-judging ability but not in prediction effect [17]. Yet it is worth noting that, even if some methods can achieve good results, these have low accuracy of the curve construction to spoil the logging interpretation, which leads to multi-solution or even wrong interpretation [18,19]. Such a status quo is caused for the following reasons: Firstly, lack of the feature information leads to loss of details when we extract the parameter eigenvalue; secondly, lack of the training data leads to model over-fitting. Only a machine learning method with high generalization performance can break this dilemma [20]. Given that the logging data of unknown depth should be close within the nearby range when reconstructing curves, we chose the depth model BI-LSTM with a facility of drawing out the context sequence information to construct the pseudo-curve.

BI-LSTM is an extension of the recurrent neural network (RNN), which is a new method based on the common LSTM. By using BI-LSTM, we could divide one-way LSTM into two directions, namely the positive timing directions and the negative timing directions, and connect two networks to the same output by putting the forward information and the reverse information as the output of the current node at the same time [21,22]. BI-LSTM compensated for the limitations of the common neural network, solved the problem of "context semantic loss" caused by the lack of operation connection between each input layers, added the reverse operation based on the RNN and achieved the final results of the training data, namely the stacking of the forward RNN and the reverse RNN [23–25]. Beyond that, owing to the presence of LSTM, the imperfection of the RNN that cannot handle the long-term dependency information was also eliminated [26,27]. Figure 4 shows the schematic diagram of BI-LSTM network structure.

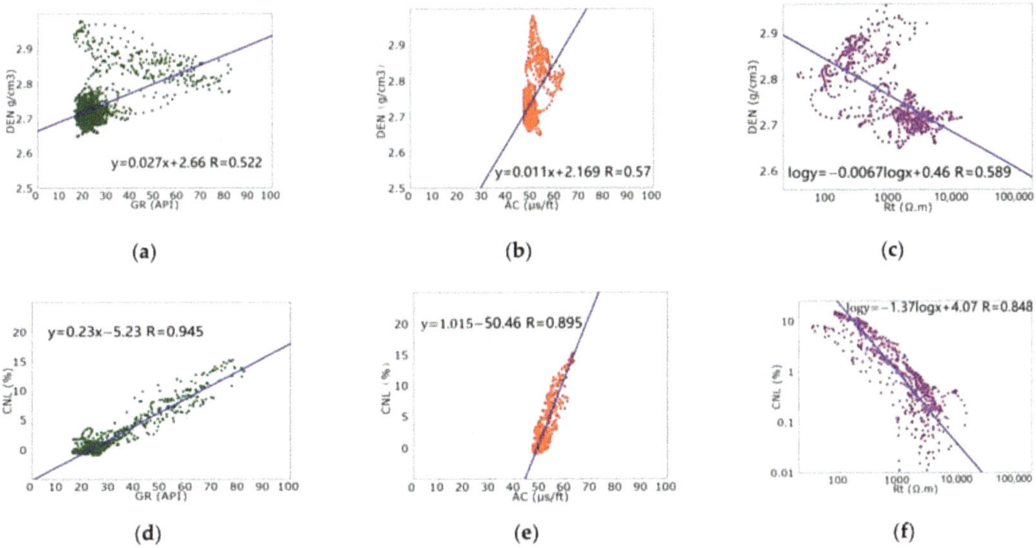

Figure 3. Cross plot of the CNL curve, the DEN curve and other curves of the Feixianguan Formation of well X1. (**a**) the DEN curve versus the GR curve; (**b**) the DEN curve versus the AC curve; (**c**) the DEN curve versus the Rt curve; (**d**) the CNL curve versus the GR curve; (**e**) the CNL curve versus the AC curve . . . and (**f**) the CNL curve versus the Rt curve.

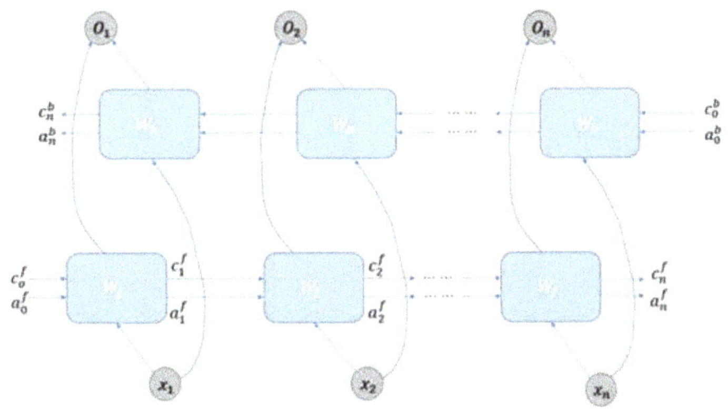

Figure 4. BI-LSTM network structure diagram.

The algorithm can be expressed as:

$$h_t = f(Ux_t + Wh_{t-1}) \tag{1}$$

$$h'_t = f(U'x_t + W'h'_{t+1}) \quad (2)$$

$$y_t = g(Vh_t + V'h'_t) \quad (3)$$

where, x_t is the input value of the BI-LSTM network at time t; y_t is the output value of the BI-LSTM network at time t; h_t is the value of the forward hidden layer at time t; h_{t-1} is the value of the forward hidden layer at time $t-1$; V is the weight of the output layer in the forward calculation; g is the corresponding activation function; U is the weight of the input layer; f is the corresponding activation function; W is the corresponding weight of the hidden layer; V' is the weight of the output layer in reverse calculation; U' is the weight of the output layer; W' is the weight of the hidden layer; h'_t is the output of LSTM memory unit in reverse calculation at time t; and h'_{t+1} is the output of LSTM memory unit in reverse calculation at time $t-1$. The output value of the BI-LSTM network at time t was obtained by the forward and reverse output.

3.3. Pseudo-Curve Reconstruction Experiment

Given the influence of the reservoir lithologies, the physical properties and the oil/gas-bearing properties on the logging curves, we used three logging curves, GR, AC and RT, as the input curves for CNL reconstruction. Taking the logging data of well X1 as the input sample and well X5 as the test sample, experiments were conducted by using multiple linear regression (MLR), BP neural network, support vector machine (SVM, and BI-LSTM [5,14,28]. According to the experiment results, the advantages of the reconstruction method for the curve can be discussed through three test parameters: the correlation coefficient (R), root mean square error (RMSE) and determination coefficient (R^2). The closer R^2 is to 1, the more accurate estimate is.

The concrete calculating methods are as follows.

SS_{res} is the sum of squares of residuals, which can be calculated by Equation (4):

$$SS_{res} = \sum_i (y_i - f_i)^2 \quad (4)$$

where, y_i is real sample data, $\overline{y}$ is sample average value and f_i is estimated data.

SS_{tot} is the total sum of squares, which can be calculated by Equation (5):

$$SS_{tot} = \sum_i (y_i - \overline{y})^2 \quad (5)$$

R^2 is the determination coefficient, which can be calculated by Equation (6):

$$R^2 = \frac{SS_{res}}{SS_{tot}} \quad (6)$$

The experimental results of well X1 and well X5 are shown in Table 3.

Table 3. Statistical comparison table of different algorithms predict accuracy.

Data Set	Evaluation Parameter	Method			
		MLR	BP	SVM	BI-LSTM
Self-judgment of Training Set X1	R	0.951	0.969	0.972	0.99
	RMSE	0.799	0.642	0.635	0.153
	R^2	0.91	0.96	0.97	0.97
Examination of Training Set X5	R	0.64	0.69	0.74	0.82
	RMSE	1.45	1.27	1.12	0.96
	R^2	0.46	0.58	0.66	0.71

Figure 5 shows the effect of regression algorithms in well X1. Combined with Table 3, the self-judging curve was found to be consistent with the measured curve in the training set. Through contrastive analysis, we drew the conclusion: To the correlation coefficient (R), R of these regression algorithms were all greater than 0.95, showing good results; to the root mean square error (RMSE), BI-LSTM was the smallest, which was only 0.153, the multiple logistic regression (MLR) was the largest, which was 0.799, BP neural network and support vector machine lay halfway in between; to the determination coefficient (R^2), SVM and BI-LSTM were the best, MLR was the worst. Briefly, all three parameters indicated that BI-LSTM was the optimal algorithm for the pseudo-curve construction.

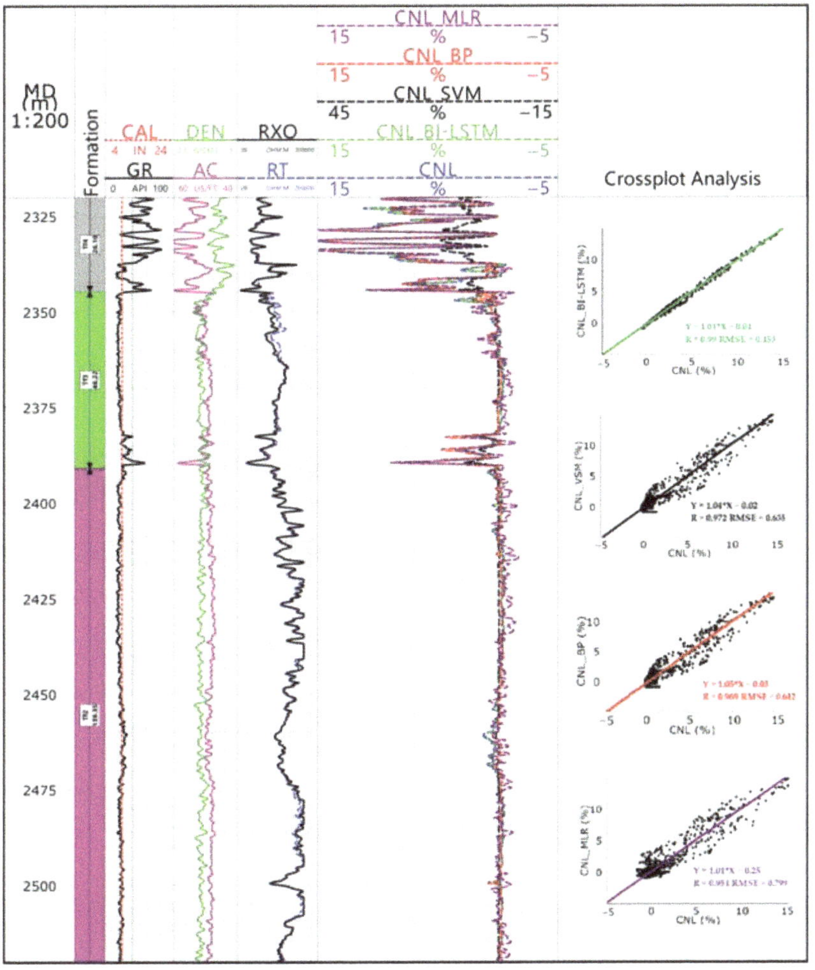

Figure 5. The back estimation sketch of well X1.

Figure 6 shows the prediction effect of different algorithms in well X5. The main contributions are as follows: For most well sections, the predicted curve was found to be consistent with the measured curve in trend and, for some well expansion sections, the predicted curve of the multiple regression and the BP algorithm was diametrically opposite to the measured curve. Three-parameter analysis showed that the prediction effect of these two algorithms was poor: R was less than 0.7, RMSE was greater than 1.2 and R^2 was less than 0.6. In comparison, though the prediction effect of the support vector was rather ideal, BI-LSTM showed greater advantages. For BI-LSTM, R was significantly higher, RMSE was only 0.96 and R^2 was up to 0.71. Through the BI-LSTM algorithm, we achieved a pseudo-neutron curve. For such a pseudo-neutron curve, the frequency band distribution range was narrower than the original curve, the high-gamma segment was smaller than the original curve and the low-gamma segment was larger than the original curve. After standardization, the pseudo-neutron curve was found to be consistent with the original curve in the low-gamma segment, but there was still a large error in the high-gamma segment. Significantly, the processing results did not interfere with the identification of the reservoirs and the calculation of the reservoir parameters, although large errors existed. The main reason for this is that the high-gamma segment was not the favorable reservoir.

Above all, the results show that the BI-LSTM network was the best method for pseudo-curve construction, which improved the prediction precision and the accuracy greatly. Through the BI-LSTM algorithm, the internal relationship of different logging curves was reconciled, the dependency relationship of the sample sequences in the logging domain was considered, the long-term dependence problem was solved and the perfect advantage of making full use of data information was reflected.

3.4. The Impact of Data Volume on Predicted Results

For the machine learning algorithm, in general, the more training samples we have, the more regularity information of the data set we achieve [4,29]. With the implementation of the training tasks and the accumulation of the experience, the model acquired a strong generalization ability. We took the data of well X1, X2, X3 and X4 as the training data in turn and took the data of well X5 as the effect test. This was done to discuss the influence of the data information on the prediction result of BI-LSTM.

It can be seen from the Figure 7 and Table 4 that, with the increase of the training data, there was a rising trend of the pseudo-curve prediction result in R and R^2 but not for RMSE, suggesting that the prediction accuracy of BI-LSTM was improving. Yet it is worth noting that BI-LSTM was not affected by the data amount solely; when we took well X1 and well X2 as the training data set, R^2 increased significantly from 0.71 to 0.76 but, with the addition of the training data of well X3 and well X4, the growth trend of R^2 slowed down gradually.

Table 4. Experimental statistic BI-LSTM algorithm in data increment.

Well Name	Evaluation Parameter	BI-LSTM			
		Training Set X1	Training Set X1X2	Training Set X1X2X3	Training Set X1X2X3X4
Test Set X5	R	0.82	0.88	0.92	0.95
	RMSE	0.96	0.72	0.67	0.37
	R^2	0.71	0.76	0.78	0.79

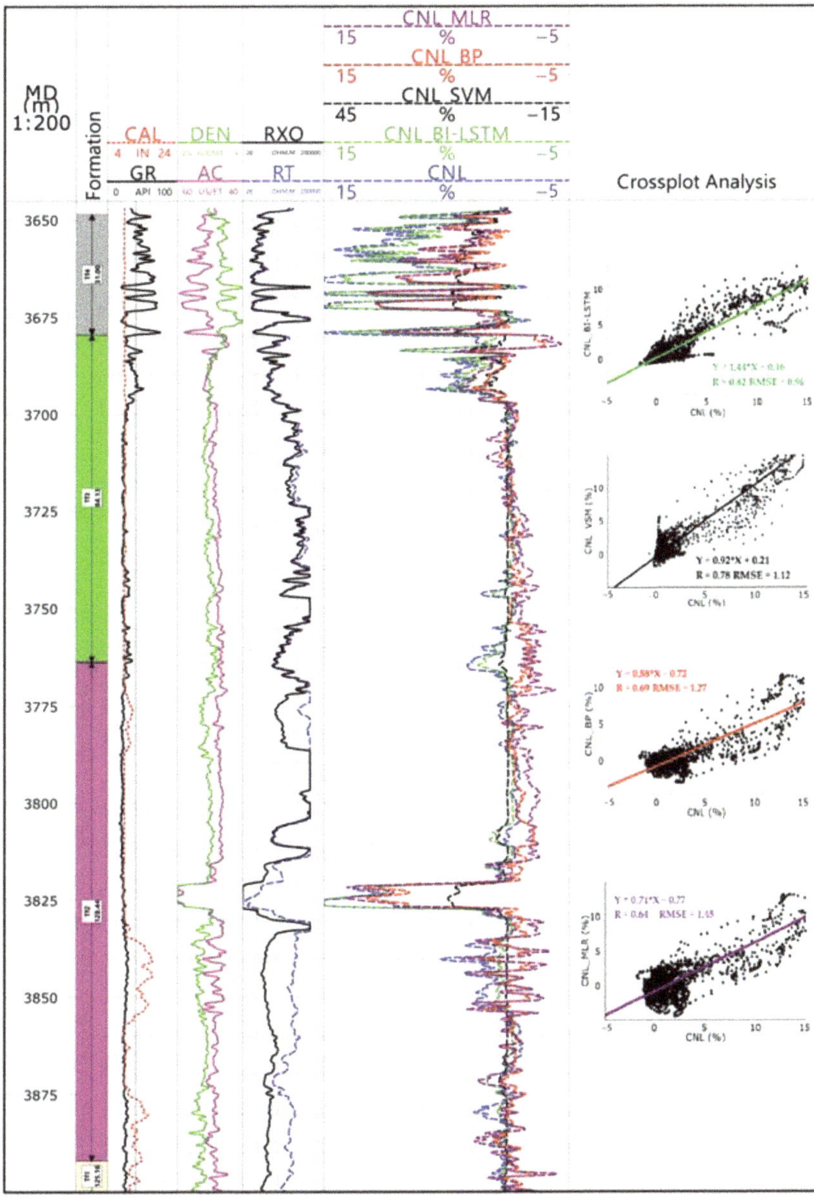

Figure 6. Forecast effect chart well X5.

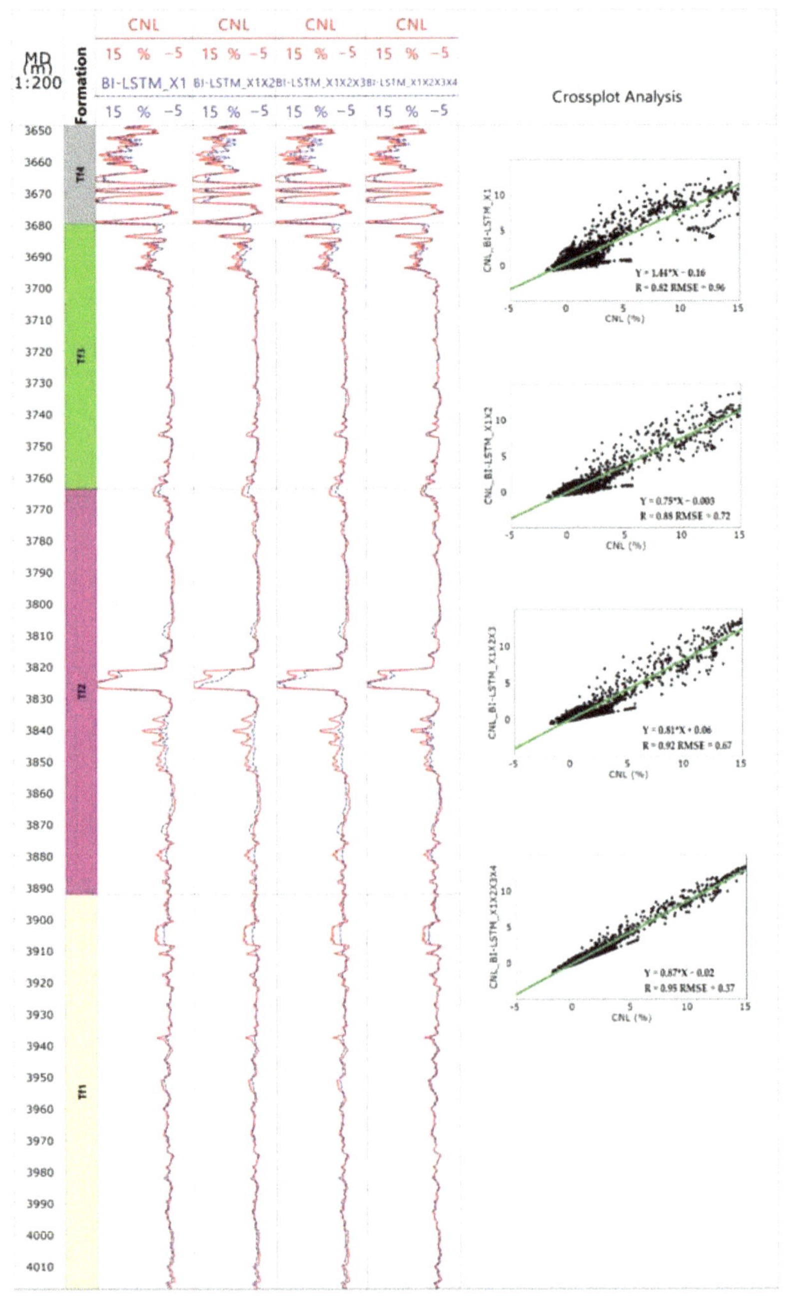

Figure 7. Prediction of BI-LSTM algorithm in well X5 data increment.

4. Application Effect Analysis

When using the optimized logging processing method for the data processing interpretation, note these points: the weight of the pseudo-neutron curve should be lower than 0.3, and the weight coefficient of the other measured curves should be set to 1. Figure 8 shows the comparison before and after the processing results of the pseudo-neutron curve construction in X6 old well. Before that, there were only four curves, GR, AC, RT and RXO. It means that the processing result can only be treated like a single limestone mineral, and the lithological, geophysical feature differs greatly from the core analysis data.

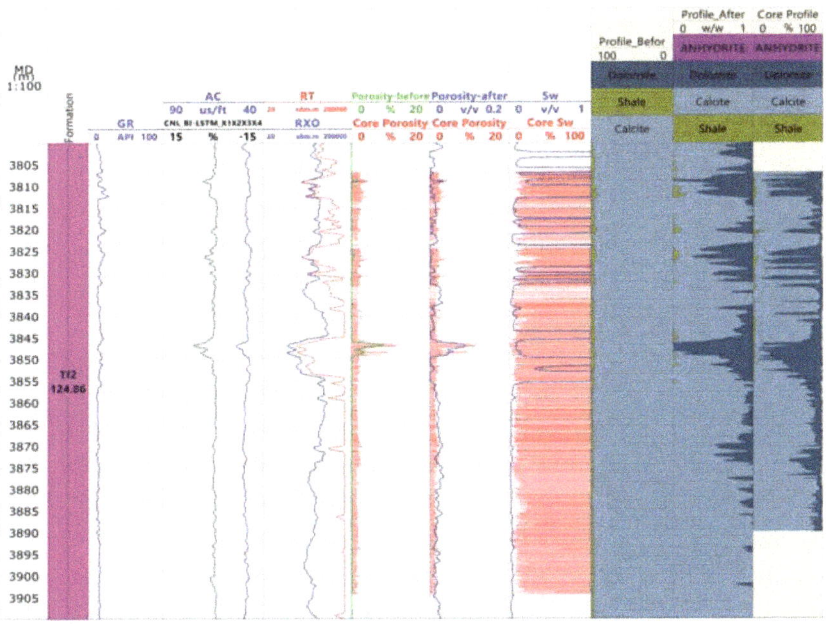

Figure 8. Comparison of well X6 logging interpretation in lithology and physical properties.

After that, the core porosity was found to be consistent with the logging porosity in trend, with a correlation coefficient of 0.896. For this kind of optimum processing result, the data points mostly distributed near the 45° line (Figure 9), the relative error was 14.7% and the absolute error was only 0.28. By comparison, the evaluation indexes were improved dramatically after curve construction (Table 5). The treated lithology profile was found to be consistent with the lithification analysis profile, basically, which demonstrates that the construction method of BI-LSTM curve is feasible.

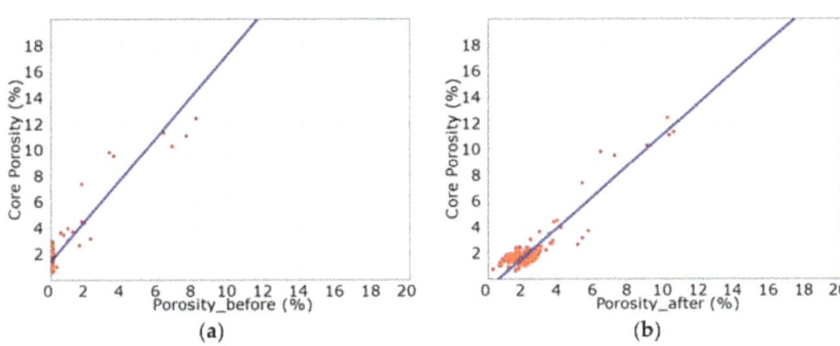

Figure 9. Cross plot of the core porosity and porosity of well X6. (**a**) the core porosity versus porosity before curve construction; (**b**) the core porosity versus log porosity after curve construction.

Table 5. Index comparison table of before and after processing results of curve construction.

Comparative Analysis of Porosity	Average (%)	Correlation Coefficient	Absolute Error (%)	Relative Error (%)
Core porosity	1.90	\	\	\
Porosity before	0.30	0.815	1.6	84.2
Porosity after	2.18	0.896	0.28	14.7

5. Conclusions

(1) According to the correlation analysis of the core physical property and the 3-porosity curve in the Feixianguan Formation, the CNL curve is most significantly correlated with the core porosity, less with the AC curve and the least with the DEN curve. The results show that, compared with the DEN curve, in constructing the pseudo-curve, the CNL curve is a more rational choice.

(2) Compared with the traditional method, BI-LSTM shows greater advantages. This method not only gives full consideration to the interdependence of the sample sequences in the logging depth domain, but also improves the construction precision and accuracy of the pseudo-curve by increasing the amount of data.

(3) With the increase of the amount of training data, the prediction accuracy of the pseudo-curve constructed by BI-LSTM is gradually improved. Yet it is worth noting that BI-LSTM is not affected by the data amount solely.

(4) By using BI-LSTM, the accuracy of the old well pseudo-curve is improved. However, it is suggested that the neutron curves should be standardized before being used for logging data processing and interpretation, and the weight coefficient should be reducing when using the curves.

Author Contributions: Conceptualization, Y.G. and C.C.; methodology, C.C.; software, S.J.; validation, J.Y.; formal analysis, L.Z.; investigation, Y.J.; resources, C.C.; data curation, Y.C.; writing original daft preparation, Y.G.; writing review and editing, C.C.; visualization, L.Z.; supervision, Y.J.; project administration, Y.C.; funding acquisition, Y.J. All authors have read and agreed to the published version of the manuscript.

Funding: This work is supported by the National Natural Science Foundation of China (no. 41430316); Funder: Jiang, Y.Q.

Institutional Review Board Statement: Not applicable.

Informed Consent Statement: Not applicable.

Data Availability Statement: The data sets used and/or analyzed during the current study are available from the corresponding author on reasonable request.

Conflicts of Interest: The authors declare no conflict of interests.

Abbreviations

AC	acoustic
GR	natural gamma ray
Rt	true formation resistivity
Rxo	flushed zone formation resistivity
CNL	compensated neutron logging
DEN	density
Tf1~Tf4	1st~4st member of Feixianguan Formation

References

1. Liu, Y.T. Reservoir characteristics of Changxing-Feixianguan Formation in northeastern Sichuan area. *Lithol. Reserv.* **2019**, *31*, 78–86.
2. Wei, G.; Yang, W.; Liu, M.; Xie, W.; Jin, H.; Wu, S.; Su, N.; Shen, J.; Hao, C. Distribution rules, main controlling factors and exploration direction of giant gas fields in the Sichuan Basin. *Nat. Gas Ind. B* **2020**, *7*, 1–12. [CrossRef]
3. Jin, Y.J.; Zhang, Q.; Wang, M.M. Well logging curve reconstruction based on genetic neural network. *Prog. Geophys.* **2021**, *36*, 1082–1087.
4. Liu, J.W.; Liu, Y.; Luo, X.L. Research and development on deep learning. *Appl. Res. Comput.* **2014**, *31*, 1921–1930.
5. Zhao, J.L.; Li, G.; Ma, P.S.; Gong, Z.W.; Meng, L.F.; Li, G. The application of network technology to petroleum logging interpretation. *Prog. Geophys.* **2010**, *25*, 1744–1751.
6. Qin, T.; Cai, J.Y.; Li, D.Y. Application of curve reconstruction technology in accurate reservoir prediction in Bohai Q oilfield. *Comput. Tech. Geophys. Geochem. Explor.* **2018**, *40*, 330–336.
7. Salehi, M.M.; Rahmatii, M.; Karimnezhad, M.; Omidvar, P. Estimation of the non records logs from existing logs using artificial neural networks. *Egypt. J. Pet.* **2016**, *26*, 957–968. [CrossRef]
8. Li, T.T.; Chen, J.R.; Guo, Y.; Pang, X.Y. A reservoir prediction application of density curve reconstruction in complex geological conditions. *China Manganese Ind.* **2019**, *37*, 19–23.
9. Yu, J. Research on reservoir prediction technology based on curve reconstruction. *Mod. Chem. Res.* **2018**, *11*, 88–89.
10. Duan, Y.X.; Li GT Sun, Q.F. Application of convolutional neural network in reservoir prediction. *J. Commun.* **2016**, *S1*, 5–13.
11. Yang, B.; Kuang, L.C.; Sun, Z.C.; Shi, J.C. *Neural Network and Its Application in Petroleum Logging*; Petroleum Industry Press: Beijing, China, 2005; pp. 94–98.
12. Li, R.J.; Cui, Y.J.; Xiong, L. Evaluating conglomeratic sandstone reservoir in block C of Bohai oilfield by reconstructing neutron log. *Well Logging Technol.* **2019**, *43*, 427–433.
13. An, P.; Cao, D.P.; Zhao, B.Y.; Yang, X.L.; Zhang, M. Reservoir physical parameters prediction based on LSTM recurrent neural network. *Prog. Geophys.* **2019**, *34*, 1849–1858. (In Chinese)
14. Yang, Z.L.; Zhou, L.; Peng, W.L.; Zheng, J.Y. Application of BP neural network technology in sonic log data rebuilding. *J. Southwest Pet. Univ. (Sci. Technol. Ed.)* **2008**, *30*, 63–66. (In Chinese)
15. Zhu, G.J. Application of acoustic curve reconstruction in reservoir prediction. *Comput. Tech. Geophys. Geochem. Explor.* **2017**, *39*, 383–387.
16. Zhou, X.; Cao, J.X.; Wang, X.J.; Wang, J.; Liao, W.P. Acoustic log reconstruction based on bidirectional gated recurrent unit neural network. *Prog. Geophys.* **2021**, 1–11. Available online: http://kns.cnki.net/kcms/detail/11.2982.P.20210209.1209.009.html (accessed on 29 December 2021).
17. Han, B.H.; Wang, F.; Liu, Q.R.; Zhang, C.A. Review of research progress on evaluation of logging reservoir classification methods. *Prog. Geophys.* **2021**, *36*, 1966–1974.
18. Hou, X.L.; Hu, Y.; Li, Y.Q.; Xu, X.H. Rational structure of multi-layer artificial neural network. *J. Northeast. Univ. (Sci. Technol. Ed.)* **2003**, *24*, 35–38.
19. Zheng, Y.Z.; Ye, Z.H.; Liu, X.A.; Zhao, L. Research on prediction method of reservoir physical properties based on deep learning. *Electron. World* **2018**, *10*, 23–26.
20. Wei, L.H.; Guo, J.Y.; Yang, Z.L.; Huang, Y.F. Analysis on key techniques of log constrained lithological inversion. *Nat. Gas Geosicience* **2006**, *17*, 731–735.
21. Cheng, X.; Cheng, Y.X.; Cheng, J.H.; Sun, Q.L. Geophysical logging system based on machine learning and big data technology. *J. Xi'an Univ. Pet. (Nat. Sci. Ed.)* **2019**, *34*, 108–116.
22. Zhou, X.Q.; Zhang, Z.S.; Zhu, L.Q.; Zhang, C.M. A new method for high-precision fluid identification in bidirectional long short-term memory network. *J. China Univ. Pet. (Ed. Nat. Sci.)* **2021**, *45*, 69–76.
23. Alizadeh, B.; Najjari, S.; Kadkhodaie-Ilkhchi, A. Artificial neural network modeling and cluster analysis for organic facies and burial history estimation using well log data: A case study of the South Pars Gas Field, Persian Gulf, Iran. *Comput. Geosci.* **2012**, *45*, 261–269. [CrossRef]
24. Bengio, Y.; Simard, P.; Frasconi, P. Learning long-term dependencies with gradient descent is difficult. *IEEE Trans. Neural Netw.* **1994**, *5*, 157–166. [CrossRef]
25. Hochreiter, S.; Schmidhuber, J. Long short-term memory. *Neural Comput.* **1997**, *9*, 1735–1780. [CrossRef] [PubMed]
26. Hornik, K. Approximation capabilities of multilayer feedforward networks. *Neural Netw.* **1991**, *4*, 251–257. [CrossRef]
27. Zhang, B.L.; Luo, D.T.; Hu, P.; Fan, J.; Jin, C. A well log curve generation method based on deep neural network. *Electron. Meas. Technol.* **2020**, *43*, 107–111.
28. Li, C.B.; Fan, J.F.; Song, X.L. A study on application of deep learning in geology. *Jiangsu Geol.* **2018**, *42*, 115–121.
29. Zhang, D.X.; Chen, Y.T.; Meng, J. Synthetic well logs generation via recurrent neural networks. *Pet. Explor. Dev.* **2018**, *45*, 598–607. [CrossRef]

coatings

Article

Dual Synergistic Effects of MgO-GO Fillers on Degradation Behavior, Biocompatibility and Antibacterial Activities of Chitosan Coated Mg Alloy

Mohammad Zolfaghari Baghbaderani [1], Somayeh Abazari [2,3], Hamid Reza Bakhsheshi-Rad [1,*], Ahmad Fauzi Ismail [4], Safian Sharif [5], Aliakbar Najafinezhad [1], Seeram Ramakrishna [6], Mohammadreza Daroonparvar [7,*] and Filippo Berto [8,*]

1 Advanced Materials Research Center, Department of Materials Engineering, Najafabad Branch, Islamic Azad University, Najafabad, Iran; Mohammadzolfaghari.bi@gmail.com (M.Z.B.); aliakbar.najafinejad@gmail.com (A.N.)
2 Department of Materials and Metallurgical Engineering, Amirkabir University of Technology, Tehran, Iran; somayeh.abazari@gmail.com
3 Iranian National Standards Organization, Tehran, Iran
4 Advanced Membrane Technology Research Center (AMTEC), University Teknologi Malaysia, Johor Bahru 81310, Johor, Malaysia; afauzi@utm.my
5 Faculty of Engineering, University Teknologi Malaysia, Johor Bahru 81310, Johor, Malaysia; safian@utm.my
6 Department of Mechanical Engineering, National University of Singapore, 9 Engineering Drive 1, Singapore 117576, Singapore; seeram@nus.edu.sg
7 Department of Chemical and Materials Engineering, University of Nevada, Reno, NV 89501, USA
8 Department of Mechanical and Industrial Engineering, Norwegian University of Science and Technology, 7491 Trondheim, Norway
* Correspondence: rezabakhsheshi@pmt.iaun.ac.ir (H.R.B.-R.); mr.daroonparvar@yahoo.com (M.D.); filippo.berto@ntnu.no (F.B.)

Citation: Baghbaderani, M.Z.; Abazari, S.; Bakhsheshi-Rad, H.R.; Ismail, A.F.; Sharif, S.; Najafinezhad, A.; Ramakrishna, S.; Daroonparvar, M.; Berto, F. Dual Synergistic Effects of MgO-GO Fillers on Degradation Behavior, Biocompatibility and Antibacterial Activities of Chitosan Coated Mg Alloy. *Coatings* **2022**, *12*, 63. https://doi.org/10.3390/coatings12010063

Academic Editors: Jian Chen, Yanxin Qiao, Fanjiang Meng and Yuxin Wang

Received: 19 November 2021
Accepted: 21 December 2021
Published: 6 January 2022

Abstract: The aim of this work was to establish and characterize chitosan/graphene oxide- magnesium oxide (CS/GO-MgO) nanocomposite coatings on biodegradable magnesium-zinc-cerium (Mg-Zn-Ce) alloy. In comparison to that of pure CS coatings, all composite coatings encapsulating GO-MgO had better adhesion strength to the Mg-Zn-Ce alloy substrate. The result depicted that the co-encapsulation of GO-MgO into the CS layer leads to diminish of contact angle value and hence escalates the hydrophilic characteristic of coated Mg alloy. The electrochemical test demonstrated that the CS/GO-MgO coatings significantly increased the corrosion resistance because of the synergistic effect of the GO and MgO inside the CS coating. The composite coating escalated cell viability and cell differentiation, according to cytocompatibility tests due to the presence of GO and MgO within the CS. The inclusion of GO-MgO in CS film, on the other hand, accelerates the formation of hydroxyapatite (HA) during 14 days immersion in SBF. Immersion results, including weight loss and hydrogen evolution tests, presented that CS/GO-MgO coating enables a considerably reduced degradation rate of Mg-Zn-Ce alloy when compared to the bare alloy. In terms of antibacterial-inhibition properties, the GO-MgO/CS coatings on Mg substrates showed antibacterial activity against *Escherichia coli* (*E. coli*), with a large inhibition area around the specimens, particularly for the coating containing a higher concentration of GO-MgO. Bacterial growth was not inhibited by the bare Mg alloy samples. The CS/GO-MgO composite coating is regarded as a great film to enhance the corrosion resistance, bioactivity, and antibacterial performance of Mg alloy implants.

Keywords: magnesium alloy; composite coating; structural characterization; mechanical properties; bioactivity; antibacterial activity; biocompatibility

1. Introduction

Magnesium (Mg) and its alloys showed great potential in the manufacture of automobile components, aviation constituents, and in the human body, owing to their outstanding mechanical properties and perfect biocompatibility. On the other hand, their current use and future development are limited due to the high degradation rate of magnesium alloys in corrosive solution containing chloride ions [1,2]. A variety of techniques were used to decrease the corrosion rate of magnesium alloys. Laser surface alloying, conversion coatings, electrochemical deposition, micro-arc oxidation, alkaline treatment, and sol-gel coatings were all proposed as surface modification techniques [2,3]. Organic polymer material composite coating via sol-gel methods were employed on the magnesium alloy surface to reduce its degradation rate and further protect Mg alloy substrate exposed to the SBF [4,5]. The bacterial infections related to implants are increasing and should not be ignored [6–8]. The mentioned infections cause high medical costs and lead to higher suffering of patients. Preparation polymer layer over the surface of Mg alloy could also increase antibacterial activity of the substrate and prevent the post infections of the body and thus expanding their use in orthopedic applications. In this respect, the antibacterial properties of chitosan (CS), a polysaccharide comprised of glucosamine and N-acetylglucosamine units, are well known. CS in bulk has a wide range of applications as an antibacterial coating layer besides its great biocompatibility, lack of toxicity, and biodegradability [6–8]. Bacterial growth was shown to be inhibited more effectively by CS–ZnO nanocomposite films [8]. Furthermore, the antibacterial performance of CS–ZnO coating (the CS coatings containing 0.05, 0.1, 0.15, and 0.2 g ZnO) against *E. coli* was demonstrated to be 1.2-fold higher than that of the CS coating without ZnO [8]. Chitosan-based composites promote the attachment of cells, cell proliferation, and differentiation due to structural similarities between chitosan and extracellular matrix structures [9]. Although CS has low mechanical strength, CS-based composites have the potential to enhance physical and chemical properties [10,11]. Furthermore, in simulated inflammatory conditions, coatings containing CS have higher adhesion strength and corrosion protection; for example, a hydroxyapatite (HA)/CS composite presented higher level of protection and greater adhesive strength than pure HA [12,13]. Hence, CS film adhesion strength, bioactivity, corrosion behavior, and antibacterial performance could be improved by adding reinforcement particles like MgO and GO [14,15].

MgO is biocompatible and can release magnesium ions, which can help human metabolism [16]. Furthermore, several studies revealed direct MgO growth on graphene due to solid interfacial bonding properties. This was suggested to be related to the strong bond between carbon and oxygen [17]. Also, MgO could construct solid interfacial bonds with the Mg matrix because of proper lattice matching, as exhibited by Y. Wang et al. [17,18]. The presence of a GO in a CS layer increases the antibacterial activity of a polymer layer containing MgO nanoparticles (NPs) [14,19,20]. Because of its exceptional mechanical, biological, chemical, and electrical characteristics, GO has found a special place in the biomedical field in addition to its applications in electronics and chemicals [21]. Because of their active functional groups, GO-modified surfaces contribute to improved protein-surface interaction, resulting in improved cell attachment and proliferation [22]. Besides, antibacterial properties and the ability to influence osteogenic differentiation are unique properties that may cause GO to become a viable material for designing advanced orthopedic devices [20]. So far, no studies were conducted on the effect of CS/MgO-GO nanocomposite coatings on corrosion behavior and antibacterial behavior of Mg alloys. In this study, CS film and CS film incorporated with MgO-GO filler were coated on the surface of Mg alloy using a dip-coating method to improve Mg alloy corrosion behavior and antibacterial performance for orthopedic applications.

2. Experimental

2.1. Materials and Method

As-cast Mg-Zn-Ce samples with 10 mm × 10 mm × 5 mm dimensions were prepared based on Ref [23]. For the coating process, CS powder with a molecular weight (MW) of 190–310 kDa, MgO NPs (both purchased from Sigma-Aldrich, Gillingham, UK), and graphene oxide (GO) powder prepared via oxidizing natural fake graphite using the modified Hummers method [24]. MgO/GO was fabricated with a simple solvothermal method according to the Ref [25]. In 1000 mL of 2.0% acetic acid aqueous solution, chitosan (1.5 g) was dissolved. After that, MgO/GO powder with various concentration (1, 2 and 3 wt.%) was added to the CS solution; to unified mixing of the powder in the chitosan solution, the mixture was stirred for three hours with a magnetic stirrer and sonicated for 2 h (Qsonica, Q125, Power Rating = 125 W, Frequency = 20 kHz). The obtained CS/MgO-GO was deposited 3 times using a dip-coating method [2] on the surface of polished Mg alloys (Scheme 1). Each layer was dried at 40 °C for 24 h before the next layer was applied. Finally, the coated samples were rinsed with a 0.1 M NaOH solution and then distilled water to reduce the effect of acetic acid on the surfaces.

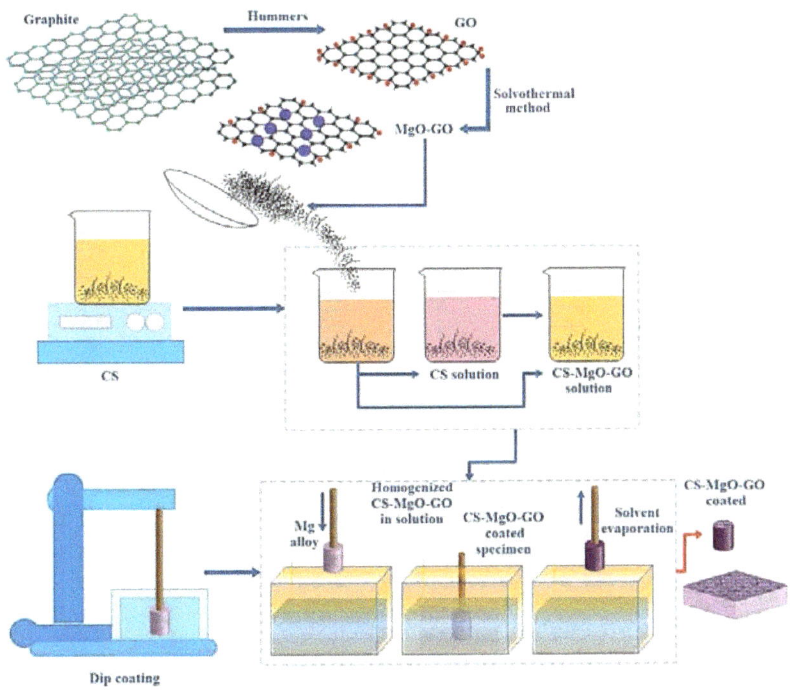

Scheme 1. Schematic of preparation process of CS/MgO-GO nanocomposite coating on Mg alloy.

2.2. Microstructural Characterization

Microstructural characterizations were performed using a field emission scanning electron microscope (FESEM, Tescan, Mira 3 Czech Republic, Prague, Czech Republic) quipped with energy dispersive spectroscopy (EDS, JSM-5910LV, JOEL Ltd., Tokyo, Japan). To reveal the phase components, X-ray diffraction (Philips PW3040, The Netherlands) with Cu-Kα radiation (30 kV, 30 mA) was used. Transmission electron microscopy (TEM; Phillips 208 m) was employed to investigate the morphology of GO nano-platelets, and MgO-GO composite powders. Fourier-transform infrared (FTIR; ALPHA-T, Bruker, Ettlingen, Germany) spectroscopy in the spectral range of 4000 to 400 cm^{-1} for detecting functional groups on the surface of coated specimens was employed. The bonding strength between the coated layer and the Mg substrate was measured at a rate of 5 mm/min using an ASTMF1044-05 universal tensile testing machine (RB 301 UNITECHM, Daejeon, Korea). A contact angle (CA) meter (GBX Digidrop, Romans-sur-Isère, France) was used to investigate the surface wettability caused by composite coating on Mg substrates. Deionized water was used in this experiment, which was carefully discharged onto the coated Mg substrates.

2.3. Corrosion Behavior

To investigate corrosion behavior, an EC-Lab machine was used to perform a potentiometric polarization test in the SBF at a voltage range of −250 to +250 mV_{SCE} an open circuit potential at a rate of 0.5 mV/s. A graphite electrode was used as the counter electrode, and a saturated calomel electrode (SCE) was used as the reference electrode, and the samples were tested with a working electrode that had a surface area of 1 cm^2 exposed to the electrolyte. The corrosion potential, the current corrosion density, and slope of the anodic and cathode curves were also determined using the EC-Lab express software. After 30 min of specimen placement in the SBF, electrochemical impedance spectrometry (EIS) was used to achieve potential stability. Using an open circuit potential and a sine signal with a potential amplitude of 10 mV, this test was performed in accordance with the ASTM G106 standard at an open circuit potential in the range of 10^5 to 10^{-2} Hz. The samples were immersed in 200 mL of Kokubo SBF solution for the immersion test, as per ASTM G1-03 (pH 7.4). The composition of the Kokubo simulated body fluid (SBF) is presented in the supporting information (Table S1). After each 24 h interval during the immersion test, the variation in the pH value of the SBF solution (PHS-3C, Shanghai Lei Ci Device Works, China) was recorded at a constant temperature of 37 °C for 3-, 7-, and 14-day durations. The surface corrosion products of composites were cleaned with a chromic acid solution (300 g/L Cr_2O_3 + 10 g/L $AgNO_3$) prior to weight loss measurements [26]. The degradation rate was calculated using the weight loss test according to the following equation:

$$C_R = \frac{W}{A \times t \times d} \tag{1}$$

where W, A, t, and d represent weight loss (g), sample area (cm^2), exposure time (s), and density (g/cm^3). The C_R unit is cm/s, which is converted to mm/year by multiplying by 3.154×10^8. To achieve repeatable results, each test was repeated three times. In addition, according to Ref. [26], the volume of H_2 released as a result of the magnesium dissolution was determined for 3-, 7-, and 14-day durations at a constant temperature of 37 °C. The coating containing MgO and GO were first immersed in SBF, then funneled, and an upside-down burette filled with SBF placed directly above them to trap the H_2 gas.

2.4. Antibacterial Activity

The antimicrobial activity against *E. coli* bacteria was determined using the disc diffusion method to study the biological behavior of each uncoated and composite coated (n = 3) sample. To achieve this, the sterile swab dipped in the microbial suspension was flushed (by pressing swabs against the pipe's side) and the cultivation environments were lawns. The specimens were incubated in an agar medium for 1 day at 37 °C in an incubator. According to Ref. [11], the inhibition area (IA) around the samples can reveal whether they have antibacterial properties.

2.5. In-Vitro Biocompatibility

The in vitro cytotoxicity of nanocomposite coatings, both uncoated and coated, was determined using an indirect 3-(4,5-dimethylthiazol-2-yl)-2,5-diphenyltetrazolium- bromide (MTT, Sigma, Saint Louis, MO, USA) assay based on the extraction technique. The nanocomposites coating were added to the culture medium and incubated for 3 and 5 days at 37 °C. Human osteogenic sarcoma MG-63 cells line were cultivated under the standard cell-culture conditions. The cell medium was refreshed with 3 and 5-day extracts after 10^4 cells/mL were refined on the 96-well plates for 24 h. After another 24 h, the medium was removed, and 100 mL of MTT agents (0.5 mg/mL in PBS) were inoculated into each well and kept in the incubator for 4 h. After 4 h, 100 μL of DMSO was injected into the well to dissolve the formazan crystals. Finally, the absorbance was measured at 545 nm using an ELISA Reader (Stat Fax-2100, Miami, FL, USA) and compared to a control group of free nanocomposites culture medium. The samples were washed twice and then fixed with paraformaldehyde for alkaline phosphatase (ALP) activity tests after being cultured for 3 and 7 days, according to Ref. [11].

2.6. Statistical Analysis

The experimental data were evaluated using the program SPSS 19.0 and expressed according to the average ± standard deviation. Student t-tests were performed be-tween various groups to assess p values that were assumed to be significant when $p < 0.05$ was used.

3. Result and Discussion

3.1. Structural Characterization GO and MgO

The TEM image GO offers a flake-like shape with a thickness of almost 3–5 nm, which is higher than that of neat graphene (0.355 nm) owing to the oxygenated groups in GO as shown in Figure 1a. However, the TEM image of MgO/GO showed that the MgO possesses a spherical shape, which merged and created an agglomerate within flake-like shape of GO along with a mean particle size of almost 80–100 nm (Figure 1b). Moreover, the images show the homogenous distribution of MgO inside the GO sheets with great attachment. The homogeneous dispersion of nanofillers in the matrix, as described in the literature, is an important requirement for preparing nanocomposite coating with unique efficiency. As presented in Figure 1c, the XRD pattern of GO powder displays characteristic peaks at $2\theta = 10.03°$ and 42°, connected to the preferable orientation (001) and (100) of GO, respectively [25,27]. The XRD patterns of the MgO/GO powder clearly show the presence of MgO, where the diffraction peaks appear at 37.50°, 42.50°, and 59.20° (JCPDS 87–0653) corresponding to the (111), (200), and (220) respectively besides to the characteristic peaks of GO (Figure 1d). The reflections from the (111), (200), (220), (311), and (222) planes of cubic MgO were recognised in the XRD peaks which presented in the supporting information (Figure S1b). Two dominant peaks were observed in the Raman spectra (Figure 1e) of the neat GO, with D and G peaks being detected between 1355 and 1582 cm^{-1}, which is aligned with the vibrational modes of the D–band caused by structural defects and the G–band of the graphic materials [25,27]. The D peak and G peak are attributed to the stretching of the in-plane C–C bond in GO. The attained I_D/I_G intensity ratio values for GO are about 0.88 [28]. FT-IR spectra of GO, and MgO–GO are shown in Figure 1f. In the curve of GO, the characteristic peaks at 3430 and 1712 cm^{-1} correspond to the stretching vibrations of O–H and C=O bands, respectively, while the peak of C–OH at 1414 cm^{-1}, the stretching peak of C–C at 1052 cm^{-1}, and the vibration peak of C=C at 1633 cm^{-1} and original peaks of C=O (1740 cm^{-1}), C–O–C (1242 cm^{-1}), and C–O (1079 cm^{-1}) is related to the GO. In the FTIR spectrum of MgO/GO, the C=C peak (1650 cm^{-1}) and C–OH peak (1401 cm^{-1}) came from GO [25,27], while the peak at 460 cm^{-1} corresponds to the vibration of the Mg–O came from MgO.

3.2. Surface Analysis

As shown in Figure 2a, the microstructure of Mg-Zn-Ce alloy consists of α-Mg and a small number of intermetallic phases, which are mainly located within grain boundaries. According to the EDS test, Mg, Zn, and Ce are the marked area constituents, possibly showing the presence of $Mg_{12}Ce$ phases (Figure 2b,c). The XRD analysis will confirm the formation of the aforementioned phases.

Figure 3 shows the FESEM micrographs of the various surfaces. Without MgO-GO, the surface of the CS-coated Mg substrate shows a smooth layer with several close pores (Figure 3a). The evaporation of organic solvent during the drying process creates these pores. The CS/(1–3)MgO-GO coating, on the other hand, has a unified surface with nanoparticle size MgO-GO dispersed throughout, as well as a bubble-like structure created on the surface film (Figure 3b,c). The CS/2MgO-GO coating produced a more uniform layer with less close porosity, which has a significant impact on the corrosion behavior of the alloy. As the MgO-GO content increases, the surfaces of CS/GO films become rougher and more porous, as shown in Figure 3d. The EDS analysis (Area A–D) revealed that these particles were mostly comprised of O and Mg elements, implying that they were MgO. Both the GO and the CS matrix were found to be encapsulated in the deposited film, which was affirmed more by the existence of carbon element.

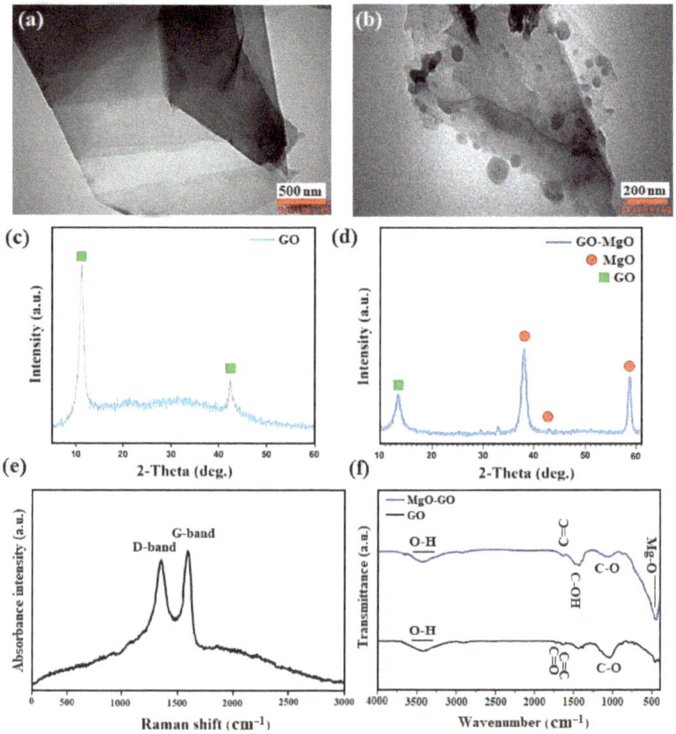

Figure 1. Transmission Electron Microscopy (TEM) images of (**a**) GO, (**b**) MgO/GO powders; X-ray Diffraction (XRD) patterns of (**c**) GO and (**d**) MgO-GO powders; and (**e**) Raman spectra of GO and (**f**) Fourier Transform Infrared (FTIR) patterns of GO and MgO-GO powders.

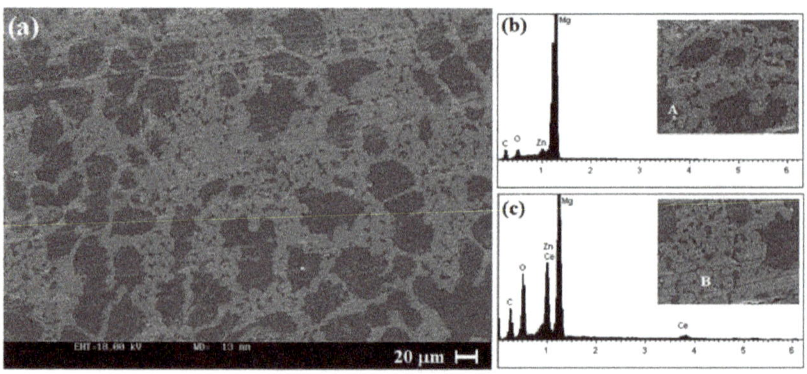

Figure 2. (**a**) Scanning Electron Microscope (SEM) micrographs of Mg-Zn-Ce alloy, and Energy Dispersive X-ray Spectroscopy (EDX) analysis of (**b**) area A and (**c**) area B.

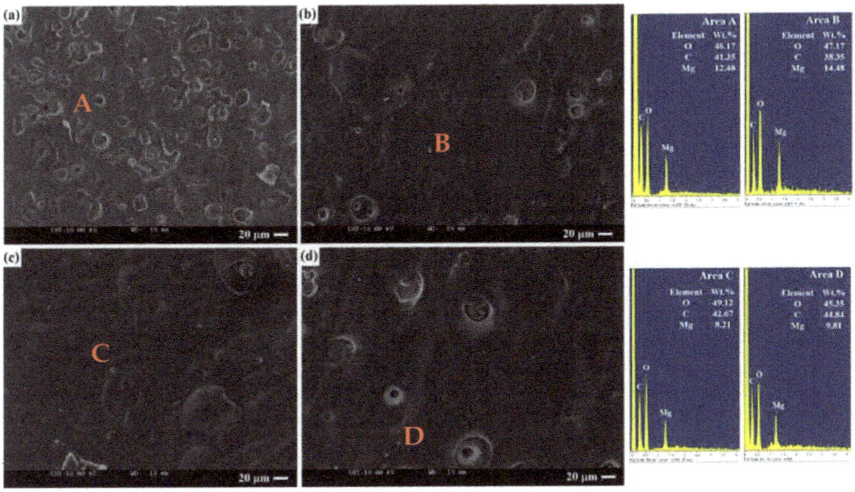

Figure 3. SEM images (**a**) CS, (**b**) CS/1MgO-GO, (**c**) CS/2MgO-GO, and (**d**) CS/3MgO-GO coated Mg alloys, and corresponding EDX analysis of Areas A, B, C, and D.

According to the XRD spectrum, the main phases in the Mg-Zn-Ce alloy are α-Mg, and $Mg_{12}Ce$ (Figure 4a). The CS coating exhibited broad peaks at $2\theta = 10°–20°$, whereas the CS/MgO-GO coating revealed the presence of an amorphous CS structure [29]. With the addition of GO and MgO, the intensity of the characteristic peak of CS decreases, indicating an increase in amorphous CS content. It was determined that the majority of the CS/MgO-GO coating was comprised of Mg and MgO. The CS/MgO-GO coating showed characteristic peaks at ~42.7° and ~62.1° related to (111) and (200) planes respectively corresponding to MgO [18,30]. Because of the coating structure, the substrate peaks were reflected in the patterns, allowing X-ray to reach the internal substrate. Commonly, the diffraction peak of GO appears around 12°, but no characteristic peak of GO was observed in this work. This could be attributed to the prepared films possess low amount of GO and also have poor crystallization [31]. In this respect, Askarnia et al. reported [12] that characteristic diffraction of graphene oxide at about 2θ = 10° was not detectable due to the low content of GO in HAp/CS/GO coatings.

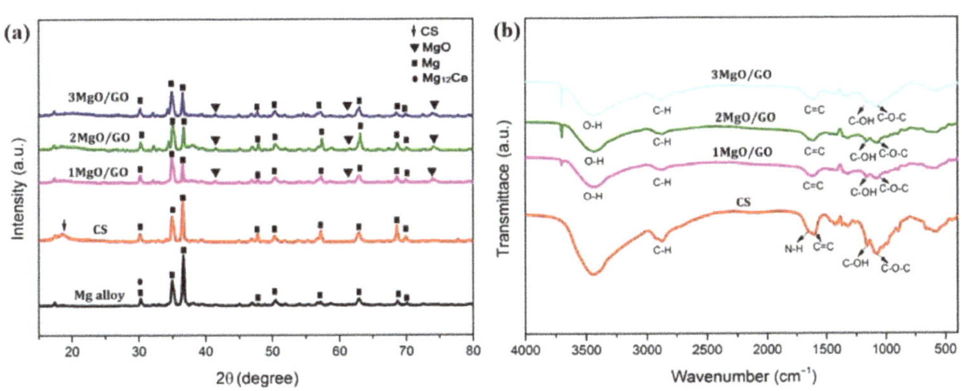

Figure 4. (**a**) XRD patterns and (**b**) FTIR analysis of CS, CS/1MgO-GO, CS/2MgO-GO, and CS/3MgO-GO coated Mg alloys.

The presence of graphene oxide, CS, and MgO in the coating was confirmed using FTIR analysis (Figure 4b). The most important bands in CS were stretching vibrations of O–H from carbohydrate rings and adsorbed water (3500–3450 cm^{-1}); N–H stretching in amine and amide (~3360 cm^{-1}); carbonyl bond (C=O) vibration in amide group at 1653 cm^{-1} and N–H bending vibration of amine group at 1580 cm^{-1} [12,32]. The peaks at 1405.8, 1153, and 1069.5 cm^{-1} correspond to the C−OH stretching vibration, C−C stretching vibration, and C−O vibration, respectively [29]. Peaks at 1047.6 and 1621.1 cm^{-1} for GO nanosheets correspond to the C−O−C stretching vibration and C−C stretching mode of the sp2 carbon skeletal network, respectively. The −OH groups in adsorbed water were responsible for the distinct peak at around 3400 cm^{-1}. Because of these oxygen-containing functional groups, GO is extremely hydrophilic [29]. As MgO increased up to 3 wt.%, a new peak at 3697 cm^{-1} was observed which is attributed to the stretching of H–O–H. It is well known that H_2O and CO_2 molecules are easily chemisorbed onto MgO surface when exposed to the atmosphere. However, as MgO concentration increased to 3 wt.%, this peak strength became stronger. The peak at 568 cm^{-1} corresponds to the vibration of the Mg–O. The FTIR spectra of CS/GO-MgO films clearly demonstrated that GO and MgO were combined in the CS matrix.

The thickness of the CS/MgO-GO coating was determined by cross-section observation to be between 38–42 µm (Figure 5a). The thickness of the CS-based films with different amounts of MgO-GO fillers did not differ significantly. Surface wettability is an important factor influencing corrosion performance and bioactivity of biomaterials [33]. Contact angle measurements were used to determine the wettability of the coated and uncoated samples (Figure 5b). Bare Mg alloy and neat CS have water contact angles (WCA) of 92.3° and 84.4°, respectively. Nevertheless, the WCA of the CS/MgO-GO coating decreased from 58.7° for 1MgO-GO to 51.2° for 3MgO-GO as the concentration of MgO-GO increased. The surface wettability of the coated samples was higher than that of the uncoated Mg alloy. This behavior can be attributed to the presence of functional groups (carboxyl group, carbonyl group, and hydroxyl group) at the edges of GO, which contributes to the hydrophilicity of the coatings, as well as the hydrophilic nature of chitosan [12,33] in the CS/MgO-GO nanocomposite coatings. Furthermore, increasing the MgO-GO content of CS-based coatings improved their hydrophilicity characteristics.

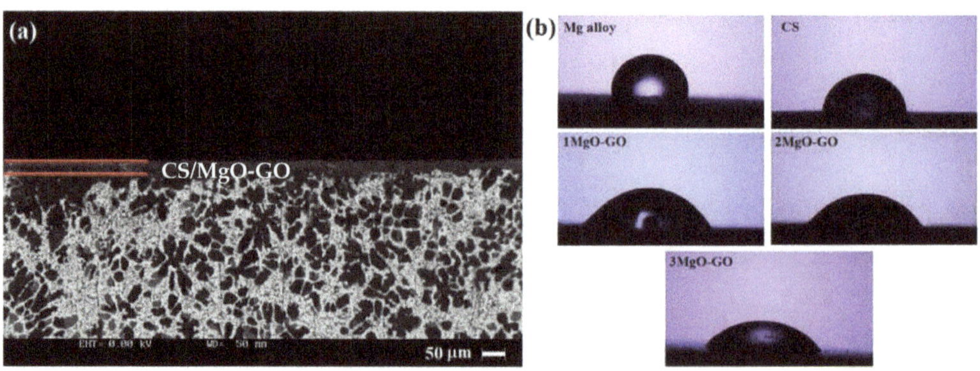

Figure 5. (**a**) Cross-section SEM image, and (**b**) images of water contact angle of uncoated, CS, CS/1MgO-GO, CS/2MgO-GO, and CS/3MgO-GO coated Mg alloys.

The adhesion test was used to evaluate the adhesion strength of the CS and CS/MgO-GO nanocomposite coatings deposited on Mg alloy. According to Figure 6, the adhesion of the CS coating on the Mg alloy was 2.64 MPa before composite coating, but this value increased to 4.42 MPa after composite coating, demonstrating the effective role of MgO-GO fillers in improving coating adhesion strength. In fact, one of the most important factors in achieving long-term implant use without premature failure *in vivo* is strong interfacial bonding between them [22]. There are oxygen functional groups at the GO sheet's edges which can conveniently form hydrogen bonds with functional oxygen groups comprising CS. At the same time, there is strong bonding between the Mg-O and CS layer sites (=O). On the other hand, the integration of MgO-GO into the CS layer, on the other hand, has a major effect on the adhesion strength between the Mg alloy specimen and the CS/MgO-GO nanocomposite coating [22,26]. In this regard, it was reported [34] that the newly formed hydroxyl groups on the Mg alloy surface may improve coating adhesion to the substrate via chemical interaction with coating molecules.

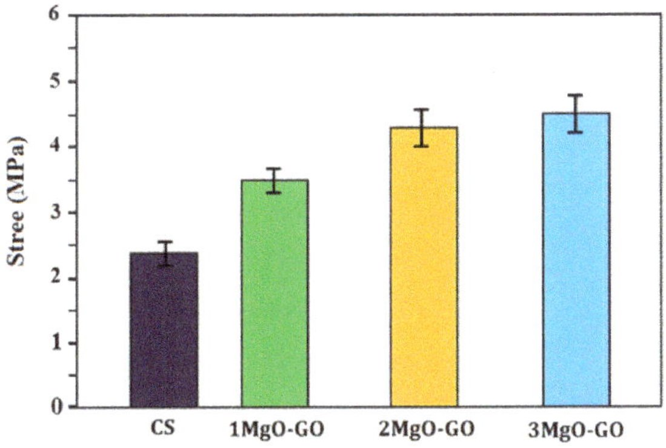

Figure 6. Bond strengths of CS, CS/1MgO-GO, CS/2MgO-GO, and CS/3MgO-GO coated Mg alloys.

3.3. *Corrosion Behavior*

The corrosion resistance of Mg alloy, CS, and CS/MgO-GO was evaluated in SBF using a potentiodynamic polarization test, and the corresponding polarization curves are shown in Figure 7a. When compared to Mg alloy (i_{corr} = 167.82 μA/cm^2), the i_{corr} value of CS shifted to a relatively negative position (50.22 μA/cm^2), whereas the i_{corr} value of CS/2MgO-GO presented a much lower corrosion current density (14.98 μA/cm^2) when compared to Mg alloy. The corrosion rate is known to be directly proportional to i_{corr}. Although the CS coating corroded more readily, this could be due to its porous surface with more defect structure, which increased the contact area between the corrosive fluid and the substrate [5]. The surface of the CS coating containing close pores was filled with MgO-GO fillers, resulting in improved corrosion resistance of the composite coating.

EIS analysis was carried out by generating Nyquist plots for the bare Mg, CS, and CS/MgO-GO coatings, as shown in Figure 7b. Based on the fitting results, the R_{ct} values for CS (1021.5 Ω.cm^2) and CS/MgO-GO were 1536.5 Ω.cm^2 (1MgO-GO), 1808.6 Ω.cm^2 (2MgO-GO), and 1257.2 Ω.cm^2 (3MgO-GO), which were significantly higher than the bare Mg alloy (401.2 Ω.cm^2). The results confirmed the higher electrochemical impedance of the CS/MgO-GO composite coating for Mg alloy, protecting substrates from over-quick degradation. The lower R_{ct} value of CS/3MgO-GO coated specimens compared to CS/2MgO-GO and CS/1MgO-GO coated samples is due to the presence of more MgO/GO aggregates in the CS/3MgO-GO film, which led to the reduction in corrosion rate. A higher value of the low-frequency impedance modulus, |Z|, indicates better corrosion protection performance [35]. As shown in Figure 7c, the value of |Z| for the CS and CS/MgO-GO coatings was greater than that of the uncoated Mg alloy. Furthermore, the phase angle (θ) at high frequency can be used to assess the completeness of the coating under the influence of corrosive electrolyte [35,36]. In Figure 7c, the phase angles of coated samples in the medium frequency range became loftier, indicating that a passivation film had formed on the surface of the samples as an inhibitor against corrosion. A complete coating has a phase angle close to 90°, whereas the bare substrate has a phase angle close to 0° [37]. The uncoated has a high frequency of phase angle of 50°, demonstrating the sample's unsatisfactory integrity. The capacitance behavior of CS/MgO-GO coating has the widest frequency range, and its phase angle is increased by approximately 60°, indicating that the formation of CS/MgO-GO coating can improve the corrosion resistance of the substrate. The equivalent is used to characterize the samples, where R_s denotes the solution resistance, CPE coat denotes the coating capacitance, R_{ct} denotes the charge transfer resistance, and CPE-dl denotes the electric double-layer capacitance of the passive film (Figure 7d). This finding indicates that the coencapsulation of MgO and GO fillers into CS film have the potential to elevate the corrosion resistance of the Mg alloy substrate.

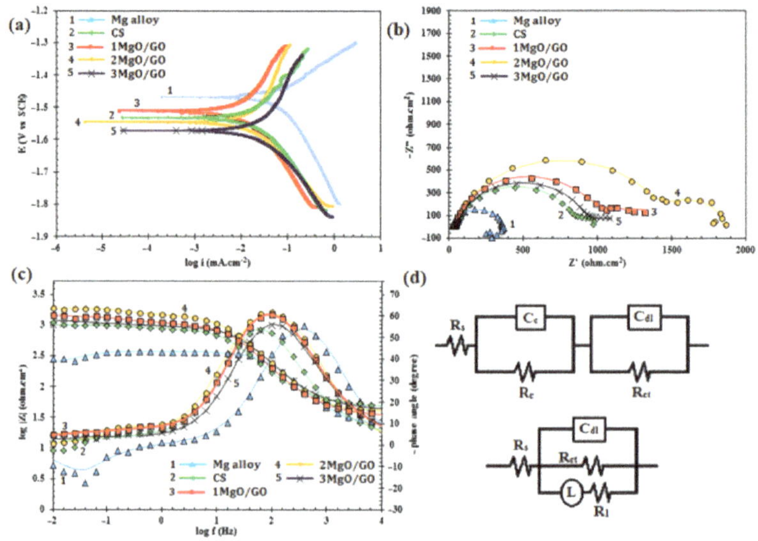

Figure 7. Electrochemical measurements of uncoated, CS, CS/1MgO-GO, CS/2MgO-GO, and CS/3MgO-GO coated Mg samples in m-SBF solution. (**a**) Potentiodynamic polarization curves; (**b**) Nyquist plots and (**c**) representative impedance (Bode magnitude and phase) spectra; and (**d**) equivalent circuit for impedance spectra.

The morphology of the surface of the various specimens after 14 days of immersion in the SBF is depicted in Figure 8a–e. After 14 days, the uncoated Mg alloy showed obvious degradation, with pitting corrosion clearly visible and a loose corrosion layer with deep cracks developing on the uncoated Mg alloy. The CS coating, on the other hand, remained intact after 14 days, with only a few surface cracks and no evidence of pitting corrosion was found. A thick layer of biomineralization was present on the surface with numerous white spherical particles, as seen in the high-magnification image. A similar biomineralization layer was observed for the CS/MgO-GO coating. The CS/MgO-GO coating's biomineralization layer was smoother with fewer defects compared to CS coating. According to the EDS spectra (Figure 8f), the precipitates layer consists of Mg, O, C, Ca, and P demonstrating the $Mg(OH)_2$ and hydroxyapatite (HAp) creation during the immersion. A large amount of mineralization layer covered the surface of the CS/MgO-GO coated film. After 14 days of immersion, CS containing GO and MgO deposited substrates show the formation of compact apatite precipitates on their surfaces. The inclusion of graphene oxide and MgO accelerate the formation of HAp on the coated surface. Surface functional groups on the surface of GO and MgO, such as –COOH and –OH, act as nucleating centers and provide a superior biomineralization process [21,22,33]. These functional groups provide negative charges on the surfaces, which encourages more Ca^{2+} cations from the SBF medium to interact electrostatically, resulting in the formation of an apatite-rich layer. Furthermore, the hydroxyl ions (OH^-) through SBF are electrically drawn to this Ca-rich positively charged surface, resulting in the precipitation of a bone apatite compound on the surfaces of the implant.

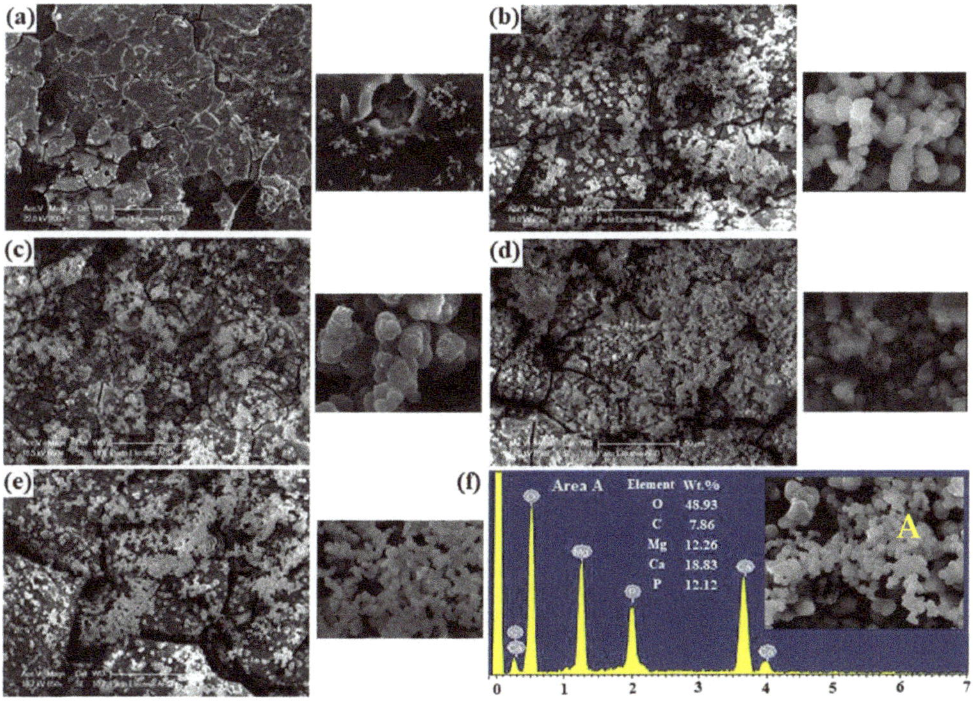

Figure 8. SEM micrographs and EDS plane-scan results of (**a**) uncoated, (**b**) CS, (**c**) CS/1MgO-GO, (**d**) CS/2MgO-GO, and (**e**) CS/3MgO-GO coated Mg samples immersed in m-SBF solution for 7 days and (**f**) EDX analysis of Area A.

During the biodegradation procedure, the coated Mg alloy had significantly lower pH values than the Mg alloy without coating. During the first 7 days, the pH value of the bare Mg alloy samples increased dramatically from 7.40 to 10.10, while the trend was reversed for the coated Mg alloy samples, particularly the CS/MgO-GO samples (Figure 9a). The pH value of the CS/MgO-GO coated Mg alloy sample increased steadily and slowly over 14 days, changing only from 7.88 to 8.76. The slope of the pH curve of the bare and coated samples decreased as the immersion time increased for all specimens. This finding may be due to the samples' reduced surface area, which delay initiation of corrosion process. During the first few days of exposure, a larger surface region was exposed to the solution, resulting in more degradation processes and the generation of hydroxide (OH) ions, increasing the pH value. However, when the exposure time was increased to 7 and 14 days, the pH value of the composites decreased and then remained stable. This phenomenon could be explained by the longer immersion time, which results in the development of protective films from corrosion products on the sample surface. These corrosion product films inhibit the specimens from coming into direct contact with the solution, lowering the pH of the specimens. Meanwhile, as shown in Figure 9b, after 14 days of immersion, the coated samples released significantly less hydrogen than the bare Mg alloy (22.8 mL/cm^2). The measured volume of hydrogen evolution for CS coated samples was 12.7 mL/cm^2, 8.8 mL/cm^2 for CS/1MgO-GO coating, and 7.9 mL/cm^2 for CS/2MgO-GO coating, all of which agreed with the pH value results. This was mainly due to the compact and protective layer of CS/MgO-GO that was effectively applied to the Mg alloy surface, which can successfully preserve the substrate from the corrosion medium and certainly decrease the hydrogen evolution rate [37]. Furthermore, the weight loss of the CS/MgO-GO sample was 0.58 mm/y, which was much lower compared to the CS coating (1.24 mm/y) and bare Mg alloy (2.47 mm/y) (Figure 9c). These findings indicated that the CS/MgO-GO coating showed the greatest corrosion resistance even after a precise 14-day immersion test. The bare sample has the highest corrosion rate, while the coated specimen has lower corrosion rates because CS/MgO-GO acts as a barrier film and reduces the infiltration rate of corrosive solution inside the substrate, and hence decline the biodegradation rate. Furthermore, the existence of MgO particles and GO within the CS film can operate as a reinforcing agent, increasing the coating layer's stability. Furthermore, these particles promote the formation of corrosion products containing hydroxyapatite, which slows the penetration of SBF to the substrate, and provide protection resulted in slowing the degradation rate. XRD spectra for the composite coating were used to confirm the precipitation of HAp on the surface of the composite coating containing GO and MgO, as shown in Figure 9d. The XRD pattern of the CS/MgO-GO composite coatings verify that the corrosion products were primarily composed of $Mg(OH)_2$ and HAp, implying that the biomineralization layer was primarily containing HAp.

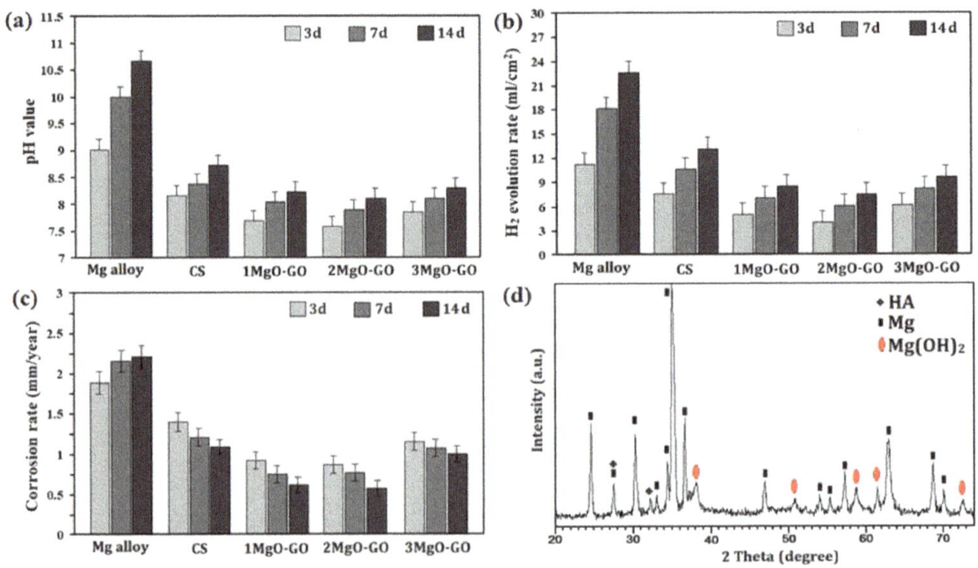

Figure 9. (**a**) pH value, (**b**) hydrogen evolution, (**c**) corrosion rate for uncoated Mg alloy, CS, CS/1MgO-GO, CS/2MgO-GO, and CS/3MgO-GO coated Mg samples coated Mg alloys versus soaking time in SBF. (**d**) X-ray diffraction patterns of CS/3MgO-GO coated Mg alloy after soaking in SBF.

3.4. Antibacterial Effect

The ability of the produced coatings to inhibit gram-negative (*Escherichia coli*) bacteria was investigated through direct contact with an agar medium. The results showed that the CS/MgO-GO coatings on Mg alloy substrates inhibited the bacterial activity of the strains (Figure 10). In comparison to the large inhibition area created for gram (−) bacteria (5.3 mm), neither the bare Mg alloy samples nor the coatings with CS coating successfully inhibited bacterial growth, even though, as reported in the literature, CS could possess antibacterial properties [8,38]. The low concentration of chitosan in the coatings may have influenced the results [38]. At the same time, the zones of antibacterial activity against *E. coli* colonies increased in the sequence of bare Mg alloy < CS < CS/1MgO-GO < CS/2MgO-GO < CS/3MgO-GO. The release of MgO and GO has a strong inhibitory effect on bacterial growth. When bacterial cells come into contact with MgO and GO, they are eventually destroyed due to various mechanisms such as oxidative stress, membrane stress, and/or wrapping isolation. Bacterial growth will be inhibited bacteriostatically or bactericidal by these mechanisms, which may operate independently or in combination [26]. Among the various suggested mechanisms for GO's antimicrobial performance, mechanical destruction to bacteria membranes through sharp edges of GO when contacting directly with GO and the formation of reactive oxygen species (ROS) occurred when oxidative stress generated by GO and charge transfer are more circumstances [7,39]. MgO as another antibacterial agent reacts with intracellular oxygen to produce ROS such as H_2O_2, $OH^\bullet$ and $O_2^{\bullet-}$, which may be released into the medium and thus inhibit bacterial proliferation and growth [19,26]. Consequently, co-encapsulation of MgO-GO into the CS layer surface significantly inhibited the bacterial activity of the strains; thus, CS/MgO-GO coating containing the highest amount of MgO-GO showed the best antibacterial activity.

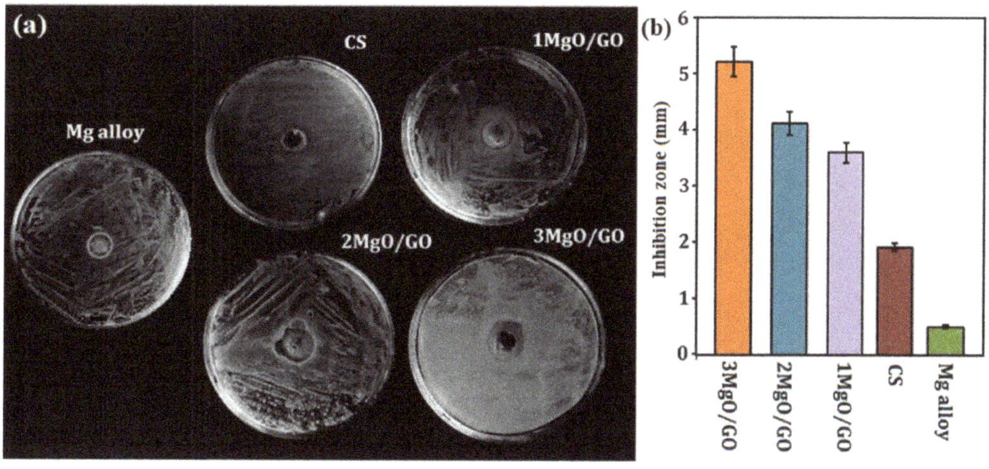

Figure 10. (**a**) Images of inhibition zones, and (**b**) values of growth-inhibition zones of uncoated Mg alloy, CS, CS/1MgO-GO, CS/2MgO-GO, and CS/3MgO-GO coated Mg samples after 24 h.

3.5. Cytotoxicity Assay

The MG63 cell response to uncoated, CS, and CS/MgO-GO coated samples was evaluated using an indirect assay. Figure 11a illustrates the changes in cell viability as a function of culture time. All extracts of CS/MgO-GO coated Mg alloy enhanced the cell viability of MG-63 cells compared to the uncoated ones. Similarly, cells contacted with all extracts of CS and CS/MgO-GO coated specimens showed low cytotoxicity after 3 and 5 days of culture, whereas that was exacerbated by extending the culture time in uncoated. Cells incubated with extract of CS/2MgO-GO coated had significantly higher viability on day 3 (91% ± 4%) and day 5 (98% ± 4%). Compared to the other three CS/MgO-GO groups. However, there was no statistical diversity ($p > 0.05$) in MG63 cell viability between the CS/MgO-GO groups after 5 days, despite a slight trend toward increased cell viability in the uncoated group. The surface oxygen-containing functional groups of GO may stimulate cell attachment [40]. According to Depan et al. [41], the hydroxyl groups on GO can serve as locations for preserving and bringing cells onto fibrous surfaces, thereby improving cell adhesion. Similarly, Shuai et al. [42] demonstrated that GO could be used to create a bio-interface for cell proliferation and growth. One of the most widely used indicators of osteoblast cell differentiation is ALP activity. The ALP activity of the MG-63 cell line cultured in media of the uncoated extract, CS, and CS/MgO-GO coated samples is shown in Figure 11b. For 3 and 7 days of culture, the CS and CS/MgO-GO coated samples demonstrated good cell differentiation than the uncoated sample. After 7 days, the ALP activity of all samples increased noticeably. Furthermore, the CS/2MgO-GO coated specimen demonstrated the greatest ALP activity at both time points tested, implying a greater potential for cell differentiation. The increased corrosion rate in uncoated Mg alloys resulted in a greater release of Mg ions [43–52], causing the culture medium to become excessively alkalinization and local hydrogen evolution. All of these were detrimental to cell differentiation. Modification with the CS/MgO-GO composite coating, on the other hand, improved corrosion resistance of the substrate while providing a quite favorable surrounding for cell differentiation.

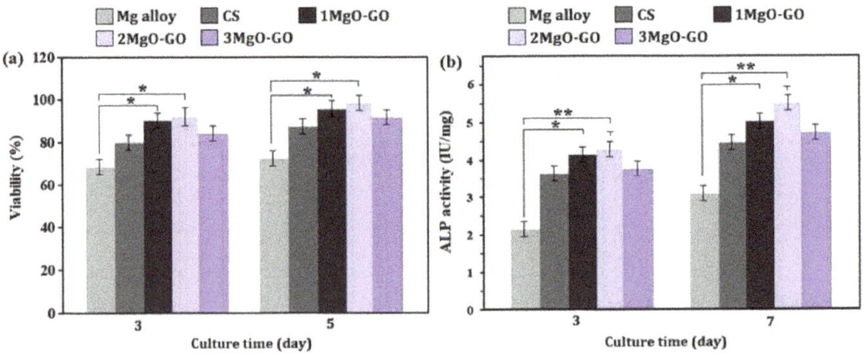

Figure 11. (**a**) Cell viability, and (**b**) ALP activity of MG63 cells cultured for various times on uncoated Mg, and Mg alloys coated with CS, CS/1MgO-GO, CS/2MgO-GO, and CS/3MgO-GO (* $p < 0.05$, and ** $p < 0.01$).

4. Conclusions

A novel CS/MgO-GO composite coating was deposited on a biodegradable Mg alloy in this study. The results of the electrochemical analysis showed that the i_{corr} values of CS (50.22 μA/cm^2) are lower compare with uncoated Mg alloy (167.82 μA/cm^2). After the addition of MgO and GO into the CS film, the i_{corr} value of CS/2MgO-GO was 14.98 μA/cm^2, further reduced relative to that of CS/3MgO-GO coating. This finding indicates that the coencapsulation of MgO and GO fillers into CS film has the potential to elevate the corrosion resistance of the Mg alloy substrate. The immersion test including H_2 evolution and weight loss indicates that the CS/MgO-GO composite coating may greatly enhance the corrosion resistance of Mg alloys. Meanwhile, after 14 days of immersion in SBF, apatite was observed on the coating surface, implying that incorporating MgO and GO in the CS film could enhance apatite formation and improve the Mg alloy bioactivity. The cytotoxicity study showed that the CS/MgO-GO composite coating improved the biocompatibility of Mg alloy substrates. Besides, the formation of apatite precipitate on the surface of the sample leads to a reduction of the degradation rate of Mg alloys. The coencapsulation of MgO and GO fillers into the CS coatings effectively enhances the antibacterial activity of Mg alloy. This research could provide a plausible and encouraging approach for preserving corrosion resistance while improving the biocompatibility and antibacterial properties of Mg alloy implants.

Supplementary Materials: The following supporting information can be downloaded at: https://www.mdpi.com/article/10.3390/coatings12010063/s1, Figure S1: XRD patterns of MgO powders; Table S1: Chemical composition of the Kokubo simulated body fluid (SBF).

Author Contributions: Methodology, formal analysis, writing—original draft preparation, M.Z.B., S.A. and A.N.; Conceptualization, supervision, visualization, writing—review and editing, A.F.I., S.S. and H.R.B.-R.; supervision, visualization, writing—review and editing, funding acquisition, S.R., M.D. and F.B. All authors have read and agreed to the published version of the manuscript.

Funding: This research received no external funding.

Institutional Review Board Statement: Not applicable.

Informed Consent Statement: Not applicable.

Data Availability Statement: All data provided in the present manuscript are available to whom it may concern.

Conflicts of Interest: The funders had no role in the design of the study; in the collection, analyses, or interpretation of data; in the writing of the manuscript, or in the decision to publish the results.

References

1. Chen, Y.; Wu, L.; Yao, W.; Chen, Y.; Zhong, Z.; Ci, W.; Wu, J.; Xie, Z.; Yuan, Y.; Pan, F. A self-healing corrosion protection coating with graphene oxide carrying 8-hydroxyquinoline doped in layered double hydroxide on a micro-arc oxidation coating. *Corros. Sci.* **2022**, *194*, 109941. [CrossRef]
2. Francis, A.; Yang, Y.; Boccaccini, A. A new strategy for developing chitosan conversion coating on magnesium substrates for orthopedic implants. *Appl. Surf. Sci.* **2019**, *466*, 854–862. [CrossRef]
3. Zhou, Z.; Zheng, B.; Gu, Y.; Shen, C.; Wen, J.; Meng, Z.; Chen, S.; Ou, J.; Qin, A. New approach for improving anticorrosion and biocompatibility of magnesium alloys via polydopamine intermediate layer-induced hydroxyapatite coating. *Surf. Interfaces* **2020**, *19*, 100501. [CrossRef]
4. Yang, Y.; Michalczyk, C.; Singer, F.; Virtanen, S.; Boccaccini, A.R. In vitro study of polycaprolactone/bioactive glass composite coatings on corrosion and bioactivity of pure Mg. *Appl. Surf. Sci.* **2015**, *355*, 832–841. [CrossRef]
5. Tian, P.; Liu, X.; Ding, C. In vitro degradation behavior and cytocompatibility of biodegradable AZ31 alloy with PEO/HT composite coating. *Colloids Surf. B Biointerfaces* **2015**, *128*, 44–54. [CrossRef] [PubMed]
6. Tiyyagura, H.R.; Rudolf, R.; Gorgieva, S.; Fuchs-Godec, R.; Boyapati, V.R.; Mantravadi, K.M.; Kokol, V. The chitosan coating and processing effect on the physiological corrosion behaviour of porous magnesium monoliths. *Prog. Org. Coat.* **2016**, *99*, 147–156. [CrossRef]
7. Pan, C.; Zhao, Y.; Yang, Y.; Yang, M.; Hong, Q.; Yang, Z.; Zhang, Q. Immobilization of bioactive complex on the surface of magnesium alloy stent material to simultaneously improve anticorrosion, hemocompatibility and antibacterial activities. *Colloids Surf. B Biointerfaces* **2021**, *199*, 111541. [CrossRef]
8. Li, J.; Zhuang, S. Antibacterial activity of chitosan and its derivatives and their interaction mechanism with bacteria: Current state and perspectives. *Eur. Polym. J.* **2020**, *138*, 109984. [CrossRef]
9. Rastegari, S.; Salahinejad, E. Surface modification of Ti-6Al-4V alloy for osseointegration by alkaline treatment and chitosan-matrix glass-reinforced nanocomposite coating. *Carbohydr. Polym.* **2019**, *205*, 302–311. [CrossRef]
10. Saadati, A.; Khiarak, B.N.; Zahraei, A.A.; Nourbakhsh, A.; Mohammadzadeh, H. Electrochemical characterization of Electrophoretically deposited Hydroxyapatite/Chitosan/Graphene Oxide composite coating on Mg substrate. *Surf. Interfaces* **2021**, *25*, 101290. [CrossRef]
11. Bakhsheshi-Rad, H.R.; Chen, X.; Ismail, A.F.; Aziz, M.; Abdolahi, E.; Mahmoodiyan, F. Improved antibacterial properties of an Mg-Zn-Ca alloy coated with chitosan nanofibers incorporating silver sulfadiazine multiwall carbon nanotubes for bone implants. *Polym. Adv. Technol.* **2019**, *30*, 1333–1339. [CrossRef]
12. Askarnia, R.; Fardi, S.R.; Sobhani, M.; Staji, H. Ternary hydroxyapatite/chitosan/graphene oxide composite coating on AZ91D magnesium alloy by electrophoretic deposition. *Ceram. Int.* **2021**, *47*, 27071–27081. [CrossRef]
13. Avcu, E.; Baştan, F.E.; Abdullah, H.Z.; Rehman, M.A.U.; Avcu, Y.Y.; Boccaccini, A.R. Electrophoretic deposition of chitosan-based composite coatings for biomedical applications: A review. *Prog. Mater. Sci.* **2019**, *103*, 69–108. [CrossRef]
14. Güler, Ö.; Bağcı, N. A short review on mechanical properties of graphene reinforced metal matrix composites. *J. Mater. Res. Technol.* **2020**, *9*, 6808–6833. [CrossRef]
15. Karimi, N.; Kharaziha, M.; Raeissi, K. Electrophoretic deposition of chitosan reinforced graphene oxide-hydroxyapatite on the anodized titanium to improve biological and electrochemical characteristics. *Mater. Sci. Eng. C* **2019**, *98*, 140–152. [CrossRef] [PubMed]
16. Khalajabadi, S.Z.; Kadir, M.R.A.; Izman, S.; Bakhsheshi-Rad, H.R.; Farahany, S. Effect of mechanical alloying on the phase evolution, microstructure and bio-corrosion properties of a Mg/HA/TiO_2/MgO nanocomposite. *Ceram. Int.* **2014**, *40*, 16743–16759. [CrossRef]
17. Shuai, C.; Wang, B.; Bin, S.; Peng, S.; Gao, C. Interfacial strengthening by reduced graphene oxide coated with MgO in biodegradable Mg composites. *Mater. Des.* **2020**, *191*, 108612. [CrossRef]
18. Shuai, C.; Zan, J.; Qi, F.; Wang, G.; Liu, Z.; Yang, Y.; Peng, S. nMgO-incorporated PLLA bone scaffolds: Enhanced crystallinity and neutralized acidic products. *Mater. Des.* **2019**, *174*, 107801. [CrossRef]
19. Raveesha, H.; Nayana, S.; Vasudha, D.; Begum, J.S.; Pratibha, S.; Ravikumara, C.; Dhananjaya, N. The electrochemical behavior, antifungal and cytotoxic activities of phytofabricated MgO nanoparticles using Withania somnifera leaf extract. *J. Sci. Adv. Mater. Devices* **2019**, *4*, 57–65. [CrossRef]
20. Rahnamaee, S.Y.; Bagheri, R.; Heidarpour, H.; Vossoughi, M.; Golizadeh, M.; Samadikuchaksaraei, A. Nanofibrillated chitosan coated highly ordered titania nanotubes array/graphene nanocomposite with improved biological characters. *Carbohydr. Polym.* **2021**, *254*, 117465. [CrossRef]

21. Munir, K.S.; Wen, C.; Li, Y. Carbon nanotubes and graphene as nanoreinforcements in metallic biomaterials: A review. *Adv. Biosyst.* **2019**, *3*, 1800212. [CrossRef]
22. Shahin, M.; Wen, C.; Munir, K.; Li, Y. Mechanical and corrosion properties of graphene nanoplatelet–reinforced Mg–Zr and Mg–Zr–Zn matrix nanocomposites for biomedical applications. *J. Magnes. Alloy.* **2021**. [CrossRef]
23. Bakhsheshi-Rad, H.R.; Hamzah, E.; Daroonparvar, M.; Saud, S.N.; Abdul-Kadir, M.R. Bi-layer nano-TiO_2/FHA composite coatings on Mg–Zn–Ce alloy prepared by combined physical vapour deposition and electrochemical deposition methods. *Vacuum* **2014**, *110*, 127–135. [CrossRef]
24. Yu, H.; Zhang, B.; Bulin, C.; Li, R.; Xing, R. High-efficient Synthesis of Graphene Oxide Based on Improved Hummers Method. *Sci. Rep.* **2016**, *6*, 36143. [CrossRef] [PubMed]
25. Guo, T.; Bulin, C. Facile preparation of MgO/graphene oxide nanocomposite for efficient removal of aqueous Congo red: Adsorption performance and interaction mechanism. *Res. Chem. Intermed.* **2021**, *47*, 945–971. [CrossRef]
26. Bakhsheshi-Rad, H.R.; Ismail, A.F.; Aziz, M.; Akbari, M.; Hadisi, Z.; Khoshnava, S.M.; Pagan, E.; Chen, X. Co-incorporation of graphene oxide/silver nanoparticle into poly-L-lactic acid fibrous: A route toward the development of cytocompatible and antibacterial coating layer on magnesium implants. *Mater. Sci. Eng. C.* **2020**, *111*, 110812. [CrossRef] [PubMed]
27. Abdel-Aal, S.K.; Ionov, A.; Mozhchil, R.N.; Naqvi, A.H. Simple synthesis of graphene nanocomposites MgO–rGO and Fe_2O_3–rGO for multifunctional applications. *Appl. Phys. A* **2018**, *124*, 365. [CrossRef]
28. Bordbar-Khiabani, A.; Ebrahimi, S.; Yarman, B. Highly corrosion protection properties of plasma electrolytic oxidized titanium using rGO nanosheets. *Appl. Surf. Sci.* **2019**, *486*, 153–165. [CrossRef]
29. Duan, J.; Liu, F.; Kong, Y.; Hao, M.; He, J.; Wang, J.; Wang, S.; Liu, H.; Sang, Y. Homogeneous Chitosan/Graphene Oxide Nanocomposite Hydrogel-Based Actuator Driven by Efficient Photothermally Induced Water Gradients. *ACS Appl. Nano Mater.* **2020**, *3*, 1002–1009. [CrossRef]
30. Li, H.; Lu, S.; Qin, W.; Wu, X. In-situ grown MgO-ZnO ceramic coating with high thermal emittance on Mg alloy by plasma electrolytic oxidation. *Acta Astronaut.* **2017**, *136*, 230–235. [CrossRef]
31. Liu, L.; Peng, F.; Zhang, D.; Li, M.; Huang, J.; Liu, X. A tightly bonded reduced graphene oxide coating on magnesium alloy with photothermal effect for tumor therapy. *J. Magnes. Alloy.* **2021**. [CrossRef]
32. Pishbin, F.; Mouriño, V.; Flor, S.; Kreppel, S.; Salih, V.; Ryan, M.P.; Boccaccini, A.R. Electrophoretic deposition of gentamicin-loaded bioactive glass/chitosan composite coatings for orthopaedic implants. *ACS Appl. Mater. Interfaces* **2014**, *6*, 8796–8806. [CrossRef]
33. Baradaran, S.; Moghaddam, E.; Basirun, W.J.; Mehrali, M.; Sookhakian, M.; Hamdi, M.; Moghaddam, M.R.N.; Alias, Y. Mechanical properties and biomedical applications of a nanotube hydroxyapatite-reduced graphene oxide composite. *Carbon* **2014**, *69*, 32–45. [CrossRef]
34. Asadi, H.; Suganthan, B.; Ghalei, S.; Handa, H.; Ramasamy, R.P. A multifunctional polymeric coating incorporating lawsone with corrosion resistance and antibacterial activity for biomedical Mg alloys. *Prog. Org. Coat.* **2021**, *153*, 106157. [CrossRef]
35. Kartsonakis, I.; Balaskas, A.; Koumoulos, E.; Charitidis, C.; Kordas, G. Evaluation of corrosion resistance of magnesium alloy ZK10 coated with hybrid organic–inorganic film including containers. *Corros. Sci.* **2012**, *65*, 481–493. [CrossRef]
36. Wang, C.; Fang, H.; Qi, X.; Hang, C.; Sun, Y.; Peng, Z.; Wei, W.; Wang, Y. Silk fibroin film-coated MgZnCa alloy with enhanced in vitro and in vivo performance prepared using surface activation. *Acta Biomater.* **2019**, *91*, 99–111. [CrossRef] [PubMed]
37. Guo, Y.; Jia, S.; Qiao, L.; Su, Y.; Gu, R.; Li, G.; Lian, J. Enhanced corrosion resistance and biocompatibility of polydopamine/dicalcium phosphate dihydrate/collagen composite coating on magnesium alloy for orthopedic applications. *J. Alloy. Compd.* **2020**, *817*, 152782. [CrossRef]
38. Ballarre, J.; Aydemir, T.; Liverani, L.; Roether, J.A.; Goldmann, W.H.; Boccaccini, A.R. Versatile bioactive and antibacterial coating system based on silica, gentamicin, and chitosan: Improving early stage performance of titanium implants. *Surf. Coat. Technol.* **2020**, *381*, 125138. [CrossRef]
39. Hegab, H.M.; ElMekawy, A.; Zou, L.; Mulcahy, D.; Saint, C.; Ginic-Markovic, M. The controversial antibacterial activity of graphene-based materials. *Carbon* **2016**, *105*, 362–376. [CrossRef]
40. Kim, J.D.; Yun, H.; Kim, G.C.; Lee, C.W.; Choi, H.C. Antibacterial activity and reusability of CNT-Ag and GO-Ag nanocomposites. *Appl. Surf. Sci.* **2013**, *283*, 227–233. [CrossRef]
41. Depan, D.; Girase, B.; Shah, J.S.; Misra, R.D.K. Structure–process–property relationship of the polar graphene oxide-mediated cellular response and stimulated growth of osteoblasts on hybrid chitosan network structure nanocomposite scaffolds. *Acta Biomater.* **2011**, *7*, 3432–3445. [CrossRef]
42. Shuai, C.; Guo, W.; Wu, P.; Yang, W.; Hu, S.; Xia, Y.; Feng, P. A graphene oxide-Ag co-dispersing nanosystem: Dual synergistic effects on antibacterial activities and mechanical properties of polymer scaffolds. *Chem. Eng. J.* **2018**, *347*, 322–333. [CrossRef]
43. Dziková, J.; Fintová, S.; Kajánek, D.; Florková, Z.; Wasserbauer, J.; Doležal, P. Characterization and corrosion properties of fluoride conversion coating prepared on AZ31 magnesium alloy. *Coatings* **2021**, *11*, 675. [CrossRef]
44. Wang, Z.; Ye, F.; Chen, L.; Lv, W.; Zhang, Z.; Zang, Q.; Peng, J.; Sun, L.; Lu, S. Preparation and Degradation Characteristics of MAO/APS Composite Bio-Coating in Simulated Body Fluid. *Coatings* **2021**, *11*, 667. [CrossRef]
45. Anawati, A.; Fitriana, M.; Gumelar, M. Improved Corrosion Resistance of Magnesium Alloy AZ31 in Ringer Lactate by Bilayer Anodic Film/Beeswax–Colophony. *Coatings* **2021**, *11*, 564. [CrossRef]
46. Wei, J.; Cai, S.; Li, Q.; Xie, Y.; Bao, X.; Xu, G. Controllable Synthesis of Nanostructured Ca-P Coating on Magnesium Alloys via Sodium Citrate Template-Assisted Hydrothermal Method and Its Corrosion Resistance. *Coatings* **2020**, *10*, 1232. [CrossRef]

47. Zhang, R.; Zhang, Z.; Zhu, Y.; Zhao, R.; Zhang, S.; Shi, X.; Li, G.; Chen, Z.; Zhao, Y. Degradation Resistance and In Vitro Cytocompatibility of Iron-Containing Coatings Developed on WE43 Magnesium Alloy by Micro-Arc Oxidation. *Coatings* **2020**, *10*, 1138. [CrossRef]
48. Toulabifard, A.; Rahmati, M.; Raeissi, K.; Hakimizad, A.; Santamaria, M. The Effect of Electrolytic Solution Composition on the Structure, Corrosion, and Wear Resistance of PEO Coatings on AZ31 Magnesium Alloy. *Coatings* **2020**, *10*, 937. [CrossRef]
49. Wang, S.; Fu, L.; Nai, Z.; Liang, J.; Cao, B. Comparison of Corrosion Resistance and Cytocompatibility of MgO and ZrO_2 Coatings on AZ31 Magnesium Alloy Formed via Plasma Electrolytic Oxidation. *Coatings* **2018**, *8*, 441. [CrossRef]
50. Hou, S.; Yu, W.; Yang, Z.; Li, Y.; Yang, L.; Lang, S. Properties of Titanium Oxide Coating on MgZn Alloy by Magnetron Sputtering for Stent Application. *Coatings* **2020**, *10*, 999. [CrossRef]
51. Li, S.; Yi, L.; Zhu, X.; Liu, T. Ultrasonic Treatment Induced Fluoride Conversion Coating without Pores for High Corrosion Re-sistance of Mg Alloy. *Coatings* **2020**, *10*, 996. [CrossRef]
52. Fu, L.; Yang, Y.; Zhang, L.; Wu, Y.; Liang, J.; Cao, B. Preparation and Characterization of Fluoride-Incorporated Plasma Electrolytic Oxidation Coatings on the AZ31 Magnesium Alloy. *Coatings* **2019**, *9*, 826. [CrossRef]

coatings

MDPI

Article

On the Effects of High and Ultra-High Rotational Speeds on the Strength, Corrosion Resistance, and Microstructure during Friction Stir Welding of Al 6061-T6 and 316L SS Alloys

Zhipeng Li, Shujin Chen *, Lingfei Meng, Yang Gao, Zhidong Yang, Mingxiao Shi, Xinyi Chen, Hao Zhang and Yuye Zhang

Jiangsu Provincial Key Laboratory of Advanced Welding Technology, Jiangsu University of Science and Technology, Zhenjiang 212000, China; 192060018@stu.just.edu.cn (Z.L.); 18452483921@163.com (L.M.); 18852894163@163.com (Y.G.); yangzhidong@just.edu.cn (Z.Y.); smx_just@163.com (M.S.); chenxinyi@just.edu.cn (X.C.); 192060003@stu.just.edu.cn (H.Z.); 199060077@stu.just.edu.cn (Y.Z.)

* Correspondence: 200800002666@just.edu.cn

Abstract: In this study, under the conditions of using tools at a high rotational speed (HRS) of 10,000 rpm and an ultra-high rotational speed (ultra-HRS) of 18,000 rpm, the produced welding heat input was utilized to weld two specimens of Al alloy 6061-T6 with 1.0 mm thickness and 316L SS with 0.8 mm thickness. The microstructural characteristics, mechanical properties, and electrochemical corrosion properties of the aluminum alloy–steel joints were analyzed. The higher tool offset forms an intermetallic compound layer of less than 1 μm at the Fe-Al interface on the advancing side (AS) at different speeds. This results in a mixed zone structure. The lower tool offset forms intermetallic compounds of only 2 μm. The formation of a composite material based on aluminum alloy in the weld nugget zone improves the hardness value. The intermetallic compounds are Fe_3Al and $FeAl_3$, respectively. It was observed that the formation of intermetallic compounds is solely related to the rotational speed, and the iron-rich intermetallic compounds produced under ultra-HRS parameters have higher corrosion resistance. When the tool offset is 0.55 mm, using the HRS parameters, the tensile strength is 220.8 MPa (about 75.9% of that of the base metal).

Keywords: friction stir welding; aluminum and steel dissimilar metals butt; high rotational speed; ultra-high rotational speed; intermetallic compound

Citation: Li, Z.; Chen, S.; Meng, L.; Gao, Y.; Yang, Z.; Shi, M.; Chen, X.; Zhang, H.; Zhang, Y. On the Effects of High and Ultra-High Rotational Speeds on the Strength, Corrosion Resistance, and Microstructure during Friction Stir Welding of Al 6061-T6 and 316L SS Alloys. *Coatings* **2021**, *11*, 1550. https://doi.org/10.3390/coatings11121550

Academic Editor: Devis Bellucci

Received: 10 November 2021
Accepted: 14 December 2021
Published: 17 December 2021

1. Introduction

To move beyond the "carbon peak", as recently proposed by the Chinese government, there has been an increased demand to meet the lightweight requirements of the automobile, rail transit, and shipbuilding industries. These industries widely use certain thin-plate materials. However, to achieve the minimized weight of material structures, the thinner sheets made of the same material fail to meet the strength requirements. Therefore, finding an economical and dependable method of connecting dissimilar materials is of the utmost importance [1–3].

A better lightweight alternative is the use of Al alloy with stainless steel [4] connections. The highly economical, lightweight 6××× series Al alloy, offering superior corrosion resistance, has often been used in various applications [5–7]. Austenitic SS is widely utilized for aircraft parts, ship parts, and auto parts due to its excellent corrosion resistance and sound weldability. The welding heat input leads to the mutual diffusion of aluminum and iron to form intermetallic compounds (IMC). At the welding interface, Al and Fe form a high-quality connection, mainly through the formation of micron-sized and uniformly distributed IMC [8,9]. As 6061-T6 Al alloy has small amounts of Si, the thickness of the IMC can be slightly reduced, thereby improving the microstructure and properties of the welding joint [10]. Therefore, the combination of the above two materials has attracted widespread attention in various application fields [11–15], such as the new energy vehicle frames, the connection of tailor-welded blanks, sealing covers, and other thin-walled structures.

There is still a crucial question to be answered about how to choose the connection method between SS and Al alloy. In the traditional fusion welding method, the higher heat input produces a thicker IMC layer. This may lead to crack propagation and reduce the mechanical properties of the joint. In comparison, friction stir welding (FSW) has numerous advantages. Primarily, the low welding-heat input will reduce sheet deformation. Furthermore, it does not produce welding defects caused by the molten state, such as pores, burn-through, cold cracks, and hydrogen embrittlement. Additionally, the thickness of the IMC layer generated at the Al-Fe interface can be precisely controlled [16]. As reported [11,17], some induction heating when using FSW technologies has achieved good results. Nevertheless, the safety risks have been heightened by adding auxiliary devices to increase the welding energy.

At present, the research on the friction stir welding of Al-Fe material only explores the use of medium and thick plates (thickness 1 mm). Ramachandran et al. [9] conducted experiments on the FSW of AA5052-H32 and hot-rolled HRSLA steel IRSM42–93 with 3 mm thickness under a rotational speed of 500 rpm, a welding speed of 50 mm/min, and a welding axial force of 7 KN. Scanning electron microscopy (SEM) was used to observe the distribution of IMC thickness produced by tapered cylindrical and straight cylindrical tools. It was observed that the IMC distribution at the Al-Fe interface of the cylindrical tool with a taper of 10° was the most uniform. In comparison to the straight cylindrical tool, the geometric shape of the tapered cylindrical tool is conducive to the generation of IMC at the bottom of the Al-Fe interface. X. Liu et al. [8,18] welded 6061-T6 aluminum alloy and phase transformation-induced plasticity (TRIP). It was found that the welding speed could not affect the type of IMC and that the tool's rotational speed increased the strain rate of the materials. The material strain rate increased static dislocations, mobile dislocations, and cracks caused by grain boundary deformation. Therefore, compared with a conventional FSW (tool rotation speed < 2000 rpm), a higher rotation speed can improve the diffusion process and produce more IMC. In the connection of welding materials, the formation of an appropriate thickness of the IMC layer at the Al-Fe interface is important. However, F. C. Liu et al. [19,20] found that at higher rotational speeds, the formation of IMC is inhibited, and nano-scale amorphous phases can form.

Micro friction stir welding (μ-FSW) refers to FSW with a thickness of less than 1 mm. For thin plates, micro friction stir welding presents a feasible connection method between Al alloy and SS. Chen et al. [21–23] pointed out that high rotation-speed FSW can improve the degree of material mixing, with less material thinning, and can reduce the machine rigidity needed. The HRS-FSW (High rotation speed-FSW) is higher than 2000 rpm and lower than 15,000 rpm, which can realize μ-FSW connection. The FSW connection can be made with a lower axial force. This reduces the rigidity requirement of the machine, which is beneficial in the case of welding robot applications.

The high rotational speed of FSW is attracting an increasing amount of attention. Mohan et al. [24] combined μ-FSW with ultra-high rotational speed friction stir welding (ultra-HRS-FSW) to weld AA1100 with a thickness of 1.0 mm. The rotation speed of ultra-HRS-FSW ranges from 15,000 to 24,000 rpm. The welding speed could attain 1000 mm/min, which improves the welding efficiency. The existing literature about HRS-FSW of dissimilar material connections focuses only on the connection between similar materials and on dissimilar materials having comparable thermomechanical properties. For example, Zhang et al. [25] conducted μ-FSW to weld brass (H62) and copper (T2), both having a thickness of 0.6 mm, a rotational speed of 12,000 rpm, and a welding speed of 280 mm/min. A relatively high tensile strength of 194 MPa was achieved, which is approximately 82.6% that of copper (T2). It was concluded that the welding process parameters have a greater effect on the quality of welding in thin plates than tool geometry. Park et al. [26] conducted tests with μ-FSW of AA6061-T6 and AA5052-H32 plates with a thickness of 1 mm, under a rotation speed of 9000 rpm and a welding speed of 300 mm/min, to reach the highest tensile strength of 226.63 MPa. It can be concluded that the welding heat input is lower at a higher speed. Additionally, using a backing plate with higher thermal conductivity has a serious impact on the material connection. However, the connection problem of dissimilar sheet materials is still unresolved. Based on the current studies, it can be concluded that high rotational speed is beneficial to μ-FSW's connections that include similar or different materials having comparable thermomechanical properties. However, for the μ-FSW of Al-Fe, the physical and mechanical properties of aluminum alloy and steel are extremely different, the binding mechanism is completely different from that of different materials with the same thermodynamic properties. In addition to the effect of rotational speed on the welding parameters, previous studies have not reported any specifics on the effect of tool offset on joint characteristics.

In the abovementioned studies, Al-Fe welding has been studied mainly in terms of medium-thick plates. Thin plates were studied mainly in terms of welding dissimilar materials or of similar materials having comparable thermomechanical properties. There are no relevant studies on Al-Fe μ-FSW, using an ideal lightweight material. In addition to the rotation speed, which is an important welding parameter of μ-FSW in the joint, the offset of the tool has a considerable influence on material fluidity and the mechanical properties of the joint. Therefore, this paper aims to investigate the feasibility of different process parameters for Al-Fe plates under HRS and super-HRS-FSW, to summarize the advantages and disadvantages of microstructures and mechanical properties, and to analyze the formation of IMC to gain insight into the intrinsic mechanism.

2. Materials and Methods

Figure 1a illustrates a schematic diagram of the welding experiment, and X is the tool offset in Figure 1b. As a result of the differences in physical properties between Al and Fe alloys, the Al alloy is placed on the retreating side (RS) to allow its strong plasticity to form a satisfactory weld [8]. In this study, a 1.0 mm-thick 6061-T6 Al alloy and a 0.8 mm-thick 316L SS were used. Table 1 shows the chemical composition of the metal as given in the material certificate provided by the supplier.

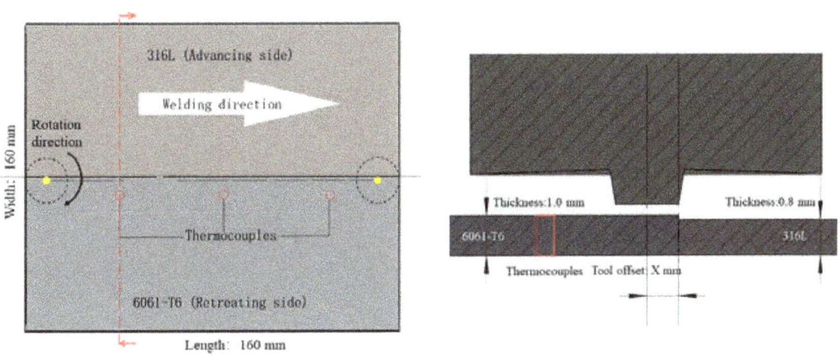

Figure 1. Schematic diagram of the FSW experiment. (**a**) Schematic diagram of tool displacement; (**b**) Cross-sectional diagram in (**a**).

Table 1. Chemical composition of the base materials.

Base	Chemical Composition (Wt.%)										
Material	C	Cu	Si	Fe	Mn	Mg	Zn	Mo	Cr	Ni	Ti	Al
6061-T6	-	0.11	0.6	0.7	0.15	0.9	0.25	-	0.09	-	0.15	Bal.
316L	0.023	-	0.45	Bal.	1.36	-	-	2.07	16.75	10.05	-	-

In Figure 1a, the circle drawn using dashed lines represents the tool of the FSW. As some of the tools will be immersed in steel and subjected to severe thermal friction conditions, high-temperature resistance and high-strength materials are required. In this study, tungsten rhenium alloy [27] was chosen for the tool, which has good high-temperature strength and a certain degree of plasticity. The geometry of the tool was chosen based on the proposal by Ramachandran et al. [9] that when the inclination of the tapered cylindrical pin reaches 10°, it facilitates the formation of a homogeneous IMC layer at the Al-Fe interface. Consequently, this improves the weld quality. On the RS, the level of resistance to material flow depends on the width of the plastic-state material around the pin and the volume of material transferred per rotation. To avoid the formation of tunneling defects, a smaller-diameter pin was chosen to reduce the volume of transferred material [28]. Figure 2 shows a physical view of the mixing head, with the exact dimensions. Austenitic SS was used as the backing plate under the workpiece because, during the welding process between the thin plate and the backing plate, using a material with high thermal conductivity would remove more heat [22]. To ensure the control accuracy of axial displacement, drill holes were drilled in specific positions on the backing plate and plate being welded. The diameter of the hole was slightly larger than that of the thermocouple to avoid the problem of insufficient material flow caused by plate consumption. The thermal conductivity of 6061-T6 aluminum alloy is much higher than that of 316L SS. A K-type thermocouple was inserted in each location marked by a red circle in Figure 1a,b, to measure the highest weld temperature. The diameter of the top of the tool pin is about 2 mm because the highest temperature point of the weld is at the edge of the tool pin. Therefore, to measure the peak temperature during welding without damaging the thermocouple, the distance between the through-hole and the Al-Fe interface must be 3 mm. For each parameter, three temperature-measuring points were selected, and the measured temperature value was taken as the average. To measure the magnitude of the axial force during the welding process, six JHBM-H3 force gauges (LYOYS, Bengbu, China) were mounted underneath the pad. The FSW welding equipment used in this study is a new type of lightweight FSW equipment developed by the research team for sheet metal welding. Its three axes can move with an accuracy of 0.01 mm, and the rotation speed ranges from 1000 to 24,000 rpm.

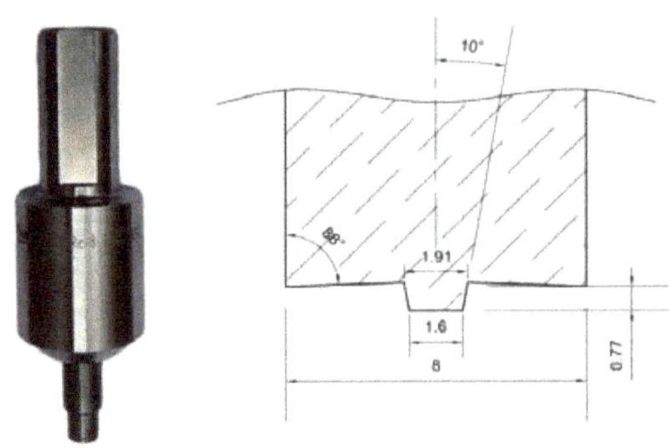

(a) (b)

Figure 2. An example of an FSW tool. (**a**) Physical drawing of the tool; (**b**) specific dimensions of the tool (unit: mm).

In this study, we mainly focus on the influence of HRS-FSW and ultra-HRS-FSW methods, taking into consideration the fact that the welding speed has a minimal effect on the composition of IMC [8], the welding heat input has a great deal of influence, but the welding heat input of FSW is a multi-factor coupling. According to Nandan et al. [29–31], the heat input formula for welding is as follows:

$$E = \left(\iint \omega r(\delta\mu p + (1-\delta)\tau_y)dA\right)/kv \quad (1)$$

E represents welding heat input, ω represents the spindle rotation speed, r represents the tool radius, δ represents the degree of slip, μ represents the coefficient of friction, p represents pressure, τ_y represents shear stress, k represents the constant, and v represents welding speed. Table 2 shows the optimal welding parameters as a result of a large number of welding tests. The deviation of the welding tool offset was within $\pm$ 0.05 mm, as a result of the error caused by the unevenness of the sheet's cross-section. The weld with the median mechanical properties was selected for analysis after each parameter experiment was repeated thrice.

Table 2. The process parameters used regarding the HFSW of Al 6061 to 316L steel.

Serial Number	Process Type	Rotational Speed (R) (rpm)	Welding Speed (FT) (mm/min)	Tool Offset (Offset) (mm)
1	HRS	10,000	80	0.70
2	HRS	10,000	80	0.55
3	Ultra-HRS	18,000	400	0.70
4	Ultra-HRS	18,000	400	0.55

A Zeiss microscope (equipment model: Imager.A2 m, PRECISE, Beijing, China) was used to observe the microstructure of the cross-section. The scanning electron microscope (QUANTA 250FEG, Thermo Fisher Scientific, Waltham Mass, MA, USA) was used to analyze the thickness and composition of the diffusion layer at the interface. The electrochemical corrosion analysis of the joint was conducted to analyze the impact of FSW on the corrosion resistance of the joint. The size of the working electrode of the heterogeneous Al-Fe-FSW material was 8 mm $\times$ 8 mm $\times$ 0.8 mm^3. The counter electrode was a Pt sheet, and the reference electrode was a saturated calomel electrode. For the polarization curve parameters, the initial potential was set to -1 V, the end potential was set to 0 V, and the scanning speed was 0.333 mV/s. Figure 3a shows the specific shape and size of the tensile specimens, as per the Chinese national standard "GB/T 2651–2008/ISO 4136:2001". All the tensile tests were conducted at a strain rate of 2 mm/min on the CMT5205 (Sinter, Jilin, China) tensile machine. As shown in Figure 3b, the hardness tests were carried out in 3 rows. A fully automatic hardness testing machine, KN30S (Sinter, Jilin, China), at an applied load of 400 g for 20 s, was used to observe the microhardness at the top, middle, and bottom in the direction of thickness across the joint. The spacing between the upper and lower rows was 0.20 mm, and the spacing between the points of each row was 0.30 mm.

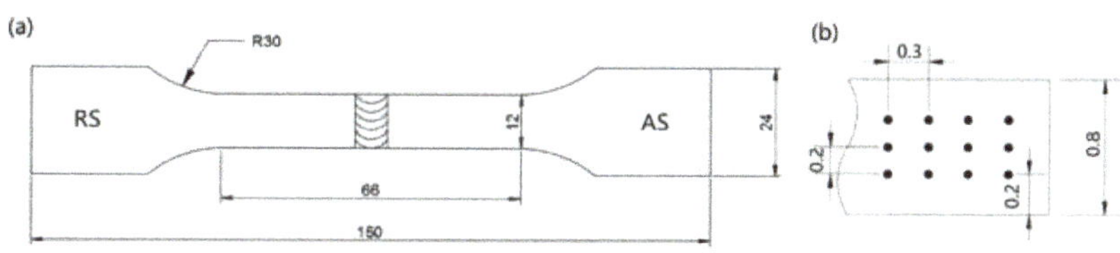

Figure 3. Tensile test and hardness test: (**a**) specific dimensions of the tensile specimen; (**b**) schematic diagram of the hardness measurement location (unit: mm).

3. Results and Discussion

3.1. Microstructure and Interface Analysis of Joints

Figure 4 shows the metallography prepared by various tool offsets, using the FSW connection method of HRS and ultra-HRS, respectively. The length of the tool pin is slightly less than the thickness of the plate; the stainless steel on the AS was driven to the aluminum side by extrusion. At a low tool offset, the extruded stainless steel was taken to the RS, creating a channel-like shape. When the axial force was measured under different parameters, it was found that the maximum axial force of FSW welding at 0.8 mm thickness was not more than 2 KN, and the minimum axial force was not less than 0.8 KN with the increase in rotational speed, indicating that the material has reached a certain degree of plasticization.

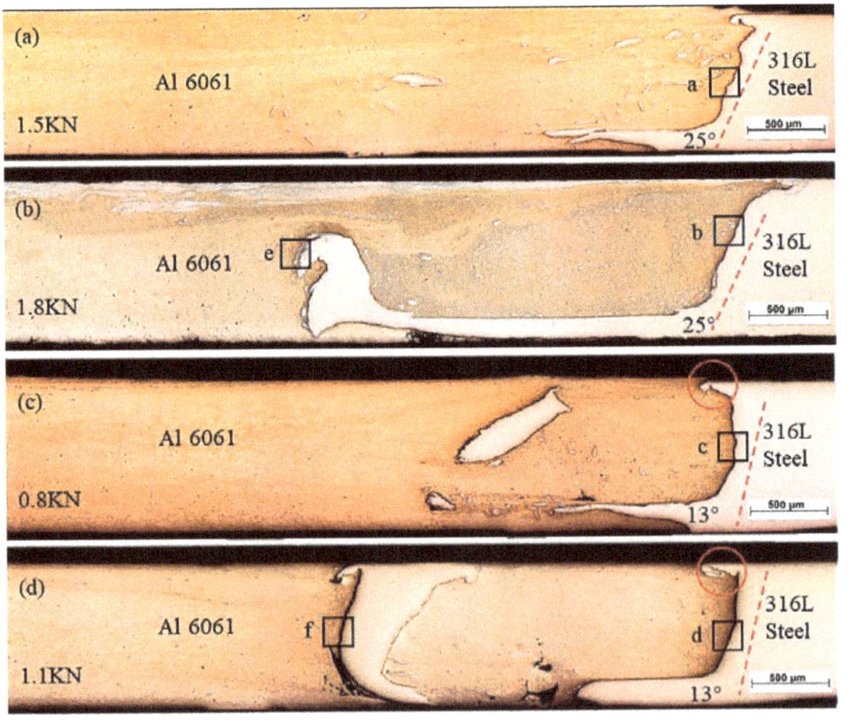

Figure 4. This picture corresponds to the Zeiss picture of the interface of (**a**) serial number 1, (**b**) serial number 2, (**c**) serial number 3, and (**d**) serial number 4 in Table 2. The red dotted line is the inclination angle of the Fe-Al interface.

As illustrated in Figure 4a,b, a flocculent gray structure or local steel particles were observed in the shoulder of the affected zone (SAZ). This occurs due to the shoulder stirring action during FSW. It does not appear in Figure 4c,d, mainly because the whole piece of SS metal that is stirred down is thrown outward at high speed to form a wall that prevents the flow of steel particles from the shaft shoulder.

Under the HRS parameters, the welding speed was slow, and the axial force was large. Consequently, the heat generation increased, and the middle part of the SS welding interface changed to a plastic state. Therefore, the local aluminum alloy was brought to the SS side and, at the Al-Fe interface, a 25° inclination angle was formed. However, under the ultra-HRS parameters, the inclination angle of the Al-Fe interface was reduced to 13°, and at the Al-Fe interface, part of the SS flowed to the aluminum alloy. This is shown in the red-circled areas of Figure 4c,d. The plastic material in the affected area of the shoulder tended to squeeze up at the tool pin because of the inclination angle of the tool shoulder. According to the Bernoulli effect, when rotation speed and welding speed increase, the linear speed of the material in the interface area reaches 306.7 mm/s, resulting in a drop in the interface pressure. The SS on the upper surface of the weld was in a plastic state under the higher rotation speed, and the plastic SS was pressed to the aluminum side. Conversely, as the welding speed increased, the welding heat input and the centrifugal force on the plate reduced, enabling the Bernoulli effect to occupy a dominant position and the inclination angle to decrease.

Figure 5 shows enlarged maps of the Al-Fe interface, corresponding to the boxes marked with black rectangles in Figure 4, and Figure 5a–d shows details of the Al-Fe interface. The steel particles on the aluminum alloy side are smashed more finely with the increase of rotational speed. In Figure 5b,d, when a small offset on the Al side was used, it is clearly observed that some small areas with gray contrast were distributed uniformly in the nugget zone. This is consistent with the phenomenon reported by X. Liu et al. [8,32] In addition, it shows that the smaller the offset, the more diffuse the distribution. On the RS near the aluminum alloy, the light gray contrast area in Figure 5e is IMC, and the corresponding SEM diagram in Figure shows that the thickness of IMC is uneven and is higher than 7 μm in Figure 5g,h.

According to the Spot EDS analysis of Figures 6–9, the compositions of spot 1 and spot 2 were consistent with those of 6061-T6 aluminum alloy and 316L aluminum alloy base metals. From Figures 6 and 8, it can be seen that the Al-Fe interface could be divided into two layers, the middle layer, and the mixed layer. After analyzing point 4 and point 5, respectively, it is clear that point 4 is mainly composed of 316L SS, with a little aluminum alloy. It can be seen that point 5 is mainly composed of 6061-T6 aluminum alloy with a portion of iron. The occurrence of the mixed layer is significantly correlated with the large offset. The smaller mixed layer generated in SS is due to the smaller transverse extrusion pressure in comparison with a lesser offset. Lee et al. [33]. demonstrated that the interface produces an island-like mixing zone. In Figures 7 and 9, at the aluminum iron interface, a more continuous interface is generated. On the steel side, a continuous and discontinuous region with small peaks is formed. This indicates that the diffusion between the atoms is dominated by iron atoms.

3.2. *Thermal History and Formation of Diffusion Layer at the Al-Fe Interface*

The key influencing factors that contribute to the formation of IMC in the welding of Al-Fe dissimilar metals are heat input and axial pressure [4,33]. Therefore, it is necessary to measure the temperature of the welding process. When using different offsets, there are slight temperature differences. Figure 10 shows the temperature measurement data for the HRS and ultra-HRS parameters. The highest temperatures measured by the HRS and ultra-HRS connection methods were 470 °C and 620 °C, respectively. The highest temperature was lower than 655 °C, the Al-Fe eutectic point. In the paper by Mahoney [34], it was found that the temperature difference along the direction of thickness was almost negligible. Since a thinner thickness was used in this experiment, the temperature can be considered to be the same in the direction of the thickness. The experimental temperature measurement point was close to the thermal mechanically affected zone on the RS. Therefore, in the welding process, the measured temperature at this point could be considered closest to the highest temperature.

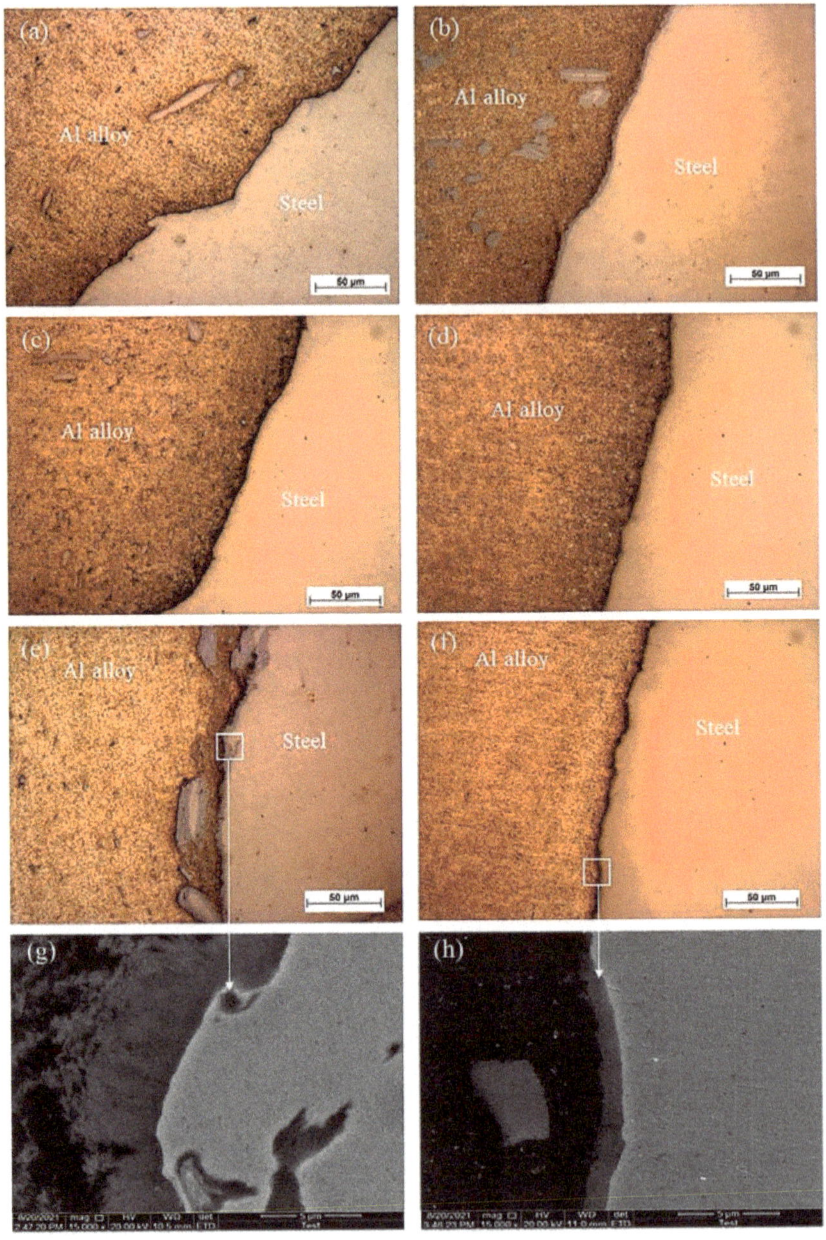

Figure 5. Details corresponding to the partially enlarged images in Figure 4 and the SEM image. (**a**) HRS-FSW Al-Fe interface under a high tool offset (**b**) HRS-FSW Al-Fe interface under a low tool offset. (**c**) Ultra-HRS-FSW Al-Fe interface under a high tool offset. (**d**) Ultra-HRS-FSW Al-Fe interface under a low tool offset. (**e**) RS side interface and SEM picture of HRS-FSW under a low tool offset. (**f**) RS side interface and SEM picture of ultra-HRS-FSW under a low tool offset. (**g**) Partial magnification in Figure e. (**h**) Partial magnification in Figure f.

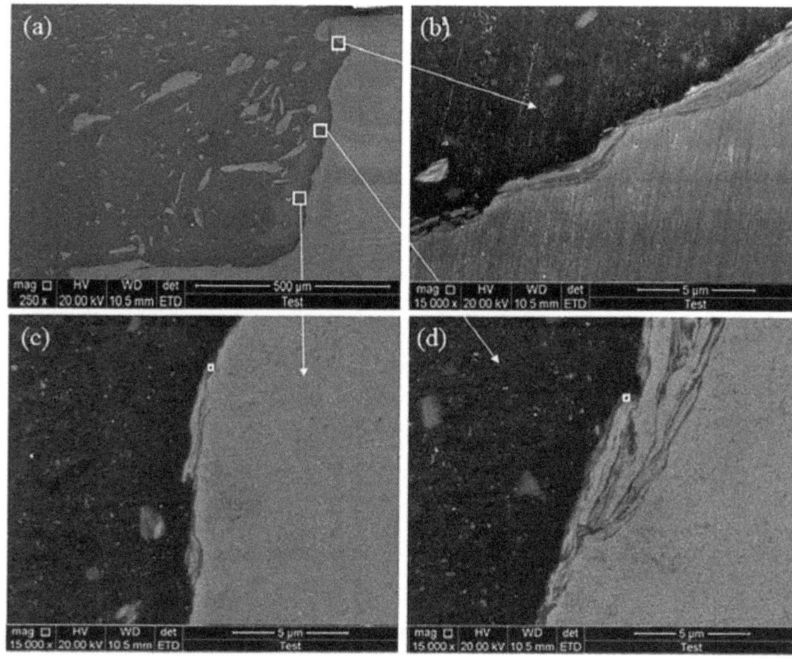

Figure 6. SEM image of the HRS-FSW interface and the EDS measuring position under a high tool offset. (**a**) Low- magnification SEM picture, (**b**) SEM Al-Fe interface picture at the top and EDS point scan position, (**c**) SEM Al-Fe interface picture in the middle, and (**d**) SEM Al-Fe interface picture at the bottom.

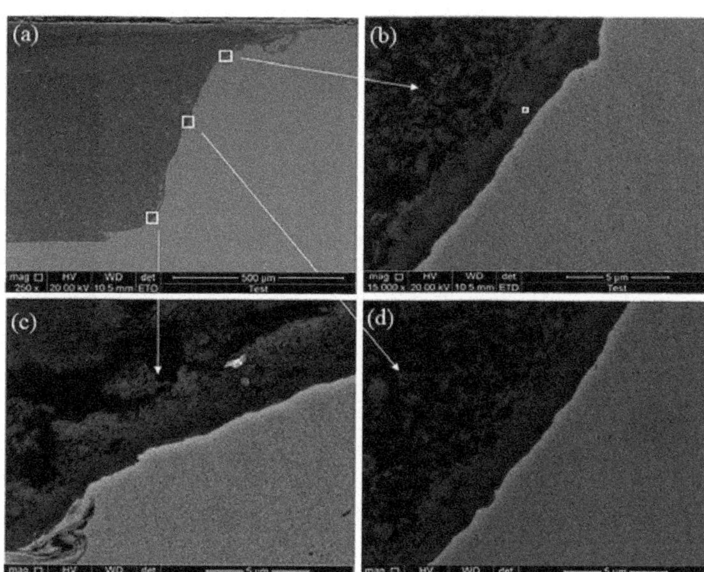

Figure 7. SEM image of HRS-FSW interface and EDS measuring position under a low tool offset. (**a**) Low magnification SEM picture, (**b**) SEM Al-Fe interface picture at the top and EDS point scan position, (**c**) SEM Al-Fe interface picture in the middle, (**d**) SEM Al-Fe interface picture at the bottom.

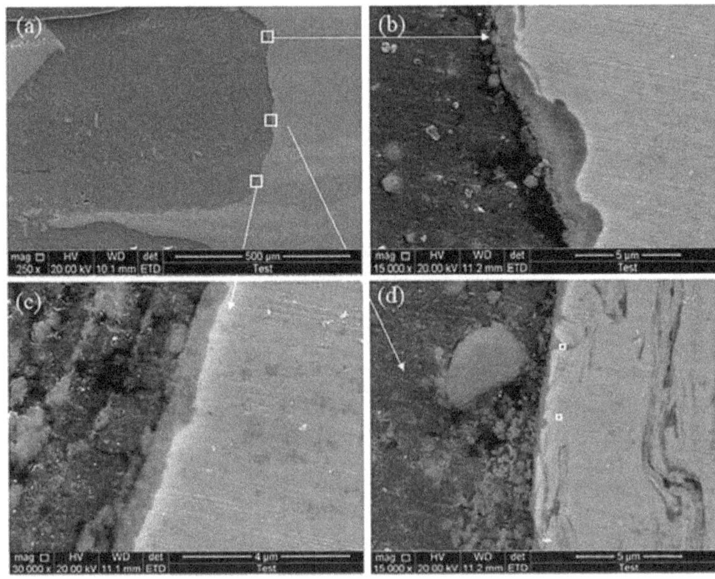

Figure 8. SEM image of ultra-HRS-FSW interface and EDS measuring position under a high tool offset. (**a**) Low magnification SEM picture, (**b**) SEM Al-Fe interface picture at the top, (**c**) SEM Al-Fe interface picture in the middle, (**d**) SEM Al-Fe interface picture at the bottom and EDS point scan position.

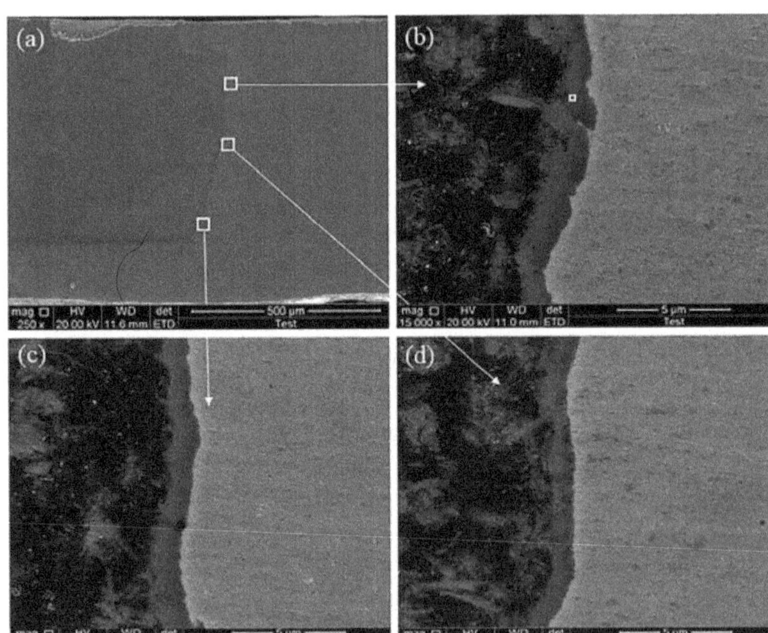

Figure 9. SEM image of ultra-HRS-FSW interface and EDS measuring position under a low tool offset. (**a**) Low magnification SEM picture, (**b**) SEM Al-Fe interface picture at the top and EDS point scan position, (**c**) SEM Al-Fe interface picture in the middle, (**d**) SEM Al-Fe interface picture at the bottom.

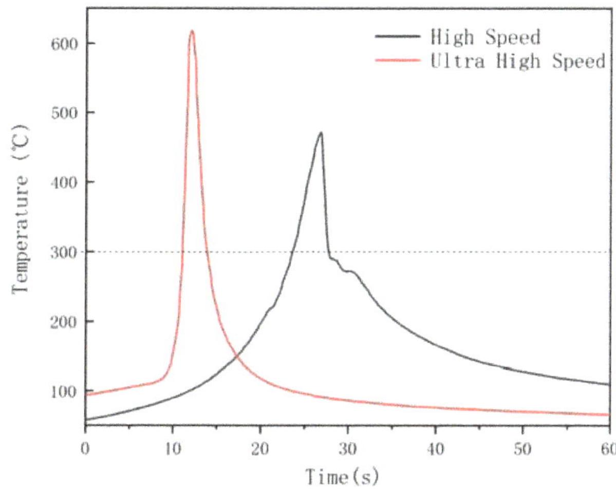

Figure 10. Thermal history of the aluminum alloy retracted side under HRS and ultra-HRS low tool offset parameters.

The atomic diffusion was related to the average shear strain rate of the interface and the diffusion coefficient of the current temperature. The average shear strain rate is defined as the ratio of the shear velocity to the shear band thickness [19]. Under HRS and ultra-HRS parameters, the shear speeds in the middle of the tool pin are 151.3 mm/s and 267.7 mm/s, respectively. Figure 11 displays the schematic diagram of the shear band thickness under HRS and ultra-HRS parameters. Without taking into account the amount of energy released by the plastic deformation at the tip of the tool pin, the middle layer of the tool pin was chosen as the minimum shear band thickness. The shear band thickness is 0.116 mm and 0.022 mm, respectively. Therefore, the average shear strain rate was 1.3×10^3 s^{-1} and 1.2×10^4 s^{-1}, respectively. The relationship between the diffusion coefficient D, the jump distance λ of the diffusion atom, and the jump frequency of the atom is D = λ^2 Γ/6. At 450 °C and 620 °C, according to the literature [35,36], the diffusion coefficients between Al and Fe are 3×10^{-15} m^2/s and 6×10^{-14} m^2/s, respectively, and the distance between Fe and Al atoms is 2 nm. The atomic jump frequencies of 1.35×10^4 s^{-1} and 9×10^4 s^{-1} were higher than the average shear strain rate, and the shear did not completely inhibit atomic diffusion. Consequently, an IMC interlayer was formed.

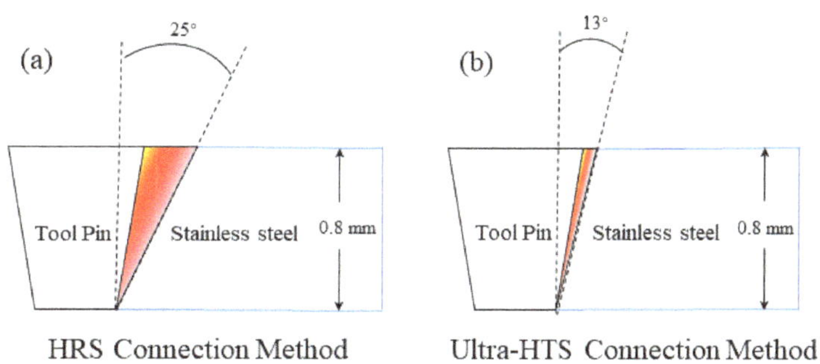

Figure 11. (**a**) Schematic diagram of shear band thickness under HRS-FSW parameter, (**b**) schematic diagram of shear band thickness under an ultra-HRS-FSW parameter.

The average values of the EDS energy spectra are obtained from three points in the IMC layer of the interface diagram. Figures 6–9 show the interface diagrams. Table 3 shows the chemical average mass composition (wt%) of the corresponding parameters. The mass fraction ratios of the chemical components of the corresponding parameters are 70.8%Al/29.2%Fe, 75.4%Al/24.6%Fe, 15.8%Al/84.2%Fe, and 17.4%Al/82.6%Fe. The corresponding IMC are $FeAl_3$, $FeAl_3$, Fe_3Al, and Fe_3Al, respectively. Yeremenko et al. [32] also obtained similar IMC components. It was found that under the HRS and ultra-HRS parameters, while the role of spindle rotation speed is dominant, the IMC type has little influence on welding temperature, welding heat input, welding speed, and tool offset. This is different from the impact of rotation speed and tool offset in the case of the IMC type described by Liu et al. [8]. The difference occurs because of the simultaneous diffusion and reaction. Atomic diffusion is inhibited by the high shear rate, and atomic diffusion makes atomic chemical reactions almost impossible. Thus, only the diffusion reaction occurs.

Table 3. The EDS results on the Al-Fe interface IMC, corresponding to the parameters of Table 2.

Serial Number	Al (wt%)	Fe (wt%)	IMC
1	70.8	29.2	$FeAl_3$
2	75.4	24.6	$FeAl_3$
3	15.8	84.2	Fe_3Al
4	17.4	82.6	Fe_3Al

It was established that the thickness of the IMC layer was about 0.7 μm when the offset was 0.7 mm, while in comparison, the thickness of the IMC layer was about 2.0 μm when the offset was 0.55 mm. The tool offset was the main factor that determined the thickness of the IMC layer at the same rotating speed and welding speed. The thickness of the IMC layer in the upper, middle, and lower parts of the Al-Fe interface decreased gradually with a larger tool offset. The upper and lower IMC layers of the Al-Fe interface were larger with a smaller tool offset, whereas the middle IMC layer was smaller. The thickness of the lower IMC layer was higher, and the upper friction heat IMC thickness of the shoulder was also higher, resulting in a thinner IMC in the middle because the axial tip of the stirring needle absorbed the energy released by the deformation of the rotating plastic backlog steel on the forward side.

3.3. Electrochemical Test and Analysis

The Al-Fe alloy FSW joint showed more IMC with two parameters of HRS and ultra-HRS under a smaller tool offset. Figure 12a shows the time-dependent curve of an open-circuit potential (OCP) in a 3.5 wt% solution. From Figure 12a, it can be seen that with the increase in immersion time, the corrosion potential (Ecorr) of the two samples moved to a positive potential, indicating that the two materials spontaneously formed a passivation film [37]. At around 800 s, the Ecorr for the HRS sample reached a stable potential of −0.74 VSCE from about −0.91 VSCE to a higher potential. However, the Ecorr of the ultra-HRS sample remained stable at −0.73 VSCE at about 700 s. In comparison with the HRS sample, the ultra-HRS sample took less time to obtain the steady potential in a 3.5 wt% NaCl solution. The Ecorr value was corrected, indicating that the Ultra-HRS sample was more corrosion-resistant [38–40]. The IMC were the main factor affecting the corrosion resistance since the corrosion interface was composed of IMC from the SS matrix and the aluminum alloy matrix.

Figure 12b shows the polarization curve of the Al-Fe alloy welded with HRS and ultra-HRS in a 3.5 wt% solution. The two samples exhibit anodic dissolution characteristics, which are caused by the anodic dissolution of the aluminum alloy at the initial stage of immersion [41]. The corrosion potential (E_{corr}) and corrosion current density (i_{corr}) of HRS and Ultra-HRS samples are −0.68 V and −0.69 V, 2.18×10^{-5} A cm^{-2}, and 9.50×10^{-6} A cm^{-2}, respectively. As compared to the HRS sample, the corrosion current density of the ultra-HRS sample has a lower value, which indicates a reduced corrosion rate and, therefore, higher corrosion resistance. In the ultra-HRS sample, the IMC layer was rich in Fe_3Al, while the HRS sample contained more $FeAl_3$. The presence of Fe_3Al improved the corrosion resistance of the alloy, which led to the improvement of the corrosion resistance of ultra-HRS samples [42]. This corrosion resistance is consistent with the results of Knibloe et al. [42,43], and the iron-rich phase has better corrosion resistance.

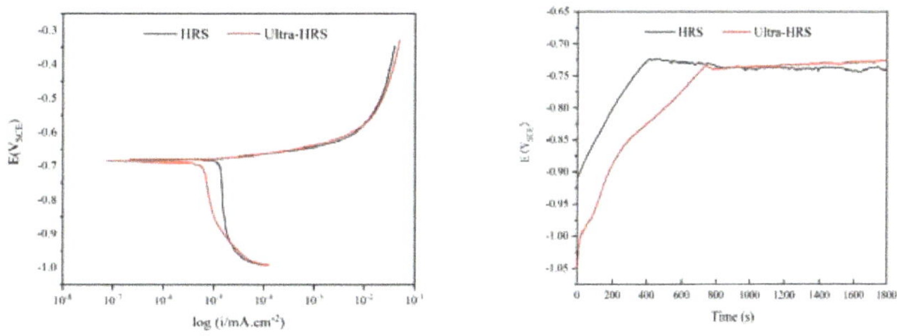

(a) (b)

Figure 12. The electrochemical corrosion behavior of HRS and ultra-HRS under low tool offset parameters. (**a**) The open-circuit potential of the connector; (**b**) the polarization curve of the connector.

3.4. Microhardness Distribution of Joint Interface

The hardness of 6061-T6 aluminum alloy is 95 ± 5 HV, and the hardness of 316L is 170 ± 5 HV. As indicated in Figure 13, the hardness of aluminum alloy gradually decreases between the aluminum alloy matrix and the heat-affected zone(HAZ) of the RS. In Figure 13, there was a peak value of hardness on the RS of the weld due to the driving effect of the tool pin. The hardness between the HAZ and SS matrix decreased gradually. The width of the HAZ of the RS was larger than that of the ultra-HRS method, as shown in Figure 13, while the HAZ of the AS of the SS was nearly the same. The main reasons for this were that the friction area of the shoulder against the SS was small, and the thermal conductivity of SS was poor. Except in the nugget area, the hardness in the upper region was observed to be different from that of the middle and lower lines. From the center stir zone to the base material, there was a noticeable decrease in hardness on the Al side, which was substantially correlated with the significantly increased grain size, along with dislocation density [44]. Interestingly, the hardness on the upper surface had a relatively lower value, which may be attributed to the coarse grains that had grown due to the resulting overheating from the rotational shoulder. On the other hand, the hardness on the top layer always exhibited a slightly higher value than that of the middle and bottom layers, since the transformation of austenite into martensite is easily induced by the heat generated during FSW. This phenomenon was identified by Lee [33], and this conjecture RS was verified by the weld's magnetic conductivity at the end of welding.

Figure 12b shows the polarization curve of the Al-Fe alloy welded with HRS and ultra-HRS in a 3.5 wt% solution. The two samples exhibit anodic dissolution characteristics, which are caused by the anodic dissolution of the aluminum alloy at the initial stage of immersion [41]. The corrosion potential (E_{corr}) and corrosion current density (i_{corr}) of HRS and Ultra-HRS samples are -0.68 V and -0.69 V, 2.18 × 10^{-5} A cm^{-2}, and 9.50 × 10^{-6} A cm^{-2}, respectively. As compared to the HRS sample, the corrosion current density of the ultra-HRS sample has a lower value, which indicates a reduced corrosion rate and, therefore, higher corrosion resistance. In the ultra-HRS sample, the IMC layer was rich in Fe_3Al, while the HRS sample contained more $FeAl_3$. The presence of Fe_3Al improved the corrosion resistance of the alloy, which led to the improvement of the corrosion resistance of ultra-HRS samples [42]. This corrosion resistance is consistent with the results of Knibloe et al. [42,43], and the iron-rich phase has better corrosion resistance.

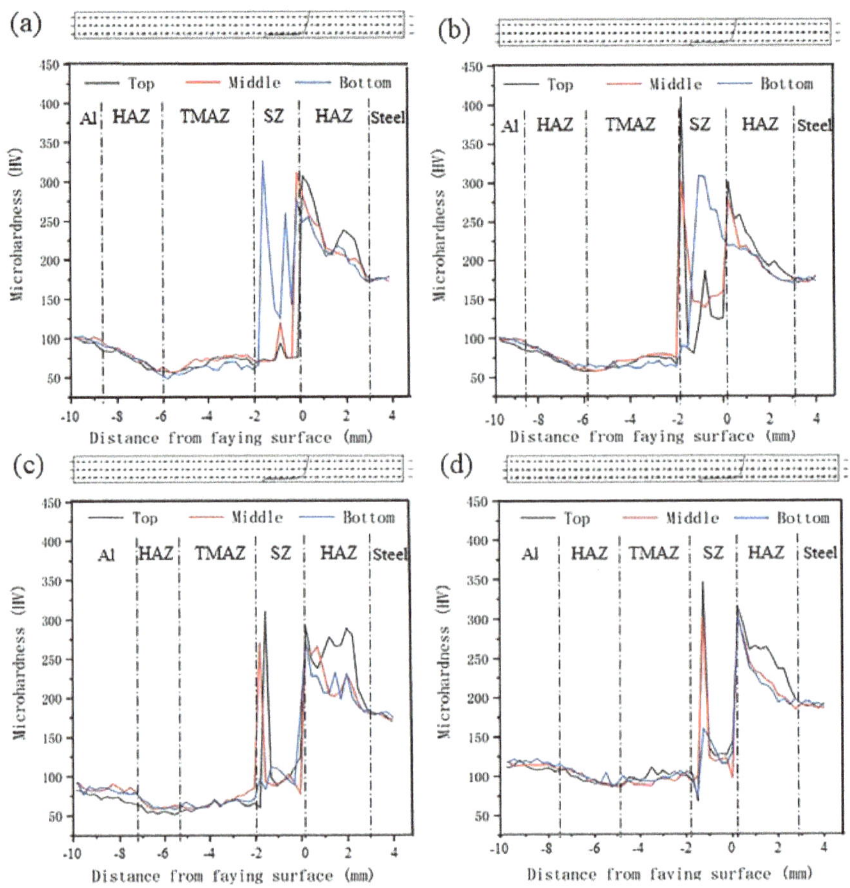

Figure 13. This figure corresponds to the microhardness values of the connector of (**a**) serial number 1, (**b**) serial number 2, (**c**) serial number 3, (**d**) and serial number 4 in the parameters of Table 2.

As shown in Figure 13b, the average hardness of the center stir zone was more than 110 HV. This phenomenon is caused by randomly distributed SS particles in the aluminum matrix, which are beneficial for the strengthening of aluminum alloy. The average hardness of the bottom region, which has the highest value, is mainly related to the finest grains, which are caused by a significant degree of dynamic recrystallization during FSW. The rotational tool drags the 316L SS from the AS to the RS [27]. Under the action of rotation, a bridge between the SS and aluminum alloy is formed at the bottom, with an average hardness of 210 HV. Some SS and IMC are pushed to the RS as a consequence of the existence of the bridge, and the hardness of the RS can reach 400 HV.

3.5. Tensile Strength Test

Figure 14 shows the stress-strain curve of the joint with the best mechanical properties after welding with HRS and ultra-HRS parameters, respectively. With a smaller tool offset, under HRS conditions, an optimum tensile strength of 220.8 MPa (about 75.9% of the base metal of aluminum alloy) was obtained. The tensile specimen broke the outer boundary of the SS involved in the HAZ on the aluminum alloy side. The decrease in strength in this area was due to the over-aging of precipitates in 6061 aluminum alloy during welding, resulting in a decrease in strength due to the growth of precipitates. Under this parameter, the hardness of the HAZ decreases, and its width increases, as shown in

Figure 13b, which is consistent with the over-aging conjecture. However, from the enlarged area of Figure 5e, it is seen that the thickness of the SS IMC layer near the aluminum alloy side was not uniform. The lowest thickness was at least 7 μm, and the mechanical properties improved when the thickness of the IMC layer was uniformly distributed at about 2 μm, at the Al-Fe interface [9]. An excessively thick IMC layer would result in crack propagation [45]. As shown in Table 4, the thickness of the IMC layer at the Al-Fe interface was uniform, and the average thickness was about 2 μm. This differs from the findings of Ramachandran et al. [9], who formed a layer of intermetallic compounds with a uniform thickness of 15 μm on the best-welded joint. This may be due to the difference in the rotational speed of the spindle. The steel particles, a kind of Al-Fe intermetallic composites distributed in the stir zone, broke at the outer boundary of SS in the heat-affected zone of the aluminum alloy side.

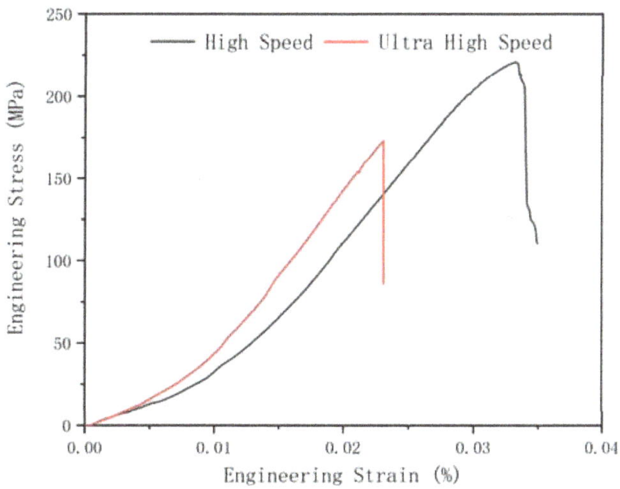

Figure 14. The stress-strain curve of the highest UTS under the parameters of HRS and ultra-HRS, respectively.

Table 4. Variables corresponding to the IMC thickness of different parts of the AlFe interface, under the parameters given in Table 2.

Serial Number	Top (μm)	Middle (μm)	Bottom (μm)	Average Value (μm)
1	1.19	0.456	0.2914	0.646
2	2.782	2.164	3.552	2.833
3	1.469	0.369	0.589	0.809
4	2.414	1.633	2.319	2.122

Under the ultra-HRS parameters, the tensile strength of the joint with a large offset was 177.1 MPa (about 61% of that of the aluminum alloy base metal). The fracture was located between the bottom of the center stir zone and the Al-Fe interface of the joint. The IMC produced at the interface is Fe_3Al, which is an iron-rich brittle compound. The strength of iron-rich compounds was higher, and their brittleness was lower than that of aluminum-rich compounds. Weld fracture was caused by three major factors: the effect of ultra-high rotational speed on the centrifugal force of welding materials, the thermal mechanically affected zone, and the upward flow of material caused by a small axial pressure. Consequently, holes were produced at the welding core, which is the source of fracture.

3.6. Fracture Analysis

Figure 15 shows the typical SEM fracture images of fractured tensile samples under the conditions of HRS with low offset and Ultra-HRS with high offset. Figure 15a shows a picture of the weld fracture with low offset under HRS parameters. The interface fracture occupied at the outside of the rolled-up SS in the nugget zone on the side of the partial aluminum alloy. The interface was mainly dimple and brittle cleavage fracture in Figure 15b.c, so it could be concluded that the fracture mode was dominated by the mixture of brittle and ductile. Figure 15d showed the picture of fracture under high offset under Ultra-HRS parameters. The interface fracture was at the Al-Fe interface, and the non-uniform IMC layer of the interface is the cause of brittle cleavage fracture. The fracture of this method was brittle fracture mechanism.

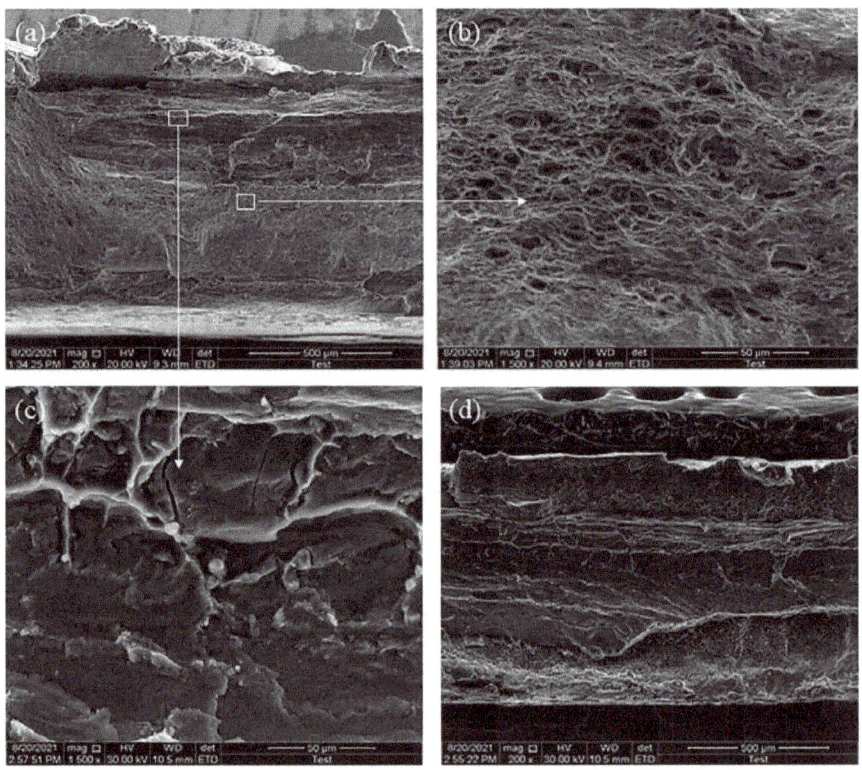

Figure 15. SEM photos of tensile fracture (**a**) fracture SEM picture under HRS-FSW. (**b**) partial enlargement in (**a**). (**c**) partial enlargement in (**a**). (**d**) fracture SEM picture under ultra-HRS-FSW.

4. Conclusions

HRS and ultra-HRS welding methods were used to perform friction stir welding experiments on 0.8 mm thick 316L SS and 1 mm thick 6061-T6 aluminum alloy under approximate welding heat input conditions. The microstructure, mechanical properties, and electrochemical corrosion properties of the joints were analyzed. This study led to the following findings.

Under the HRS parameters, when the tool offset is 0.55 mm (about 75.9% of the base metal), a tensile strength of 220.8 MPa is obtained. The IMC compounds are uniformly distributed at the Al-Fe interface, and the weld breaks at the Al-Fe adhesion interface outside the stirring pin on the RS.

Under high rotation speed and ultra-high rotation speed, the welding temperature, welding speed, and tool offset have little influence on the IMC type. High rotation speed means a high shear rate, which directly affects the interatomic diffusion, thereby affecting the type of intermetallic compounds produced. Therefore, the rotation speed plays a crucial role in the type of IMC.

The tensile strength of the iron-rich phase (Fe_3Al) of the joint in the interlayer is not as high as that of the aluminum-rich phase ($FeAl_3$) of the joint in the interlayer. In the case of the joint obtained by ultra-HRS-FSW, the SS block cut by high stirring leads to the formation of loose holes between the joints and a decrease in connection strength. The iron-rich phase (Fe_3Al) has better corrosion resistance.

According to the hardness analysis, the hardness of the top layer of the joint is found to be substantially different from that of the middle and lower layers. The aluminum side has lower hardness levels than the steel side. This difference in hardness is related to the high tool rotation speed on the aluminum alloy side, higher friction, and plastic deformation caused by high-speed rotation of the shoulder. The grain on the side of aluminum alloy absorbs energy, and the material softens, while SS is transformed into martensite as a result of temperature changes and plastic deformation.

When the tool offset is 0.7 mm under HRS-FSW and ultra-HRS-FSW welding methods, along with an intermetallic compound at the Al-Fe interface, a mixed zone of iron and aluminum is formed. When the tool offset is 0.55 mm, only intermetallic compounds are formed at the Al-Fe interface.

From the above findings, it can be concluded that although the mechanical properties of ultra-HRS-FSW joints are not as high as those of HRS-FSW joints, the iron-rich phases produced by ultra-HRS-FSW have better strength, toughness and corrosion resistance. Future research can be focused on improving the ultra-HRS-FSW connection modes, such as improving the shape of the stirring head or adding auxiliary pulses or induction heating to vary the microstructure distribution of joints. Finally, improved methods and processes can lead to the formation of excellent welds.

Author Contributions: Data curation, Y.Z., H.Z., X.C., M.S., Z.Y., Y.G. and L.M.; writing—original draft, Y.G., L.M., S.C. and Z.L.; writing—review and editing, Z.L. and S.C.; supervision, S.C.; validation, Y.Z., H.Z., X.C., M.S. and Z.L. All authors have read and agreed to the published version of the manuscript.

Funding: This research received no external funding.

Institutional Review Board Statement: Not applicable.

Informed Consent Statement: Not applicable.

Conflicts of Interest: The authors declare no conflict of interest.

References

1. Li, K.; Liu, X.; Zhao, Y. Research Status and Prospect of Friction Stir Processing Technology. *Coatings* **2019**, *9*, 129. [CrossRef]
2. Riahi, M.; Amini, A. Effect of different combinations of tailor-welded blank coupled with change in weld location on mechanical properties by laser welding. *Int. J. Adv. Manuf. Technol.* **2013**, *67*, 1937–1945. [CrossRef]
3. Wan, L.; Huang, Y. Friction stir welding of dissimilar aluminum alloys and steels: A review. *Int. J. Adv. Manuf. Technol.* **2018**, *99*, 1781–1811. [CrossRef]
4. Springer, H.; Kostka, A.; Payton, E.; Raabe, D.; Kaysser-Pyzalla, A.; Eggeler, G. On the formation and growth of intermetallic phases during interdiffusion between low-carbon steel and aluminum alloys. *Acta Mater.* **2011**, *59*, 1586–1600. [CrossRef]
5. Wang, Z.; Bi, X.; Liu, B.; Xu, M.; Dong, Z. Adhesion enhancement of PEEK/6161-T6 FLJ joints via laser surface modification. *Compos. Part. B Eng.* **2021**, *216*, 108797. [CrossRef]
6. Verma, R.P.; Pandey, K.; Sharma, Y. Effect of ER4043 and ER5356 filler wire on mechanical properties and microstructure of dissimilar aluminium alloys, 5083-O and 6061-T6 joint, welded by the metal inert gas welding. *Proc. Inst. Mech. Eng. Part B J. Eng. Manuf.* **2014**, *229*, 1021–1028. [CrossRef]
7. Cabral-Miramontes, J.; Gaona-Tiburcio, C.; Estupinán-López, F.; Lara-Banda, M.; Zambrano-Robledo, P.; Nieves-Mendoza, D.; Maldonado-Bandala, E.; Chacón-Nava, J.; Almeraya-Calderón, F. Corrosion Resistance of Hard Coat Anodized AA 6061 in Citric–Sulfuric Solutions. *Coatings* **2020**, *10*, 601. [CrossRef]

8. Liu, X.; Lan, S.; Ni, J. Analysis of process parameters effects on friction stir welding of dissimilar aluminum alloy to advanced high strength steel. *Mater. Des.* **2014**, *59*, 50–62. [CrossRef]
9. Ramachandran, K.K.; Murugan, N.; Kumar, S.S. Effect of tool axis offset and geometry of tool pin profile on the characteristics of friction stir welded dissimilar joints of aluminum alloy AA5052 and HSLA steel. *Mater. Sci. Eng. A* **2015**, *639*, 219–233. [CrossRef]
10. An, J.; Liu, Y.B.; Zhang, M.Z.; Yang, B. Effect of Si on the interfacial bonding strength of Al–Pb alloy strips and hot-dip aluminized steel sheets by hot rolling. *J. Mater. Process. Technol.* **2002**, *120*, 30–36. [CrossRef]
11. Kaushik, P.; Dwivedi, D.K. Induction preheating in FSW of Al-Steel combination. *Mater. Today Proc.* **2021**, *46*, 1091–1095. [CrossRef]
12. Sisan, M.M.; Sereshki, M.A.; Khorsand, H.; Siadati, M. Carbon coating for corrosion protection of SS-316L and AA-6061 as bipolar plates of PEM fuel cells. *J. Alloy. Compd.* **2014**, *613*, 288–291. [CrossRef]
13. Öztürk, B. Energy consumption model for the pipe threading process using 10 wt.% Cu and 316L stainless steel powder-reinforced aluminum 6061 fittings. *Mater. Test.* **2019**, *61*, 797–805. [CrossRef]
14. Kap, I.; Starostin, M.; Shter, G.E.; Grader, G.S. Corrosion of aluminium, stainless steels and AISI 680 nickel alloy in nitrogen-based fuels. *Mater. Corros.* **2011**, *63*, 571–579. [CrossRef]
15. Chen, C.S.; Peng, X.L.; Li, Y.L.; Hu, X.W.; Wang, W.Q. Influence of Cu Coating on the Microstructure and Mechanical Properties of Laser Welded Joints in Dissimilar Metals. *Lasers Eng.* **2019**, *43*, 131–144.
16. Guan, Q.; Long, J.; Yu, P.; Jiang, S.; Huang, W.; Zhou, J. Effect of steel to aluminum laser welding parameters on mechanical properties of weld beads. *Opt. Laser Technol.* **2019**, *111*, 387–394. [CrossRef]
17. Mohan, D.G.; Tomków, J.; Gopi, S. Induction Assisted Hybrid Friction Stir Welding of Dissimilar Materials AA5052 Aluminium Alloy and X12Cr13 Stainless Steel. *Adv. Mater. Sci.* **2021**, *21*, 17–30. [CrossRef]
18. Tylecote, R.F. *The Solid Phase Welding of Metals*; Hodder & Stoughton Educational: London, UK, 1968.
19. Liu, F.; Dong, P.; Zhang, J.; Lu, W.; Taub, A.; Sun, K. Alloy amorphization through nanoscale shear localization at Al-Fe interface. *Mater. Today Phys.* **2020**, *15*, 100252. [CrossRef]
20. Liu, F.; Dong, P. From thick intermetallic to nanoscale amorphous phase at Al-Fe joint interface: Roles of friction stir welding conditions. *Scr. Mater.* **2021**, *191*, 167–172. [CrossRef]
21. Chen, S.; Zhou, Y.; Xue, J.; Ni, R.; Guo, Y.; Dong, J. High Rotation Speed Friction Stir Welding for 2014 Aluminum Alloy Thin Sheets. *J. Mater. Eng. Perform.* **2017**, *26*, 1337–1345. [CrossRef]
22. Liu, F.J.; Fu, L.; Chen, H.Y. Effect of high rotational speed on temperature distribution, microstructure evolution, and mechanical properties of friction stir welded 6061-T6 thin plate joints. *Int. J. Adv. Manuf. Technol.* **2018**, *96*, 1823–1833. [CrossRef]
23. Zhang, H.J.; Wang, M.; Qi, R.L.; Zhu, Z.; Zhang, X.; Yu, T.; Wu, Z.Q. Effect of rotation speed on nugget structure and property of high rotation speed friction stir welded Al-Mn aluminum alloy. *Int. J. Adv. Manuf. Technol.* **2017**, *92*, 2401–2410. [CrossRef]
24. Mohan, R.; Jayadeep, U.B.; Manu, R. CFD modelling of ultra-high rotational speed micro friction stir welding. *J. Manuf. Process.* **2021**, *64*, 1377–1386. [CrossRef]
25. Zhang, C.; Qin, Z.; Rong, C.; Shi, W.; Wang, S. The Preliminary Exploration of Micro-Friction Stir Welding Process and Material Flow of Copper and Brass Ultra-Thin Sheets. *Materials* **2020**, *13*, 2401. [CrossRef] [PubMed]
26. Park, S.; Joo, Y.; Kang, M. Effect of Backing Plate Materials in Micro-Friction Stir Butt Welding of Dissimilar AA6061-T6 and AA5052-H32 Aluminum Alloys. *Metals* **2020**, *10*, 933. [CrossRef]
27. Barnes, S.J.; Bhatti, A.R.; Steuwer, A.; Johnson, R.M.; Altenkirch, J.; Withers, P. Friction Stir Welding in HSLA-65 Steel: Part I. Influence of Weld Speed and Tool Material on Microstructural Development. *Met. Mater. Trans. A* **2012**, *43*, 2342–2355. [CrossRef]
28. Kumar, K.; Kailas, S.V. The role of friction stir welding tool on material flow and weld formation. *Mater. Sci. Eng. A* **2008**, *485*, 367–374. [CrossRef]
29. Kumar, R.; Singh, K.; Pandey, S. Process forces and heat input as function of process parameters in AA5083 friction stir welds. *Trans. Nonferrous Met. Soc. China* **2012**, *22*, 288–298. [CrossRef]
30. Nandan, R.; Debroy, T.; Bhadeshia, H. Recent advances in friction-stir welding–Process, weldment structure and properties. *Prog. Mater. Sci.* **2008**, *53*, 980–1023. [CrossRef]
31. Pew, J.W.; Nelson, T.W.; Sorensen, C.D. Torque based weld power model for friction stir welding. *Sci. Technol. Weld. Join.* **2007**, *12*, 341–347. [CrossRef]
32. Yeremenko, V.N.; Natanzon, Y.V.; Dybkov, V.I. The effect of dissolution on the growth of the Fe2Al5 interlayer in the solid iron -liquid aluminium system. *J. Mater. Sci.* **1981**, *16*, 1748–1756. [CrossRef]
33. Lee, W.-B.; Schmuecker, M.; Mercardo, U.A.; Biallas, G.; Jung, S.-B. Interfacial reaction in steel–aluminum joints made by friction stir welding. *Scr. Mater.* **2006**, *55*, 355–358. [CrossRef]
34. Mahoney, M.W.; Rhodes, C.G.; Flintoff, J.G.; Bingel, W.H.; Spurling, R.A. Properties of friction-stir-welded 7075 T651 aluminum. *Metall. Mater. Trans. A* **1998**, *29*, 1955–1964. [CrossRef]
35. Hirano, K.-I.; Agarwala, R.; Cohen, M. Diffusion of iron, nickel and cobalt in aluminum. *Acta Met.* **1962**, *10*, 857–863. [CrossRef]
36. Hood, G.M. The diffusion of iron in aluminium. *Philos. Mag.* **1970**, *21*, 305–328. [CrossRef]
37. Qin, T.; Lin, X.; Yu, J.; Wang, M.; Guo, P.; Li, J.; Zhang, Y.; Liu, J.; Zhang, S.; Huang, W. Performance of different microstructure on electrochemical behaviors of laser solid formed Ti–6Al–4V alloy in NaCl solution. *Corros. Sci.* **2021**, *185*, 109392. [CrossRef]
38. Tang, Y.; Shen, X.; Qiao, Y.; Yang, L.; Chen, J.; Lu, D.; Zhang, Z. Corrosion Behavior of a Selective Laser Melted Inconel 718 Alloy in a 3.5 wt.% NaCl Solution. *J. Mater. Eng. Perform.* **2021**, *30*, 5506–5514. [CrossRef]

39. Qiao, Y.; Wang, X.; Yang, L.; Wang, X.; Chen, J.; Wang, Z.; Zhou, H.; Zou, J.; Wang, F. Effect of aging treatment on microstructure and corrosion behavior of a Fe-18Cr-15Mn-0.66N stainless steel. *J. Mater. Sci. Technol.* **2021**, *107*, 197–206. [CrossRef]
40. Chen, Z.; Hu, H.; Guo, X.; Zheng, Y. Effect of groove depth on the slurry erosion of V-shaped grooved surfaces. *Wear* **2022**, *488–489*, 204133. [CrossRef]
41. Gupta, R.K.; Das, H.; Pal, T.K. Influence of Processing Parameters on Induced Energy, Mechanical and Corrosion Properties of FSW Butt Joint of 7475 AA. *J. Mater. Eng. Perform.* **2012**, *21*, 1645–1654. [CrossRef]
42. Knibloe, J.R.; Wright, R.N.; Trybus, C.L. Microstructure and mechanical properties of Fe3Al alloys with chromium. *J. Mater. Sci.* **1993**, *28*, 2040–2048. [CrossRef]
43. Johnson, M.; Mikkola, D.E.; March, P.A.; Wright, R.N. The resistance of nickel and iron aluminides to cavitation erosion and abrasive wear-ScienceDirect. *Wear* **1990**, *140*, 279–289. [CrossRef]
44. Jin, H.O.; Ko, C.; Saimoto, S.; Threadgill, P.L. Microstructure of Friction Stir Welded Joints in AA5182. *Mater. Sci. Forum* **2000**, *331–337*, 1725–1730. [CrossRef]
45. Miles, M.P.; Pew, J.; Nelson, T.W.; Li, M. Comparison of formability of friction stir welded and laser welded dual phase 590 steel sheets. *Sci. Technol. Weld. Join.* **2006**, *11*, 384–388. [CrossRef]

coatings

MDPI

Article

Laboratory Experimental Study on Influencing Factors of Drainage Pipe Crystallization in Highway Tunnel in Karst Area

Huaming Li [1], Shiyang Liu [2,3,*], Shuai Xiong [4], Hao Leng [1], Huiqiang Chen [2,3], Bin Zhang [2] and Zhen Liu [2,3]

1 Sichuan Lehan Expressway Co., Ltd., Leshan 614000, China; huam_li@126.com (H.L.); hao_leng@126.com (H.L.)
2 State Key Laboratory of Mountain Bridge and Tunnel Engineering, Chongqing Jiaotong University, Chongqing 400074, China; chqlxj@126.com (H.C.); 611150086011@mails.cqjtu.edu.cn (B.Z.); zhen_ll@126.com (Z.L.)
3 College of Civil Engineering, Chongqing Jiaotong University, Chongqing 400074, China
4 College of Architectural Engineering, Neijiang Normal University, Neijiang 641112, China; shuai_xx@126.com
* Correspondence: liushiyang@mails.cqjtu.edu.cn

Abstract: The crystalline blockage of tunnel drainage pipes in a karst area seriously affects the normal operation of drainage system and buries hidden dangers for the normal operation of the tunnel. In order to obtain the influencing factors and laws of tunnel drainage pipe crystallization in a karst area, based on the field investigation of crystallization pipe plugging, the effects of groundwater velocity, drainage pipe diameter, drainage pipe material, and structure on the crystallization law of tunnel drainage pipe in karst area are studied by indoor model test. The results show that: (1) With the increase of drainage pipe diameter (20–32 mm), the crystallinity of drainage pipes first increases and then decreases. (2) With the increase of water velocity in the drainage pipe (22.0–63.5 $cm \cdot s^{-1}$), the crystallinity of the drainage pipes gradually decreases from 1.20 g to 0.70 g. (3) The crystallinity of existing material drainage pipe is: M3 (poly tetra fluoroethylene) > M2 (pentatricopeptide repeats) > M4 (high density polyethylene) > M1 (polyvinyl chloride); M8 (polyvinyl chloride + coil magnetic field) is used to change the crystallinity of drain pipe wall material. (4) When the groundwater flow rate is 34.5 $cm \cdot s^{-1}$, M1 (polyvinyl chloride) and M8 (polyvinyl chloride + coil magnetic field) can be selected for the tunnel drainage pipe. The research on the influencing factors of tunnel drainage pipe crystallization plugging fills a gap in the research of tunnel drainage pipe crystallization plugging. The research results can provide a basis for the prevention and treatment technology of tunnel drainage pipe crystallization plugging.

Keywords: tunnel; crystallization and clogging in pipe; flow velocity; pipe diameter; pipe material

Citation: Li, H.; Liu, S.; Xiong, S.; Leng, H.; Chen, H.; Zhang, B.; Liu, Z. Laboratory Experimental Study on Influencing Factors of Drainage Pipe Crystallization in Highway Tunnel in Karst Area. *Coatings* **2021**, *11*, 1493. https://doi.org/10.3390/coatings11121493

Academic Editor: Francesco Di Quarto

Received: 29 October 2021
Accepted: 1 December 2021
Published: 3 December 2021

1. Introduction

With the accelerated development and modernization of China, the construction of railways and expressways in China has obviously developed by leaps and bounds. There are numerous tunnels in hard water regions in Southwest China. Water in this geological environment contains nearly saturated bicarbonate and calcium ions, which flow with groundwater into drainage pipes, slowly form calcium carbonate ($CaCO_3$) crystals, and precipitate to clog tunnel drainage systems (Figure 1). The clogging of tunnel drainage pipes may seem to be a minor problem, but it can lead to substantial hidden dangers to the safety of the entire tunnel lining structure and the normal operation of the drainage system.

The problem of tunnel damage due to crystallization has long been recognized around the world. When SI_c (calcite saturation index) is greater than 1.0, calcium carbonate deposition begins to occur rapidly. Then, in the downstream direction, the pH value and SiC of water no longer increase, but decrease [1]. The presence of phosphate ions in the precipitating solution stabilizes the initially formed vaterite by significantly reducing the rate of transformation of vaterite to calcite [2]. Within a certain concentration range of Ca^{2+} and CO_3^{2-}, the magnetic field has a significant inhibition effect on the growth and precipitation of $CaCO_3$ crystals [3]. The magnetic field affects the phase equilibrium of $CaCO_3$ polymorphs by influencing the CO_2/water interface or through the hydration of CO_3^{2-} ions [4]. Magnetic treatment can significantly lower the pH of the solution and, in the meantime, reduce the deposition of scale by 48% [5]. The activation energy of water molecules at high temperatures is 3.16 kcal·mol^{-1} higher than that at low temperatures and is close to the hydrogen bond energy between water and polyacrylate [6]. The diffusion coefficient of water molecules increases continuously as the temperature of aqueous solutions increases, while the diffusion coefficients of Ca^{2+} and CO_3^{2-} are relatively small at 293 K [7]. Ions in aqueous solutions most readily react at 353 K to form ion pairs, which then grow into crystallites, and are least likely to crystallize at 343 K [8]. The orders of binding energies of polymers with two calcite crystal surfaces are polyacrylic acid (PAA) > acrylic acid-methyl acrylate copolymer (AA-MA) > polymethylacrylic acid (PMAA) [9]. The solubility of $CaCO_3$ in water increases as the water activity increases [10]. The scale inhibition effect of magnetic fields is related to the magnetic flux density, and the relationship between the two is not a monotonic but rather a multi-extreme function [11]. With the increase of temperature (20–80 °C), the carbonate mineral crystal mainly changes from granular or acicular single crystal to granular and needle crystal aggregate, and the crystallization degree changes from heteromorphic or semi automorphic to automorphic [12]. Scaling quantity increased with increasing the initial concentration of Ca^{2+} and SO_4^{2-}, the influence of HCO_3^- and pH on scaling quantity was little, and the scale would dissolve with increasing Cl^- [13]. Bacteria and fungi can promote or lead to karst erosion, and some bacteria in water can promote the formation of $CaCO_3$ precipitation [14]. Bacteria are the main producers of CO_2 in soil [15] and can lead to $CaCO_3$ crystallization [16,17]. The terpolymer β-CD-MA-SSS has a scale inhibition rate of nearly 99.9% for $CaCO_3$ [18]. Currently, this substance is the most widely used scale inhibitor [19]. The crystal of the tunnel drainage system is mainly insoluble calcium carbonate [20–23], which is mainly calcite crystal form. When the pH value is 8~10, the crystallinity of the semi filled pipe is higher. When the pH value is 11, the crystallinity of the full filled pipe is gradually higher than that of the semi filled pipe 23 days after the start of the test, and the crystallinity is greatly affected by the pH value; the crystals are calcite. The higher the pH value when the main crystal form is spindle shape, the closer the accumulation, and the smaller and uniform the grain size [24]. The main influencing factors of tunnel drainage pipe crystallization blockage are groundwater type, shotcrete characteristics, physicochemical conditions of drainage pipe solution crystallization and drainage pipe deposition conditions [25].

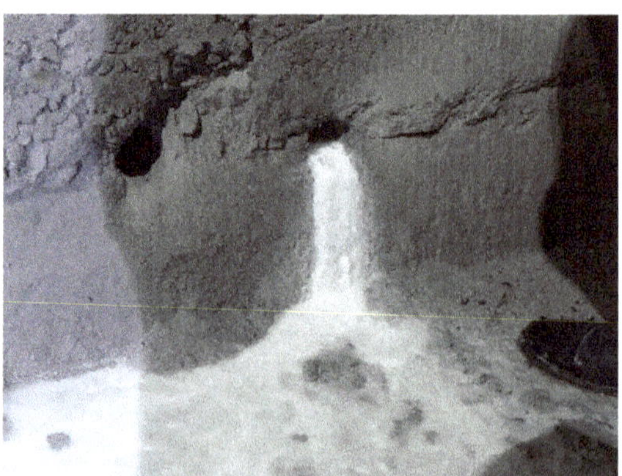

Figure 1. Crystallization and clogging in drainage pipes in the Baoligang Tunnel of the Emeishan-Hanyuan Expressway.

It can be seen from the above research literature that there has been some research on the blockage of tunnel drainage pipe, but it is still in its infancy. For example, a lot of research work needs to be carried out on the influencing factors of crystallization blockage of tunnel drainage pipe. Therefore, aiming at the problem of crystalline plugging of tunnel drainage pipe, based on the field investigation, this paper analyzes the effects of groundwater velocity, drainage pipe diameter and drainage pipe material on the crystallization law of drainage pipe using the methods of indoor orthogonal model experiment and quantitative and qualitative analysis to determine what materials and parameters of drainage pipe that can prevent crystallization blockage suitable for tunnel field application. The overall structure of the paper is as follows: Section 2 explores the indoor model test design, while Section 3 provides the results of the indoor model test. Section 4 is a quantitative and qualitative analysis of the results and discussion of the indoor model test. Section 5 indicates the research conclusions.

2. Theoretical Tests

2.1. Research Route

Based on the field investigation, this paper analyzes the influence of three factors on the crystallization of drain pipe: the flow rate of groundwater in the drain pipe, the diameter of drain pipe and the material of drain pipe. The main research route is shown in Figure 2.

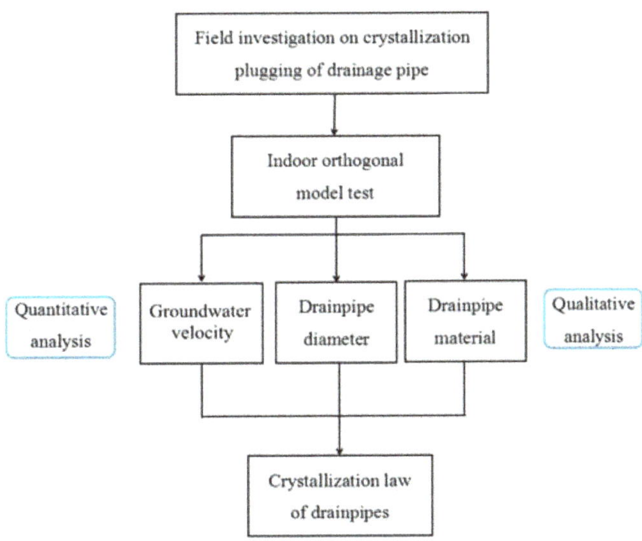

Figure 2. Research technology route.

2.2. Test Scheme

Three sets of laboratory tests were designed to assess the drainage pipe diameter, pipe material, and flow velocity. Based on the test contents, a test apparatus was designed and custom built (Figure 3) and mainly consisted of a water tank, a pump, a ball valve, and test pipes. Table 1 lists the variable values set for the main influencing factors of the tests. The indoor test run is shown in Figure 4. The drainage pipe used in the test is the drainage pipe sold in the market, and its durability meets relevant standards.

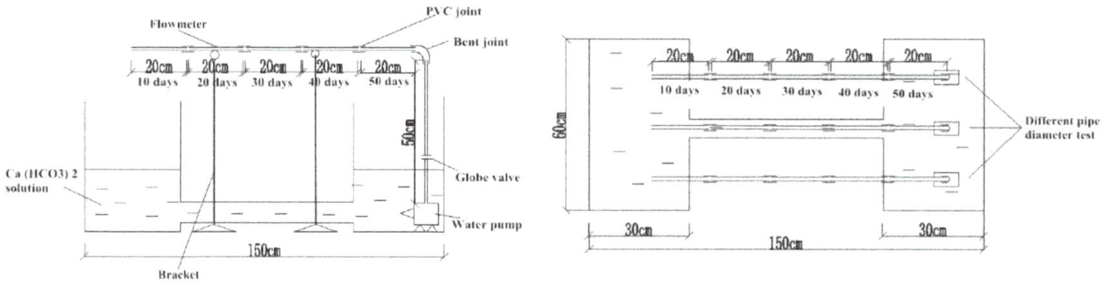

(a) (b)

Figure 3. Test device diagram. (**a**) Elevation. (**b**) Top view.

Table 1. Influencing factors of the tests.

Influencing Factor	Variable Value
Drainage pipe diameter	D1 = 20 mm; D2 = 25 mm; D3 = 32 mm
Flow velocity	V1 = 22.0 cm·s^{-1}, V2 = 26.5 cm·s^{-1}, V3 = 34.5 cm·s^{-1}, V4 = 44.5 cm·s^{-1}, V5 = 63.5 cm·s^{-1}
Drainage pipe material	M1 = PVC, M2 = PPR, M3 = PTFE, M4 = HDPE, M5 = PVC + hydrophobic antistatic self-cleaning agent, M6 = PVC + flocking film, M7 = PVC + silicone, M8 = PVC + magnetic field coil, M9 = PVC + PE powder

In the table: PVC (Polyvinyl chloride); PPR (pentatricopeptide repeats); PTFE (Poly tetra fluoroethylene); HDPE (High Density Polyethylene).

Figure 4. Indoor test operation.

The tests were conducted at room temperature. The test solution was prepared from anhydrous $CaCl_2$, $NaHCO_3$, and deionized water. The aqueous solution was circulated for a total of 50 days, which was divided into five 10-day test cycles. After the completion of each cycle, the test was stopped, and the pipe section under study was dried and weighed.

2.3. Test Solution

A field investigation showed that the groundwater in the region where the engineering project is located is composed of clastic rock pore fracture water, carbonate rock fracture karst water, and loose rock pore water. Water samples were taken at K108+435 of the right line of the Dayan Tunnel of the Emeishan-Hanyuan Expressway for testing and analysis. The composition and content of various major anions and cations in water sample solutions were determined, and the corresponding test results are shown in Table 2.

Table 2. Concentration of main ions in tunnel drains.

Ion Type	Ion	Concentration *c* (mmol·L^{-1})	Percentage (%)	Ion Type	Ion	Concentration *c* (mmol·L^{-1})	Percentage (%)
Cation	K^++Na^+	3.68	5.17	Anion	Cl^-	1.8	0.69
	Ca^{2+}	45.2	63.50		SO_4^{2-}	51.9	19.83
	Mg^{2+}	22.3	31.33		HCO_3^-	208	79.48

The selected water samples were rich in a variety of ions, including cations such as Ca^{2+}, Mg^{2+}, K^+, and Na^+. Ca^{2+} was the most abundant cation, accounting for 63.50% of the total. The most common anions were HCO_3^-, SO_4^{2-}, and Cl^-, and HCO_3^- was the most abundant, accounting for 79.48% of the total.

The analysis of the data in Table 2 indicated that the presence of a wide range of anions and cations in the natural water samples would increase the number of factors affecting crystallization in tunnel drainage pipes and the sensitivity of these factors. To achieve controllability and operability of the factors influencing crystallization and to minimize repetition in the laboratory tests, representative types of anions and cations in the water samples were determined for the tests. Ca^{2+} and HCO_3^- were selected, and accordingly, a saturated $CaHCO_3$ solution was used as the test solution.

3. Results

3.1. Effect of Pipe Diameter

To investigate the effect of the diameter of drainage pipes on crystallization, three pipe diameters (20 mm, 25 mm, and 32 mm) were used, and the flow velocity was controlled at 34.5 cm·s^{-1}. Each type of pipe was completely filled during the test. The variation in the mass of crystals with time is shown in Figure 5, and the variation in the increase in the mass of crystals with time is shown in Figure 6.

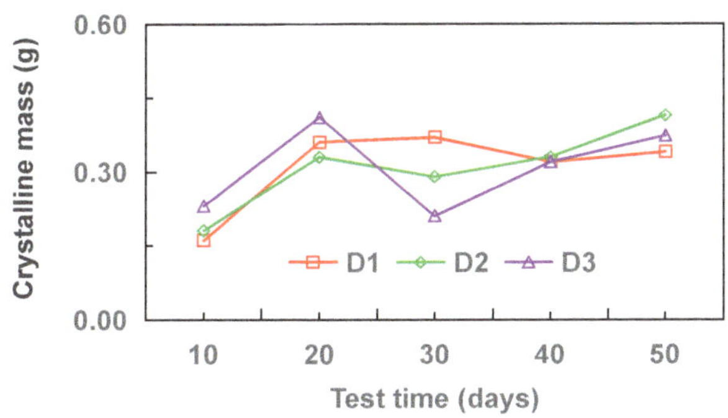

Figure 5. Variation of crystal quality with time.

Figure 5 shows that (1) the 25-mm pipe (D2) was the most sensitive to $CaCO_3$ crystallization and had the largest change in the mass of crystals, with the maximum and minimum being 0.415 g and 0.18 g, respectively; in other words, the difference between the maximum and minimum was 0.235 g, and the maximum was more than 2.3 times the minimum; (2) the difference between the maximum and the minimum masses of crystals in the 20-mm pipe (D1) was 0.21 g, and the maximum was 2.3 times the minimum, exhibiting a pattern that was the same as that in the case of the 25-mm pipe (D2); and (3) the 32-mm pipe (D3) was the most stable, where the difference between the maximum and minimum masses of crystals was 0.2 g, and the maximum was 1.95 times the minimum.

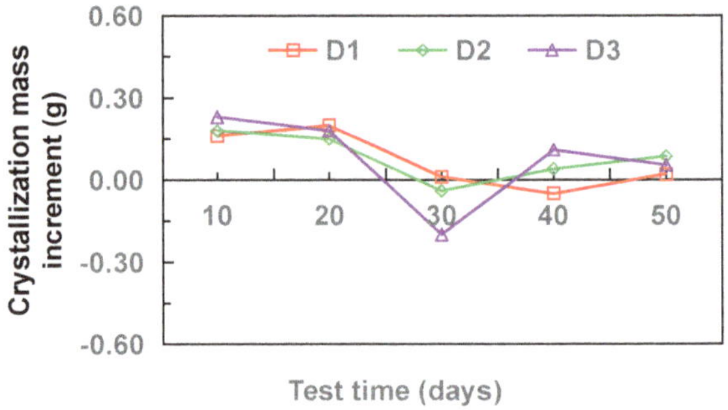

Figure 6. Variation of crystal mass increment with time.

Figure 6 indicates that the variation in the mass of crystals in the 20-mm pipe (D1) was characterized by a fast increase in the early phase followed by gradual stabilization with time. The curves of all three types of pipe showed a negative growth, indicating that the pipe wall played a role in weakening the crystallization for a certain period of time. The mass of crystals in the 32-mm pipe (D3) experienced the largest negative growth, reaching 0.2 g, which was 20 and 5 times the negative growth in the 20-mm and 25-mm pipes, respectively. In addition, Figure 6 shows that for all types of pipe, the mass of crystals started to increase abruptly after reaching the minimum negative growth. During the test, there was initially a large interaction between the crystals and the pipe wall, which was favorable to the formation of crystals. In a later phase, the interaction between the $CaCO_3$ crystals and the pipe wall was surpassed by the interaction between the $CaCO_3$ crystals themselves, which mainly played a repulsive role. Under the continued action of water flow, the presence of friction reduced the originally formed crystals, thereby resulting in the phenomenon of negative growth. As the circulation test proceeded, the surface of crystals was covered continuously by other tiny particles and impurities in the solution, thereby resulting in growth again in the later phase. The above process occurred cyclically during the test.

3.2. Effect of Flow Velocity

The effect of the flow velocity on crystallization in drainage pipes was investigated using a 32-mm polyvinyl chloride (PVC) pipe and considering five different flow velocities (22 $cm{\cdot}s^{-1}$, 26.5 $cm{\cdot}s^{-1}$, 34.5 $cm{\cdot}s^{-1}$, 44.5 $cm{\cdot}s^{-1}$, and 63.5 $cm{\cdot}s^{-1}$). The variation in the mass of crystals with time is shown in Figure 7, and the variation in the increase in the mass of crystals with time is shown in Figure 8.

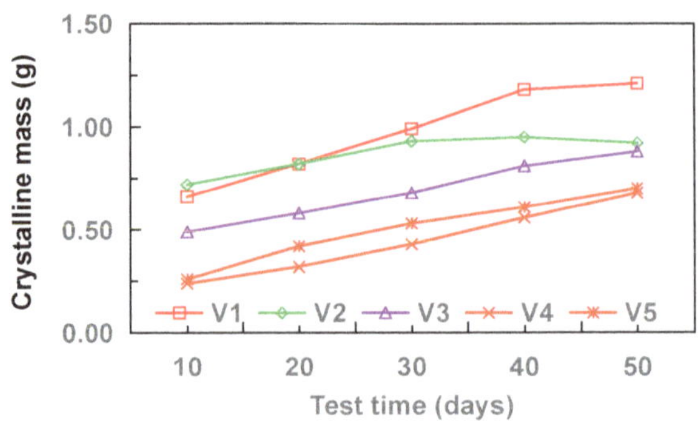

Figure 7. Variation of crystal quality with time.

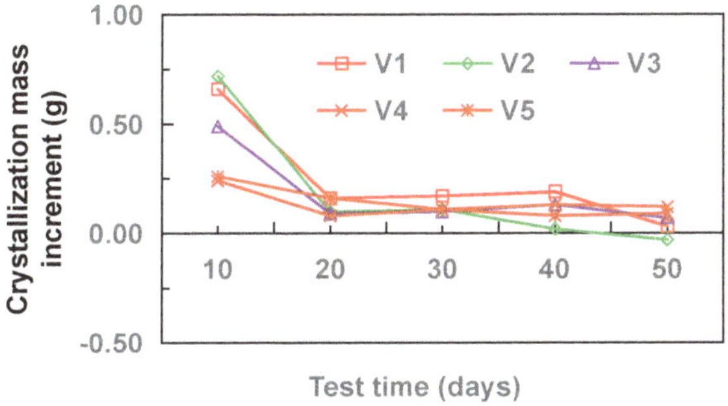

Figure 8. Variation of crystal mass increment with time.

Figure 7 shows that, overall, the rate of crystallization was faster in the pipes with a low flow velocity than in the pipes with a high flow velocity. Specifically, the results reveal the following: (1) Under a flow velocity of 22 cm·s^{-1}, the difference between the maximum and minimum crystal mass was 0.55 g; and the maximum was 1.83, 1.32, 1.8, 2.83, and 2.69 times the minimum for flow velocities of 26.5 cm·s^{-1}, 34.5 cm·s^{-1}, 44.5 cm·s^{-1}, and 63.5 cm·s^{-1}, respectively. (2) Under the influence of each flow velocity, the mass of crystals first increased, then reached the saturated state in the later phase, and remained stable, with the crystallization rate slowing down gradually. In the early phase, there was a high interaction force between the pipe wall and the $CaCO_3$ crystals, resulting in the attachment of a large mass of crystals to the pipe wall and the gradual accumulation of impurities in the solution. In the later phase, the interaction between the $CaCO_3$ crystals and the pipe wall was no longer dominant, and its effect gradually weakened. Meanwhile, the precipitates on the wall were rounded under the continuous impact of the water flow, so the change in mass was very small. Figure 8 shows that the phenomenon of negative growth in the mass of crystals occurred only under a flow velocity of 26.5 cm·s^{-1}, exhibiting a negative growth of 0.03 g, i.e., on the order of magnitude of 0.01 g, which was very small compared to the order of magnitude of 1 g under a flow velocity of 26.5 cm·s^{-1}. There was a large fluctuation in the mass of crystals in the pipes with a high flow velocity.

3.3. Effect of Pipe Material

This group of experiments included nine kinds of pipe experiments. The pipe diameter of the drainage pipe was 32 mm and the flow rate was controlled to 34.5 cm·s^{-1}. Among the pipes, M1–M4 were mainly the existing pipes, and M5–M9 mainly included the addition of other materials on the basis of the existing pipes, which changed the inner wall characteristics of the drainage pipe and the internal physical field of the drainage pipe, and analyzed the influence law of the drainage pipe material and structure on the crystallization of the drainage pipe. The variation law of crystallization quality of M1–M4 drainage pipes with test time is shown in Figure 9, and the variation curve of crystallization quality increment with test time is shown in Figure 10. The variation law of crystallization quality of M5–M9 drainage pipe with test time is shown in Figure 11. The variation curve of crystallization quality increment with test time is shown in Figure 12.

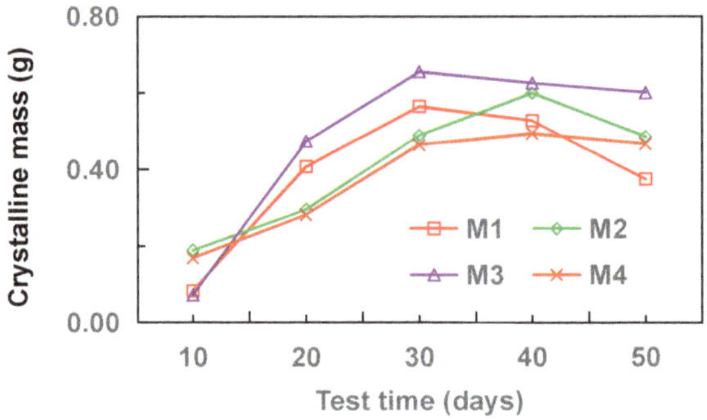

Figure 9. Variation of crystal quality with time.

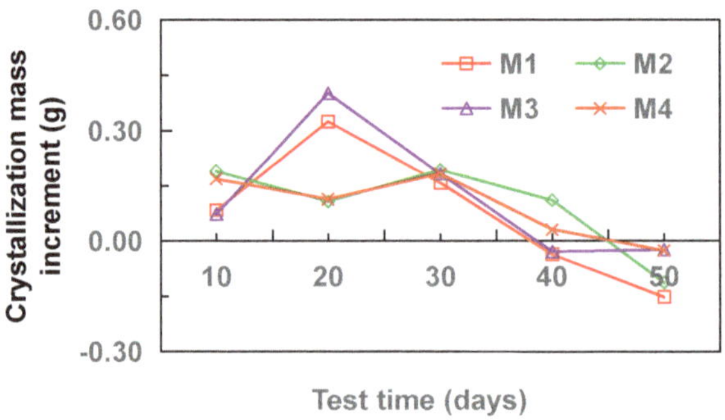

Figure 10. Variation of crystal mass increment with time.

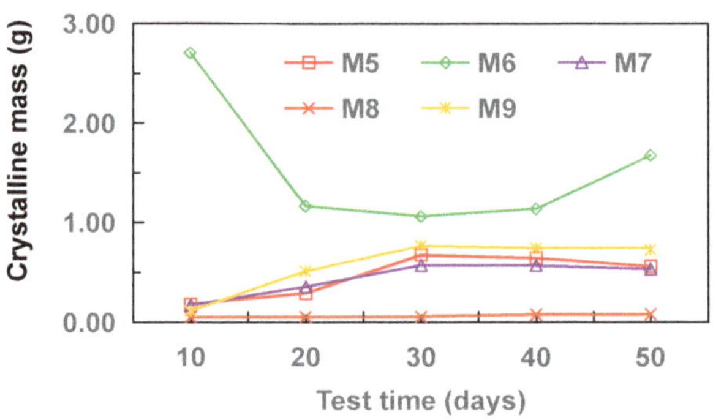

Figure 11. Variation of crystal quality with time.

As shown in Figure 9, the trends of the effects of the different pipe materials on $CaCO_3$ crystallization were approximately the same, i.e., there was always an initial stage of increase followed by a stage of decrease. In the increasing stage, the interaction between the $CaCO_3$ crystals and the pipe wall was dominated by attraction in the early phase, and there was an accumulation of other impurities in the solution, resulting in an increase in the mass of crystals. Specifically, in the case of the PVC (M1) pipe, the mass of crystals reached the maximum of 0.564 g on day 30, which was 6.8 times the corresponding minimum of 0.083 g; for the polypropylene random (PPR) (M2) pipe, the mass of crystals attained its maximum of 0.599 g on day 40, which was 3.2 times its minimum of 0.189 g; for the polytetrafluoroethylene (PTFE) (M3) pipe, the mass of crystals reached the maximum of 0.655 g at day 30, which was 9.1 times the minimum of 0.072 g; and for the HDPE (M4) pipe, the mass of crystals reached the maximum of 0.494 g on day 40, which was 2.94 times the minimum of 0.168 g. After running the test for a period of time, the mass of crystals reached the maximum, and then repulsion became dominant instead of attraction. However, as the test continued, the impurities in the solution and the previously formed crystals interacted with each other through extrusion and friction, which rounded them and resulted in a decreasing process. Overall, the mass of crystals in the PTFE (M3) pipe was higher than that in the other types of pipe and thus did not meet our goal of reducing $CaCO_3$ crystallization. Considering the economic cost, the use of PVC (M1) pipe is recommended. In addition, Figure 10 shows that the PVC (M1) pipe achieved the largest negative growth, with a steep slope of change. During the period of day 10 to day 20, there was a positive increase in the variable value for all pipes, with the largest increase (0.401 g) occurring in the PTFE (M3) pipe, and the smallest increase (0.107 g) occurring in the PVC (M1) pipe, indicating that there was a process of aggregation in each pipe in the early phase. Overall, the largest negative growth (0.153 g) occurred in the PVC (M1) pipe between day 40 and day 50, during which the increase in each parent material was less than 0 g.

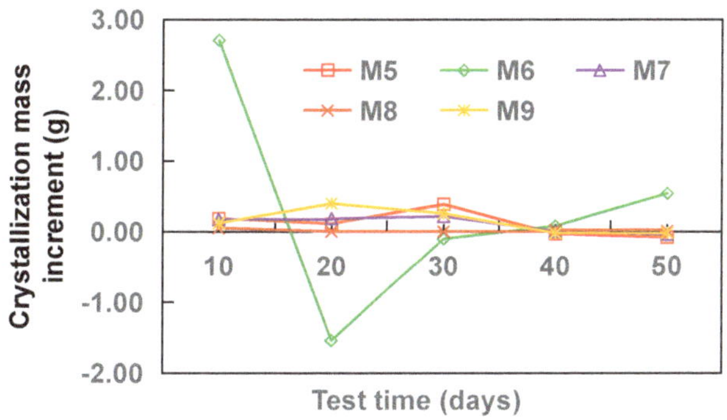

Figure 12. Variation of crystal mass increment with time.

Figure 11 reflects the variation trend in the mass of crystals in pipes made of different modified materials with time. The mass of crystals in the pipe using a magnetic field coil (M8) was always smaller than the mass in the pipes using different modified materials, indicating a significant modification effect; the maximum mass of crystals (0.075 g) was 1.5 times the minimum (0.05 g). The minimum mass of crystals was mainly due to the directional movement of Ca^{2+} and HCO_3^- in the circulating water under the action of the electromagnetic field generated by the current, which thus interfered with their mutual binding. The curve of the flocking film (M6) pipe is apparently above the curves of other pipes, indicating that the mass of crystals in this pipe was much larger than that in the pipes using other types of modified materials; the maximum mass of crystals reached 2702 g, which was 2.55 times the minimum of 1058 g. The mass of crystals in this pipe was the largest mainly because the villi were negatively charged and interacted strongly with the substances in solution. In addition, the presence of villi greatly increased the contact area, so other substances in solution continuously accumulated as the water flowed through the pipe. For these two reasons, the mass of crystals in the flocking film (M6) pipe was much larger than that in the pipes using other modified materials. The other three modified materials had very similar modification effects, with the effect of polyethylene (PE) powder (M9) being slightly less than that of the hydrophobic material (M5) and silicone (M7).

Figure 12 shows the variation trend in the increase in the mass of crystals with time in the pipes using different types of modified materials. The variation in the mass of crystals in the flocking film (M6) pipe was the largest, with the maximum increase in the mass of crystals reaching 1539 g, which was 40.5 times the minimum (0.038), mainly due to its own negative charge and large contact area. The increase in the mass of crystals produced in the pipe using the magnetic field coil (M8) was relatively stable (almost in a straight line), with an average increase of 0.0075 g, indicating that the magnetic field generated by a current of 1 A in the test had a satisfactory effect on the directional movement of ions in solution and thus achieved the purpose of reducing the crystals on the pipe wall. Except for the stable mass of crystals in the pipe using the magnetic field coil (M8), the mass of crystals in the other four pipes using modified materials all showed a negative growth stage, indicating that with the circulation of water flow, the binding force between the $CaCO_3$ crystals and each modified material had an obvious stage of change between attraction and repulsion.

4. Discussion

It can be seen from the influence experiment of drainage pipe diameter that when the drainage pipe is full of groundwater, the crystallinity also increases gradually in the initial stage. However, after it increases to a certain extent, due to the decrease of drainage pipe cross-section, when the groundwater flow is certain, the groundwater velocity in the pipe will increase, resulting in the increase of flow velocity pressure at the cross-section. Finally, the crystals on the pipe wall fall off. It can be seen that when the groundwater flow is certain, the change of pipe diameter also reflects the change of groundwater velocity in the pipe; that is, the groundwater velocity has a strong impact on the crystallization of the drainage pipe. It can be seen from the flow velocity influence experiment that the crystals in the drainage pipe increase gradually under different flow velocities, but the larger the flow velocity is, the smaller the crystals as a whole, which also verifies that the flow velocity of groundwater in the drainage pipe has a significant impact on the crystallization of the drainage pipe. The test results are consistent with literature [26,27].

In the pipe influence experiment, the change curve of crystallinity with time of the existing pipes in M1~M4 markets is basically the same, and the total crystallinity has little difference. Therefore, the effect of these four kinds of pipes on preventing crystal plugging is not ideal. Among the five composite pipes M5~M9, although the crystallization laws of M6 and M8 are different, the crystallization laws of the other three pipes are basically the same. Because fluff is pasted on the pipe wall of M6, the contact opportunity between the crystal and fluff is increased due to the existence of a large number of fluff at the initial stage of the test, so the attached crystal is lower than that of other materials. In the later stage, when the local water force is greater than the bonding force between the crystal and fluff, the crystal will fall off, and then the crystal will increase. The difference of M8 was that a magnetic field was added. Under the action of magnetic field, calcium carbonate crystals in groundwater are mainly unstable aragonite crystals, which are easily washed away by groundwater. The modification results are consistent with the research results in literature [5,28,29].

5. Conclusions

Aiming at the problem of crystalline plugging of tunnel drainage pipe in karst area, based on the field investigation, this paper studied the effects of groundwater velocity, diameter, material, and structure of drainage pipe on the crystallization law of a tunnel drainage pipe in karst area using the methods of indoor orthogonal model test and quantitative and qualitative analysis. The main conclusions are as follows:

(1) With the increase of drainage pipe diameter (20–32 mm), the crystallinity of drainage pipe increases first and then decreases gradually. With the increase of water velocity in the drainage pipe, the crystallinity of the drainage pipe decreases gradually. At different flow rates, the growth rate of crystallization increased first and then gradually stabilized. Without adding other materials, the crystallinity of drainage pipe is: M3 > M2 > M4 > M1. When other materials are added, the crystallinity of drainage pipe is M6 > M9 > M5 > M7 > M8. When the groundwater flow rate is 34.5 $cm \cdot s^{-1}$, M1 and M8 drainage pipes can be used in the tunnel to ensure that the drainage pipe will not be blocked by crystallization.

(2) There is a 5-fold relationship between the diameter of indoor test (20–32 mm) and the diameter of field drainage pipe (80–160 mm). Therefore, whether there is a 5-fold

relationship between the law obtained from indoor test and the actual crystallization change law needs to be further determined. Ca^{2+} and HCO_3^- were mainly selected as the ions in the test solution, while the actual groundwater on site also contains other ions that may affect crystallization, such as Mg^{2+}, requiring a lot of further research work.

Author Contributions: Investigation, H.L. (Huaming Li) and H.L. (Hao Leng); writing—original draft preparation, S.X.; writing—review and editing, S.L.; formal analysis, H.C. and B.Z.; data curation, Z.L. All authors have read and agreed to the published version of the manuscript.

Funding: This research was financially supported by the Scientific Research Project of the Emei Hanyuan Expressway Project (Grant No. LH-HT-45), and the State Key Laboratory of Mountain Bridge and Tunnel Engineering (Grant No. SKLBT-2110).

Institutional Review Board Statement: Not applicable.

Informed Consent Statement: Not applicable.

Data Availability Statement: The datasets used and/or analyzed during the current study are available from the corresponding author on reasonable request.

Conflicts of Interest: The authors declare no conflict of interest.

References

1. Liu, Z.; Zhang, M.; You, S.; Qiang, L.; Sun, H.; Wang, J.; Kongyun, W. Spatial and diurnal variations of geochemical indicators in a calcite-precipitating stream-Case study of Baishuitai. Yunnan. *Geochimica* **2004**, *33*, 269–278.
2. Katsifaras, A.; Spanos, N. Effect of inorganic phosphate ions on the spontaneous precipitation of vaterite and the transformation of vaterite to calcite. *J. Cryst. Growth* **1999**, *204*, 183–190. [CrossRef]
3. Lin, Y.; Fang, J.; Li, J. Quantitive study of the effect of a magnetic field on the precipitation of calcium carbonate aqueous solutions. *Ind. Water Treat.* **2002**, *22*, 16–18.
4. Knez, S.; Pohar, C. The magnetic field influence on the polymorph composition of CaCO3 precipitated from carbonized aqueous solution. *J. Colloid Interface Sci.* **2005**, *281*, 377–388. [CrossRef]
5. Parsons, S.; Wang, B.; Judd, S.; Stephenson, T. Magnetic treatment of Calcium Carbonate scale-effect of PH control. *Water Res.* **1997**, *31*, 339–342. [CrossRef]
6. Dermitzaki, E.; Bauer, J.; Wunderle, B.; Michel, B. Diffusion of water in amorphous polymers at different temperatures using molecular dynamics simulation. In Proceedings of the 2006 1st Electronic Systemintegration Technology Conference, Dresden, Germany, 5–7 September 2006; Volume 2, pp. 762–772.
7. You, X.; Zhang, X. Molecular dynamics simulation on an aqueous solution of calcite. *J. North Univ. China* **2013**, 170–173.
8. Wang, D.; Huang, L. Molecular dynamics simulation on an acIueous solution of calcium carbonate. *Sci. Technol. Eng.* **2009**, *9*, 6619–6623.
9. Zhang, S.; Shi, W.; Lei, W.; Xia, M.; Wang, F. Molecular dynamics simulation of interaction between calcite crystal and water-soluble polymers. *Acta Phys. Chim. Sin.* **2005**, *21*, 1198–1204.
10. Song, J. *Experimental Study on the Impact of Magnetic Field on Calcium Carbonate Crystallization Process*; Shandong University: Jinan, China, 2015.
11. Chen, J. *Manufacturing Water Treatment Devices Generating Alternating Magnetic Field and Experimental Study on Its Antiscaling Effect*; Chongqing University: Chongqing, China, 2014.
12. Huang, S.; Song, H. Study on the crystallization and precipitation of carbonate at different temperature conditions. *Geoscience* **1991**, 442–449 and 475–476.
13. Jiang, S.; Yu, H.; Liu, C. Research on the scaling dynamic model of the high salinity system. *Appl. Chem. Ind.* **2011**, *40*, 1623–1628.
14. Zhang, J.; Li, S.; Zhou, Y. On bacterial and fun gal effects on karst process and its application. *Carsologica Sin.* **1997**, *16*, 362–369.
15. Jakucs, L. *Morphogenetics of Karst Regions: Variants of Karst Evolution*; Adam Hilger: New York, NY, USA, 1977.
16. Greenfield, L. Metabolism and concentration of calcium and magnesium and precipitation of calcium carbonate by a marine bacteria. *Ann. N. Y. Acad. Sci.* **1963**, *109*, 23–45. [CrossRef]
17. Krumbein, W. Photolithotopic and chemoorganotrophic activity of bacteria and algae as related to beachrock formation and degradation (Gulf of Aqabu, Sinai). *Deep. Sea Res. Part B Oceanogr. Lit. Rev.* **1979**, *1*, 139–203.
18. Gu, X.; Qiu, F.; Zhou, X.; Yang, D.; Guo, Q.; Guo, X. Synthesis and application of terpolymer scale inhibitor in the presence of β-cyclodextrins. *J. Pet. Sci. Eng.* **2013**, *109*, 177–186. [CrossRef]
19. Wang, H.; Zhou, Y.; Liu, G.; Huang, J.; Yao, Q.; Ma, S.; Cao, K.; Liu, Y.; Wu, W.; Sun, W.; et al. Investigation of calcium carbonate precipitation in the presence of fluorescent-tagged scale inhibitor for cooling water systems. *Desalination Water Treat.* **2015**, *53*, 3491–3498. [CrossRef]

20. Yoon, S.; Park, E.; Lee, J.; Chun, B. Laboratory test of molecular vibration for preventing drainage pipe blockage in deteriorated tunnel. *J. Korean Geotech. Soc.* **2012**, *28*, 69–77. [CrossRef]
21. Zhang, X.; Zhou, Y.; Zhang, B.; Zhou, Y.; Liu, S. Investigation and Analysis on Crystallization of Tunnel Drainage Pipes in Chongqing. *Adv. Mater. Sci. Eng.* **2018**, *2018*, 1–6. [CrossRef]
22. Guo, X. Crystallization mechanism and countermeasures of drainage system for railway tunnel. *China Railw. Sci.* **2020**, *41*, 71–77.
23. Ye, F.; Tian, C.; He, B.; Zhao, M.; Wang, J.; Han, X.; Song, G. Experimental study on scaling and cloggingin drainage system of tunnels under construction. *China J. Highw. Transp.* **2021**, *34*, 159–170.
24. Xiang, K.; Zhou, J.; Zhang, X.; Huang, C.; Song, L.; Liu, S. Experimental Study on Crystallization Rule of Tunnel Drainpipe in Alkaline Environment. *Tunn. Constr.* **2019**, *39*, 207–212.
25. Tian, C.; Ye, F.; Song, G.; Wang, Q.; Zhao, M.; He, B.; Wang, J.; Han, X. On Mechanism of crystal blockage of tunnel drainage system and preventive countermeasures. *Mod. Tunn. Technol.* **2020**, *57*, 66–76.
26. Lu, G.; Wang, P.; Yang, Y.; Mao, C.; Wu, Y.; Wu, J.; Dong, P.; Wu, J. Advance in mechanism of groundwater crystallization blockage in tunnel drainage pipe and scale inhibiting techniques in karst area. *Mod. Tunn. Technol.* **2021**, 1–16. Available online: http://kns.cnki.net/kcms/detail/51.1600.U.20210818.1810.006.html (accessed on 25 November 2021).
27. Mao, C.; Yang, Y.; Wu, J.; Dong, P.; Wu, J. Numerical simulation of crystal blockage in tunnel drainage pipe based on dynamic grid and level set. *Carsologica Sin.* **2021**, 1–18. Available online: http://kns.cnki.net/kcms/detail/45.1157.P.20211025.1840.002.html (accessed on 25 November 2021).
28. Higashitani, K.; Kage, A.; Katamura, S.; Imai, K.; Hatade, S. Effects of a Magnetic Field on the Formation of $CaCO_3$ Particles. *J. Colloid Interface Sci.* **1993**, *156*, 90–95. [CrossRef]
29. Alimi, F.; Tlili, M.; Amor, M.B.; Maurin, G.; Gabrielli, C. Influence of magnetic field on calcium carbonate precipitation in the presence of foreign ions. *Surf. Eng. Appl. Electrochem.* **2009**, *45*, 56–62. [CrossRef]

coatings

Article

Optimization Model of Engineering Specifications Based on Grey Quality Gain-Loss Function

Bo Wang [1,2,3], Qi Yang [1], Chen Liu [4,*], Qikai Li [1] and Xiangtian Nie [1,2,3,4]

1 School of Water Conservancy, North China University of Water Resources and Electric Power, Zhengzhou 450046, China; wangbo@ncwu.edu.cn (B.W.); 201703219@stu.ncwu.edu.cn (Q.Y.); 201403516@stu.ncwu.edu.cn (Q.L.); niexiangtian@ncwu.edu.cn (X.N.)
2 Henan Collaborative Innovation Center for Water Efficient Utilization and Guarantee Engineering, Zhengzhou 450046, China
3 Henan Province Key Laboratory of Water Environment Simulation and Treatment, Zhengzhou 450046, China
4 School of Management and Economics, North China University of Water Resources and Electric Power, Zhengzhou 450046, China
* Correspondence: x201810105116@stu.ncwu.edu.cn

Abstract: In view of the fact that the target values of some quality characteristics are grey, the grey quality gain-loss function model was applied in the analysis of the quality characteristics. At the same time, based on the analysis of engineering specifications and process capability, an optimization model of engineering specifications was proposed to minimize the expected total loss of each product and maximize the expected compensation with inspection costs, scrap costs and grey quality gain-loss into consideration. The optimal engineering specification can be obtained by using the optimization model. Through the example analysis and its application in dam concrete construction, the practicability of the model is verified, which provides an important reference for the research of the new theory of dam concrete construction quality control.

Keywords: grey quality gain-loss; engineering specifications; process capability

Citation: Wang, B.; Yang, Q.; Liu, C.; Li, Q.; Nie, X. Optimization Model of Engineering Specifications Based on Grey Quality Gain-Loss Function. *Coatings* **2021**, *11*, 1327. https://doi.org/10.3390/coatings11111327

Academic Editor: Chi-wai Kan

Received: 9 September 2021
Accepted: 25 October 2021
Published: 29 October 2021

1. Introduction

According to Shewhart, one of the most important tasks in quality control is to understand and evaluate the needs and expectations of customers, and ultimately to provide products and services that meet or exceed those needs and expectations [1]. Therefore, the first step in meeting the needs of customers is to make the physical characteristics of the products meet the satisfied standards. The second step is to determine the way and the method of setting the standards for the physical characteristics of products, understand and evaluate the needs and expectations of the customers, and to provide products and services that meet or exceed those needs and expectations.

A quality loss function is proposed by Taguchi to determine the ideal target value (parameter design) and to evaluate the loss caused by deviation from the target value [2]. Because parameter design reduces performance variation by reducing the influence of the sources of variation rather than by controlling them, it is a very cost-effective technique for improving product quality. Kackar introduces the concepts of off-line quality control and parameter design and then discusses the Taguchi Method for conducting parameter design experiments [3,4]. Taguchi's philosophy recognizes the importance of both off-line and on-line quality control in reducing variation and its effects to the minimum, and covers the entire engineering and manufacturing process. Barker looks at key topics such as loss function, performance variations, signal to noise, system design and parameter and tolerance design [5].

The concept of quality loss function focuses on the evaluation of quality losses caused by deviation of quality characteristics from the target value, while the goal of quality planning is to minimize the total loss of producers and consumers for society; Kapur and Wang have described the use of the normal and log-normal quality characteristics to design the speciation limits based on Taguchi's quality loss function [6]. On this basis, Kapur proposed an optimization model to develop the acceptable specifications or tolerance limits. Using this model, the total quality costs caused by variation (which is always present within any acceptable limits), inspection (measurement to determine the nonconforming units) and costs related to nonconforming units (scrap or rework) can be minimized [7].

Considering the multivariate normal distribution of quality characteristics. The specification region is given by truncating the multivariate normal distribution. Kapur and Cho present the optimization model to develop the specification region for multiple quality characteristics based on the framework of multivariate quality loss function [8]. Chatterjee and Chakraborty have studied the expression of process capability index of some unilateral specification limits and the relationship between the expression and the defect rate [9]. The control chart of process capability is designed and used to evaluate the process capability in the entire production cycle.

Chen proposes the topic of selecting lower specification limits under Taguchi quality loss function [10]. Given that product quality characteristics comply with exponential distribution, an improved method is proposed based on Kapur and Wang for the economic design of specification limits; Pan and Pan have modified the Kapur's model so that the economic specification limits of symmetric loss and asymmetric loss can be established [6,7,11].

Since Taguchi quality loss function cannot describe the quality compensation effect in production practice, the concept of quality gain-loss function is proposed, and the quality gain-loss transfer model and the tolerance optimization method of quality characteristics are studied with the foundation that the constant term in Taylor series expansion is given the meaning of quality compensation. When the loss of linear term cannot be ignored and the compensation amount is constant, the quality gain-loss function model with larger-the-better and smaller-the-better expectation is designed. Considering the compensation effect of quality characteristics in dam concrete construction, the tolerance optimization model of dam concrete construction quality is constructed, and the optimal tolerance adjustment rate of the characteristic per unit quality in each stage of the dam concrete construction is studied, so as to improve the overall construction quality to the greatest extent [12–14]. On this basis, an inverted normal quality gain-loss function based on functional characteristics of the inverted normal function is proposed, aiming at the issue that the existing quality gain-loss function cannot describe the unbounded problem in engineering practice, and the multivariate quality gain-loss function and quality gain-loss function with larger-the-better, smaller-the-better characteristics are designed. In this paper, fuzziness of quality characteristics was analyzed, the fuzzy quality gain-loss function model was constructed and the optimal process mean for the asymmetric fuzzy quality gain-loss function was designed [15,16].

2. Grey Quality Gain-Loss Function

Intuition and judgment play an important role when determining the target value of some quality characteristics, and some factors related to personal needs are also relevant. For this kind of quality characteristics, the determination of the target value is grey, so the application of Grey System Theory is adopted to expand the inverted normal quality gain-loss model. With the following methods, the grey quality gain-loss function is constructed.

Assuming that the target value y_0 of a quality characteristic y is grey, it can be expressed as grey number $\otimes \in [T_1, T_2]$, where $[T_1, T_2]$ is the interval range of the grey number, in which the quality loss can be considered as 0. In this case, the grey quality gain-loss function is defined as:

$$\widetilde{L}(y) = \begin{cases} \widetilde{g}(y) + A\left[1 - exp\left(-\frac{(y-T_1)^2}{2\sigma^2}\right)\right] & , y \in [-\infty, T_1) \\ \widetilde{g}(y) & , y \in [T_1, T_2] \\ \widetilde{g}(y) + A\left[1 - exp\left(-\frac{(y-T_2)^2}{2\sigma^2}\right)\right] & , y \in (T_2, +\infty] \end{cases} \quad (1)$$

Many quality gain-losses deviating from the target value are asymmetric in engineering practice, and therefore in order to make the grey quality gain-loss function more universal and closer to the actual situation, the piecewise function theory is applied to modify it, and the asymmetric grey quality gain-loss function model is constructed. The modified asymmetric grey quality gain-loss function is

$$\widetilde{L}(y) = \begin{cases} \widetilde{g}(y) + A_1\left[1 - exp\left(-\frac{(y-T_1)^2}{2\sigma_1^2}\right)\right] & , y \in [-\infty, T_1) \\ \widetilde{g}(y) & , y \in [T_1, T_2] \\ \widetilde{g}(y) + A_2\left[1 - exp\left(-\frac{(y-T_2)^2}{2\sigma_2^2}\right)\right] & , y \in (T_2, +\infty] \end{cases} \quad (2)$$

where T_1 and T_2 are the upper and lower limits of interval grey number respectively, A_1 and A_2 are the maximum possible loss caused by deviation from the target value on both sides, and σ_1^2 and σ_2^2 are the shape parameters of quality gain-loss function on both sides.

2.1. Constant Compensation

The quality compensation is constant, i.e., $\widetilde{g}(y) = a$, and the grey quality gain-loss function of quality characteristic y is

$$\widetilde{L}(y) = \begin{cases} a + A_1\left[1 - exp\left(-\frac{(y-T_1)^2}{2\sigma_1^2}\right)\right] & , y \in [-\infty, T_1) \\ a & , y \in [T_1, T_2] \\ a + A_2\left[1 - exp\left(-\frac{(y-T_2)^2}{2\sigma_2^2}\right)\right] & , y \in (T_2, +\infty] \end{cases} \quad (3)$$

when $\widetilde{g}(y) = a$ (constant), the curve of grey quality gain-loss function y is shown in Figure 1.

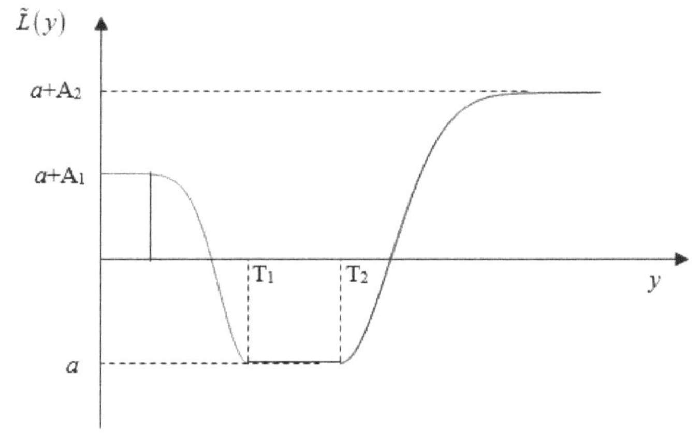

Figure 1. Curve of constant compensation.

Therefore, when the compensation is constant, the maximum grey quality compensation amount in the target value range is a. As the quality characteristics deviate from the target value, the grey quality loss increases.

2.2. Hyperbolic Tangent Compensation

Furthermore, considering the variable quality compensation and the grey quality characteristic target value, if it is assumed that the target value y_0 can be expressed as the grey number $\otimes \in [T_1, T_2]$, the quality compensation will be the largest in this interval. In this case, let the quality compensation term be a function of the quality characteristic value y, and the grey hyperbolic tangent compensation function is constructed according to the properties of hyperbolic tangent function. At this moment, the grey quality compensation function is:

$$\tilde{g}(y) = \begin{cases} -\alpha\left\{1 - \frac{2}{\exp[2(y-T_1)]+1}\right\} + \beta \text{ , } y < T_1 \\ \beta & \text{, } T_1 \leq y \leq T_2,\ \alpha > 0 \\ \alpha\left\{1 - \frac{2}{\exp[2(y-T_2)]+1}\right\} + \beta \text{ , } y > T_2 \end{cases} \tag{4}$$

In the case of asymmetric quality compensation, the asymmetric hyperbolic tangent grey quality compensation function can be expressed as:

$$\tilde{g}(y) = \begin{cases} -\alpha_1\left\{1 - \frac{2}{\exp[2(y-T_1)]+1}\right\} + \beta \text{ , } y < T_1 \\ \beta & \text{, } T_1 \leq y \leq T_2,\ \alpha > 0 \\ \alpha_2\left\{1 - \frac{2}{\exp[2(y-T_2)]+1}\right\} + \beta \text{ , } y > T_2 \end{cases} \tag{5}$$

The curve of the asymmetrical hyperbolic tangent grey quality compensation function is shown in Figure 2.

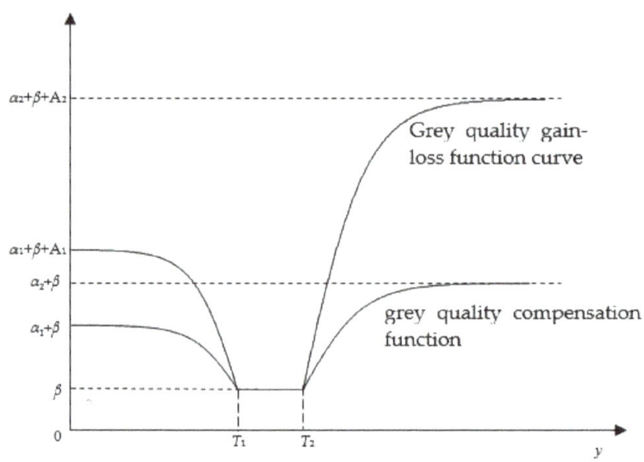

Figure 2. Comparison of grey quality function and grey gain-loss function under asymmetric hyperbolic tangent compensation ($\beta > 0$).

Therefore, the grey quality gain-loss function of asymmetric hyperbolic tangent compensation is:

$$\tilde{L}(y) = \begin{cases} -\alpha_1\left\{1 - \frac{2}{\exp[2(y-T_1)]+1}\right\} + \beta + A_1\left[1 - exp\left(-\frac{(y-T_1)^2}{2\sigma_1^2}\right)\right] \text{ , } y < T_1 \\ \beta & \text{, } T_1 \leq y \leq T_2 \\ \alpha_2\left\{1 - \frac{2}{\exp[2(y-T_2)]+1}\right\} + \beta + A_2\left[1 - exp\left(-\frac{(y-T_2)^2}{2\sigma_2^2}\right)\right] \text{ , } y > T_2 \end{cases} \tag{6}$$

The curve of grey quality gain-loss function of asymmetric hyperbolic tangent compensation is shown in Figure 2.

Therefore, when the compensation is hyperbolic tangent compensation, the maximum grey quality compensation in the target value interval is β. With the quality characteristic deviating from the target value, the grey quality loss increases.

3. Process Capability Analysis

3.1. Process Capability

Process capability refers to the ability of a process to ensure product quality in a stable state. Usually, process capability is expressed by six times the standard deviation (6σ), that is, when the process is in a stable state, 99.73% of its quality characteristic value falls within the interval of $\mu \pm 3\sigma$, where μ is the overall mean of the quality characteristic value, and σ is the overall standard deviation of the quality characteristic value, that is, 99.73% of the quality characteristic value falls within the 6σ range above. Generally speaking, the smaller the value of the standard deviation is, the better it will be.

Process capability analysis is to analyze and evaluate the degree to which the process output meets the engineering standard or the technical specification proposed by customers. First, the determination and analysis of process capability is the basic work to ensure the product quality. Only when the process capability is known, can satisfied quality in the manufacturing process be controlled. If the process capability cannot meet the requirements of product design, then quality control will be impossible. Second, the measurement and analysis of process capability is an effective means to improve process capability. Through the process capability analysis, we can find the dominant factors that affect the process capability. Therefore, process capability can be increased by improving the process, equipment, operation level as well as environmental conditions. Besides, effective process methods and operation procedures, as well as strict process discipline, can be formulated to this end. Third, measurement and analysis of process capability can find out the direction for quality improvement. The actual quality status of process capability processing is the objective basis of guaranteeing product quality. Through process capability analysis, key data of process capability can be provided for designers and processes. At the same time, main problems of process capability can help find a direction to improve processing capability and product quality.

3.2. Process Capability Index

Process capability index is a value to measure the capacity of a process, which has been widely used in many industries. It is the main measure index to evaluate process capacity, representing the production process level and the rate of finished products in a stable production state. As for the indicators of the satisfaction degree with technical requirements, the greater the process capability index is, the better the process capability will satisfy the technical requirements, and there will even be a certain capacity reserve.

The process capability index was originally proposed by Juran [17]. If the product quality characteristics are stable and comply with normal distribution, the process capability index is defined as:

$$C_P = \frac{USL - LSL}{6\sigma} = \frac{\Delta}{3\sigma} \tag{7}$$

Process capability index C_p is the ratio of specification width to natural tolerance. USL is the upper limit of engineering specifications, LSL is the lower limit, $\Delta = (USL - LSL)/2$ and σ is the standard deviation of the process.

Since C_p only reflects the relative relationship between the tolerance zone width and the process capability 6σ, instead of reflecting the positional relationship of the mean value μ of the process quality characteristic value with the tolerance zone, it only reflects the potential capability of the process quality to meet the technical standards.

In order to improve the shortcomings of C_p, Kane considers the positional relationship between the mean of the process quality characteristic value and the tolerance zone, and proposes the process capability index C_{pk}, whose expression is as follows [18]

$$C_{pk} = \min\left\{C_{pl}, C_{pu}\right\} = \min\left\{\frac{USL-\mu}{3\sigma}, \frac{\mu-LSL}{3\sigma}\right\} = \frac{\Delta-|\mu-M|}{3\sigma} \quad (8)$$

where $M = (USL + LSL)/2$ is the center of the specifications and μ is the mean value of the process. C_{pk} takes the average deviation of the process from the specification center into account, but it cannot reflect the deviation of the process from the target value.

Aiming at the shortcomings of C_p and C_{pk}, Chan, Cheng and Spring proposes a new indicator C_{pm} [19]

$$C_{pm} = \frac{USL-LSL}{6\left\{E(Y-T)^2\right\}^{\frac{1}{2}}} = \frac{\Delta}{3\sqrt{\sigma^2+(\mu-T)^2}} \quad (9)$$

where T is the target value. This index is similar to Taguchi loss function. When the mean value of the process deviates from the target value, the process will face square loss.

Pearn, Kotz, and Johnson integrate C_p and C_{pm} into the index C_{pmk} [20]:

$$C_{pmk} = \frac{\min(USL-\mu, \mu-LSL)}{3\sqrt{E\left[(Y-T)^2\right]}} = \frac{\Delta-|\mu-M|}{3\sqrt{\sigma^2+(\mu-T)^2}} \quad (10)$$

Obviously, when $\mu = T = M$, the above four process capability indexes are equal. In addition, there are also process capability indicators with larger-the-better and smaller-the-better characteristics. The summary of process capability indicators with various quality characteristics is shown in Table 1.

Table 1. Summary of process capability index of nominal-the-best, larger-the-better and smaller-the-better characteristics.

Nominal-the-Best Characteristic		Larger-the-Better Characteristic		Smaller-the-Better Characteristic	
Index	**Formula**	**Index**	**Formula**	**Index**	**Formula**
C_p	$\frac{USL-LSL}{6\sigma}$	C_p	$\frac{USL-T}{6\sigma}$	C_p	$\frac{T-LSL}{6\sigma}$
C_{pk}	$\min\left\{\frac{USL-\mu}{3\sigma}, \frac{\mu-LSL}{3\sigma}\right\}$	C_{Pu}	$\frac{USL-\mu}{3\sigma}$	C_{pl}	$\frac{\mu-LSL}{3\sigma}$
C_{pm}	$\frac{USL-LSL}{3\sqrt{\sigma^2+(\mu-T)^2}}$	C_{Pt}	$\frac{\mu-T}{3\sigma}$	C_{Pt}	$\frac{T-\mu}{3\sigma}$

It can be seen from Table 1 that after the specification limit (*USL*, *LSL*) is determined, the smaller the standard deviation σ is, the larger the process capability index and the better the process capability will be. Taking C_p as an example, the quality level of process can be divided into five levels according to the C_p value.

Based on the levels, corresponding judgment and disposal methods can be developed in terms of management shown in Table 2. The levels, judgment and disposal in this table are also applicable to C_{pk}, C_{pm} and C_{pmk}.

Table 2. Grading judgment and disposal of process capability index.

C_p	Level	Range	Bilateral Tolerance Range (T)	Disposal Measures
$C_p > 1.67$	Special grade	High capacity	$T \geq 10\sigma$	(1) Narrow the tolerance range appropriately; (2) Large external fluctuations are allowed to improve efficiency; (3) Lower the accuracy of the equipment appropriately to reduce the cost; (4) Make a brief test.
$1.67 \geq C_p > 1.33$	First grade	Full capacity	$T = 8\sigma \sim 10\sigma$	(1) If it is not a key product, a certain degree of external fluctuations is allowed; (2) Simplified test; (3) Control by a control chart.
$1.33 \geq C_p > 1$	Second grade	Acceptable capacity	$T = 6\sigma \sim 8\sigma$	(1) Control by a control chart to prevent external fluctuations. (2) For product sampling inspection, pay attention to the sampling method and the interval. (3) Inspection cannot be loosened.
$1 \geq C_p > 0.67$	Third grade	Insufficient capacity	$T = 4\sigma \sim 6\sigma$	(1) Analyze the causes of excessive range and take corresponding remedial measures; (2) If the final quality and assembly of the product is not affected, the tolerance range can be appropriately expanded; (3) All products are inspected or screened by grades.
$C_p \leq 0.67$	Fourth grade	Severely insufficient capacity	$T < 4\sigma$	The unqualified rate is too high, so the cause should be traced immediately and remedial measures should be taken to improve the process capability.

3.3. Improvement of Process Capability

It can be seen from Table 1 that if the product quality needs to be improved, we must enhance the process capability. In actual production, there are usually two solutions for insufficient process capacity: the first one is to find a way of reducing the process variation and improve the stability of the process. In most cases, however, to reduce the process variation means that the production technology or construction technology needs to be greatly improved, or the production equipment needs to be updated. This method requires a lot of money, which may increase the total quality cost, and thus it has great limitations; the second method is to optimize the engineering specifications. That is, develop specification limits for the process and truncate the distribution of the quality characteristic by screening to sort out the products that do not meet the prescribed limits. From the viewpoint of manufacturers, they must decide whether screening should be performed on their products before shipping to the customer. If they decide not to screen, all the products they produce would be sent directly to the customer, and then the loss due to variability is sustained by the customer. If they decide to screen, their key concern would be how to determine the specifications for the products [21]. The formulation of engineering specifications is directly related to the qualification rate of products, costs of obsolescence and rework, and the loss of customers caused by the deviation of quality characteristics from the target value. Therefore, it is of great significance to formulate an appropriate specification limits in the total quality cost reduction. The improvement of engineering specifications can enhance the product quality, reduce the total quality cost and raise productivity without increasing capital investment.

Therefore, when the current process cannot be improved (i.e., decrease σ^2), one way to decrease variance of the units delivered to the customers is to add specification limits on the process and truncate the distribution by inspection. The question is how these specification limits can be developed. In quality management, engineering specifications of products are the standard to determine whether the products are qualified or not. In order to formulate the engineering specifications of products from the perspective of social and economic losses, the principle of "minimum social and economic loss" must be followed. Social and economic losses include the total cost (inspection cost, scrap (rework) cost, etc.) of an enterprise for ensuring and improving the product quality and the quality loss of the customers in the process of using the products. Specification limits are developed on the premise of minimizing the total quality cost. If the specification limits are set arbitrarily, it is likely to increase the total cost in the form of inspection. Therefore, in the product manufacturing process, improvement of engineering specifications can not only enhance the process capability, but also have an important impact on the estimation of quality loss.

4. Improvement of the Formulation Method of Engineering Specifications

4.1. Shortcomings of Traditional Formulation Method of Engineering Specifications

In the traditional quality evaluation system, if the quality characteristics of a product do not conform to the engineering specifications, the product will be deemed as unqualified, which will result in quality loss. Moreover, Taguchi believes that the society suffers losses from the initial design stage to the stage of product use due to improper product design. Kapur and Wang put forward an optimization method to determine the specification limits on the basis of minimizing the total cost or the loss of customers and manufacturers [5,6]. This method aims to reduce the loss borne by the customers through the inspection of online quality management, with the main idea of dividing the product quality loss into three categories, i.e., the loss suffered by customers, scrap cost or rework cost and inspection cost. Suppose that the quality characteristic y conforms to the normal distribution, and the target value of quality characteristic is y_0. If the product is fully inspected, the loss costs of the quality characteristic value lower than LSL and higher than USL will be scrap costs. The total quality cost under the full inspection mode is

$$T = H(y) + (1 - p)SC + IC \tag{11}$$

where $H(y)$ is the expected loss per unit product suffered by the customers, SC is the scrap cost per unit product, IC is the inspection and p is the qualification rate. When μ and σ are known, the total expected loss of the product can be minimized, and the optimal solution η can be obtained, therefore the optimal $LSL = \mu - \eta\sigma$ and the optimal $USL = \mu + \eta\sigma$ can be calculated. With the established engineering specification limits, the products are fully inspected to screen out the products that meet the engineering specifications, and customers' loss can be reduced by truncating the normal distribution.

Traditionally, the 3σ principle is used to determine the upper and lower limits of engineering specifications with the target value y_0 plus or minus three times the standard deviation into consideration, that is, $(LSL, USL) = (y_0 - 3\sigma, y_0 + 3\sigma)$. This is because when the distribution range is $y_0 \pm 3\sigma$, the process achieves the maximum quality loss at the boundary of $y_0 \pm 3\sigma$, and the qualification probability of product quality can reach 99.73%, which is close to 1. Therefore, $\pm 3\sigma$ or 6σ is taken as the standard to measure whether the process capability is accurate enough and economic at the same time. However, if the engineering specifications (LSL, USL) formulated by 6σ standard are used for the full inspection of products, the qualified rate of the products is 99.73%, and thus almost all the inspected products are qualified. At this moment, the inspection cost caused by full inspection increases the total cost of products, which causes a waste of resources, and therefore full inspection does not exert its deserved effect. In addition, before putting forward the concept of quality compensation, measurement of quality only rests with the size of quality losses, thus the formulation of traditional optimal engineering specifications is only targeted at the quality loss function, instead of considering the role of quality compensation. As for the quality gain-loss function, inclusion of the compensation part inevitably has an impact on the engineering specifications, thus affecting the judgment on quality and the estimation of benefits. Therefore, formulation of optimal engineering specifications also requires analysis of the quality compensation part, so as to minimize the average total quality loss and maximize the average quality compensation. Therefore, formulation of engineering specifications should combine the quality loss part with the quality compensation part to optimize the formulation method of engineering specifications on the premise of minimizing the total quality loss and maximizing the quality compensation.

4.2. Optimal Engineering Specifications under Grey Quality Loss

In many existing literatures on formulation of engineering specifications, it is all assumed that the population distribution is normal. However, after the products are fully inspected, the quality characteristic value of the inspected products, i.e., the product delivered to the customers, is within the range of (LSL, USL). Therefore, this section assumes that the quality characteristics conform to the truncated normal distribution, and the process average is equal to the target value, that is, $(LSL, USL) = (\mu - \eta\sigma, \mu + \eta\sigma)$. With the combination of the loss part of the grey quality gain-loss function, engineering specifications that minimize the average quality loss are formulated.

Suppose that the product quality characteristic y conforms to normal distribution $Y \sim N(\mu, \sigma^2)$, $F(y)$ and $f(y)$ are its distribution function and probability density function respectively. The probability density function is

$$f(y) = \frac{1}{\sqrt{2\pi}\sigma} exp[-\frac{(y - y_0)^2}{2\sigma^2}],\ -\infty \leq y \leq \infty \tag{12}$$

where μ is the process mean and σ^2 is the variance, $-\infty \leq \mu \leq \infty, \sigma^2 \geq 0$.

It is considered that y is not less than LSL and not more than USL, that is, y is a random variable in the interval (LSL, USL). The distribution function of the quality characteristic y is:

$$F_T(y) = P(Y \leq y | LSL \leq y \leq USL) = \frac{P(LSL \leq Y \leq y)}{P(LSL \leq Y \leq USL)} = \frac{F(y) - F(LSL)}{F(USL) - F(LSL)} \tag{13}$$

The probability density is

$$f_T(y) = \frac{f(y)}{F(USL) - F(LSL)} \tag{14}$$

where $LSL = \mu - \eta\sigma, USL = \mu + \eta\sigma, F(x) = \Phi\left(\frac{x-\mu}{\sigma}\right), F(USL) - F(LSL) = \Phi(\eta) - \Phi(-\eta) = p$.
Therefore, the loss of customers caused by product quality characteristics falling within the specification limits is as follows:

$$\begin{aligned} L_1 &= \int_{\mu-\eta\sigma}^{T_1} A_1\left[1 - exp\left(-\frac{(y-T_1)^2}{2\sigma_1^2}\right)\right] f_T(y)dy + \int_{T_2}^{\mu+\eta\sigma} A_2\left[1 - exp\left(-\frac{(y-T_2)^2}{2\sigma_2^2}\right)\right] f_T(y)dy \\ &= \frac{A_1}{p}\left\{ \begin{array}{l} \Phi\left(\frac{T_1-\mu}{\sigma}\right) + \Phi(\eta) - 1 - \frac{\sigma_1}{\sqrt{\sigma^2+\sigma_1^2}} \times \left[exp\left(-\frac{(\mu-T_1)^2}{2(\sigma^2+\sigma_1^2)}\right)\right] \times \\ \left[\Phi\left(\frac{\sigma_1(T_1-\mu)}{\sigma\sqrt{\sigma^2+\sigma_1^2}}\right) - \Phi\left(\frac{\sigma(\mu-T_1)-\eta(\sigma^2+\sigma_1^2)}{\sigma_1\sqrt{\sigma^2+\sigma_1^2}}\right)\right] \end{array} \right\} \\ &+ \frac{A_2}{p}\left\{ \begin{array}{l} \Phi(\eta) - \Phi\left(\frac{T_2-\mu}{\sigma}\right) - \frac{\sigma_2}{\sqrt{\sigma^2+\sigma_2^2}} \times \left[exp\left(-\frac{(\mu-T_2)^2}{2(\sigma^2+\sigma_2^2)}\right)\right] \times \\ \left[\Phi\left(\frac{\sigma(\mu-T_2)+\eta(\sigma^2+\sigma_2^2)}{\sigma_2\sqrt{\sigma^2+\sigma_2^2}}\right) - \Phi\left(\frac{\sigma_2(T_2-\mu)}{\sigma\sqrt{\sigma^2+\sigma_2^2}}\right)\right] \end{array} \right\} \end{aligned} \tag{15}$$

Among which, when the quality characteristics of products exceed the specification limits, the per unit scrap cost of production is:

$$L_2 = (1-q)SC \tag{16}$$

The per unit inspection cost is IC. The total expected loss is

$$L = L_1 + L_2 + IC \tag{17}$$

where $\eta \geq (\mu - T_1)/\sigma$ and $\eta \geq (T_2 - \mu)/\sigma$, $\Phi(x)$ is the function of standard normal distribution.

The Relationship between average expected loss L and optimal solution η is shown in Figure 3.

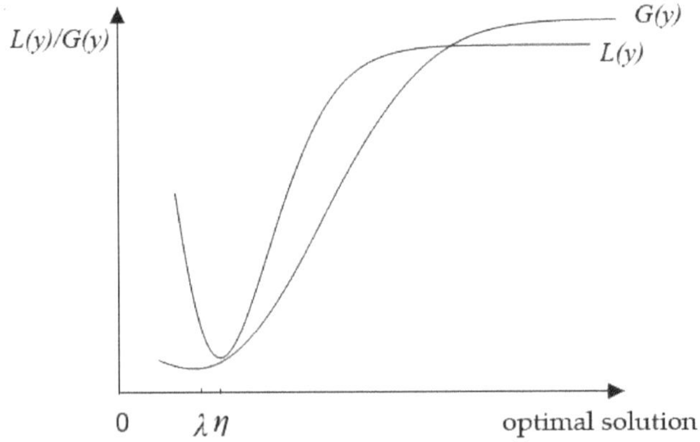

Figure 3. Relationship between average expected loss L, average expected compensation G and optimal solution η, λ.

4.3. Optimal Engineering Specifications under Grey Quality Compensation

For the quality gain-loss function, formulation of engineering specifications should consider not only the loss function, but also the compensation function. If the engineering specifications are developed with the traditional method, it is likely to miss the interval that maximizes the quality compensation. Therefore, formulation of the optimal engineering specifications with quality compensation must be carried out on the premise of minimizing the loss and maximizing the compensation. Optimal specifications of compensation items are collectively referred to as compensation specifications below, and the method for formulating optimal compensation specifications is explained as follows.

Calculation of the expected compensation by the hyperbolic tangent compensation function of Equation (5) is relatively complicated, therefore a function with the same variation trend can be selected to replace it when formulating the compensation specifications. It is known that the independent variable x of hyperbolic tangent function is continuous on $(-\infty, +\infty)$, and $th(x)$ has a third derivative at $x = 0$, which can be expanded by Taylor series:

$$th(x) = th(0) + \frac{th'(0)}{1!}x + \frac{th''(0)}{2!}x^2 + \frac{th'''(0)}{3!}x^3 + o\left(x^3\right) \tag{18}$$

Because $th(0) = 0$, $th'(0) = 1$, $th''(0) = 0$, $th'''(0) = -2$, $th(x)$ can be estimated by $th^*(x)$:

$$th^*(x) = x - 1/3x^3 \tag{19}$$

However, there is an estimation deviation between $th^*(x)$ and $th(x)$, and the variation trend between the two is consistent, that is, the interval between the maximum and minimum values is also consistent, and thus $th^*(x)$ can be used to replace $th(x)$ to construct the compensation function, so as to develop the compensation specifications. Therefore, quality compensation function of modified grey hyperbolic tangent can be estimated as:

$$\widetilde{g}_s^*(y) = \begin{cases} -\alpha_1\left[(y-T_1)-(y-T_1)^3/3\right]+\beta\,,\ y \le T_1 \\ \beta & ,\ T_1 \le y \le T_2,\ \alpha > 0 \\ \alpha_2\left[(y-T_2)-(y-T_2)^3/3\right]+\beta\,,\ y > T_2 \end{cases} \tag{20}$$

Let the lower bound of compensation specifications be $LSL^* = \mu - \lambda_1\sigma$ and the upper bound be $USL^* = \mu + \lambda_2\sigma$, then the average expected compensation within the limits of compensation specifications is:

$$\begin{aligned} G &= E[\widetilde{g}^*(y)] = \int_{\mu-\lambda\sigma}^{\mu+\lambda\sigma} \widetilde{g}^*(y) f_T(y)dy \\ &= \frac{1}{P}\left\{ \begin{aligned} &\int_{\mu-\lambda\sigma}^{T_1}\left\{-\alpha_1\left[(y-T_1)-(y-T_1)^3/3\right]+\beta\right\}f(y)dy + \int_{T_1}^{T_2}\beta f(y)dy \\ &+\int_{T_2}^{\mu+\lambda\sigma}\left\{\alpha_2\left[(y-T_2)-(y-T_2)^3/3\right]+\beta\right\}f(y)dy \end{aligned} \right\} \\ &= \frac{1}{P}\left\{ \begin{aligned} &\beta\int_{\mu-\lambda\sigma}^{\mu+\lambda\sigma} f(y)dy - \alpha_1\left[\begin{aligned} &\left(\tfrac{1}{3}T_1^3 - T_1\right)\int_{\mu-\lambda\sigma}^{T_1} f(y)dy + (1-T_1)^2\int_{\mu-\lambda\sigma}^{T_1} yf(y)dy \\ &+T_1\int_{\mu-\lambda\sigma}^{T_1} y^2 f(y)dy - \tfrac{1}{3}\int_{\mu-\lambda\sigma}^{T_1} y^3 f(y)dy \end{aligned} \right] \\ &+\alpha_2\left[\begin{aligned} &\left(\tfrac{1}{3}T_2^3 - T_2\right)\int_{T_2}^{\mu+\lambda\sigma} f(y)dy + (1-T_2)^2\int_{T_2}^{\mu+\lambda\sigma} yf(y)dy \\ &+T_2\int_{T_2}^{\mu+\lambda\sigma} y^2 f(y)dy - \tfrac{1}{3}\int_{T_2}^{\mu+\lambda\sigma} y^3 f(y)dy \end{aligned} \right] \end{aligned} \right\} \end{aligned} \tag{21}$$

where:

$$F(a) = \Phi\left(\frac{a-\mu}{\sigma}\right)$$

$$\int_a^b f(y)dy = \Phi\left(\frac{b-\mu}{\sigma}\right) - \Phi\left(\frac{a-\mu}{\sigma}\right)$$

$$\int_a^b yf(y)dy = \mu\left[\Phi\left(\frac{b-\mu}{\sigma}\right) - \Phi\left(\frac{a-\mu}{\sigma}\right)\right] + \sigma\left[\phi\left(\frac{a-\mu}{\sigma}\right) - \phi\left(\frac{b-\mu}{\sigma}\right)\right]$$

$$\begin{aligned} \int_a^b y^2 f(y)dy =& (\sigma^2+\mu^2)\left[\Phi\left(\tfrac{b-\mu}{\sigma}\right) - \Phi\left(\tfrac{a-\mu}{\sigma}\right)\right] \\ &+[\sigma(a+\mu)+2\mu]\phi\left(\tfrac{a-\mu}{\sigma}\right) - [\sigma(b+\mu)+2\mu]\phi\left(\tfrac{b-\mu}{\sigma}\right) \end{aligned}$$

$$\begin{aligned} \int_a^b y^3 f(y)dy =& (\mu^3+3\mu\sigma^2)\left[\Phi\left(\tfrac{b-\mu}{\sigma}\right) - \Phi\left(\tfrac{a-\mu}{\sigma}\right)\right] \\ &+\left\{\sigma^2\left[(a-\mu)^2+2\sigma^2\right]+3\mu\sigma^2(a-\mu)+3\mu^2\sigma^2\right\}\phi\left(\tfrac{a-\mu}{\sigma}\right) \\ &-\left\{\sigma^2\left[(b-\mu)^2+2\sigma^2\right]+3\mu\sigma^2(b-\mu)+3\mu^2\sigma^2\right\}\phi\left(\tfrac{b-\mu}{\sigma}\right) \end{aligned}$$

where $\Phi(x)$ and $\phi(x)$ are the distribution function and probability density of standard normal distribution, respectively. Obviously, when μ and σ are known, G is a function of λ. When $\beta < 0$, the quality compensation will be positive compensation, which means that the quality compensation is opposite to the quality loss. If the quality loss is always greater than 0, the quality positive compensation will always be less than 0, and the smaller the quality compensation is, the greater its absolute value will be, indicating that the more the positive compensation is, the better the quality will be. When $\beta > 0$, it is negative compensation, and like the quality loss, it will always be positive. The smaller the value is, the less the negative compensation and the better the quality will be. In conclusion, an economic engineering specification which minimizes the average total expected compensation G per unit product should be developed, i.e., $Min\{G(\lambda)\}$. Through MATLAB programming calculation, the optimum solution corresponding to $Min\{G(\lambda)\}$ is λ, and the optimal compensation specification $(LSL^*, USL^*) = (\mu - \lambda\sigma, \mu + \lambda\sigma)$ can be obtained.

The Relationship between average expected compensation G and optimal solution λ is shown in Figure 3.

If the minimum expected loss and the maximum expected compensation are required at the same time, the intersection of the optimal loss specification and the optimal compensation specification should be taken, that is $(LSL, USL) = \{max(LSL, LSL^*), min(USL, USL^*)\}$.

5. Example Calculation

In the heightening project of Danjiangkou dam, the heightening construction scope of overflow dam section includes the heightening construction of 11 dam sections from overflow 14 to 24. The construction process involves the concrete pouring of the weir surface, the second-phase concrete construction of the gate slot, the planting of steel bars and the installation of steel bars. This study took the second-phase concrete construction of the gate slot as an example to conduct research on the formulation of optimal engineering specifications based on the acceptance evaluation form of construction quality through the inspection process of general concrete appearance quality from the Construction Quality Acceptance and Evaluation Form and Formfilling Instructions for Unit Works of Water Conservancy and Hydropower Projects, edited by the Construction and Management Department of the Ministry of Water Resources, with the combination of design specifications and design requirements. In the construction quality acceptance and evaluation of the general concrete appearance quality inspection process, the surface flatness of general items was inspected. The maximum deviation per 3 m was 10.75 mm and the minimum deviation was 9.25 mm, which met the quality requirements in the Construction Quality Acceptance and Evaluation Form and Formfilling Instructions for Unit Works of Water Conservancy and Hydropower Projects.

Therefore, $[9.25, 10.75]$ can be taken as the range of target value and denoted as interval grey number $\otimes \in [T_1, T_2]$, where $T_1 = 9.25$ and $T_2 = 10.75$; according to the design requirements and engineering practice, the surface evenness Y is a continuous random variable and it conforms to normal distribution, with mean value $\mu = 10$ and variance $\sigma^2 = 2^2$, i.e., $Y \sim N(10, 2^2)$. When the measured value of quality characteristics deviates from the target value, the maximum loss can be $A = 11.25$, the per unit scrap cost $SC = 3$, the per unit inspection cost $IC = 0.2$, compensation coefficient $\alpha = 2$ and maximum compensation $\beta = -15$.

The process causes the largest loss at $T_1 - 3\sigma$ and $T_2 + 3\sigma$, where $T_1 - 3\sigma = 9.25 - 3 \times 2 = 3.25$, $T_2 + 3\sigma = 10.75 + 3 \times 2 = 16.75$, adjustment coefficient $\sigma_1 = \sigma_2 = \sigma_L$, $\sigma_1 = \frac{(9.25-3.25)}{4} = 1.5$, $\sigma_2 = \frac{(16.75-10.75)}{4} = 1.5$.

Substitute $A = 11.25, \sigma = 0.5, \sigma_L = 0.375$ and $(T_1, T_2) \in (9.25, 10.75)$ into Equation (11). The total expected loss of quality per unit product is:

$$\begin{aligned} L = {} & L_1 + L_2 + IC \\ = {} & \frac{11.25}{2\Phi(\eta)-1}\left\{\begin{array}{l} [-2\Phi(0.375) + 2\Phi(2.5)] \\ -0.6 \times \left\{\begin{array}{l} [exp(-0.045)] \times [\Phi(-0.225) - \Phi(0.4 - 1.6667\eta)] \\ -[exp(-0.045)] \times [\Phi(-0.4 + 1.6667\eta) - \Phi(0.225)] \end{array}\right\} \end{array}\right\} \\ & +(2 - 2\Phi(\eta)) \times 3 + 0.2 \end{aligned}$$

where:

$$\begin{aligned} L_1 = {} & \frac{A}{p}\left\{\begin{array}{l} \left[\Phi\left(\frac{T_1-\mu}{\sigma}\right) - \Phi\left(\frac{T_2-\mu}{\sigma}\right) - \Phi\left(\frac{\omega_1-\mu}{\sigma}\right) + \Phi\left(\frac{\omega_2-\mu}{\sigma}\right)\right] - \frac{\sigma_L}{\sqrt{\sigma^2+\sigma_L^2}} \times \\ \left\{\begin{array}{l} \left[exp\left(-\frac{(\mu-T_1)^2}{2(\sigma^2+\sigma_L^2)}\right)\right] \times \left[\Phi\left(\frac{\sigma_L(T_1-\mu)}{\sigma\sqrt{\sigma^2+\sigma_L^2}}\right) - \Phi\left(\frac{\sigma(\mu-T_1)-\eta(\sigma^2+\sigma_L^2)}{\sigma_L\sqrt{\sigma^2+\sigma_L^2}}\right)\right] \\ +\left[exp\left(-\frac{(\mu-T_2)^2}{2(\sigma^2+\sigma_L^2)}\right)\right] \times \left[\Phi\left(\frac{\sigma(\mu-T_2)+\eta(\sigma^2+\sigma_L^2)}{\sigma_L\sqrt{\sigma^2+\sigma_L^2}}\right) - \Phi\left(\frac{\sigma_L(T_2-\mu)}{\sigma\sqrt{\sigma^2+\sigma_L^2}}\right)\right] \end{array}\right\} \end{array}\right\} \\ = {} & \frac{11.25}{2\Phi(\eta)-1}\left\{\begin{array}{l} [-2\Phi(0.375) + 2\Phi(2.5)] \\ -0.6 \times \left\{\begin{array}{l} [exp(-0.045)] \times [\Phi(-0.225) - \Phi(0.4 - 1.6667\eta)] \\ -[exp(-0.045)] \times [\Phi(-0.4 + 1.6667\eta) - \Phi(0.225)] \end{array}\right\} \end{array}\right\} \end{aligned}$$

$$L_2 = (1 - p)SC = (2 - 2\Phi(\eta)) \times 3$$

Obviously, it is known that L is a function of η. Next, the economic engineering specification that minimizes the average total expected loss L per unit product is formulated, that is:

$$Min\left\{\begin{array}{l} L=\frac{11.25}{2\Phi(\eta)-1}\left\{\begin{array}{l}[-2\Phi(0.375)+2\Phi(2.5)] \\ -0.6\times\left\{\begin{array}{l}[exp(-0.045)]\times[\Phi(-0.225)-\Phi(0.4-1.6667\eta)] \\ -[exp(-0.045)]\times[\Phi(-0.4+1.6667\eta)-\Phi(0.225)]\end{array}\right\}\end{array}\right\} \\ +(2-2\Phi(\eta))\times 3+0.2\end{array}\right\}$$

With charting and calculation by MATLAB programming, the optimal solution corresponding to $Min\{L(\eta)\}$ is $\eta = 0.8750$, as shown in Figure 4.

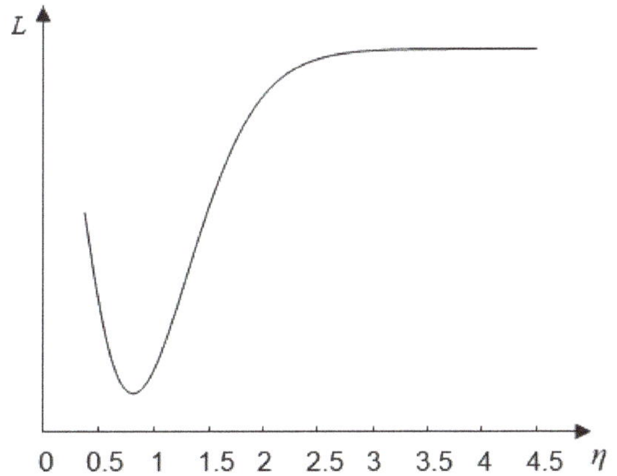

Figure 4. Relationship between average expected loss l and optimal solution η.

Substitute the optimal solution into the formula $(USL, LSL) = (\mu - \eta\sigma, \mu + \eta\sigma)$ for the upper and lower limits of the optimal engineering specification, it can be obtained that:

The lower limit of the optimal specification for the quality loss part is $LSL = \mu - \eta\sigma = 10 - 0.745 \times 2 = 8.51$.

The upper limit of the optimal specification for the quality loss part is $USL = \mu + \eta\sigma = 10 + 0.745 \times 2 = 11.49$.

Substitute $\alpha = 2$, $\beta = -15$, $A = 11.25$, $(T_1, T_2) \in (9.25, 10.75)$, $\mu = 10$, $\sigma = 2$, $\sigma_L = 1.5$, into Equation (15), we have:

$$\begin{aligned} G = & E[g^*(y)] = -15[1-2\Phi(\lambda)] \\ & -2\left[880.0521 - 1523.8594\Phi(-\lambda) + \left(483.7916 - 63.6667\lambda - \tfrac{16}{3}\lambda^2\right)\phi(-\lambda)\right] \\ & +2\left[2098.6407[\Phi(\lambda)-\Phi(0.375)] - 880.1667\phi(0.375) - \left(432.4583 - 37\lambda - \tfrac{4}{3}\lambda^2\right)\phi(\lambda)\right] \end{aligned} \quad (22)$$

Obviously, G is a function of λ. $\beta < 0$, so the quality compensation is positive, which means that the quality compensation value is always less than 0, and the smaller the quality compensation value is, the larger its absolute value will be, indicating more compensation. Therefore, the economic engineering specification that minimizes the average total expected compensation G per unit product is formulated, that is, $Min\{G(\eta)\}$. With charting and calculation by MATLAB programming, the optimal solution corresponding to $Min\{G(\eta)\}$ is $\lambda = 0.69$, as shown in Figure 5.

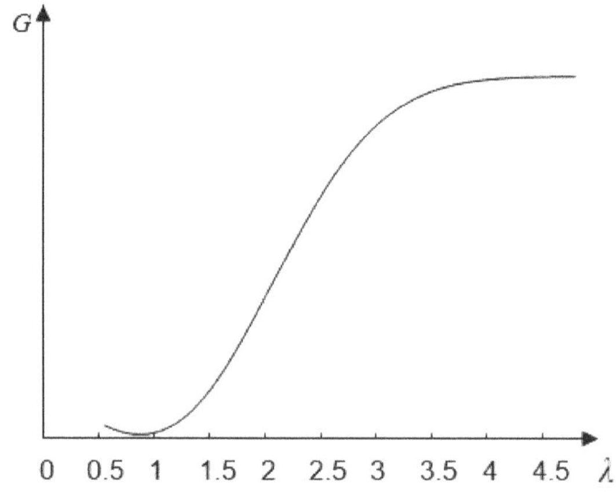

Figure 5. Relationship between average expected compensation g and optimal solution λ.

By substituting the optimal solution into the formula $(LSL', USL') = (\mu - \lambda\sigma, \mu + \lambda\sigma)$ of the upper and lower limit of the optimal compensation specification, we can get:

The lower limit of the optimal compensation specification $LSL^* = \mu - \lambda\sigma = 10 - 0.69 \times 2 = 8.62$.

The upper limit of the optimal compensation specification $USL^* = \mu + \lambda\sigma = 10 + 0.69 \times 2 = 11.38$.

In conclusion, in order to minimize the expected loss and maximize the expected compensation, the intersection of the optimal loss specification and the optimal compensation specification shall be taken as the optimal engineering specification, i.e., $(LSL, USL) = \{max(8.51, 8.62), min(11.49, 11.38)\} = (8.62, 11.38)$. The specifications of the project meet the quality requirements in the construction quality acceptance evaluation form and filling instructions of unit works of water resources and hydropower projects. Taking this section as the specification limit of the ordinary concrete appearance quality inspection process, the distribution can be cut off through inspection, which can improve the concrete appearance quality, reduce the total quality cost and improve the concrete construction efficiency without increasing capital investment.

6. Conclusions

Because the quality evaluation standard is grey in some cases, the calculation of quality profit and loss is also grey. The definition of some quality indicators in the actual construction and production process is also grey. Intuition and judgment play an important role in determining the target value of some quality characteristics, and some factors related to personal needs also have an effect. For this kind of quality characteristics, determination of the target value has a certain degree of greyness, therefore the grey quality gain-loss function model was established. In view of insufficient process capability, specification bounds can be imposed on the process and the distribution can be truncated with inspection if the current process cannot be improved, so as to lessen the differences of projects delivered to customers. However, the formulation of specification limits should take the minimization of the total quality cost of producers and consumers as the basis. Losses borne by producers include costs related to inspection (tests to determine defective products) and defective products (obsolescence or rework). At the same time, for the products with compensation functions, the effect of quality compensation should be considered, and the maximum quality compensation should be achieved on the premise of minimizing the total quality cost. With the optimization model proposed in this paper, optimal engineering specification limits can be obtained. Finally, taking the Danjiangkou dam heightening project as an example, in the construction quality acceptance and evaluation of the general concrete appearance quality inspection process, for the surface flatness of general items, the optimal engineering specification limits can be obtained by using the optimization model proposed in this paper. Without increasing capital investment, the appearance quality of concrete is improved, the total quality cost is reduced and the concrete construction efficiency is improved.

In this paper, the research is carried out on the premise that the quality characteristics obey the normal distribution. However, in engineering practice, the distribution characteristics of each quality characteristic often have a variety of situations, such as uniform distribution, exponential distribution and Weber distribution. Therefore, subsequent researchers can extend the case of non-normal distribution. For the part of compensation function, this paper only considers two forms: constant compensation and hyperbolic tangent compensation. Subsequent researchers can develop other compensation functions in line with the actual compensation situation. The multivariate grey quality gain-loss function can also be deeply discussed and analyzed.

Author Contributions: Conceptualization, B.W. and Q.Y.; methodology, B.W.; software, C.L.; validation, Q.Y. and Q.L.; formal analysis, X.N.; investigation, Q.Y.; resources, C.L.; data curation, B.W.; writing—original draft preparation, Q.Y.; writing—review and editing, C.L.; visualization, X.N.; supervision, Q.L.; project administration, B.W.; funding acquisition, B.W. All authors have read and agreed to the published version of the manuscript.

Funding: Natural Science Foundation of China (No. 51709116).

Institutional Review Board Statement: Not applicable.

Informed Consent Statement: Not applicable.

Data Availability Statement: No data supporting results.

Conflicts of Interest: The authors declare no conflict of interest.

References

1. Shewhart, W.A. *Economic Control. of Quality Manufactured Product*; Van Nostrand Company: New York, NY, USA, 1931.
2. Taguchi, G. Quality engineering in Japan. Communication in Statistics. *Theory Methods* **1985**, *14*, 2785–2801. [CrossRef]
3. Kackar, R.N. Off-line quality control, parameter design and the Taguchi method. *J. Qual. Technol.* **1985**, *17*, 176–188. [CrossRef]
4. Kackar, R.N. Taguchi's Quality Philosophy: Analysis and Commentary. *Qual. Prog.* **1989**, *2*, 3–21. [CrossRef]
5. Barker, T.R. *Quality Engineering by Design: Taguchi's Philosophy*; Taylor & Francis: Oxfordshire, UK, 1986; pp. 37–42.
6. Kapur, K.C.; Wang, C.J. *Economic Design of Specifications Based on Taguchi's Concept of Quality Loss Function. Quality Design, Planning, and Control*; The American Society of Mechanical Engineers: Boston, MA, USA, 1987; pp. 23–36.
7. Kapur, K.C. An approach for development of specifications for quality improvement. *Qual. Eng.* **1988**, *1*, 63–77. [CrossRef]
8. Kapur, K.C.; Cho, B.-R. Economic design of the specification region for multiple quality characteristics. *IIE Trans.* **1996**, *28*, 237–248. [CrossRef]
9. Chatterjee, M.; Chakraborty, A.K. Some process capability indices for unilateral specification limits—Their properties and the process capability control charts. *Commun. Stat. Theory Methods* **2014**, *45*, 7130–7160. [CrossRef]
10. Chen, C.-H. Specification limit under a quality loss function. *J. Appl. Stat.* **1999**, *26*, 903–908. [CrossRef]
11. Pan, J.N.; Pan, J.B. A comparative study of various loss functions in the economic tolerance design. In Proceedings of the IEEE International Conference on Management of Innovation and Technology, Singapore, 21–23 June 2006; pp. 783–787.
12. Wang, B.; Li, Z.; Gao, J.; Vaso, H. Critical Quality Source Diagnosis for Dam Concrete Construction Based on Quality Gain loss Function. *J. Eng. Sci. Technol. Rev.* **2014**, *7*, 137–151. [CrossRef]
13. Wang, B.; Zhou, H.G.; Li, Z.Y.; Nie, X.T. Tolerance optimization for quality characteristics of the dam concrete construction based on quality gain-loss function. *Math. Pract. Theory* **2016**, *46*, 90–99.
14. Wang, B.; Fan, T.Y.; Tian, J.; Liu, M.Q.; Nie, X.T. Designing a quality gain-loss function for larger-the-better char-acteristic and smaller-the-better characteristic under not neglecting the linear term loss and keeping compensation amount constant. *Math. Pract. Theory* **2019**, *49*, 153–160.
15. Nie, X.T.; Liu, C.; Wang, B. Inverted normal quality gain-loss function and its application in water project con-struction. *J. Coast. Res.* **2020**, *104*, 415–420. [CrossRef]
16. Nie, X.T.; Liu, C.; Guo, W.J.; Liu, M.Q.; Wang, B. Fuzzy quality gain-loss function model and process mean design optimization. *Henan Sci.* **2020**, *38*, 1377–1386.
17. Ireson, W.G.; Juran, J.M. Quality-Control Handbook. *J. Am. Stat. Assoc.* **1952**, *47*, 317. [CrossRef]
18. Kane, V.E. Process capability indices. *J. Qual. Technol.* **1986**, *18*, 41–52. [CrossRef]

19. Chan, L.K.; Cheng, S.W.; Spiring, F.A. A New Measure of Process Capability:Cpm. *J. Qual. Technol.* **1988**, *20*, 162–175. [CrossRef]
20. Pearn, W.L.; Kotz, S.; Johnson, N.L. Distributional and Inferential Properties of Process Capability Indices. *J. Qual. Technol.* **1992**, *24*, 216–231. [CrossRef]
21. Kapur, K.; Cho, B.-R. Economic Design and Development of Specifications. *Qual. Eng.* **1994**, *6*, 401–417. [CrossRef]

coatings

MDPI

Article

Adsorption Features of Loess Calcareous Nodules to Heavy-Metal Ions in Aqueous Solution

Qi Li [1,2,3,*] and Yanan Li [1,2,3]

1 Shaanxi Provincial Land Engineering Construction Group Co., Ltd., Shaanxi Key Laboratory of Land Consolidation, Xi'an 710075, China; 2015127001@chd.edu.cn
2 Shaanxi Provincial Land Engineering Construction Group Co., Ltd., Institute of Land Engineering and Technology, Xi'an 710075, China
3 Shaanxi Provincial Land Consolidation Engineering Technology Research Center, Xi'an 710075, China
* Correspondence: 2015126049@chd.edu.cn

Abstract: This paper explores the use of calcareous tuberculosis as an adsorbent and heavy-metal ions (Cu^{2+}, Zn^{2+}, Cd^{2+}, and Pb^{2+}) as adsorbates, and the influence of varying levels of particle size, adsorption time, pH, adsorbent dosage, and initial concentration of heavy metals is studied through an experiment of single heavy-metal adsorption. In addition, the impact of the temperature and other factors on the adsorption of heavy-metal ions by calcareous nodules is analyzed to identify the optimal conditions for the adsorption of heavy-metal ions by calcareous nodules. As shown by the research findings, the adsorption rates of Cu^{2+}, Zn^{2+}, and Pb^{2+} gradually declined with the increase in particle size, with no evident effect on Cd^{2+}. In the meantime, with further increases in factors such as the adsorption time, adsorbent dosage, and temperature, the adsorption rates of Cu^{2+}, Zn^{2+}, Cd^{2+}, and Pb^{2+} experienced gradual increases. The adsorption rates of Cu^{2+}, Zn^{2+}, and Cd^{2+} gradually declined with the increase in initial concentration of heavy-metal ions, whereas the adsorption rate of Pb^{2+} experience increased first and then declined. As the pH increased, the adsorption rate of Cd^{2+} experience increased first and then declined at a slow pace. The adsorption rates of Cu^{2+}, Zn^{2+}, and Pb^{2+} increased first and then decreased. The adsorption capacity of calcareous nodules toward the four heavy-metal ions was in the order of $Pb^{2+} > Zn^{2+} > Cu^{2+} > Cd^{2+}$. When the particle size was set to 0.25 mm, the adsorption time was set to 120 min, and the dosage was set to 0.6 g, the calcareous nodules included Pb^{2+}, Zn^{2+}, and Cu^{2+}. Moreover, Cd^{2+} was able to achieve stronger adsorption capacity, with the adsorption rate able to reach 83.33%, 77.78%, 73.81%, and 81.93% of its maximum level. Therefore, as the particle size of the heavy-metal ions decreased, the adsorption capacity generally became stronger. As the adsorption time increased, the temperature and the amount of adsorbent also increased. The optimal pH value for the adsorption of calcareous nodules toward Pb^{2+}, Zn^{2+}, Cu^{2+}, and Cd^{2+} was found to be 7, 6, 5, and 8, respectively, and the optimal temperature was 50 °C. In summary, calcareous nodules are a natural, low-cost, and effective adsorbent.

Keywords: loess calcareous nodules; heavy-metal ions; single adsorption; adsorption rate

Citation: Li, Q.; Li, Y. Adsorption Features of Loess Calcareous Nodules to Heavy-Metal Ions in Aqueous Solution. *Coatings* **2021**, *11*, 1314. https://doi.org/10.3390/coatings11111314

Academic Editor: Artur P. Terzyk

Received: 16 September 2021
Accepted: 27 October 2021
Published: 28 October 2021

1. Introduction

Over recent years, with the rapid progress made in industrialization and urbanization, a large amount of wastewater containing heavy-metal ions such as Cu^{2+}, Zn^{2+}, Cd^{2+}, and Pb^{2+} has been discharged into water ecosystems. As a result, the contamination of water bodies caused by heavy metals has become increasingly severe [1]. Heavy metals feature strong toxicity and can easily migrate but are hardly degraded. Subsequent to entrance into the water body, they can impose severe damage to human health. Therefore, identifying ways of purifying water and removing metal ions has become a research hotspot within the topic of environmental pollution [2]. Given that the treatment of water and metal pollution is deemed a long-term and arduous task, the cost and environmental friendliness must be taken into full account during selection of the treatment methodology. The adsorption method is more applicable to various water bodies with a low concentration of heavy-metal pollution, mainly due to the advantages of this method, such as fast treatment rate, simple operation, and optimal environmental coordination [3,4]. During the application of adsorption methods, the commonly used adsorbents include activated carbon, chitosan, resin, zeolite, and sepiolite, in addition to clay minerals such as kaolinite, vermiculite, and montmorillonite [5]. According to the research findings, vermiculite features a strong adsorption capacity for Pb^{2+}, Cd^{2+}, and Zn^{2+} in water, and sepiolite features an optimal adsorption effect for Ni^{2+}, Cd^{2+}, Zn^{2+}, and Cu^{2+} in water [6–8]. In addition, illite, kaolinite, and montmorillonite feature an optimal adsorption effect on Cu^{2+}, Zn^{2+}, Cr^{3+}, Cd^{2+}, and Pb^{2+} in water [9].

Loess calcareous nodules are also referred to as ginger stones. They are aggregates of soils of varying sizes and shapes, which are subject to the action of leaching, sedimentation, alteration, and human activities under the circumstances of alternating wet and dry conditions and freeze–thaw conditions in the soil on the Loess Plateau [10]. Calcareous nodules are widely distributed in the heavily eroded Loess Plateau, and they are mainly composed of primary minerals such as calcite, quartz, and feldspar, in addition to layered silicate minerals such as kaolinite, montmorillonite, and illite. In particular, illite and montmorillonite feature a large specific surface area, high activity, and optimal porosity; thus, they are able to effectively adsorb heavy-metal ions in water [11]. Previous studies mainly focused on the origin, distribution, structure, and mechanical properties of loess calcareous nodules, but few of them investigated the adsorption features of heavy-metal ions on the surface of calcareous nodules [10,12–14]. As natural mineral aggregates, calcareous nodules feature advantages such as a wide range of sources, low cost, prevention of secondary pollution when added to water, and ease of application. In addition, the approach of treating waste with waste not only improves the soil body itself, but also lowers the content of heavy-metal ions in the water body. They are, thus, regarded as materials with an optimal effect of environmental coordination. For this reason, this paper elaborates on the adsorption features of loess calcareous nodules toward water and heavy-metal ions, and it is expected to lay a theoretical foundation for the application of this adsorbent in eliminating heavy metals from polluted water.

2. Materials and Methods

2.1. Materials

The loess calcareous nodules used for the experiments were collected from forest land in Tongchuan City, Shaanxi Province, China (34°99′ N; 108°92′ E). Calcium nodules of similar size and shape were mainly collected in the surface soil at 0–20 cm depth and air-dried for later use. $Cu(NO_3)_2$, $Zn(NO_3)_2$, $Cd(NO_3)_2$, $Pb(NO_3)_2$, HNO_3, and NaOH were all analytical reagent-grade. The water adopted in the experiments was distilled water, which was used to prepare standard solutions of metal ions.

2.2. Methods

2.2.1. Scanning Electron Microscopy and Microwave Digestion of Calcareous Tuberculosis

The calcareous tuberculosis powder was prepared into a suspension, which was then evenly coated on a copper table with a pipette. Subsequent to natural air-drying, the powder was scanned using a high-resolution scanning electron microscope (SU8010, Hitachi, Tokyo, Japan). The chemical composition of calcareous nodules was extracted through the method of microwave digestion using aqua regia–hydrofluoric acid as the reagent, and the extract was measured by an inductively coupled plasma mass spectrometer (ICP-MS, Agilent Technologies Inc., Tokyo, Japan) [15,16].

2.2.2. Adsorption Kinetics

Calcareous nodules (0.4 g) were added to 100 mL of a 30 mg/L solution containing Cu^{2+}, Zn^{2+}, Cd^{2+}, and Pb^{2+}. Continuous stirring was carried out at 25 °C. Samples of 0.5 mL were subjected to 1, 3, 5, 15, 30, 60, 90, and 120 min of shaking, and the contents of heavy-metal ions (Cu^{2+}, Zn^{2+}, Cd^{2+}, and Pb^{2+}) were determined in the supernatant. Each group of experiments was conducted in triplicate.

2.2.3. Calcium Nodule Adsorption to Heavy-Metal Ions

Calcareous nodules were used as adsorbents. Aqueous solutions containing heavy-metal ions, including Cu^{2+}, Zn^{2+}, Cd^{2+}, and Pb^{2+}, were adopted as the objects of adsorption. The experiment on single heavy-metal adsorption was carried out under varying conditions.

The experiment on the size of particles involved taking 0.2 g of calcium nodules sieved through 1, 0.5, 0.25, 0.18, and 0.149 mm meshes, respectively. Subsequently, 50 mL of Cu^{2+}, Zn^{2+}, Cd^{2+}, or Pb^{2+} was added at a concentration of 30 mg/L and shaken at a constant temperature (25 ± 1 °C) for 120 min, whereas the supernatant was centrifuged, using ICP-MS to identify the content of heavy-metal ions.

The experiment on the time of adsorption involved taking 0.2 g of calcic nodules sieved through a 0.25 mm mesh and 50 mL of Cu^{2+}, Zn^{2+}, Cd^{2+}, or Pb^{2+} added at a concentration of 30 mg/L. Samples were taken at shaking times of 1, 3, 5, 15, 30, 60, 90, and 120 min, and then the content of heavy-metal ions in the supernatant was measured accordingly.

The experiment on the pH value involved taking 0.2 g of calcareous nodules sieved through a 0.25 mm mesh, which were then added to 50 mL of solutions at a concentration of 30 mg/L of Cu^{2+}, Zn^{2+}, Cd^{2+}, or Pb^{2+}, with the pH values reaching 4, 5, 6, 7, and 8. Other experimental conditions were the same as for the experiment on the size of particles.

The experiment on the sorbent dosage involved taking 0.2, 0.4, 0.6, 0.8, and 1 g of calcareous nodules sieved through a 0.25 mm mesh, which were added to 50 mL solutions containing 30 mg/L Cu^{2+}, Zn^{2+}, Cd^{2+}, or Pb^{2+}. Other experimental conditions were the same as above.

The experiment on the initial concentration of heavy metals involved taking 0.2 g of calcareous nodules sieved through a 0.25 mm mesh, which were added to 50 mL solutions of Cu^{2+}, Zn^{2+}, Cd^{2+}, or Pb^{2+} at concentrations of 10, 20, 30, 50, or 100 mg/L. Other experimental conditions were the same as above.

The experiment on the adsorption temperature involved taking 0.2 g of calcareous nodules sieved through a 0.25 mm mesh, which were added to 50 mL solutions of Cu^{2+}, Zn^{2+}, Cd^{2+}, and Pb^{2+} at a concentration of 30 mg/L. After shaking for 120 min at temperatures of 20 °C, 30 °C, 40 °C, 50 °C, and 60 °C, the supernatant was centrifuged to identify the content of heavy-metal ions.

The experiment on the adsorption isotherm involved adding the calcareous nodules (0.2 g) to 50 mL solutions of Cu^{2+}, Zn^{2+}, Cd^{2+}, or Pb^{2+} at concentrations of 10, 20, 30, 50, or 100 mg/L, respectively. After shaking at 25 °C for 5 h, the content of Cu^{2+}, Zn^{2+}, Cd^{2+}, or Pb^{2+} in the supernatant was measured. Each group of experiments was conducted in triplicate.

2.3. Data Processing

The adsorption rate was adopted to assess the adsorption capacity of calcium nodules to heavy metals, calculated as follows:

$$Q = (C_i - C_e)/C_i \times 100\% \quad (1)$$

where Q (%) is the adsorption rate, C_i (mg/kg) is the initial concentration of heavy metal, and C_e (mg/kg) is the equilibrium concentration of heavy metal added to calcareous nodules.

SPSS 18.0 software was adopted for statistical analysis, and Origin 8.0 software was adopted for plotting.

3. Results and Analysis

3.1. Adsorption Isotherm Experiment

The fitting parameters of isothermal adsorption are specified in Table 1. The fitting results of the Langmuir model (Equation (2)) and the Freundlich model bas (Equation (3)) are illustrated in Figure 1. It can be seen from Table 1 that the adsorption of calcareous nodules to Cu^{2+}, Zn^{2+}, Cd^{2+}, and Pb^{2+} conformed to the Langmuir model overall, whereas the fitting result of the Freundlich model was better for Zn^{2+}. As can be seen from the Langmuir isotherm model, during the adsorption of Cu^{2+}, Cd^{2+}, and Pb^{2+} to calcium nodules, monolayer adsorption and heterogeneous surface adsorption were found to coexist, but monolayer adsorption played the dominant role. Moreover, in the Freundlich isotherm model, nonuniform surface adsorption played the dominant role in the adsorption of Zn^{2+} to calcareous nodules, mainly composed of silica. The results of our isotherm adsorption experiment on calcareous nodules are consistent with those of other adsorbents whose main component is silica. For example, the research results of Lee et al. in a study of adsorption of Pb(II) and Cu(II) metal ions to functionalized large-pore mesoporous silica revealed that the correlation coefficient of the Langmuir isotherm (R^2 = 0.983 – 0.999) was higher than that of the Freundlich isotherm (R^2 = 0.826 – 0.983) for adsorption to copper ions [17]. Similarly, the research results of Melnyk et al. are consistent with ours, whereby a better fit was obtained with the Langmuir isotherm [18].

$$\text{Langmuir: } C/Q = 1/Kq_m + C/q_m \tag{2}$$

$$\text{Freundlich: } \lg Q = \lg KF + n\lg C \tag{3}$$

Table 1. Fitness of isotherm models and corresponding parameters.

Heavy Metal	Langmuir Model			Freundlich Model		
	K (L/mg)	Q_m (mg·g^{-1})	R^2	K_F (mg/g)/(mg/L)$^{1/n}$	n	R^2
Cu^{2+}	−2.1111111	52.6315789	0.9908	78.072844	−0.1023	0.677
Zn^{2+}	0.53947368	121.95122	0.988	45.593192	0.378	0.9906
Cd^{2+}	−1.7845304	30.9597523	0.9924	55.309534	−0.2322	0.9921
Pb^{2+}	−1.5471014	23.4192037	0.9867	76.155277	−0.4468	0.5419

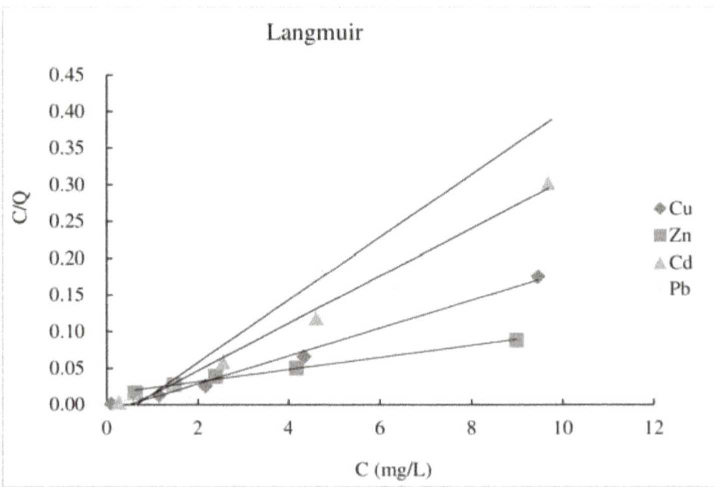

Figure 1. Fitting curves of the Langmuir adsorption isotherm model.

3.2. Adsorption Kinetics Experiment

The fitting parameters of the adsorption kinetics are specified in Table 2. The fitting results of the quasi-first-order kinetic model (Equation (4)), the quasi-second-order kinetic model (Equation (5)), and the Elovich kinetic model (Equation (6)) are illustrated in Figure 2. As can be seen from Table 2, the correlation coefficients of the Elovich kinetic model describing the adsorption of the calcareous nodules to Cu^{2+}, Zn^{2+}, Cd^{2+}, and Pb^{2+} exceeded those of the quasi-first-order kinetic model and the quasi-second-order kinetic model. The research findings indicate that the adsorption behavior of calcium nodules to Cu^{2+}, Zn^{2+}, Cd^{2+}, and Pb^{2+} was more consistent with the Elovich kinetic model. The reason is that the calcareous nodules are mainly formed by evaporation or leaching in nodular authigenic sediments, which are composed of calcium carbonate in semiarid plains or lowland soils. The Elovich kinetic model is quite applicable to processes with large changes in activation energy during the reaction, such as the processes taking place at the interface of soil and sediment.

$$\text{Elovich kinetic model: } Q_t = a + b\ln t \quad (4)$$

$$\text{First-order kinetic model: } \ln(C_0/C_t) = K_1 t \quad (5)$$

$$\text{Second-order kinetic model: } 1/C_t - 1/C_0 = K_2 t \quad (6)$$

Table 2. Kinetic fitting results.

Heavy Metal	Elovich			First-Order Dynamics		Second-Order Kinetics	
	A	b	R^2	K_1 (min^{-1})	R^2	K_2 ($10^{-5} g \cdot mg^{-1} \cdot min^{-1}$)	R^2
Cu^{2+}	17.817	12.634	0.938	0.0017	0.6995	0.0007	0.7299
Zn^{2+}	65.842	4.9672	0.9042	0.0007	0.6062	0.0003	0.6242
Cd^{2+}	6.7267	8.1169	0.9817	0.001	0.7941	0.0004	0.8068
Pb^{2+}	8.1764	23.856	0.8958	0.0039	0.8061	0.0018	0.8268

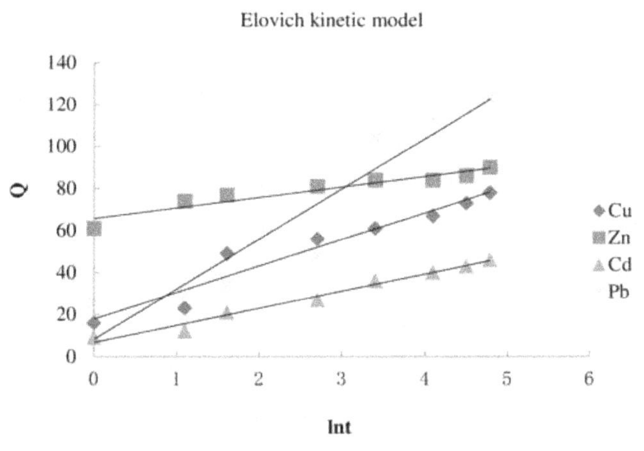

Figure 2. *Cont.*

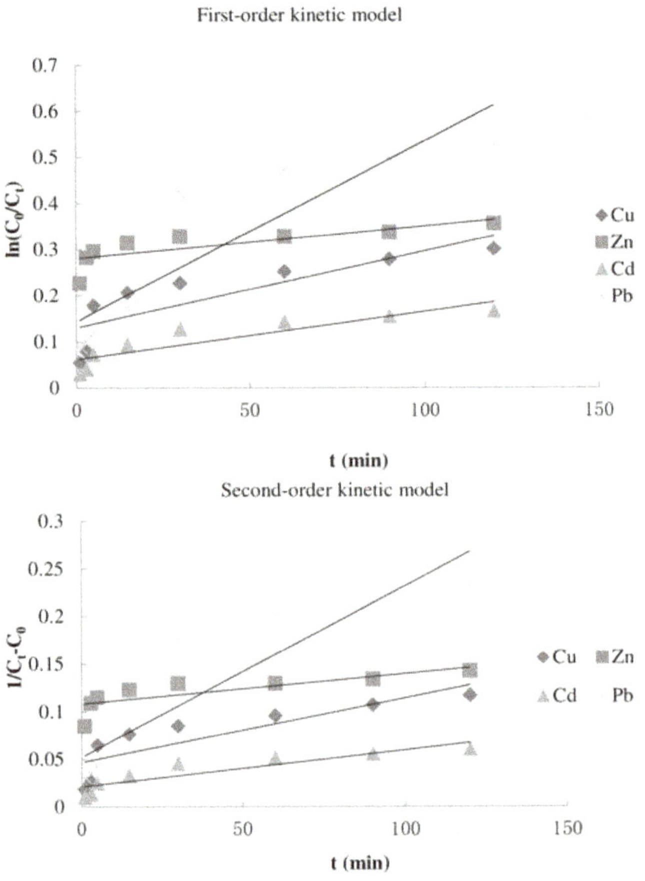

Figure 2. Fitting curves of the adsorption kinetics models.

3.3. Micromorphology and Chemical Composition of Calcareous Nodules

As can be seen from the scanning electron micrograph (illustrated in Figure 3), the calcareous nodules featured a compact texture, a rough surface, and a stepped shape, whereas the particles were smooth and had no evident rhomboids. Some of the particles were found to have large pores. The particles were of varying sizes and shapes, and they primarily existed in the form of rods, ellipses, flakes, thin strips, etc. The thin strips and rods are likely to be illite, the ellipses are likely to be iron oxide, and the flakes are likely to be kaolinite and montmorillonite [19]. The adsorption of calcium nodules to heavy-metal ions in water is linked to the type and content of minerals. With respect to the chemical composition, the major component of calcareous nodules was SiO_2, followed by CaO and Al_2O_3. The sum of the three components was roughly 90%. In contrast, K_2O, NaO, MgO, and Fe_2O_3 are easily leached out with water, and their content was relatively low (as specified in Table 3). The ratio of silicon to aluminum in calcareous nodules (SiO_2/Al_2O_3) was found to be 2.89. Combined with the percentage of each oxide, it can be inferred that the calcareous nodules mainly included 2:1 collision type minerals such as illite and montmorillonite. The 2:1 collision type minerals feature a large specific surface area and fine particle size, with an optimal effect toward the adsorption and precipitation of heavy-metal ions. Therefore, calcium nodules can be adopted to adsorb and eliminate heavy-metal ions in water [20,21].

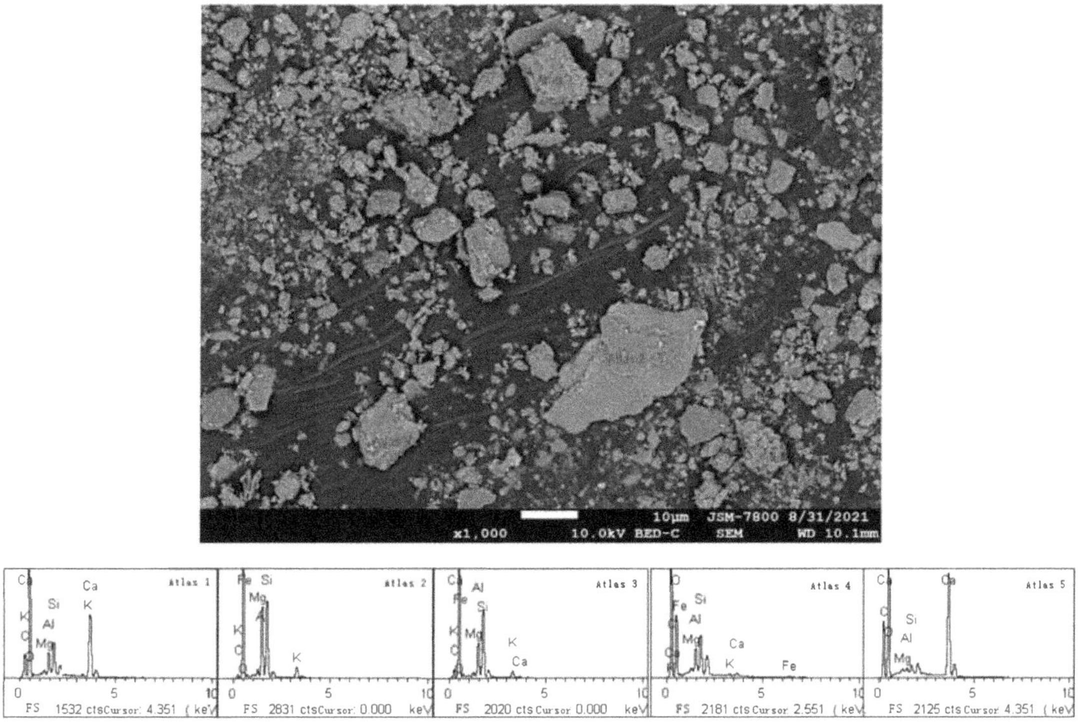

Figure 3. Scanning electron micrograph of calcareous nodules. 1. Silicate (calcium-based); 2. Silicate + a small amount of iron oxide; 3. Silicate + a small amount of iron oxide; 4. Silicate + a small amount of iron oxide; 5. Calcium oxide + a small amount of silicate.

Table 3. Chemical composition of calcareous modules.

Chemical Components	K_2O	CaO	NaO	MgO	Al_2O_3	Fe_2O_3	SiO_2	SiO_2/Al_2O_3
Content (%)	2.16	34.57	1.23	1.78	14.36	4.34	41.56	2.89

3.4. *Impact of the Particle Size on the Adsorption of Heavy-Metal Ions by Calcium Nodules*

The adsorption rate of heavy-metal ions was subject to significant changes in accordance with the size of the calcium nodules (as illustrated in Figure 4). When the particle size ranged between 0.149 mm and 0.25 mm, the adsorption rate of Cu^{2+}, Zn^{2+}, and Pb^{2+} declined at a slow pace. For particle sizes smaller than 0.25 mm, the adsorption rate declined to a significant extent. The adsorption rates of Cu^{2+}, Zn^{2+}, and Pb^{2+} ranged from 20.33% to 85.00%, 28.67% to 89.33%, and 43.67% to 82.00%, respectively. When the particle size of Cd^{2+} ranged between 0.149 mm and 0.25 mm, the adsorption rate declined to a relatively significant extent (by 14.67%). For particle sizes exceeding 0.25 mm, the adsorption rate was generally stable. As the particle size of calcium nodules further expanded, the adsorption rate of heavy-metal ions gradually declined. The reason behind is perhaps that, as the particle size of the calcareous nodules of the same mass increased, parameters such as the specific surface area and the relative content of layered silicate minerals decreased, including montmorillonite and illite, thereby lowering the adsorption rate of heavy-metal ions. Furthermore, the discrepancy in the content of carbonate, sulfate, and hydrated oxides of iron, aluminum, and magnesium in calcium nodules with varying particle sizes would also have an impact on the adsorption effect of heavy-metal ions. It was observed that, as the particle size decreased, the solution likely became more turbid, with a longer time required before settling. Therefore, it was better to opt for calcareous nodules sieved through 0.25 mm mesh when conducting the adsorption of heavy-metal ions.

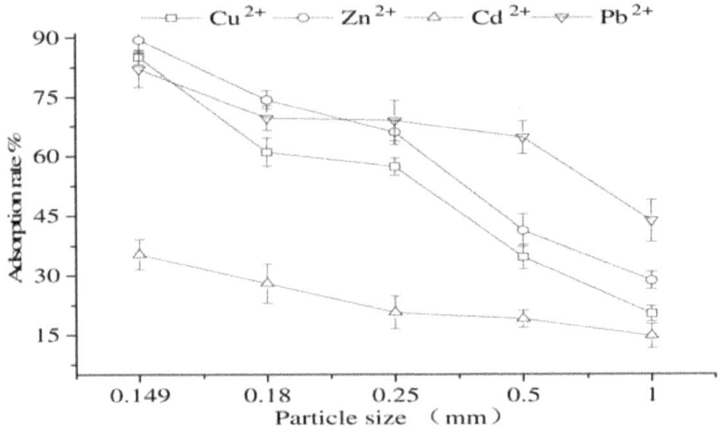

Figure 4. Effects imposed by calcium nodules of varying particle size on the adsorption of Cu^{2+}, Zn^{2+}, Cd^{2+}, and Pb^{2+} in solution (25 °C, pH 7).

3.5. Impact of the Adsorption Time on the Adsorption of Heavy-Metal Ions by Calcium Nodules

With an increase in experimental time, the adsorption rate of heavy-metal ions gradually increased, but the adsorption curve showed a different trend (as illustrated in Figure 5). The adsorption rates of Cd^{2+} and Pb^{2+} experienced significant increases within the adsorption time of 1 to 60 min, whereas the increase in adsorption rates slowed with a further increase in adsorption time. Compared with Cd^{2+} and Pb^{2+}, Cu^{2+} and Zn^{2+} were able to attain a higher adsorption rate in a shorter time (within 15 min). Cu^{2+}, Zn^{2+}, Cd^{2+}, and Pb^{2+} were able to attain the high adsorption rates within 30 min, eventually reaching maximum rates of 26%, 30.00%, 15.33%, and 41.00%, respectively. Cu^{2+}, Zn^{2+}, Cd^{2+}, and Pb^{2+} exhibited better adsorption effects within 60 min. This was likely caused by the combination of mineral types and their contents in calcareous nodules. Due to the large concentration gradient existing between the calcium nodules and the solutions for heavy-metal adsorption, adsorption occurred at a relatively faster pace. The adsorption rate of calcium nodules to the four heavy-metal ions increased smoothly after 60 min. According to the adsorption time, the degree of adsorption of heavy-metal ions can be determined to a certain extent. At adsorption times longer than 60 min, the adsorption rates of the four heavy-metal ions experienced slight increases with time. The research findings indicate that the adsorption time is not the primary contributing factor to the adsorption rate of heavy-metal ions. Accordingly, a time of 120 min was selected for comparative analysis.

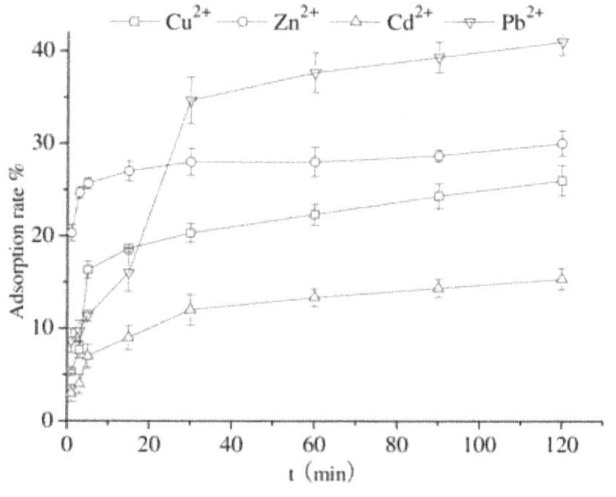

Figure 5. Effects imposed by the adsorption time on the adsorption of Cu^{2+}, Zn^{2+}, Cd^{2+}, and Pb^{2+} in solution (25 °C, pH 7).

3.6. Impact of pH Value on the Adsorption of Heavy-Metal Ions by Calcium Nodules

The pH value is a vital factor mediating the adsorption rate, which is linked to the chemical properties of various heavy-metal ions and their existence in solution. As illustrated in Figure 6, when pH = 4, the adsorption rates of calcium nodules to Cu^{2+}, Zn^{2+}, Cd^{2+}, and Pb^{2+} reached the minimum levels of 16.00%, 19.67%, 9.67%, and 10.33%, respectively. At this pH, calcium carbonate and the oxides of iron, aluminum, and magnesium in the calcareous nodules were subject to decomposition under the action of H^+, thereby damaging their structure and properties. In addition, the concentrations of Ca^{2+}, Fe^{2+}, Al^{3+}, and Mg^{2+} in the solutions constantly increased. As a result, a competitive relationship among Cu^{2+}, Zn^{2+}, Cd^{2+}, and Pb^{2+} in the solutions was formed, thus impacting the adsorption of heavy-metal ions. As the pH value gradually increased, the adsorption rate of Cd^{2+} first increased rapidly and then slowly, reaching a maximum of 30.33% (pH = 8). The adsorption rates of Zn^{2+} and Pb^{2+} first increased and then declined, reaching a maximum of 40.67% (pH = 6) and 49.67% (pH = 7), respectively. When the pH value was set to 5, the adsorption rate of Cu^{2+} reached its maximum. When the pH exceeded 5, the adsorption rate declined to a significant extent, whereas it remained basically stable in the pH value range of 6 to 8.

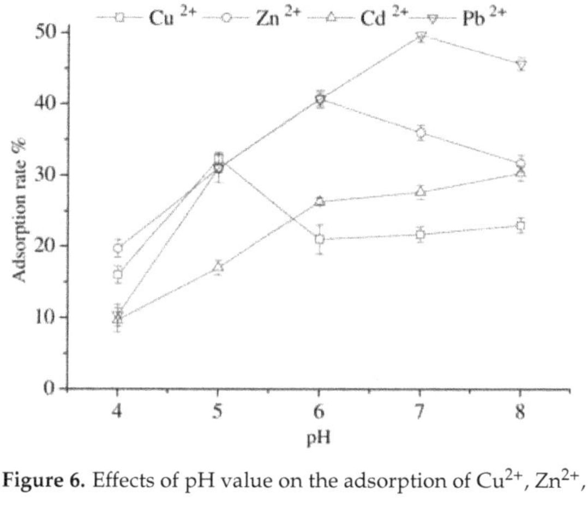

Figure 6. Effects of pH value on the adsorption of Cu^{2+}, Zn^{2+}, Cd^{2+}, and Pb^{2+} in solution (25 °C).

As shown by the research findings, the metal ions showed different patterns as a function of pH. Cd^{2+} adsorption mainly exhibits characteristics of complicated and potential adsorption [4]. Under such circumstances, a higher pH value is more conducive to the adsorption of Cd^{2+}. On the other hand, Cu^{2+}, Zn^{2+}, and Pb^{2+} are primarily adsorbed through precipitation. As the pH value increases, they are gradually transformed into hydroxide complex ions. In particular, Cu^{2+} starts to precipitate as the pH value exceeds 5, whereas Zn^{2+} starts to precipitate as the pH value exceeds 6, and Pb^{2+} starts to precipitate as the pH value exceeds 7. Therefore, with an increase in the pH value, the changes in the adsorption rates of the four heavy-metal ions varied [22–25].

3.7. Impact of the Amount of Adsorbent on the Adsorption of Heavy-Metal Ions by Calcium Nodules

The adsorption rates of calcium nodules to Cu^{2+}, Zn^{2+}, Cd^{2+}, and Pb^{2+} in solution gradually increased with the increasing amount of calcium nodules (illustrated in Figure 7). The adsorption rate increased rapidly between dosages of 0.2 g and 0.6 g, before slowing down. The amount of calcareous nodules weakly impacted the adsorption rate of Cd^{2+}, only increasing from 16.00% and 27.67%. On the other hand, the amount of calcareous nodules most evidently impacted the adsorption rate of Pb^{2+}, increasing from 43.00% and 76.00%, a nearly twofold increase. The amount of calcareous nodules increased the adsorption rate according to the following order: $Pb^{2+} > Zn^{2+} > Cu^{2+} > Cd^{2+}$. Specifically, the maximum adsorption rates of Cu^{2+} and Cd^{2+} were lower than 40%, whereas the adsorption rate of Pb^{2+} reached a level as high as 80%. This is likely linked to the discrepancy in the electronegativity of Cu^{2+}, Zn^{2+}, Cd^{2+}, and Pb^{2+} [26,27]. Although a high dosage of calcium nodules was conducive to increasing the adsorption rate of heavy-metal ions, it also led to turbid water and a poor visual effect. Therefore, 0.6 g was deemed most suitable for practical application.

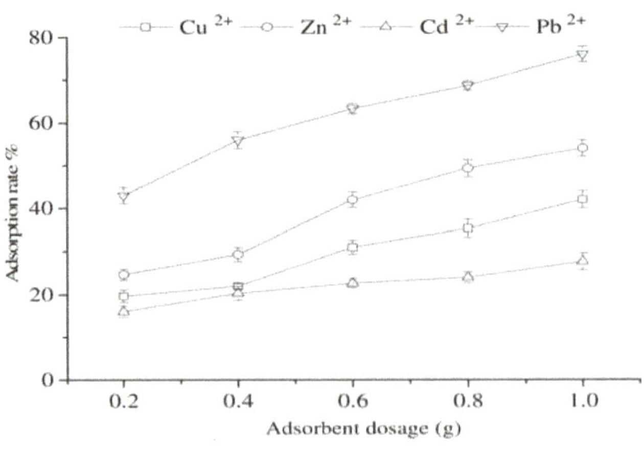

Figure 7. Effects of the amount of adsorbent on the adsorption of Cu^{2+}, Zn^{2+}, Cd^{2+}, and Pb^{2+} in solution (25 °C, pH 7).

3.8. Impact of the Initial Concentration of Heavy Metals on the Adsorption of Heavy-Metal Ions by Calcium Nodules

As the initial concentration of heavy-metal ions increased, the adsorption rates of calcium nodules to Cu^{2+}, Zn^{2+}, and Cd^{2+} showed a significant declining trend, whereas that of Pb^{2+} first increased and then decreased (as illustrated in Figure 8). This is likely due to the fact that, at a low initial concentration, there are numerous adsorption sites on the surface of calcareous nodules; thus, the adsorption rate decreased with the concentration of heavy-metal ions in solution as the adsorption sites became saturated. The initial concentration of heavy-metal ions had a much greater impact on the adsorption rates of Cu^{2+}, Cd^{2+}, and Pb^{2+} than Zn^{2+}. At concentrations up to 100 mg/L, the adsorption rate of Zn^{2+} changed most slowly, declining by merely 27.80%. On the other hand, the adsorption rates of Cu^{2+}, Cd^{2+}, and Pb^{2+} declined to a significant extent by 84.60%, 70.80%, and 69.00% respectively. The adsorption rates of the four heavy-metal ions were optimal at lower concentrations. Up to a concentration of 30 mg/L, the adsorption rate of Pb^{2+} remained above 40%, despite its declining trend. A similar observation was made for Cu^{2+} up to a concentration of 20 mg/L. The adsorption rates of Cd^{2+} and Zn^{2+} were lower than 30% up to a concentration of 20 mg/L, reaching the maximum levels of 74.00% and 38.00%, respectively, at 10 mg/L. Therefore, the calcium nodules can be used for the treatment of low-concentration wastewater.

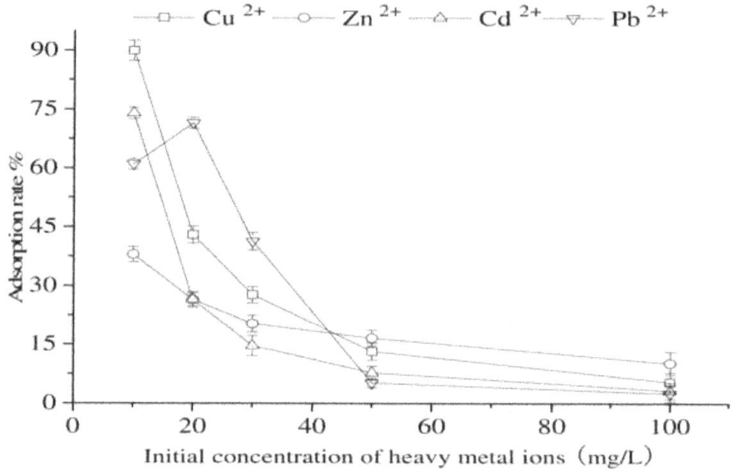

Figure 8. Effects of the initial concentration of heavy-metal ions on the adsorption of Cu^{2+}, Zn^{2+}, Cd^{2+}, and Pb^{2+} in solution (25 °C, pH 7).

3.9. Impact of the Temperature on the Adsorption of Heavy-Metal Ions by Calcium Nodules

As the temperature increased, the adsorption rates of calcium nodules to Cu^{2+}, Zn^{2+}, Cd^{2+}, and Pb^{2+} all increased (as illustrated in Figure 9), indicating that heat is conducive to the adsorption process. When the temperature ranged between 20 °C and 40 °C, the adsorption rates experienced a slow increase. When the temperature reached 40 °C, the adsorption rates increased sharply. The contributing factor is likely that, as the temperature increased further, the adsorption and exchange rates of heavy-metal ions also increased. The order of impact of the changing temperature on the adsorption of heavy-metal ions to calcium nodules was as follows: $Pb^{2+} < Cu^{2+} < Zn^{2+} < Cd^{2+}$. Specifically, within the range of 20–40 °C, temperature had a weak effect on the adsorption rate of Pb^{2+}, with the adsorption rate increasing by merely 5.67%. When the temperature increased to 60 °C, the adsorption rate increased by 46.00%. The adsorption rates of Cu^{2+}, Zn^{2+} and Cd^{2+} increased linearly with increasing temperature. When the temperature increased to 60 °C, the adsorption rates increased by 55.67%, 66.33%, and 73.67%, respectively. When the temperature was set to 50 °C, the adsorption rates of Cu^{2+}, Zn^{2+}, Cd^{2+}, and Pb^{2+} were found to reach 79.00%, 76.00%, 72.00%, and 70.00%, respectively, of their maximum adsorption rates. Although an increase in temperature enhanced the adsorption effect of heavy-metal ions, an excessively high temperature risks lowering the survival rate of aerobic microorganisms in the water body, thus imposing a negative impact on the decomposition of other contaminants by the microorganisms in the water body. Accordingly, 50 °C was selected as the ideal temperature.

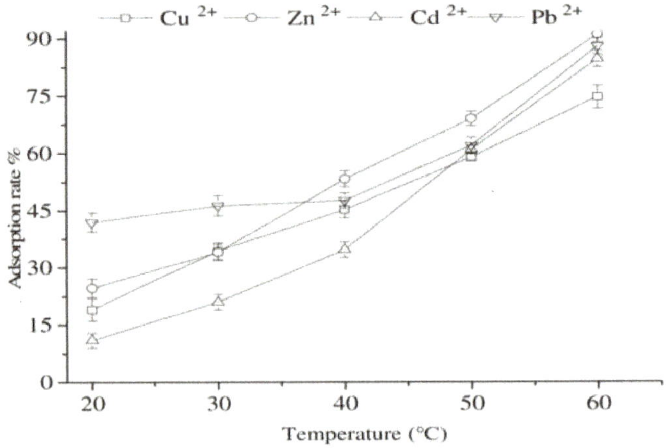

Figure 9. Effects of temperature on the adsorption of Cu^{2+}, Zn^{2+}, Cd^{2+}, and Pb^{2+} in solution.

4. Discussion

Since heavy metals exhibit significant toxicity to humans, animals, plants, and microorganisms and are not able to be degraded by microorganisms, wastewater containing heavy metals has become a major environmental concern around the world [28]. Compared with other approaches, the adsorption method features the advantages of simple operation, high efficiency, and low cost; thus, it has been widely adopted in the treatment of heavy-metal wastewater [4]. The adsorbent is the core element during the application of the adsorption method, and its performance impacts both the quality and the efficiency of separation. At present, the adsorbents adopted in the treatment of heavy metals can be divided into natural adsorbents, biological adsorbents, and synthetic adsorbents according to their source and chemical structure [29].

Loess calcium is categorized as a natural adsorbent, but few studies have been carried out on this adsorbent so far. Nevertheless, some scholars have carried out research on the adsorption effect of other adsorbents toward water containing heavy metals. As shown by previous results, the pH value, time, adsorbent dosage, temperature, and initial concentration of heavy metals are crucial factors affecting the adsorption effects of varying adsorbents, consistent with our research findings [1,2,4–6,29–33]. For instance, Sheikhhosseini et al. elaborated on the competitive sorption of nickel, cadmium, zinc, and copper on palygorskite and sepiolite silicate clay minerals [8]. Their research findings indicated that palygorskite and sepiolite can effectively eliminate Cu from solution regardless of the presence of other metals. Zou et al. adopted corn stalks to prepare biochar, conducted experiments on the adsorption effects of heavy metals including Cd^{2+} and Pb^{2+}, and analyzed the capacity and efficiency of the biochar's adsorption to heavy metals [3]. Their experimental results showed that the optimal adsorption condition for Cd^{2+} was reached when the pH value was set to 5, and the adsorption equilibrium amounted to 120 min. The optimal adsorption condition for Pb^{2+} was reached when the pH value was set to 1, and the adsorption equilibrium amounted to 60 min. These findings are basically consistent with the results in this paper. Li et al. studied the adsorption mechanism of corn stover biochar toward Cd(II) [4]. Their experimental results indicated that, with an increase in the pH value, the adsorption rate constantly increased, which is consistent with our research findings. However, upon reaching a certain threshold, the adsorption rates of various heavy metals increase slowly or even decline. According to Peng's research, the optimal pH value for the adsorption of various heavy-metal ions may differ due to their discrepancy in metal electronegativity, in addition to the standard reduction potential of heavy-metal ions and the first stability constant of the associated metal hydroxide [29]. Furthermore, a proper amount of adsorbent helps achieve an optimal effect. On the other hand, as the adsorption time is lengthened, the adsorption effect is also enhanced, but this effect slows over time; therefore, an optimal time should be selected. Lastly, an increase in temperature leads to an enhanced adsorption effect, but the most suitable temperature varies according to the target heavy metal. The aforementioned research conclusions are overall consistent with our research findings.

5. Conclusions

Different adsorption conditions impose varying effects on the adsorption of heavy-metal ions by calcium nodules. According to our results, we reached four conclusions, as outlined below.

First, it is better to opt for calcium nodules that pass through a 0.25 mm sieve when promoting the adsorption of heavy-metal ions.

Second, as the adsorption time and the amount of adsorbent increase, the adsorption rates of Cu^{2+}, Zn^{2+}, Cd^{2+}, and Pb^{2+} increase rapidly before slowing down, with quite similar kinetics. The adsorption time is not a major contributing factor to the adsorption rate of heavy-metal ions. Accordingly, a duration of 120 min is deemed the best option. Although a high amount of calcium nodules is conducive to increasing the adsorption rates of heavy-metal ions, a high dosage is likely to lead to turbid water and a poor visual effect. Thus, a dosage of 0.6 g is considered appropriate for practical application.

Third, both the pH value and the temperature impose a great impact on the adsorption of heavy-metal ions to calcium nodules. However, proper levels should be selected as a function of the heavy metal. In particular, the optimal pH values for the adsorption of Pb^{2+}, Zn^{2+}, Cu^{2+}, and Cd^{2+} were 7, 6, 5, and 8, respectively. The optimal temperature for adsorption was found to be 50 °C.

Fourth, judging by the impact of the initial concentration of heavy metals on their adsorption to calcium nodules, calcium nodules are more applicable for the treatment of low-concentration wastewater.

Generally, as the particle size of heavy-metal ions decreases, the adsorption capacity increases, whereas a longer adsorption time, higher temperature, and greater amount of adsorbent improve the adsorption effect. In summary, calcareous nodules can be used as pure natural adsorbents that are convenient to obtain and low in cost, with a good adsorption effect.

Author Contributions: Q.L. designed and developed the research idea; Y.L. collected and processed data; Q.L. and Y.L. participated in drafting the manuscript; Q.L. contributed to the scientific contentand editing of the article; and all the authors revised the manuscript. All authors have read and agreed to the published version of the manuscript.

Funding: The Fund Project of Shaanxi Key Laboratory of Land Consolidation (Program No.2019-ZD05).

Institutional Review Board Statement: Not applicable.

Informed Consent Statement: Not applicable.

Data Availability Statement: The data presented in this study are available on request from the corresponding author. The data are not publicly available due to the data author belongs to the unit of its own experiment.

Acknowledgments: The Research Square Platform LLC is thanked for improving the article.

Conflicts of Interest: The authors declare no conflict of interest.

References

1. Zhang, Z.; Bao, L.I.; Liang, R. Comparison of sediment heavy metal fractions at estuary and center of Nanyang Zone from Nansi Lake, China. *Huanjing Kexue Xuebao/Acta Sci. Circumstantiae* **2015**, *35*, 1408–1416.
2. Wang, Y.Y.; Liu, Y.X.; Lu, H.H.; Yang, R.Q.; Yang, S.M. Competitive adsorption of Pb (II), Cu (II), and Zn (II) ions onto hydroxyapatite-biochar nano composite in aqueous solutions. *J. Solid State Chem.* **2018**, *261*, 53–61. [CrossRef]

3. Zou, J.; Sun, D.; Zhao, J.; Wang, S.; Li, S.; Li, W.; Lv, Y.; Dou, H. Adsorption of heavy metal Cd, Pb by maize straw Biochar. *J. Beihua Univ.* **2018**, *281*, 96–99.
4. Li, L.; Lu, Y.; Liu, Y.; Sun, H.W.; Liang, Z.Y. Adsorption mechanisms of cadmium (II) on biochars derived from corn straw. *J. Agro-Environ. Sci.* **2012**, *31*, 2277–2283.
5. Liu, J.; Liu, L.; Xue, J. Research progress on treatment of heavy metal wastewater by adsorption. *Environ. Chem.* **2018**, *37*, 2016–2024.
6. Hu, H.; Liu, G. The trend of heavy metal wastewater treatment by adsorption method. In Proceedings of the 2012 2nd International Conference on Remote Sensing, Environment and Transportation Engineering, Nanjing, China, 1–3 June 2012.
7. Ji, Z.; Wu, X.F.; Li, Y.; Feng, C. Kinetic adsorption and change in chemical potential of heavy metal ions in aqueous solutions. *Environ. Chem.* **2015**, *11*, 53–61.
8. Sheikhhosseini, A.; Shirvani, M.; Shariatmadari, H. Competitive sorption of nickel, cadmium, zinc and copper on palygorskite and sepiolite silicate clay minerals. *Geoderma* **2013**, *192*, 249–253. [CrossRef]
9. He, H.P.; Guo, J.G.; Zhu, J.X.; Yang, D. An experimental study of adsorption capacity of montmorillonite, kaolinite and illite for heavy metals. *Acta Mineral. Petrol.* **2001**, *20*, 573–578.
10. Gong, T.; Zhu, Y. Spatial pattern of caliche nodule in surface soil of the hillslopes in Liudaogou catchment in the wind-water erosion crisscross zone of the Loess Plateau. *Sci. Soil Water Conserv.* **2016**, *14*, 42–49.
11. Teng, Z.H.; Liu, R.M. Research on calcareous nodules in the loess strata of China. *Chin. Sci. Bull.* **1990**, *192*, 1008–1011.
12. Wang, S.Y. Study on the Petrological Characteristics of N_2 Red Clay in Northern Shanxi. Master's Thesis, China University of Mining and Technology, Xuzhou, China, 2017.
13. Gong, T.X. Study on the Distribution of Calcareous Nodules and Their Water Characteristics on Slopes in Loess Area. Master's Thesis, University of Chinese Academy of Sciences, Beijing, China, 2017.
14. Li, C.A.; Wu, J.P.; Cao, J.X. The morphology of loess calcareous nodules in northwest Hebei and their genesis dynamic characteristics and stratigraphic environmental significance. *Earth Sci.* **1995**, *38*, 511–514.
15. Yang, L.G.; Qiao, D.Y.; Hu, P.Y. Determination of thirteen elements in soil by inductively coupled plasma mass spectrometry with automatic digestion instrument. *Soil Fertil. Sci. China* **2019**, *38*, 89–93.
16. Wu, H.Z.; Meng, L.F. Liquid chromatography-UV determination of heavy metal ions in environmental samples using dispersive liquid-liquid microextraction coupled with magnetic nanoparticles. *Appl. Ecol. Environ. Res.* **2019**, *17*, 1571–1584. [CrossRef]
17. Lee, J.Y.; Chen, C.H.; Cheng, S.; Li, H.Y. Adsorption of Pb(II) and Cu(II) metal ions on functionalized large-pore mesoporous silica. *Int. J. Environ. Sci. Technol.* **2016**, *13*, 65–76. [CrossRef]
18. Melnyk, I.V.; Gdula, K.; Dbrowski, A.; Zub, Y.L. Magneto-sensitive adsorbents modified by functional nitrogen-containing groups. *Nanoscale Res. Lett.* **2016**, *11*, 61. [CrossRef] [PubMed]
19. Gao, B.Y.; Qiu, T.; Li, R.H.; Qin, R.; Zhang, G.; Li, X.; Zhang, Z. Investigation of heavy metal ions adsorption ability by thiol-modified corn stalk powder. *J. Northwest A F Univ. -Nat. Sci. Ed.* **2012**, *40*, 185–190.
20. Qin, H.F.; Zhang, W.M. Adsorption of lead and cadmium by hydroxyapatite coated quartz sands. *Sci. Technol. Eng.* **2018**, *20*, 183–189.
21. Kang, J.; Kim, T.; Park, J.; Lee, K.Y.; Park, D.H.; Park, S.; Kim, S.; Jung, Y. A mesoporous chelating polymer-carbon composite for the hyper-efficient separation of heavy metal ions. *J. Nanosci. Nanotechnol.* **2020**, *20*, 3042–3046. [CrossRef]
22. Tsao, T.M.; Chen, Y.M.; Sheu, H.S.; Zhuang, S.Y.; Shao, P.H.; Chen, H.W.; Shea, K.S.; Wang, M.K.; Shau, Y.H.; Chiang, K.Y. Red soil chemistry and mineralogy reflect uniform weathering environments in fluvial sediments, Taiwan. *J. Soils Sediments* **2012**, *12*, 1054–1065. [CrossRef]
23. Zhu, X.; Chen, H.; Li, W.; He, Y.; Brookes, P.C.; Xu, J. Aggregation kinetics of natural soil nanoparticles in different electrolytes. *Eur. J. Soil Sci.* **2014**, *65*, 206–217. [CrossRef]
24. Zhang, Z.Y.; Huang, L.; Liu, F.; Kuang Wang, M.; Ling Fu, Q.; Zhu, J. The properties of clay minerals in soil particles from two Ultisols, China. *Clays Clay Miner.* **2017**, *65*, 273–285. [CrossRef]
25. Chou, Y.M.; Song, S.R.; Tsao, M.T.; Lin, C.S.; Wang, M.K.; Lee, J.J.; Chen, F.J. Identification and tectonic implications of nano-particle quartz (<50 nm) by synchrotron X-ray diffraction in the Chelungpu fault gouge, Taiwan. *Tectonophysics* **2014**, *619*, 36–43. [CrossRef]
26. Ye, T.; Huang, L.; Zhang, K.Q.; Zhang, B.; Chang, H.; Liu, Z.J.; Du, L.Z. Evaluation of the combined removal of heavy metals by saponin and citric acid from municipal sewage sludges and metal stability features. *Huanjing Kexue* **2017**, *38*, 4850–4859. [CrossRef]
27. Wang, W.; Han, N.; Yang, C.; Zhang, W.; Miao, J.; Li, W.; Zhang, X. Fabrication of P(AN-MA)/rGO-g-PAO superhydrophilic nanofiber membrane for removal of heavy metal ions. *J. Nanosci. Nanotechnol.* **2020**, *20*, 1685–1696. [CrossRef]
28. Tofighy, M.A.; Mohammadi, T. Adsorption of divalent heavy metal ions from water using carbon nanotube sheets. *J. Hazard. Mater.* **2011**, *185*, 140–147. [CrossRef] [PubMed]
29. Peng, W.; Li, H.; Liu, Y.; Song, S. A review on heavy metal ions adsorption from water by graphene oxide and its composites. *J. Mol. Liq.* **2017**, *230*, 496–504. [CrossRef]
30. Gu, D.G.; Fein, J.B. Adsorption of metals onto graphene oxide: Surface complexation modeling and linear free energy relationships. *Colloids Surf. A Physicochem. Eng. Asp.* **2015**, *481*, 319–327. [CrossRef]

31. Rahnama, E.; Bazrafshan, O.; Asadollahfardi, G. Application of data-driven methods to predict the sodium adsorption rate (SAR) in different climates in Iran. *Arab. J. Geosci.* **2020**, *13*, 1160. [CrossRef]
32. Zhu, L.; Wang, J.; Liu, J.; Nasir, M.S.; Zhu, J.; Li, S.; Liang, J.; Yan, W. Smart formaldehyde detection enabled by metal organic framework-derived doped electrospun hollow nanofibers. *Sens. Actuators B Chem.* **2021**, *326*, 128819. [CrossRef]
33. Liu, J.W.; Wang, J.N.; Zhu, L.; Chen, X.; Ma, Q.; Wang, L.; Wang, X.; Yan, W. A high-safety and multifunctional MOFs modified aramid nanofiber separator for lithium-sulfur batteries. *Chem. Eng. J.* **2021**, *411*, 128540. [CrossRef]

coatings

MDPI

Article

Friction and Wear Behaviors of Fe-19Cr-15Mn-0.66N Steel at High Temperature

Shaolong Sheng [1], Huiling Zhou [1], Xiaojing Wang [1], Yanxin Qiao [1,2,*], Hongtao Yuan [3], Jian Chen [1,4], Lanlan Yang [1], Dongpeng Wang [1], Zhenguang Liu [1], Jiasheng Zou [1], Zhibin Zheng [2,*] and Jingyong Li [1,*]

1 School of Materials Science and Engineering, Jiangsu University of Science and Technology, Zhenjiang 212003, China; 152210602420@stu.just.edu.cn (S.S.); zhouhl@just.edu.cn (H.Z.); wxj@just.edu.cn (X.W.); jchen496@uwo.cn (J.C.); lanlanyang@just.edu.cn (L.Y.); dpwang@just.edu.cn (D.W.); zgliu@just.edu.cn (Z.L.); zjzoujs@126.com (J.Z.)
2 Institute of New Materials, Guangdong Academy of Sciences, Guangzhou 510650, China
3 Shanghai Waigaoqiao Shipbuilding Co., Ltd., Shanghai 200137, China; yuanht@chinasws.com
4 Department of Chemistry, Western University, 1151 Richmond St., London, ON N6A 5B7, Canada
* Correspondence: yxqiao@just.edu.cn (Y.Q.); zbzheng712003@163.com (Z.Z.); jingyong_li@126.com (J.L.)

Abstract: The friction and wear behaviors of Fe-19Cr-15Mn-0.66N steel were investigated under applied loads of 5 N and 15 N at the wear-testing temperatures of 300 °C and 500 °C using a ball-on-disc tribometer. The wear tracks were evaluated by scanning electron microscopy (SEM) and laser scanning confocal microscopy (LSCM) to reveal the variation in morphologies. Energy-dispersive X-ray spectroscopy (EDS) and X-ray photoelectron spectroscopy (XPS) were used to determine the components of oxide layers formed on wear surfaces. The results demonstrated that the oxide layers are favorable for obtaining a low friction coefficient under all conditions. The average friction coefficient decreased with increasing load at 300 °C, while it increased with the increase in applied load at 500 °C. At 300 °C, severe abrasive wear characterized by grooves resulted in a high friction coefficient with 5 N applied, whereas the formation of a denser oxide layer consisting of Cr_2O_3, $FeCr_2O_4$, Fe_2O_3, etc., and the increased hardness caused by work hardening led to a decrease in friction characterized by mild adhesive wear. At 500 °C, the transformation of Fe_2O_3 to the relatively softer Fe_3O_4 and the high production of lubricating Mn_2O_3 resulted in a minimum average friction coefficient (0.34) when 5 N was applied. However, the softening caused by high temperature weakened the hardening effect, and thus the friction coefficient increased with 15 N applied at 500 °C.

Keywords: high-nitrogen austenitic stainless steel; wear; friction coefficient; high temperature

Citation: Sheng, S.; Zhou, H.; Wang, X.; Qiao, Y.; Yuan, H.; Chen, J.; Yang, L.; Wang, D.; Liu, Z.; Zou, J.; et al. Friction and Wear Behaviors of Fe-19Cr-15Mn-0.66N Steel at High Temperature. *Coatings* **2021**, *11*, 1285. https://doi.org/10.3390/coatings11111285

Academic Editor: Alina Vladescu

Received: 28 August 2021
Accepted: 19 October 2021
Published: 22 October 2021

1. Introduction

High-nitrogen austenitic stainless steel (HNSS) can be used for non-magnetic drill collars in oil exploitation, bearing steels for automobile industry, and many other applications due to its excellent properties, i.e., high strength, good ductility, and superior corrosion resistance [1–6]. However, because of harsh service conditions, such as high temperature, heavy load, and high velocity, the demand for surface hardness and wear resistance is enhanced [7,8]. Investigations on preparation methods, mechanical properties, and corrosion resistance of HNSS have been widely carried out [9–12]. However, few reports about its tribological properties, especially at high temperatures, are available. Therefore, it is of great significance to investigate the friction and wear behaviors of HNSS at high temperature.

Wear at high temperature is a serious problem in many industrial applications, such as power generation, materials processing, and high-temperature bearing [13–15]. Metallic materials can be oxidized or softened at high temperature, which plays a significant role, causing a change in overall friction behavior [16]. This is because oxides can reduce metal wear by reducing or eliminating metal–metal contact, allowing a transition from severe to mild wear [17]. In addition, at elevated temperatures, the increasing dominance of dislocation climb, aided by increased lattice diffusion at high temperatures, results in a significant softening of metals [18,19]. Thus, the wear mechanism at high temperature is different from that at ambient temperature [20,21]. Hemant et al. [22] investigated the sliding friction and wear behaviors of austenitic stainless steel at a temperature of 473 K and 823 K, respectively, and found serious damage at 823 K due to the softening of the rubbing surfaces. Wang et al. [23] evaluated the tribological behaviors of sodium carbonate coating on stainless steel at high temperatures, and the results suggested that the lubrication capability of NCO is partially due to the reconstruction of the oxide scale. The high-temperature wear mechanism of 445 stainless steel was studied, and the Cr-rich oxide scale was able to stabilize the friction coefficient and reduce the wear rate [24]. In addition to temperature, the applied load also affects friction and wear behaviors at high temperature. Wang et al. [25] studied the effect of load on wear behavior at high temperature and found that the oxide layer at a higher load is easier to transfer onto the counterpart surface, leading to higher wear loss. Torres et al. [26] concluded that a higher applied load at high temperatures could lead to higher wear rates for the steels. Stott et al. [27] studied the effects of load on wear-protective layers during sliding at elevated temperatures, and the results showed that the layers were broken down, leading to enhanced wear damage, particularly at higher loads at 550 and 600 °C. Additionally, Razali et al. [28] used a combination of numerical and experimental methods to reveal the relationship between tribological conditions and microstructural evolution. In addition, Lee et al. [29] used similar methods that revealed that the friction coefficient suddenly changes at critical surface strain, and the lubricant film loses its function when the surface strain of the material exceeds the critical surface strain.

At present, there is a lack of research on the tribological properties of HNSS at high temperatures, and the synergistic effects of load and temperature on the wear mechanisms of HNSS need to be further studied. In this work, high-temperature friction and wear tests under the conditions of 5 N at 300 °C, 15 N at 300 °C, 5 N at 500 °C, and 15 N at 500 °C were applied on HNSS of Fe-19Cr-15Mn-0.66N. The microstructure and tribological properties of the samples were systematically characterized and analyzed.

2. Experimental Details

The material used in this study was Fe-19Cr-15Mn-0.66N HNSS. Its chemical composition is listed in Table 1. The material was cut into coupons with dimensions of 20 mm × 20 mm × 5 mm. The samples were gradually ground using 2000# SiC papers, then polished with a diamond paste of 2.5 μm, cleaned in ethanol, and finally dried using cool air.

Table 1. The chemical composition of the HNSS (wt%).

C	Si	Mn	P	S	Cr	Mo	N	Fe
0.044	0.24	15.80	0.017	0.005	18.40	2.19	0.66	bal.

The wear tests were evaluated using a ball-on-disc tribometer (HTC-2, Lanzhou Zhongke Kaihua Technology Co., Ltd., Lanzhou, China). The ball and the disc were Si_3N_4 and the HNSS, respectively. After installing the ball and disc in the holder on the tribometer, they were heated to the target temperatures of 300 and 500 °C. The sliding tests were conducted under loads of 5 N and 15 N (yielding an initial Hertz peak contact pressure of 370–630 MPa) and were conducted continuously for a duration of 1800 s in air without lubrication. The samples were taken out immediately when the sliding tests ceased to prevent further oxidation. The morphologies of the wear tracks were observed using laser scanning confocal microscopy (LSCM, LEXT OLS400, Tokyo, Japan) and scanning electron microscopy (SEM, JEOL, JSM-6480, Takeno, Japan) under secondary electron image (SEI) mode at an accelerating voltage of 20 kV and a probe current of 50 pA. The oxide layers formed on the wear tracks were characterized using energy-dispersive X-ray spectroscopy (EDS, X-Max 20, Oxford, UK) and X-ray photoelectron spectroscopy (XPS, ESCALAB 250Xi T, ThermoFisher Scientific, Waltham, MA, USA) using an Al $K\alpha$ (1486.6 eV) radiation source. The analysis of the spectra was performed using CasaXPS software version 2.3.15 (CasaXPS Software Ltd.). The background baseline of all the signals was subtracted using a standard Shirley background line type. Before conducting curve fitting, all the peaks were calibrated by shifting C 1s to 284.6 eV.

3. Results

3.1. Friction Coefficient

Figure 1 depicts the friction coefficient as a function of time for the HNSS under different test conditions. It is obvious that almost all the friction coefficient curves exhibited a similar tendency under all the four test conditions with the increase in time. Indeed, the curves could be divided into two stages, as can be seen from Figure 1. The curves of the friction coefficient dropped dramatically in the initial stage for the first 400 s and then rose up slowly without significant fluctuations or maintained a relatively stable trend as the time increased.

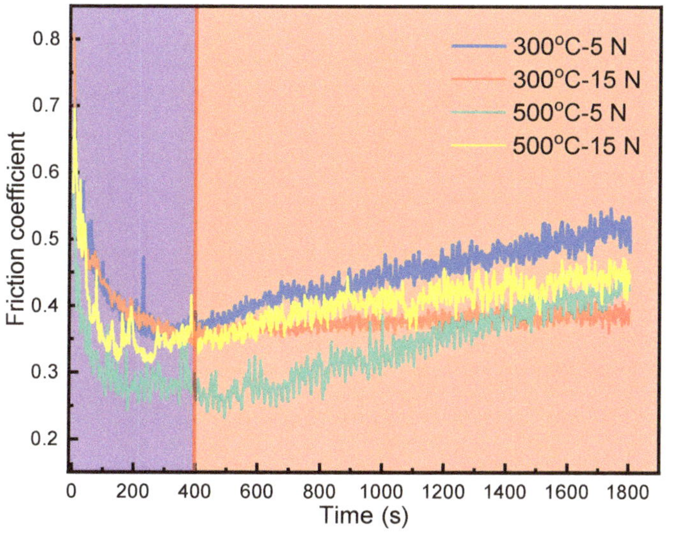

Figure 1. Friction coefficients of the HNSS under different conditions.

To observe the magnitude of the friction coefficient under different conditions more visually, the average friction coefficients of all four test conditions are shown in Figure 2. When the samples were tested at 300 °C, the average friction coefficients obtained by applying 5 N and 15 N were 0.44 and 0.38, respectively, indicating that the change in the friction coefficient is directly related to the applied load [30]. When the test temperature was 500 °C, the average friction coefficients obtained by applying 5 N and 15 N were 0.34 and 0.40, respectively. The variation in the friction coefficient shows an interesting phenomenon. When the HNSS was tested at 300 °C, its friction coefficient decreased with increasing load but showed an inverse trend when the temperature was 500 °C. This will be further explained in the Discussion section.

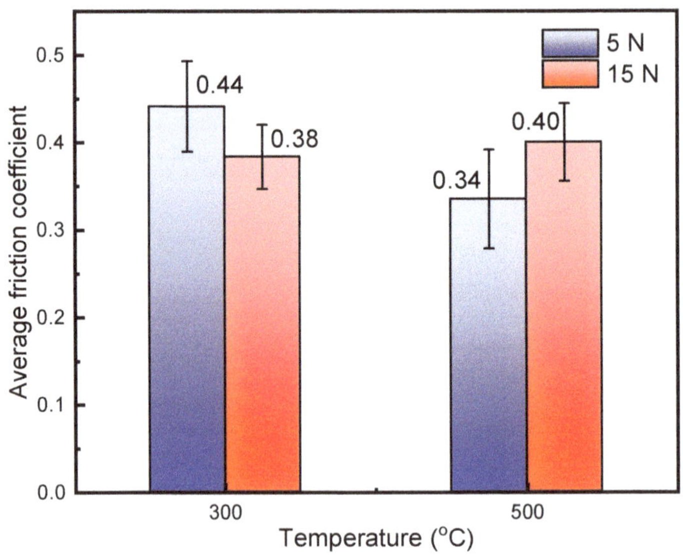

Figure 2. Average friction coefficients of the HNSS under different conditions.

3.2. LSCM of the Wear Tracks

Figure 3 shows the LSCM diagrams of the wear tracks. It can be seen that the surfaces of the wear tracks were seriously damaged. Grooves and oxide layers (dark zone) were observed on the surfaces of the wear tracks under all four conditions. As shown in Figure 3a, the width of the wear track was about 480 μm, and the oxide layer on the surface of the wear track was discontinuous when 5 N was applied at 300 °C. When the load was changed to 15 N, the morphology of the wear track was similar to that at 5 N, except that the width of the wear track increased to 682 μm, as shown in Figure 3b. At 500 °C, the worn surface was almost completely covered by the oxide layer, and the width of the wear track was about 423 μm with 5 N applied. When 15 N was applied at 500 °C (Figure 3d), the width of the wear track increased to 683 μm, and there were a large number of grooves as well as delamination of oxide on the surface.

3.3. Morphologies of the Wear Tracks

Figure 4 shows the SEM morphologies of the wear tracks for the HNSS under different conditions. It could be found that the surface was rough when 5 N was applied at 300 °C. The micrograph of the wear track showed a variety of superficial damages, such as grooves and stripping areas. The material near the grooves was subjected to severe plastic deformation and plastic damage, which subsequently formed the wear debris [31], which potentially acted as a third-body abrasive source, resulting in severe abrasive wear. When the load was increased to 15 N, the morphology of the wear track was as can be seen in Figure 4b. Some small particles and stripping areas could be found. Meanwhile, work hardening caused by dislocation accumulation occurred due to the increased load, resulting in an increase in hardness [5,32–35]. Thus, the wear resistance of the sample was better than that of the sample with 5 N applied at the same temperature.

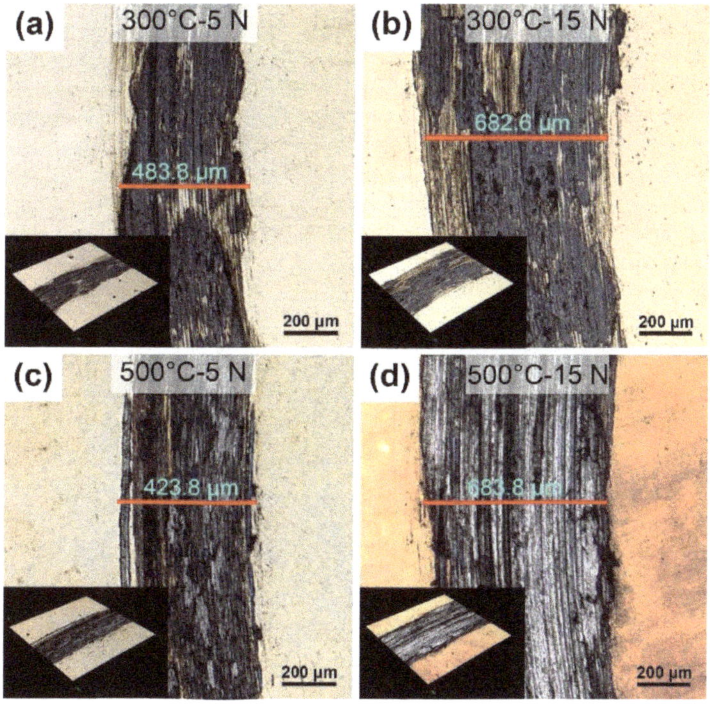

Figure 3. Morphologies of wear tracks for the HNSS under different conditions: (**a**) 300 °C–5 N, (**b**) 300 °C–15 N, (**c**) 500 °C–5 N, and (**d**) 500 °C–15 N.

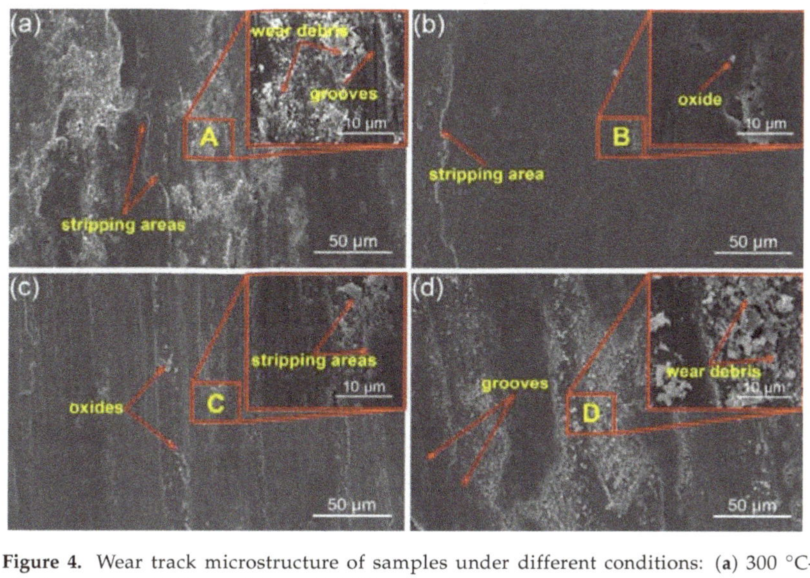

Figure 4. Wear track microstructure of samples under different conditions: (**a**) 300 °C–5 N, (**b**) 300 °C–15 N, (**c**) 500 °C–5 N, and (**d**) 500 °C–15 N.

Figure 4c shows the micrograph of the wear track when 5 N was applied at 500 °C. It is shown that the worn surface was relatively smooth. However, some shallow grooves along the sliding direction could still be seen, which was mainly caused by wear debris under dry-friction conditions. Local stripping areas could also be seen on the surface. Figure 4d shows the surface of the wear track when 15 N was applied at 500 °C; a large number of wear debris and some grooves could obviously be seen on the surface, which showed relatively serious wear behavior.

3.4. Microstructure Characteristics of the Wear Tracks

The SEM morphologies and corresponding distributions of elements on the worn surfaces of the HNSS are shown in Figure 5. It is obvious that the O element was distributed in the wear tracks of all conditions, indicating that the samples were oxidized at both 300 °C and 500 °C. It can be found from the EDS mapping that the oxide layers mainly consisted of the elements Fe, Cr, Mn, and O. Although there was no obvious change in the distribution of elements, a more serious discontinuous distribution could be observed on the oxide layer at 300 °C, which was consistent with the results of LSCM, indicating an oxidation–scratch–reoxidation mechanism in the process of friction [36]. It could also be found that oxidation was more severe with increasing temperature, and this was demonstrated by the distribution of the O element, which was distributed from being mainly present in the wear track at 300 °C to also being significantly distributed on the substrate at 500 °C.

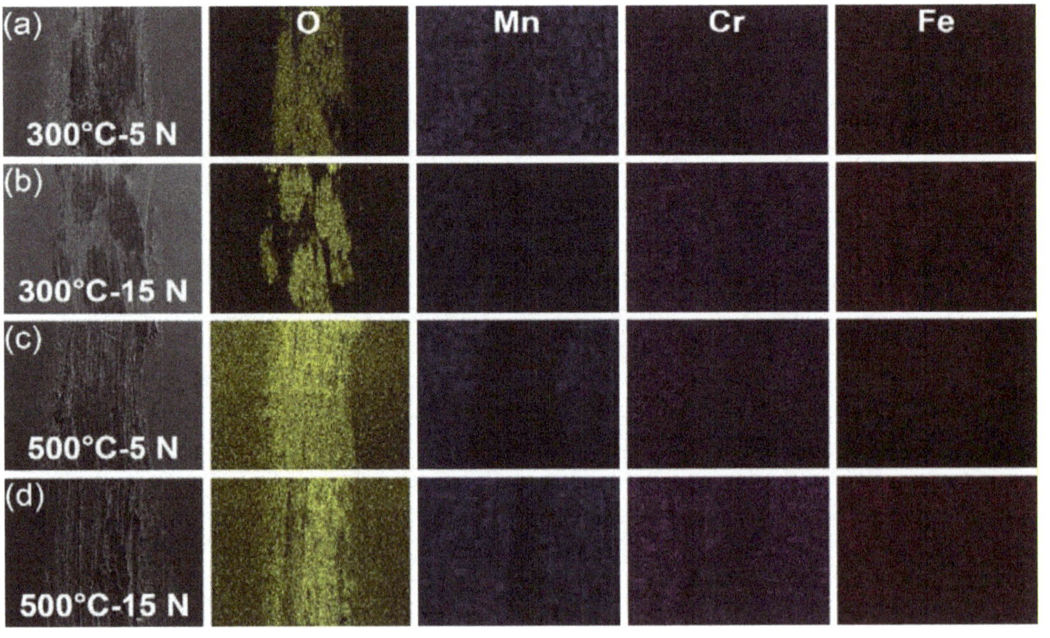

Figure 5. SEM morphologies of worn surfaces and corresponding element mapping: (**a**) 300 °C–5 N, (**b**) 300 °C–15 N, (**c**) 500 °C–5 N, and (**d**) 500 °C–15 N.

XPS was used to investigate the components of oxide layers on the worn surfaces under different conditions. Figure 6 presents the XPS spectra of the oxide layers. The peaks of Fe 2p, Mn 2p, Cr 2p, and O 1s were found in all the survey spectra. In addition, the intensity of Cr 2p dropped obviously when the temperature increased to 500 °C, while the intensity of Mn 2p showed a slight increase.

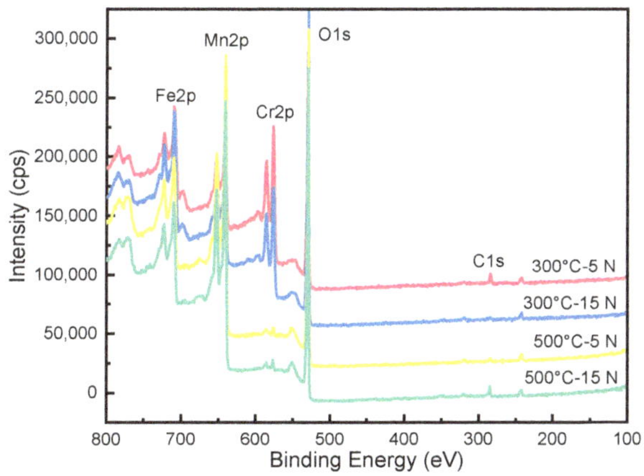

Figure 6. XPS spectra of oxide layers at wear tracks for the HNSS under different conditions.

The high-resolution XPS spectra of Fe 2p of the wear tracks under different conditions are shown in Figure 7. The element of Fe has two common oxidation states, ferrous (Fe^{2+}) and ferric (Fe^{3+}) [37], and the two oxidation states exist under all conditions. The binding energy (BE) of Fe^{3+} (2p3/2) in Fe_2O_3 and Fe^{2+} (2p3/2) in Fe_3O_4 was 711.6 eV and 709.2 eV, respectively [38,39]. The peak of metallic Fe^0 (2p3/2) was located at 706.6 eV [40]. It can be concluded from Figure 7 that Fe is mainly present in the above three forms under different conditions, with Fe_2O_3 and Fe_3O_4 in all cases. However, metallic Fe could only be detected at 300 °C, indicating more severe oxidation with increasing temperature, which is also consistent with the result in Figure 5.

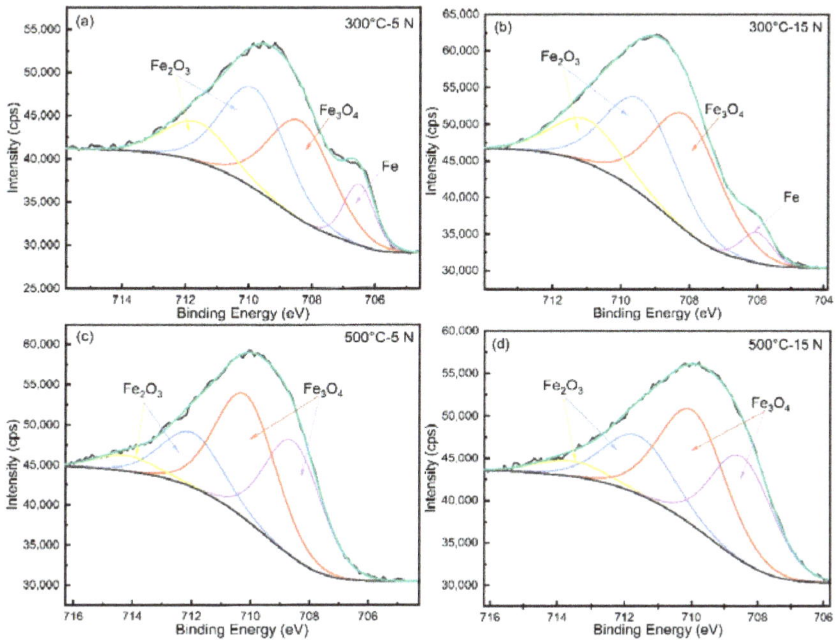

Figure 7. Spectra of Fe 2p of wear tracks under different conditions. (**a**) 300 °C–5 N, (**b**) 300 °C–15 N, (**c**) 500 °C–5 N, and (**d**) 500 °C–15 N.

The high-resolution XPS spectra of Cr 2p of the wear tracks under different conditions are shown in Figure 8. The BE of Cr^{3+} (2p3/2) in Cr_2O_3 was 576.4 eV [41], and a peak at ~2.6 eV lower than the Cr^{3+} (2p3/2) peak was observed, which was attributed to the BE of metallic Cr^0 [42,43]. $FeCr_2O_4$ with a BE of 577.9 eV could also be detected at 300 °C [38]. However, there only existed Cr_2O_3 on the wear tracks at 500 °C, as seen in Figure 8.

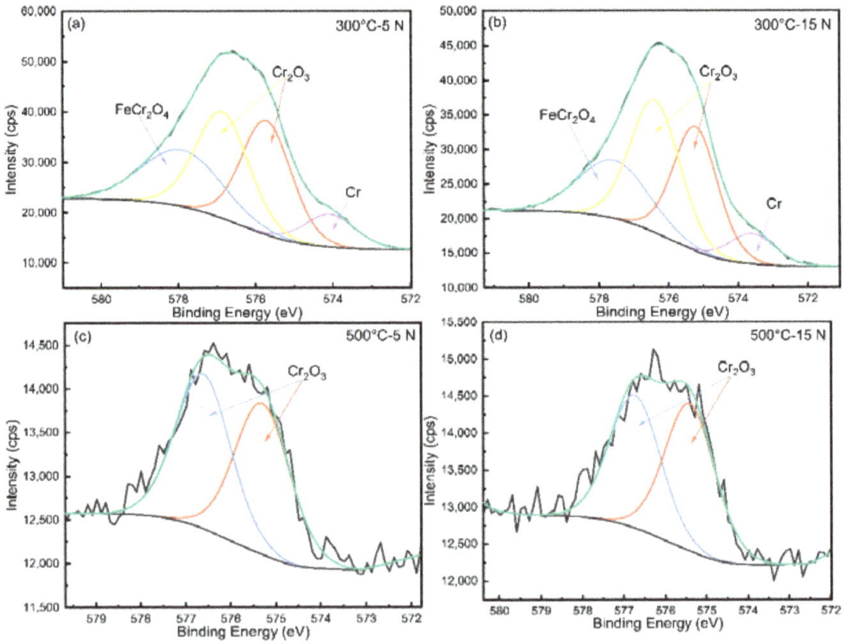

Figure 8. Spectra of Cr 2p of wear tracks under different conditions. (**a**) 300 °C–5 N, (**b**) 300 °C–15 N, (**c**) 500 °C–5 N, and (**d**) 500 °C–15 N.

The high-resolution XPS spectra of Mn 2p of the wear tracks under different conditions are shown in Figure 9. The binding energy (BE) of Mn^{3+} (2p3/2) in Mn_2O_3 and Mn^{2+} (2p3/2) in MnO was 641.9 eV and 643.2 eV, respectively [38,39]. The peak of metallic Mn^0 (2p3/2) was located at 639.6 eV [38]. In addition, the above three states of Mn were present under all four conditions.

Figure 10 shows the contents of Fe, Mn, and Cr oxides under different conditions based on the results of XPS. It could be observed that the composition of the oxides changed significantly with the change in the test conditions. When the temperature was 300 °C, the oxides of Cr were the most abundant among the oxides, while they decreased significantly when the temperature increased to 500 °C, which could also be proved by the survey spectra. In addition, the oxides content of Mn increased significantly with increasing temperature. Although the oxide content of Fe also increased at high temperatures, the increase was not as great as that of Mn.

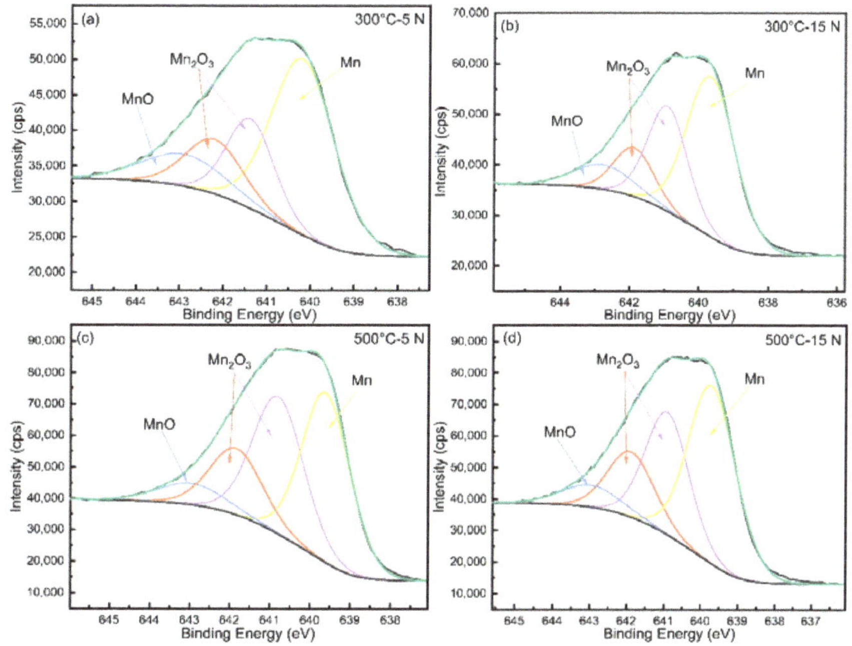

Figure 9. Mn 2p spectra of wear tracks under different conditions. (**a**) 300 °C–5 N, (**b**) 300 °C–15 N, (**c**) 500 °C–5 N, and (**d**) 500 °C–15 N.

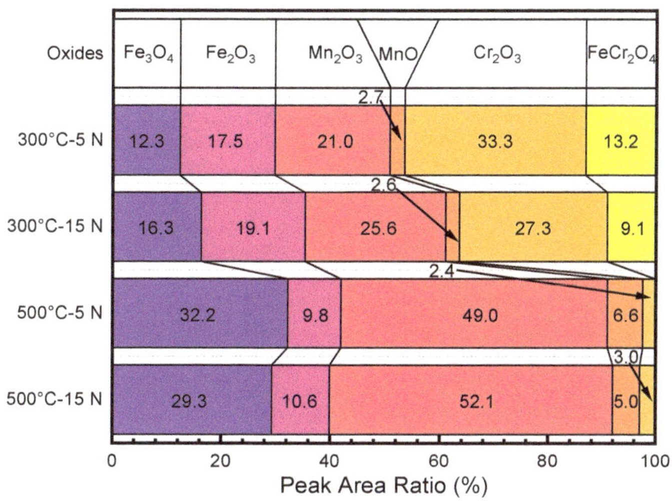

Figure 10. Contents of oxides under different conditions.

4. Discussion

The friction coefficient curves show some significant variations in trends and values under different conditions. Detailed studies of the wear tracks have shown that these changes are associated with the oxide layers formed on the wear surfaces. Oxidation wear is a common phenomenon in a high-temperature environment, characterized by the presence of an oxide layer on the sliding surface [44,45]. The surfaces of the samples are easily oxidized under high temperature due to the high affinity of Cr, Mn, and Fe to

O [23,24]. These oxides usually became wear particles due to the normal stress and shear force from the friction couple during the friction test, and the wear particles, which act as a third body, often plough the wear tracks as the sliding develops, resulting in a fluctuation of the friction coefficient and a relatively high value [16]. After the generation of debris particles, some are lost from between the surfaces, while most are retained and they are involved in the development of a compact oxide layer [23]. This layer can provide wear protection for long periods due to its high smooth and hardness, resulting in a decrease in the friction coefficient. The development of this layer from the agglomeration and compaction of the wear debris particles is described elsewhere [46].

As is seen, oxide layers were formed on the surfaces of all wear tracks (Figures 3 and 4), but the stability of the oxide layers was different under different conditions. In general, this compact oxide layer can form at a temperature higher than 150 °C [16]. In addition, the rate of establishment of this layer usually increases with the increase in temperature [17], which can explain the phenomenon that the friction coefficients at 500 °C were lower than those at 300 °C for the first 400 s. It can be seen from the results of the friction coefficient that there was no significant fluctuation after 400 s (Figure 1), indicating that the oxides formed on the wear track surface play a crucial role in the wear process. However, there existed a little difference in the friction coefficient trend after 400 s. Most of them increased slowly as the time increased, except for the curve of 300 °C–15 N, which maintained a relative constant at 0.38. After the establishment of the compact oxide layer, there were two competitive processes, breakdown and reconsolidation of this layer, indicating that the layer established at 300 °C with 15 N applied may be the most stable one.

When a load of 5 N was applied at 300 °C, an oxide layer consisting of Fe_2O_3, Fe_3O_4, Cr_2O_3, $FeCr_2O_4$, MnO, and Mn_2O_3 was formed on the wear track, which can be concluded from the results of EDS and XPS. It can be revealed that the content of Cr_2O_3 is higher than XPS (Figures 6 and 10), and it is usually used as the third body to grind the surface of the wear tracks due to its high hardness [24]. Thus, severe abrasive wear occurred at this time. In addition, part of the Cr_2O_3 can react easily with Fe to form $FeCr_2O_4$ at 300 °C [47], which has high hardness and can be easily removed from the surface during the process of friction, thus greatly reducing the protective ability of the oxide layer [37,48]. With the plowing of Cr_2O_3, the transformation of Cr_2O_3 to $FeCr_2O_4$ and the long-term shear action of the friction couple, the layer was destroyed, which resulted in the exposure of the matrix and the formation of a large amount of wear debris on the surface of the wear track (Figure 4a) [37]. The HNSS matrix was ploughed by hard debris under the extrusion of the friction couple, forming some grooves parallel to the sliding direction on the surface. Thus, the friction coefficient increased continuously and the average friction coefficient (0.44) was the highest under this condition with the wear mechanism of oxidation wear and severe abrasive wear [49].

The addition of the N element could further reduce the stacking fault energy of austenitic stainless steel, which increased the ability of work hardening. When 15 N was applied at 300 °C, the increased load facilitated plastic deformation of the HNSS, which led to an increase in the dislocation density and an increment in the dislocation, resulting in the effect of work hardening and the improvement of the hardness of the matrix [5]. A material with higher hardness usually shows a lower friction coefficient to that of a material with lower hardness [50–53]. The higher load also makes it easier to crush oxide particles, such as Cr_2O_3, which are usually squeezed into previously formed grooves, which in turn makes the formed compact layer denser and also enhances the load-bearing capacity [17,54]. During the period of 0~400 s of the test, the curve of the friction coefficient was almost coincident with that at 300 °C–5 N, but it became much more stable after 400 s due to the increased hardness and compact oxide layer (Figure 1). At this time, a significant reduction in the content of Cr_2O_3 and $FeCr_2O_4$ also occurred, and the grinding effect on the surface was reduced. However, a partial oxide layer may peel off and adhere to the surface of the friction couple during friction due to shear action and the accumulation of plastic deformation in the contact area. The peeling oxide layer was then gradually removed from the surface, and the exposed metal matrix under the stripping area was oxidized to form a new oxide layer (Figure 5b). The exfoliated oxidation products hindered direct contact between the friction couple and the matrix. The effects of increased hardness, the formation of a denser oxide layer, and the exfoliated oxidation products under this condition make the HNSS have better wear resistance than that of 300 °C–5 N, showing the wear mechanism of oxidation wear and mild adhesive wear and forming a stable friction state (Figure 1).

When a load of 5 N was applied at 500 °C, only a few oxide particles, shallow grooves, and sporadic stripping areas could be observed, as seen in Figure 4c. It could also be found that the content of Cr oxides was significantly reduced, and Cr_2O_3 had the effect of inhibiting the diffusion of O to the inside. Thus, a large amount of Fe and Mn oxides were formed when Cr_2O_3 was reduced. Among them, Mn has a stronger affinity with O compared to Fe, so the oxides of Mn, i.e., MnO and Mn_2O_3, are more abundant [23,24]. The latter, in particular, is generated in large quantities and is beneficial for lubrication during friction [55]. In addition, at 500 °C, Fe_2O_3 is easily transformed into Fe_3O_4, which has lower hardness than Fe_2O_3, and also plays a certain protective effect [24,56]. Additionally, the oxide layer is softened due to the high temperature, which can reduce friction to some extent [38,57,58]. Due to the accumulation of plastic deformation, the partial layer peeled off and adhered to the friction couple. Therefore, the friction process mainly occurred between the oxides and the friction couple, which could play a better role in tribological properties [59–61], and the average friction coefficient (0.34) was the smallest under this condition in all cases (Figure 2) [62,63].

When a load of 15 N was applied at 500 °C, the composition of the oxide layer did not change much compared to that at 500 °C–5 N, while the increased temperature softened the matrix, resulting in the weakening of work hardening [64]. In addition, the wear of the oxide layer was aggravated and the layer was destroyed due to the increased load, leading to large amounts of wear debris and some grooves (Figures 3d and 4d), showing typical abrasive wear. The increase in Mn content on the wear track of EDS indicated that the compacted oxide layer on the surface was slightly damaged, which led to an increase in the friction coefficient (0.40) [26], so the wear resistance weakened. However, the oxide layer still played a certain role in protection, so the wear resistance at this time was better than that at 300 °C with 5 N applied. The wear mechanism of the HNSS under this condition was mainly oxidation wear and severe abrasive wear.

5. Summary

The friction and wear behaviors of the Fe-19Cr-15Mn-0.66N HNSS at high temperature were tested under different conditions with different loads and temperatures, and the conclusions can be drawn as follows:

(1). The friction coefficient of HNSS decreased with time up to 400 s but rose up slowly or maintained a relatively stable trend as the time increased. The average friction coefficient decreased with increasing load at 300 °C, while it showed a reverse trend at 500 °C, which depended on the wear mechanism that reflects the competitive processes, including breakdown and reconsolidation of the oxide layer formed on wear tracks.

(2). At 300 °C, the friction coefficient curve of HNSS with 15 N applied showed a stable friction state after 400 s due to the formation of a denser oxide layer consisting of Cr_2O_3, Fe_2O_3, and Mn_2O_3 and increased hardness caused by work hardening, showing the wear mechanisms of oxidation wear and adhesive wear.

(3). Under the conditions of 500 °C–5 N, softening at high temperature and the production of large amounts of Mn_2O_3, which has a lubricating effect, minimized the friction of the sample, showing the lowest friction coefficient with a value of 0.34. In addition, the increased temperature also weakened the work hardening, resulting in severe abrasive wear at 500 °C–15 N.

Author Contributions: Conceptualization, Y.Q. and Z.Z.; methodology, J.C.; software, X.W.; validation, S.S., H.Z. and L.Y.; formal analysis, S.S.; investigation, H.Z.; data curation, S.S.; writing—original draft preparation, S.S., H.Z. and H.Y.; writing—review and editing, Y.Q. and Z.Z.; supervision, J.Z.; and funding acquisition, D.W., Z.L. and J.L. All authors have read and agreed to the published version of the manuscript.

Funding: This research received no external funding.

Institutional Review Board Statement: Not applicable.

Informed Consent Statement: Not applicable.

Data Availability Statement: The data used to support the findings of this study are available from the corresponding author upon request.

Acknowledgments: This work is funded by the National Natural Science Foundation of China (nos. 51905110 and 52005228), the Guangdong Province Key Area R&D Program (2020B010184001), and the Science Foundation of Jiangsu Province (BK 20180984).

Conflicts of Interest: The authors declare no conflict of interest.

References

1. Zhang, X.R.; Sun, S.C.; Sun, G.X.; Han, S.; Jiang, Z.H.; Lian, J.S. Nanoindentation creep deformation behaviour of high nitrogen nickel-free austenitic stainless steel. *Mater. Sci. Technol.* **2019**, *35*, 1592–1599. [CrossRef]
2. Li, J.G.; Li, H.; Peng, W.; Xiang, T.; Xu, Z.Y.; Yang, J.C. Effect of simulated welding thermal cycles on microstructure and mechanical properties of coarse-grain heat-affected zone of high nitrogen austenitic stainless steel. *Mater. Charact.* **2019**, *149*, 206–217. [CrossRef]
3. Vats, V.; Baskaran, T.; Arya, S.B. Tribo-corrosion study of nickel-free, high nitrogen and high manganese austenitic stainless steel. *Tribol. Int.* **2018**, *119*, 659–666. [CrossRef]
4. Qiao, Y.X.; Wang, X.Y.; Chen, J.; Yang, L.L.; Wang, X.J.; Zhou, H.L.; Zou, J.S. Electrochemical behavior and passive film composition of a high-nitrogen nickel-free austenitic stainless steel. *Arab. J. Sci. Eng.* **2021**, 1–8. [CrossRef]
5. Qiao, Y.X.; Sheng, S.L.; Zhang, L.M.; Chen, J.; Yang, L.L.; Zhou, H.L.; Wang, Y.X.; Li, H.B.; Zheng, Z.B. Friction and wear behaviors of a high nitrogen austenitic stainless steel Fe-19Cr-15Mn-0.66N. *J. Min. Metall. Sect. B Metall.* **2021**, *57*, 285–293. [CrossRef]
6. Zhao, H.; Ren, Y.; Dong, J.; Yang, K. The microstructure and tribological behavior of a pre-cold-deformed 0.90 % nitrogen containing stainless steel. *Mater. Werkst.* **2018**, *49*, 1439–1448. [CrossRef]
7. Lin, H.; Yang, M.S.; Shu, B.P. Fretting wear behaviour of high-nitrogen stainless bearing steel under lubrication condition. *J. Iron Steel Res. Int.* **2020**, *27*, 849–866. [CrossRef]
8. Chen, Z.X.; Hu, H.X.; Guo, X.M.; Zheng, Y.G. Effect of groove depth on the slurry erosion of V-shaped grooved surfaces. *Wear* **2021**, *488–489*, 204133. [CrossRef]
9. Qiao, Y.X.; Tian, Z.H.; Cai, X.; Chen, J.; Wang, Y.X.; Song, Q.N.; Li, H.B. Cavitation erosion behaviors of a nickel-free high-nitrogen stainless steel. *Tribol. Lett.* **2019**, *67*, 1. [CrossRef]
10. Qiao, Y.X.; Chen, J.; Zhou, H.L.; Wang, Y.X.; Song, Q.N.; Li, H.B.; Zheng, Z.B. Effect of solution treatment on cavitation erosion behavior of high-nitrogen austenitic stainless steel. *Wear* **2019**, *424–425*, 70–77. [CrossRef]
11. Qiao, Y.X.; Wang, X.Y.; Yang, L.L.; Wang, X.J.; Chen, J.; Wang, Z.B.; Zhou, H.L.; Zou, J.S.; Wang, F.H. Effect of aging treatment on microstructure and corrosion behavior of a Fe-18Cr-15Mn-0.66N stainless steel. *J. Mater. Sci. Technol.* **2021**. [CrossRef]
12. Li, J.; Yang, Y.X.; Ren, Y.B.; Dong, J.H.; Yang, K. Effect of cold deformation on corrosion fatigue behavior of nickel-free high nitrogen austenitic stainless steel for coronary stent application. *J. Mater. Sci. Technol.* **2018**, *34*, 660–665. [CrossRef]
13. Korashy, A.; Attia, H.; Thomson, V.; Oskooei, S. Fretting wear behavior of cobalt—Based superalloys at high temperature—A comparative study. *Tribol. Int.* **2020**, *145*, 106155. [CrossRef]
14. Lu, Y.; Zhu, S.Y.; Wang, X.Y.; Sui, X.D.; Zhang, S.T.; Hao, J.Y.; Zhao, T. High temperature tribological behavior of polymer-derived Ta4HfC5 nanoceramics. *Tribol. Int.* **2021**, *156*, 106859. [CrossRef]
15. Xue, Y.W.; Wu, C.H.; Shi, X.L.; Zhang, K.P.; Huang, Q.P. High temperature tribological behavior of textured CSS-42L bearing steel filled with Sn-Ag-Cu-Ti_3C_2. *Tribol. Int.* **2021**, *164*, 107205. [CrossRef]
16. Pauschitz, A.; Roy, M.; Franek, F. Mechanisms of sliding wear of metals and alloys at elevated temperatures. *Tribol. Int.* **2008**, *41*, 584–602. [CrossRef]
17. Stott, F.H. High-temperature sliding wear of metals. *Tribol. Int.* **2002**, *35*, 489–495. [CrossRef]
18. Eder, S.J.; Grutzmacher, P.G.; Ripoll, M.R.; Dini, D.; Gachot, C. Effect of temperature on the deformation behavior of copper nickel alloys under sliding. *Materials* **2020**, *14*, 60. [CrossRef]
19. Torres, H.; Varga, M.; Ripoll, M.R. High temperature hardness of steels and iron-based alloys. *Mater. Sci. Eng. A-Struct. Mater. Prop. Microstruct. Process.* **2016**, *671*, 170–181. [CrossRef]
20. Li, L.; He, K.; Sun, S.Y.; Yang, W.Z.; Yue, Z.F.; Wan, H. High-temperature friction and wear features of nickel-based single crystal superalloy. *Tribol. Lett.* **2020**, *68*, 26. [CrossRef]

21. Pei, Y.C.; Xia, D.X.; Wang, S.R.; Cong, L.; Wang, X.L.; Wang, D.Y. Effects of temperature on the tribological properties of NM600 under sliding wear. *Materials* **2019**, *12*, 4009. [CrossRef] [PubMed]
22. Kumar, H.; Ramakrishnan, V.; Albert, S.K.; Meikandamurthy, C.; Tata, B.V.R.; Bhaduri, A.K. High temperature wear and friction behaviour of 15Cr–15Ni–2Mo titanium-modified austenitic stainless steel in liquid sodium. *Wear* **2010**, *270*, 1–4. [CrossRef]
23. Wang, L.; Tieu, A.K.; Wang, J.; Sang, P.T.; Xia, C.Y.; Zhu, H.T.; Deng, G.Y. High load capability, sticking scale inhabitation and promising lubrication of sodium carbonate coating for steel/steel contact at high temperature. *Tribol. Int.* **2021**, *153*, 106594. [CrossRef]
24. Cheng, X.W.; Jiang, Z.Y.; Kosasih, B.; Wu, H.; Luo, S.Z.; Jiang, L.Z. Influence of Cr-Rich oxide scale on sliding wear mechanism of ferritic stainless steel at high temperature. *Tribol. Lett.* **2016**, *63*, 28. [CrossRef]
25. Wang, M.J.; Wang, Y.X.; Liu, H.; Wang, J.Q.; Yan, F.Y. Interrelated effects of temperature and load on fretting behavior of SAF 2507 super duplex stainless steel. *Tribol. Int.* **2019**, *136*, 140–147. [CrossRef]
26. Torres, H.; Varga, M.; Adam, K.; Ripoll, M.R. The role of load on wear mechanisms in high temperature sliding contacts. *Wear* **2016**, *364–365*, 73–83. [CrossRef]
27. Stott, F.H.; Jordan, M.P. The effects of load and substrate hardness on the development and maintenance of wear-protective layers during sliding at elevated temperatures. *Wear* **2001**, *250*, 391–400. [CrossRef]
28. Razali, M.K.; Kim, S.W.; Irani, M.; Kim, M.C.; Joun, M.S. Practical quantification of the effects of flow stress, friction, microstructural properties, and the tribological environment on macro- and micro-structure formation during hot forging. *Tribol. Int.* **2021**, *164*, 107226. [CrossRef]
29. Lee, S.; Lee, J.; Joun, M. On critical surface strain during hot forging of lubricated aluminum alloy. *Tribol. Int.* **2020**, *141*, 105855. [CrossRef]
30. Cao, Y.J.; Sun, J.Q.; Ma, F.; Chen, Y.Y.; Cheng, X.Z.; Gao, X.; Xie, K. Effect of the microstructure and residual stress on tribological behavior of induction hardened GCr15 steel. *Tribol. Int.* **2017**, *115*, 108–115. [CrossRef]
31. Berns, H.; Koch, S. Influence of abrasive particles on wear mechanism and wear resistance in sliding abrasion tests at elevated temperatures. *Wear* **1999**, *233–235*, 424–430. [CrossRef]
32. Kumar, S.S.S.; Raghu, T.; Bhattacharjee, P.P.; Rao, G.A.; Borah, U. Work hardening characteristics and microstructural evolution during hot deformation of a nickel superalloy at moderate strain rates. *J. Alloy Compd.* **2017**, *709*, 394–409. [CrossRef]
33. Ball, A. On the Importance of work hardenning in the design of wear-resistant materials. *Wear* **1983**, *91*, 201–207. [CrossRef]
34. Llewellyn, D.T. Work hardening effects in austenitic stainless steels. *Mater. Sci. Technol.* **2013**, *13*, 389–400. [CrossRef]
35. Wang, D.P.; Zhang, H.T.; Guo, P.Y.; Sun, B.A.; Wang, Y.X. Nanoscale periodic distribution of energy dissipation at the shear band plane in a Zr-based metallic glass. *Scr. Mater.* **2021**, *197*, 4. [CrossRef]
36. Chen, Y.J.; Wang, S.H.; Hao, Y.; Pu, J.B.; Jiang, X.; Huang, L.F.; Wang, L.P. Friction and wear behavior of CrN coating on 316L stainless steel in liquid sodium at elevated temperature. *Tribol. Int.* **2020**, *143*, 106079. [CrossRef]
37. Xu, Y.W.; Jing, H.Y.; Xu, L.Y.; Han, Y.D.; Zhao, L. Effect of overload on the oxidation behavior of CF8A austenitic stainless steel in a high-temperature water environment. *Corros. Sci.* **2020**, *162*, 108219. [CrossRef]
38. Biesinger, M.C.; Payne, B.P.; Grosvenor, A.P.; Lau, L.W.M.; Gerson, A.R.; Smart, R.S.C. Resolving surface chemical states in XPS analysis of first row transition metals, oxides and hydroxides: Cr, Mn, Fe, Co and Ni. *Appl. Surf. Sci.* **2011**, *257*, 2717–2730. [CrossRef]
39. Mills, P.; Sullivan, J.L. A study of the core level electrons in iron and its three oxides by means of X-ray photoelectron spectroscopy. *J. Phys. D Appl. Phys.* **1983**, *16*, 723–732. [CrossRef]
40. Duhamel, C.; Sennour, M.; Georgi, F.; Guerre, C.; Chaumun, E.; Crépin, J.; Héripré, E.; Curières, I.D. Characterization of oxide scales formed on alloy 82 in nominal PWR Primary water at 340 °C and in hydrogenated steam at 400 °C. *Corros. Sci.* **2018**, *131*, 386–403. [CrossRef]
41. Agostinelli, E.; Battistoni, C.; Fiorani, D.; Mattogno, G.; Nogues, M. An XPS study of the electronic structure of the ZnxCd1-xCr2X4 (X = S, Se) spinel system. *J. Phys. Chem. Solids* **1989**, *50*, 269–272. [CrossRef]
42. Allen, G.C.; Tucker, P.M. Multiplet splitting of X-Ray photoelectron lines of chromium complexes. the effect of covalency on the 2p core level spin-orbit separation. *Inorg. Chim. Acta* **1976**, *16*, 41–45. [CrossRef]
43. Wang, D.P.; Shen, J.W.; Chen, Z.; Chen, F.G.; Guo, P.Y.; Geng, Y.X.; Wang, Y.X. Relationship of corrosion behavior between single-phase equiatomic CoCrNi, CoCrNiFe, CoCrNiFeMn alloys and their constituents in NaCl solution. *Acta Metall. Sin. (Engl. Lett.)* **2021**, 1–11. [CrossRef]
44. Quinn, T.F.J.; Sullivan, J.L.; Rowson, D.M. Origins and development of oxidational ambient temperatures wear at low ambient temperatures. *Wear* **1984**, *94*, 175–191. [CrossRef]
45. Stott, F.H. The role of oxidation in the wear of alloys. *Tribol. Int.* **1998**, *31*, 61–71. [CrossRef]
46. Jiang, J.; Stott, F.H.; Stack, M.M. The role of triboparticulates in dry sliding wear. *Tribol. Int.* **1998**, *31*, 245–256. [CrossRef]
47. Young, D.J. *High Temperature Oxidation and Corrosion of Metals*; Elsevier Ltd.: Amsterdam, The Netherlands, 2008.
48. Jin, X.Y.; Zhang, Y.F.; Chen, L.; Yu, J.H.; Xue, W.B. Preparation and tribological behaviors of DLC/spinel composite film on 304 stainless steel formed by cathodic plasma electrolytic oxidation. *Surf. Coat. Technol.* **2018**, *338*, 38–44. [CrossRef]
49. Lai, Z.W.; Wu, J.L.; Wen, Y.H.; Li, N.; Wang, Z.C. Study of a new type austenitic stainless steel with better wear resistance. *Tribology* **2009**, *29*, 128–133.

50. Song, Y.L.; Yu, C.; Miao, X.; Han, X.H.; Qian, D.S.; Chen, X. Tribological performance improvement of bearing steel GCr15 by an Alternating magnetic treatment. *Acta Metall. Sin. (Engl. Lett.)* **2017**, *30*, 957–964. [CrossRef]
51. Choi, B.; Cho, I.S.; Jung, D.H.; Lee, M.G.; Jeon, Y. Friction and wear behavior of direct metal deposition on SUH3. *Arch. Metall. Mater.* **2019**, *64*, 841–844.
52. Sourani, F.; Enayati, M.H.; Taghipour, M. High-temperature oxidation and wear behavior of (Fe,Cr)Al intermetallic compound and (Fe,Cr)Al-Al_2O_3 nanocomposites. *J. Mater. Eng. Perform.* **2021**, *30*, 3654–3669. [CrossRef]
53. Song, Q.N.; Tong, Y.; Li, H.L.; Zhang, H.N.; Xu, N.; Zhang, G.Y.; Bao, Y.F.; Liu, W.; Liu, Z.G.; Qiao, Y.X. Corrosion and cavitation erosion resistance enhancement of cast Ni-Al bronze by laser surface melting. *J. Iron Steel Res. Int.* **2021**. [CrossRef]
54. Blau, P.J.; Brummett, T.M. High-temperature oxide regrowth on mechanically damaged surfaces. *Tribol. Lett.* **2008**, *32*, 153–157. [CrossRef]
55. Jin, B.; Chen, G.Y.; Zhao, J.; He, Y.Y.; Huang, Y.Y.; Luo, J.B. Improvement of the lubrication properties of grease with Mn_3O_4/graphene (Mn_3O_4#G) nanocomposite additive. *Friction* **2021**, *9*, 1361–1377.
56. Qu, J.; Blau, P.J.; Jolly, B.C. Tribological properties of stainless steels treated by colossal carbon supersaturation. *Wear* **2007**, *263*, 719–726. [CrossRef]
57. Chandra-Ambhorn, S.; Chueaprakha, S.; Siripongsakul, T. High-temperature oxidation of the dissimilar weld between AISI 304L and Fe-15.6Cr-8.5Mn using Ar and Ar-4%N2 shielding gas. *Anti-Corros. Method. Mater.* **2019**, *66*, 210–214. [CrossRef]
58. Li, L.L.; Wang, Z.B.; He, S.Y.; Zheng, Y.G. Correlation between depassivation and repassivation processes determined by single particle impingement: Its crucial role in the phenomenon of critical flow velocity for erosion-corrosion. *J. Mater. Sci. Technol.* **2021**, *89*, 158–166. [CrossRef]
59. Heras, E.D.L.; Santamaría, D.G.; García-Luis, A.; Cabo, A.; Brizuela, M.; Ybarra, G.; Mingolo, N.; Brühl, S.; Corengia, P. Microstructure and wear behavior of DC-pulsed plasma nitrided AISI 316L austenitic stainless steel. *Plasma Processe. Polym.* **2007**, *4*, S741–S745. [CrossRef]
60. Devaraju, A.; Perumal, A.E.; Alphonsa, J.; Kailas, S.V.; Venugopal, S. Sliding wear behavior of plasma nitrided austenitic stainless steel type AISI 316LN in the temperature range from 25 to 400 °C at 10^{-4} bar. *Wear* **2012**, *288*, 17–26. [CrossRef]
61. Duarte, M.C.S.; Godoy, C.; Wilson, J.A.B. Analysis of sliding wear tests of plasma processed AISI 316L steel. *Surf. Coat. Technol.* **2014**, *260*, 316–325. [CrossRef]
62. Wilson, S.; Alpas, A.T. Effect of temperature on the sliding wear performance of A1 alloys and A1 matrix composites. *Wear* **1996**, *196*, 270–278. [CrossRef]
63. Pearson, S.R.; Shipway, P.H.; Abere, J.O.; Hewitt, R.A.A. The effect of temperature on wear and friction of a high strength steel in fretting. *Wear* **2013**, *303*, 622–631. [CrossRef]
64. Mitchell, D.R.G.; Stott, F.H. The friction and wear of thin titanium nitride and silicon nitride coatings on stainless steel at temperatures to 500 °C. *Surf. Coat. Technol.* **1992**, *50*, 151–160. [CrossRef]

coatings

Article

Numerical Simulation of the Effect of Freeze–Thaw Cycles on the Durability of Concrete in a Salt Frost Environment

Hao Li [1,*], Yuan Zhang [2] and Haolong Guo [1]

1 Water Conservancy and Civil Engineering College, Inner Mongolia Agricultural University, Huhhot 010018, China; guohl@emails.imau.edu.cn
2 School of Economics and Management, Inner Mongolia Technical College of Construction, Huhhot 010070, China; 20020100127@imaa.edu.cn
* Correspondence: hao.li@imau.edu.cn

Abstract: In order to improve the accuracy of the analysis of the impact of freeze–thaw cycle on concrete durability in a salt freezing environment, the numerical simulation of the impact of the freeze–thaw cycle on concrete durability in a salt freezing erosion environment is studied in this paper. Firstly, considering the influence of axial force and bending moment on the relationship between bending moment and curvature, a concrete fiber beam column model is established. Then, according to the joint influence of temperature field, stress field and seepage field on concrete in the process of freezing and thawing, the control differential equation of the freezing and thawing cycle is established. The freeze–thaw damage section is divided, the non-uniform distribution of freeze–thaw damage is determined, and the division of the freeze–thaw damage section is completed. According to the linear relationship between freeze–thaw damage degree, relative dynamic elastic modulus, freeze–thaw cycle times and position variables, the durability of concrete is numerically simulated, and the attenuation law of bond strength at different section depths after freeze–thaw is determined. The results show that the temperature curve simulated by the design method is consistent with the actually measured temperature curve, which can better reduce the temperature change of the inner core of the test block during freezing and thawing, and the relative dynamic elastic modulus is in good agreement with the actual value, which can prove that the method in this paper has certain practical application value. It is expected to provide some reference for solving the durability problem of concrete in a salt frost erosion environment and the optimal design of concrete structures.

Keywords: salt freezing; erosion; freeze thaw cycle; concrete; durability; numerical simulation

Citation: Li, H.; Zhang, Y.; Guo, H. Numerical Simulation of the Effect of Freeze–Thaw Cycles on the Durability of Concrete in a Salt Frost Environment. *Coatings* **2021**, *11*, 1198. https://doi.org/10.3390/coatings11101198

Academic Editor: Paolo Castaldo

Received: 1 September 2021
Accepted: 21 September 2021
Published: 30 September 2021

1. Introduction

Various causes have led concrete structures to suffer from destruction, collapse and other material durability problems in the past 40 or 50 years, seriously affecting the normal benefits and safe use of these structures [1]. Many engineering structures have to end their service ahead of time, and countries all over the world have paid a huge price for this. In practical engineering, the durability failure of or damage to concrete structures are common, causing huge economic losses [2]. Over a long period, under the combined action of salt erosion and freeze–thaw and other environmental factors, the damage to concrete structures in irrigation areas is very serious, and a considerable number of concrete structures lose their functions prematurely before reaching the designed lifetime, which has a big impact on the safety and benefits of water projects. A salt frost environment is an environment where crystals such as frost precipitate on the concrete surface, making it more prone to freeze–thaw. The mechanism of freezing and thawing is that the water in the pores of the soil and its matrix metal or rock expands in the freezing process, resulting in the continuous increase in cracks and the overall fragmentation of soil or rock. After ablation, its anti-corrosion stability is greatly reduced. Under the action of gravity, the rock and soil begin to move downward along the slope. In cold and dry areas with lakeside sedimentary rocks as the main foundation conditions, the sulfate and chloride ions abundant in groundwater and soil become the main media of erosion damage to concrete buildings. When the concrete structure is in a working environment with freeze–thaw in winter, it will suffer from the combined action and linkage damage of two or more elements. Concrete buildings suffer from the combined action of salt erosion and freeze–thaw over a long period. The damage and performance deterioration of concrete structures in these areas are serious, and a considerable number of concrete structures are damaged prematurely before reaching the designed lifetime. At present, in order to alleviate freeze–thaw damage, scholars have proposed various mitigation methods, such as adding creto material, changing the mix proportion parameters and the connection type of foundation parts, and using freeze–thaw- and corrosion-resistant materials [3–7].

Although these mitigation methods have been proposed, in order to improve the pertinence of mitigation methods, it is necessary to analyze the impact of the melting cycle on concrete durability. Therefore, some scholars have studied this. For example, Wang et al. examined a durability analysis method of shotcrete lining under the coupling action of nitric acid erosion and the freeze–thaw cycle. Taking a long highway tunnel with a shotcrete single-layer permanent lining in a cold area as the engineering background, the pore structure of concrete under the combined action of nitric acid erosion and the freeze–thaw cycle was characterized and analyzed by means of the straight-line conductor method [8]. Peng et al. predicted and analyzed the corrosion life of reinforced concrete structures in a salt freeze–thaw environment. Based on Fick's second law, a chloride ion transport model considering time, temperature, humidity, chloride ion binding and freeze–thaw damage effects in concrete was established to study the durability and life of concrete [9]. Tian, Y. G also studied the durability of highway C50 high performance concrete, combined with the rich local industrial wastes (fly ash and mineral powder), and prepared C50 concrete with excellent performance, so as to study the influence of mineral admixtures on the durability of concrete [10]. Yang et al. conducted experimental research on the freeze–thaw damage depth of concrete in a deicing salt environment. By comparing the change law of concrete's appearance, the detection results of ultrasonic equipment and the penetration of concrete contact surface, the damage depth of concrete under different freeze–thaw cycles were obtained. It was concluded that after 20 freeze–thaw cycles, the freeze–thaw damage of concrete begins to expand into the interior of concrete. In order to alleviate the impact of freeze–thaw, relevant measures can be put forward according to the above research results [11]; in order to alleviate the impact of freeze–thaw, Huang, J. B and others also analyzed the freeze–thaw deterioration and damage characteristics of existing cracked concrete, used the prefabricated crack method to simulate the concrete with cracks in actual projects, studied the damage deterioration process and mechanical characteristics of cracked concrete under the conditions of the freeze–thaw cycle, and obtained the calculation formula of the peak stress of the prefabricated crack specimen and the number of freeze–thaw cycles. Relevant personnel can take necessary mitigation measures to solve relevant freeze–thaw problems according to the calculated formula [12]. Although the above scholars have studied this, there is a problem of a poor numerical analysis effect. In order to solve the above problems and improve the accuracy of the impact analysis of the freeze–thaw cycle on concrete durability in a salt freezing environment, this paper puts forward a new research method, namely the following innovative research:

(1) In the analysis process, considering the influence of the axial force and bending moment on the relationship between bending moment and curvature, a concrete fiber beam column model is established to divide the section of concrete members into several discrete small elements and improve the characteristics of different small elements of concrete;
(2) According to the joint influence of temperature field, stress field and seepage field on concrete in the process of freeze–thaw, the control differential equation of freeze–thaw

cycle is established and substituted into the concrete fibers at different positions of the section, so as to analyze the fiber section in the process of freeze–thaw damage evolution;

(3) Based on the coupling algorithm, the freeze–thaw damage section is divided, the non-uniform distribution of freeze–thaw damage is determined, and the division of freeze–thaw damage section is completed;

(4) According to the linear relationship between the freeze–thaw damage degree, the relative dynamic elastic modulus, number of freeze–thaw cycles and location variables, the durability of concrete is numerically simulated. On this basis, the attenuation law of bond strength at different section depths after freeze–thaw is determined.

Through the above four innovations, this paper completes the numerical simulation of the effect of freeze–thaw cycle on concrete durability in a salt frost environment.

2. Numerical Simulation of the Effect of Freeze–Thaw Cycles on the Durability of Concrete in Salt Frost Environment

2.1. Establishment of Concrete Fiber Beam Column Model

The fiber model divides the cross section of the concrete member into a certain number of discrete small elements, and the mechanical properties of each small element are expressed by the axial stress–strain relationship of the steel bar and concrete. Furthermore, the lateral displacement of the column can be obtained along the longitudinal direction of the member, which has high accuracy. The fiber model is established in OPENSEES Windows version 3.2.2 (as shown in Figure 1), and the element type is based on a nonlinear beam column element.

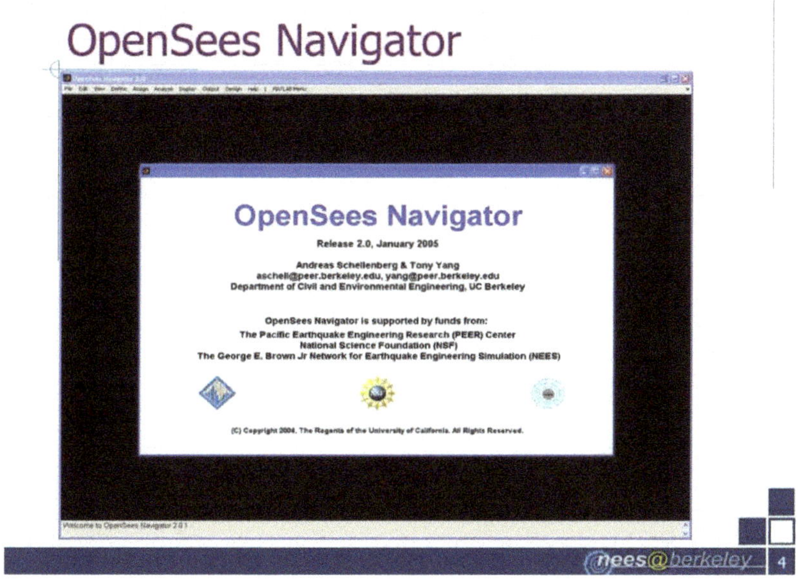

Figure 1. OPENSEES interface.

In the displacement-based model, the element displacement field is expressed as the function of the node displacement field, which is the approximation of the real structural element displacement field. In order to obtain a good approximation result, it is necessary to divide a member into several elements. In order to determine the number of fibers needed for cross-section analysis, sensitivity analysis is carried out, and 50 × 50 fibers are obtained, which can basically achieve the balance. The concrete model of OPENSEES is selected as Concrete0l. The model is based on the uniaxial constitutive relation of concrete, and does not consider the tensile strength of concrete. The relationship is determined by the Karsan-Jirsa unloading rule [13]. The parameters include the axial compressive strength and corresponding strain, the residual strength and corresponding strain. The formula for calculating the axial strength and peak strain of unfrozen concrete is as follows:

$$\begin{cases} w = 0.76p \\ u = \frac{w}{2500\sqrt{w'}} \end{cases} \quad (1)$$

In Formula (1), w is the axial strength of concrete; p is the residual strength of concrete; u is the peak strain; and w' is the peak intensity. The constitutive model of concrete in the core area is expressed in sections. The first part selects the Concrete04 model in OPENSEES. The parameters needed to be defined in the model are: the axial compressive strength and corresponding strain of concrete, the elastic modulus of concrete and the ultimate compressive strain of concrete. For restrained concrete, the value of the elastic modulus is the same as that of unconstrained concrete. Considering the influence of stirrups on the strength and ductility of concrete in the core area, the compressive strength and peak strain of unconstrained concrete are modified to obtain the corresponding value of constrained concrete. The second part is the Minmax material in OPENSEES, which can be used to obtain the strain threshold. In this paper, the third part is set as the Mander constitutive model failing when the concrete strain exceeds the ultimate compressive strain, and the stress–strain relationship enters the linear degradation [14]. The Steel02 model in OPENSEES is selected as the constitutive model of reinforcement, which can consider the isotropic strain hardening phenomenon and Bauschinger effect of reinforcement, and the calculation efficiency of the model is high, which is in good agreement with the results of the repeated loading test of reinforcement. The parameters to be defined in the Steel02 model mainly include: the yield strength, initial elastic modulus, strain hardening rate, the control parameter of curve curvature from the elastic transition to the plastic phase, the isotropic hardening parameter and initial stress value. Because the freeze–thaw cycle mainly affects the mechanical properties of concrete, the same selection method is adopted for the constitutive model of steel bars of freeze–thaw and non-freeze–thaw specimens.

2.2. Governing Differential Equations of Freeze Thaw Cycles

In this section, the relative dynamic elastic modulus of concrete is used as the physical damage index to measure the degree of freeze–thaw damage. The relative dynamic elastic modulus refers to the ratio of stress and strain of an object under dynamic load. It is generally measured by a dynamic resistance strain gauge and an acoustic instrument. Based on the material property test data in the existing research, the uneven distribution and development law of damage are established, and then the damage evolution law of concrete mechanical properties with the change is formed. Thus, under the condition of given cycles, different constitutive parameters can be given to the fibers at different positions of the member section, and the non-uniformity of freeze–thaw damage can be considered. The freeze–thaw process of concrete is affected by the joint action of the temperature field, stress field and seepage field. Firstly, it is controlled by the temperature field. The freezing and thawing cycle means that the temperature decreases and increases continuously. The second is the effect of the stress field, which is due to the pore pressure caused by the freezing of water in the pores. Finally, it is also affected by the seepage field, which is due to the influence of water migration and phase transformation in pores. In order to describe the influence of the three fields, there are respective governing equations. In the process of freeze–thaw cycle, the water in the pores of porous materials will produce the transformation of the liquid phase and the gas phase. The heat conduction equation of water phase's latent heat is as follows [15]:

$$mc\frac{T}{t} = \nabla(\alpha\nabla T) + \lambda\frac{p}{t} \quad (2)$$

In Formula (2), m is the density of the system; c is the specific heat capacity of the system; T is temperature; t represents time; p is the thermal conductivity of the system; and α is the thermal conductivity of the system. In the porous system, the relationship between effective stress and strain is as follows [16]:

$$\chi = B\gamma - \left(1 - \frac{\varphi_1}{\varphi_2}\right)\eta E \tag{3}$$

In Formula (3), χ is the stress; B is the stiffness matrix; γ is the total strain; φ_1 and φ_2 are the elastic modulus of the system and aggregate; η is the average pore pressure; and E is the identity matrix. In the absence of external load, there is an equilibrium differential equation of mechanical field. The migration obeys Darcy's law, and the water in the pores also conforms to the law of mass conservation in the process of liquid and solid phase changes. According to the constitutive equation of each phase, the Darcy seepage field equation in a porous media system can be deduced as follows [17]:

$$S = \nabla\left(\frac{\theta}{\nu}\nabla S\right) + V - \beta\delta \tag{4}$$

In Formula (4), S represents the permeation field of the system; θ is the permeability coefficient; ν is the dynamic viscosity coefficient of water; V is the expansion volume of the system; β is porosity; and δ is volume strain. In the process of failure, the temperature, stress and seepage field are coupled with each other, jointly promoting the damage. Therefore, the failure process of concrete in the freeze–thaw cycle can be described by the above-mentioned differential equations. When the number of freeze–thaw cycles does not reach a certain value, the change in the relative dynamic elastic modulus is small. Therefore, the critical value of freeze–thaw cycles is proposed to characterize this trend, and the critical value increases linearly with section depth. When the critical number is exceeded, the relative dynamic elastic modulus decreases linearly with the increase in the number of cycles, and the degradation rate is similar at different section depths. Therefore, a consistent degradation rate of the relative dynamic elastic modulus at different positions of the cross section is proposed to characterize this trend. On this basis, the relative dynamic elastic modulus of concrete is used as the bridge between the constitutive parameter eigenvalue of concrete before and after freeze–thaw and the section depth and freeze–thaw cycles. For the fiber section, the above-mentioned modified freeze–thaw concrete constitutive equation can be substituted into the concrete fibers at different positions of the section, so as to form the fiber section which can consider the evolution process of freeze–thaw damage.

2.3. Division of Freeze–Thaw Damage Section Based on Coupling Algorithm

During the freeze–thaw cycle, the freeze–thaw damage on the cross section of the specimen is not evenly distributed, but gradually decreases with the increase in the distance from the freeze–thaw surface. If the non-uniform distribution of freeze–thaw damage is ignored and the degree of freeze–thaw damage is assumed to be consistent in the whole section of the specimen, there will be large errors in the calculation of bearing capacity and deformation of the specimen in the process of simulation analysis. Therefore, it is very important to the damage distribution of the specimen that we ensure the accuracy of the analysis results. In this paper, the freeze–thaw damage section is divided based on the coupling algorithm. In this coupling algorithm, the whole computational domain is discretized into LBM grids by means of the partition alternating solution. Among them, the grid occupied by structure is marked as domain, and the rest of the grid is marked as watershed. The fluid movement is solved by the fluid solver, and the displacement of the structure is solved by the structure solver. At each time step, the algorithm solves the flow field and the structure domain in turn, and exchanges the force and displacement information of the fluid structure coupling surface. After obtaining the structure displacement, it is input into the fluid solver to determine the position of the fluid structure coupling surface to update the basin grid. In the whole solution time domain, the structure coupling problem can be solved alternately by performing the process. According to the variation of the relative dynamic elastic modulus at different distances from contact surface, the relationship between the critical influence depth and the number of cycles is proposed as follows:

$$h_o = 11.7\left(\frac{M}{2.5}\right)^{0.9} \tag{5}$$

In Formula (5), h_o is the critical depth of freeze–thaw—that is, the maximum depth of the cross-section of the specimen affected by the freeze–thaw cycle—and M is the number of freeze–thaw cycles corresponding to the freeze–thaw cycle test. Based on the damage equivalence principle, the relative dynamic elastic modulus is selected as the quantitative index of freeze–thaw damage. Using the linear interpolation method, 80 freeze–thaw cycles of NAC, RAC50 and RAC100 can be obtained under the non-standard freeze–thaw test, which are equivalent to 200 cycles of the standard freeze–thaw test, respectively. NAC, RAC50 and rac100 refer to different types of concrete. NAC is small particle size asphalt concrete and RAC50 and RAC100 refer to recycled concrete with a 50% and 100% replacement rate of coarse aggregate. Based on this, combined with Formula (5), the critical influence depth of freeze–thaw of concrete specimen in this paper can be obtained:

$$h_o = 0.15\sqrt[0.9]{\frac{n_1 M'}{n_2}} \tag{6}$$

In Formula (6), n_1 and n_2 represent the number of freeze–thaw cycles under the standard and non-standard freeze–thaw tests, and M' represents the number of freeze–thaw cycles corresponding to the non-standard freeze–thaw test in this paper. After determining the depth of the freeze–thaw effect, it is necessary to further estimate the freeze–thaw damage distribution on the section. After the freeze–thaw cycle, the relative dynamic elastic modulus of the specimen section basically obeys the linear distribution. In order to simplify the analysis, it is assumed that the freeze–thaw damage degree of the section decreases linearly from the outside to the inside, the outermost edge of the section is the relative dynamic elastic modulus of the prism in the freeze–thaw test, and the innermost edge of the freeze–thaw influence depth in the section is 1 (the freeze–thaw damage degree is zero). Therefore, the constitutive relationship of a recycled concrete pier at different positions of the section can be determined according to the values of the relative dynamic elastic modulus and replacement rate of recycled coarse aggregate, and then the section division of the specimen can be completed. In this paper, the concrete layer is divided into 15 mm grids.

2.4. Numerical Simulation of Concrete Durability

According to the linear relationship between the relative dynamic elastic modulus of damage degree and location variables, the durability is numerically simulated to determine the degradation law of bond strength at different section depths after freeze–thaw. The critical number of freeze–thaw cycles is used to represent the number of freeze–thaw cycles required for concrete to begin to experience freeze–thaw damage at a certain section depth. On this basis, the degradation law of bond strength at different section depths after freeze–thaw can be established only by establishing the change in relative bond strength with the freeze–thaw damage index. The relative bond strength and the corresponding freeze–thaw damage degree data obtained from the pull-out test of freeze–thaw specimens are analyzed, and these data have a linear regression. Considering the boundary conditions (damage index = 1, relative bond strength = 1 without freeze–thaw damage), the regression formula coefficient is obtained according to the least square method, and the damage model of the freeze–thaw damage index with respect to cycle number and location variable is obtained. The concrete expression is as follows:

$$I = \begin{cases} 1, & K \leq 1.05h - 0.23 \\ 1 - 0.0112[h - (1.05h - 0.23)], & K > 1.05h - 0.23 \end{cases} \tag{7}$$

In Formula (7), I is the freeze–thaw damage index; K is the number of cycles; h is the position variable of section depth. The relationship between bond strength and position variables can be expressed as follows:

$$Y = \begin{cases} 1, & K \leq 1.05h - 0.23 \\ 1.05 - 0.0104[h - (1.05h - 0.23)], & K > 1.05h - 0.23 \end{cases} \tag{8}$$

In Formula (8), Y is the freeze–thaw bond strength. The relative dynamic elastic modulus of each mix proportion model decreases continuously. Additionally, with the increase in time, the amount of ice increases gradually, resulting in greater frost heaving pressure. The porosity inside the concrete increases, and micro-cracks are gradually generated, which eventually leads to the connection between the pores, and micro cracks gradually increase, causing greater damage to the concrete [18]. In this vicious circle, the freeze–thaw damage of concrete is more serious, which is reflected in the dynamic modulus, and the value of dynamic modulus decreases faster. According to the established bond strength model of freeze–thaw damage, within a certain range of anchorage length, the bond stress after freeze–thaw has a linear relationship with the position variable. In order to facilitate the subsequent derivation, firstly, the coordinate system is established according to the boundary conditions of the anchorage end of the steel bar; that is, the stress and strain of the steel bar are zero, and the required stress penetration length after freeze–thaw damage can be obtained. This length indicates that the bond stress degenerates only within this depth range [19]. In the elastic stage, the model still satisfies the force balance condition, and the bond stress is trapezoidal in the range of stress penetration length. Under the condition of given reinforcement stress, the unique solution regarding the length of seepage stress is obtained. At this time, the stress increment at both ends of the reinforcement in the micro section has the surrounding bond stress balance. When entering the area not affected by freezing and thawing, the average stress calculation method is used to obtain the corresponding slip. In the post yield stage, the position coordinates of the steel bar are obtained according to the equilibrium equation. The total slip can be divided into yield slip and new slip after the steel bar enters the plastic stage. In conclusion, regardless of whether the reinforcement yield exists or not, under the condition of given reinforcement stress, the slip value can be calculated by solving the stress penetration length, and so it can be directly applied to the analysis at the component level.

3. Experiment

3.1. Experimental Preparation

First, raw materials of NAC natural aggregate concrete, RAC50 recycled concrete and RAC100, mainly including cement, coarse aggregate, fine aggregate, water and the mixture, were prepared and combined into a uniform mixture as required to obtain high-strength concrete performance. Relevant standards for the mix proportion design of ordinary concrete (JGJ55–2011) were followed in the experiment [20].

In this paper, the numerical simulation of concrete under salt frost erosion environment was studied. In order to verify the effect of this method, the test results were compared with the temperature shock field test results. The temperature impact field experiment was carried out with reference to the actual environment of the concrete structure. Material parameters are shown in Table 1.

Table 1. Material parameters.

Parameter	Value	Parameter	Value
Conductivity (W/M·K)	1.29	Specific heat (J/kg·°C)	0.2
Density (kg/m^3)	2.5	Expansion angle	40
Dynamic modulus of elasticity (Pa)	2.486	Eccentricity	0.15
Poisson's ratio	0.2	Strain loading rate	e^{-2}
Expand (1/°C)	e^{-5}	Viscosity parameter	e^{-5}

The grid type was a linear order three-dimensional stress type—that is, eight nodes in a linear hexahedral element c3d8r—which was calculated by means of the reduced integral and hourglass control method. Because there were many temperature points in the test block and the amount of temperature sampling and measurement was large, the key points of the structure were selected for temperature curve measurement. The center of the concrete corresponded to the temperature sensor. The key points of the selected structure in this experiment were located at 50 mm from the lower surface, 50 mm from the sample's center and 50 mm from the upper surface, marked as A, B and C, respectively. Based on the above tests, the temperature impact field test was carried out, and the numerical simulation was carried out by using the method in this paper.

3.2. Experimental Analysis of Temperature Shock Field

Firstly, the temperature impact field test was carried out on the concrete specimen. The temperature measurement results of the key points of each structure are shown in Table 2.

Table 2. Temperature measurement results of key points of the specimen structure.

Time (min)	Point A Temperature (°C)	Point B Temperature (°C)	Point C Temperature (°C)
0	12.26	13.65	12.66
100	−22.55	−23.57	−22.55
200	20.88	21.82	20.87
300	−9.62	−9.26	−10.24
400	−18.29	−19.32	−19.32
500	23.55	24.95	24.23
600	−15.88	−15.28	−15.55
700	−24.26	−24.55	−24.78
800	18.33	19.12	18.52
900	−17.62	−18.43	−18.24

According to the measured results in Table 2, the temperature fluctuation curve could be obtained. Because A, B and C were on the outer surface of the structure, the peak point was the largest, close to ±25 °C. The measured temperatures of the three key points were processed to obtain the average measured temperature curve. Compared with the measured temperature curve, the experimental results are shown in Figure 2.

As can be seen from Figure 2, compared with the other three methods, the temperature curve of the temperature impact field obtained by numerical simulation in this method was similar to the temperature curve measured by the actual test block, and had higher accuracy. Because the heat transfer process of concrete is time-dependent, it takes a period of time from the outer surface to the center point, so the temperature change of the inner center point and the time taken to reach the peak value are slightly delayed. In the process of heat conduction, with the continuous loss of heat, the temperature reaching the temperature sensor is far lower than the temperature of the outer surface. In the numerical simulation, the concrete material was set to be uniform, and the temperature curve obtained by heat conduction was consistent with the set curve. However, the internal material of the test block used in the test process was uneven. The addition of aggregates such as the superplasticizer increases the thermal conductivity of the test block, and the temperature curve lags behind relatively. This method can obtain the structure's center's temperature through numerical simulation, so as to better reduce the temperature change of the core in the test block during freezing and thawing. Therefore, the numerical simulation method proposed in this paper can simulate the influence of the temperature field on concrete durability, and has certain feasibility.

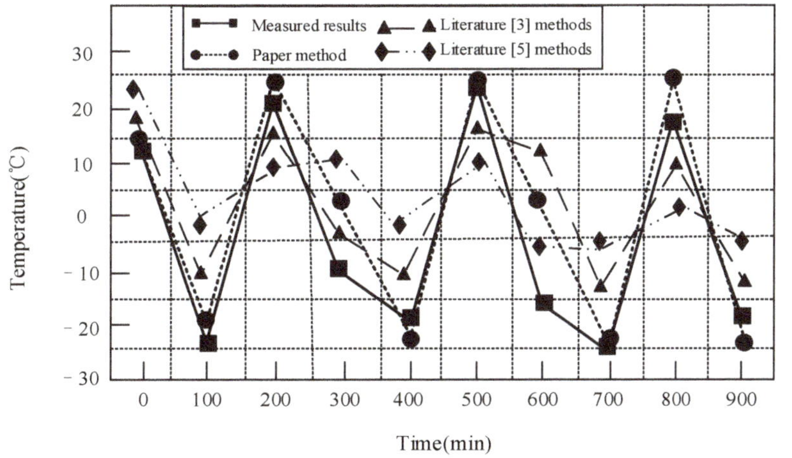

Figure 2. Experimental comparison results.

3.3. *Experimental Analysis of Relative Dynamic Elastic Modulus*

Then, the relative dynamic elastic modulus of concrete was tested, which was used as the physical damage index to measure the degree of freeze–thaw damage. The relative dynamic elastic modulus of different concrete types under different freeze–thaw cycles is shown in Table 3.

Table 3. Relative dynamic elastic modulus of different concrete specimens under different freeze–thaw cycles.

Freeze Thaw Cycle Times	NAC (%)	RAC50 (%)	RAC100 (%)
0	100	100	100
100	70.72	91.53	97.04
200	69.38	86.91	94.4
300	61	79.56	90.42
400	–	75.03	87.47
500	–	60	84.17
600	–	54.61	82.8

According to the data in Table 3, under the action of 300 freeze–thaw cycles of NAC concrete, the dynamic elastic modulus of the specimen decreases to 61.3% of the initial dynamic elastic modulus, and under the action of 310 freeze–thaw cycles, the dynamic elastic modulus of the specimen reaches 60% of the initial dynamic elastic modulus; under 500 freeze–thaw cycles, the dynamic elastic modulus of RAC50 concrete is 60% of the initial dynamic elastic modulus; after 600 freeze–thaw cycles, the dynamic elastic modulus of RAC100 concrete is always greater than 60% of the initial dynamic elastic modulus. Comparing the three numerical simulation methods with the actual values in Table 2, the fitting degree between the three numerical simulation methods and the actual values can be obtained, as shown in Figures 3–5.

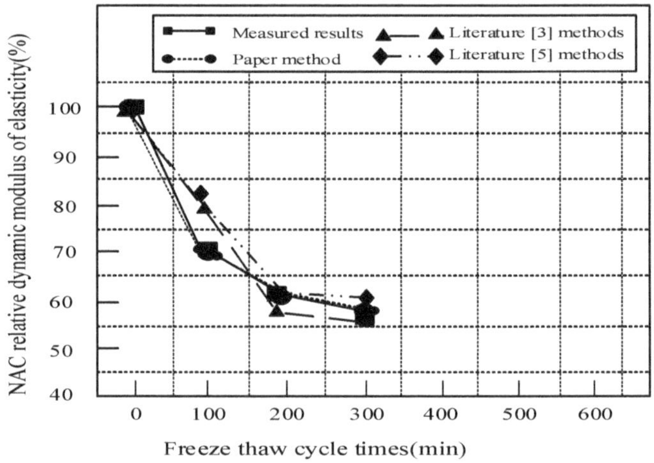

Figure 3. Relative dynamic elastic modulus of NAC specimens with different methods.

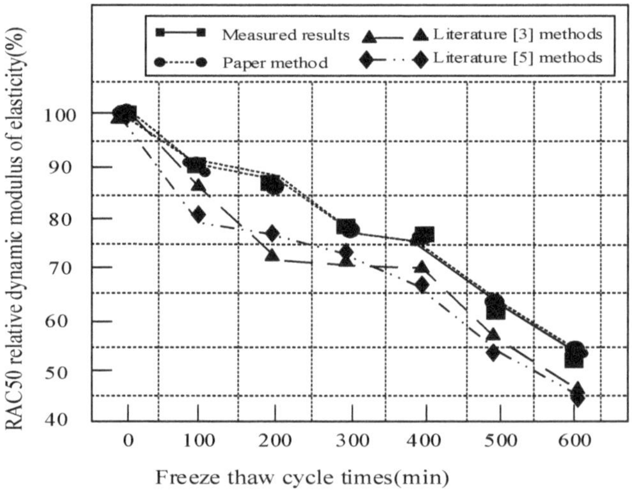

Figure 4. Relative dynamic elastic modulus of RAC50 specimens with different methods.

From Figure 2 to Figure 4, compared with the other two numerical simulation methods, the design method in this paper has a higher fitting degree with the actual value, and the maximum error is no more than 3%. The NAC sample can reach 60% within 310 min because of the characteristics of the NAC sample. Therefore, it can be seen that the design method in this paper can improve the accuracy of the analysis of the impact of the freeze–thaw cycle on concrete durability in a salt frost environment, and has a certain practical application value.

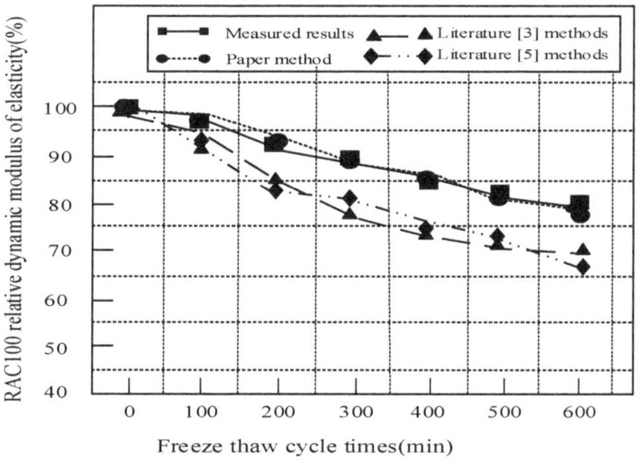

Figure 5. Relative dynamic elastic modulus of RAC100 specimens with different methods.

4. Conclusions

In order to improve the accuracy of the analysis of the influence of the freeze–thaw cycle on the durability of concrete in a salt freezing environment, the influence of the freeze–thaw cycle on the durability of concrete in a salt freezing erosion environment is numerically simulated in this paper. Firstly, considering the influence of the axial force and bending moment on the relationship between the bending moment and curvature, a concrete fiber beam column model is established. Then, according to the joint influence of the temperature field, stress field and seepage field on concrete in the process of freeze–thaw, the governing differential equation of the freeze–thaw cycle is established. Based on the coupling algorithm, the freeze–thaw damage section is divided, the non-uniform distribution of freeze–thaw damage is determined, and the division of the freeze–thaw damage section is completed. According to the linear relationship between freeze–thaw damage degree, relative dynamic elastic modulus, freeze–thaw cycle times and position variables, the durability of concrete is numerically simulated, and the attenuation law of bond strength at different section depths after freeze–thaw is determined. The results show that the temperature curve obtained by numerical simulation is consistent with the temperature curve measured by the actual test block, which can better reduce the temperature change of the inner core during freezing and thawing, the relative dynamic elastic modulus is in good agreement with the actual value, and the maximum error is no more than 3%. This provides a theoretical basis for solving the durability problem of concrete in a salt frost erosion environment and the optimal design of concrete structure. In order to better improve the accuracy of the impact analysis of the freeze–thaw cycle on concrete durability in a salt freezing environment, this paper will also improve the damage evaluation index of concrete under the influence of the freeze–thaw cycle in future research, so as to make the research more in-depth.

Author Contributions: Conceptualization, H.L. and H.G.; methodology, Y.Z.; validation, H.L.; formal analysis, H.G.; investigation, Y.Z.; resources, H.G.; data curation, Y.Z.; writing–original draft preparation, H.L.; writing–review and editing, H.L.; supervision, H.G.; project administration, H.L.; funding acquisition, H.L., H.G., Y.Z. All authors have read and agreed to the published version of the manuscript.

Funding: This research was funded by the Study on the Effect and Mechanism of Freeze–thaw Cycle on Concrete Durability in Saline Environment [NJZZ18044]; the Study on the Effect and Mechanism of Dry-Wet Freeze–thaw Cycle on Concrete in Saline Environment in Wuliangsuhai Area (BJ2015-2) and the Research on Concrete Durability in Saline Environment in Cold and Arid Regions of Inner Mongolia.

Institutional Review Board Statement: Not applicable.

Informed Consent Statement: Not applicable.

Data Availability Statement: Not applicable.

Acknowledgments: Thank you for all editors and reviewers.

Conflicts of Interest: The authors declare no conflict of interest.

References

1. Zhang, Y. Seismic performance prediction simulation of bridge steel-concrete composite structure. *Comput. Simul.* **2020**, *37*, 480–483.
2. Ji, Y.; Zhang, J.; Song, Q.; Zhu, H.; Shang, C. Effect of Nano-SiO_2 on concrete performance in bittern corrosion environment. *Bull. Chin. Ceram. Soc.* **2019**, *38*, 1425–1432.
3. Jiang, Z.W.; Li, L.J.; Zhang, Z.Y.; Yan, F.; He, P.; Niu, L. Analysis and application improvement of concrete freeze-thaw failure. *Jiangsu Build. Mater.* **2020**, *174*, 34–36.
4. Huang, J.C.; Yang, K.K.; Hou, C.Y.; Ren, Z.J. Application of creto material in freeze-thaw damage repair of diversion channel lining slab concrete of Wannianzha pump station. *Haihe Water Resour.* **2020**, *224*, 66–67, 72.
5. Fang, X.W.; Yao, R.F.; Yu, F.; Yu, J.W.; Yan, N.X. Effect of mix proportion parameters on concrete sulfate freeze-thaw damage. *Int. J. Hydroelectr. Energy* **2020**, *233*, 109, 129–132.
6. Liu, N.; Chen, X.F.; Zhang, K. Cause analysis and countermeasures of freeze-thaw failure of thin-walled concrete members. *Zhihuai* **2019**, *8*, 42–43.
7. Hou, Z.G. Repair and treatment of freeze-thaw damage of pier in tailrace of unit 5 of Wanjiazhai Water Control Project. *Hydropower Autom. Dam Monit.* **2019**, *5*, 93–96.
8. Wang, J.B.; Niu, D.T. Study on the durability performance of shotcrete lining under coupling effect of nitric acid attack and freeze-thaw cycles (Part I): Deterioration law of performance and air-void structure. *Mater. Rev.* **2019**, *33*, 1340–1347.
9. Peng, Y.Z.; Gao, J.; Xu, G.; Li, X. Prediction for corrosion life of reinforced concrete structures under salt freeze-thaw environment. *Adv. Sci. Technol. Water Resour.* **2019**, *39*, 44–49.
10. Tian, Y.G.; Lu, D.; Wang, S.F.; Jia, K.; Wang, Z. Study on durability of C50 High performance concrete in the highway from Tongchuan to Huangling. *Constr. Technol.* **2019**, *48*, 20–24.
11. Yang, X.M.; Sun, G.J. Experimental study on freeze-thaw damage depth of concrete in deicing salt environment. *J. Nat. Disaster* **2020**, *29*, 52–59.
12. Huang, J.B.; Ning, B.K.; Li, M.S.; Xia, X. Analysis of freeze-thaw deterioration and damage characteristics of existing cracked concrete. *Ind. Constr.* **2020**, *564*, 146–151.
13. Uur, B.; Toklu, H.Z. Effect of multi-cycle freeze-thaw tests on the physico-mechanical and thermal properties of some highly porous natural stones. *Bull. Eng. Geol. Environ.* **2020**, *79*, 255–267. [CrossRef]
14. Xiaolai, Y.U.; Tian, R.; Xing, L.; Liu, Y. Modification of constrained concrete mechanics calculation model based on mander model. *J. Yangzhou Univ. Nat. Sci. Ed.* **2018**, *21*, 61–65.
15. Truong, G.T.; Son, M.K.; Choi, K.K. Mechanical performance and durability of latex-modified fiber-reinforced concrete. *J. Adv. Concr. Technol.* **2019**, *17*, 79–92. [CrossRef]
16. Brostow, W.; Chetuya, N.; Gencel, O.; Hong, H.J.; Menard, N.; Sayana, S. Durability of portland concrete containing polymeric fillers and fly ash. *Mater. Sci.* **2019**, *26*, 103–108. [CrossRef]
17. Kandasami, S. Performance-based specification of concrete for durability: Carbonation-induced corrosion. *Indian Concr. J.* **2019**, *4*, 61–79.
18. Raveesha, P.; Prakash, K.E.; Babu, B. Impact of salinity in Netravali estuary sand on durability of reinforced cement concrete. *Indian Concr. J.* **2019**, *93*, 27–33.
19. Feo, L.; Ascione, F.; Penna, R.; Lau, D.; Lamberti, M. An experimental investigation on freezing and thawing durability of high performance fiber reinforced concrete (HPFRC). *Compos. Struct.* **2019**, *234*, 111673. [CrossRef]
20. Ding, W.; Leng, L.; Wei, Q.; Zhang, X.; Zhou, Y.; Tian, G.; He, X.; Ji, X.; Wang, J. Introduction to code for mix proportion design of ordinary concrete (JGJ55–2011). *Concr. World* **2011**, *12*, 76–79.

coatings MDPI

Article

Wear and Corrosion Resistance of CoCrFeNiSiMoW Medium-Entropy Alloy Coatings on Q235 Steel

Qingxian Hu *, Xiaoli Wang *, Xinwang Shen, Fanglian Fu and Zemin Tan

School of Materials Science and Engineering, Jiangsu University of Science and Technology, Zhenjiang 212003, China; 1440602222@stu.just.edu.cn (X.S.); flfu1997@stu.just.edu.cn (F.F.); 199060046@stu.just.edu.cn (Z.T.)
* Correspondence: huqingxian@just.edu.cn (Q.H.); xlwang@just.edu.cn (X.W.)

Abstract: CoCrFeNiSiMoW medium-entropy alloy coatings (MEACs) were fabricated by plasma-arc surfacing welding on Q235 steel. The microstructures and mechanical properties of CoCrFeNiSiMoW MEACs were studied. CoCrFeNiSiMoW MEACs are made from a mixture of NiCrCoMo cubic (FCC) solid solution phase, (Fe, Ni), Mo1.24Ni0.76, and CoCx phases by XRD analysis. The average hardness values of the one- and two-layer CoCrFeNiSiMoW MEACs obtained were 186 ± 1.56 and 198 ± 1.78 HV, respectively. Compared with the one-layer CoCrFeNiSiMoW coating, the two-layer coating has a better wear performance due to its higher hardness. Its corrosion resistance is better because of its higher Ni content.

Keywords: CoCrFeNiSiMoW; medium-entropy alloy coatings; microstructure; wear

Citation: Hu, Q.; Wang, X.; Shen, X.; Fu, F.; Tan, Z. Wear and Corrosion Resistance of CoCrFeNiSiMoW Medium-Entropy Alloy Coatings on Q235 Steel. *Coatings* **2021**, *11*, 1053. https://doi.org/10.3390/coatings11091053

Academic Editors: Ludmila B. Boinovich and Diego Martinez-Martinez

Received: 7 July 2021
Accepted: 28 August 2021
Published: 31 August 2021

1. Introduction

Every year, the economic losses caused by the wear and corrosion of metal materials are high. At the same time, the wear and corrosion of metal materials also affect the quality of industrial products, which contribute to a poor market image for enterprises, and will inevitably bring challenges to the realization of economic benefits for enterprises. It is particularly important to prepare wear- and corrosion-resistant coatings on the surface of the material. Due to the increasing demand of modern industry for wear- and corrosion-resistant coatings, researchers continue to explore new coating materials.

High- and medium-entropy alloys (HEAs and MEAs, respectively) possess a list of excellent properties, such as excellent strength and hardness [1,2], outstanding wear performance [3], and good corrosion resistance [4,5]. When HEA and MEA were initially proposed, they were considered to be composed of at least five elements. After an in-depth study of HEAs and MEAs, some quaternary alloys not equal to or close to equal atomic percentages, ranging from 5% to 35%, were also determined to be considered HEAs and MEAs [6]. It provided the new idea to fabricate alloy coatings with good wear and corrosion resistance due to their high content of multiple corrosion-resistant elements [7]. Many types of research about high- and medium-entropy alloy coatings (HEACs and MEACs, respectively) have been conducted. Feng et al. [8] researched the corrosion properties of laser cladding CrCoNi MEACs and found they had better corrosion resistance than SS304. Zhao et al. [9] studied the fabrication and friction properties of an AlFeCrCo MEAC on a magnesium alloy using a resistance seam-welding method. The results showed that the AlFeCrCo MEAC has a better wear performance than the substrate. Tian et al. [10] synthesized a CrMnFeCoNi HEAC with a thickness of 180 μm on Q235 steel using a mechanical alloying method and found it had better corrosion resistance than Q235 steel in a 3.5% NaCl solution. A Al2CrFeNiMox HEAC was fabricated by laser cladding, and its wear resistance was improved because of the properties of Mo [11]. Xu et al. [12] fabricated a CoCrFeNiTiMo HEAC and studied its corrosion resistance. Lloyd et al. [13] studied the effects of the elements Cr, Mo, and W, on the passive film of Ni-Cr-Mo (W) alloys in acidic solutions. Their results showed that high Cr-alloys form thicker oxides with a core-shell structure consisting of an inner Cr-Ni oxide layer and an outer Mo/W oxide in a 0.1 M H_2SO_4 + 1.0 M NaCl solution. Some studies have shown that Mo can also alter mechanical properties and the corrosion resistance of HEAs because it can change the morphology, the segregation behavior of elements, and the phase structure [14,15]. In addition, Cr can improve the wear and corrosion resistance of materials, and Ni can help improve the corrosion performance in acidic solutions.

The methods of fabricating HEACs include laser cladding [16], magnetron sputtering [17], plasma spraying [18], and more. Laser cladding technology is a common technique to fabricate wear-resistant coatings. In the process of laser cladding, the heat source is a laser, and the cladding material is a powder material. The cladding efficiency is relatively low and the price of power is high. Compared to laser cladding, plasma-arc surfacing technology uses welding wire as a cladding material and a plasma arc as a heat source, which has a relatively higher efficiency and is a more economical preparation method of cladding coating. In this process, the dilution effect of the base metal affects the mechanical properties of coatings. The Fe element from the base metal can change the proportion of elements in coatings so that the HEA/MEA coatings can be obtained by making full use of the dilution and selecting the appropriate welding wire. To study the effects of this change on the microstructure and properties, one- and two-layer coatings were prepared.

Hastelloy X (HX) alloy contains Cr, Ni, and Mo, and it was selected to fabricate HEACs/MEACs by plasma-arc surfacing welding in this study. HX has been widely studied in high temperature conditions because it has extraordinary properties, such as formability, oxidation resistance, and mechanical properties in the temperature range of 540–1000 °C [19–21]. However, the literature on fabricating HEACs/MEACs with HX and about its wear and corrosion resistance at room temperature is plentiful.

In this work, one- and two-layer CoCrFeNiSiMoW MEACs were fabricated with HX wire by plasma-arc surfacing welding, and its wear and corrosion resistance were studied.

2. Experimental Details

2.1. CoCrFeNiSiMoW Coating Fabrication

Q235 steel was taken as the base metal in this study. Before plasma-arc surfacing welding, the surface of Q235 was ground with 200# sandpaper and cleaned with acetone. The HX wire was used as filling wire to fabricate the coatings. The main elemental compositions of Q235 and HX wire are shown in Table 1.

Table 1. The main elemental compositions of Q235 steel and HX wire (wt.%).

Elements	Mn	Si	W	Co	Mo	Cr	Fe	Ni
Q235	0.3	0.15	/	/	/	/	Bal.	/
HX	0.247	0.264	0.201	0.875	9.16	22.34	18.15	Bal.

A Trans Tig5000 Series Digital automatic welding machine (Fronius, Pettenbach, Austria) was adopted to fabricate the coatings. Figure 1 shows a schematic diagram of plasma-arc surfacing welding. First, the non-arc power supply causes the non-arc between the tungsten electrode and the nozzle. Once the non-arc is produced, the plasma gas between the tungsten electrode and the nozzle causes the formation of the path. When the main arc power supply is turned on, the arc will start between the base metal and the tungsten electrode. The main arc is then generated, also known as the transfer arc. Because the current of the main arc is generally large, the energy density of the main arc is also relatively high, which can form a molten pool on the surface of the base metal. Welding wire enters the plasma flame, is melted by the plasma flame, and is deposited on the base metal to form a coating. Protective gas protects the plasma flame from oxidation of the molten pool by oxygen during surfacing welding through a protective gas channel. The direction of the arc movement is the direction of deposition. Due to the influence of dilution, different surfacing layers may affect the performance of the coatings. Two coatings were deposited by HX wire: one was HX1 coating with one layer of HX; the other was HX2 coating with two layers of HX. Table 2 shows the surfacing welding process parameters. Argon, with a purity of 99.99%, was used as a protective gas throughout these processes.

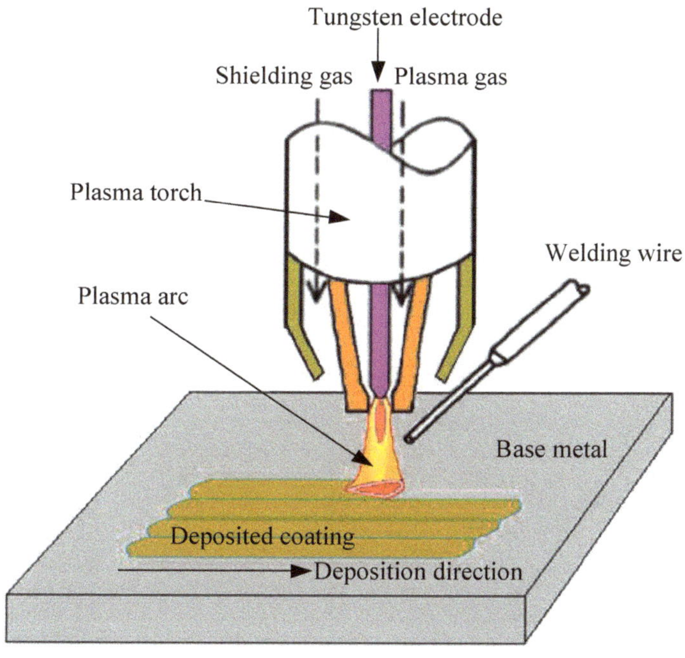

Figure 1. A schematic diagram of plasma-arc surfacing welding.

Table 2. The parameters of the plasma-arc welding process.

Parameters	Welding Voltage (V)	Welding Current (A)	Welding Speed (cm/min)	Welding Torch Height (mm)	The Plasma Gas Flow (L/min)
Value	20.1	120	24	7	2.5

2.2. *Microstructure Observation and Hardness Measurement*

Square coatings with dimensions of 10 × 10 mm^2 were cut and polished for the XRD test, microstructure observation, and hardness measurement.

The main chemical compositions of the HX1 and HX2 coatings were tested by a direct-reading spectrograph with a PAD5500II (Shimadzu, Kyoto, Japan). The crystal structures of CoCrFeNiSiMoW coatings were tested by X-ray diffractometer (XRD, Bruker D8 Advance, Karlsruhe, Germany). The diffraction angle (2θ) ranged from 20° to 90°. The scan rate was 5° per minute with a step size of 0.02°. The microstructures of the coatings were characterized by an optical microscope (Nikon Epiphot 300 model, Tokyo, Japan) and a scanning electron microscope (SU-70 model, Tokyo, Japan) after the coatings were mechanically polished and etched in a 3 wt % HNO_3 solution.

A DHV-1000 hardness tester (Shanghai Shangcai Tester Machine Co., LTD., Shanghai, China) was used to measure the surface hardness of the HX1 and HX2 coatings every 0.25 mm with a test load of 500 g and a dwell time of 15 s. These experiments were repeated five times. The test result was the average value of five experiments. The ranges between plus and minus were obtained by calculating the standard deviation of five experimental data. The horizontal axis was the distance, and the vertical axis was the hardness value.

2.3. Wear Measurements

Square coatings with dimensions of 30 × 30 mm^2 were cut and polished for a wear resistance test. A dry reciprocating sliding wear test was carried out by the HSR-2M friction tester (Lanzhou Institute of Chemical Physics, Lanzhou, China) at room temperature. A Si_3N_4 ceramic ball was used as the friction couple, and the working disk was coated in HX1 and HX2. The Si_3N_4 ball was reciprocated with a stroke length of 5 mm at a sliding speed of 25 mm/s for a duration of 30 min. The loads were chosen to be 20, 30, and 40 N. During the friction testing, the data were transmitted to a computer by the sensor on the loading rod, and the corresponding friction coefficient was calculated. The friction tests were repeated at least five times. The test results are the average of five experiments. After the friction tests, images of wear scars were observed by a laser scanning confocal microscope (LEXT, OLS400 LSCM, Tokyo, Japan) and a scanning electron microscope (SEM, JEOL, JSM-6480, Tokyo, Japan). The wear rates ($mm^3\ N^{-1}{\cdot}m^{-1}$), represented by W_S, can be calculated by the following formula [22]:

$$W_S = \frac{CA}{FL} \tag{1}$$

where parameters C, A, F and L, correspond to the width of the wear scar (mm), average wear area of wear loss (mm^2), applied load (N), and the distance of sliding friction (mm), respectively.

2.4. Electrochemical Corrosion

An electrochemical corrosion analysis of CoCrFeNiSiMoW coatings was conducted in 3.5% NaCl solution by a CHI600E electrochemical workstation (Wuhan Corrtest Instruments, Corp., Wuhan, China) at room temperature. A three-electrode electrochemical cell, including a saturated calomel reference electrode (SCE), a platinum counter electrode, and a working electrode of coatings of HX1 and HX2, was adopted [23]. The scan rate of the tests was 0.5 mV/s, and the test range was −0.5~1.5 V.

3. Results and Discussion

3.1. Morphology Observation

The surface appearance of CoCrFeNiSiMoW coatings after plasma-arc surfacing welding is shown in Figure 2. It is evident that the HX wires were melted sufficiently, and continuous coatings were formed on the Q235 base metal. Each coating had a uniform geometric appearance, and the surface of the HX2 coating was smoother than that of HX1 coating. The overlaps between adjacent welding passes were almost parallel to each other.

Figure 3 shows the optical microstructures of the HX1 and HX2 coatings. From Figure 3a, it can be seen that the microstructure of the HX1 coating contains various forms of equiaxed microstructures, epitaxial dendrites, and cellular crystals. Interface also can be seen in the overlap area between welding passes, which can be attributed to different growth velocities and different growth directions. Figure 3b shows a typical equiaxed microstructure on the surface of the HX2 coating. Figure 3c–e is the microstructure of the cross section of the HX2 coating: the bottom, medium and top, respectively. It can be seen that from the bottom to the top, i.e., the deposition direction, dendrites grew in different directions. There are tiny sedimentary facies between dendrites. These smaller deposits are evenly distributed between dendrites at the bottom and middle. At the top, the dendrites are coarse, and there are more small precipitates of particles between the dendrites. This is due to the sectioning of the epitaxial dendrites growing upward along the deposition direction. Compared with the HX1 coating shown in Figure 3a, the microstructure of the HX2 coating shown in Figure 3b is obviously smaller than that of the HX1 coating, which may lead to the higher hardness of the HX2 coating.

Figure 2. The surface appearance of CoCrFeNiSiMoW coating after plasma-arc surfacing welding. HX1 is a one-layer coating, and HX2 is a two-layer coating.

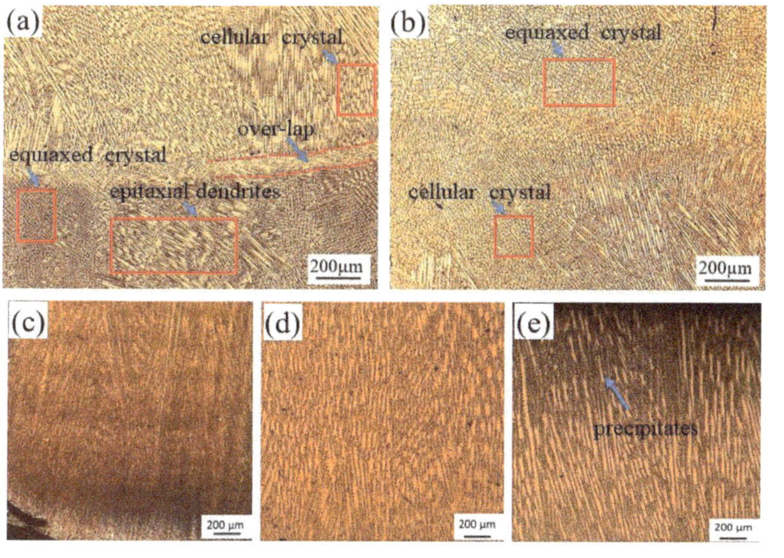

Figure 3. Optical microstructures of the HX1 and HX2 coatings: (**a**) the microstructure of the surface of the HX1 coating; (**b**) the microstructure of the surface of HX2 coating; (**c**–**e**) the microstructures of the bottom, medium, and top cross-sections of the HX2 coating, respectively.

3.2. XRD Analysis

Table 3 shows the main elemental compositions of the HX1 and HX2 coatings. It was shown that they were the CoCrFeNiSiMoW coatings. The entropy value (ΔS) of the coatings can be obtained by Boltzmann's hypothesis [24]:

$$\Delta S = -R[X_1 lnX_1 + X_2 lnX_2 + \ldots X_n lnX_n] = -R\sum_{i=1}^{n} X_i lnX_i \quad (2)$$

where R is a gas constant, X_i is the molar ratio of the principal element, and n is the number of mixing elements. According to the value of ΔS, alloys are divided into high-, medium-, and low-entropy alloys. When the value of ΔS is greater than or equal to 1.5 R, it is considered to be a high-entropy alloy. When the value of ΔS is between R and 1.5 R, it is considered to be a medium-entropy alloy. If the value of ΔS is less than R, it is identified as a low-entropy alloy [24].

According to Formula (2), the mixing entropy of the alloying layers' system can be calculated by the molar fraction of each element. The mixed entropy of the HX1 and HX2 coatings was 1.30 R and 1.28 R, respectively. Therefore, the HX1 and HX2 coatings can be considered CoCrFeNiSiMoW medium-entropy coatings.

Table 3. Analysis results of the elemental composition on the surface of the HX1 and HX2 coatings (mole fraction, %).

Elements	Ni	Cr	Fe	Mo	Co	W	Si
Coating HX1	42.21	18.85	31.85	4.43	1.28	0.15	1.23
Coating HX2	44.37	19.95	28.53	4.81	2.02	0.16	0.90

From Table 3, it can be seen that the Ni, Cr, Co, W, and Mo contents in the HX2 coating are higher than in the HX1 coating. Ni, Cr, and Co can be formed in solid solution. The elements of W and Mo can solubilize in solid solution, which contributes to solid solution strengthening. The atomic radii of Mo and W are larger than that of the other elements involved, and the solid solubility of the W and Mo in the Ni-Cr-Co solid solution matrix is large. They can cause lattice deformations, resulting in a solid solution, as has been reported previously [25,26]. This may lead to the higher hardness of the HX2 coating compared to the HX1 coating.

The XRD patterns of the HX1 and HX2 coatings are shown in Figure 4. It can be confirmed that the CoCrFeNiSiMoW coatings of HX1 and HX2 display a solution matrix of CoCrFeNiSiMoW cubic phase (FCC), according to JCPDS card number 35-1489. The high mixing entropy effect can effectively reduce the mixing Gibbs free energy, thus promoting the formation of a FCC solid solution during solidification [14]. In addition to the FCC matrix, some weak diffraction peaks were detected. These diffraction peaks correspond to some secondary phases. Their structures are similar to (Fe, Ni), Mo1.24Ni0.76, and CoCx phases, according to JCPDS card numbers 47-1417, 47-1129, and 44-0962, respectively. Compared with the XRD diffraction peak of HX, the peak of the HX2 coating is shifted to the left. The stronger lattice distortion is caused by the higher content of W and Mo in the HX2 coating [27].

3.3. Hardness of Coatings

Figure 5 shows the surface hardness profile of the HX1 and HX2 coatings. The hardness values of the HX1 and HX2 coatings fluctuate from 165 to 195 HV and from 192 to 200 HV, with average hardness values of 186 ± 1.56 and 198 ± 1.78 HV, respectively. The hardness of the HX2 coating is higher than that of the HX1 coating. The hardness tests are consistent with the analysis results in Figure 3 and Table 3. Hardness represents the ability of materials to resist residual deformation and anti-failure. In general, the higher the hardness of the material, the better the wear resistance.

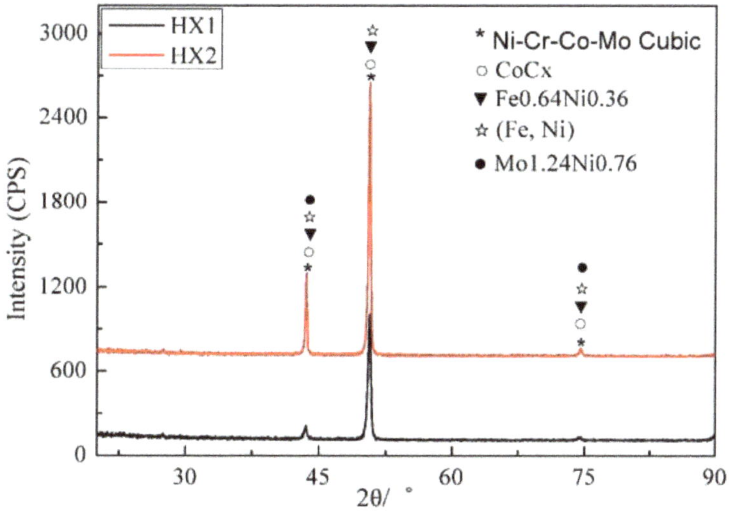

Figure 4. XRD patterns of the HX1 and HX2 coatings.

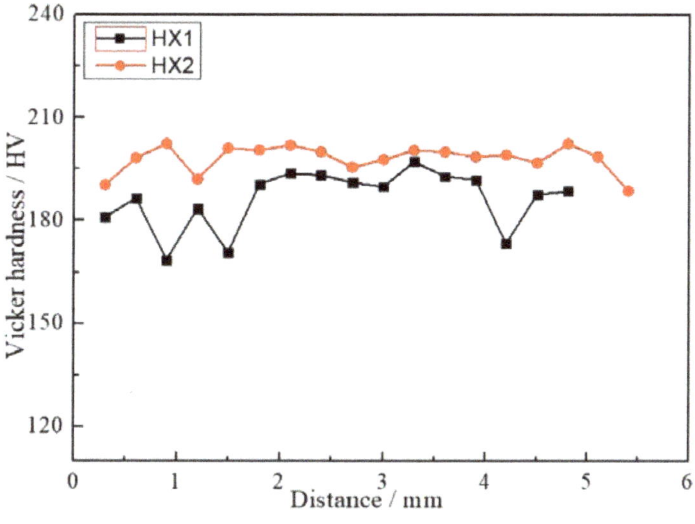

Figure 5. The surface hardness of the HX1 and HX2 coatings.

3.4. Wear Analysis

Figure 6 represents the change in the friction coefficients of CoCrFeNiSiMoW MEA coatings as a function of wear test time under different loads. During test periods of 150 s, the friction coefficients increased over time, then declined and remained fluctuating. The increase indicates the transition from material intact to material fracture [28]. It also can be seen that the friction coefficient decreased with the increased load. Under the same load, the friction coefficient of the HX2 coating was less than that of the HX1 coating. It indicates that the change of the friction coefficient might relate to the wear mechanism caused by the changing load. The wear mechanism will be further analyzed later in detail.

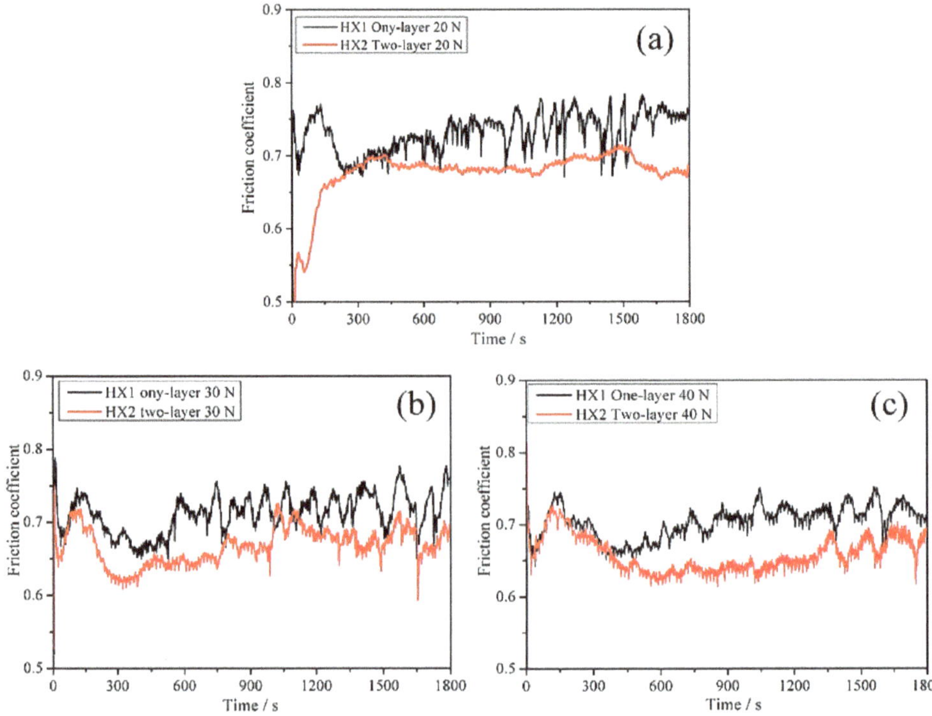

Figure 6. The change of friction coefficient of the HX1 and HX2 coatings as a function of time under different loads. 20 N (**a**), 30 N (**b**), 40 N (**c**).

Figure 7 represents the wear rate of the HX1 and HX2 coatings with increasing loads. The wear rate of the HX1 coating increased from 0.348 to 0.489 $mm^3 \cdot N^{-1} \cdot m^{-1}$, while the wear rate of the HX2 coating increased from 0.341 to 0.439 $mm^3 \cdot N^{-1} \cdot m^{-1}$ when the loads increased from 20 to 40 N. This demonstrates that the wear rate increased with increasing loads. Comparing the HX1 and HX2 coatings at a load of 20 N, the wear rates of HX1 and HX2 are 0.348 and 0.341 $mm^3 \cdot N^{-1} \cdot m^{-1}$, respectively, which is a statistically insignificant difference. This means that the performance of both coatings is similar at a load of 20 N. At a load of 30 or 40 N, the wear rate of HX2 is lower than that of HX1. Wear belongs to the category of contact surface stress, that is, the wear rate under pressure due to the interaction between the surface of the coating and the Si_3N_4 ball. In this case, the wear resistance of a high hardness coating is better than that of low hardness coating. The test results are consistent with the analysis of the hardness test results.

Figure 8 presents the wear scars of the HX1 and HX2 coatings under different loads. When the applied loads change from 20 to 40 N, it can be seen that the widths of the wear scars of the HX1 and HX2 coatings increase from 1007.60 to 1264.86 μm, and from 982.10 to 1192.85 μm, respectively. The width of the wear scars increases with the increase in applied load. In Figure 8a–c, there is a visibly large quantity of continuous spalling surfaces and some ploughs on the worn surface of the HX1 coating. It indicates that the main wear mechanism of the HX1 coating is adhesion wear, followed by abrasive wear. For the HX2 coating, as shown in Figure 8d–f, there is a large number of grooves and some discontinuous spalling surfaces. It means that the main wear mechanism of the HX2 coating is abrasive wear, followed by adhesion wear.

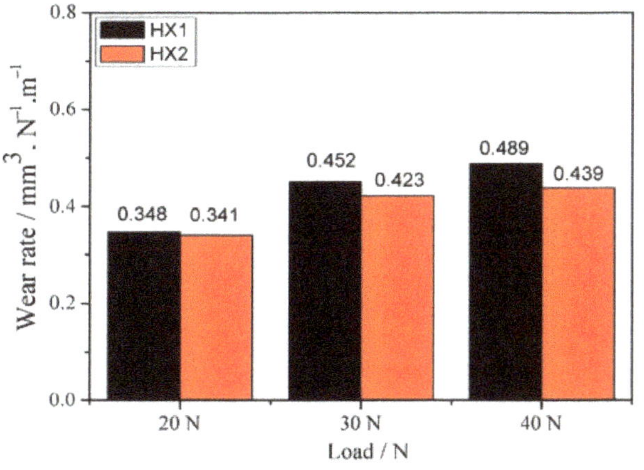

Figure 7. Wear rate of the HX1 and HX2 coatings under different applied loads.

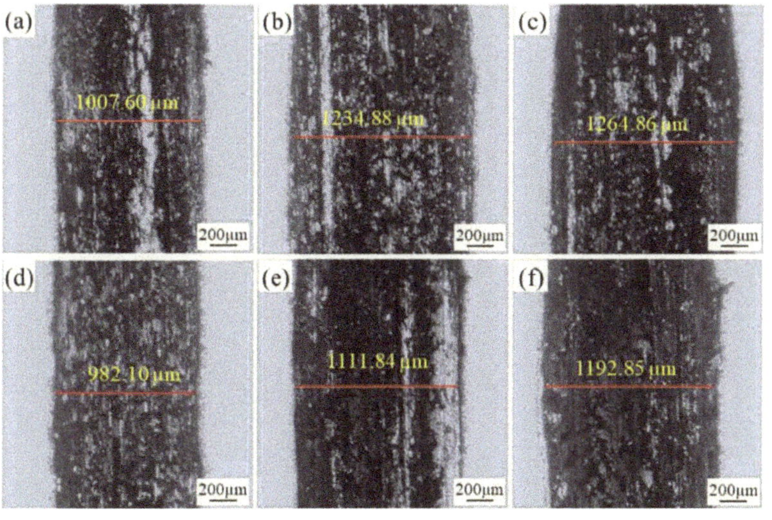

Figure 8. LSCM images of the wear scars under different loads: 20 N (**a**,**d**); 30 N (**b**,**e**); 40 N (**c**,**f**). (**a**–**c**) The images of the HX1 coating; (**d**–**f**) the images of the HX2 coating.

Figure 9 plots the two-dimensional (2D) profiles of the cross-section of the wear scars under different loads. When the load increased from 20 to 40 N, the depths of the wear scars on the HX1 and HX2 coatings increased from 11.5 to 21.5 μm, and from 10.5 to 20.8 μm, respectively. The width of the wear scar also increased with the increasing load. The wear track depth of the HX1 coating was slightly larger than that of the HX2 coating after sliding wear under the same load. The edges of the HX1 and HX2 coatings became evident due to extrusion with the load increases.

Figure 10a–f shows the worn surface morphology of the HX1 and HX2 coatings examined in detail using SEM after being subjected to sliding wear under different loads. From Figure 10a, it can be seen there were flake debris, spalling from the surface of the HX1 coating, and broken debris under a load of 20 N. When the load increased to 30 N, flake debris, spalling, and broken debris were aggravated, and cracks appeared, as shown in Figure 10b. When the load reached 40 N, the flake debris and surface appalling shown in Figure 10c changed little compared with that shown in Figure 10b. This morphology is consistent with the analysis results in Figure 8a–c. The main wear mechanism of the HX1 coating is adhesion wear, followed by abrasive wear. From Figure 10d, it can be seen there were wear debris and a discontinuous glaze layer on the worn surface of the HX2 coating. Cracks and broken debris appeared, and wear debris was visible on the surface of the HX2 coating, as shown in Figure 10e,f. It indicates that the wear mechanisms of the HX2 coating is adhesion and abrasive wear. The significant fluctuations in the friction coefficients, as shown in Figure 6, are attributable to inhomogeneous distributions of the glaze layer and wear debris.

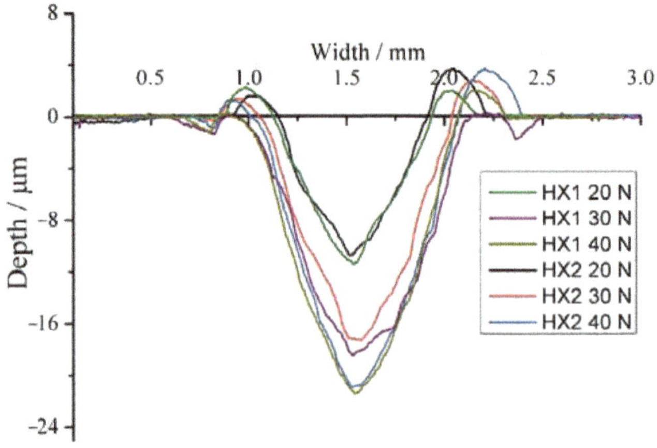

Figure 9. Two-dimensional (2D) profiles of the cross-section of the wear scars under different loads.

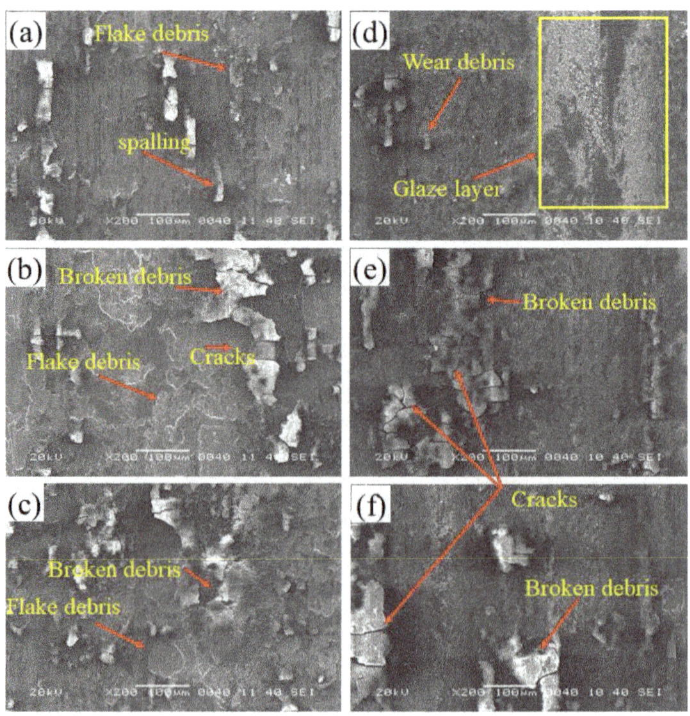

Figure 10. SEM micrographs of the worn surface under different loads: 20 N (**a**,**d**); 30 N (**b**,**e**); 40 N (**c**,**f**). (**a**–**c**) Images of the HX1 coating; (**d**–**f**) images of the HX2 coating.

3.5. Electrochemical Corrosion

Figure 11 presents the polarization curve of the CoCrFeNiSiMoW MEACs in a 3.5% NaCl solution. Both MEACs underwent passivation, even though the two MEACs exhibit the same corrosion potential. The corrosion current density of the two coatings increased rapidly when the potential was about 1.0 V and passive corrosion occurred, which can obstruct the corrosion of the coating surface by the 3.5% NaCl solution.

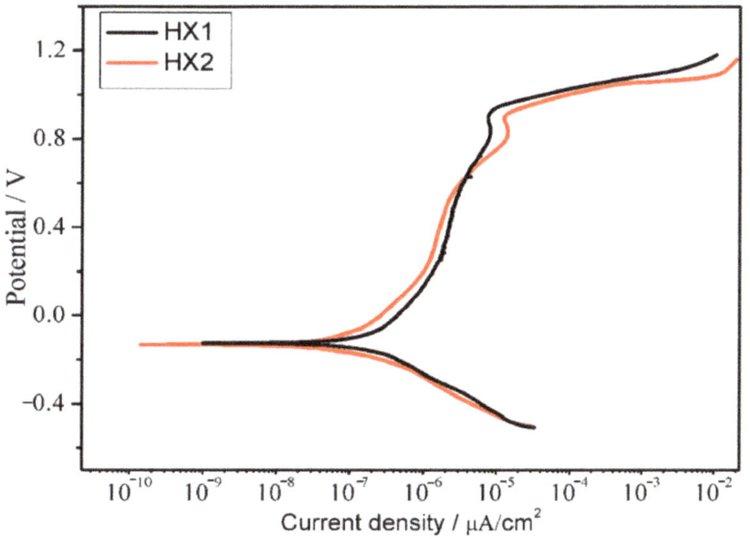

Figure 11. Polarization curve of the CoCrFeNiSiMoW MEACs in a 3.5% NaCl solution.

Table 4 shows the free corrosion potential (Ecorr) and the current density (icorr) value, which are determined from the linear portion of the polarization curve by Tafel extrapolation. According to classical electrochemical theory, the free corrosion potential (Ecorr) only represents the thermodynamic trend in the coatings in the corrosion process, while the free corrosion density (Icorr) is caused by the dissolution of the coatings. Therefore, Icorr is more accurate for evaluating corrosion resistance of coatings [29,30]. Compared to the HX1 coating, the HX2 coating has a lower Ecorr and smaller Icorr. It can be concluded that the HX2 coating has better corrosion resistance than the HX1 coating.

Table 4. Corrosion potential and corrosion current density of the coatings in a 3.5% NaCl solution.

Coating	Ecorr (VSCE)	Icorr ($\mu A/cm^2$)
HX1	−0.12468	4.0998×10^{-7}
HX2	−0.13313	9.3562×10^{-8}

Figure 12 shows SEM images of corrosion of the two HX coatings. In Figure 12a, it can be seen that the dendrite core of the HX1 coating and the deposited particles between cellular crystals were corroded, while the dendrite core of the HX2 coating remained intact, and the deposited particles between dendrites were visible, as shown in Figure 12b. On the one hand, this result can be attributed to some large dendrites on the surface of the HX1 coating, as shown in Figure 3a, and the finding that the boundaries between the dendrites maximized the interaction area between the surface of the HX1 coating and the corrosive solution. On the other hand, it can be seen in Table 3 that the HX2 coating contained a higher amount of Ni than the HX1 coating, and Ni has a higher electrochemical potential, which is also beneficial to improving the corrosion resistance of the HX2 coating. The SEM observation results are consistent with the analysis in Figure 11.

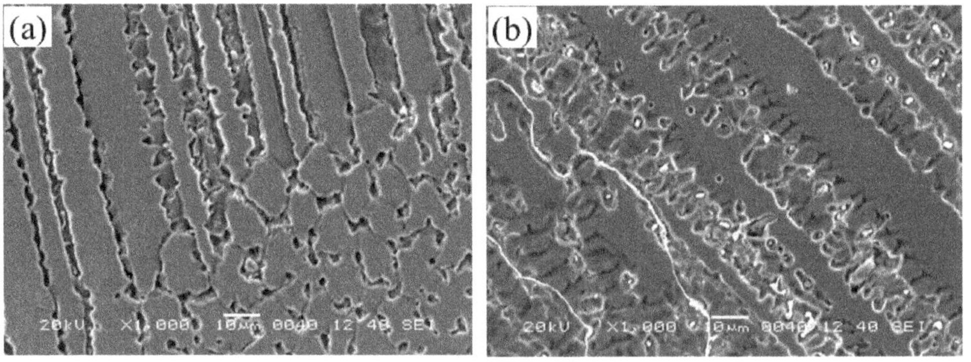

Figure 12. SEM images of corrosion: (**a**) image of the one-layer HX1 coating; (**b**) image of the two- layer HX2 coating.

4. Conclusions

The CoCrFeNiSiMoW medium-entropy alloy coating was successfully fabricated by plasma-arc surfacing welding.

The CoCrFeNiSiMoW MEACs are a mixture of a NiCrCoMo cubic (FCC) solid solution phase, (Fe, Ni), Mo1.24Ni0.76, and CoCx phases.

The average hardness values of the one-layer and two-layer CoCrFeNiSiMoW MEACs are 186 ± 1.56 and 198 ± 1.78 HV, respectively. The two-layer coating has a higher hardness because of a higher content of W and Mo on the surface of the coating. The radii of W and Mo atoms are larger than those of the other elements in the coating, and they solubilize in the Ni-Cr-Co solid solution and cause lattice deformation, resulting in solid solution strengthening.

At a load of 20 N, the one-lay coating and two-layer coatings have similar wear resistance. At loads of 30 and 40 N, the wear resistance of the two-layer coating is better than that of a one-layer coating because of the higher hardness of the two-layer coating.

The corrosion resistance of the two-layer CoCrFeNiSiMoW coating is better than that of a one-layer coating because the two-layer coating contains a higher amount of Ni than a one-layer coating, and Ni has a higher electrochemical potential, which is also beneficial to improving the corrosion resistance, which is evident in the two-layer coating.

Author Contributions: Project administration, Q.H.; data curation, X.W.; formal analysis, X.S., F.F. and Z.T. All authors have read and agreed to the published version of the manuscript.

Funding: The National Science Foundation of China (Nos. 51205176 and 51675249) supported this research.

Institutional Review Board Statement: Not applicable.

Informed Consent Statement: Not applicable.

Data Availability Statement: The data used to support the findings of this study are available from the corresponding authors upon request.

Conflicts of Interest: The authors declare no conflict of interest.

References

1. Zhang, Z.T.; Axinte, E.; Ge, W.J.; Shang, C.Y.; Wang, Y. Microstructure, mechanical properties and corrosion resistance of CuZrY/Al, Ti, Hf series high-entropy alloys. *Mater. Des.* **2016**, *108*, 106–113. [CrossRef]
2. Wu, P.H.; Peng, Z.; Liu, N.; Niu, M.Y.; Zhu, Z.X.; Wang, X.J. The effect of Mn content on the microstructure and properties of CoCrCu0.1Fe0.15Mo1.5MnxNi. *Mater. Trans.* **2016**, *57*, 5–8. [CrossRef]
3. Chuang, M.H.; Tsai, M.H.; Wang, W.R.; Lin, S.J.; Yeh, J.W. Microstructure and wear behavior of AlxCo1.5CrFeNi1.5Tiy high-entropy alloys. *Acta Mater.* **2011**, *59*, 6308–6317. [CrossRef]
4. Ye, Q.F.; Feng, K.; Li, Z.G.; Lu, F.G.; Li, R.F.; Huang, J.; Wu, Y.X. Microstructure and corrosion properties of CrMnFeCoNi high entropy alloy coating. *Appl. Surf. Sci.* **2017**, *396*, 1420–1426. [CrossRef]
5. Nair, R.B.; Arora, H.S.; Mukherjee, S.; Singh, S.; Singh, H.; Grewal, H.S. Exceptionally high cavitation erosion and corrosion resistance of a high entropy alloy. *Ultrason. Sonochem.* **2018**, *41*, 252–260. [CrossRef] [PubMed]
6. Li, Z.M.; Pradeep, K.G.; Deng, Y.; Raabe, D.; Tasan, C.C. Metastable high-entropy dual-phase alloys overcome the strength-ductility trade-off. *Nature* **2016**, *534*, 227–230. [CrossRef] [PubMed]
7. Qiu, Y.; Thomas, S.; Gibson, M.A.; Fraser, H.L.; Pohl, K.; Birbilis, N. Microstructure and corrosion properties of the low-density single- phase compositionally complex alloy AlTiVCr. *Corros. Sci.* **2018**, *133*, 386–396. [CrossRef]
8. Feng, K.; Zhang, Y.; Li, Z.G.; Yao, C.W.; Yao, L.; Fan, C.Y. Corrosion properties of laser cladded CrCoNi medium entropy alloy coating. *Surf. Coat. Technol.* **2020**, *397*, 126004. [CrossRef]
9. Zhao, D.C.; Yamaguchi, T.; Tusbasa, D.J.; Wang, W.Q. Fabrication and friction properties of the AlFeCrCo medium-entropy alloy coating on magnesium alloy. *Mater. Des.* **2020**, *193*, 108872. [CrossRef]
10. Tian, Y.; Lu, C.Y.; Shen, Y.F.; Feng, X.M. Microstructure and corrosion property of CrMnFeCoNi high entropy alloy coating on Q235 base metal via mechanical alloying method. *Surf. Interfaces.* **2019**, *15*, 135–140. [CrossRef]
11. Wu, W.; Jiang, L.; Jiang, H.; Pan, X.M.; Cao, Z.Q. Phase evolution and properties of Al2CrFeNiMox high-entropy alloys coatings by laser cladding. *J. Therm. Spray Technol.* **2015**, *24*, 1333–1340. [CrossRef]
12. Xu, J.; Peng, S.; Li, Z.Y.; Jiang, S.Y.; Xie, Z.H.; Munroe, P.; Lu, H. Remarkable cavitation erosion-corrosion resistance of CoCrFeNiTiMo high-entropy alloy coatings. *Corros. Sci.* **2021**, *190*, 109663. [CrossRef]
13. Lloyd, A.C.; Noël, J.J.; Mclntyre, S.; Shoesmith, D.W. Cr, Mo and W alloying additions in Ni and their effect on passivity. *Electrochim. Acta* **2004**, *49*, 3015–3027. [CrossRef]
14. Shun, T.T.; Chang, L.Y.; Shiu, M.H. Microstructure and mechanical properties of multiprincipal component CoCrFeNiMox alloys. *Mater. Charact.* **2012**, *70*, 63–67. [CrossRef]
15. Hsu, Y.J.; Chiang, W.C.; Wu, J.K. Corrosion behavior of FeCoNiCrCux high-entropy alloys in 3.5% sodium chloride solution. *Mater. Chem. Phys.* **2005**, *92*, 112–117. [CrossRef]
16. Li, Y.Z.; Shi, Y. Microhardness, wear resistance, and corrosion resistance of Al0.8CrFeCoNiCu high-entropy alloy coatings on aluminum by laser cladding. *Mater. Res. Express.* **2021**, *7*, 026504.
17. Chen, T.K.; Shun, T.T.; Yeh, J.W.; Wong, M.S. Nanostructured nitride films of multi-element high-entropy alloys by reactive DC sputtering. *Surf. Coat. Technol.* **2004**, *188–189*, 193–200. [CrossRef]
18. Zeng, Q.F.; Xu, Y.T. A comparative study on the tribocorrosion behaviors of AlFeCrNiMo high entropy alloy coating and 304 stainless steel. *Mater. Today Commun.* **2020**, *24*, 101261. [CrossRef]
19. Wang, F. Mechanical property study on rapid additive layer manufacture Hastelloy X alloy by selective laser melting technology. *Int. J. Adv. Manuf. Technol.* **2011**, *58*, 545–551. [CrossRef]
20. Zhao, J.C.; Larsen, M.; Ravikumar, V. Phase precipitation and time temperature-transformation diagram of Hastelloy X. *Mater. Sci. Eng. A.* **2000**, *293*, 112–119. [CrossRef]
21. Blue, C.A.; Blue, R.A.; Lin, R.Y.; Lei, J.F.; Williams, W.D. Joining of Hastelloy X to Inconel 718 using an infrared process. *J. Mater. Process. Technol.* **1996**, *58*, 32–38. [CrossRef]
22. Qiao, Y.X.; Sheng, S.L.; Zhang, L.M.; Chen, J.; Zheng, Z.B. Friction and wear behaviors of a high nitrogen austenitic stainless steel Fe-19Cr-15Mn-0.66N. *J. Min. Metall. Sect. B Metall.* **2021**, *57*, 285–293.
23. Qiao, Y.X.; Xu, D.K.; Wang, S.; Ma, Y.J.; Chen, J.; Wang, Y.X.; Zhou, H.L. Effect of hydrogen charging on microstructural evolution and corrosion behavior of Ti-4Al-2V-1Mo-1Fe alloy. *J. Mater. Sci. Technol.* **2021**, *60*, 168–176. [CrossRef]
24. Sharma, P.; Dwivedi, V.K.; Dwivedi, S.P. Development of high entropy alloys: A review. *Mater. Today Proc.* **2021**, *43*, 502–509. [CrossRef]
25. Liu, N.; Ding, W.; Wang, X.J.; Du, J.J.; Liu, L.X. Microstructure evolution and phase formation of Fe25Ni25CoxMoy Multi-principal component alloys. *Metall. Mater. Trans. A* **2020**, *51*, 2990–2997. [CrossRef]
26. Ding, W.; Liu, N.; Fan, J.C.; Cao, J.; Wang, X.J. Diffusion bonding of copper and titanium with an interlayer of CoCrFeMnNi high-entropy alloy. *Intermetallics* **2021**, *129*, 107027. [CrossRef]
27. Lin, K.H.; Tseng, C.M.; Chueh, C.C.; Chang, S.Y.; Lo, Y.C.; Wang, C.C.; Lin, S.J.; Yeh, J.W. Different lattice distortion effects on the tensile properties of Ni-W dilute solutions and CrFeNi and CoCrFeMnNi concentrated solutions. *SSRN Electron. J.* **2021**. [CrossRef]

28. Xi, Y.T.; Bai, Y.Y.; Gao, K.W.; Pang, X.L.; Yang, H.S.; Yan, L.C.; Volinsky, A.A. Residual stress and microstructure effects on mechanical, tribological and electrical properties of TiN coating on 304 stainless steel. *Ceram. Int.* **2018**, *44*, 15851–15858. [CrossRef]
29. Li, Y.Z.; Shi, Y. Microhardness, wear resistance, and corrosion resistance of AlxCrFeCoNiCu high-entropy alloy coatings on aluminum y laser cladding. *Opt. Laser Technol.* **2021**, *134*, 106632. [CrossRef]
30. Qiao, Y.X.; Chen, Y.; Li, L.L.; Chen, J.; Emori, W.; Wang, X.J.; Yang, L.; Zhou, H.L.; Song, G.; Naik, N.; et al. Corrosion behavior of a nickel-free high-nitrogen stainless steel with hydrogen charging. *JOM* **2021**, *73*, 1165–1172. [CrossRef]

coatings

Article

Effect of Intermediate Principal Stress on the Bearing Capacity of Footings in Soft Rock

Zongyuan Ma [1,*], Faning Dang [1] and Hongjian Liao [2]

1 Institute of Geotechnical Engineering, Xi'an University of Technology, Xi'an 710048, China; dangfn@mail.xaut.edu.cn
2 Department of Civil Engineering, Xi'an Jiaotong University, Xi'an 710049, China; hjliao@mail.xjtu.edu.cn
* Correspondence: mzy_gogo@sohu.com

Abstract: The bearing capacity for footings is a fundamental scientific problem in civil engineering. The evaluation of the bearing capacity of footings usually does not take into account the effect of the intermediate principal stress. In practice, the intermediate principal stress has certain influences on the strength of geomaterials (e.g., rock and soil) or concrete. In this paper, a series of numerical solutions are presented to evaluate the bearing capacity of footings in a soft rock foundation via a two-dimensional finite difference code (FLAC) with a strain hardening/softening constitutive model based on the unified strength theory (UST). The values of the bearing capacity factor N_c and N_γ for strip, circular and square footings in a soft rock foundation were evaluated using the strain hardening/softening constitutive model. The effect of the intermediate principal stress on the bearing capacity of strip, circular and square footings in a soft rock foundation was analyzed. The results of the numerical computation show that the intermediate principal stress has a significant influence on the bearing capacity and failure mechanisms of a soft rock medium. The influence of the intermediate principal stress on the peak and residual values of the bearing capacity for a strip footing is much greater than for circular and square footings. Research works for the reasonable estimation of the bearing capacity of footings in soft rock are facilitated by this study.

Keywords: soft rock; strain hardening/softening; bearing capacity of footing; numerical solution; effect of intermediate principal stress

Citation: Ma, Z.; Dang, F.; Liao, H. Effect of Intermediate Principal Stress on the Bearing Capacity of Footings in Soft Rock. *Coatings* **2021**, *11*, 1019. https://doi.org/10.3390/coatings11091019

Academic Editor: Manuel Miguel Jordan-Vidal

Received: 3 August 2021
Accepted: 21 August 2021
Published: 25 August 2021

1. Introduction

A rock material can be classified as soft rock if the uniaxial compressive strength (USC) is below 30 MPa [1,2]. The deformation process of soft rock often has a remarkable strain-softening characteristic with a peak and residual strength [3–6]. Many researchers investigate the problems of soft rock in geotechnical engineering. For example, the stability of tunnels constructed in soft rock was analyzed by Zhu and Tokiwa [7,8]. The evaluation of the bearing capacity for footings is an essential problem in geotechnical engineering. In recent years, the bearing capacity of footings in clay or sand is often researched by many investigators using analytical or numerical methods with a linearly elastic-perfectly plastic constitutive model [9–15]. However, the bearing capacity of footings in strain-softening materials (e.g., soft rock) is seldom studied [16]. Using the linearly elastic, perfectly plastic constitutive model, Ma investigated the influence of the intermediate principal stress on the bearing capacity of strip or circular footings [17]. The intermediate principal stress has more influence on the bearing capacity of a strip footing than that of circular footing, and the intermediate principal stress has much more influence on the bearing capacity factor N_γ than that of factor N_c [17].

Many complex stress tests (e.g., true-triaxial tests, plane strain tests, or torsion shear tests) have verified that the intermediate principal stress has specific influences on the mechanical behavior of geomaterials [18–26]. The polyaxial test data of sand obtained by Sutherland and Mesdary suggest that the intermediate principal stress has a marked effect on the strength of sand and that the Mohr–Coulomb theory will therefore underestimate the strength over most of the stress state range [20]. Mogi found that in true triaxial testing of many rock samples, the effect of the intermediate principal stress must be considered for rock [21]. The results of complex stress tests for concrete indicate that the intermediate principal stress also has a marked effect on the strength of concrete [27,28]. We can conclude that the intermediate principal stress effect is an inherent mechanical behavior of many materials. Several strength criteria that take the effect of the intermediate principal stress and nonlinear yield surface into account were proposed for geomaterials [29]. These criteria include the Zienkiewicz–Pande criterion, the Lade–Duncan criterion, and the Matsuoka–Nakai criterion [30,31].

In this study, a numerical solution of the bearing capacity of footings in a soft rock foundation was proposed taking the strain hardening/softening material behavior and the intermediate principal stress effect into account. The influence of the intermediate principal stress on the bearing capacity of strip, circular and square footings in a soft rock foundation will be analyzed.

2. Materials, Theories and Methods

The unified strength theory (UST), which takes the influence of the intermediate principal shear stress (or intermediate principal stress) into account, was proposed by Yu [32]. The expressions of the UST can be written in terms of the principal stresses as follows:

$$\begin{cases} f = \frac{b\sigma_2+\sigma_3}{(1+b)(1+\sin\varphi)/(1-\sin\varphi)} - \sigma_1 + \frac{2c}{\sqrt{(1+\sin\varphi)/(1-\sin\varphi)}}, & \text{when } \sigma_2 \le \frac{1+\sin\varphi}{2}\sigma_1 + \frac{1-\sin\varphi}{2}\sigma_3 \\ f' = \frac{\sigma_3}{(1+\sin\varphi)/(1-\sin\varphi)} - \frac{\sigma_1+b\sigma_2}{1+b} + \frac{2c}{\sqrt{(1+\sin\varphi)/(1-\sin\varphi)}}, & \text{when } \sigma_2 \ge \frac{1+\sin\varphi}{2}\sigma_1 + \frac{1-\sin\varphi}{2}\sigma_3 \end{cases} \quad (1)$$

where b is a coefficient reflecting the effect of the intermediate principal stress on the strength of geomaterials, c and φ are the cohesion and friction angle of geomaterials, respectively. The order of the three principal stresses follows $\sigma_1 \ge \sigma_2 \ge \sigma_3$. For the non-associated flow rule, the plastic potential function g can be written as follows:

$$\begin{cases} g = \frac{b\sigma_2+\sigma_3}{(1+b)(1+\sin\psi)/(1-\sin\psi)} - \sigma_1, & \text{when } \sigma_2 \le \frac{1+\sin\varphi}{2}\sigma_1 + \frac{1-\sin\varphi}{2}\sigma_3 \\ g' = \frac{\sigma_3}{(1+\sin\psi)/(1-\sin\psi)} - \frac{\sigma_1+b\sigma_2}{1+b}, & \text{when } \sigma_2 \ge \frac{1+\sin\varphi}{2}\sigma_1 + \frac{1-\sin\varphi}{2}\sigma_3 \end{cases} \quad (2)$$

where ψ is the dilation angle for geomaterials (the plastic flow rule is non-associated if $\psi < \varphi$, and the plastic flow rule is associated if $\psi = \varphi$). The limit loci of UST on the deviatoric plane or principal stress space are shown in Figure 1a,b, where θ_b is the stress angle for the junction of two yield surfaces and depends only on the friction angle of the material. Comparisons between the yield surfaces of the UST and test data for granite, marble, and sand under a complex stress state are plotted in Figure 2, the intermediate principal stress has a substantial influence on the strengths of geomaterials under complex stress states. UST can predict the potential strengths of different geomaterials with various values of the parameter b. The yield surfaces of the UST cover the entire region of convex theory from the lower bound (UST b = 0.0) to the upper bound (UST b = 1.0). The Mohr–Coulomb strength criterion is the special case of the UST when the coefficient b equals zero.

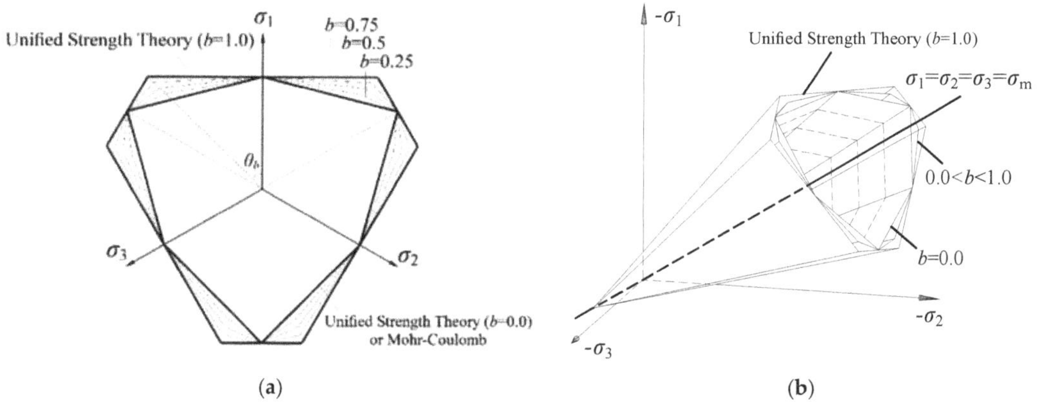

Figure 1. Limit loci of UST: (**a**) deviatoric plane, (**b**) principal stress space.

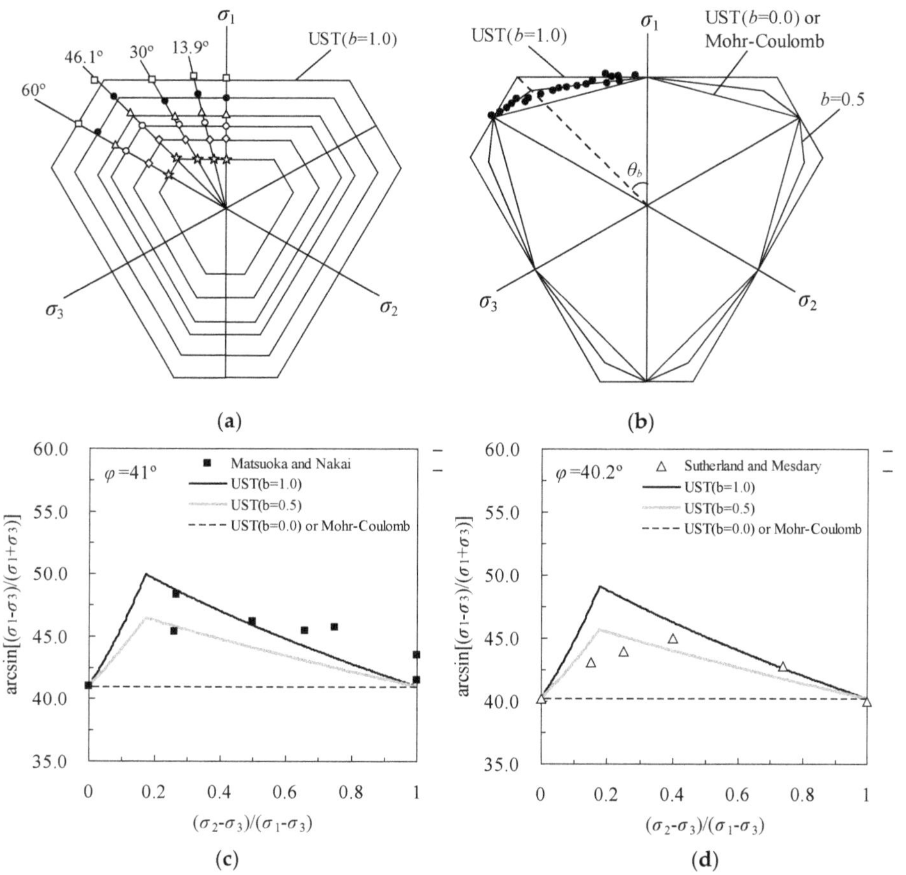

Figure 2. Limit surfaces of rock and sand under complex stress states: (**a**) Granite, (**b**) Marble, (**c**) Toyoura sand, (**d**) Loch Aline sand.

In this study, the vertical bearing capacity solutions for strip, circular and square footings in a soft rock foundation were evaluated using a finite difference code FLAC/FLAC3D and strain hardening/softening constitutive model based on the UST. FLAC/FLAC3D (fast Lagrangian analysis of continua) is a two/three-dimensional (2D/3D) finite-difference code that uses an explicit Lagrangian computation scheme. The behavior of structures built of soil, rock, or other materials that may undergo plastic flow when their yield limits are reached can be simulated easily by FLAC/FLAC3D [33,34]. The derivation and verification of the UST elastoplastic constitutive model in FLAC/FLAC3D were discussed by Ma [35,36]. In this study, the strain hardening/softening constitutive model based on the UST is written in C++ as a user-written constitutive model and compiled as a DLL file (dynamic link library), and it can be loaded into the FLAC/FLAC3D code. The elastic relations between the elastic principal strain increments and principal stress increments in FLAC/FLAC3D are as follows:

$$\Delta\sigma_i = S_i(\Delta\varepsilon_n^e)\ i = 1, n \text{ and } n = 1, 3 \tag{3}$$

where S_i is a linear function of the elastic principal strain increments $\Delta\varepsilon_n^e$. The plastic principal strain increments can be written as follows:

$$\Delta\varepsilon_i^p = \lambda\partial g/\partial\sigma_i \tag{4}$$

where λ is a non-negative multiplier if plastic loading occurs. The expression of the elastoplastic constitutive model for FLAC/FLAC3D can be formulated as follows:

$$\sigma_i^N = \sigma_i^I - \lambda \cdot S_i\left(\frac{\partial g}{\partial\sigma_n}\right) \tag{5}$$

where $S_i(\partial g/\partial\sigma_n)$ is the matrix of the constitutive model composed by the plastic principal strain component, g is the plastic potential function, σ_i^I are the stress components obtained from the elastic Hooke's law (elastic trial stress), and σ_i^N are the new stress components obtained from the plastic flow rule (if the elastic trial stress exceed the Equation (1)). The expression of the plastic multiplier λ can be written as follows:

$$\lambda = f\left(\sigma_n^I\right)/\left[f\left(S_n\left(\frac{\partial g}{\partial\sigma_n}\right)\right) - f(0)\right] \tag{6}$$

Equation (6) is accurate only for an elastic perfectly plastic material, for which the strength parameters are constant. Theoretically, Equation (6) is non-rigorous for hardening/softening material because it lacks the hardening modulus. In this study, the hardening parameter e^{ps} is used to characterize the hardening/softening behavior via the plastic shear strain [37], and e^{ps} is defined as follows:

$$e^{ps} = \left\{\frac{1}{2}\left[\left(\varepsilon_1^{ps} - \varepsilon_m^{ps}\right)^2 + \left(\varepsilon_2^{ps} - \varepsilon_m^{ps}\right)^2 + \left(\varepsilon_3^{ps} - \varepsilon_m^{ps}\right)^2\right]\right\}^{1/2} \tag{7}$$

where ε_1^{ps}, ε_2^{ps} and ε_3^{ps} are the three plastic principal strains of the shear strength envelope and $\varepsilon_m^{ps} = \left(\varepsilon_1^{ps} + \varepsilon_2^{ps} + \varepsilon_3^{ps}\right)/3$. It is observed from Equation (7) that the hardening parameter e^{ps} equals the square root of the second invariant of the strain. For the plastic hardening/softening stage, the plastic potential function g can be written as follows:

$$g[\sigma_n, c(e^{ps}), \varphi(e^{ps})] = 0\ n = 1, 3 \tag{8}$$

Liao et al. and Li conducted consolidated-undrained (CU) triaxial shear tests on saturated diatomaceous soft rock and mudstone, respectively [38,39]. The effective stress is used to calculate the shear strength parameters using the test data of the CU test for diatomaceous rock and mudstone. The triaxial test results on the diatomaceous soft rock and mudstone show that the effective strength parameter φ for the two types of soft rock has peak and residual values with the value of axial strain ε_1 increased (shown in Figure 3a). The effective strength parameter c (cohesion) of the two types of soft rock remains constant during the deformation progress of the soft rock sample. In this study, the friction angle φ in the plastic potential function g varies with the values of the hardening parameter e^{ps}. In contrast, cohesion c does not depend on the hardening parameter e^{ps}. The hump curve function is used as the hardening/softening function to simulate the strain-softening behavior of soft rock in this study. The relationship between the friction angle φ and the hardening parameter e^{ps} predicted by the hump curve function is shown in Figure 4, and the hump curve function is described as follows:

$$\varphi(e^{ps}) = \frac{e^{ps}(H + Re^{ps})}{(H + Pe^{ps})^2} \tag{9}$$

where the parameter H controls the slope of the hump curve, and the parameters P and R control the peak and residual value of the hump curve. The values of the parameters H, P, and R in Equation (9) can be determined by the variation in the strength parameter φ yielded by a triaxial test of the soft rock. The UST elastoplastic constitutive model can be established via the substitution of Equations (4), (6), and (9) into Equation (5). The plastic shear strain e^{ps} are updated by each time step (time increment Δt) in the numerical computation, and the values of friction angle φ will be changed with the variation of the plastic shear strain e^{ps} after each time step. Thus, the type of hardening rule is isotropic. The loading surface of the strain hardening/softening model is expanded with the variation of the values of friction angle φ, and the hardening/softening behavior of the soft rock can be predicted.

A 2D axisymmetric element was used to simulate a test sample of soft rock, and the consolidated-undrained triaxial test was simulated using FLAC code. The confining pressure σ_3 was applied by the stress boundary condition on all sides of the test sample element to generate the consolidated stress. For the conventional triaxial test situation (simple stress state), the three principal stresses follow $\sigma_1 \geq \sigma_2 = \sigma_3$, a vertical velocity load (1×10^{-6} m/step) was applied to the top of the test sample element to simulate the maximum principal stress σ_1. The bottom of the test sample element was fixed in the vertical direction. The values of the material property parameters for numerical analysis are presented in Table 1. Because the test sample of soft rock had slight volume dilation after the consolidated-undrained triaxial test, the dilation angle ψ was set to zero (non-associated flow rule) throughout the whole calculation process of the triaxial test simulation. Figure 3a,b show the relationship of stress (σ_1–σ_3) versus axial strain (ε_1) for diatomaceous soft rock and mudstone, respectively, measured by the consolidated-undrained triaxial test or simulated by FLAC using the hump curve function (Equation (9)). The comparison between the model prediction and measured results shows that the stress–strain relationship of soft rock under the triaxial test state ($\sigma_1 \geq \sigma_2 = \sigma_3$) can be simulated by the elastoplastic model with the hump curve function. Then, the values of the bearing capacity factor N_c and N_γ for strip, circular and square footings in diatomaceous soft rock were evaluated by FLAC/FLAC3D using the strain hardening/softening constitutive model following the UST.

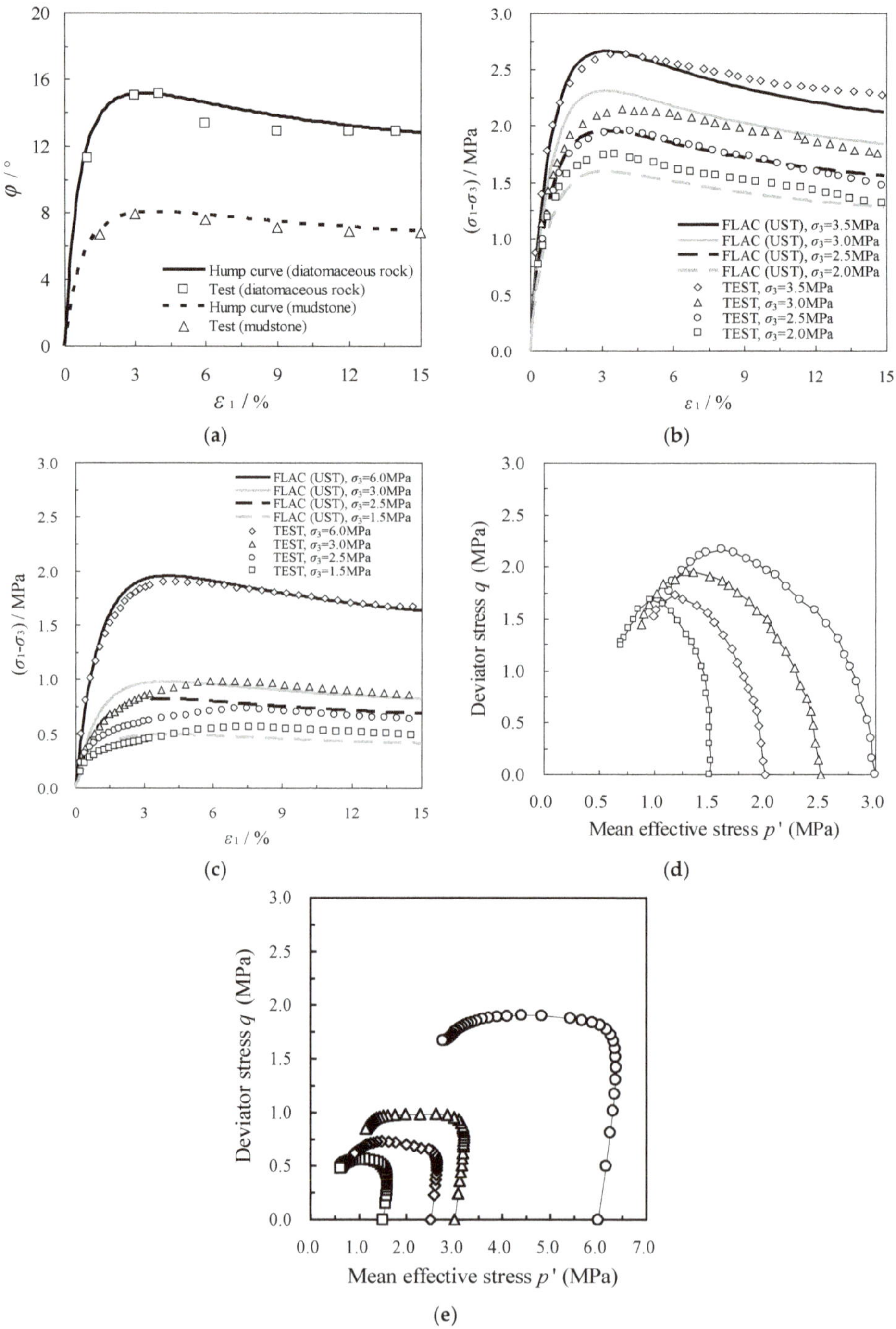

Figure 3. Test and simulated results for soft rock measured by CU triaxial test [38,39] or computed by FLAC: (**a**) Variation of friction angle φ, (**b**) Stress–strain behavior for diatomaceous soft rock, (**c**) Stress–strain behavior for mudstone, (**d**) Effective stress path for diatomaceous soft rock, (**e**) Effective stress path for mudstone.

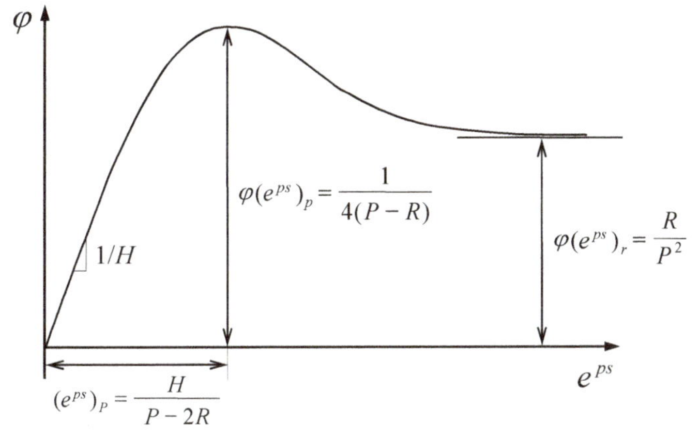

Figure 4. Determination of parameters for hump curve function.

Table 1. Parameters and values for numerical analysis.

Name	Density ρ (kg/m^3)	Elastic Modulus E (MPa)	Poisson's Ratio ν	Cohesion c (kPa)	Peak Friction Angle (°)	Residual Friction Angle (°)	P	H	R
Diatomaceous soft rock	2183	500	0.25	40	15.1	12.8	0.021	0.04	0.0045
Mudstone	2032	500	0.25	1.2	8.0	6.8	0.039	0.09	0.0079

3. Calculations and Results

The 2D problem of the plane strain and axisymmetry is also under the complex stress state; three principal stress are independent and not equals to each other. Thus, the intermediate principal stress will have a specific influence on the 2D problem. Strip and circular footings can be considered as plane strain and axisymmetric problems, respectively. The square footing must be analyzed as a 3D problem. Vesić suggested an equation of the ultimate bearing capacity for a strip footing on a soil foundation. The formula is expressed as follows [40]:

$$P_u = c \cdot N_c + \frac{\gamma \cdot B}{2} N_\gamma \tag{10}$$

$$N_c = \cot\varphi \cdot \left[\frac{1+\sin\varphi}{1-\sin\varphi}\exp(\pi \cdot \tan\varphi) - 1\right] \tag{11}$$

$$N_\gamma = 2\left[\frac{1+\sin\varphi}{1-\sin\varphi}\exp(\pi \cdot \tan\varphi) + 1\right]\tan\varphi \tag{12}$$

$$P_u = c \cdot N_c + qN_q + 0.4BN_\gamma \text{ (for square foundation)} \tag{13}$$

where P_u is the ultimate bearing capacity of strip footing, N_c is the bearing capacity factor of cohesion c, γ is the soil unit weight, B is the footing width, and N_γ is the soil self-weight bearing capacity. Terzaghi also proposed the formulas of factors N_c and N_γ for rough rigid and strip footings [41]. Circular footings, which belong to the class of axisymmetric problems, were considered by Eason and Shield and Cox et al. using the characteristic method [42,43]. The bearing capacity of a square footing has been investigated by some researchers using numerical and testing methods [44–47].

In this study, the foundation was considered a strain-softening and homogeneous material (soft rock) and discretized into several finite-difference element meshes. The footings are considered to be perfectly rigid and rough. The influence of the intermediate principal stress on the bearing capacity of strip, circular and square footings was analyzed, taking the strain hardening/softening behavior of soft rock into account. The residual strength of soft rock is practically irrelevant, and it is only of the theoretical value. In this study, the peak and residual bearing capacity of soft rock foundations are both considered for analysis. The values of the factors N_c and N_γ for strip, circular and square footings in diatomaceous soft rock were evaluated by FLAC and FLAC3D code with strain hardening/softening constitutive model following the UST and dilation angle ψ = 0. Because the problem domain is symmetric, only half of the problem domain was considered, and the footing width B was held constant at 20 m. Because the left vertical boundary was the symmetry plane, the horizontal displacement was fixed, and the vertical displacement was free. The right vertical boundary was only constrained in the horizontal direction. The displacement of the bottom boundary was fixed in both the vertical and horizontal directions. The boundary condition and mesh for numerical analysis are shown in Figure 5. The analyses were performed by applying a vertical velocity (1×10^{-6} m/step) to simulate the load from the rigid footing base. The contact stress P beneath the strip footing was calculated as the sum of the vertical nodal forces beneath the footing divided by the half-width of the footing [10]. The contact stress P beneath the circular footing was calculated by dividing the sum of the vertical footing nodal forces by the footing area, the radius of which is equal to the distance to the center of the first element outside the footing [9]. The horizontal velocity at the surface nodes beneath the footing was set to zero to simulate the rough interface between the footing and soft rock foundation. Previous studies observed that the value of Poisson's ratio ν and elastic modulus E do not influence the value of the bearing capacity of footings [11]. The tensile failure at the free surface near the footing base has little influence on the ultimate bearing capacity of footings [9,10]. Shear failure is placed in most elements beneath the footing under the limit state, and the ultimate bearing capacity of footing mainly depends on the shear strength parameters (c and φ) of soft rock.

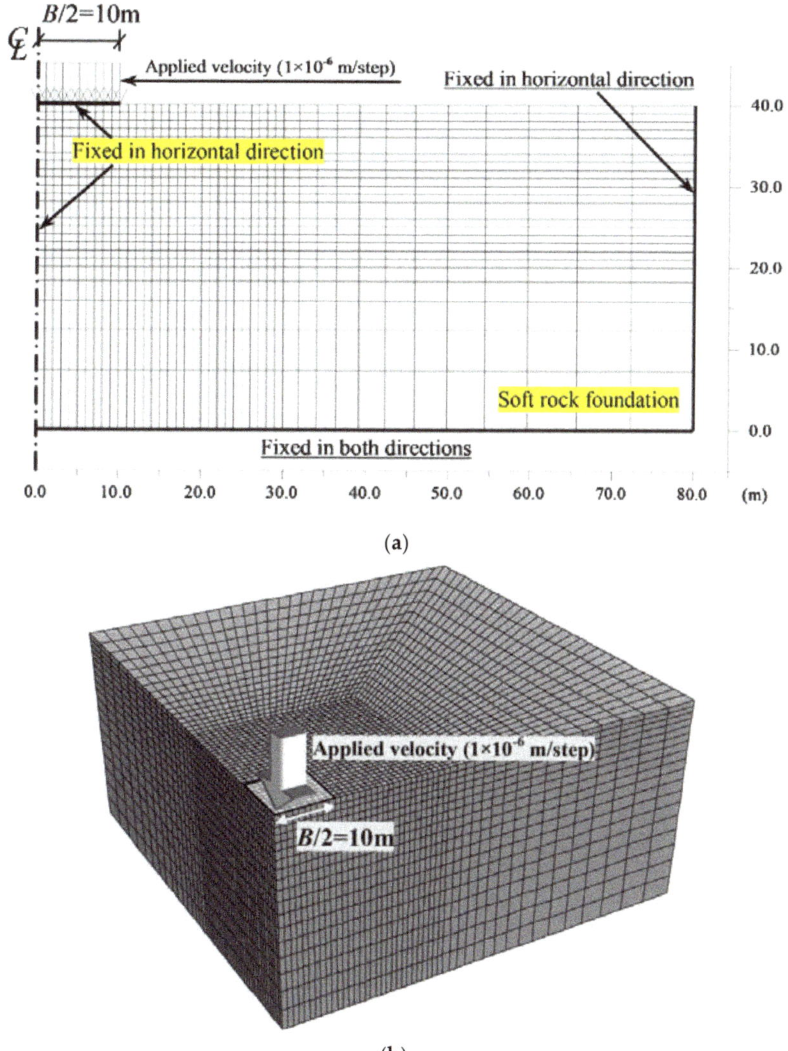

Figure 5. Finite difference meshes with boundary conditions used for analysis: (**a**) Strip or circular footing (symmetrical model), (**b**) Square footing (quarter model).

The two bearing capacity factors (N_c and N_γ) can be obtained individually using the numerical method [10,11] and can be expressed by the following equations:

$$\begin{cases} N_c = P/c, \text{ when } \gamma = 0 \\ N_\gamma = 2P/\gamma B, \text{ when } c = 0 \end{cases} \tag{14}$$

where, P is the contact stress beneath the footing base, B is the width of the footing. The initial soil stress state must be established before the numerical computation of the bearing capacity factor N_γ. The initial geostress state for foundation soil is calculated with gravity, and a load of footing base is not applied to the surface of the foundation. Secondly, the vertical load is applied to the nodes underneath the footing base after the initial geostress of the foundation is established. The peak and residual values of bearing capacity factor N_c

and N_γ yield by FLAC and UST are shown in Tables 2 and 3, respectively. Two parameters ξ_c and ξ_γ are given to estimate the influence of the intermediate principal stress and are described as follows:

$$\xi_c = \frac{N_c(\text{UST},\ b \geq 0)}{N_c(\text{Mohr}-\text{Coulomb})},\ \xi_\gamma = \frac{N_\gamma(\text{UST},\ b \geq 0)}{N_\gamma(\text{Mohr}-\text{Coulomb})} \tag{15}$$

where ξ_c and ξ_γ are the efficiency factors of the intermediate principal stress effect for the bearing capacity factors N_c and N_γ, respectively. Figure 6 shows the relationship of $N_c = P/c$ and $N_\gamma = 2P/\gamma B$ versus vertical displacement s/w for strip, circular and square footings in diatomaceous soft rock as calculated by the finite-difference code FLAC. The relationship of the peak and residual values of the bearing capacity factors N_c and N_γ for strip, circular and square footings versus the parameter b of UST are shown in Figure 7. The results from Figure 6 indicate that the strain-softening characteristic for N_γ is more remarkable than that of N_c. The results from Figure 7 suggest that the peak and residual values of the factors N_c and N_γ for strip, circular and square footings are increased with increasing values of the parameter b in UST. Figure 8 shows the relationship of the efficiency factors ξ_c and ξ_γ versus the values of parameter b of UST. The results from Figure 8 indicate that the values of the efficiency factor ξ_γ for N_γ are generally higher than the efficiency factor ξ_c with the values of the parameter b increased. The influence of the intermediate principal stress on the peak and residual values of N_γ is more apparent than that of N_c. The influence of the intermediate principal stress on the peak and residual values of the bearing capacity for a strip footing is much more significant than for circular and square footings. The intermediate principal stress has a similar influence on the peak and residual values of the bearing capacity of circular and square footings. The difference between the influence of the intermediate principal stress on the peak and residual values of the bearing capacity of a soft rock foundation appears relatively small. Figure 9 shows the maximum shear strain rate contours for a rough strip footing when the soft rock foundation's peak or residual bearing capacity has been reached.

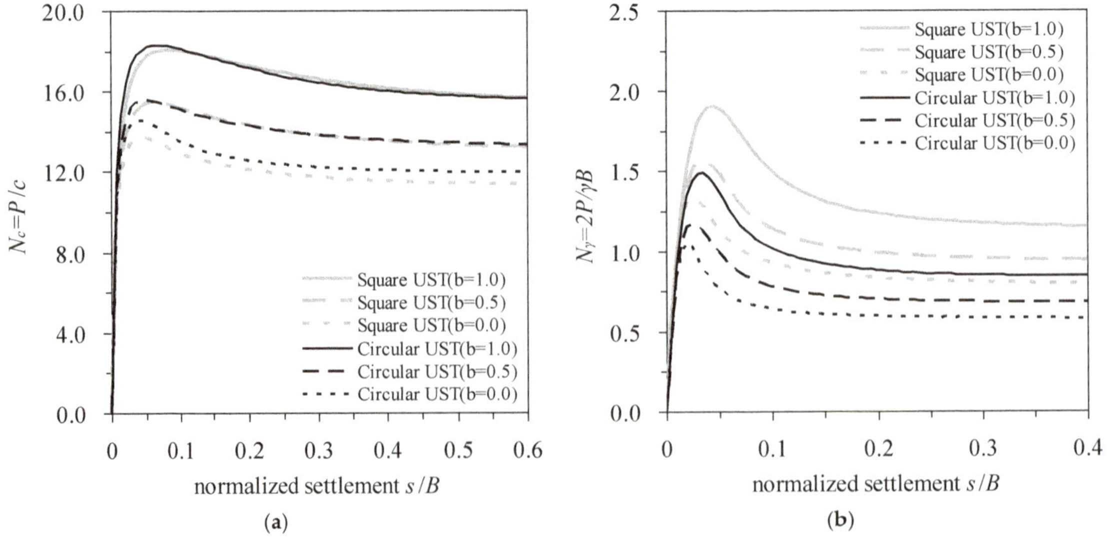

Figure 6. Relationship between bearing capacity factors and vertical displacement: (**a**) N_c, (**b**) N_γ.

Table 2. Peak and residual values of bearing capacity factor N_c yield by FLAC and UST.

Geometry	Strip					Circular					Square				
UST	$b = 0.0$	$b = 0.25$	$b = 0.5$	$b = 0.75$	$b = 1.0$	$b = 0.75$	$b = 1.0$	$b = 0.5$	$b = 0.75$	$b = 1.0$	$b = 0.0$	$b = 0.25$	$b = 0.5$	$b = 0.75$	$b = 1.0$
Peak values	10.27	12.27	14.22	15.77	17.20	14.56	14.88	15.57	16.69	18.28	13.68	14.45	15.43	16.55	18.05
Residual values	8.72	10.30	11.90	13.10	14.16	11.94	12.46	13.32	14.17	15.60	11.34	12.21	13.17	14.31	15.65

Table 3. Peak and residual values of bearing capacity factor N_γ yield by FLAC and UST.

Geometry	Strip					Circular					Square				
UST	$b = 0.0$	$b = 0.25$	$b = 0.5$	$b = 0.75$	$b = 1.0$	$b = 0.0$	$b = 0.25$	$b = 0.5$	$b = 0.75$	$b = 1.0$	$b = 0.0$	$b = 0.25$	$b = 0.5$	$b = 0.75$	$b = 1.0$
Peak values	1.17	1.51	1.83	2.09	2.35	1.05	1.11	1.18	1.30	1.49	1.33	1.44	1.57	1.73	1.90
Residual values	0.67	0.86	0.99	1.14	1.28	0.58	0.63	0.68	0.75	0.84	0.79	0.85	0.94	1.03	1.15

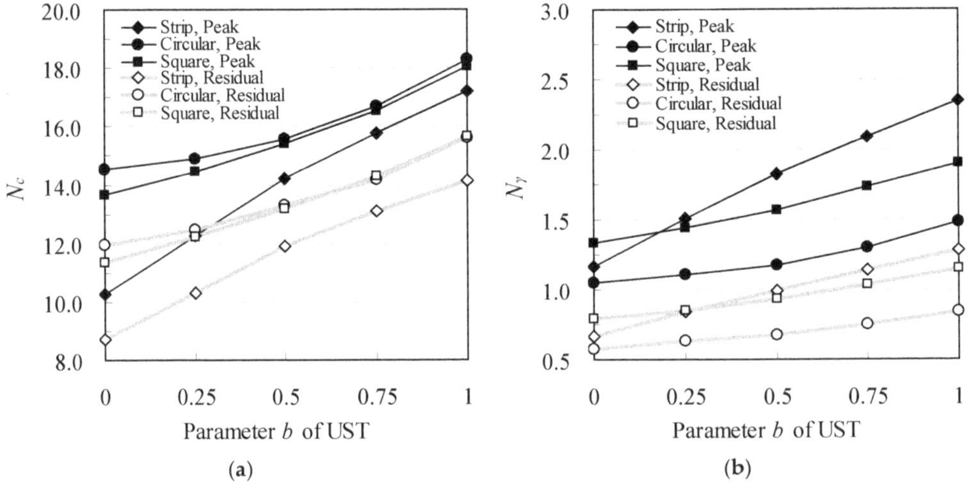

Figure 7. Peak and residual values of bearing capacity factor of strip, circular and square footings vary with UST parameter b: (**a**) N_c, (**b**) N_γ.

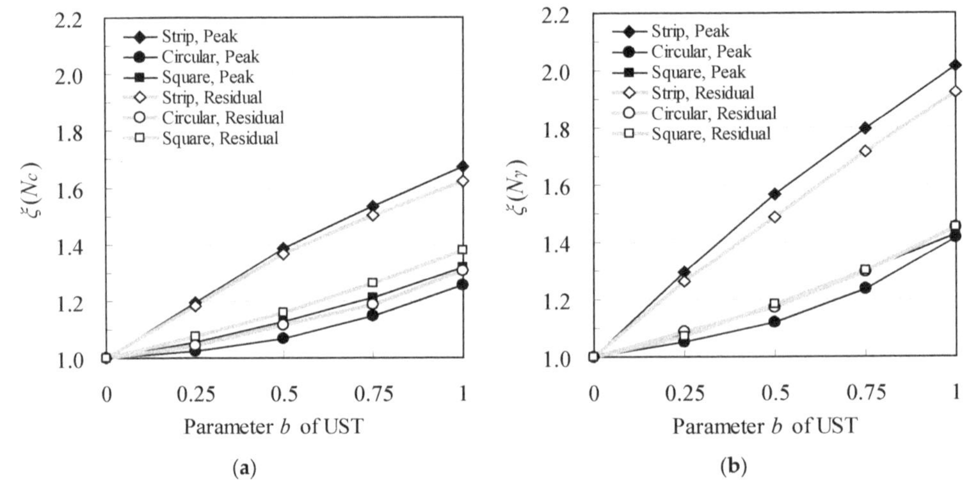

Figure 8. Efficiency factors of intermediate principal stress effect for bearing capacity of strip, circular and square footing: (**a**) N_c, (**b**) N_γ.

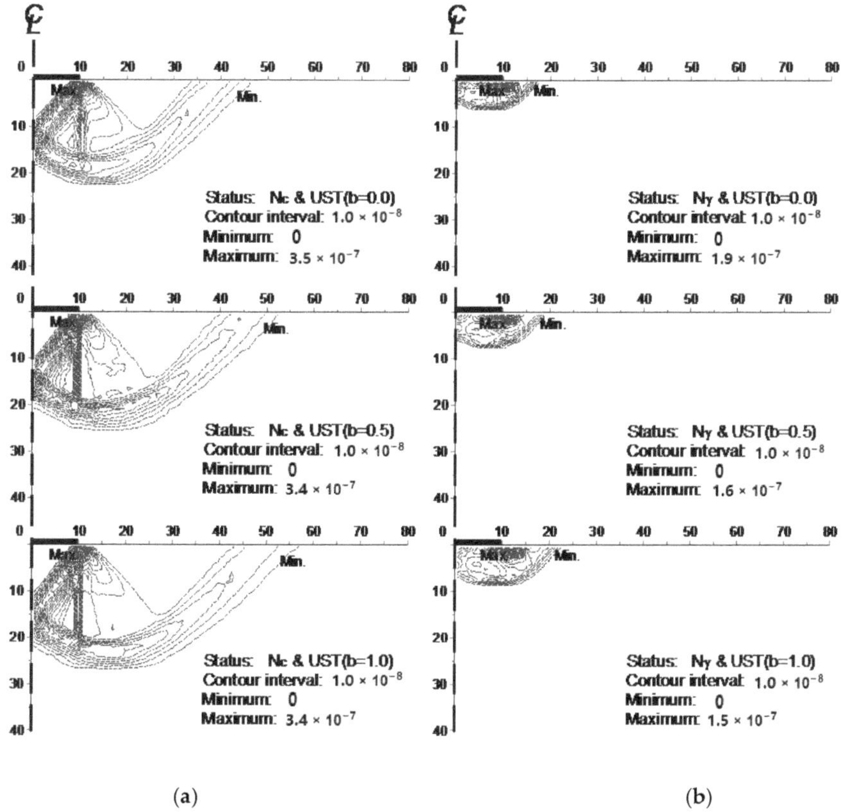

Figure 9. Isolines of maximum shear strain rate of foundation beneath strip footing under ultimate state: (**a**) N_c, (**b**) N_γ.

4. Discussion

The results of numerical computation show that the influence of the intermediate principal stress on the peak and residual values of N_γ is more apparent than on the bearing capacity factor N_c. The influence of the intermediate principal stress on the peak and residual values of the bearing capacity for a strip footing is much greater than for circular and square footings. The strain-softening behavior of soft rock foundations beneath strip, circular and square footings for factor N_γ evaluations are more apparent than that of the N_c evaluations. The influence of the intermediate principal stress on the peak and residual values of N_γ is more apparent than on the bearing capacity factor N_c. The influence of the intermediate principal stress on the peak and residual values of the bearing capacity for a strip footing is much greater than for circular and square footings. The difference between the influence of the intermediate principal stress on the peak and residual values of the bearing capacity of a soft rock foundation is relatively small. The size of the shear zone in the soft rock foundation increases as the value of the UST parameter b increases. The shear zone of the soft rock foundation under peak values of the bearing capacity is larger than that of the residual values of the bearing capacity.

The results obtained from FLAC/FLAC3D code using the UST indicate that a clear difference exists between the failure mechanisms of the foundation when the influence of the intermediate principal stress is taken into account versus when it is not. The size of the shear zone increases as the value of parameter b increases, and more soft rock beneath the footing contributes to the bearing capacity of the foundation. The size of the shear zone in the peak bearing capacity of soft rock foundations is larger than that of the residual bearing capacity, and the size of the shear zone for N_c evaluation is larger than that of the N_γ.

5. Conclusions

This paper presents a series of numerical solutions for the bearing capacity of strip, circular, and square footings in a soft rock foundation. Based on the results of this study, the following conclusions can be drawn:

(1) A strain hardening/softening constitutive model which takes the influence of the intermediate principal stress into account was established in this study. The mechanical behavior of the strain-softening material under the complex stress state can be analyzed using this model.

(2) The intermediate principal stress significantly influences a soft rock foundation's bearing capacity and failure mechanisms. The intermediate principal stress has less influence on the bearing capacity of footing when gravity is neglected. The influence of the intermediate principal stress on the bearing capacity for strip footing is much more significant than that of circular and square footings. The size of the failure area in the soft rock foundation increases with the increased effect of the intermediate principal stress.

(3) The research works of our study would benefit the reasonable estimation of the bearing capacity of footings in a soft rock medium. In our future research, taking the nonlinear failure envelope on the meridian plane into account, the influence of the intermediate principal stress effect on soft rock material will be investigated.

Author Contributions: Conceptualization, Z.M.; Methodology, Z.M.; Investigation, Z.M.; Resources, H.L.; Data Curation, H.L.; Writing-Original Draft Preparation, Z.M.; Writing-Review and Editing, F.D. All authors have read and agreed to the published version of the manuscript.

Funding: This research received the funding of the National Nature Science Foundation of China (No. 51879212, 51979225).

Institutional Review Board Statement: Not applicable.

Informed Consent Statement: Not applicable.

Data Availability Statement: All data, models, and code generated or used during the study appear in the submitted article.

Conflicts of Interest: The authors declare no conflict of interest.

References

1. British Standards Institution. *Geotechnical Investigation and Testing—Identification and Classification of Rock*; ISO14689: 2017; British Standards Institution: London, UK, 2018.
2. International Society for Rock Mechanics (ISRM). *The ISRM Suggested Methods for Rock Characterization, Testing and Monitoring: 2007–2014*; Ulusay, R., Ed.; Pergamon Press: Oxford, UK, 2015.
3. Berre, T. Triaxial Testing of Soft Rocks. *Geotech. Test. J.* **2011**, *34*, 61–75. [CrossRef]
4. Liao, H.J.; Su, L.J.; Pu, W.C.; Yin, J.H. Test and numerical analysis of the constitutive relation of a diatomaceous soft rock. *Mar. Geotechnol.* **2003**, *21*, 183–200. [CrossRef]
5. Jovičić, V.; Lapčević, R.; Bogdanović, S. Preservation of Historical Underground Sites in Soft Rock: A Case Example. *Geosciences* **2020**, *10*, 256. [CrossRef]
6. Liu, B.; Jiang, X. Consolidation and deformation characteristics of soft rock foundation in hydrological wetland environment. *Earth Sci. Res. J.* **2020**, *24*, 183–190. [CrossRef]
7. Zhu, H.; Ye, B.; Cai, Y.; Feng, Z. An elasto-viscoplastic model for soft rock around tunnels considering overconsolidation and structure effects. *Comput. Geotech.* **2013**, *50*, 6–16. [CrossRef]
8. Tokiwa, T.; Tsusaka, K.; Matsubara, M.; Ishikawa, T. Fracture characterization around a gallery in soft sedimentary rock in horonobe url of japan. *Int. J. Rock Mech. Min. Sci.* **2014**, *65*, 1–7. [CrossRef]
9. Erickson, H.L.; Drescher, A. Bearing capacity of circular footings. *J. Geotech. Geoenvironmental Eng.* **2002**, *128*, 38–43. [CrossRef]
10. Frydman, S.; Burd, H. Numerical studies of bearing-capacity factor ngamma. *J. Geotech. Geoenvironmental Eng.* **1997**, *123*, 20–29. [CrossRef]

11. Loukidis, D.; Salgado, R. Bearing capacity of strip and circular footings in sand using finite elements. *Comput. Geotech.* **2009**, *36*, 871–879. [CrossRef]
12. Salgado, R.; Lyamin, A.V.; Sloan, S.W.; Yu, H.S. Two and three-dimensional bearing capacity of foundations in clay. *Géotechnique* **2004**, *54*, 297–306. [CrossRef]
13. Sukmak, G.; Sukmak, P.; Horpibulsuk, S.; Hoy, M.; Arulrajah, A. Load Bearing Capacity of Cohesive-Frictional Soils Reinforced with Full-Wraparound Geotextiles: Experimental and Numerical Investigation. *Appl. Sci.* **2021**, *11*, 2973. [CrossRef]
14. Hosamo, H.; Sliteen, I.; Ding, S. Numerical Analysis of Bearing Capacity of a Ring Footing on Geogrid Reinforced Sand. *Buildings* **2021**, *11*, 68. [CrossRef]
15. Galindo, R.; Alencar, A.; Isik, N.S.; Olalla Marañón, C. Assessment of the Bearing Capacity of Foundations on Rock Masses Subjected to Seismic and Seepage Loads. *Sustainability* **2020**, *12*, 10063. [CrossRef]
16. Ma, Z.Y.; Liao, H.J.; Dang, F.N. Effect of intermediate principal stress on strength of soft rock under complex stress states. *J. Cent. South Univ.* **2014**, *21*, 1583–1593. [CrossRef]
17. Ma, Z.Y.; Liao, H.J.; Dang, F.N. Influence of intermediate principal stress on the bearing capacity of strip and circular footings. *J. Eng. Mech. ASCE* **2014**, *140*, 04014041. [CrossRef]
18. Lade, P.V. Assessment of test data for selection of 3-d failure criterion for sand. *Int. J. Numer. Anal. Methods Geomech.* **2006**, *30*, 307–333. [CrossRef]
19. You, M. True-triaxial strength criteria for rock. *Int. J. Rock Mech. Min. Sci.* **2009**, *46*, 115–127. [CrossRef]
20. Sutherland, H.B.; Mesdary, M.S. The influence of the intermediate principal stress on the strength of sand. In Proceedings of the 7th International Conference on Soil Mechanics and Foundation Engineering, Mexico City, Mexico, 12–16 January 1969; pp. 391–399.
21. Mogi, K. Effect of the intermediate principal stress on rock failure. *J. Geophys. Res.* **1967**, *72*, 5117–5131. [CrossRef]
22. Xue, Y.; Liu, J.; Ranjith, P.G.; Liang, X.; Wang, S. Investigation of the influence of gas fracturing on fracturing characteristics of coal mass and gas extraction efficiency based on a multi-physical field model. *J. Pet. Sci. Eng.* **2021**, *206*, 109018. [CrossRef]
23. Kulatilake, P.H.S.W. 3-D Rock Mass Strength Criteria—A Review of the Current Status. *Geotechnics* **2021**, *1*, 7. [CrossRef]
24. Xu, P.; Sun, Z.; Shao, S.; Fang, L. Comparative Analysis of Common Strength Criteria of Soil Materials. *Materials* **2021**, *14*, 4302. [CrossRef]
25. Liu, Y.; Liao, X.; Li, L.; Mao, H. Discrete Element Modelling of the Mechanical Behavior of Sand–Rubber Mixtures under True Triaxial Tests. *Materials* **2020**, *13*, 5716. [CrossRef]
26. Song, Z.; Li, M.; Yin, G.; Ranjith, P.G.; Zhang, D.; Liu, C. Effect of Intermediate Principal Stress on the Strength, Deformation, and Permeability of Sandstone. *Energies* **2018**, *11*, 2694. [CrossRef]
27. Guan, H.-X.; Wang, H.-Q.; Liu, H.; Yan, J.-J.; Lin, M. The Effect of Intermediate Principal Stress on Compressive Strength of Different Cement Content of Cement-Stabilized Macadam and Different Gradation of AC-13 Mixture. *Appl. Sci.* **2018**, *8*, 2000. [CrossRef]
28. Rukhaiyar, S.; Sajwan, G.; Samadhiya, N.K. Strength behavior of plain cement concrete subjected to true triaxial compression. *Can. J. Civ. Eng.* **2018**, *45*, 179–196. [CrossRef]
29. Yang, Q.; Zan, Y.; Xie, L.G. Comparative analysis of the nonlinear unified strength criterion for rocks and other three-dimensional Hoek–Brown strength criteria. *Geomech. Geophys. Geo-Energy Geo-Resour.* **2018**, *4*, 29–37. [CrossRef]
30. Zienkiewicz, O.C.; Pande, G.N. *Finite Elements in Geomechanics*; Wiley: New York, NY, USA, 1977.
31. Matsuoka, H.; Nakai, T. Relationship among tresca, mises, mohr-coulomb and matsuoka-nakai failure criteria. *Soils Found.* **1985**, *25*, 123–128. [CrossRef]
32. Yu, M.H. *Unified Strength Theory and Its Applications*; Springer: Berlin/Heidelberg, Germany, 2004.
33. Itasca Consulting Group Inc. *FLAC-Fast Lagrangian Analysis of Continua (Version 8.1) User's Manual*; Itasca Consulting Group Inc.: Minneapolis, MN, USA, 2019.
34. Itasca Consulting Group Inc. *FLAC3D-Fast Lagrangian Analysis of Continua in 3 Dimensions (Version 7.0) User's Manual*; Itasca Consulting Group Inc.: Minneapolis, MN, USA, 2019.
35. Ma, Z.Y.; Liao, H.J.; Dang, F.N. Unified elastoplastic finite difference and its application. *Appl. Math. Mech.* **2013**, *34*, 457–474. [CrossRef]
36. Ma, Z.Y.; Dang, F.N.; Liao, H.J. Numerical study of the dynamic compaction of gravel soil ground using the discrete element method. *Granul. Matter* **2014**, *16*, 881–889. [CrossRef]
37. Vermeer, P.A.; Borst, D. Non-associated plasticity for soils, concrete and rocks. *Heron* **1984**, *29*, 3–64. [CrossRef]
38. Li, H.Z.; Liao, H.J.; Kong, L.W.; Leng, X.L. Experimental study on stress-strain relationship of expansive mud-stone. *Yantu Lixue/Rock Soil Mech.* **2007**, *28*, 107–110. (In Chinese)
39. Kang, X.S.; Liao, H.J. A Bounding Surface Plasticity Model Considering Spacing Ratio for Overconsolidated Jointed Soft Rocks. *Rock Mech. Rock Eng.* **2020**, *53*, 59–69. [CrossRef]
40. Vesic, A.S. Analysis of ultimate loads of shallow foundations. *J. Soil Mech. Found. Div. ASCE* **1973**, *99*, 45–73. [CrossRef]
41. Terzaghi, K. *Theoretical Soil Mechanics*; Wiley: New York, NY, USA, 1943.
42. Eason, G.; Shield, R.T. The plastic indentation of a semi-infinite solid by a perfectly rough circular punch. *Z. Angew. Math. Phys. Zamp* **1960**, *11*, 33–43. [CrossRef]
43. Cox, A.D. Axially-symmetric plastic deformation in soils—II. indentation of ponderable soils. *Int. J. Mech. Sci.* **1962**, *4*, 371–380. [CrossRef]

44. Dixit, M.S.; Patil, K.A. Experimental estimate of nγ values and corresponding settlements for square footings on finite layer of sand. *Geomech. Eng.* **2013**, *5*, 363–377. [CrossRef]
45. Puzakov, V.; Drescher, A.; Michalowski, R.L. Shape factor for shallow footings. *Geomech. Eng.* **2009**, *1*, 113–120. [CrossRef]
46. Long, Y.; Liu, J.; Kong, X.J.; Hu, Y. Three-dimensional large deformation fe analysis of square footings in two-layered clays. *J. Geotech. Geoenviron. Eng. ASCE* **2011**, *137*, 52–58. [CrossRef]
47. Zhu, M.; Radoslaw, L.; Michalowski, R.L. Shape factors for limit loads on square and rectangular footings. *J. Geotech. Geoenviron. Eng.* **2005**, *131*, 223–231. [CrossRef]

coatings

MDPI

Article

Anti-Blocking Mechanism of Flocking Drainage Pipes in Tunnels Based on Mathematical Modeling Theory

Shiyang Liu [1,2,*], Xuefu Zhang [1,2] and Feng Gao [1,2]

1 College of Civil Engineering, Chongqing Jiaotong University, Chongqing 400074, China; zhangxuefu400074@126.com (X.Z.); 990020701005@cqjtu.edu.cn (F.G.)
2 State Key Laboratory of Mountain Bridge and Tunnel Engineering, Chongqing Jiaotong University, Chongqing 400074, China
* Correspondence: cqjtulsy@163.com

Abstract: Crystalline pipe plugging in tunnel drainage systems is one of the causes of tunnel lining cracking and water leakage. Therefore, effective prevention of crystalline pipe blockage in tunnel drainage systems is very important to ensure the safety and stability of the lining structure during tunnel operation. Combined with the theories of fluid mechanics, structural mechanics and basic physics, the flocking and anti-blocking mechanism of drainage pipe was comprehensively analyzed by using the method of mathematical modeling, including: (1) the calculation expression of average velocity of the flocked section of a flocked drainage pipe $v = Q/(C_1 - C_2(r + r'))$ and the calculation formula of flowing water pressure under the action of groundwater $F_w = KA\gamma Q^2/(2g(C_1 - C_2(r + r'))^2)$; (2) the flow velocity v_0 in the flocked drainage pipe shall meet $v^2 < 4g\pi rl\tau_1/\gamma KA$, crystals will be attached to the fluff and the crystals will maintain dynamic balance; (3) the flow velocity v_0 in the flocked drainage pipe shall meet $v^2 \geq 4g\pi rl\tau_1/\gamma KA$, crystals will not adhere to the fluff and the flocked drainage pipe will remain unobstructed. The research on the mechanism of preventing blockage of flocking drainage pipes fills the gap in research theory in this regard, contributes to the popularization and application of blocking prevention technology of flocking drainage pipes, reduces the maintenance cost during operation of tunnel drainage systems and ensures the safe and normal operation of tunnels.

Keywords: anti-blocking of flocking drainage pipe; drainage pipe blockage by crystals; mechanism; mathematical modeling; tunnel

Citation: Liu, S.; Zhang, X.; Gao, F. Anti-Blocking Mechanism of Flocking Drainage Pipes in Tunnels Based on Mathematical Modeling Theory. *Coatings* **2021**, *11*, 961. https://doi.org/10.3390/coatings11080961

Received: 5 July 2021
Accepted: 11 August 2021
Published: 13 August 2021

1. Introduction

As of the end of 2020, there were 21,316 highway tunnels in China, including 1394 extra-long tunnels and 5541 long tunnels [1]. However, with the operation of the tunnels, various types of damage have gradually appeared, especially lining leakage (Figure 1) and cracking (Figure 2) caused by the crystallization-induced blockage of the drainage system of the tunnel in the karst area (Figure 3) [2], which seriously affects the safety of the tunnel lining. At present, a tunnel drainage system blockage caused by crystals can only be dredged by high-pressure gas (water) [3] or excavation, which is costly and affects the normal passage of the tunnel.

To make the lining structure of the tunnel safe and stable, it is necessary to ensure the effective and smooth operation of the drainage system of the tunnel. Therefore, it is necessary to prevent the occurrence of crystallization-induced pipe blockage in the drainage system in advance. In view of the problem of crystallization blockage in the tunnel drainage system, the current focus is mainly on three aspects: the analysis of crystal composition, the analysis of crystallization influencing factors and the technology to prevent crystallization. The main component of the crystalline material of the tunnel drain pipes is insoluble rectangular calcium carbonate crystals [4–6] such as calcite crystal, and the calcium element in the crystal is derived from the groundwater [4,7] in the tunnel and the cement [8] in the shotcrete. The main influencing factors of crystallization-induced blockage include CO_2 partial pressure, flow velocity, temperature, pH, ion type and concentration [9,10]. In addition to the thickness of the diffusion boundary layer (DBL), hydrodynamic factors such as liquid flow velocity and liquid level height as well as the friction coefficient of the interior wall of drainage pipes also have an important impact on the precipitation and crystallization rate of drainage pipes in the karst area [11]. The amount of crystal in the drain pipe increases with the increase in pH value, but is affected by the coupling of the pH and the water filling state of the drainage pipe. The calcium carbonate crystals are mainly spindle shaped. The higher the pH value, the smaller the grain size, and the denser the accumulation [12]. The use of protective coatings to hydrophobicize the concrete base surface and the wall surface of the polyvinyl chloride (PVC) pipe can reduce the attachment of calcium carbonate crystals [13,14]; the generation of crystals can be effectively reduced by optimizing the concrete and concrete mix ratio, reducing the contact between groundwater and concrete, preventing CO_2 from entering the tunnel drainage pipe, and adding appropriate fly ash to shot concrete [10,15]. PEG-b-PAA-b-PS, (poly(ethylene glycol)-block-poly(acrylic acid)-block-poly(styrene)) can prevent the phase transition from vaterite to calcite [16]. In the presence of biopolymer, the relative content of vaterite increases with the application of ultrasonic treatment [17]; ultrasonic treatment makes the aggregated calcium carbonate crystals more fragile [18]. RS1600, a green corrosion inhibitor, changes the crystal structure of calcium carbonate from calcite to vaterite [19]. A cleaning solvent of organic acid reagents of single molecule carboxylic acid with a concentration of 2000 ppm and a dichchromate index of 17.71% and polymerized carboxylic acid can effectively remove karst crystals in a drainage system under the premise of ensuring environmental protection [20]. Through a large number of 1:1 indoor model tests and numerical simulation analyses, Liu et al. studied the feasibility and reliability of the anti-blocking mechanism of flocking drainage pipes from the macro perspective, and obtained some good flocking parameters [21–25]. However, the anti-blocking mechanism of flocking drainage pipes is not very mature, and further research is needed.

Figure 1. Water receiver at lining leakage.

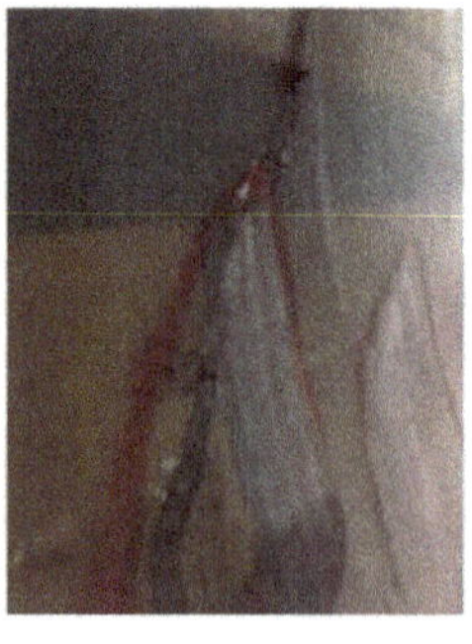

Figure 2. Cracking of the tunnel lining.

Figure 3. Tunnel drainage pipe blockage by crystals.

Based on the above research results, combined with the relevant knowledge of basic physics and mechanics, this paper discusses the mathematical expression of the anti-blocking mechanism of flocking drainage pipes using the mathematical modeling method, which provides theoretical support for the anti-blocking mechanism of flocking drainage pipes.

2. Method

2.1. Average Flow Velocity of Flocking Drainage Pipe Section

Due to the existence of fluff in the flocking drain pipes, the local velocity of the flocking section differs from that of the non-flocking section, but it still follows the most basic physical law. The schematic diagram of the average flow velocity of a flocking drainage pipe section is shown in Figure 4.

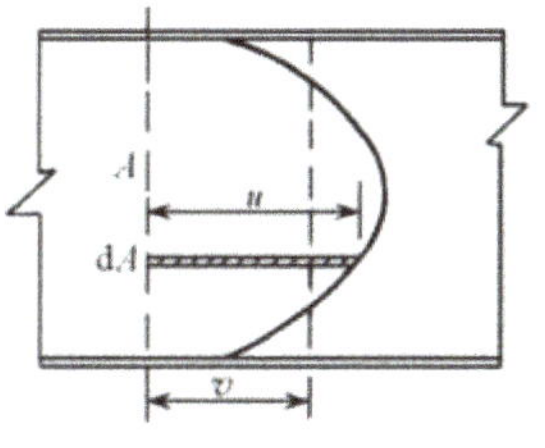

Figure 4. The average flow velocity of a flocking drainage pipe section.

According to the definition of the average flow velocity of a flocking drainage pipe section, the equation is:

$$Q = \int_A u dA = \int_A v dA = v \int_A dA = vA, \tag{1}$$

or

$$v = \frac{Q}{A}, \tag{2}$$

where

- Q—Rate of flow;
- u—Flow velocity at a point in the pipe;
- Sectional area of water in the pipe;
- v—Average flow velocity in the pipe section.

2.2. Rule of the Flow Velocity Distribution of Flocking Drainage Pipe

It is generally believed that the flow velocity presents a parabolic distribution on the vertical surface, and its curve equation is as follows [26]:

$$V = V_{max}(h/H)^n, \tag{3}$$

where

- V—flow velocity at any depth of vertical surface;
- V_{max}—flow velocity of water surface;
- H—water depth;
- h—water depth at any point;
- n—constant, which was determined by the nature of water flow and the interior wall of the drainage pipe.

Figure 5 shows the flow velocity distribution curve and the joint force focus position of flowing water pressure at different n values (shown by the arrow).

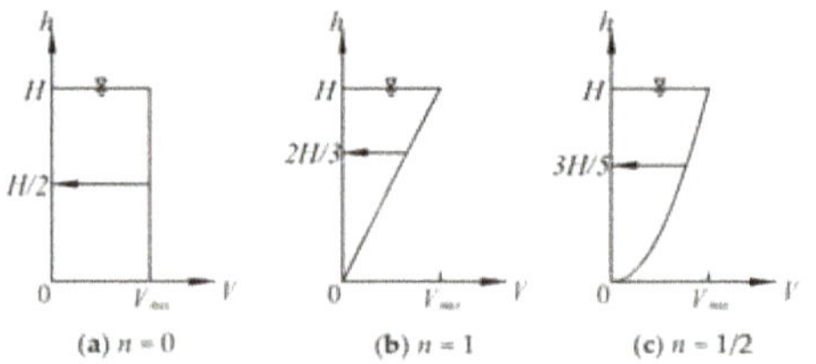

Figure 5. The flow velocity distribution curve and the joint force focus position of flowing water pressure: (**a**) $n = 0$; (**b**) $n = 1$; (**c**) $n = 1/2$. [26].

3. Results and Discussion

3.1. Analysis of the Flow Velocity of Flocking Drainage Pipes

Assuming mountain size A, tunnel size B and drainage pipe size C, then C < B < A and C << A. The groundwater level in the mountain is stable in most cases, and the water flow in the tunnel drainage pipe is very small compared with the groundwater in the entire mountain. According to the definition of the flow field, the groundwater flow field in the tunnel drainage pipe can be regarded as a constant flow. The existence of fluff changes the local water crossing section dimension of the flocking drainage pipe, so the groundwater flow field in the tunnel drainage pipe is non-uniform flow. In conclusion, the groundwater flow field in the tunnel drainage pipe is a constant and non-uniform flow.

For flocking drainage pipes, the average flow velocity of the section of the non-flocked part can be calculated according to Formula (2). The existence of fluff changes the local water crossing section dimension of flocking drainage pipe, so the equation for its average flow speed needed to be deduced. The average flow velocity of a flocking drainage pipe section was deduced according to the basic principle of Equation (2), and the calculation diagram is shown in Figure 6.

Area of △OAB:

$$A_1 = \sqrt{R^2 - (R-H)^2} \cdot (R-H) \tag{4}$$

Area of the fluff of the water crossing section:

$$A_2 = 2r \cdot l \cdot N, \tag{5}$$

$$N = \frac{\theta}{\alpha} - 1 \tag{6}$$

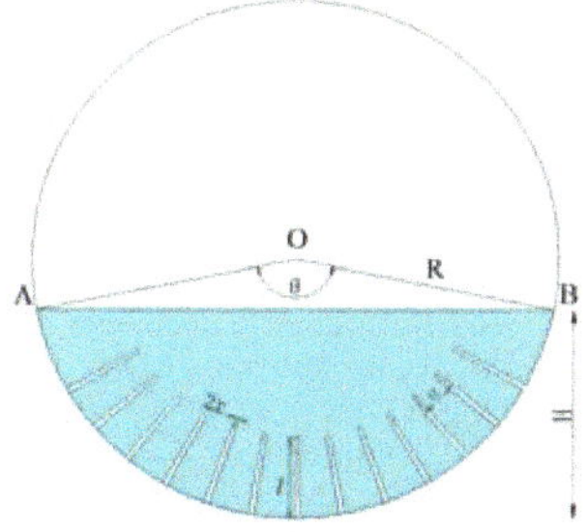

Figure 6. The average flow velocity of the flocking drainage pipe section.

Area of arc corresponding to θ:

$$A = \frac{\pi R^2 \theta}{360} \tag{7}$$

Water crossing area of the flocking drainage pipe section:

$$A_3 = A - A_1 - A_2 \tag{8}$$

The following equation was derived from the geometric relationship in the calculation diagram:

$$\theta = 2\arccos(\frac{R-H}{R}) \tag{9}$$

Equations (3)–(6) were substituted into Equation (8) to obtain:

$$A_3 = \frac{\pi R^2}{180}\arccos(\frac{R-H}{R}) - \sqrt{R^2-(R-H)^2}\cdot(R-H) - \left[\frac{2}{\alpha}\arccos(\frac{R-H}{R}) - 1\right]\cdot 2rl \tag{10}$$

Then, Equation (10) was substituted into Equation (2), and the average flow velocity of the flocking drainage pipe section was obtained:

$$v = \frac{Q}{\frac{\pi R^2}{180}\arccos(\frac{R-H}{R}) - \sqrt{R^2-(R-H)^2}\cdot(R-H) - \left[\frac{2}{\alpha}\arccos(\frac{R-H}{R}) - 1\right]\cdot 2rl}, \tag{11}$$

where

- v—average flow velocity of flocking drainage pipe section (mm/s);
- Q—groundwater flow in flocking drainage pipe (mL/s);
- R—radius of flocking drainage pipe (mm);
- H—cross-section height of groundwater in flocked drainage pipe (mm);
- α—flocking circular spacing of flocked drainage pipe (°);
- r—radius of fluff of flocking drainage pipe (mm);
- l—length of fluff of flocking drainage pipe (mm); N—amount of fluff of flocking drainage pipe section.

It was assumed that the crystal fluff under water was covered by crystals during the movement of groundwater, and the thickness of crystal was r′. Therefore, we only needed to replace the r in the above Equation (11) with (r + r'), and simplify the constant term in the equation according to the mathematical modeling theory to obtain:

$$v = \frac{Q}{A_3} = \frac{Q}{C_1 - C_2(r + r')} \tag{12}$$

As can be seen from the above Equation (12), when the groundwater flow was constant, the greater the r', the smaller the area of the water crossing section, and the greater the v. It was consistent with the distribution law of a 2D flow field of flocking drainage pipes by Liu et al. [24]. Through a simple mathematical transformation of Equation (12), the equation between the thickness of crystal and the flow velocity of the flocking drainage pipe section was obtained.

$$r' = \frac{(C_1 - Q/v)}{C_2} - r \tag{13}$$

3.2. Analysis of the Fluff Stress of Flocking Drainage Pipe

Referring to the General Specifications for Design of Highway Bridges and Culverts issued by the Ministry of Transport of the People's Republic of China in 2015 [27], the equation of the flowing water pressure in the flocking drainage pipe was:

$$F_w = KA\frac{\gamma v^2}{2g}, \tag{14}$$

where

- F_W—standard value of flowing water pressure (kN);
- γ—unit weight of groundwater (kN/m^3);
- v—average flow velocity of flocking drainage pipe section (m/s);
- water resistance area by fluff (m^2),
- $A = 2rl$, g—acceleration of gravity, $g = 9.81$ m/s^2;
- K—coefficient of fluff shape, which was 0.8 for round fluff, 1.3 for rectangular fluff.

Equation (12) was substituted into Equation (14) to obtain the equation of flowing water pressure at the fluff in the flocking drainage pipe:

$$F_w = KA\frac{\gamma Q^2}{2g(C_1 - C_2(r + r'))^2} \tag{15}$$

It can be seen from the above formula that the amount of crystal on the fluff of the flocking drainage pipe is inversely proportional to the flowing water pressure of the fluff. When the crystals increase, the flowing water pressure on the fluff becomes greater; when the crystals decrease, the flowing water pressure on the fluff decreases.

3.3. Analysis of the Interaction between the Fluff, the Crystal and the Groundwater

During the movement of groundwater in the flocking drainage pipe, the easily crystallized anions and cations in the water form crystals, which gradually adhere to the pipe wall and the fluff. When the water pressure is greater than the adhesion force between the crystals and the fluff, the crystals are shed from the fluff. When the water pressure is greater than the adhesion force between the crystals and less than the adhesion between the crystals and the fluff, some of the crystals fall off.

The adhesion force between the crystal and the fluff:

$$F_1 = 2\pi rl \cdot \tau_1 \tag{16}$$

The adhesion force between the crystals:

$$F_2 = 2\pi(r + r')l \cdot \tau_2, \tag{17}$$

where

- τ_1—adhesion force between the crystal and the fluff (kN/m^2);
- τ_2—adhesion force between the crystals (kN/m^2);
- r'—fixed thickness of crystal on the fluff (m).

(1) When the flowing water pressure was greater than the adhesion force between the crystals and the fluff, namely:

$$KA\frac{\gamma v^2}{2g} > 2\pi rl \cdot \tau_1, \tag{18}$$

and through simple transformation, we obtained:

$$v^2 > \frac{4g\pi rl \cdot \tau_1}{\gamma KA} \tag{19}$$

That is, when the squared value of test flow velocity in the field was greater than the calculated value of the right side of Equation (19), the crystals will not adhere to the fluff, but will flow out of the pipe with groundwater or adhere to the pipe wall.

(2) When the water pressure is greater than the adhesion force between the crystals and less than the adhesion between the crystals and the fluff, namely:

$$2\pi(r + r')l \cdot \tau_2 < KA\frac{\gamma v^2}{2g} < 2\pi rl \cdot \tau_1 \tag{20}$$

and through simple transformation, we obtained:

$$\frac{4g\pi(r + r')l \cdot \tau_2}{\gamma KA} < v^2 < \frac{4g\pi rl \cdot \tau_1}{\gamma KA} \tag{21}$$

That is to say, when the square value of the flow velocity is between the calculated value on the left side of Equation (19) and that of the right side of Equation (21), some of the crystals will be attached to the fluff, and but the remaining crystals would fall from the fluff under the effect of groundwater. By substituting the calculation results of Equation (13) into Equations (19) and (21), we determined the status of crystal on a single piece of fluff in the flocking drainage pipe.

3.4. Analysis of the Changing Law of Crystals in Flocking Drainage Pipes over Time

By using Equation (21), the changing law of r' over time is obtained.

(1) When the initial velocity v_0 met $v^2 \le \frac{4g\pi(r + r')l \cdot \tau_2}{\gamma KA}$, the crystals began to increase. When the r' increased, the v gradually increased to $v^2 = \frac{4g\pi(r + r')l \cdot \tau_2}{\gamma KA}$, and the amount of crystal on the fluff remained unchanged.

Assuming that the crystallization velocity of crystal on the fluff was constant, namely

$$\frac{dr'}{dt} = \alpha_1$$

and by finding the derivative of both sides of Equation (12) to t, we obtained:

$$\frac{dv}{dt} = \frac{Q}{(C_1 - C_2(r + r'))^2 C_2 \frac{dr'}{dt}} = \frac{QC_2\alpha_1}{(C_1 - C_2(r + r'))^2} = \frac{C_2\alpha_1 v^2}{Q}$$

Through transformation, we obtained:

$$\frac{dv}{v^2} = \frac{C_2\alpha_1 dt}{Q}$$

By finding the integral of the both sides of the above equation simultaneously, we obtained:

$$\int \frac{dv}{v^2} = \int \frac{C_2\alpha_1 dt}{Q},$$
$$-(\frac{1}{v} - \frac{1}{v_0}) = \frac{C_2\alpha_1 t}{Q},$$
$$-\frac{1}{v} = \frac{C_2\alpha_1 t}{Q} - \frac{1}{v_0}$$

Finally, we obtained:

$$v(t) = \frac{1}{(\frac{1}{v_0} - \frac{C_2\alpha_1 t}{Q})} \tag{22}$$

By substituting Equation (13) into the above equation, we obtained the equation between the amount of crystal on the fluff and the time:

$$r'(t) = \frac{(C_1 - \frac{Q}{v(t)})}{C_2} - r = \frac{(C_1 - Q(\frac{1}{v_0} - \frac{C_2\alpha_1 t}{Q}))}{C_2} - r \tag{23}$$

To solve the t_f in the steady state, make $v^2 = \frac{4g\pi\,(r+r')l \cdot \tau_2}{\gamma KA}$.

Then, $v = \sqrt{\frac{4g\pi(r+r')l \cdot \tau_2}{\gamma KA}} = \frac{1}{(\frac{1}{v_0} - \frac{C_2\alpha_1 t}{Q})}$.

We obtained $t_f = \frac{Q}{C_2\alpha_1}\left(\frac{1}{v_0} - \sqrt{\frac{\gamma KA}{4g\pi(r+r')l \cdot \tau_2}}\right)$.

In summary:

When $t < t_f$, $r'(t) = \frac{(C_1 - Q(\frac{1}{v_0} - \frac{C_2\alpha_1 t}{Q}))}{C_2} - r$.

When $t \geq t_f$, $r'(t) = \frac{(C_1 - Q(\frac{1}{v_0} - \frac{C_2\alpha_1 t_f}{Q}))}{C_2} - r$.

(2) When the initial velocity v_0 met $\frac{4g\pi\,(r+r')l \cdot \tau_2}{\gamma KA} < v^2 < \frac{4g\pi rl \cdot \tau_1}{\gamma KA}$, some of the crystals fell from the fluff under the effect of groundwater. When the r' decreased, the v gradually decreased to $v^2 = \frac{4g\pi(r+r')l \cdot \tau_2}{\gamma KA}$, and the amount of crystal on the fluff remained unchanged.

Assuming that the falling off velocity of crystals on the fluff was constant, namely

$$\frac{dr'}{dt} = -\alpha_2$$

and by finding the derivative of both sides of Equation (12), we obtained:

$$\frac{dv}{dt} = \frac{Q}{(C_1 - C_2(r+r'))^2 C_2 \frac{dr'}{dt}} = \frac{-QC_2\alpha_2}{(C_1 - C_2(r+r'))^2} = \frac{-C_2\alpha_2 v^2}{Q}$$

Through transformation, we obtained:

$$\frac{dv}{v^2} = \frac{-C_2\alpha_2 \mathrm{d}t}{Q}$$

By finding the integral of both sides of the above equation simultaneously, we obtained:

$$\int \frac{dv}{v^2} = \int \frac{-C_2\alpha_2 dt}{Q},$$
$$-(\frac{1}{v} - \frac{1}{v_0}) = \frac{-C_2\alpha_2 t}{Q},$$
$$\frac{1}{v} = \frac{C_2\alpha_2 t}{Q} + \frac{1}{v_0}$$

Finally, we obtained:

$$r'(t) = \frac{(C_1 - \frac{Q}{v(t)})}{C_2} - r = \frac{(C_1 - Q(\frac{1}{v_0} - \frac{C_2\alpha_1 t}{Q}))}{C_2} - r \tag{24}$$

By substituting Equation (13) into the above equation, we obtained the equation between the number of crystals on the fluff and the time:

$$r'(t) = \frac{(C_1 - \frac{Q}{v(t)})}{C_2} - r = \frac{(C_1 - Q(\frac{1}{v_0} + \frac{C_2\alpha_2 t}{Q}))}{C_2} - r \tag{25}$$

To solve the t_f in the steady state, we make $v^2 = \frac{4g\pi(r+r')l\cdot\tau_2}{\gamma KA}$.

Then, $v = \sqrt{\frac{4g\pi(r+r')l\cdot\tau_2}{\gamma KA}} = \frac{1}{(\frac{1}{v_0} + \frac{C_2\alpha_2 t}{Q})}$.

We can obtain $t_f = \frac{Q}{C_2\alpha_2}\left(-\frac{1}{v_0} + \sqrt{\frac{\gamma KA}{4g\pi(r+r')l\cdot\tau_2}}\right)$.

In summary:

When $t < t_f$, $r'(t) = \frac{(C_1 - Q(\frac{1}{v_0} + \frac{C_2\alpha_2 t}{Q}))}{C_2} - r$.

When $t \geq t_f$, $r'(t) = \frac{(C_1 - Q(\frac{1}{v_0} + \frac{C_2\alpha_2 t_f}{Q}))}{C_2} - r$.

(3) When the initial velocity v_0 met $v^2 \geq \frac{4g\pi rl\cdot\tau_1}{\gamma KA}$, all crystals fell from the fluff. When $t > 0$,

$$r'(t) = 0$$

That is, no crystal would attach to the fluff.

4. Conclusions

Through theoretical analysis and derivation, the conditions and operation time of crystal attachment and falling off in the flocking drainage pipe were obtained. The main conclusions were as follows:

(1) When the groundwater velocity v_0 in the pipe met $v^2 \leq 4g\pi(r+r')l\tau_2/\gamma KA$, crystals were attached to the fluff. When the running time of groundwater in the drainage pipe was $t > (Q/C_2\alpha_1)((1/v_0) - (\gamma KA/(4g\pi(r+r')l\cdot\tau_2))^{1/2})$, the crystals attached to the fluff, maintain dynamic balance.

(2) When the groundwater velocity v_0 in the pipe met $4g\pi(r+r')l\tau_2/\gamma KA < v^2 < 4g\pi rl\tau_1/\gamma KA$, crystals were still attached to the fluff. When the running time of groundwater in the drainage pipe was $t > (Q/C_2\alpha_2)((-1/v_0) + (\gamma KA/(4g\pi(r+r')l\cdot\tau_2))^{1/2})$, the crystals attached to the fluff, maintaining dynamic balance.

(3) When the groundwater velocity v_0 in the pipe met $v^2 \geq 4g\pi rl\tau_1/\gamma KA$, crystals did not adhere to the fluff.
(4) The adhesion between crystals and fluff and between crystals of the flocking drainage pipe needs to be determined by more in-depth experimental research, so as to improve the calculation model of the flocking anti-crystallization blocking mechanism of drainage pipes.

Author Contributions: Conceptualization, S.L., X.Z. and F.G.; Data curation, X.Z. and F.G.; Formal analysis, S.L.; Writing—original draft, S.L. and X.Z.; Writing—review & editing, S.L. All authors have read and agreed to the published version of the manuscript.

Funding: This research was funded by the National Natural Science Foundation of China (Grant No.51778095), Scientific Research Project of Emei Hanyuan Expressway Project (Grant No. LH-HT-45).

Institutional Review Board Statement: Not applicable.

Informed Consent Statement: Not applicable.

Data Availability Statement: The study did not report any data.

Acknowledgments: This research was financially supported by the National Natural Science Foundation of China (Grant No. 51778095) and the Scientific Research Project of the Emei Hanyuan Expressway Project (Grant No. LH-HT-45).

Conflicts of Interest: The authors declare no conflict of interest.

References

1. Ministry of Transport of the People's Republic of China. Statistical Bulletin on the Development of Transportation Industry in 2020 [EB/OL]. Available online: https://xxgk.mot.gov.cn/2020/jigou/zhghs/202105/t20210517_3593412.html (accessed on 19 May 2021).
2. Duau, H.; Chen, F.; Yao, C.; Wang, B.; Shu, W. The phenomenon of crystal pipe blocking in karst tunnel and its induced risk. *China Civ. Eng. J.* **2020**, *53*, 332–335.
3. Zhao, P.; Guo, X.; Ma, W. Research and application of dredge equipment for drainage pipeline in railway tunnel. *Railw. Eng.* **2018**, *58*, 30–32+66.
4. Zhang, X.; Zhou, Y.; Zhang, B.; Zhou, Y.; Liu, S.; Li, Y. Investigation and analysis on crystallization of tunnel drainage pipes in chongqing. *Adv. Mater. Sci. Eng.* **2018**, *2018*, 1–6. [CrossRef]
5. Guo, X. Crystallization mechanism and countermeasures of drainage system for railway tunnel. *China Railw. Sci.* **2020**, *41*, 71–77.
6. Lei, L.; Hua, Y.; Li, G.; Qi, J.; Li, D. Analysis and treatment of blind pipe blockage in humaling tunnel on lanzhou-chongqing railway. *Mod. Tunn. Technol.* **2020**, *57*, 149–153.
7. Wu, Y.; Li, G.; Jia, Q.; Qi, H.; Li, D. Study on characteristics and sources of crystal components of leakage water in the loess tunnel. *Mod. Tunn. Technol.* **2020**, *57*, 154–159.
8. Ye, F.; Tian, C.; He, B.; Zhao, M.; Wang, J.; Han, X.; Song, G. Experimental study on scaling and cloggingin drainage system of tunnels under construction. *China J. Highw. Transp.* **2021**, *34*, 159–170.
9. Jiang, Y.; Du, K.; Tao, L.; Zhao Jumei Xiao, H. Investigation and discussion on blocking mechanism of drainage system in karst tunnels. *Railw. Stand. Des.* **2019**, *63*, 131–135.
10. Tian, C.; Ye, F.; Song, G.; Wang, Q.; Zhao, M.; He, B.; Wang, J.; Han, X. On mechanism of crystal blockage of tunnel drainage system and preventive countermeasures. *Mod. Tunn. Technol.* **2020**, *57*, 77–83.
11. Zhou, W.; Wang, Y.; Li, Y.; Wang, L.; Jia, X. Research on the influence factors of crystallization and sedimentation of tunnel drainage pipeline in karst area. *Highway* **2021**, *66*, 347–352.
12. Xiang, K.; Zhou, J.; Zhang, X.; Huang, C.; Song, L.; Liu, S. Experimental Study on Crystallization Rule of Tunnel Drainpipe in Alkaline Environment. *Tunn. Constr.* **2019**, *39*, 207–212.
13. Zhou, Y.; Zhang, X.; Wei, L.; Liu, S.; Zhang, B.; Zhou, C. Experimental Study on Prevention of Calcium Carbonate Crystallizing in Drainage Pipe of Tunnel Engineering. *Adv. Civ. Eng.* **2018**, *2018*, 1–11. [CrossRef]
14. Jiang, Y.; Du, K.; Liao, J.; Chen, X.; Xiao, H. Experimental research on maintainability of drainage facilities in lining construction joints of karst tunnel. *Railw. Stand. Des.* **2019**, *63*, 91–96.
15. Ye, F.; Tian, C.; Zhao Meng He, B.; Wang, J.; Han, X. The disease of scaling and cloging in the drainage pipes of a tunnel under construction in Yunnan. *China Civ. Eng. J.* **2020**, *53*, 336–341.
16. Su, Y.; Yang, H.; Shi, W.; Guo, H.; Zhao, Y.; Wang, D. Crystallization and morphological control of calcium carbonate by functionalized triblock copolymers. *Colloids Surf. A Physicochem. Eng. Asp.* **2009**, *355*, 158–162. [CrossRef]
17. Semra, K.; Mualla, O.; Emel, A. The effect of ultrasonication on calcium carbonate crystallization in the presence of biopolymer. *J. Cryst. Growth* **2014**, *401*, 266–270.

18. Su, M.; Han, J.; Li, Y.; Chen, J.; Zhao, Y.; Keith, C. Ultrasonic crystallization of calcium carbonate in presence of seawater ions. *Desalination* **2015**, *369*, 85–90. [CrossRef]
19. Rayane, M.; Samira, G.; Mohamed, T. Calcium carbonate inhibition by green inhibitors: Thiamine and pyridoxine. *Desalination* **2017**, *404*, 147–154.
20. Hong, Y.; Qian, X.; Li, J.; Yang, H.; Zhang, P. On scavenging performances of cleaning solvents for the clogging in the drainage system of karst tunnels. *Mod. Tunn. Technol.* **2020**, *57*, 160–170.
21. Liu, S.; Zhang, X.; Lü, H.; Liu, Q.; Wang, B. The effect of flocking pvc pipe on the prevention and crystallization of tunnel drains. *Sci. Technol. Eng.* **2018**, *18*, 313–319.
22. Liu, S.; Zhang, X.; Lü, H.; Liu, Q.; Wang, B. The effect of anti-crystallization of tunnel plumage drain pipe under different water filling state. *Sci. Technol. Eng.* **2018**, *18*, 156–163.
23. Liu, S.; Gao, F.; Zhou, Y.; Liu, Q.; Lü, H.; Wang, B.; Xiang, K.; Xiao, D. Effect of fuzz length on the prevention of crystallization of tunnel flocking drainpipes. *Sci. Technol. Eng.* **2019**, *19*, 234–239.
24. Liu, S.; Gao, F.; Zhang, X.; Han, F.; Zhou, Y.; Xiang, K.; Xiao, D. Experimental study on anti-crystallization law of tunnel transverse flocking drainpipe at different velocities. *Asia-Pac. J. Chem. Eng.* **2020**, *15*, 1–9. [CrossRef]
25. Liu, S.; Zhang, X.; Gao, F.; Wei, L.; Liu, Q.; Lü, H.; Wang, B. Two-dimensional flow field distribution characteristics of flocking drainage pipes in tunnel. *Open Phys.* **2020**, *18*, 139–148. [CrossRef]
26. Zhou, Y. Calculation of bridge water pressure. *East China Highw.* **1992**, *5*, 24–25.
27. *JTG D60-2015 General Specifications for Design of Highway Bridges and Culverts*; People's Communications Publishing House Co., Ltd.: Beijing, China, 1 December 2015. Available online: https://xxgk.mot.gov.cn/2020/jigou/glj/202006/t20200623_3312312.html (accessed on 10 October 2015).

coatings

Article

Multi-Damage Identification of Multi-Span Bridges Based on Influence Lines

Yunkai Zhang, Qingli Xie, Guohua Li * and Yali Liu

School of Civil and Transportation Engineering, Beijing University of Engineering and Architecture, Beijing 102616, China; 201802040124@stu.bucea.edu.cn (Y.Z.); 201802010113@stu.bucea.edu.cn (Q.X.); 2108140619003@stu.bucea.edu.cn (Y.L.)
* Correspondence: liguohua@bucea.edu.cn

Abstract: The framework security of a bridge is essential as a critical component of traffic engineering. Even though the bridge structure is damaged to various degrees due to various reasons, the bridge will be wrecked when the damage reaches a particular level, suggesting a negative influence on people's lives. Based on the current situation and existing problems of structural damage identification of bridges, a structural damage identification technology of continuous beam bridges based on deflection influence lines is proposed in this paper in order to keep track of and always detect broken bridge elements, thereby extending the bridge's service life and reducing the risk of catastrophic accidents. The line function expression of deflection impact on a multi-span continuous beam bridge was first obtained using Graphic Multiplication theory. From the theoretical level, the influence line function of the continuous beam bridge without extensive damage was computed, and a graph was generated. The photographs of the DIL as well as the first and second derivatives, the deflection influence line distinction and its first and second derivatives, and the DIL distinction and its first and second derivatives of a continuous beam bridge in a single position and multi-position destruction were fitted in this paper. Finally, after comparing multiple work conditions and multiple measuring points, it was found that the first derivative of deflection influence line difference had the best damage identification effect. The design was completed and tested, which had verified the feasibility of this theory.

Keywords: traffic engineering; damage identification; deflection influence line; continuous beam bridge; multi-span bridges; structural damage

Citation: Zhang, Y.; Xie, Q.; Li, G.; Liu, Y. Multi-Damage Identification of Multi-Span Bridges Based on Influence Lines. *Coatings* **2021**, *11*, 905. https://doi.org/10.3390/coatings11080905

Academic Editor: Paolo Castaldo

Received: 16 June 2021
Accepted: 23 July 2021
Published: 28 July 2021

1. Introduction

The economy of any nation is heavily reliant on transportation infrastructure. It is critical to ensure that this infrastructure has dependable mobility and serviceability. Bridges are an important element of this infrastructure, and they need to be inspected and maintained regularly to stay in good working order [1]. Appropriate maintenance helps to prevent costly bridge repair and replacement by extending the structural life. The timing of maintenance activity, the activity itself, and the scope of such an activity are all important components of effective maintenance programs. Bridge maintenance is an important part of the operations necessary to maintain bridge infrastructure, maintain essential safe operational levels with minimal disruptions, and cost-effectively extend the life of structures [2]. As a result, bridge maintenance is an essential component of any bridge management strategy. It is widely acknowledged that appropriate maintenance operations will extend the bridge's operational life [1].

As a vital transport system, a bridge has a long service life, at a minimum of more than 50 years, and the life span is continuously moving on [3]. Due to comprehensive actions of numerous aspects, for example environs, raw ingredients, and lassitude possessions, the bridge's functional structure deteriorates due to such aspects and catastrophic accidents occur when it is more serious [4–6]. By 2016, a total of 830,000 highway bridges were found in China, so it urgently needed to identify the damages of domestic bridges with firm and precise fundamental impairments credentials [7]. Other countries are also under the same conditions [8–10]. There are about 690,000 bridges in the United States, more than 50% of which have been used for a minimum of 50 years [11,12]. Simply, fast and precise operational detection methods could resolve this unmet problem of bridge detection [13–16].

To analyze and evaluate the bridge's deterioration detection, Kou Xiaona first practically used the effectual deflection contour (comprising its first derivative and second derivative) to detect deteriorations in the structure [17]. She functionalized the theory of mechanics to the numerical analysis of the purely maintained beam model of the reinforced concrete, and justification was made to achieve the deflection influence line (DIL) of the simply supported beam and its first and second derivatives. Afterwards, she performed a finite element modeling analysis by ANSYS. Finally, through the experimental verification, she learned that the DIL could detect the presence of damage and the severity of the damage. Liu Yunshuai selected the deflection difference influence line (DDIL) as a foundation of destruction identification [18]. He assumed the simply supported beam for illustration, the derivatization of the analytical formula for the deflection difference influence line previously, and, later, single-point destruction of the simply supported beam, and at that moment additionally resolved the curvature of the influence line (IL). In addition, he devised a symmetrical loading strategy to remove the finite element model's inaccuracy. The deflection difference influences line curvature, and it is an excellent identification index and is far more vulnerable to damage identification. Jia Yaping recognized the index of structural damage identification of the continuous beam bridge based on DIL: DDIL, the first and the second derivative of DDIL, and the second derivative of DIL [19]. Formerly, an aluminum alloy plate with a three-span simulated coupling beam was utilized to clarify that the four indexes can recognize if there any destruction and the site of destruction. In the current study, based on DIL and calculation basis theory, the calculation method of DIL of the multi-span continuous beam was proposed by theoretical derivation and experimental verification through MIDAS/Civil modeling. A relevant study was performed on whether there is any damage to the DIL (and its first and second derivatives), the damage position, and damage degree, and the damage location and the damage degree is deduced with the IL (and its first and second derivatives). A structural damage identification (SDI) system based on the function of the DIL (and its first and second derivatives) was obtained, and the theory was confirmed by a simulation test of a plexiglass plate. Besides the structural elements, previous studies showed that non-structural elements also play a significant role in bridge deflection. For instance, previous authors demonstrated that the non-structural elements, including the pedestrian sidewalk and overlay of joint granite block, could significantly cause deflection in the historical Boco bridge [20].

2. Numerical Methods and Materials

Here we explained the theoretical deflection influence line equations of the continuous bean under damage.

2.1. Relevant Parameters of Calculation Model

The two-span continuous beam (as publicized in Figure 1) was taken as the theoretical study subject, the span was 160 cm, the left end support was immovable hinge support, and the other supports are sliding or movable hinge supports. The DIL equations of the model under the condition of health and damage were deduced.

2.2. Deflection Influences Line Equation of Structure without Loss

Next, captivating the midspan section of the second span of the laboratory model as a specimen, the theoretical derivation of non-destructive deflection IL of the structure of the continuous beam bridge was carried out. As shown in Figure 2, for two equal-span continuous beams, the calculated section of the influence line was in the midspan of the second span, and one unit force F = 1 was acting on the structure, F was x far from the t left side of the structure.

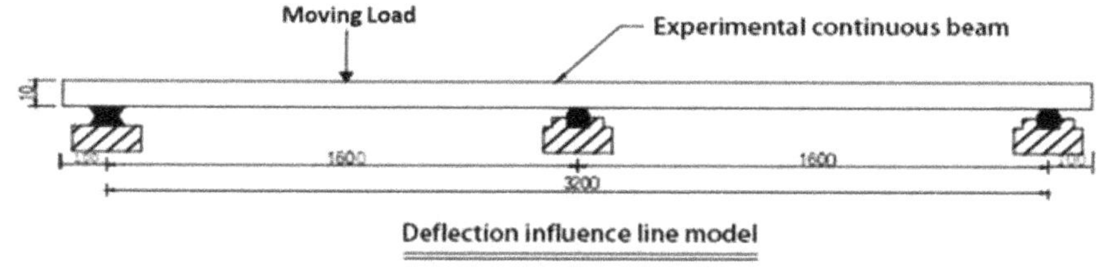

Figure 1. Experimental model.

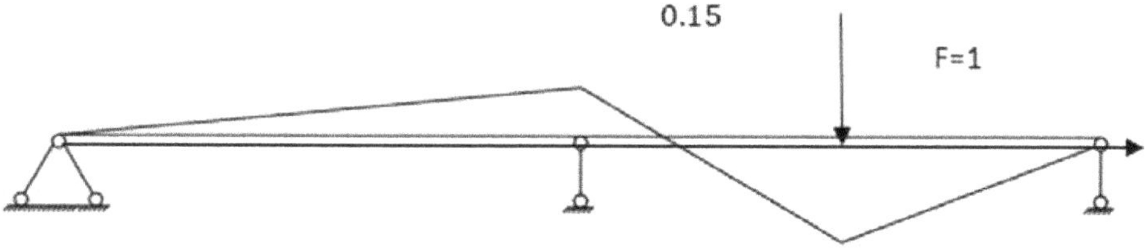

Figure 2. Figure *M*.

The fulcrum on the left side of the Figure *M* was considered as the coordinate origin, and the image equation of *M* is:

$$\begin{cases} y = \frac{1.5}{16}x & x \le 1.6 \\ y = -0.59375x + 1.1 & 1.6 < x \le 2.4 \\ y = 0.40625x - 1.3 & 2.4 < x \le 3.2 \end{cases} \quad (1)$$

When the unit load acts on the 1st span, the bending moment diagram of the basic structure is shown in Figure 3.

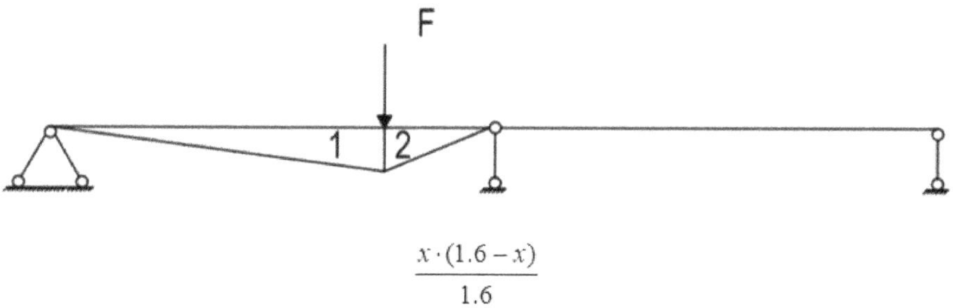

Figure 3. Basic structure $\overline{M}_K$ figure.

The area of area 1 is: $S_1 = \frac{x^2 \cdot (1.6 - x)}{3.2}$.
The gravity is:

$$\frac{2}{3}x \quad (2)$$

The area of Area 2 is:
$S_2 = \frac{x \cdot (1.6 - x)^2}{3.2}$.
The gravity is:

$$\frac{2 \cdot x}{3} + \frac{1.6}{3} \quad (3)$$

When the unit load acts on the 2nd span, the bending moment diagram of the basic mechanism is shown in Figure 4.

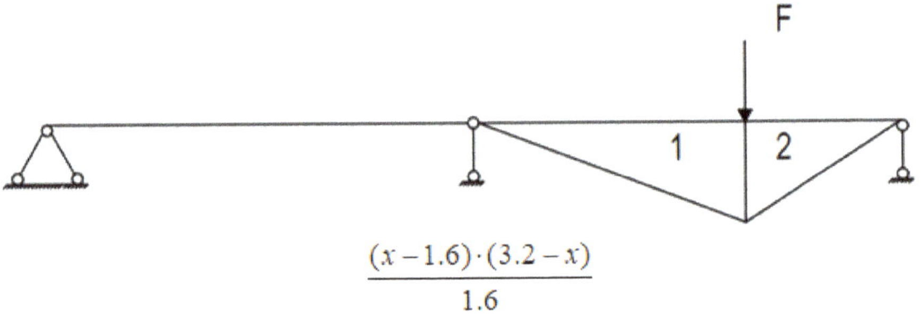

Figure 4. Basic structure $\overline{M}_K$ figure.

The area of Area 1 is:
$S_1 = \frac{(x-1.6)^2\cdot(3.2-x)}{3.2}$.
The gravity is:

$$\frac{2}{3}x + \frac{1.6}{3} \tag{4}$$

The area of Area 2 is:
$S_2 = \frac{(x-1.6)\cdot(3.2-x)^2}{3.2}$.
The gravity is:

$$\frac{2x}{3} + \frac{3.2}{3} \tag{5}$$

The following is the multiplication of graphs.

When the unit load (UL) acts on the 1st span, the center of gravity (Cg) of Area 1 and Area 2 is in the first equation. Where $0 \le x \le 1.6$, the deflection is:

$$\Delta = \frac{1}{EI}\cdot\left[\frac{x^2\cdot(1.6-x)}{3.2}\cdot\frac{1.5}{16}\cdot\frac{2}{3}x + \frac{x\cdot(1.6-x)^2}{3.2}\cdot\frac{1.5}{16}\cdot(\frac{2}{3}x+\frac{1.6}{3})\right] \tag{6}$$

When the UL acts on the 2nd span, the Cg position of Area 1 and Area 2 is discussed again.

(1) When the Cg is all in the 2nd equation: $y = -0.59375x + 1.1$

It shall meet:

$$\begin{cases} 1.6 < \frac{2}{3}x + \frac{1.6}{3} \le 2.4 \\ 1.6 < \frac{2}{3}x + \frac{3.2}{3} \le 2.4 \end{cases} \tag{7}$$

Then, $1.6 < x \le 2$
The deflection is:

$$\Delta = \frac{1}{EI}\cdot\left[\frac{(x-1.6)^2\cdot(3.2-x)}{3.2}\cdot\left(1.1-0.59375\cdot(\frac{2}{3}x+\frac{1.6}{3})\right)+ \frac{(x-1.6)\cdot(3.2-x)^2}{3.2}\cdot(1.1-0.59375\cdot(\frac{2}{3}x+\frac{3.2}{3}))\right] \tag{8}$$

(2) When the Cg of Area 1 is in the 2nd equation: $y = -0.59375x + 1.1$;

When the Cg of Area 2 is in the 3rd equation: $y = 0.40625x - 1.3$, then

$$\begin{cases} 1.6 < \frac{2}{3}x + \frac{1.6}{3} \le 2.4 \\ 2.4 < \frac{2}{3}x + \frac{3.2}{3} \le 3.2 \end{cases} \tag{9}$$

i.e., $2 < x \le 2.8$

The deflection is:

$$\Delta = \frac{1}{EI}\left[\frac{(x-1.6)^2\cdot(3.2-x)}{3.2}\cdot(11-0.59375\cdot(\tfrac{2}{3}\cdot x+\tfrac{1.6}{3}))+\frac{(x-1.6)\cdot(3.2-x)^2}{3.2}\cdot(0.40625\cdot(\tfrac{2}{3}\cdot x+\tfrac{3.2}{3})-1.3)\right] \tag{10}$$

(3) When the Cg of Area 1 and Area 2 is in the 3rd equation:

$y = 0.40625x - 1.3 \quad 2.8 < x \leq 3.2.$

The deflection is:

$$\Delta = \frac{1}{EI}\left[\frac{(x-1.6)^2\cdot(3.2-x)}{3.2}\cdot(0.40625\cdot\left(\tfrac{2}{3}x+\tfrac{1.6}{3}\right)-1.3)+\frac{(x-1.6)\cdot(3.2-x)^2}{3.2}\cdot(0.40625\cdot(\tfrac{2}{3}x+\tfrac{3.2}{3})-1.3)\right] \tag{11}$$

Figure 5 shows the results of drawing the function image of the theoretical deflection effect line using MATLAB. When calculating the DIL of the damaged continuous beam, the bending stiffness of the deteriorated section is set EI Reduced to $EI\prime$. The calculation process is the same; when the damage section is involved, the graph multiplication is used separately. The deflection influence line is deduced theoretically and the following rules are obtained:

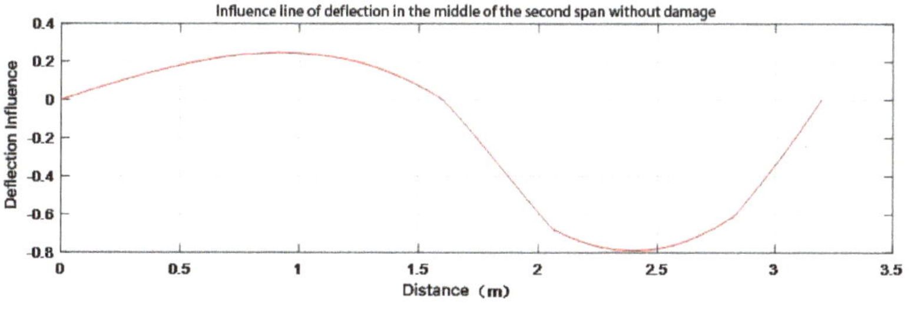

Figure 5. Fitting image of no loss deflection influence line.

(1) The location x of the moving load has a cubic connection with the DIL in any part of the continuous beam, and the DIL is continuous throughout the continuous beam structure.

(2) The bending stiffness EI changes significantly impact the DIL of a continuous beam, and the difference in the DIL will directly represent bending stiffness variations.

3. Study on Deflection Influence Line Identification of Continuous Beam with Multiple Damage under Ideal Condition

Using the continuous two-span beam as the opposite direction, Midas Civil was utilized to analyze the DIL of continuous beam bridge under normal condition and damage conditions of important portions. The determining points were respectively organized at 25/100 point of the 1st span and the 2nd span, and at 75/100 point in the middle of the span. The deflection data of key positions were collected and processed (25/100 of the 1st span, the first span and 75/100 of the 1st span) to evidence the precision of the theory and applied use. First, the ideal model simulation was carried out (that is, a single point dynamic load was defined to act on the continuous beam), and the continuous beam was shown in Figure 6. The span of the two spans was 1.6 m, and the local structure was shown in the figure. The detailed research was divided into single location damage identification and multi-location destruction identification.

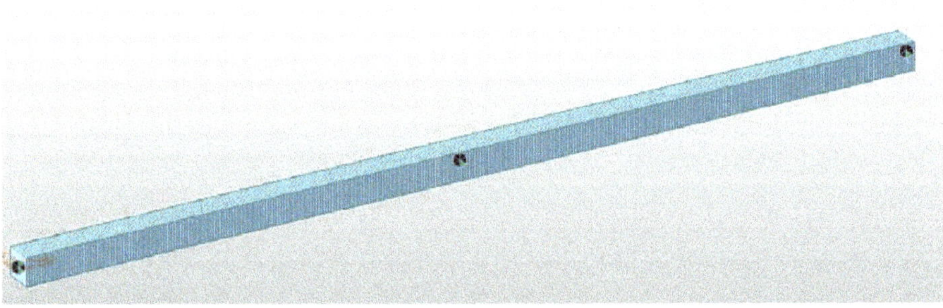

Figure 6. Midas civil model of test continuous beam.

3.1. Establishment of Multiple Damage Model

As displayed in Figure 6, the idealized model of the continuous beam structure was that the UL acted on the central line of the beam. Locating the structure type to the X-Z plane could shorten the design process. To make the model easier to understand, C50 concrete under the JTG3362-18(RC) code was directly selected as the material, including the elastic modulus of concrete $E = 3.45 \times 10^4$ kN/mm^2, Poisson's ratio $\mu = 0.2$, and the coefficient of linear expansion $\alpha = 1 \times 10^{-5}$ 1/°C. Two sections were added and named "normal" and "damage", respectively. The solid web rectangular section was selected, and the section size was 100 mm × 100 mm. The numerical model had 321 nodes, 320 elements, and boundary conditions and a schematic diagram are evaluated.

The bending stiffness of normal material is EI, and that of damaged material is EI*α (α is the reduction factor). In the Midas Civil model, only the damaged part changes I_{yy} to simulate the loss.

The boundary conditions were applied at Nodes 1, 161, and 321, respectively. The fixed hinge bearing was used at Node 1 to limit the displacement in x and z directions, and the sliding hinge bearing is used at nodes 161 and 321 to limit the displacement in z direction.

The loads were organized once the boundary conditions were specified. Only one lane was set up to make the computation easier. Because the track surface was not involved, only the lane line had to be set up. After setting the load, i.e., moving the load, observe the DIL.

In view of the multi-position damage of continuous beam bridges in practical engineering, 10 kinds of work conditions were simulated. The damage simulation of the key position of the continuous beam was carried out respectively. Because the two-span continuous beam of the test model was symmetrical structure, only half-span damage simulation was carried out.

The damage degree was divided into four grades: 10%, 20%, 30%, and 50%. The specific work conditions are shown in Figures 7 and 8 and Tables 1 and 2. Each work condition was further subdivided according to the damage degree.

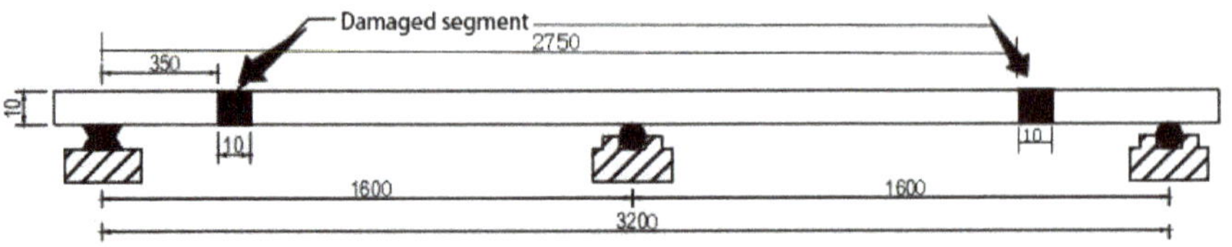

Figure 7. Diagram of work condition Model 1.

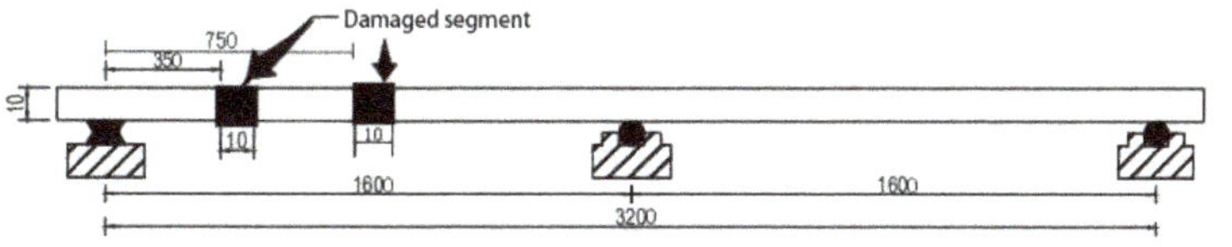

Figure 8. Diagram of work condition Model 2.

Table 1. Damage information of Condition 1.

Damage Location	**35–45 cm and 275–285 cm**				
Work condition No.	Work condition No.1-1	Work condition No.1-2	Work condition No.1-3	Work condition No.1-4	Work condition No.1-5
Number of injuries	2	2	2	2	2
Degree of damage	0	10%	20%	30%	50%

Table 2. Work condition 2 damage information.

Damage Location	**35–45 cm and 75–85 cm**				
Work condition No.	Work condition 2-1	Work condition 2-2	Work condition 2-3	Work condition 2-4	Work condition 2-5
Number of injuries	2	2	2	2	2
Degree of damage	0	10%	20%	30%	50%

3.2. Multi-Point Destruction Recognition of Deflection Influence Line

Draw the deflection influence line data obtained from work condition 1 (1–5) and work condition 2 (1–5) of the 1st span 1/4 determining point. As shown in Figures 9 and 10, there is no significant difference in the trend and value of the image under different damage locations, so the location of multi-point damage cannot be identified by the deflection influence line.

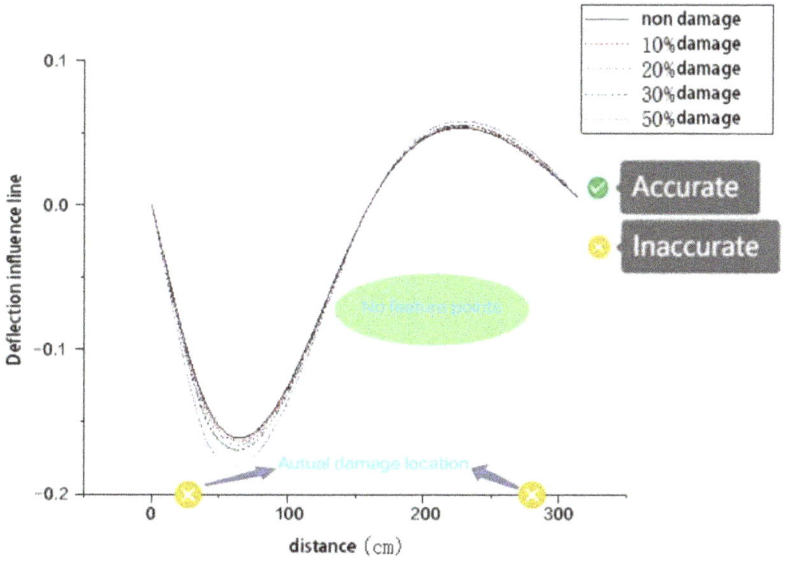

Figure 9. Deflection influence of 1/4 determining point of the 1st span following work condition 1 (1-5).

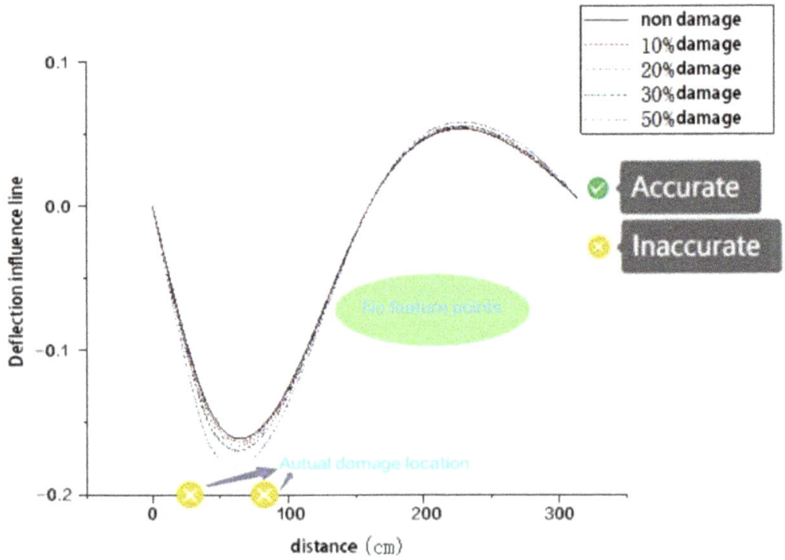

Figure 10. Deflection influence line of 1/4 determining point of the 1st span following work condition 2 (1-5).

3.3. *Multi-Point Damage Identification Based on 1st Derivative of Deflection Influence Line*

Under condition 1 (1-5) and condition 2 (1-5), the 1st derivative of deflection influence line obtained from 1/4 determining point of the first span is shown in Figures 11 and 12. It was found that the destruction site could not be recognized by image.

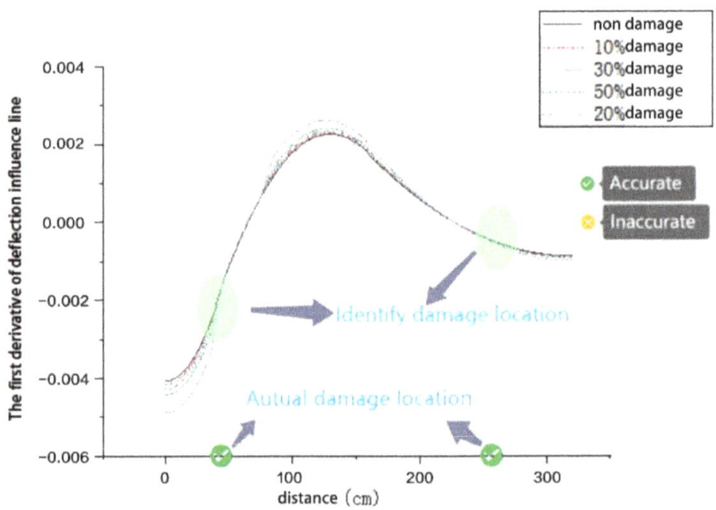

Figure 11. The first derivative of deflection influence line at 1/4 determining point of the 1st span following work condition 1 (1-5).

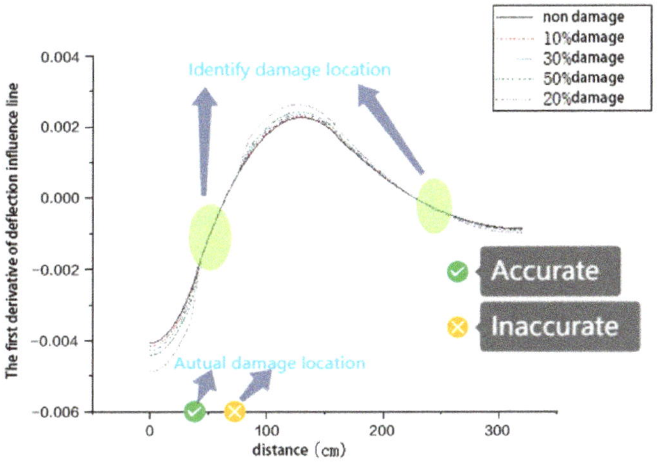

Figure 12. The 1st span1/4First derivative of deflection influences the point following work condition 2 (1-5).

3.4. Multi-Point Damage Identification Based on the 2nd Derivative of Deflection Influence Line

As shown in the figures, in the case of multi-point damage of continuous beam bridge, the deflection sensor was used to collect data and process the 2nd derivative of DIL. According to the rule of the 2nd derivative of DIL in single-point damage identification, the image was analyzed. On the basic line, the position where the large mutation and deviation occur was the damage location. Compared with Figures 13 and 14, it was found that this method could also reflect the damage location well when there was only one measuring point. Compared with Figures 14 and 15, it was found that the effect of different measuring points on structural damage identification was the same under the same work condition.

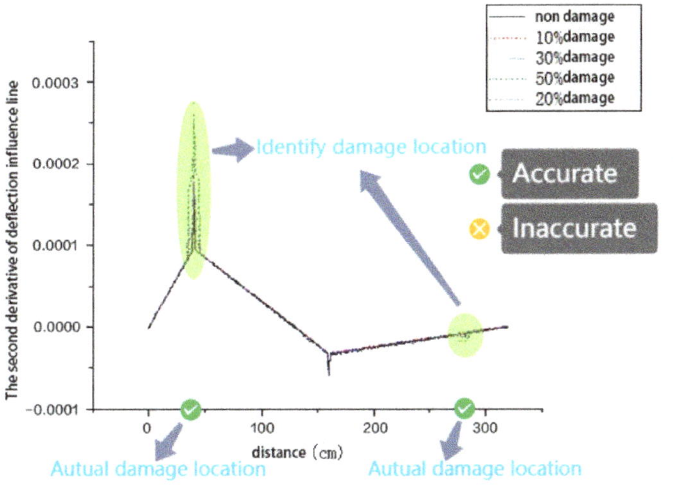

Figure 13. The 2nd derivative of deflection influence line at 1/4 determining point of the 1st span under work condition 1 (2-5).

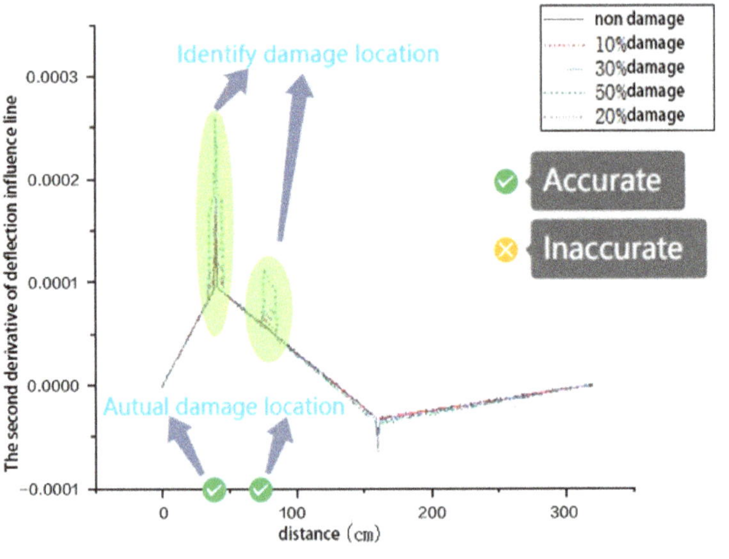

Figure 14. The 2nd derivative of deflection influence line at 1/4 determining point of the 1st span following work condition 2 (2-5).

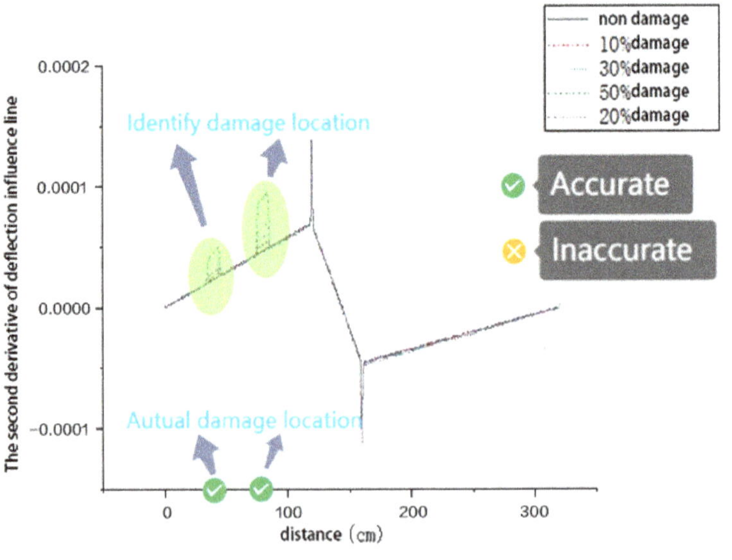

Figure 15. The 2nd derivative of deflection influence line at 3/4 determining point of the 1st span under work condition 2 (2-5).

At the same time, it was found that when the damage location coincides with the observation point, the observed effect was greatly weakened. Because the quadratic function of the DIL of the continuous beam structure was linear, the peak value appeared at the measuring point and the upper fulcrum of the beam. It was difficult to observe when the deviation position coincides with the peak value.

3.5. Multi-Point Damage Identification Based on Deflection Influence Line Difference

Under the condition of multi-point damage simulated by work condition 1 and work condition 2, the image of the DIL difference of each measuring point was extracted (as shown in Figures 16–18). According to the relationship between the image change and the damage location, it was found that the image had a small peak value in the structural damage section, but it did not have a great impact on the trend of the whole curve. Therefore, it was theoretically feasible to identify the damage location only through the DIL difference image, but the identification process was very difficult. In order to locate the greatest and excellent parameter index of multi-point destruction detection, the 1st derivative of DIL difference was studied.

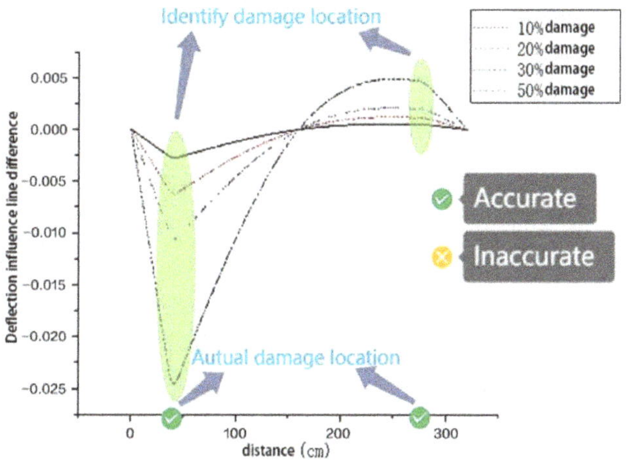

Figure 16. Deflection influence line difference of 25/100 determining point of the 1st span following work condition 1 (2-5).

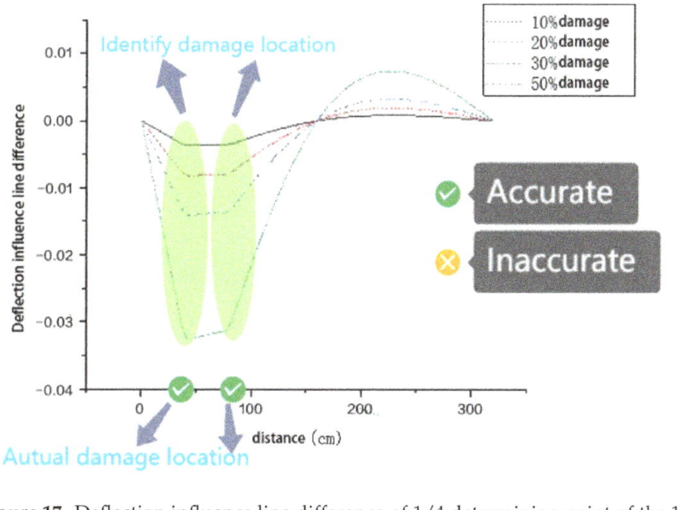

Figure 17. Deflection influence line difference of 1/4 determining point of the 1st span following work condition 2 (2-5).

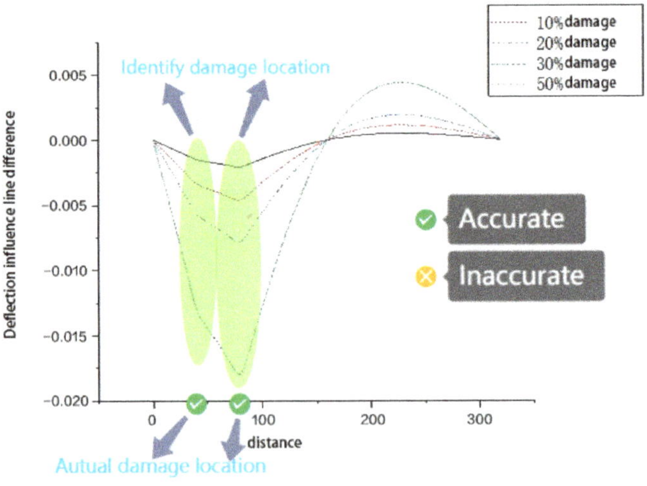

Figure 18. Deflection influence line difference of 3/4 determining points of the 1st span under work condition 2 (2-5).

3.6. Multi-Point Damage Identification Based on 1st Derivative of Deflection Influence Line Difference

Draw the 1st derivative image of DIL difference under multi-point damage of work condition 1 and work condition 2, as shown in Figures 19–21. Based on the first derivative of the DIL difference in the case of single-point damage, the position of the image mutation segment was the corresponding position of the damaged segment. The identification results are shown in the figure. The method had obvious characteristics for the damaged section, and could accurately reflect the damage location in different positions and different working conditions. Consequently, the 1st derivative of the DIL difference was a suitable parameter for structural damage identification.

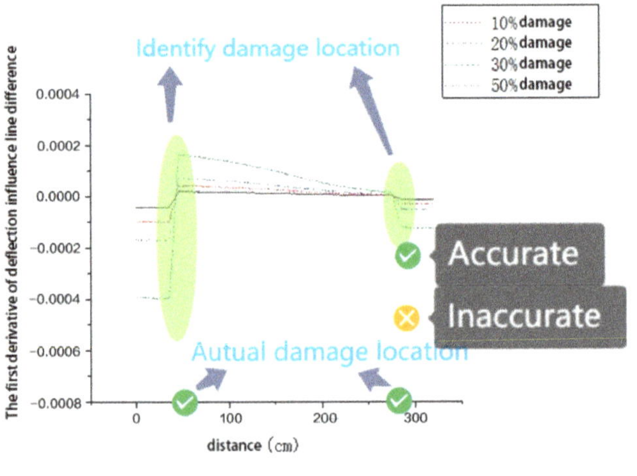

Figure 19. The 1st derivative of deflection influences line difference at 1/4 determining point of the 1st span under work condition 1 (2-5).

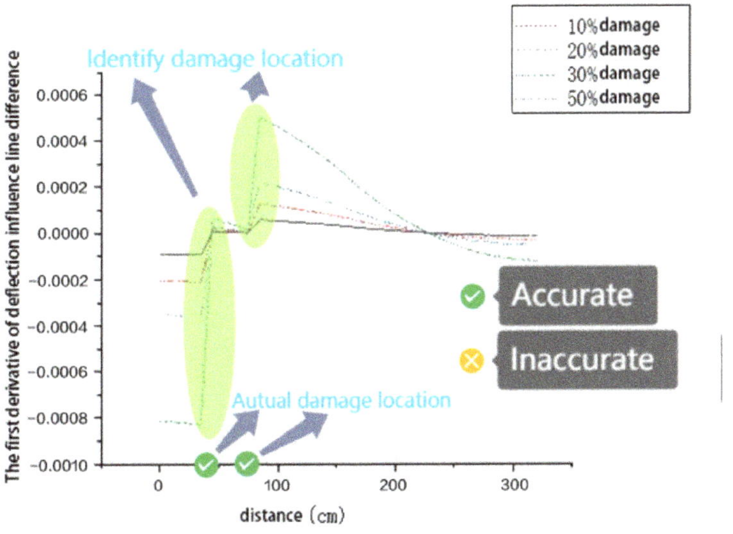

Figure 20. The first derivative of deflection influences line difference at 1/4 determining point of the first span under work condition 2 (2-5).

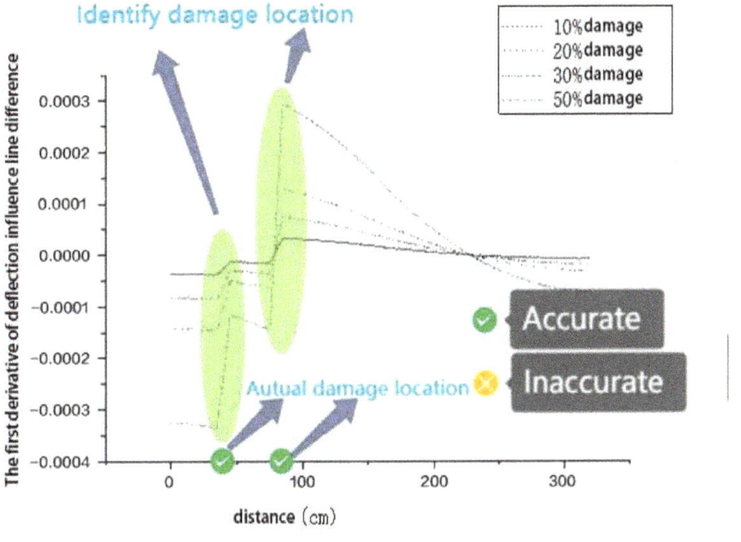

Figure 21. The 1st derivative of deflection influence line difference at 3/4 determining points of the first span under work condition 2 (2-5).

In order to choose the most appropriate settings, the 1st derivative of the deflection influence line difference was derived again to verify the identification effect of the second derivative for multi-point damage of the continuous beam bridge structure.

3.7. Multi-Point Damage Identification Based on the Second Derivative of Deflection Influence Line Difference

Gather and progress the data of each measuring point under work condition 1 and work condition 2, and acquire the second derivative photograph of the DIL difference of each measuring point (as shown in Figures 22–24). According to the second derivative identification law of the influence line difference under single-point damage of the continuous beam bridge structure, the location of the image peak was the location of the damage section. However, due to the influence of measurement accuracy, the 2nd derivative image had great fluctuation, so it had high requirements for measuring instruments to use the 2nd derivative of the IL difference as the identification parameter.

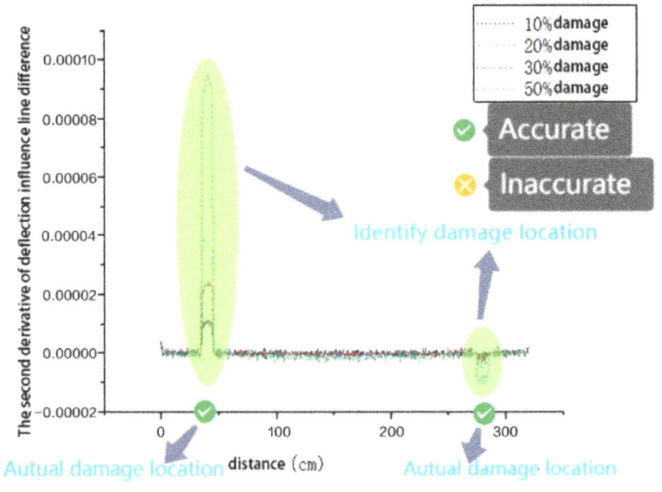

Figure 22. The 2nd derivative of deflection influences line difference at 1/4 determining point of the 1st span following work condition 1 (2-5).

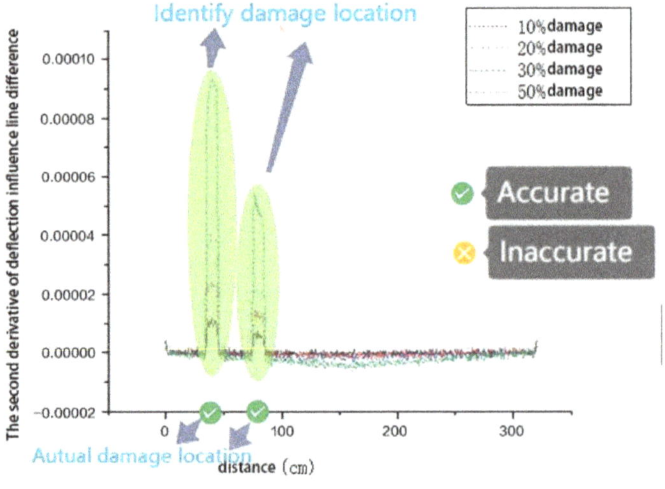

Figure 23. The 2nd derivative of deflection influences line difference at 1/4 determining point of the 1st span following work condition 2 (2-5).

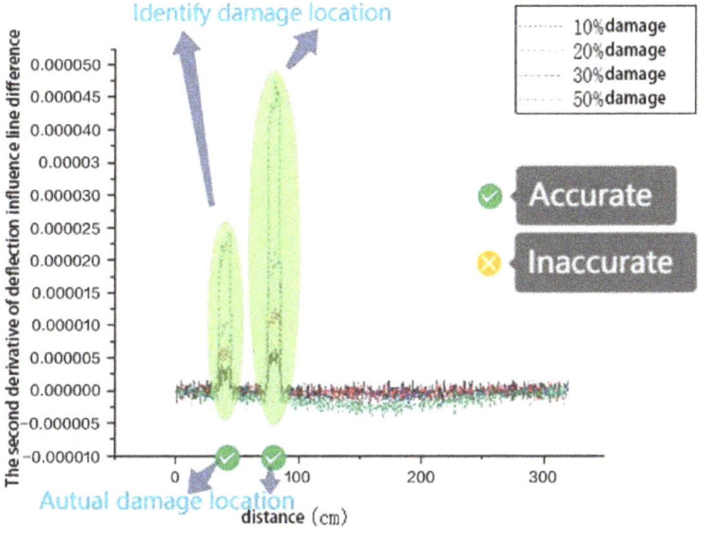

Figure 24. The 2nd derivative of deflection influence line difference at 3/4 determining point of the 1st span following work condition 2 (2-5).

4. Laboratory Validation of the Theoretical Model

4.1. Test Plan

The model is shown in Figure 25, and the bridge was simulated with the acrylic plate. The layout of fixed hinge bearing and sliding bearing is shown in Figure 26, and the bridge deck used a trolley with controllable speed as the moving load, as shown in Figure 27. Two stages of loading were designed in the experiment. The first stage loading was an empty car (mass 1 kg), and the second stage loading was car loading with a 1 kg weight (total mass is 2 kg). In order to avoid vehicle bridge coupling, the vehicle speed was set at 0.128 m/s [15]. Based on the theoretical derivation of each working condition, the non-destructive state and 25% damage state are set, and the thickness of the damaged acrylic plate was 75% of the non-destructive state.

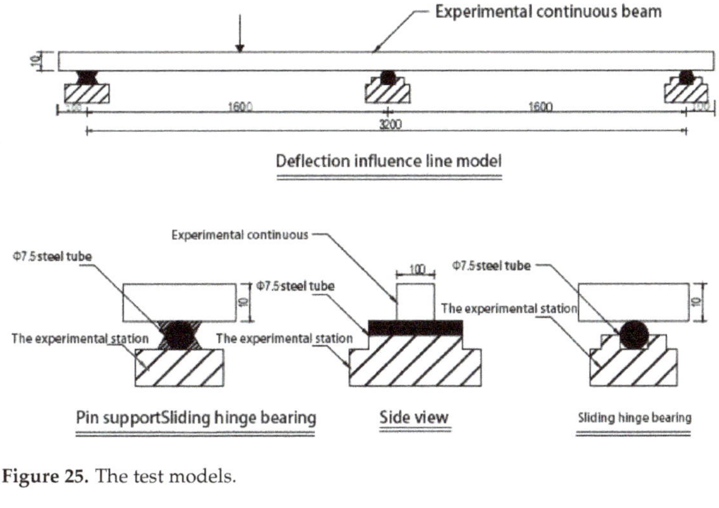

Figure 25. The test models.

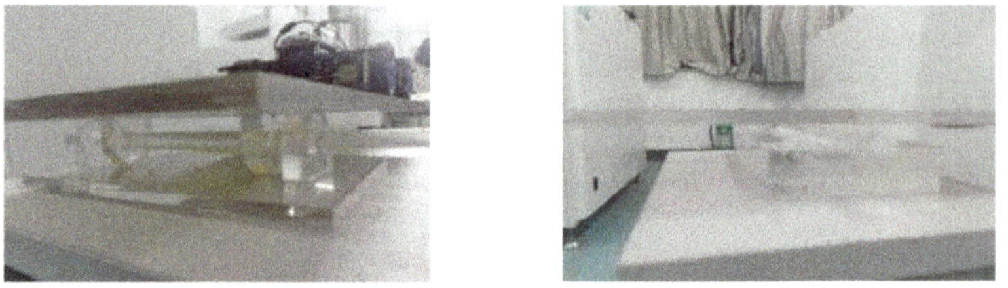

Figure 26. Fixed hinge bearing and sliding hinge bearing.

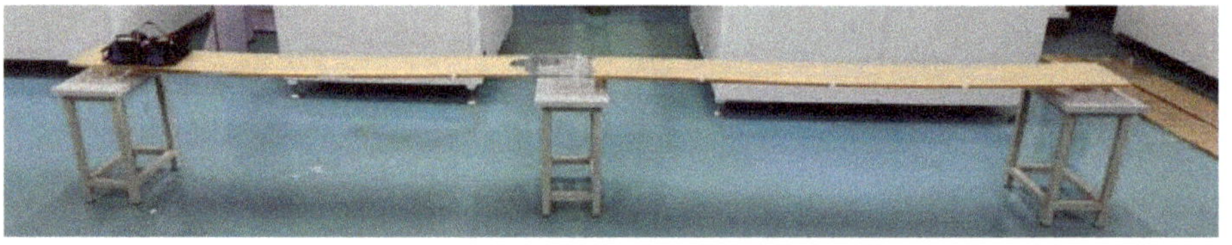

Figure 27. Dynamic load layout.

4.2. Testing Equipment

SMTN-X pro-multi-point dynamic video measurement was used to monitor the 1/4 1/2 3/4 position of each span (Figure 28), and the DIL of the measuring point was obtained.

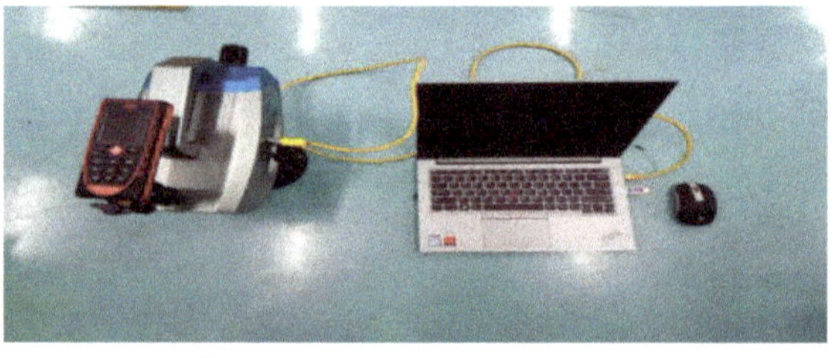

Figure 28. SMTN-X Pro multi-point dynamic video measurement system.

4.3. Identification of Damage Location by Deflection Influence Line and Its First and Second Derivatives

Because of the small deflection of the first level loading, the damage location could not be well reflected, so the second level loading was analyzed.

According to the curve obtained in work condition 5 (Figures 29–31), the first derivative of DIL could not effectively detect the destruction site, but the second derivative of DIL could effectively identify the specific destruction site. The results were consistent with the data.

According to the curve obtained in case 4 (Figure 32), it could be seen that the second derivative of the deflection difference IL was universal to identify the destruction site, and it could effectively identify the damage location when it was extended to any damage location.

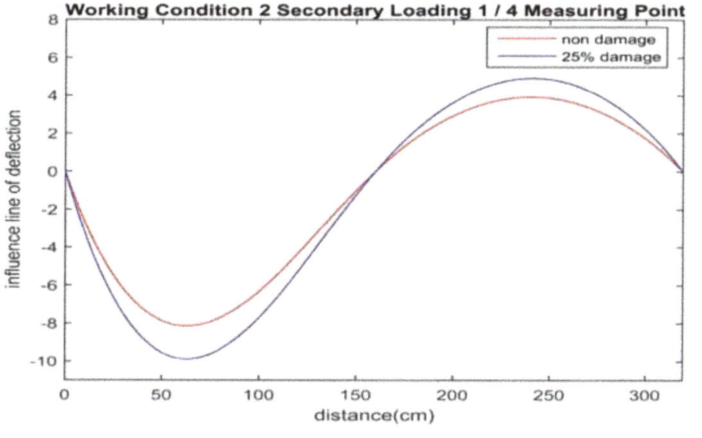

Figure 29. Deflection influence line at 25/100 determining point of the first span in work condition 5.

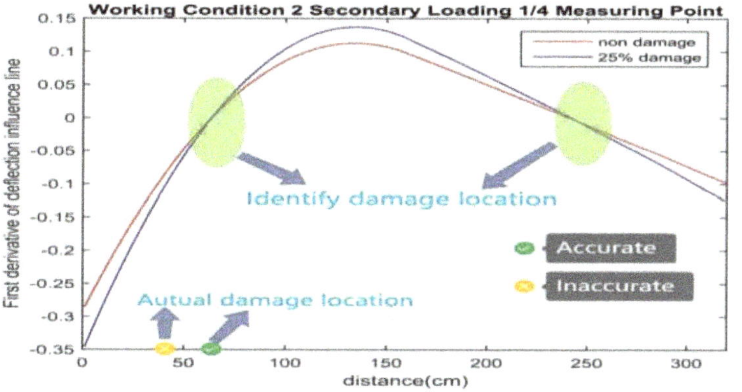

Figure 30. The first derivative of deflection influence line at 25/100 determining point of the first span following condition 5.

4.4. Identification of Damage Location by Deflection Influence Line Difference and Its First and Second Derivatives

According to the DIL difference and its first-order derivative and second-order derivative obtained from work condition 4 (Figures 33–35), the first-order derivative and second-order derivative of DIL could effectively detect the destruction site. However, for the first-order derivative of the DIL difference, the second-order derivative data processing was more difficult, so it was more convenient to identify the damage location by using the first-order derivative of the DIL difference.

According to the curve obtained in work condition 5 (Figures 36 and 37), it could be extended to any case to recognize the damaged location by using the first-order and second-order derivatives of the DIL difference.

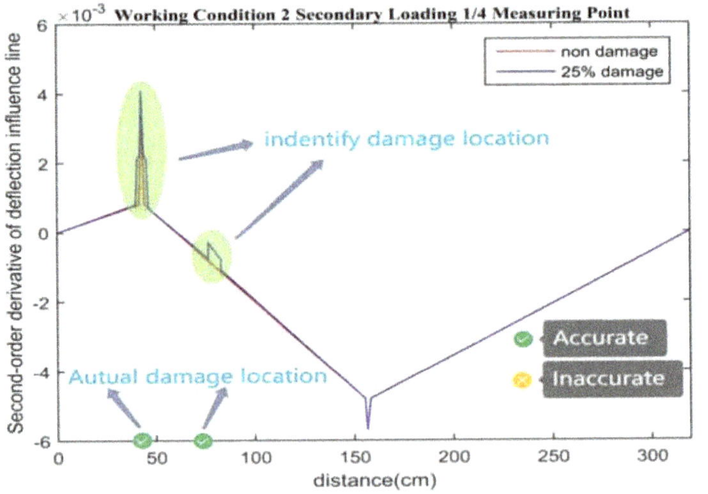

Figure 31. The second derivative of deflection influence line at 25/100 measuring point of the first span following work condition 5.

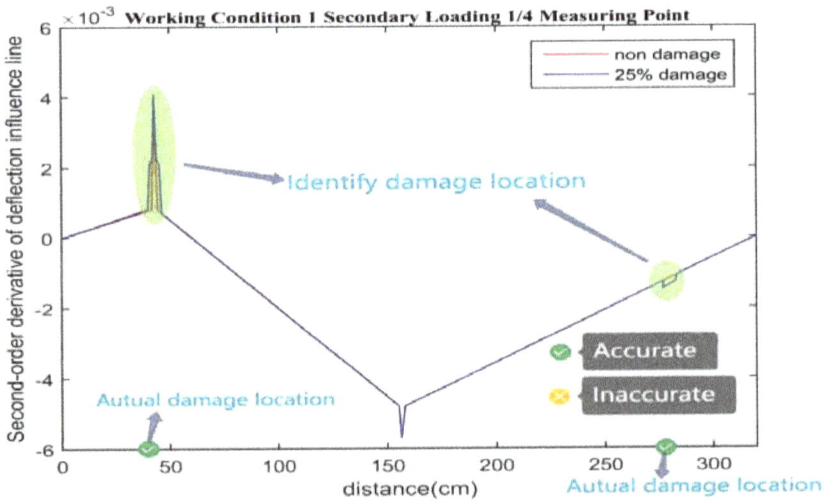

Figure 32. The second derivative of deflection influence the line at 1/4 determining point of the first span following work condition 4.

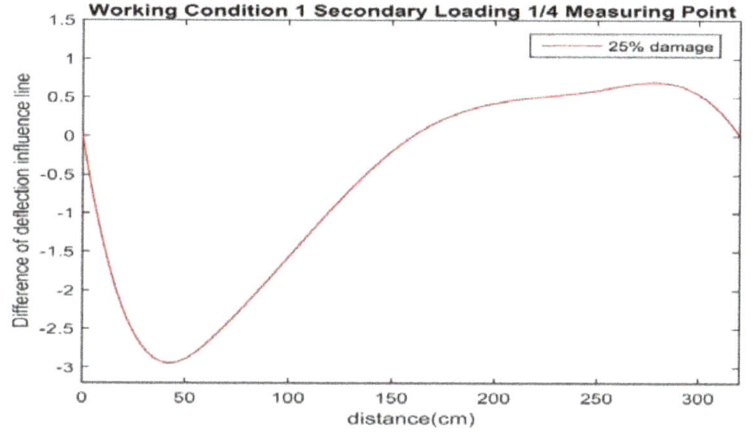

Figure 33. The deflection influence line difference of the first span 25/100 determining point in work condition 4.

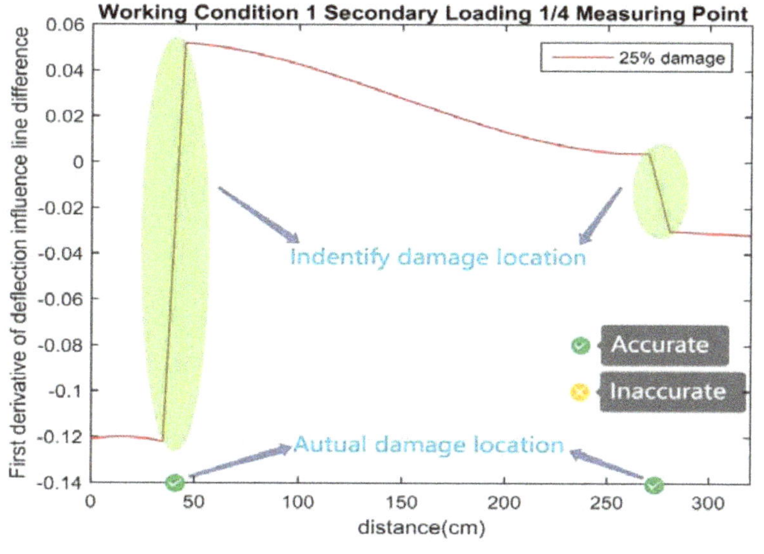

Figure 34. The first-order derivation of deflection influences line difference at 25/100 calculating point of the first span following work condition 4.

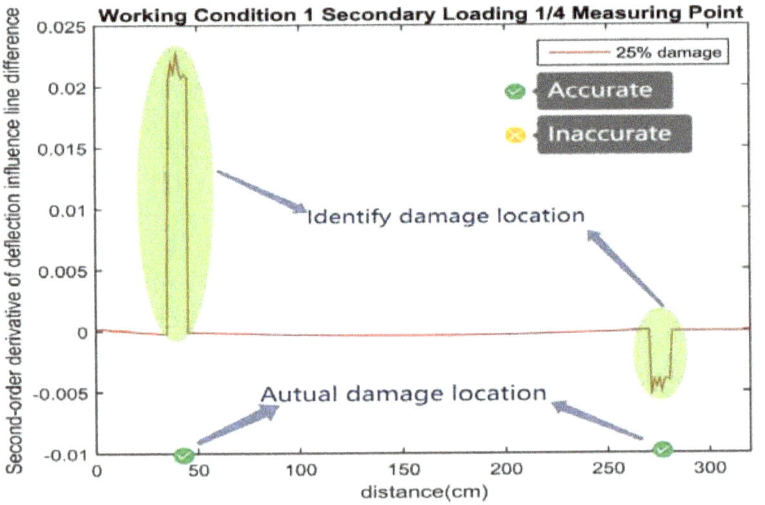

Figure 35. The second-order derivation of deflection influence line difference at 25/100 measuring point of the first span under work condition 4.

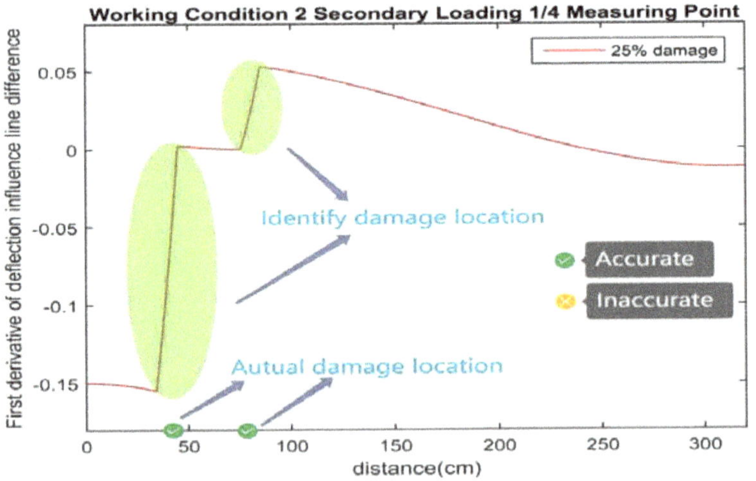

Figure 36. The first-order derivation of deflection influences line difference at 25/100 measuring point of the first span following work condition 5.

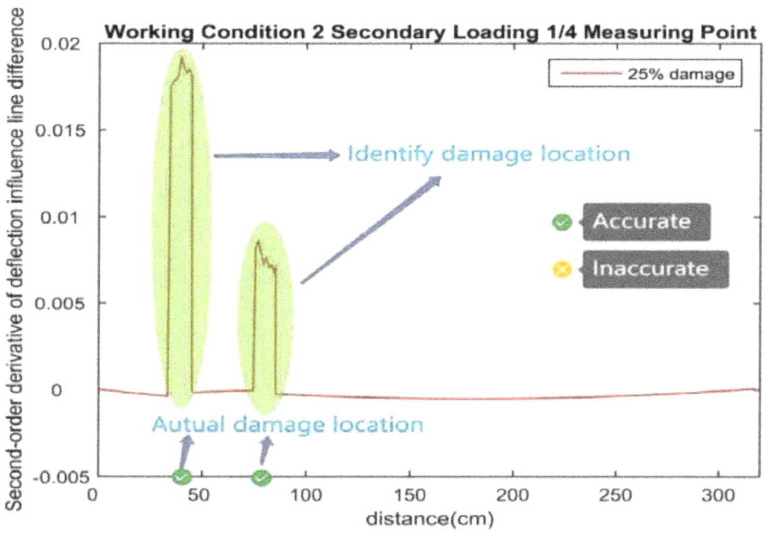

Figure 37. The second-order derivation of deflection influences line difference at 25/100 measuring point of the first span under work condition 5.

5. Discussion

Bridge nondestructive testing procedures can reveal defects in structures without affecting their functionality. When choosing a nondestructive testing approach, accuracy, accessibility, cost, and the repercussions of detection failures or false indications must be considered. There is now many studies being done on how to test bridges quickly and accurately. For many years, researchers have been looking into the idea of utilising vibration methods to identify bridge collapse [21–24]. Damping ratios are sensitive to structure damage, but they have become quite difficult to calculate. Vibration frequency is frequently used to determine the degree of damage in a structure. It is simple to measure, but it is not sensitive to changes in stiffness. As a result, developing a method of nondestructive testing to detect damage from a global viewpoint is crucial.

Based on the current situation and existing problems of structural damage identification of bridges, a structural damage identification technology of continuous beam bridges based on deflection influence lines is proposed in this paper in order to keep track of and detect broken bridge elements at all times, thereby extending the bridge's service life and reducing the risk of catastrophic accidents. The line function expression of deflection influence on a multi-span continuous beam bridge was first obtained using the Graphic Multiplication theory. The influence line function of the continuous beam bridge without severe damage was estimated at the theoretical level, and a graph was created. This article included pictures of the DIL and its first and second derivatives, the deflection influence line distinction and its first and second derivatives, and the DIL distinction and its first and second derivatives of a continuous beam bridge in single and multi-position destruction. According to our theoretical model, the position x of the moving load has a cubic link with the DIL in any portion of the continuous beam, according to our theoretical model, and the DIL is continuous throughout the continuous beam structure. Furthermore, changes in bending stiffness have a substantial influence on a continuous beam's DIL, and the difference in the DIL will directly indicate bending stiffness variations. Finally, it was discovered that the first derivative of deflection influence line difference had the best damage identification impact after evaluating different work circumstances and multiple measuring sites. The design was completed and tested, which had verified the feasibility of this theory. The design was completed and tested, and we validated the experiments.

The deflection sensor was utilized to gather data and calculate the second derivative of DIL in the event of multi-point damage of a continuous beam bridge, according to our findings. We examined the picture using the rule of the second derivative of DIL in single-point damage identification. The damage site on the basic line was where the major mutation and deviation occurred. Meanwhile, our analysis shows that the observed effect is substantially diminished when the damage site coincides with the observation point. The peak value appeared at the measuring point and the top fulcrum of the beam because the quadratic function of the DIL of the continuous beam structure was linear. When the deviation location coincided with the peak value, it was difficult to see.

The image of the DIL difference of each measuring point was retrieved under the conditions of multi-point damage simulated by work conditions 1 and 2 (as shown in Figures 16–18). We discovered that the image had a small peak value in the structural damage region, but it had limited influence on the overall curve's trend, based on the link between image change and damage location. As a result, identifying the damage location just using the DIL difference picture was theoretically possible, but the identification procedure was highly challenging. The first derivative of the DIL difference was investigated in order to find the best and most accurate parameter index for multi-point destruction detection.

The first derivative of the deflection influence line difference at 3/4 determination points of the first span under work condition 2 exhibited clear characteristics for the damaged section and could properly indicate the damage site in various locations and working circumstances. As a result, the first derivative of the DIL difference proved to be a useful measure for detecting structural deterioration. The current study suggests that the first derivative of the deflection influence line difference be computed again to validate the identified impact of the second derivative for multi-point damage of continuous beam bridge construction to pick the most appropriate settings.

Flooding occurrences were demonstrated in Italian suburbs by Sasuu et al. whose construction and regulations do not adequately cover maintenance [25]. Several failure scenarios led to this, including vehicle dragging when crossing bridges (A), erosion induced by bridge overtopping, and erosion and floating produced by upward buoyant force from Archimedes' principle on the bridge slab. The failure mechanisms presented here can be used to identify not just structural failures but also service failures. As a result, for identifying and forecasting multiple damage, especially in multi-span bridges, a suitable model and numerical computation are essential. Crocea et al. tested several techniques and concluded that the suggested procedure is not only appropriate for the intended applications, but also sufficiently "robust" [26]. By further verifying the approach suggested in this work under various experimental settings, it may be able to circumvent such eventualities.

There are several damage identification methods for bridge structures currently available, including intelligent algorithm-based, Bayesian theory-based, time-domain signal processing-based damage identification methods, sparsity information and sparse recovery theory-based, neural network-based damage identification methods, and various model-based methods [27–33]. These approaches are described conceptually and then proven in actual bridges through tests and experimentation. All of them have benefits and drawbacks; nevertheless, we feel that the first derivative of the DIL divergence is the greatest damage detection indicator, exceeding other parameters in terms of simplicity and accuracy of observation. Damage detection approaches based on displacement influence lines may be successfully utilised to detect and locate damage based on numerical studies and experimental verification. This will allow it to be used in a more suitable and useful way in the real bridge.

6. Conclusions

To sum up, the current study identifies the multi-damage of multi-span bridges based on influence lines. This paper fits the DIL and the first and second derivatives of the continuous beam bridge when the damage occurs at multiple positions, and the difference between the DIL and the first and second derivatives of the image. The photographs of the DIL and the first and second derivatives, the deflection influence line distinction and its first and second derivatives, and the DIL distinction and its first and second derivatives of a continuous beam bridge in a single position and multi-position destruction were fitted in this paper. The ideal identification parameters were determined as the basis of visual tracking systems by evaluating the detection methods and effectiveness under various locations, degrees, and measurement locations. This was the crucial criterion for identifying deterioration in a continuous beam building frame that has been in use, and the deterioration was found. The second derivative of the DIL, the first derivative of the DIL discrepancy, and the second derivative of the DIL distinction were the basic design damage detection indexes. The photographic characteristics were not visible and hard to recognize because there were still mutation spots in the second derivative of the deflection effect line in addition to the site of damage. The second derivative of the deflection effect line differences had to be determined using a high-precision device that was hard to fit, but it may also approximate the damage site. As a result, employing the second derivative of the deflection effect line difference posed a significant challenge to the instrument's reliability. To summarize, the first derivative of the DIL divergence was the best damage detection indicator, surpassing other parameters in terms of observation ease and accuracy, which will enable this to be more appropriately and beneficially applied to real bridges. We hope that our study will be of great interest for the readers of this journal.

Author Contributions: Conceptualization, Y.Z. and Q.X.; methodology, Y.Z.; software, Y.Z.; validation, Y.Z., Q.X., G.L. and Y.L.; formal analysis, Y.Z., Q.X., G.L. and Y.L.; investigation, Y.Z., Q.X., G.L. and Y.L.; resources, G.L. and Y.L.; writing—original draft preparation, Y.Z.; writing—review and editing, Y.Z. and Y.L.; supervision, Y.L.; project administration, Y.L.; funding acquisition, Y.L. All authors have read and agreed to the published version of the manuscript.

Funding: This research was funded by National College Students' Innovation and Entrepreneurship Training Project, grant number 202010016018.

Institutional Review Board Statement: Not Applicable.

Informed Consent Statement: Not Applicable.

Acknowledgments: This research was supported by National College Students' Innovation and Entrepreneurship Training Project (No. 202010016018).

Conflicts of Interest: The authors declare no conflict of interest.

References

1. Alampalli, S.; Ettouney, M.M.; Agrawal, A.K. Structural health monitoring for bridge maintenance. *Bridge Struct.* **2005**, *1*, 345–354. [CrossRef]
2. Fujino, Y.; Siringoringo, D.M.; Ikeda, Y.; Nagayama, T.; Mizutani, T. Research and Implementations of Structural Monitoring for Bridges and Buildings in Japan. *Engineering* **2019**, *5*, 1093–1119. [CrossRef]
3. An, Y.; Chatzi, E.; Sim, S.-H.; Laflamme, S.; Blachowski, B.; Ou, J. Recent progress and future trends on damage identification methods for bridge structures. *Struct. Control. Health Monit.* **2019**, *26*, e2416. [CrossRef]
4. Zhou, Y.; Di, S.; Xiang, C.; Li, W.; Wang, L. Damage identification in simply supported bridge based on rotational-angle influence lines method. *Trans. Tianjin Univ.* **2018**, *24*, 587–601. [CrossRef]
5. Deng, L.; Wang, W.; Yu, Y. State-of-the-art review on the causes and mechanisms of bridge collapse. *J. Perform. Constr. Facil.* **2016**, *30*, 04015005. [CrossRef]
6. Gehl, P.; D'Ayala, D. Development of Bayesian Networks for the multi-hazard fragility assessment of bridge systems. *Struct. Saf.* **2016**, *60*, 37–46. [CrossRef]
7. Xu, F.Y.; Zhang, M.J.; Wang, L.; Zhang, J.R. Recent highway bridge collapses in China: Review and discussion. *J. Perform. Constr. Facil.* **2016**, *30*, 04016030. [CrossRef]
8. Acar, E.; Izgi, A.; Serenbay, S.K. Note on Jakimovski-Leviatan Operators Preserving e–x. *Appl. Math. Nonlinear Sci.* **2019**, *4*, 543–550. [CrossRef]
9. Sahin, R.; Yagcı, O. Fractional calculus of the extended hypergeometric function. *Appl. Math. Nonlinear Sci.* **2020**, *5*, 369–384. [CrossRef]

10. Tan, J.-S.; Elbaz, K.; Wang, Z.-F.; Shen, J.S.; Chen, J. Lessons learnt from bridge collapse: A view of sustainable management. *Sustainability* **2020**, *12*, 1205. [CrossRef]
11. Gidaris, I.; Padgett, J.E.; Barbosa, A.R.; Chen, S.; Cox, D.; Webb, B.; Cerato, A. Multiple-hazard fragility and restoration models of highway bridges for regional risk and resilience assessment in the United States: State-of-the-art review. *J. Struct. Eng.* **2017**, *143*, 04016188. [CrossRef]
12. Omar, T.; Nehdi, M.L. Condition assessment of reinforced concrete bridges: Current practice and research challenges. *Infrastructures* **2018**, *3*, 36. [CrossRef]
13. Sun, L.; Shang, Z.; Xia, Y.; Bhowmick, S.; Nagarajaiah, S. Review of bridge structural health monitoring aided by big data and artificial intelligence: From condition assessment to damage detection. *J. Struct. Eng.* **2020**, *146*, 04020073. [CrossRef]
14. Shokravi, H.; Shokravi, H.; Bakhary, N.; Heidarrezaei, M.; Rahimian Koloor, S.S.; Petrů, M. Application of the subspace-based methods in health monitoring of civil structures: A systematic review and meta-analysis. *Appl. Sci.* **2020**, *10*, 3607. [CrossRef]
15. Shang, Z.; Sun, L.; Xia, Y.; Zhang, W. Vibration-based damage detection for bridges by deep convolutional denoising autoencoder. *Struct. Health Monit.* **2020**, 1475921720942836. [CrossRef]
16. Wang, T.B. Damper performance requirements for improving durability of bridge structures. *Bridge Construction.* **2016**, *46*, 29–34.
17. Kang, X.N. Preliminary Study on Damage Identification Method of Bridge Structure Based on Deflection Influence Line. Ph.D. Thesis, Chongqing Jiaotong University, Chongqing, China, 2008.
18. Liu, Y.S. Study on Damage Identification of Simply Supported Beam Bridge Based on Deflection Difference Influence Line. Ph.D. Thesis, Lanzhou University of Technology, Lanzhou, China, 2009.
19. Jia, Y. Research on Structural Damage Identification of Continuous Beam Bridge Based on Deflection Influence Line. Ph.D. Thesis, Guangzhou University, Guangzhou, China, 2014.
20. Bautista-De Castro, Á.; Sánchez-Aparicio, L.J.; Ramos, L.F.; Sena-Cruz, J.; González-Aguilera, D. Integrating geomatic approaches, Operational Modal Analysis, advanced numerical and updating methods to evaluate the current safety conditions of the historical Bôco Bridge. *Constr. Build. Mater.* **2018**, *158*, 961–984. [CrossRef]
21. Neild, S.A.; Williams, M.S.; McFadden, P.D. Nonlinear vibration characteristics of damaged concrete beams. *J. Struct. Eng.* **2003**, *129*, 260–268. [CrossRef]
22. Unger, J.F.; Teughels, A.; De Roeck, G. System identification and damage detection of a prestressed concrete beam. *J. Struct. Eng.* **2006**, *132*, 1691–1698. [CrossRef]
23. Zhang, R.F.; Zhang, X.M.; Qi, C.X. Application of the Influence Line on the Bridge Testing. *Adv. Mater. Res.* **2012**, *594*, 1586–1589. [CrossRef]
24. Kato, M.; Shimada, S. Vibration of PC bridge during failure process. *J. Struct. Eng.* **1986**, *112*, 1692–1703. [CrossRef]
25. Sassu, M.; Giresini, L.; Puppio, M.L. Failure scenarios of small bridges in case of extreme rainstorms. *Sustain. Resilient Infrastruct.* **2017**, *2*, 108–116. [CrossRef]
26. Croce, P.; Marsili, F.; Klawonn, F.; Formichi, P.; Landi, F. Evaluation of statistical parameters of concrete strength from secondary experimental test data. *Constr. Build. Mater.* **2018**, *163*, 343–359. [CrossRef]
27. Wang, F.L.; Chan, T.H.T.; Thambiratnam, D.P.; Tan, A.C.C. Damage diagnosis for complex steel truss bridges using multi-layer genetic algorithm. *J. Civ. Struct. Health Monit.* **2013**, *3*, 117–127. [CrossRef]
28. Zheng, W.; Shen, J.; Wang, J. Improved computational framework for efficient bayesian probabilistic inference of damage in truss structures based on vibration measurements. *Transp. Res. Rec.* **2014**, *2460*, 117–127. [CrossRef]
29. Behmanesh, I.; Moaveni, B. Probabilistic identification of simulated damage on the Dowling Hall footbridge through Bayesian finite element model updating. *Struct. Control. Health Monit.* **2015**, *22*, 463–483. [CrossRef]
30. Nair, K.K.; Kiremidjian, A.S.; Law, K.H. Time series-based damage detection and localization algorithm with application to the ASCE benchmark structure. *J. Sound Vib.* **2006**, *291*, 349–368. [CrossRef]
31. Zhu, L.; Wang, J.; Liu, J.; Nasir, M.S.; Zhu, J.; Li, S.; Liang, J.; Yan, W. Smart Formaldehyde Detection Enabled by Metal Organic Framework-Derived Doped Electrospun Hollow Nanofibers. *Sensor Actuat. B-Chem.* **2020**, *326*, 128819. [CrossRef]
32. Zhou, S.; Bao, Y.; Li, H. *Structural Damage Identification Based on Substructure Sensitivity and l1 Sparse Regularization*; International Society for Optics and Photonics: Washington, WA, USA, 2013; p. 86923N.
33. Zhang, Q.C.; Sun, Q.S. Damage detection of self-anchored suspension bridge based on neural network model and genetic-simulated annealing algorithm. *Adv. Mater. Res.* **2011**, *243*, 1963–1967. [CrossRef]

coatings

MDPI

Article

Effects of Mechanical Activation on Physical and Chemical Characteristics of Coal-Gasification Slag

Feng Wu [1,2], Hui Li [1,2,*] and Kang Yang [1,2]

[1] College of Materials Science and Engineering, Xi'an University of Architecture and Technology, Xi'an 710055, China; wufeng@xauat.edu.cn (F.W.); yangkang@xauat.edu.cn (K.Y.)

[2] State Key Laboratory of Green Building in Western China, Xi'an University of Architecture and Technology, Xi'an 710055, China

[*] Correspondence: sunshineli@vip.sina.com

Abstract: Coal-gasification slag (CGS) was subjected to mechanical grinding by three different methods. We studied the effects of mechanical activation on various physical and chemical characteristics of the CGS, including particle-size distribution, specific surface area, mineral composition, degree of crystallinity, particle morphology, chemical bonding, surface activity and binding energy, anionic-polymerization degree and hydration properties. The results show that there are different effects on CGS characteristics depending on the type of activation applied. Mechanical activation also can increase the specific surface area and the dissolution rates of activated SiO_2 and Al_2O_3, and the major elements (O, Si, Al, Ca) in CGS, whereas the degree of crystallinity and of polymerization of $[SiO_4]$ and $[AlO_6]$ are reduced by mechanical activation. We also found that the effects of different mechanical-activation methods on the compressive strength and activity were similar and could accelerate the hydration process.

Keywords: mechanochemical activation; coal-gasification slag; particle characteristics; cementitious material; reactivity

Citation: Wu, F.; Li, H.; Yang, K. Effects of Mechanical Activation on Physical and Chemical Characteristics of Coal-Gasification Slag. *Coatings* **2021**, *11*, 902. https://doi.org/10.3390/coatings11080902

Academic Editors: Andrea Nobili, Filippo Berto and Frederic Sanchette

Received: 26 May 2021
Accepted: 20 July 2021
Published: 28 July 2021

1. Introduction

In recent years, coal-gasification technology (CGT)—such as used by Texaco, Shell and Grup Servicii Petroliere—has become one of the leading clean coal technologies. A new technology, CGT can not only make the coal attain high value-added utilization, but it also will not produce sulfur dioxide, carbon dioxide and other harmful gases [1,2]. Thus, it effectively supports environmental protection. Coal-gasification slag (CGS), a byproduct of CGT, could reduce the advantages of the CGT as a clean technology [3,4]. To highlight the advantages of CGT, such as 'zero emissions', the environmental safety problems of CGT byproducts must be addressed and all-encompassing solutions developed and sustainable development achieved [5,6]. There are two types of CGS, fine-grained and coarse-grained. The fine-grained slag, 10% of the total CGS output, has a relatively high carbon content (as high as 15–30%) [7]. Inhomogeneous and porous, it can be used as an adsorbent to purify water. The coarse-grained slag, 90% of the total CGS output, has a relatively low carbon content and is dense, hard, inhomogeneous and amorphous [8]. It has many uses, such as a polishing compound, a cement additive, a component of concrete aggregate and road-surface coating [5,7,9–11].

As is widely known, much of the CGS produced by CGT has been harmful to the local ecological environment. Due to similar serious environmental problems are known to have been caused by CGS abroad (such as in Spain, Poland and Japan), a lot of research on uses of CGS has been carried out [12–17]. They found that the CGS could be used to fabricate building bricks, but the mechanical properties of such bricks could be improved. They also confirmed the potential for lightweight aggregates by using integrated-gasification combined cycle slag and demonstrated that their properties rivalled those of existing commercial lightweight aggregates. In China, gasification slags have been used mostly in civil construction such as raw materials for pavement or building bricks and pavement-base/subbase [18–22]. However, these research results do not make full use of the cementitious characteristics of CGS.

It is also well known that the physical and chemical characteristics of CGS determine its suitability for various uses [23]. Mechanical activation is a method that can induce a series of physical and chemical changes in solid materials by applying shear, compression, impact, bending and extension, which also induce solid–liquid, solid–gas and solid–solid chemical reactions [24]. In the process of mechanical grinding, the structure of solid materials will be greatly changed: The particle size of inorganic nonmetallic materials will be rapidly refined, the surface area of particles will rapidly increase, crystal-lattice distortion will occur and lattice defects will be produced and the amorphous (non-crystalline) character will be intensified [25,26]. At the same time, the chemical activity will increase, and the reaction conditions between inorganic nonmetallic materials and other materials will be reduced, thus resulting in the so-called mechanization-learning effect [27,28]. In the process of mechanical grinding, if the chemical composition of the solid material does not change, it is called mechanical activation, but if the chemical composition or structure of the solid material also changes, it is called mechanochemical activation [29,30]. Under the action of mechanochemistry, solid materials have more active points per unit mass, which can greatly improve the reactivity; it has been used widely in practical applications [31]. In addition, the most commonly used mechanical activation instruments in the laboratory are ball mill, vibration mill and prototype machine. However, there are few experimental studies about improving activity by mechanical activation, the effect of mechanical-activation methods on basic physical and chemical CGS properties. Too, the reaction mechanism of preparing cementitious materials from CGS have not been reported [32,33].

As CGS is generally coarse-grained, reactivity is minimal without grinding. In this study, we focused our research on CGS produced by CGT. The differences in physical and chemical characteristics of CGS subjected to various mechanical activations were analyzed from different perspectives, such as particle-size distribution, mineral composition, degree of crystallinity, particle morphology, chemical bond, surface activity and binding energy, anionic-polymerization degree and hydration properties. To understand the physical and chemical changes induced by various mechanical activations, the activated CGS was characterized by particle-size analysis, X-ray diffraction (XRD), scanning electron microscopy (SEM), Fourier transform infrared (FTIR) spectroscopy and X-ray photoelectron spectrometry (XPS). Furthermore, we assessed the hydration characteristics of cementitious material prepared from activated slag.

2. Raw Material and Experimental Methods

2.1. *Raw Material*

The CGS used in this research was provided by Shenhua Ningmei Group. It was ground to a specific surface area by applying three grinding methods: ball mills (QM), vibration mills (ZD) and a type of prototype machine (ZY), ZY is a kind of small equipment similar to the refiner, it is ground by friction and rubbing between the rotating disc and the material. The yielded average particle sizes of ~20 μm and ~7 μm, respectively. Cement, provided by Sheng wei cement company, had a specific surface area of 370 m^2/kg. The chemical composition of the CGS and of the cement are shown in Table 1, and a photomicrograph of the CGS is shown in Figure 1.

Table 1. Composition of Coal-gasification Slag and Cement.

Composition	SiO_2	Al_2O_3	CaO	Fe_2O_3	MgO	Na_2O	K_2O	P_2O_5	SO_3	Other	LOI
CGS	48.75	20.05	10.69	9.67	2.84	1.65	2.11	0.17	0.56	1.88	1.52
Cement	18.21	3.37	64.59	2.88	3.89	0.51	1.02	0.48	1.37	0.48	3.19

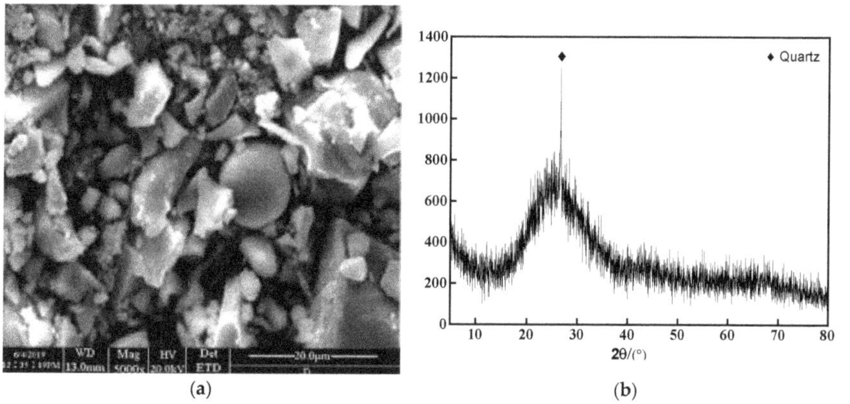

(**a**) (**b**)

Figure 1. Microscopic morphology (**a**) and mineral components (**b**) of coal-gasification slag.

The CGS contains mainly silica (SiO_2), aluminum oxide (Al_2O_3), calcium oxide (CaO) and iron oxide (Fe_2O_3) with small amounts of various RO and R_2O phases; the total content of SiO_2, Al_2O_3 and CaO is >70%, and the loss on ignition is 1.52%. The CGS is characterized by heterogeneous morphology; its irregular forms include flakes and rods, and a small amount of spherical particles form an included powder. The surface of the particles is very dense, and CGS porosity is low. The CGS contains mainly an amorphous glass phase and small amounts of silica (quartz phase). The relatively high glass-phase content is favorable for the dissolution of the activated SiO_2 and Al_2O_3.

2.2. Mechanical-Activation-Material Sample Preparation

CGS grinding experiments were carried out by using the three methods QM (Φ 300 mm × 300 mm), ZD (MZS-3 in laboratory) and ZY (a type of prototype machine), resulting in particle sizes of ~20 um and ~7 um, respectively, all the grinding methods are dry grinding. When the QM was used for grinding and using the most compact packing theory to match the ball, the grinding material was steel ball, the ratio of ball to material was 10:1. When the ZD was used for grinding, the grinding material was steel ball, the grinding material was zirconia ball, and the ratio of ball to material was 10:1, the filling rate of zirconia ball was 80%. There are three particle sizes of zirconia ball: Φ 5 mm, Φ 10 mm and Φ 15 mm, the total weight of zirconia balls was 6.0 kg, and the ratio was 3:5:7. When the ZY was used for grinding, the addition amount of coal gasification slag was 60 g.

2.3. Experimental Methods

1. The mineral composition of the activated CGS was analyzed by XRD, the X-ray diffraction method (D/max 2200, Rigakul Corporation, Tokyo, Japan). The measuring speed was at 10° (2θ)/min in the 2θ range of 5–75°. the chemical composition was measured by X-ray fluorescence (XRF, Brooke Instruments Co., Ltd., Brooke, Germany), the particle morphology of the activated slag was observed by SEM (FEI Company, Oregon, OR, USA), the anionic-polymerization degree of the CGS was analyzed by FTIR (Shimadzu Corporation, Kyoto, Tokyo), and the surface activity and elements were analyzed by chemical analysis and XPS (Rigakul Corporation, Tokyo, Japan).
2. The CGS paste was prepared as follows: First, it was prepared from 70 wt% cement and 30 wt% CGS, resulting in a water–binder ratio of 0.5 for the paste. The specimens were cured at 20 ± 1 °C and >95% humidity for 3 d, 7 d and 28 d, respectively. Then the strength of CGS was measured according to Chinese National Standard GB/T17671-1999.
3. We used three types of equipment to grind CGS in this study: ball and vibration mills and a prototype machine. During the grinding process, CGS underwent a series

of physical and chemical changes, and its microstructure also changed accordingly. Changing the structure modified the performance, thereby improving its reactivity. For each grinding mode, a parallel group was set by changing the duration of mechanical grinding to form two CGS samples, one fine-grained and one coarse-grained. The performance of the CGS of the same particle size (fine- or coarse-grained) produced by different grinding methods was compared, and the effects of different grinding methods were analyzed. The influence of CGS gelation activity and the grinding parameters of different equipment are shown in Table 2.

Table 2. Grinding Parameters.

Grinding Equipment	QM	ZD	ZY
Grinding Time	2.5 h 1 h	2 h 1 h	2 min 1 min

3. Properties of Mechanically Activated Coal-Gasification Slag

3.1. Particle-Size Distribution of Activated Slag

The particle-size distribution of CGS activated by different grinding methods was measured by a laser particle-size analyzer (Figure 2). The particle size of by different grinding methods are shown in Table 3, The CGS particle size is significantly reduced by controlled extension of the grinding time by a particular method. This may be due to the natural defects in the CGS subjected to the action of mechanical force. When the slag is being destroyed, its particle size is rapidly refined along the defect interface, showing a macroscopic decrease in particle size and an increase in specific surface area.

The particle-size and frequency distribution curves for CGS activated by the three grinding methods are similar. The particle size averages ~7 μm, and <20 μm particles account for >90%. The median particle size and volume average particle size of the CGS are similar for the three different grinding methods (QM, ZD and ZY), and their particle-size and frequency distribution curves are very similar. The particle size is concentrated mainly ~20 μm, and the particle size of >90% particles is <50 μm. Thus, it is apparent that the different grinding methods have little effect on the CGS particle-size distribution. The purpose of extending the activation time of mechanical activation is to optimize the particle size distribution of samples. Previous studies have shown that the geopolymer synthesized by the samples has a low strength value, which is closely related to the particle size distribution of samples [34].

The specific surface area of the material output by different grinding methods applied for different grinding times is shown in Table 4.

After applying any of the different grinding methods, the specific surface area of the activated slag significantly increases owing to structural damage to the material under the action of mechanical force, which leads to the reduction of particle size. Comparing the different grinding methods resulting in the same particle size, the specific surface area of the CGS produced by mechanical activation by the prototype machine is the largest, and that generated by the vibration mill is the smallest. This is caused by the different working principles of different equipment for grinding materials. However, when applying the same grinding method, the specific surface area of the material increases significantly with the extension of grinding time. This phenomenon is related to the particle-size distribution of the material, to which in turn the change in surface area corresponds [35]. With the appropriate increase in grinding time, the material particle size decreases and its specific surface area increases, and the phase composition will change.

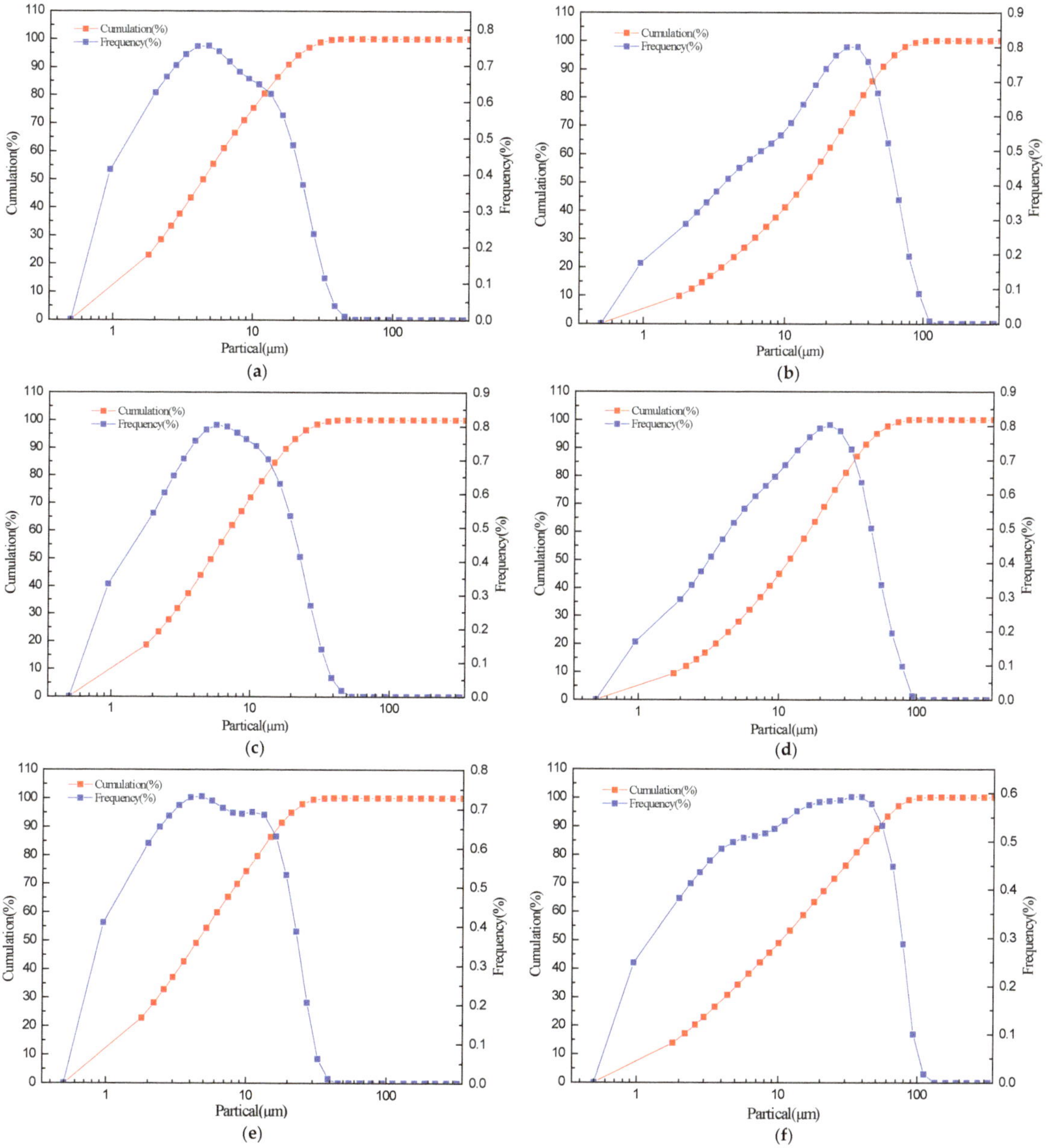

Figure 2. Influence of different mechanical-activation methods on powder particle-size distribution. (**a**) QM-2.5 h; (**b**) QM-1 h; (**c**) ZD-2 h; (**d**) ZD-1 h; (**e**) ZY-2 min; (**f**) ZY-1 min. (QM, ball milling; ZD, vibration mill; ZY, prototype machine.).

Table 3. Characteristic Particle Size of Activated Slag.

Grinding Method	QM		ZD		ZY	
Grinding Time	2.5 h	1 h	2 h	1 h	2 min	1 min
Median Particle Size D_{50} (μm)	5.41	14.25	5.27	11.86	4.55	10.52
D_{90} (μm)	18.40	48.66	18.36	40.35	17.20	52.04
Average Diameter (μm)	8.79	19.84	7.74	17.17	7.08	19.28

Table 4. Specific Surface Area of Activated Slag.

Grinding Method	QM		ZD		ZY	
Grinding Time	2.5 h	1 h	2 h	1 h	2 min	1 min
Specific Surface Area (m^2/kg)	864.55	471.29	703.77	409.55	921.20	671.56

3.2. Mineral-Composition Analysis

The diffraction pattern of the cinder (Figure 3) is a disordered pattern. With no obvious crystal peak, it represents mainly amorphous material. The mechanochemical effect of the CGS was produced by different mechanical treatments. A narrow and sharp crystal peak and a non-crystal (amorphous) bulge peak are apparent at 20~30° (2θ); the crystal peak represented the quartz phase, produced by all three methods in a short grinding time. In contrast, a crystal peak at 47° represents an iron phase that appeared in the ball-milling diffraction pattern but was not produced by the other two methods. This may be due to the mechanical damage to the CGS structure, particle-size reduction, being encased in the inner quartz-crystal structure later exposed, or the formation of the iron phase may be due to the extrusion, impact and collision between grinding balls, which results in the mixing of some iron phases in the powder. With increased grinding time, the intensity of the diffraction peak of the original crystal decreases, the intensity of the amorphous bulge peak increases, and the diffraction peak is more diffused. With the simultaneous quartz-crystal-lattice distortion, the structure changes to amorphous as the crystallinity of the slag decreases, the inert glass structure depolymerizes and the dissolution rate of active Si and Al in the CGS increases.

It is known that mechanical activation can cause the reduction in crystallinity degree of mineral phases, i.e., and active sites increase in the amorphous content. The transformation of crystalline material into amorphous enhances its reactivity. Mechanical activation influenced enhanced availability of activated SiO_2 and Al_2O_3 during the process of geopolymerization [36,37].

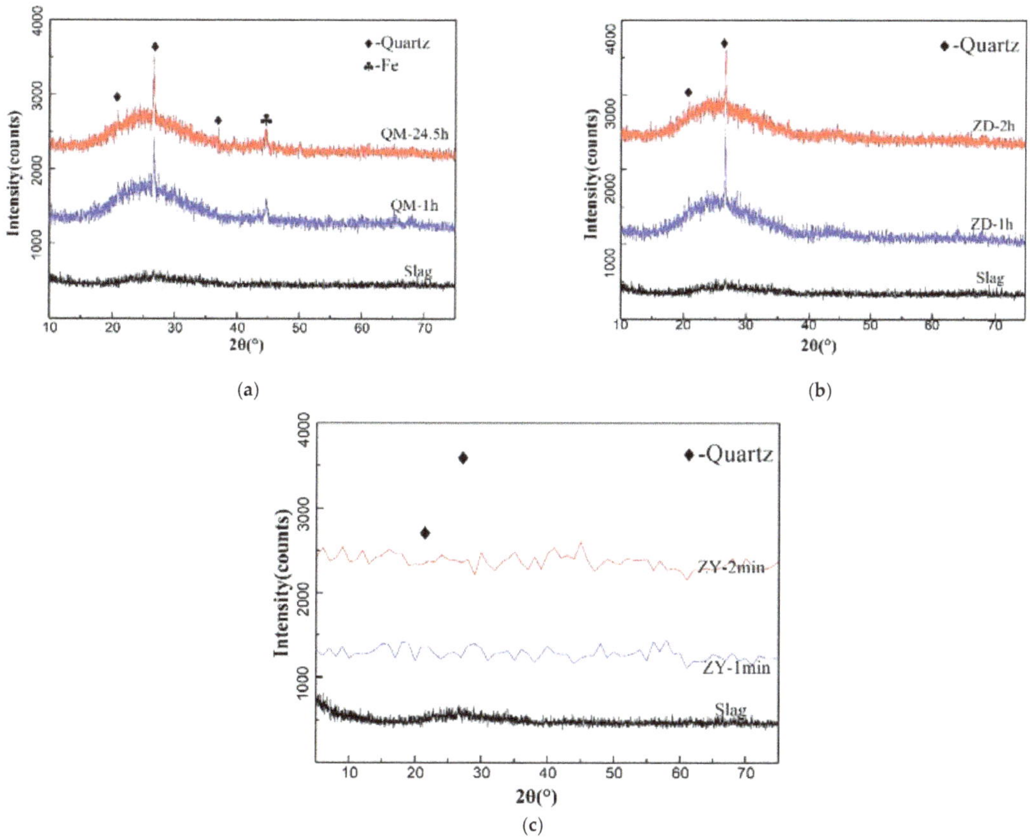

Figure 3. Mineral composition of slag produced by different mechanical-activation methods. (**a**) QM; (**b**) ZD; (**c**) ZY.

3.3. Particle Morphology

CGS micromorphology has changed greatly after being ground by various grinding equipment. Under the action of mechanical force, spherical particles and dense irregular larger particles form sheetlike or blocklike configurations to varying degrees; the particles are rapidly ground finer, the original spherical particles giving way to compact irregularly shaped particles with sharp edges and corners and showing good dispersion [37].

By comparing the output from the same grinding method at different grinding times, we found that (Figure 4) over time, the irregular large particles collided and impacted each other under the action of mechanical force, whereby they were further refined into smaller irregular particles showing no evidence of the structure of the original spherical particles. A good particle-size distribution is established as a small number of large particles and a large number of small particles become thoroughly intermixed, while also a large number of spherical particles in the slag become almost invisible and a large number of irregular polygonal particles form. We also found that the CGS particles formed by ball and vibration mills have sharp edges and corners, whereas the CGS formed by the prototype sample-making machine has relatively smooth edges and corners. A photomicrograph also revealed that the particle size of the CGS sample after grinding by the prototype machine is smallest and after grinding by the vibration mill is the largest; this outcome is consistent with the specific surface area of resulting particles given different grinding methods. The sample-making machine yields the largest specific surface area, while the vibration mill yields the smallest.

Figure 4. Photomicrograph of coal-gasification slag with different grinding methods. (**a**) QM-2.5 h; (**b**) ZD-2 h; (**c**) ZY-2 min; (**d**) QM-1 h; (**e**) ZD-1 h; (**f**) ZY-1 min.

3.4. Chemical-Bond Analysis

The CGS records changes in a physical-chemistry series under the action of mechanical force. We used FTIR to analyze the internal structural change of the CGS activated by mechanical treatment.

The presence of quartz causes a series of bands at about 1150, 1080, 796–778 (double band) and 480 cm^{-1} [38–40]. There are stretching-vibration peaks in the slag (Figure 5); the strong absorption peak at 3441 cm^{-1} and 1593 cm^{-1} are caused by stretching vibration of the O–H bond, and the absorption peak at 1380 cm^{-1} and 1352 cm^{-1} are caused by stretching vibration of C–O–C, the absorption peaks at 1087 cm^{-1} and 989 cm^{-1} are caused by the asymmetric stretching vibration of the Si–O–Si bond. The wave band at 761 cm^{-1} may be related to the symmetric stretching vibration of Si–O–Si and the AlO_4 vibration, and the absorption peak at 481 cm^{-1} is caused by the bending vibration of the Si–O–Si bond.

The FTIR spectra of CGS activated by different grinding methods are similar, and the absorption peaks of CGS activated by ball or vibration milling and by prototype-machine grinding at 1087 cm^{-1} and 989 cm^{-1} shift to lower wavenumbers to different degrees, and with the increase in grinding time, the absorption peak at 989 cm^{-1} decreases gradually until it disappears. The shift from the absorption peak to the low wavenumber can reflect the change of the structure of the silicon–oxygen tetrahedron in the activated CGS; the change of the degree of polymerization from the high to the low state leads to the increase of the content and activity of the low state vitreous. The bending vibration of the Si–O–Si bond at 481 cm^{-1} also confirms this phenomenon.

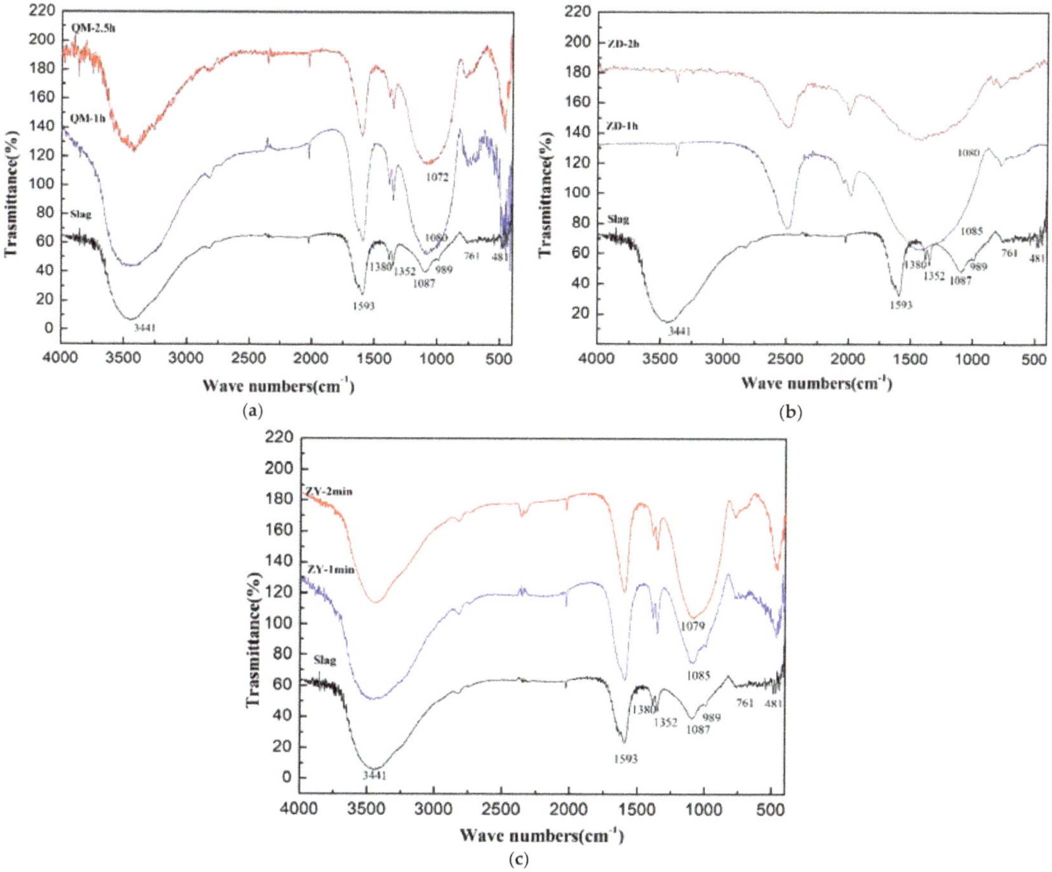

Figure 5. FTIR spectra of activated coal-gasification slag under different grinding methods. (**a**) QM; (**b**) ZD; (**c**) ZY.

Given the same grinding method, at shorter grinding times, the peak intensity of the spectral lines increases and the absorption-band intensity increases because of the quartz-crystal phase in the CGS after grinding. However, as the grinding continues, the band of the absorption spectrum becomes wider and weaker owing to the destruction of the crystal lattice and the depolymerization of the glass structure of the high polymer state. From the above analysis, it can be concluded that the CGS bond structure can be changed by using different mechanical-grinding methods; the CGS crystallinity is reduced, the structure becoming amorphous as the content of activated glass increases.

3.5. *Surface-Activity Analysis*

From these results (Table 5), it can be seen that three different grinding methods can effectively improve the digestion rate of activated SiO_2 and Al_2O_3 in CGS. Furthermore, the total amount of SiO_2 and Al_2O_3 in activated CGS from the use of a vibration mill or a prototype machine was similar, while the total amount of SiO_2 and Al_2O_3 dissolved from activated CGS was the lowest when ball milling was used. Given the same grinding method, the dissolution rate of active SiO_2 and Al_2O_3 increases with the increase in grinding time, because with the extension of grinding time, the mechanical force acting on the material destroys its structure, crystalline material becomes amorphous, the structure of the inert vitreous body is dispersed and depolymerized and the SiO_2 and Al_2O_3 in the amorphous mineral are separated to form active SiO_2 and Al_2O_3. With higher dissolution rates of

SiO_2 and Al_2O_3 in activated CGS, both the activity of activated SiO_2 and Al_2O_3 and the degree of participation of activated SiO_2 and Al_2O_3 as cementitious material in hydration reactions increase.

Table 5. Dissolution Rate of Activated SiO_2 and Al_2O_3 in Different Activated Coal-gasification Slags.

Grinding Mode	Dissolution Rate of Active Silicon/%	Leaching Rate of Active Aluminum/%	Total Dissolution of Silicon and Aluminum/%	Activity Rate K_a/%
CGS	6.789	3.105	9.894	14.38
QM-2.5 h	20.544	10.383	30.927	44.95
QM-1 h	15.525	7.506	23.031	33.48
ZD-2 h	19.154	11.876	31.03	45.10
ZD-1 h	16.142	8.398	24.540	35.67
ZY-2 min	21.058	11.010	32.068	46.61
ZY-1 min	15.006	9.027	24.033	34.93

3.6. Surface Binding Energy

XPS analysis was used to further understand the chemical-state changes of several typical elements in CGS particles before and after mechanochemical activation. Full-spectrum scanning indicated that the elements O, Si, Al and Ca were mainly near the surface of slag (especially in ball-milled slag). Then we applied fine-spectra scanning of O_{1s}, Si_{2p}, Al_{2p} and Ca_{2p}.

That the electron-binding energy of elements changes with mineral structure enables qualitative analysis of the structure and activity of minerals using XPS analysis (Figure 6). By analyzing cement, fly ash and slag, we found that the binding energy of Si_{2p} is related to the activity of cementitious materials; accordingly, lower Si_{2p} binding energy means higher cementitious activity of materials. The maximum binding energy of Si_{2p} is 103.23 eV, and the minimum binding energy of 4.5 h ball-milled slag is 102.67 eV, which shows that the activity of 4.5 h ball-milled slag is the highest, that of 2.5 h ball-milled slag is second and the original slag is lowest (Table 6).

In the Al_{2p} XPS spectra, the peak position shifts in the direction of small-electron-binding energy with the increase in ball-milling time. The electron-binding energy of ball milling 2.5 h slag is the smallest and that of raw slag is the largest; thus, from the binding-energy value, it can be concluded that the aluminum in the raw slag is mainly in the state of six-coordination, and the aluminum in the ball-milled CGS is mainly in the form of the aluminum–oxygen tetrahedron, which shows that the bond structure in the CGS has been changed by the action of mechanical force, and the change of binding energy increases the activity of slag. In minerals, silicon and aluminum exist mainly in the structure of silicon–oxygen and aluminum–oxygen tetrahedra. The change in the binding energy of silicon and aluminum results mainly from the polarization of silicon and aluminum by oxygen, during which the binding energy of oxygen also changes. The changing trend of the binding energy of oxygen, which is the same as that of silicon and aluminum, may reflect CGS activity enhanced by mechanical action, and that activity increases with the duration of mechanical action (Table 6).

The reactivity of material attributes to its high surface [40–42], The activity of cementitious materials also is higher with higher Ca content on the surface of the samples. According to the XPS spectra, the peak area of Ca_{2p} and the content of Ca_{2p} on the surface increased with grinding time, which indicated that the degree of Ca^{2+} participating in hydration reactions had increased. Apparently, mechanical activation can promote CGS gelling activity. The regularity of the surface electron-binding energy of O_{1s}, Si_{2p}, Al_{2p} and Ca_{2p} orbitals before and after CGS activation shows that the activation can change the CGS mineral structure. A decrease in binding energy indicates a lower degree of aggregation of minerals, presumably owing to the mineral structure's becoming unstable, with an increase in surface reactivity. This interpretation is consistent with XRD and FTIR analysis before and after CGS activation.

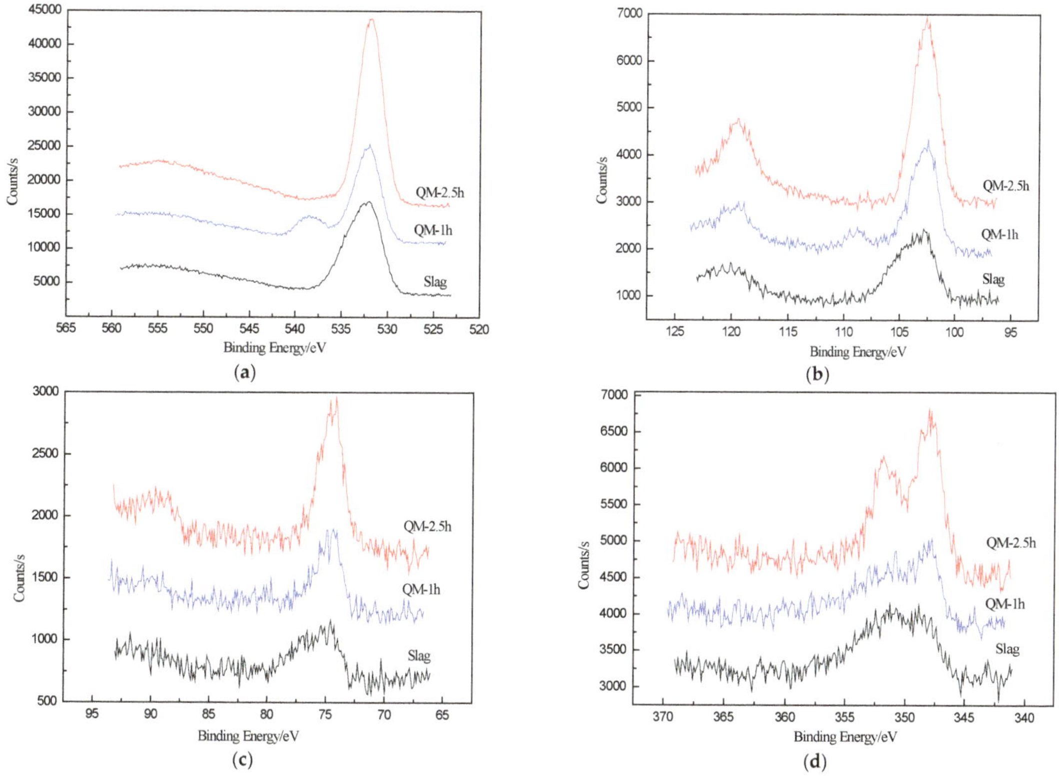

Figure 6. X-ray photoelectron spectrometry diagram of four elements of slag activated by QM methods: O_{1s}, Si_{2p}, Al_{2p} and Ca_{2p}. (**a**) XPS Spectra of O_{1s}; (**b**) XPS Spectra of Si_{2p}; (**c**) XPS Spectra of Al_{2p}; (**d**) XPS Spectra of Ca_{2p}.

Table 6. Electron-binding Energies of Different Elements in Studied Samples.

Electronic Binding Energies	Element			
	O_{1s}	Si_{2p}	Al_{2p}	Ca_{2p}
QM-2.5 h	531.91	102.67	74.48	348.04
QM-1 h	532.12	102.71	74.49	348.06
Slag	532.31	103.23	74.86	350.56

4. Activity Index of Cementitious Materials Prepared from Activated Slag

The composite cementitious materials used in this study were prepared by mixing activated slag and cement in a 7:3 ratio and using a water–cement ratio of 2:5. By analyzing and characterizing the mechanical and microcosmic properties of the composite cementitious block, we explored the effect of activated CGS on the hydration process of the composite system and determined the optimum process of generating composite cementitious material, thereby providing a reliable way for the application of CGS.

According to the comparative analysis of the strength of samples subjected to different grinding methods (Figure 7), the hydration effect of the system with ball-milled CGS is the worst, while the hydration effect of the system with vibration-milled CGS is the best (and its activity index is higher). For the QM-2.5 h block, the compressive strength of the block had decreased by 35.16%, 24.69% and 18.89% by the ages 3 d, 7 d and 28 d, respectively, and for QM-1 h specimen, the compressive strength was 43.23%, 36.90% and 17.67% lower by the ages 3 d, 7 d and 28 d, respectively, than that of cement paste. The difference in compressive strength between the composite cementitious block and the pure cement block was decreasing; the strength of ZD-2 h and ZY-2 min samples even exceeded that of pure-cement samples after hydration for 28 d. The activity index of ZD-2 h and ZY-2 min samples were 123.2% and 114.6% (Table 7), respectively.

Previous studies have shown that the reactivity of materials increases (characterized by increasing the mechanical strength of geopolymers) [43–46]. The results of this study confirmed that the reactivity of CGS was greatly enhanced in particle size. When the median size (D_{50}) decreases to less than 10 μm, the geopolymer strength tends to increase. The early low strength and late high strength of the composite cementitious block may be due to the activity effect and microaggregate effect of CGS. As the activated CGS was used as cementitious material to replace part of the cement, the clinker mineral content of the composite cementitious system decreased. Early hydration of the system was less than that of a pure-cement system. The pozzolanic activity of activated slag was fully stimulated in the alkaline environment formed by hydration of the cement, and the reaction between the calcium dihydroxide so formed and the activated SiO_2 and Al_2O_3 dissolved by activated CGS promoted the secondary hydration of the cement. However, because the particle size of the activated CSG was smaller than that of cement, it could develop good particle gradation with cement and fill the gap between cement particles. This improved the densification of the composite cementitious system before hydration, and, at a later stage, the strength of the block was also improved.

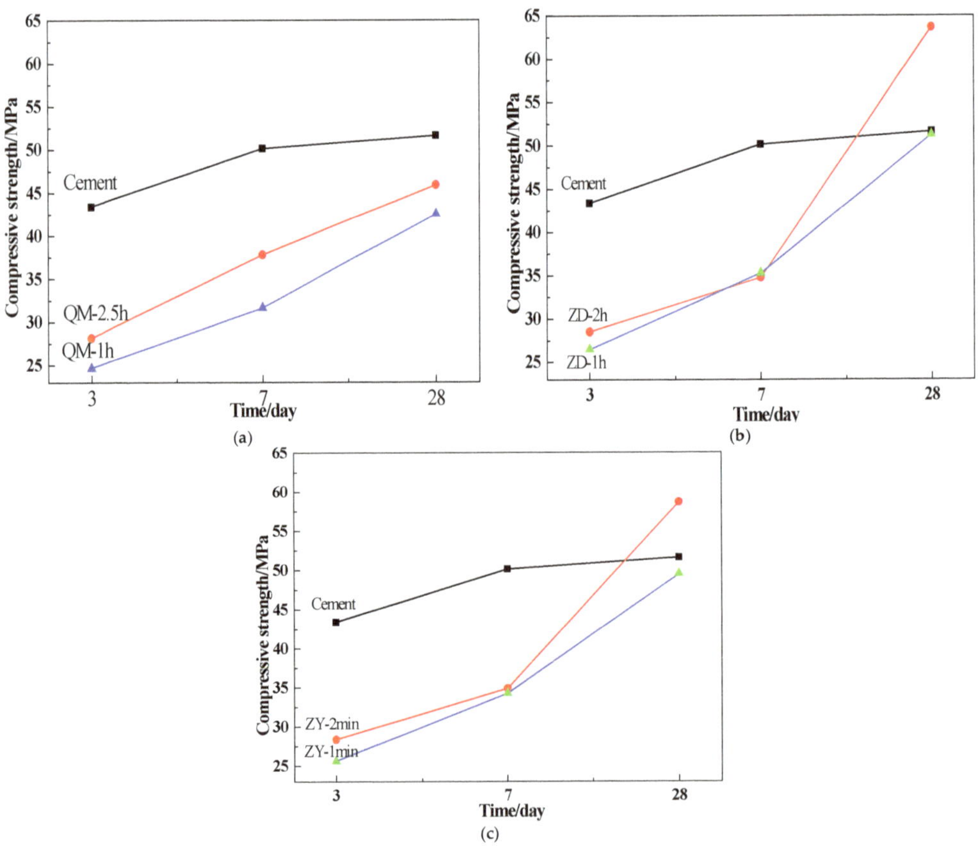

Figure 7. Mechanical properties of composite cementitious materials activated by different grinding methods. (**a**) QM; (**b**) ZD; (**c**) ZY.

Table 7. Dissolution Rate of Activated Silica and Aluminum in Different Activated Slags.

Grinding Mode	Activity Index/%		
	3 d	7 d	28 d
Cement	100	100	100
QM-2.5 h	64.8	75.3	88.9
QM-1 h	56.8	63.1	82.3
ZD-2 h	65.7	69.3	123.2
ZD-1 h	61.1	70.3	99.3
ZY-2 min	65.4	69.6	114.6
ZY-1 min	59.1	68.3	95.9

Thus overall, the specific surface area of powder particles can be improved, the particle size reduced, the complete crystal structure of powder particles was broken and the degree of crystallization of slag reduced. Furthermore, the surface energy of Si, Al, Ca and other effective elements can be improved and the dissolution of active SiO_2 and Al_2O_3 enhanced through mechanical activation. In summary, the CGS can be used as cement admixture.

5. Conclusions

In this study, the properties of CGS activated by three different mechanical methods were investigated, and the feasibility of using mechanically activated slag as cementitious material was researched by experiment. Our findings support the application of CGS in cement and concrete. The results of our experiments are as follows:

1. There are significant differences in particle-size distribution, mineral composition, degree of crystallinity, particle morphology, chemical bond, surface activity and binding energy, anionic-polymerization degree and hydration properties resulting from three types of grinding.
2. We found that three different mechanical-activation methods could change the internal structure of CGS and improve its gelling activity to different degrees at the same level of particle fineness. We also confirmed that CGS activated by vibration milling and by a prototype grinding machine have the same activity as that by ball milling, and in all three, CGS activity increases with duration of grinding.
3. Mechanical-grinding action reduced the degree of polymerization of [SiO_4] and [AlO_6], and with longer grinding time, the action effect became more significant. The 3-d, 7-d and 28-d strength of CGS paste after mechanical grinding improved significantly, and 28-d activity index of the CGS reached 88.9%, 123.2% and 114.6%, respectively.
4. To sum up, although all three kinds of mechanically activated slags can be used as secondary cementitious materials in cement and concrete, considering the activity index, the grinding power and consumption of materials, the best plan appears to be to replace 30% cement with activated slag as a secondary cementitious material. It can meet the requirement of strength, reduce the amount of cement, reduce the cost and improve the comprehensive utilization ratio of CGS.

Author Contributions: Conceptualization, Funding acquisition, F.W.; Methodology, Writing—review & editing, H.L.; Formal analysis, Investigation, K.Y. All authors have read and agreed to the published version of the manuscript.

Funding: This research was funded by National Natural Science Foundation of China, grant number Z20180222, Independent Research and Development project of State Key Laboratory of Green Building in Western China, grant number LSZZ202021 and Natural Science Foundation of Shaanxi Provincial Department of Education, grant number 20JY041 and Qingyuan Science and Technology Plan Project, grant number 2020KJJH040.

Institutional Review Board Statement: Not applicable.

Informed Consent Statement: Not applicable.

Data Availability Statement: All data included in this study are available upon request by contact with the corresponding author.

Acknowledgments: We acknowledge financial supports by National Natural Science Foundation of China (Z20180222), Independent Research and Development project of State Key Laboratory of Green Building in Western China (LSZZ202021) and Natural Science Foundation of Shaanxi Provincial Department of Education (20JY041) and Qingyuan Science and Technology Plan Project (2020KJJH040).

Conflicts of Interest: The authors declare that there is no conflict of interest in this manuscript.

References

1. Pan, C.; Liang, Q.; Guo, X.; Dai, Z.; Liu, H.; Gong, X. Characteristics of different sized slag particles from entrained-flow coal gasification. *Energy Fuel* **2016**, *30*, 1487–1495. [CrossRef]
2. Ratafia-Brown, J.A.; Manfredo, L.M.; Hoffmann, J.W.; Ramezan, M.; Stiegel, G.J. In an environmental assessment of IGCC power systems. In *Nineteenth Pittsburgh Coal Conference*; University of Pittsburgh: Pittsburgh, PA, USA, 2002; p. 200216.
3. *Gasification Markets and Technologies-Present and Future an Industry Perspective*; US Department of Energy: Washington, DC, USA, 2002; p. 100.
4. Wu, T.; Gong, M.; Lester, E.; Wang, F.; Zhou, Z.; Yu, Z. Characterisation of residual carbon from entrained-bed coal water slurry gasifiers. *Fuel* **2007**, *86*, 972–982. [CrossRef]
5. Wu, S.Y.; Huang, S.; Ji, L.Y.; Wu, Y.Q.; Gao, J.S. Structure characteristics and gasification activity of residual carbon from entrained-flow coal gasification slag. *Fuel* **2014**, *122*, 67–75. [CrossRef]
6. Li, Z.Z.; Zhang, Y.Y.; Zhao, H.Y.; Chen, H.X.; He, R. Structure characteristics and composition of hydration products of coal gasification slag mixed cement and lime. *Constr. Build. Mater.* **2019**, *213*, 265–274. [CrossRef]
7. Wu, S.; Huang, S.; Wu, Y.; Gao, J. Characteristics and catalytic actions of inorganic constituents from entrained-flow coal gasification slag. *J. Energy Ins.* **2015**, *88*, 93–103. [CrossRef]
8. Duan, W.; Yu, Q.; Liu, J.; Hou, L.; Xie, H.; Wang, K.; Qin, Q. Characterizations of the hot blast furnace slag on coal gasification reaction. *Appl. Therm. Eng.* **2016**, *98*, 936–943. [CrossRef]
9. Zhao, X.; Zeng, C.; Mao, Y.; Li, W.; Peng, Y.; Wang, T.; Eiteneer, B.; Zamansky, V.; Fletcher, T. The surface characteristics and reactivity of residual carbon in coal gasification slag. *Energy Fuels* **2010**, *24*, 91–94. [CrossRef]
10. Wagner, N.J.; Matjie, R.H.; Slaghuis, J.H.; Van Heerden, J.H.P. Characterization of unburned carbon present in coarse gasification ash. *Fuel* **2008**, *87*, 683–691. [CrossRef]
11. Xu, S.Q.; Zhou, Z.J.; Gao, X.X.; Yu, G.S.; Gong, X. The gasification reactivity of unburned carbon present in gasification slag from entrained-flow gasifier. *Fuel Process. Technol* **2009**, *90*, 1062–1070. [CrossRef]
12. Acosta, A.; Iglesiasa, I.; Ainetoa, M.; Romero, M.; Rincón, J.M. Utilisation of IGCC slag and clay steriles in soft mud bricks (by pressing) for use in building bricksmanufacturing. *Waste Manag.* **2002**, *22*, 887–891. [CrossRef]
13. Aineto, M.; Acosta, A.; Rincón, J.M.; Romero, M. Production of Lightweight Aggregates from Coal Gasification Fly Ash and Slag. *World Coal Ash* **2005**, 11–15.
14. Yoshitaka, I. Utilization of coal gasification slag collected from IGCC as fine aggregate for concrete. In Proceedings of the EUROCOALASH Conference, Thessaloniki, Greece, 25–27 September 2012.
15. Martin, I.; Echeveria, A.A.; Garcia-Romero, E. Recycling of residual IGCC slags and their benefits as degreasers in ceramics. *J. Environ. Manag.* **2013**, *129*, 1–8. [CrossRef]
16. Masafumi, K.; Reiji, T.; Takao, E.; Shigemi, N. Utilization of the gasification-melting slag from municipal solid waste as fine aggregate for concrete, Concr. *Res. Technol.* **2002**, *13*, 89–98.
17. Pomykala, R. The mechanical properties of coal gasification slag as a component of concrete and binding mixtures. *Pol. J. Environ. Stud.* **2014**, *23*, 1403–1406.
18. Maghool, F.; Arulrajah, A.; Horpibulsuk, S.; Du, Y. Laboratory evaluation of ladle furnace slag in unbound pavement-base/subbase applications. *J. Mater. Civ. Eng.* **2017**, *29*, 1–9. [CrossRef]
19. Maghool FArulrajah, A.; Du, Y.; Horpibulsuk, S.; Chinkulkijniwat, A. Environmental impacts of utilizing waste steel slag aggregates as recycled road construction materials. *Clean Technol. Environ.* **2016**, *19*, 949–958. [CrossRef]
20. Arulrajah, A.; Mohammadinia, A.; Phummiphan, I.; Horpibulsuk, S.; Samingthong, W. Stabilization of recycled demolition aggregates by geopolymers comprising calcium carbide residue, fly ash and slag precursors. *Constr. Build. Mater.* **2016**, *114*, 864–873. [CrossRef]
21. Mohammadinia, A.; Arulrajah, A.; Sanjayan, J.; Disfani, M.M.; Bo, M.W.; Darmawan, S. Stabilisation of demolition materials for pavement base/subbase applications using fly-ash and slag geopolymers: Laboratory investigation. *J. Mater. Civ. Eng.* **2016**, *28*, 1–9. [CrossRef]
22. Heidrich, C.; Ward, C.; Chalmers, D.; Heeley, P.; Ness, J.; Williams, R. Production and handling of coal combustion products. In *Coal Combustion Products Handbook*, 2nd ed.; Ward, C., Heidrich, C., Yeatman, O., Eds.; Ash Development Association of Australia: Port Kembla, Australia, 2014; pp. 1–33.
23. Suryanarayana, C. Mechanical alloying and milling. *Prog Mater. Sci.* **2001**, *46*, 1–184. [CrossRef]

24. Hewitt, S.A.; Kibble, K.A. Effects of ball milling time on the synthesis and consolidation of nanostructured WC–Co composites. *Int J. Refract. Met. Hard Mater.* **2009**, *27*, 937–948. [CrossRef]
25. Sharafi, S.; Gomari, S. Effects of milling and subsequent consolidation treatment on the microstructural properties and hardness of the nanocrystalline chromium carbide powders. *Int. J. Refract. Met. Hard Mater.* **2012**, *30*, 57–63. [CrossRef]
26. Lee, J.S.; Choi, M.S.; Hung, N.V.; Kim, Y.S.; Kim, I.W.; Park, E.C. Effects of high energy ball-milling on the sintering behavior and piezoelectric properties of PZT based ceramics. *Ceram. Int.* **2007**, *33*, 1283–1286. [CrossRef]
27. Razmjou, A.; Mansouri, J.; Chen, V. The effects of mechanical and chemical modification of TiO_2 nanoparticles on the surface chemistry, structure and fouling performance of PES ultrafiltration membranes. *J. Membr. Sci.* **2011**, *378*, 73–84. [CrossRef]
28. Patil, A.G.; Anandhan, S. Ball milling of class-F Indian Fly ash obtained froma thermal power station. *Int. J. Energy Eng.* **2012**, *2*, 57–62.
29. Sharma, A.; Srivastava, K.; Devra, V.; Rani, A. Modification in properties of fly ash through mechanical and chemical activation. *Am. Chem. Sci. J.* **2012**, *2*, 177–187. [CrossRef]
30. Akcay KSirkecioğlu, A.; Tatlıer, M.; Savaşcı, Ö.T.; Erdem-Şenatalar, A. Wet ball milling of zeolite HY. *Powder Technol.* **2004**, *142*, 121–128. [CrossRef]
31. Watanabe, H. Critical rotation speed for ball-milling. *Powder Technol.* **1999**, *104*, 95–99. [CrossRef]
32. Patil, A.G.; Shanmugharaj, A.M.; Anandhana, S. Interparticle interactions and lacunarity of mechano-chemically activated fly ash. *Powder Technol.* **2015**, *272*, 241–249. [CrossRef]
33. Zhao, J.H.; Wang, D.M.; Liao, S.C. Effect of mechanical grinding on physical and chemical characteristics of circulating fluidized bed fly ash from coal gangue power plant. *Constr. Build. Mater.* **2015**, *101*, 851–860. [CrossRef]
34. Kumar, S.; Kumar, R.; Alex, T.C.; Bandopadhyay, A.; Mehrotra, S.P. Effect of mechanically activated fly ash on the properties of geopolymer cement. In Proceedings of the 4th world congress geopolymer, Saint-Quentin, France, 28 June–1 July 2005; pp. 113–116.
35. Arjuan, P.; Silbee, M.R.; Roy, D.M. Quantitative determination of the crystalline and amorphous phases in low calcium fly ashes. In Proceedings of the 10th International Congress of the Chemistry of Cement, Gothenburg, Sweden, 2–6 June 1997; pp. 2–6.
36. Zhang, Q.; Saito, F. A review on mechanochemical synthesis of functional materials. *Adv. Powder Technol.* **2012**, *23*, 523–531. [CrossRef]
37. Fu, X.; Li, Q.; Zhai, J.; Sheng, G.; Li, F. The physical–chemical characterization of mechanically-treated CFBC fly ash. *Cem Concr Compos.* **2008**, *30*, 220–226. [CrossRef]
38. Ding, P.; Orwa, M.G.; Pacek, A.W. De-agglomeration of hydrophobic and hydrophilic silica nano-powders in a high shear mixer. *Powder Technol* **2009**, *195*, 221–226. [CrossRef]
39. Sakthivel SKrishnan, V.V.; Pitchumani, B. Influence of suspension stability on wet grinding for production of mineral nanoparticles. *Particuology* **2008**, *6*, 120–124. [CrossRef]
40. Zhang, J.; Cui, H.; Wang, B.; Li, C.; Zhai, J.; Li, Q. Preparation and characterization of fly ash cenospheres supported CuO–$BiVO_4$ heterojunction composite. *Appl. Surf. Sci.* **2014**, *300*, 51–57. [CrossRef]
41. Oliveira, R.C.; Hammer, P.; Guibal, E.; Taulemesse, J.M.; Garcia, O., Jr. Characterization of metal–biomass interactions in the lanthanum (III) biosorption on Sargassum sp. using SEM/EDX, FTIR, and XPS: Preliminary studies. *Chem. Eng. J.* **2014**, *239*, 381–391. [CrossRef]
42. Huang, Z.Q.; Yi, S.H.; Chen, H.X.; He, X.Q. Parameter analysis of damaged region for laminates with matrix defects. *J. Sandw. Sructures Mater.* **2019**, *23*, 580–620. [CrossRef]
43. Kumar, S.; Kumar, R. Mechanical activation of fly ash: Effect on reaction, structure and properties of resulting geopolymer. *Ceram. Int.* **2011**, *37*, 533–541. [CrossRef]
44. Kumar, S.; Kumar, R.; Alex, T.C.; Bandopadhyay, A.; Mehrotra, S.P. Influence of reactivity of fly ash on geopolymerization. *Adv. Appl. Ceram.* **2007**, *106*, 120–127. [CrossRef]
45. Komljenović, M.; Baščarević, Z.; Bradić, V. Mechanical and microstructural properties of alkali-activated fly ash geopolymers. *J. Hazard. Mater.* **2010**, *181*, 35–42. [CrossRef]
46. Gholipour, G.; Zhang, C.; Mousavi, A.A. Numerical analysis of axially loaded RC columns subjected to the combination of impact and blast loads. *Eng. Struct.* **2020**, *219*, 110924. [CrossRef]

coatings

Article

Optimization Study of Fluffy Materials Flocking Drainage Pipes to Resist Blockage Based on MD Binding Energy

Shiyang Liu [1,2,*], Xuefu Zhang [1,2], Yuanfu Zhou [1,2] and Feng Gao [1,2]

1 State Key Laboratory of Mountain Bridge and Tunnel Engineering, Chongqing Jiaotong University, Chongqing 400074, China; zhangxuefu@cqjtu.edu.cn (X.Z.); 990201500027@cqjtu.edu.cn (Y.Z.); 990020701005@cqjtu.edu.cn (F.G.)

2 College of Civil Engineering, Chongqing Jiaotong University, Chongqing 400074, China

* Correspondence: liushiyang@mails.cqjtu.edu.cn

Abstract: Drainage pipe blockage resulting from crystals is one of the causes for cracking and leakage of tunnel lining. Therefore, effective prevention from drainage pipe blockage caused by crystals is crucial to ensure the safety and stability of lining structures during the operation of tunnel drainage system. Based on a large number of indoor model tests and numerical simulation analyses, binding energy between four materials and the calcium carbonate aqueous solution ("solid + liquid" system) and that between the four materials and the two typical growth crystals of calcium carbonate ("solid + solid" system) were studied. The research results indicated that: (1) The four materials all had an adsorption effect on the calcium carbonate aqueous solution system, and the PA6 had the greatest adsorption effect while the PP had the smallest adsorption effect; (2) There was spontaneous adsorption between the PVC or PA6 and the two typical growth crystals of calcium carbonate and no adsorption between the PP or SiC and the two typical growth crystals of calcium carbonate unless external energy was in place; (3) The PP and SiC can be used as the materials for drainage pipe flocking, but it shall be ensured that the fluffy material has a good geometrical property. The prevention technology for crystallization that causes drainage pipe blockage fills the gap in the research of drainage pipe blockage caused by crystals, which can reduce the maintenance cost for the operation of the tunnel drainage system and ensure safe and normal operation of the tunnel.

Keywords: molecular dynamics; binding energy; crystallization prevention; flocking for resisting blockage; drainage pipe

Citation: Liu, S.; Zhang, X.; Zhou, Y.; Gao, F. Optimization Study of Fluffy Materials Flocking Drainage Pipes to Resist Blockage Based on MD Binding Energy. *Coatings* **2021**, *11*, 853. https://doi.org/10.3390/coatings11070853

Received: 25 June 2021
Accepted: 13 July 2021
Published: 15 July 2021

1. Introduction

With the operation of tunnel projects, tunnel defects gradually emerge, among which drainage pipe blockage caused by crystals is a major factor affecting the service life of a tunnel. Carbonates formed by the inter-reaction of various ions of groundwater calcify into crystals over time, and these crystals accumulate in drainage pipes and cause blockage (Figure 1). Improper treatment of the crystals can affect the smooth operation of a tunnel drainage system and further lead to the cracking and leakage of the tunnel lining (Figure 2). Even worse, it may affect traffic safety and cause imponderable losses.

There are two kinds of causes for the blockage of tunnel drainage pipes. One is the environmental factor [1–3]: the concentration of ions such as calcium and magnesium in groundwater, the concentration of carbon dioxide in the air, and pH value of groundwater; the other is the construction factor [2,4]: concrete composition ratio, the form of the drainage system, etc. At present, research on the prevention technology of drainage pipe blockage caused by crystals is at the initial stage. The attachment of calcium carbonate crystals can be reduced through hydrophobic treatment on concrete base surfaces and PVC pipe walls with protective coating [5,6]; generation of crystals can be effectively lowered by optimizing the concrete materials and the concrete composition ratio, reducing the contact between groundwater and concrete, preventing CO_2 from entering the tunnel drainage pipe, as well as adding appropriate fly ash to shotcrete [2,4]. Drainage pipe crystals mainly include insoluble calcite crystals, and the PEG-b-PAA-b-PS, poly(ethylene glycol)-block-poly(acrylic acid)-block-poly(styrene) can prevent the phase transition from vaterite to calcite [7]; in the presence of biopolymer, the relative content of vaterite increases with the application of ultrasonic treatment [8]; ultrasonic treatment makes the gathered calcium carbonate crystals more fragile [9]; RS1600, a green corrosion inhibitor, makes the crystal structure of calcium carbonate change from calcite to vaterite [10]; the cleaning solvent of organic acid reagents of single molecule carboxylic acids with a concentration of 2000 ppm and a dichromate index of 17.71% and that of polymerized carboxylic acids can effectively remove the karst crystal of a drainage pipe system while ensuring environmental protection [11]. Through a large number of 1:1 indoor model tests and numerical simulation analyses, Liu Shiyang, et al. [12–16] studied the feasibility and reliability of drainage pipe flocking for resisting blockage from a macro perspective, and some good flocking parameters were obtained. The real solution to the attachment of crystals of drainage pipes depends on the micro binding energy between the crystals and the drainage pipe. The presence of Fe^{2+} and Mg^{2+} can inhibit the growth of $CaCO_3$, and the greater the concentration of Fe^{2+} or Mg^{2+} is, the stronger the inhibitory effect [17] will be. High voltage electric field can hinder the synthesis of calcium ions and carbonate ions and reduce the binding action of calcium ions or carbonate ions with the calcite growth crystal surface, and it also promotes the dissolution of scale [18].

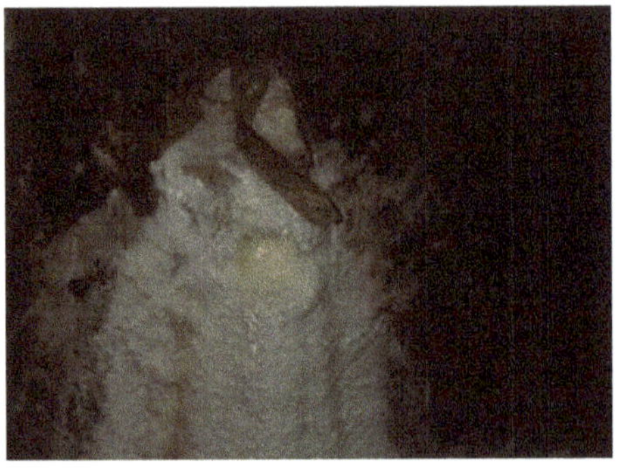

Figure 1. Blockage of a tunnel drainage pipe caused by crystals.

Figure 2. Cracking and leakage of the tunnel lining.

From the above analysis, we can see that at present, many researches on the prevention technology of drainage pipe blockage caused by crystals are at a macro level, and there are few researches at a micro level. Therefore, based on the above indoor model tests and numerical simulation analyses, binding energy between the materials and crystals was analyzed by the molecular dynamics software to find the best flocking materials for drainage pipes to resist blockage, which can provide a theoretical basis for the mechanism of resisting blockage by drainage pipe flocking.

2. Methods

2.1. Molecular Dynamics Software

Molecular Dynamics (MD) simulation has a history of about 50 years, whose success depends on the selection of an appropriate force field and a correct calculation method. The widely used MD simulation is becoming more and more important with the rapid development of computers. There are some commercial molecular dynamics computing software designed for the MD simulation, represented by the Materials Studio software of Accelary, which is a molecular dynamics simulation software featuring powerful functions, easy usage and clear images.

MD simulation considers the system to be studied as a collection of a large number of interacting particles whose motions follow the classical equation of motion (Newtonian equation, Hamiltonian equation or Lagrangian equation). By analyzing the force of each particle, equations of motion of various particles that constitute the system were numerically solved directly to obtain theses particles' coordinates and momentum at every moment. Then, the microscopic state consisting of the coordinates and momentum was averaged against time to calculate the macroscopic properties such as multisystem pressures, energy and temperatures.

2.2. Model Building

From a micro level, different materials produce different binding energy with the groundwater solution system or the crystals' microscopic crystal surface. The stronger the binding energy is, the stronger the adsorption between the materials and the ions or the crystals' crystal surface in the solution will be. Thus, the anti-crystallization effect of fluffy materials was analyzed based on the intensity of binding energy. The characteristics of fluffy materials (Figure 3) and drainage pipe materials are shown in Table 1 below. The "solid–solid" model and the "solid–liquid" model were constructed with the amorphous cell of Materials Studio 8.0. The "solid–solid" model mainly included the double-layer model of different materials and the crystal surface of calcium carbonate crystals, and the "solid–liquid" model mainly included the double-layer model of different materials and calcium carbonate aqueous solution.

Figure 3. Three types of fluffy materials.

Table 1. The characteristics of fluffy materials and drainage pipe materials.

Material Category	Density	Chemical Composition	Chemical Formula	Structural Formula	Schematic Diagram
M1	1.13 g/cm^3	PA6	$(C_6H_{11}NO)_n$	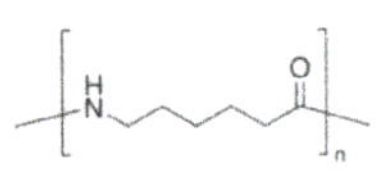	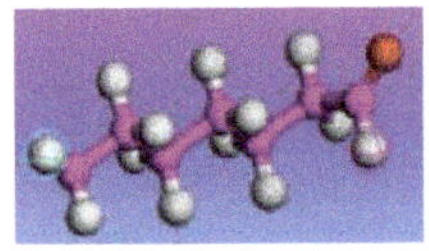
M2	0.90 g/cm^3	PP	$(C_3H_6)_n$	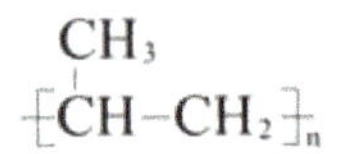	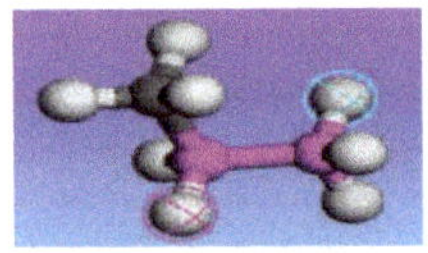
M3	3.20 g/cm^3	SiC	SiC	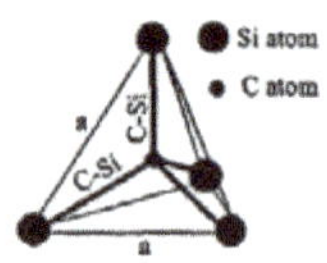	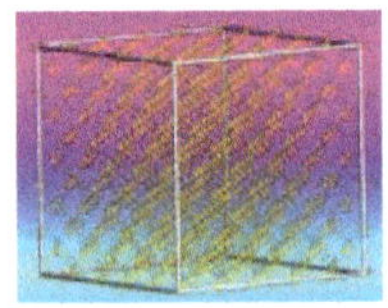
PVC	1.38 g/cm^3	VCM	$(C_2H_3Cl)_n$	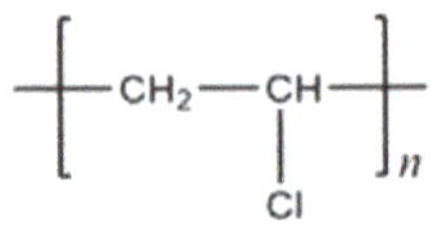	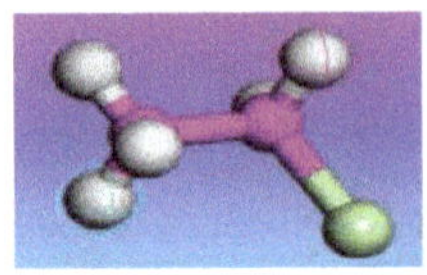

According to the Inorganic Crystal Structure Database (ICSD), calcite belongs to the R-3cH space group, with the spatial parameters of a = b = 4.983 Å, c = 17.078 Å, α = 90°, β = 90°, γ = 120° (Figure 4). The study in [19] shows that the growth faces of calcite were (1 −1 0) and (1 0 4) crystal surfaces. The former was positively charged, while the latter was not charged. To obtain the binding energy between the materials and the calcite crystals, the models between different materials and the (1 −1 0) and (1 0 4) crystal surfaces were built respectively in the "solid–solid" model. The "solid–liquid" model mainly included the double-layer model of different materials and calcium carbonate aqueous solution. Given the low solubility of calcium carbonate, the calcium carbonate aqueous solution was formed with 350 water molecules, $3Ca^{2+}$ and $3CO_3^{2-}$ (Figure 5), and the solution volume was 21.74 × 21.74 × 19.34 Å^3.

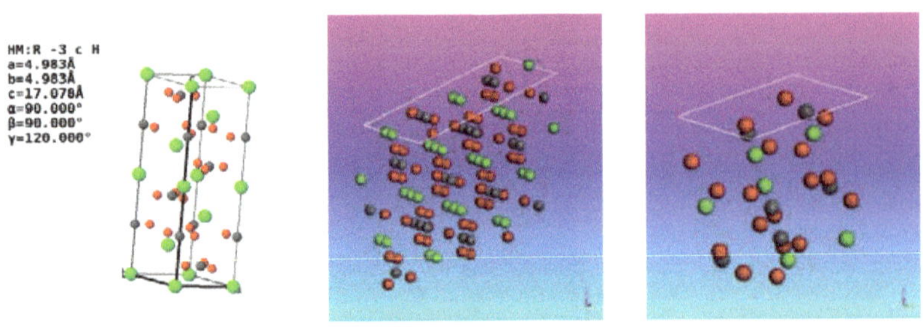

(a) (b) (c)

Figure 4. Details of the calcium carbonate calcite crystals: (**a**) calcite cell configuration; (**b**) (1 −1 0) crystal surface; (**c**) (1 0 4) crystal surface.

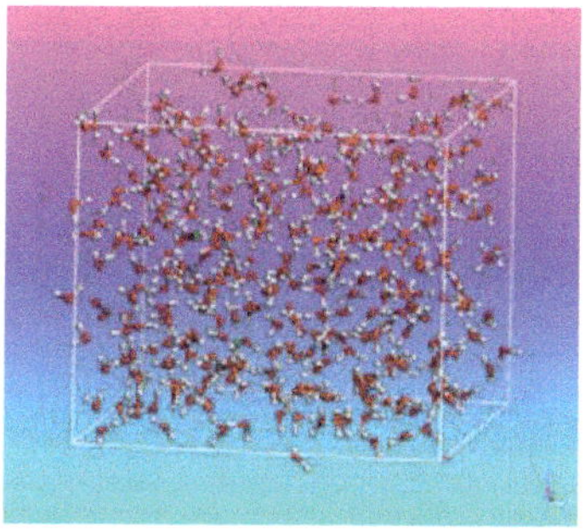

Figure 5. Calcium carbonate aqueous solution.

The supercell models of (1 −1 0) and (1 0 4) crystal surfaces were built respectively through the command of Build→Symmetry→Supercell with a size of 21.74 Å × 21.74 Å × 17.60 Å; the polymer cells of the materials were built through the command of Modules→Amorphous cell→Calculation with a size of 21.74 Å × 21.74 Å × 19.34 Å. PVC is amorphous polymer, PA6, PP and SiC are crystalline polymer. PVC, PA6 and PP all construct amorphous polymer chain unit cells through single chain molecules, while SiC unit cells are obtained by expanding the unit cell structure of the software. First, geometric and energy optimization of all cells were performed. Then, the built upper and lower models were used to build the "solid + solid" model and the "solid + liquid" model through the command of Build→Build layers in MS. The volume of the "solid + solid" model and that of the "solid + liquid" model were both 21.74 × 21.74 × 60 $Å^3$, where the upper layer was calcium carbonate aqueous solution (supercells of (1 −1 0) or (1 0 4) crystal surfaces) and the lower layer was polymer. A 5 Å vacuum layer is set between the upper and lower layers as the contact surface of the two materials, and a 20 Å vacuum layer is set at the top of the upper layer as the interface. The combination model of the materials and the supercells of (1 −1 0) or (1 0 4) crystal surfaces is shown in Figures 6 and 7, and the combination model of the materials and the calcium carbonate aqueous solution is shown in Figure 8.

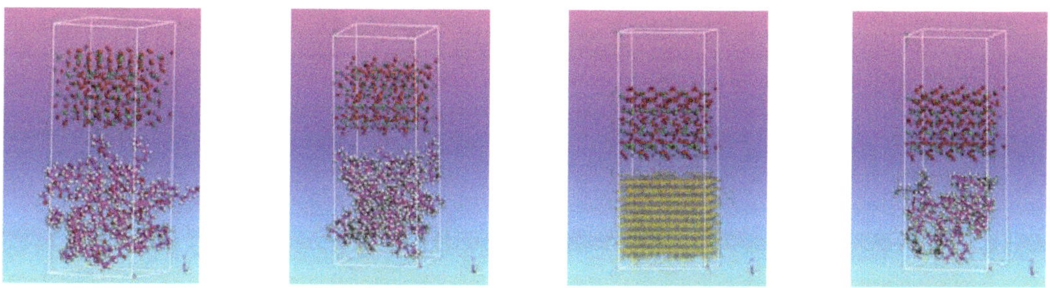

(a) (b) (c) (d)

Figure 6. Combination model of the materials and the supercells of (1 −1 0) crystal surface. (**a**) PA6-(1 −1 0); (**b**) PP-(1 −1 0); (**c**) SiC—(1 −1 0); (**d**) PVC-(1 −1 0).

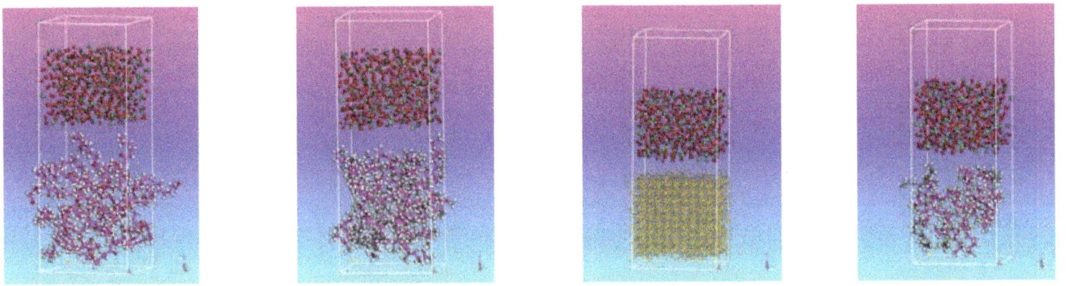

(a) (b) (c) (d)

Figure 7. Combination model of the materials and the supercells of (1 0 4) crystal surface. (**a**) PA6-(1 0 4); (**b**) PP-(1 0 4); (**c**) SiC-(1 0 4); (**d**) PVC-(1 0 4).

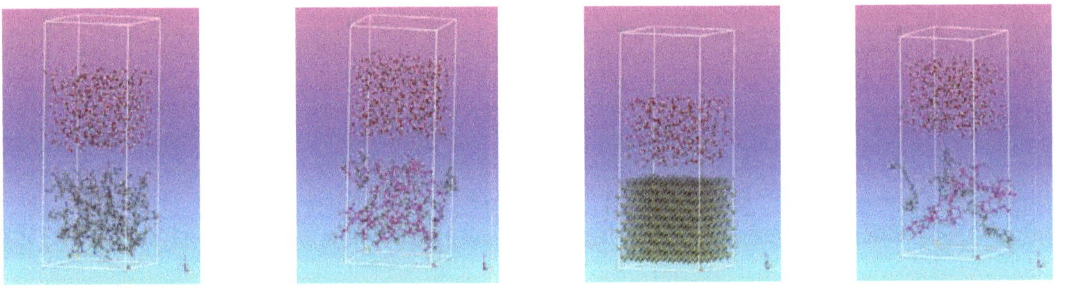

(a) (b) (c) (d)

Figure 8. Combination model of the materials and the calcium carbonate aqueous solution. (**a**) PA6-solution; (**b**) PP-solution; (**c**) SiC-solution; (**d**) PVC-solution.

2.3. Parameter Setting

The MD simulation was performed by the Forcite module in Materials Studio. First, the positions were assigned, and all atomic coordinates of the material layer were fixed through the universal COMPASS force field of high precision. The NVT was adopted because the system pressure was not a key factor. The 100 ps MD simulation under the NVT and velocity scale was first conducted to allow the system to reach an equilibrium state. Then, the MD simulation was performed under the NVT and Andersen thermostatic heat bath, with a time step of 1 fs, a simulation time of 200 ps, the system track being recorded every 1000 steps, a simulation temperature of 298 K, and a cutoff radius of 12.5 Å.

3. Results and Discussion

3.1. System Equilibrium

The equilibrium of the system was determined by the temperature and energy. The accuracy of the simulation was characterized by the ratio of energy convergence parameter ($\Delta E_{converge}$), the total energy fluctuation value rms (E_t), and the kinetic energy fluctuation value rms (E_k), as shown in Formulas (1) and (2) where $E(0)$ and the $E(i)$ were the initial total energy and the total energy when the iteration reached the ith step respectively, and N_{nm} was the times of simulation. When $\Delta E_{converge} \leq 0.001$, $R \leq 0.001$, the calculation results were reliable. After calculation, the $\Delta E_{converge}$ and R of the simulation system at each temperature conformed to the above value range, indicating that the system reached equilibrium and the simulated calculation results were reliable.

$$\Delta E_{coverge} = \frac{1}{N_{nm}} \sum_i \left| \frac{E(0) - E(i)}{E(0)} \right| \quad (1)$$

$$R = \frac{rms(E_t)}{rms(E_k)} \quad (2)$$

Figure 9 shows the energy output curve of the equilibrium process, and Figure 10 shows the temperature output curve of the equilibrium process. From Figure 9, we can see that the potential energy, kinetic energy, non-bond energy and total energy flattened over time, indicating that the various energy of the system reached the equilibrium. From Figure 10, we can see that the temperature fluctuated 10% around 298 K, indicating that the temperature of the system also reached the equilibrium.

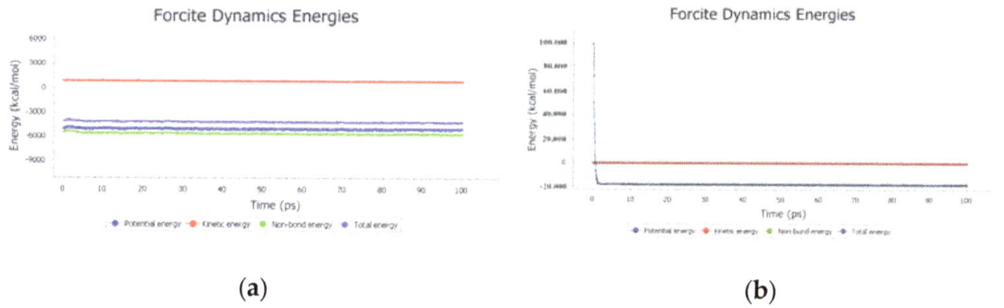

Figure 9. Energy output curve of the system after MD simulation at T = 298 K. (**a**) The "solid-liquid" model; (**b**) The "solid-solid" model.

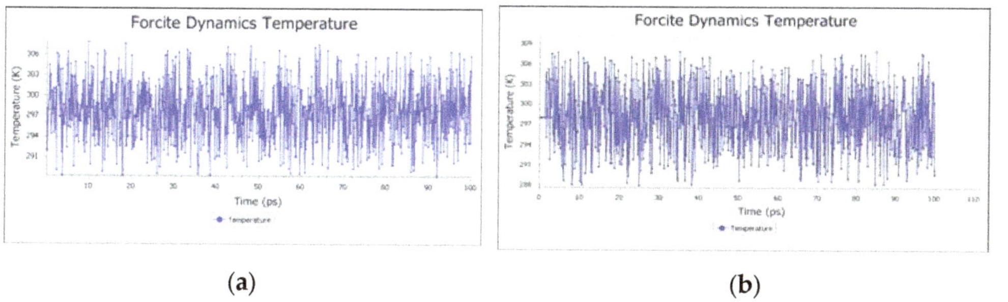

Figure 10. Temperature output curve of the system after MD simulation at T = 298 K. (**a**) The "solid-liquid" model; (**b**) The "solid-solid" model.

3.2. Binding Energy Analysis

Interaction between the materials and the crystallized ion solution (two typical growth crystal surfaces of calcium carbonate) was simulated by the molecular dynamics software. If the interaction was very strong, the crystallized ion aqueous solution (two typical growth crystal surfaces of calcium carbonate) would easily attach to the material layer, which meant that the contact area between the pipe wall and the fluffy material was prone to crystallization.

When using the double-layer model for simulation, the data of the fully balanced double-layer structure was collected at an appropriate temperature and a proper ensemble to obtain a series of equilibrium configurations. Then each possible equilibrium configuration was treated as follows: (1) Restore the lower fixed atoms to allow them to move freely, copy three backups, and calculate the total energy (E_{total}) of the system with backup 1; (2) Keep the lower layer of the backup 2 only, delete the upper layer, and calculate the energy (E_{lower}) of the lower layer; (3) Keep the upper layer of the backup 3 only, delete the lower layer, calculate the energy (E_{upper}) of the upper layer, and finally calculate the interaction energy by Formula (3). To calculate the adsorption energy of each material layer, let the interaction energy of the system be ΔE, and the binding energy ($E_{binding}$) be the opposite number of the interaction energy ΔE (Formula (4)). The details are as follows:

$$\Delta E = E_{total} - \left(E_{upper} + E_{lower}\right) \quad (3)$$

$$E_{binding} = -\Delta E \quad (4)$$

where E_{total} was the total energy of the system, E_{lower} was the single-point energy of the material layer, E_{upper} was the single-point energy of the crystallized ion aqueous solution system (two typical growth crystal surfaces of calcium carbonate) after interaction. Through simulation calculation, the energy value of the system after the interaction between the "solid + solid" model and the "solid + liquid" model is shown in Table 2, and the changing trend of the binding energy with the materials is shown in Figures 11 and 12.

Table 2. Energy value of the system after interaction between the "solid-solid" model and the "solid-liquid" model (unit: kcal/mol).

Model System			E_{total}	E_{lower}	E_{upper}	ΔE	$E_{binding}$
solid-solid	$CaCO_3$-PVC	1 −1 0	−20,132.466	−574.596	−18,487.159	−1070.711	1070.711
		1 0 4	−38,045.005	−542.662	−36,704.997	−797.346	797.346
	$CaCO_3$-PA6	1 −1 0	−21,054.710	−937.917	−18,467.027	−1649.766	1649.766
		1 0 4	−38,758.331	−928.115	−36,671.616	−1158.600	1158.600
	$CaCO_3$-PP	1 −1 0	−17,540.010	676.839	−18,487.308	270.459	−270.459
		1 0 4	−35,527.521	616.249	−36,683.798	540.028	−540.028
	$CaCO_3$-SiC	1 −1 0	−117,279.169	−107,275.987	−18,473.126	8469.944	−8469.944
		1 0 4	−133,290.641	−107,264.983	−36,691.537	10,665.879	−10,665.879
solid-liquid	Solution-PVC		−4703.853	−577.787	−4028.226	−97.840	97.840
	Solution-PA6		−3849.500	269.641	−3992.464	−126.677	126.677
	Solution-PP		−3470.688	625.414	−4043.291	−52.811	52.811
	Solution-SiC		−111,111.148	−107,276.559	−3757.73	−76.859	76.859

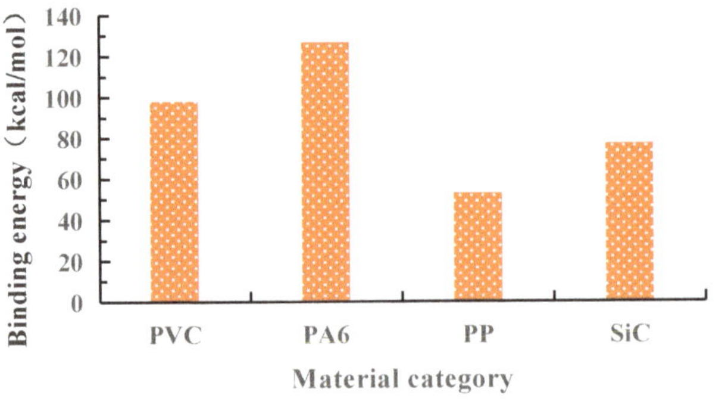

Figure 11. The "solid-liquid" binding energy.

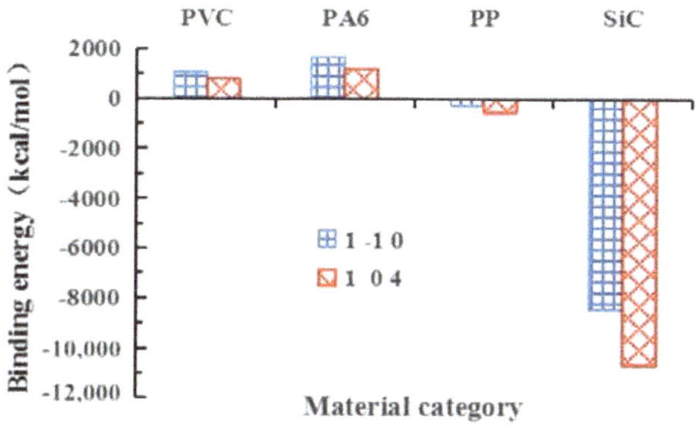

Figure 12. The "solid-and solid" binding energy.

The interaction energy was negative, suggesting that the adsorption of the crystallized ion aqueous solution (two typical growth crystal surfaces of calcium carbonate) on each material surface was a spontaneous process, and a relatively stable system could be formed [20]. As it can be seen from Table 2 and Figure 8, the interaction energy between the four materials and the calcium carbonate aqueous solution was all negative in the "solid + liquid" model, and the binding energy was all positive, indicating that all the four materials had an adsorption effect on the calcium carbonate aqueous solution, and the PA6 had the greatest adsorption effect while the PP had the smallest adsorption effect, with the former being about 2.5 times of the latter. As it can be seen from Table 2 and Figure 9, the interaction energy between the materials of PVC or PA6 and the two typical growth crystals of calcium carbonate was all negative, and the binding energy was all positive, indicating that there was spontaneous adsorption between PVC or PA6 and the two typical growth crystals of calcium carbonate, and the binding energy between either of the two materials and (1 −1 0) crystal surface was greater than that between (1 0 4) crystal surface; the interaction energy between the materials of PP or SiC and the two typical growth crystals of calcium carbonate was all positive, and the binding energy was all negative, indicating that the adsorption between the PP or SiC and the two typical growth crystals of calcium carbonate was impossible unless there was external energy, and the adsorption of (1 0 4) crystal surface was greater than that of (1 −1 0) crystal surface.

Through the above analysis, given the binding energy between the four materials and the crystallized ion aqueous solution, we can select the PP with low binding energy as the materials for drainage pipes; given the binding energy between the four materials and the two typical growth crystals of calcium carbonate, we can select the PP and SiC as the materials for drainage pipes. From the analysis of MD simulation results, we speculated that SiC had the best anti-crystallization effect in drainage pipe flocking, however, the result of the actual indoor macroscopic test [15] was contrary. Thus, the geometric property of the fluffy material (Figure 3) played a major role in the anti-crystallization effect of drainage pipes flocking. Therefore, for better anti-crystallization effect, the PP and SiC can be used as the materials for flocking drainage pipes, but the fluffy material shall have a good geometrical property (smooth surface and straight in the lengthwise direction) so as to maximize the anti-crystallization effect of the drainage pipe flocking.

4. Conclusions

In this paper, the binding energies of PA6, PVC, SiC, PP and calcium carbonate aqueous solution and calcium carbonate were studied by molecular dynamics numerical simulation method.

(1) PA6, PVC, SiC and PP all have adsorption effect on calcium carbonate solution, and the order of binding energy is PA6 > PVC > SiC > PP.

(2) The results show that PVC, PA6 and $CaCO_3$ can spontaneously adsorb on each other, while PP and SiC can only adsorb on each other with the help of external energy. The energy absorbed by (1 0 4) crystal face is greater than that absorbed by (1 −1 0) crystal face.
(3) The follow-up research can start from the energy of the solution system and the crystal itself and find the technology to make the energy of the system or crystal in a low state, so that the crystal and the pipe are not combined.
(4) From the point of view of anti-crystallization effect, while considering the binding energy between materials, it is also necessary to ensure the excellent geometric characteristics of pile (smooth surface and straight length direction), so as to maximize the anti-crystallization effect of flocking drainage pipe. In addition to the binding energy, the molecular weight and production process of polymer should also be considered in the follow-up study.

Author Contributions: Conceptualization, S.L.; Data curation, X.Z., Y.Z. and F.G.; Formal analysis, Y.Z. and F.G.; Writing—original draft, S.L. and X.Z.; Writing—review & editing, S.L. All authors have read and agreed to the published version of the manuscript.

Funding: This research was funded by the National Natural Science Foundation of China (Grant No.51778095), Scientific Research Project of Emei Hanyuan Expressway Project (Grant No. LH-HT-45).

Institutional Review Board Statement: Not applicable.

Informed Consent Statement: Not applicable.

Data Availability Statement: The study did not report any data.

Conflicts of Interest: The authors declare no conflict of interest.

References

1. Jiang, Y.; Du, K.; Tao, L.; Zhao, J.; Xiao, H. Investigation and discussion on blocking mechanism of drainage system in karst tunnels. *Railw. Stand. Des.* **2019**, *63*, 131–135.
2. Tian, C.; Ye, F.; Song, G.; Wang, Q.; Zhao, M.; He, B.; Wang, J.; Han, X. On Mechanism of crystal blockage of tunnel drainage system and preventive countermeasures. *Mod. Tunn. Technol.* **2020**, *57*, 66–76.
3. Xiang, K.; Zhou, J.; Zhang, X.; Huang, C.; Song, L.; Liu, S. Experimental study on crystallization rule of tunnel drainpipe in alkaline environment. *Tunn. Constr.* **2019**, *39*, 207–212.
4. Ye, F.; Tian, C.; Zhao, M.; He, B.; Wang, J.; Han, X. The disease of scaling and cloging in the drainage pipes of a tunnel under construction in yunnan. *China Civ. Eng. J.* **2020**, *53*, 336–341.
5. Zhou, Y.; Zhang, X.; Wei, L.; Liu, S.; Zhang, B.; Zhou, C. Experimental study on prevention of calcium carbonate crystallizing in drainage pipe of tunnel engineering. *Adv. Civ. Eng.* **2018**, *2018*, 1–11. [CrossRef]
6. Jiang, Y.; Du, K.; Liao, J.; Chen, X.; Xiao, H. Experimental research on maintainability of drainage facilities in lining construction joints of karst tunnel. *Railw. Stand. Des.* **2019**, *63*, 91–96.
7. Su, Y.; Yang, H.; Shi, W.; Guo, H.; Zhao, Y.; Wang, D. Crystallization and morphological control of calcium carbonate by functionalized triblock copolymers. *Colloids Surf. Physicochem. Eng. Asp.* **2009**, *355*, 158–162. [CrossRef]
8. Kirboga, S.; Oner, M.; Akyol, E. The effect of ultrasonication on calcium carbonate crystallization in the presence of biopolymer. *J. Cryst. Growth* **2014**, *401*, 266–270. [CrossRef]
9. Su, M.; Han, J.; Li, Y.; Chen, J.; Zhao, Y.; Chadwick, K. Ultrasonic crystallization of calcium carbonate in presence of seawater Ions. *Desalination* **2015**, *369*, 85–90. [CrossRef]
10. Menzri, R.; Ghizellaoui, S.; Tlili, M. Calcium carbonate inhibition by green inhibitors: Thiamine and pyridoxine. *Desalination* **2017**, *404*, 147–154. [CrossRef]
11. Hong, Y.; Qian, X.; Li, J.; Yang, H.; Zhang, P. On scavenging performances of cleaning solvents for the clogging in the drainage system of karst tunnels. *Mod. Tunn. Technol.* **2020**, *57*, 160–170.
12. Liu, S.; Zhang, X.; Lü, H.; Liu, Q.; Wang, B. The effect of flocking PVC pipe on the prevention and crystallization of tunnel drains. *Sci. Technol. Eng.* **2018**, *18*, 313–319.
13. Liu, S.; Zhang, X.; Lü, H.; Liu, Q.; Wang, B. The effect of anti-crystallization of tunnel plumage drain pipe under different water filling state. *Sci. Technol. Eng.* **2018**, *18*, 156–163.
14. Liu, S.; Gao, F.; Zhou, Y.; Liu, Q.; Lü, H.; Wang, B.; Xiang, K.; Xiao, D. Effect of fuzz length on the prevention of crystallization of tunnel flocking drainpipes. *Sci. Technol. Eng.* **2019**, *19*, 234–239.
15. Liu, S.; Gao, F.; Zhang, X.; Han, F.; Zhou, Y.; Xiang, K.; Xiao, D. Experimental study on anti-crystallization law of tunnel transverse flocking drainpipe at different velocities. *Asia-Pac. J. Chem. Eng.* **2020**, *15*, 1–9. [CrossRef]

16. Liu, S.; Zhang, X.; Gao, F.; Wei, L.; Liu, Q.; Lü, H.; Wang, B. Two-dimensional flow field distribution characteristics of flocking drainage pipes in tunnel. *Open Phys.* **2020**, *18*, 139–148. [CrossRef]
17. Leeuw, N.H.D. Molecular dynamics simulation of the growth inhibiting effect of Fe^{2+}, Mg^{2+}, Cd^{2+},and Sr^{2+} on calcite crystal growth. *J. Phys. Chem. B* **2002**, *106*, 5241–5249. [CrossRef]
18. Wang, D.; Huang, L.; Wei, M.; Wang, T.; Tian, Y. MD simulation of influence of high voltage static electric field on crystallization of calcium carbonate. *Ind. Water Wastewater* **2010**, *41*, 76–79.
19. Hädicke, E.; Rieger, J.; Rau, I.U.; Boeckh, D. Molecular dynamics simulations of the incrustation inhibition by polymeric additives. *Phys. Chem. Chem. Phys.* **1999**, *1*, 3891–3898. [CrossRef]
20. Zhao, W.; Zhang, Q.; Chen, T.; Zhang, J.; Lu, T.; Zhang, H. Molecular dynamics simulation of adsorption of polyelectrolyte on the polystyrene microspheres. *J. Mater. Sci. Eng.* **2009**, *27*, 329–331.

coatings

MDPI

Article

Facile Synthesis of Copper(I) Oxide Nanochains and the Photo-Thermal Conversion Performance of Its Nanofluids

Zhongjin Ni [1], Xiaohai Cao [2], Xinyi Wang [1], Shiyu Zhou [1], Caixia Zhang [2], Bin Xu [2,*] and Yihua Ni [1,*]

[1] College of Engineering, Zhejiang A&F University, Lin'an 311300, China; neezj@zafu.edu.cn (Z.N.); wangxinyi9219@sina.com (X.W.); shiyuzhou98@sina.com (S.Z.)
[2] College of Materials Science and Engineering, Zhejiang University of Technology, Hangzhou 310014, China; xiaohaicao_zjut@sina.com (X.C.); caixiazhang_zjut@sina.com (C.Z.)
* Correspondence: xub@zjut.edu.cn (B.X.); 20120006@zafu.edu.cn (Y.N.)

Abstract: In this thesis, Cu_2O nanochains were synthesized by thermal decomposition with copper formate-octylamine as the precursor, oleic acid and oleylamine as the catalyst stabilizer agent and paraffin as the solvent. The phase structure and micromorphology of Cu_2O nanochains were characterized by X-ray diffraction and transmission electron microscopy. The effect of reaction time and concentration of the precursor on the Cu_2O nanochains were discussed, and the formation mechanism of the Cu_2O nanochains was analyzed. The results show that Cu_2O nanochains were self-assembled by Cu_2O nanocrystals; with the extension of the reaction time, Cu_2O nanochains gradually become granular; increasing the concentration of the precursor will increase the entanglement degree of the nanochains. Oleic acid contributes to the formation of Cu_2O, and oleylamine plays a directional role in the formation of nanochains. On the basis of those phenomenon, a comparison of the Cu_2O nanochain-water nanofluids with that of a water-based liquid showed that after irradiating for 3000 s, the temperature of nanofluids reached 91.1 °C while the water was only 75.7 °C. This demonstrates the better performance of the Cu_2O nanochain-water nanofluid in the ability of light absorption, thermal conductivity and photothermal conversion.

Keywords: cuprous oxide nanochains; thermal decomposition; nanofluids; photo-thermal conversion performance

Citation: Ni, Z.; Cao, X.; Wang, X.; Zhou, S.; Zhang, C.; Xu, B.; Ni, Y. Facile Synthesis of Copper(I) Oxide Nanochains and the Photo-Thermal Conversion Performance of Its Nanofluids. *Coatings* **2021**, *11*, 749. https://doi.org/10.3390/coatings11070749

Academic Editor: Ivan Jerman

Received: 21 May 2021
Accepted: 17 June 2021
Published: 22 June 2021

1. Introduction

Solar thermal utilization is a technology that can utilize the sun's radiation efficiently. Meanwhile, the solar collector is a device that can be used to absorb the sun's radiation and convert it into thermal energy, then the thermal energy is transferred to the heat transfer medium. Currently, all of the traditional flat plate and vacuum tube collectors are almost using indirect heat collection. After the solar radiation is absorbed by selective coating and converted into thermal energy, only by going through the thermal conductivity process of the coated metal plate or glass tube wall can the thermal energy be transferred to the working fluid. Therefore, with the increasing surface temperature of the heat collector, the radiation heat loss from the high-temperature surface to the environment is increasing as well, consequently, the conversion efficiency will be reduced. In the 1975s, Minardi proposed a conception about direct absorption solar collectors (DASC) in order to reduce the high-temperature radiant heat loss and improve the thermal efficiency [1]. Different from the indirect heat collector, this kind of solar heat collector does not have a heat-absorbing surface and the solar radiation will be absorbed directly by the working fluid so it can reduce the radiant heat loss and avoid the conducted thermal resistance between the endothermic surface and the working fluid. However, the working fluid for the DASC has usually been a kind of black liquid which is prepared by adding black ink, black dye and other common micron particles into the medium, such as water or ethylene glycol. However, this black liquid has some defects such as instability of the light, heat-induced degradation, ease clogging and low thermal conductivity. Therefore, the DASC based on the black liquid can not achieve the desired high efficiency and obstructed the development and application of this kind of new type collector. With the development of nanotechnology and the application of nanomaterial, in 1995, Choi et al. proposed a concept called "Nanofluids" in the international community for the first time, which is a kind of liquid adding nanoscale metal and non-metallic oxide particles in a certain manner and ratio, which will form a new class of heat transfer medium [2]. Soon after, numerous researches for the thermal physical properties, enhancing heat transfer performance et al. of the nanofluids have been performed by scientists; the results indicated that nanofluids have significant heat transfer performance [3,4]. The binary CuO/ZnO nanofluids were prepared and the thermal conductivity, as well as the photothermal conversion, were investigated by Fang [5]. The optical absorption properties and the thermal conductivity were both enhanced due to the presence of the CuO/ZnO nanoparticles in the nanofluids.

In the meantime, with the existence of nanoparticles, the absorption, transmission, and scattering of solar radiation by nanofluids has changed and its optical properties are different from the common fluids which mean light can be strongly absorbed or selectively absorbed by the nanofluids at some bands in solar radiation [6]. In view of the special light absorption performance and the good heat transport performance of nanofluids, in recent years, some scholars proposed that in order to enhance the efficiency of the heat collector, nanofluids can be used as the working medium for the DASC, which can absorb the sun's radiation energy directly. It can be demonstrated from the research results that the efficiency of heat collection for DASC based on the nanofluids had obviously been enhanced [7–9]. It showed that nanofluids have a large application potential as the working media for DASC.

Cuprous oxide nanoparticles are known as a kind of typical p-type semiconductor with a bandgap of 2.0–2.3 eV [10], which has a high absorption coefficient for visible light. On account of its excellent electrical [11] and magnetic properties [12], Cu_2O has become a hot research topic in the field of optoelectronics, catalysis and diluted magnetic semiconductors (DMSs). It is an important material to manufacture low-cost solar cell and gas sensors. However, there are two factors impeding its development in the field of photoelectrochemistry. One is the inconsistency between the diffusion length and the absorption depth of the light near the bandgap of Cu_2O intrinsic carrier [13–15], and the other one is the instability [16–18] of optical absorption of Cu_2O in solution. The Cu_2O has a huge surface and a high aspect ratio, and it can collect and separate the effect carrier from the radial in the nanometer scale while it is absorbing the light along the long axis, so it has a higher ability of light absorption than the Cu_2O nanoparticles [19]. Additionally, the chain (liner) structure of the Cu_2O nanochains makes it easier to form the thermal network chain, which can improve the thermal conductivity of the nanofluids, thereby its heat transfer efficiency will be improved.

The nanoparticles tend to agglomerate in the base fluid due to the effect of the Van der Waals force and hydrogen bonding. During the experiment to determine the thermal conductivity of Fe_3O_4-water nanofluids by Zhu [20], they found that the nanoparticles had been aligned to nanochains caused by the agglomeration accidentally. The number and length of alignments increase with the increase in the nanoparticle's concentration, leading to the sharp increases of the thermal conductivities. The results are consistent with the simulation results of the effective thermal conductivity of fluid-particle mixture by Kumar and Murthy [21]. The enhancement from the nanochains to the nanofluids can be explained from two sides.

First, the single nanochains have a thermal conductivity enhancement compared to the single nanoparticles. The heat transfer process in the nanochains is a non-Fourier phenomenon, therefore the thermal conductivity changes with the shape and the size of the materials. According to the result from the experiment about the thermal conductivity of monatomic nanochains by Chen [22], an atomic nanochain can be thought of as a finite one-dimensional lattice in which lattice vibrational waves must be reflected at the ends.

The transmission direction of the energy will be changed as well so that the energy flux will be reduced as will the thermal conductivity. The shorter the nanochains are, the smaller the attenuation of a lattice vibration wave will be before it reaches the end. The whole energy flux will reduce which will result in the reduction of the thermal conductivity.

The result demonstrates the relationship between the length of the nanochains and the thermal conductivity from the microscope point with the view of phonon and lattice wave transmission. On the other side, Moran et al. compared the conduction of heat in the nanochains to the flow of gas in the nanochannels by theoretical analysis and molecular dynamics (MD) simulation from the macroscope point when studying the thermal conductivity of silicon nanochains [23]. The thermal resistance caused by the end effect in this way of heat transfer was:

$$f_T = \frac{2\gamma\rho C^2 T}{c^2 k\left(1 - e^{-D/l}\right)e^{-ND/l}} q \tag{1}$$

where q is the heat flux, T the local thermodynamic equilibrium temperature, ρ the solid density, C the solid specific heat capacity, γ the Grüneisen constant, l the length of nanochain and N can be understood as the dimensionless length of ends effect region normalized by the channel width. For a straight channel, an empirical value of N is 5. From this formula, the heat resistance caused by the end effect will be negligible when the length of nanochain is long enough. That is, the thermal conductivity will be improved when the length of nanochain increases.

Secondly, the synergistic effect among the nanochains in the nanofluids will also lead to the enhancement of the thermal conductivity of nanofluids. Figure 1a shows the network structure formed by the bridge-like overlap of the nanochains which are parallel to the direction of the heat transfer and the randomly aligned nanochains in the nanofluids. This phenomenon of bridge-like overlap is mainly caused by the high ratio aspect of the nanochains. These networks provide a low-resistance pathway to heat conduction and increase the overall thermal conductivity of the composite. Compared to the nanochains, the nanoparticles are unable to join to form an effective thermal conduction path in the nanofluid, shown in Figure 1b. So the thermal conductivity of the nanofluids with nanoparticles will be lower than that of the nanochains [24].

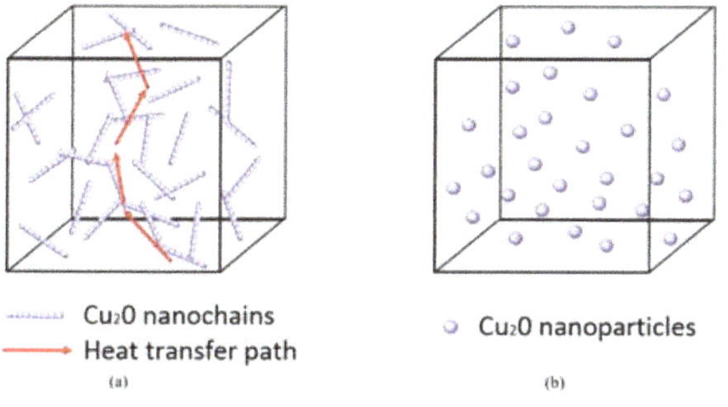

Figure 1. Schematics diagram of the heat transfer process of (**a**) Cu_2O nanochains and (**b**) Cu_2O nanoparticles in the nanofluids.

Due to the large specific surface area and high aspect ratio, Cu_2O nanochains can absorb light along the long axis and the effective carrier collection and separation can be carried out at the same time in the nanometer scale [25], which is expected to solve the problem above. However, the research about the photo-thermal conversion performance of Cu_2O nanochains-water nanofluids has been barely developed by now. In this paper, the innovative mixing of Cu_2O nanochains with water as the working media can prepare nanofluids to improve the heat collection efficiency of DASC with the photothermal conversion properties it possesses. Cu_2O nanochains are mainly synthesized by the template method [26] and the chemical reduction method [27], both of which have relatively obvious disadvantages. It is difficult to separate using the template method, and the chemical reduction method depends on the environmentally unfriendly reductant. Here we report a method to facile synthesis of the Cu_2O nanochains, and several types of research about the formation mechanism of Cu_2O nanochains and the optical performance of Cu_2O nanochains-water nanofluids was performed. Section 2 details the preparation process and characterisation scheme of Cu_2O nanochains and their nanofluids. Section 3 analyses and discusses the influencing factors and formation mechanism of Cu_2O nanochains and the properties of Cu_2O nanofluids. Section 4 summarises the experimental results.

2. Materials and Methods

2.1. Materials and Reagents

Copper(II) formate tetrahydrate (AR) was purchased from Guanghua Sci-Tech Co., Ltd. Octylamine (AR, alladin, Shanghai, China); Efficient sliced paraffin (melting point 60~62 °C) was purchased from Shanghai Hualing Rehabilitation Equipment Company, (Shanghai, China); Oleic acid (CP) was purchased from Hangzhou Shuanglin Chemical Reagent Company(Hangzhou, China); Oleylamine (AR, alladin); n-hexane (AR) was purchased from Yonghua Fine Chemistry Co., Ltd. (Suzhou, China).

2.2. Synthesis

The synthesis process of Cu_2O nanochains is shown in Figure 2. Copper(II) formate tetrahydrate and octylamine were mixed with a mole ratio of 1:2. After stirring for 1 h at 35.0 °C, the precursor (the coordination complex of copper formate and octylamine) was prepared. The precursor (0.02 mol) was added to paraffin (6 g) with oleic acid (0.005 mol) and oleylamine (0.02 mol) stirring for 1 h at 70 °C, the mixture was added to the 50 mL flask with three necks with 6 g of the melted paraffin. After the addition process, the reaction mixture was heated to 130 °C in a temperature-controlled oil bath. A constant amount of nitrogen is introduced into the three-necked flask to ensure that the reaction apparatus is airtight and free of gas leaks. The precursor began to decompose under a constant stream of nitrogen and the color of the solution turned to dark brown from light blue. After heating for 1 h, the resulting products were centrifugally washed by n-hexane 3~4 times. Finally, the target products were obtained.

2.3. Characterization

The micro morphology of the Cu_2O nanochains was characterized by Tecnai G2 F30 Transmission electron microscope; the phase analysis was determined by using X'PERT PROX shooting line diffractometer over the scan range from 10° to 80° at a rate of 5°/min, using a cooper target as the diffraction source (Cu Ka, λ = 0.15406 nm); the light absorption performance of the samples was determined by Lamda 750 s UV-visible-near-infrared spectrophotometer with the wavelength from 380 to 800 nm; the laboratory self-assembled light-heat conversion tester (Figure 3) was used to analyze the optical thermal conversion performance of the nanofluids which is prepared by spreading 4.8 wt.% Cu_2O nanochains into deionized water and going through the CTAB and ultrasonic process; Thermal conductivity of the suspension was determined by TC-3000 thermal conductivity meter in the temperature range from 30 to 80 °C at intervals of 10 °C, and the temperature was measured repeatedly and averaged for 10 times. The thermal conductivity enhancement ratio of nanofluids was calculated by the following equation:

$$E = (K_1 - K_0)/K_0 \quad (2)$$

where E is the thermal conductivity enhancement ratio; K_1 is the thermal conductivity of nanofluids, $W{\cdot}m^{-1}{\cdot}K^{-1}$; K_0 is the thermal conductivity of water, $W{\cdot}m^{-1}{\cdot}K^{-1}$.

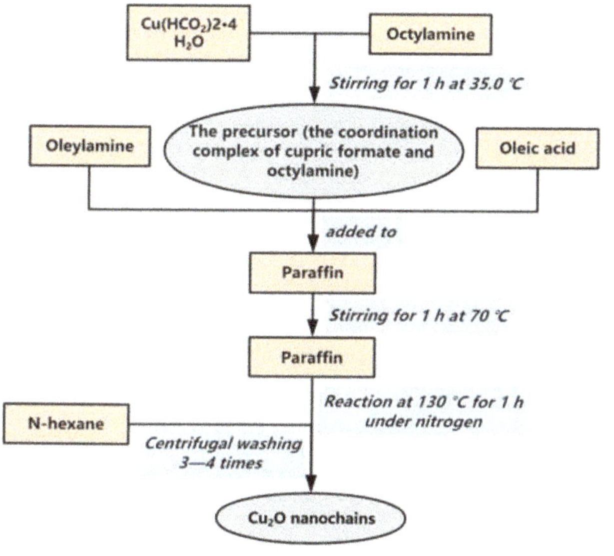

Figure 2. Synthesis process flow chart of Cu_2O nanochains.

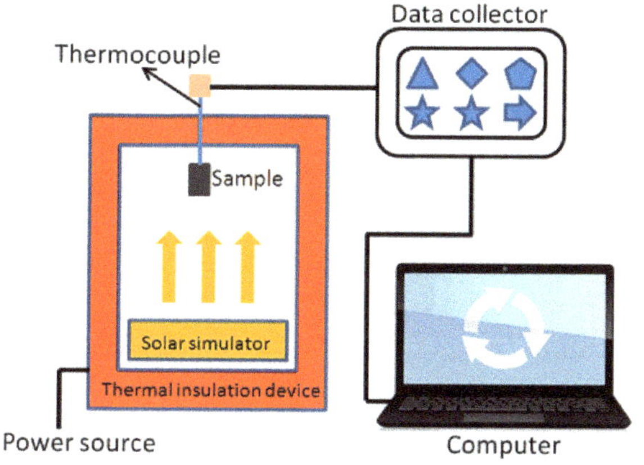

Figure 3. The apparatus for evaluating the light-thermal conversion performance [28].

3. Results and Discussion

3.1. The Characterization of Cu_2O Nanochains

Figures 4 and 5 show the XRD images and TEM images of Cu_2O nanochains, respectively. In the XRD pattern (Figure 4), the diffraction peaks are consistent with standard diffraction data of Cu_2O (No. JCPDF 040836). The diffraction peaks at 2θ values of 36.50°, 42.50°, 61.33° and 74.09° are corresponding to the diffractions of (111), (200) and (311) crystal planes of Cu_2O, respectively. The diffraction peaks are very sharp, which indicates the good crystallinity of the microcrystals. There are no other peaks that can be observed, implying that the nanochains prepared by thermal decomposition have high purity.

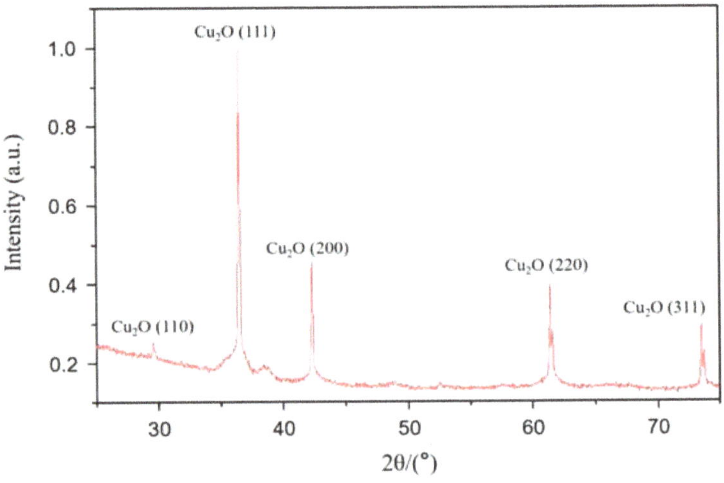

Figure 4. XRD pattern of Cu_2O nanochains.

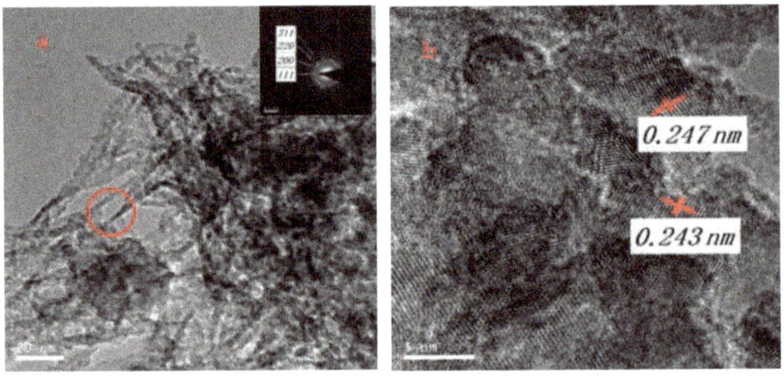

Figure 5. TEM photos of Cu_2O nanochains and (**b**) is an enlarged view of the red zone of (**a**).

The morphology of the nanofluids by TEM is presented in Figure 5. A large number of entangled nanochains can be observed in the TEM image shown in Figure 5a, the diameter is about 2~3 nm. The selected area electron diffraction (SAED) of the upper right corner of Figure 5a shows the multi-crystal diffraction ring, it is corresponding to the (111), (200) and (220) crystal planes of Cu_2O from the inner space to the outside in turn. The different lattice fringe directions in this picture indicate that there are many Cu_2O micro-crystallizes with different sizes, and Cu_2O nanochains are composed of a large number of Cu_2O micro-crystallizes by self-assembly.

3.2. The Effect Factors of the Formation of Cu_2O Nanochains

3.2.1. The Effect of Reaction Time on the Morphology of the Product

In order to study the effect of reaction time on the morphology of the product, we set the time to 0.5 and 2 h and keep the other parameters constant. The morphology of the product is shown in Figure 6. When the reaction time is 0.5 h, the products are small particles with a diameter of about 10 nm (shown in Figure 6a); when the reaction time prolonged from 0.5 h to 1 h, the nanochains appeared (shown in Figure 5); when the reaction time reached 2 h, the diameter of larger particles with whiskers around the surface is about 100 nm (shown in Figure 6b). That is due to the particles entering the intraparticle ripening stage with the extension of the reaction time during the growing of nano-crystallizes. In the reaction system where the Cu_2O was prepared, the precursor produces plenty of molecules of Cu_2O by thermal decomposition; these molecules nucleate and polymerize into nanocrystalline and then self-assemble into nanochains within a short time period. Afterward, the concentration of Cu_2O decreases rapidly and the Cu_2O nanochains will enter the Ostwald ripening stage [29]. The small particles with larger solubility tend to be dissolved, and the monomers will recrystallize on the surface of larger particles. These phenomena all follow the Ostwald ripening mechanism. Oswald ripening is a phenomenon that can be observed in solid or liquid solutes and describes the change of a non-uniform structure with the passage of time: smaller crystalline or solute particles in the solute dissolve and are deposited again on larger crystalline or solute particles. Consequently, when the Cu_2O nanochains appear in the reaction system, the low concentration of Cu_2O results in the dissolving of nano-crystallizes progressively, and the larger particles of Cu_2O will grow on by adsorbing the small particles of Cu_2O. Finally, a large crystal of Cu_2O is obtained.

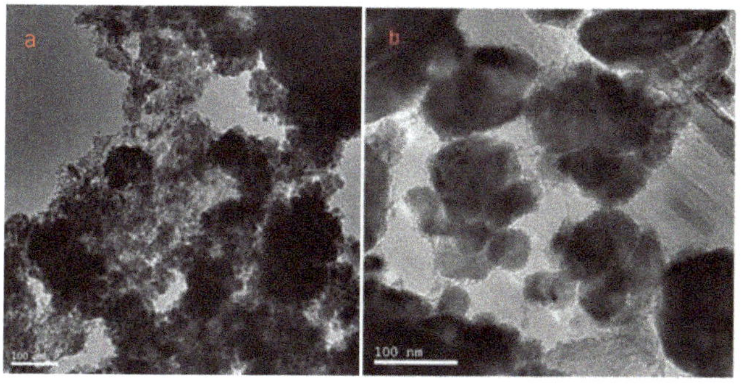

Figure 6. TEM photos of the products with different reaction times (**a**) 0.5 h; (**b**) 2 h.

3.2.2. The Effect of Precursor Concentration on the Morphology of the Product

In order to investigate the influence of precursor concentration on the morphology of the product, the precursor concentration was adjusted to 0.005 mol and 0.04 mol and keep the other parameters constant. The morphology of the resulting product is shown in Figure 7. It can be seen from the picture that when the precursor concentration is 0.005 mol, less Cu_2O is generated, the particle is mainly formed by nucleation, therefore the morphology of the final product is mainly small particles (Figure 7a), those small particles have connected together, the condition of forming chains was initially formed. With the increase in the concentration of the precursor to 0.04 mol, the number of nanochains increased, and the entanglement degree of the nanochains also increased (Figure 7b). In this paper, the Cu_2O nanochains reaction system is a homogeneous reaction system, and the nucleation process and growth process are all dependent on the supersaturation of Cu_2O, that is the difference between the solubility of Cu_2O concentration and Cu_2O nanoparticles [30]. When Cu_2O has a lower supersaturation, Cu_2O nanoparticles will grow faster than the nucleation, and it is conducive to the growth of Cu_2O nanoparticles; when Cu_2O supersaturation is higher, the nucleation of Cu_2O nanoparticles will be the dominant part in the growing process. Therefore, increasing the precursor concentration can increase the number of Cu_2O molecules, which also contributes to improving the supersaturation of the system. This will help the nucleation of Cu_2O nanoparticles to form more nanocrystals, which then self-assemble into Cu_2O nanochains. Consequently, too many Cu_2O nanochains lead to an increasing degree of entanglement.

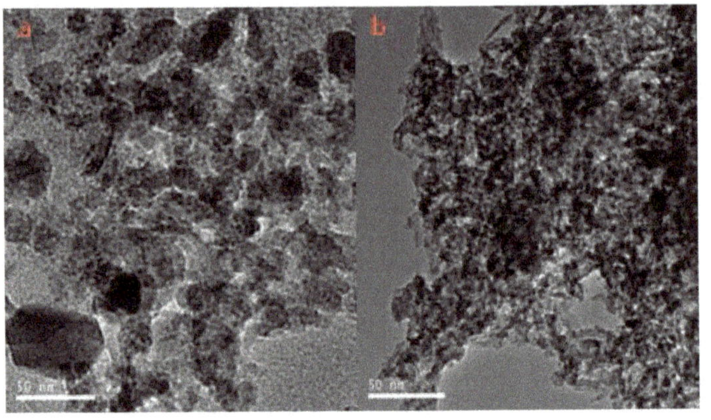

Figure 7. TEM photos of the products with different precursor concentration (**a**) 0.005 mol; (**b**) 0.04 mol.

3.3. The Formation Mechanism of Cu_2O Nanochains

The complexes made up of the copper formate and octylamine can increase its solubility in the paraffin, which can reduce the thermal decomposition temperature of copper formate, and inhibit the growth of the nanoparticles [31]. The addition of oleic acid is beneficial to the formation of Cu_2O by the final thermal decomposition of the complex. The formation mechanism of Cu_2O is that the coordination complex of copper formate and octylamine ($Cu(HCOO)_2{\cdot}4H_2O$) can produce H_2O and $Cu(HCOO)_2$ by thermal decomposition, for the water is surrounded by the liquid paraffin and forming a W/O structure in the liquid paraffin. $Cu(HCOO)_2$ is a polar molecule, being dissolved in water can promote Cu $(HCOO)_2$ partial ionization and producing Cu^{2+} and $HCOO^-$. Additionally, organic acid has reducibility [32], while $HCOO^-$ does not have reducibility, when oleic acid is added into the reactant, due to the weak acidity of oleic acid, a small amount of hydrogen ions are captured by $HCOO^-$, resulting in the $HCOO^-$ having reducibility. Therefore, in the presence of hydrogen ions, $HCOO^-$ reduced the Cu^{2+} to Cu^+, the reaction equation is as follows:

$$2HCOO^- + 2Cu^{2+} \rightarrow H_2\uparrow + 2CO_2\uparrow + 2Cu^+ \quad (3)$$

Cu^+ combined with H_2O to generate CuOH, CuOH is heated to generate Cu_2O and H_2O, the reaction equation is described as follows:

$$Cu^+ + H_2O + HCOO^- \rightarrow CuOH\downarrow + HCOOH \quad (4)$$

$$2CuOH \rightarrow Cu_2O + H_2O \quad (5)$$

The total reaction equation as follows [33]:

$$4HCOO^- + 2Cu^{2+} + H_2O \rightarrow H_2\uparrow + 2CO_2\uparrow + Cu_2O\downarrow + 2HCOOH \quad (6)$$

The formation mechanism of Cu_2O nanochains is shown in Figure 8. Compared to the electron-based carboxyl group of oleic acid, the amino group of the oleylamine with a stronger nucleophilicity reacts with the copper atoms in Cu_2O more easily to compensate unbalance of the charge so that the surface of the crystal will be stabilized. The crystal plane of Cu_2O (111) contains an unsaturated dangling bond of copper, therefore it reacts easier with oleylamine and reduces the surface energy of the crystal plane (111) compared to that of the crystal plane (100) or (110) [34]. In addition, a clear steric hindrance, formed by the long carbon chain of oleylamine on the surface of the nanoparticles, increases the repulsion of the connection of the Cu_2O crystal plane (111). Therefore, a long-range order structure with high leak-proof ability is formed on the surface of Cu_2O covered by oleylamine, while the rest of the Cu_2O surface still is highly active [35]. The interacted Cu_2O nano-crystallizes with special structure self-assemble into a thread and all the produced surfaces are crystal planes (111), as shown in Figure 8.

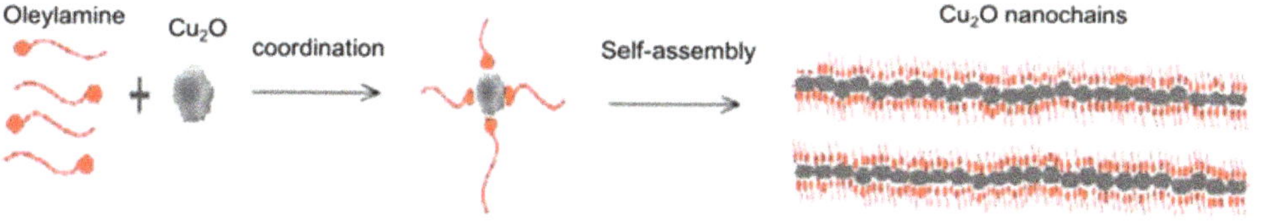

Figure 8. The schematic diagram of the formation mechanism of Cu_2O nanochains.

3.4. Thermo-Physical Properties and Photo-Thermal Performance of Nanofluids

Figure 9A shows the temperature-thermal conductivity curve of the Cu_2O nanochains-water nanofluids and water. It can be seen from the curve that the thermal conductivity of nanofluids is all higher than that of water. This is mainly due to the fact that the thermal conductivity of Cu_2O nanochains is much higher than that of the water, and there is the micro convection heat transfer caused by the irregular Brownian motion of the Cu_2O nanochains [36] inside the fluids in the meantime. The thermal conductivity of nanofluids increases from 0.6330 $W \cdot m^{-1} \cdot K^{-1}$ to 0.7137 $W \cdot m^{-1} \cdot K^{-1}$ as the temperature increases from 30 to 80 °C, that is, the thermal conductivity increases almost linearly, which is consistent with the results of Zamzamian et al. [37]. As is shown in Figure 9B, the thermal conductivity enhancement ratio of nanofluids was 3.16% at 30 °C. When the temperature is raised to 80 °C, the thermal conductivity enhancement ratio of nanofluids reaches 6.68%. The enhancement of the thermal conductivity is mainly due to the fact that the viscosity of the nanofluids decreases as the temperature is raised up, with the irregular Brownian motion increases, so that the micro-convection heat transfer inside the fluids will be enhanced [38–41].

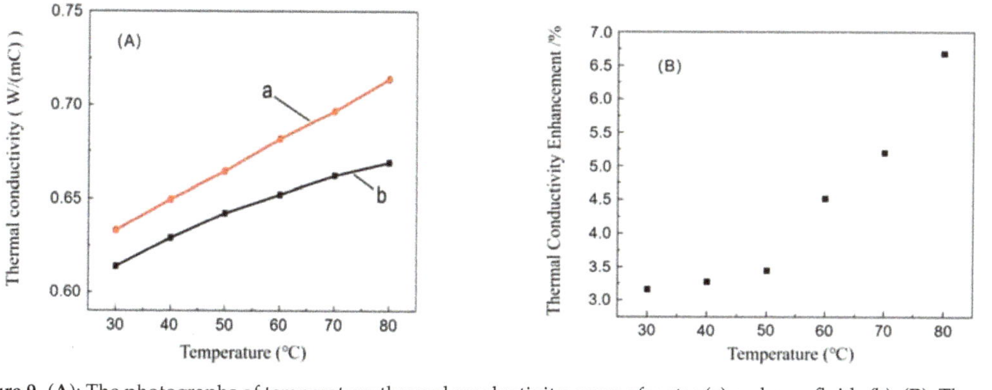

Figure 9. (**A**): The photographs of temperature-thermal conductivity curve of water (a) and nanofluids (b); (**B**): Thermal conductivity enhancement ratio of nanofluids in the different temperatures.

As is shown in this spectrum of nanofluids, the light absorption performance of nanofluids is obviously better than that of water. Attributes to the high ability of light absorption which the nanochains possessed, the addition of the nanochains will be able to enhance the light absorption capacity of nanofluids significantly (Figure 10).

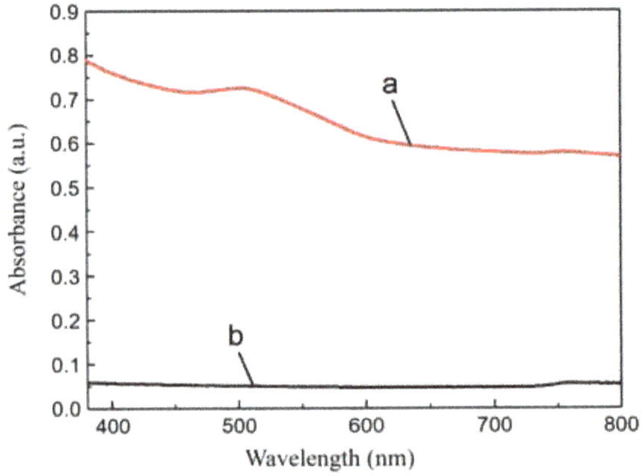

Figure 10. UV-vis pattern of water (a) and Cu_2O nanochains-water nanofluids (b).

Figure 11 shows the time-temperature curve of Cu_2O nanochains-water nanofluids and water under the irradiation of a self-made analog light source. From these pictures, we can figure out that both curves are in a monotonous upward trend while the rate of rising is both decreasing under the irradiation of the same light intensity (1000 $W{\cdot}m^{-2}$). Additionally, the temperature of nanofluids was always higher than that of water. After irradiating for 3000 s, the temperature of nanofluids reached 91.1 °C while the water was only 75.7 °C. Obviously, the photothermal conversion performance of Cu_2O nanochains-water nanofluids was enhanced significantly compared to that of water. That is mainly due to the fact that the solar spectral absorption characteristics have changed and the scattering and the absorption of light have been enhanced when Cu_2O nanochains are added into the water. The above results show that the Cu_2O nanochains-water nanofluids possess significant thermal-physical and photo-thermal conversion performance, and they can be the heat transfer medium for DASC.

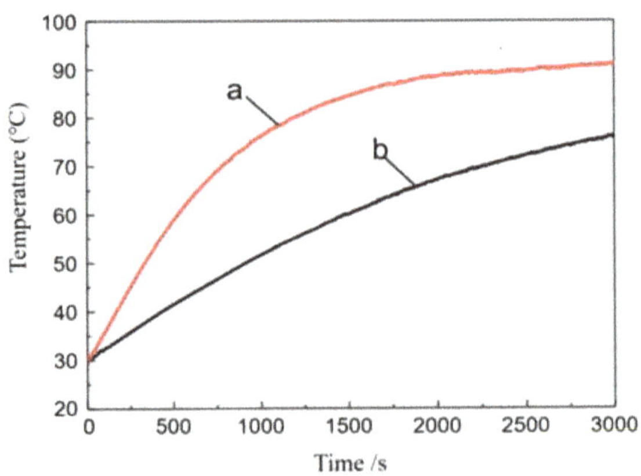

Figure 11. The time–temperature curve of water (a) and Cu_2O nanochains-water nanofluids (b) under the irradiation of self-made analog light source.

4. Conclusions

In melted paraffin, the Cu_2O nanochains (2–3 nm in diameter), self-assembled from Cu_2O nanocrystals, are prepared by thermal decomposition of a precursor with coordination compound with copper formate tetrahydrate and octylame. Cu_2O nanochains become granular gradually as reaction time prolongs. The increasing concentration of the precursor accelerates the agglomeration of the nanochains. Oleic acid is conducive to form Cu_2O, while oleylamine plays a directional role in forming the nanochains. The comparison between the Cu_2O nanochains, water nanofluids and the water base fluids shows that the former has better light absorption ability, thermal conductivity and photo-thermal conversion ability, which makes it qualify as a heat transfer medium for DASC. Future research can be conducted in several ways. Firstly, the composition of the nanochains in nanofluids can have a significant impact on the performance of nanofluids, and therefore research into optimal nanochain materials is one of the priorities for future research. Secondly, previous experiments have demonstrated that there is a correlation between nanofluids efficiency and concentration, but an increase in nanofluids concentration leads to an increase in viscosity, which affects the nanofluids efficiency. Therefore, an optimal concentration range may exist. In addition, previous studies have not considered the correlation between the properties of nanofluids and the size of the nanostrands within it, so this may also be a direction of interest for future research.

Author Contributions: Conceptualization and writing-original draft preparation, Z.N.; data collection and data analysis, X.C.; data interpretation, X.W.; literature search, S.Z.; data collection, C.Z.; study design and writing, B.X.; study design, Y.N. All authors have read and agreed to the published version of the manuscript.

Funding: This research was funded by the Planning Project of Application Research for Public Service Technology of Zhejiang Province (LGG18E060002) and Zhejiang Provincial Collaborative Innovation Center for Bamboo Resources and High-efficiency Utilization (2017ZZY2-15).

Institutional Review Board Statement: Not applicable.

Informed Consent Statement: Informed consent was obtained from all subjects involved in the study.

Data Availability Statement: The data used to support the findings of this study are obtained directly from the simulation by the authors.

Conflicts of Interest: The authors declare no conflict of interest.

References

1. Goel, N.; Taylor, R.A.; Otanicar, T. A review of nanofluid-based direct absorption solar collectors: Design considerations and experiments with hybrid PV/Thermal and direct steam generation collectors. *Renew. Energy* **2020**, *145*, 903–913. [CrossRef]
2. Wang, K.; He, Y.; Liu, P.; Kan, A.; Yu, W. Highly-efficient nanofluid-based direct absorption solar collector enhanced by re-verse-irradiation for medium temperature applications. *Renew. Energy* **2020**, *159*, 652–662. [CrossRef]
3. Hong, S.W.; Kang, Y.-T.; Kleinstreuer, C.; Koo, J. Impact analysis of natural convection on thermal conductivity measurements of nanofluids using the transient hot-wire method. *Int. J. Heat Mass Transf.* **2011**, *54*, 3448–3456. [CrossRef]
4. Saidur, R.; Leong, K.; Mohammed, H. A review on applications and challenges of nanofluids. *Renew. Sustain. Energy Rev.* **2011**, *15*, 1646–1668. [CrossRef]
5. Fang, J.; Xuan, Y. Investigation of optical absorption and photothermal conversion characteristics of binary CuO/ZnO nanofluids. *RSC Adv.* **2017**, *7*, 56023–56033. [CrossRef]
6. Lenert, A.; Wang, E.N. Optimization of nanofluid volumetric receivers for solar thermal energy conversion. *Sol. Energy* **2012**, *86*, 253–265. [CrossRef]
7. Otanicar, T.P.; Phelan, P.E.; Prasher, R.S.; Rosengarten, G.; Taylor, R.A. Nanofluid-based direct absorption solar collector. *J. Renew. Sustain. Energy* **2010**, *2*, 033102. [CrossRef]
8. Saidur, R.; Meng, T.C.; Said, Z.; Hasanuzzaman, M.; Kamyar, A. Evaluation of the effect of nanofluid-based absorbers on direct solar collector. *Int. J. Heat Mass Transf.* **2012**, *55*, 5899–5907. [CrossRef]
9. Tyagi, H.; Phelan, P.; Prasher, R. Predicted Efficiency of a Low-Temperature Nanofluid-Based Direct Absorption Solar Col-lector. *J. Sol. Energy Eng.* **2009**, *131*, 143–155. [CrossRef]
10. Xu, B.; Chen, C.; Zhou, J.; Ni, Z.; Ma, X. Preparation of novel microencapsulated phase change material with Cu-Cu_2O/CNTs as the shell and their dispersed slurry for direct absorption solar collectors. *Sol. Energy Mater. Sol. Cells* **2019**, *200*, 109980. [CrossRef]
11. Dai, P.; Li, W.; Xie, J.; He, Y.; Thorne, J.; McMahon, G.; Zhan, J.; Wang, D. Forming buried junctions to enhance the photovoltage generated by cuprous oxide in aqueous solutions. *Angew. Chem. Int. Ed.* **2014**, *53*, 13493–13497. [CrossRef]
12. Wang, J.; Ma, J.; Li, X.; Li, Y.; Zhang, G.; Zhang, F.; Fan, X. Cu_2O mesoporous spheres with a high internal diffusion capacity and improved catalytic ability for the aza-Henry reaction driven by visible light. *Chem. Commun.* **2014**, *50*, 14237–14240. [CrossRef] [PubMed]

13. Shah, J.; Kumar, S.; Ranjan, M.; Sonvane, Y.; Thareja, P.; Gupta, S.K. The effect of filler geometry on thermo-optical and rheo-logical properties of CuO nanofluid. *J. Mol. Liq.* **2018**, *272*, 668–675. [CrossRef]
14. Hacialioglu, S.; Meng, F.; Jin, S. Facile and mild solution synthesis of Cu_2O nanochains and nanotubes driven by screw dis-locations. *Caryologia* **2012**, *31*, 449–456.
15. Zhao, L.; Dong, W.; Zheng, F.; Fang, L.; Shen, M. Interrupted growth and photoelectrochemistry of Cu_2O and Cu particles on TiO_2. *Electrochim. Acta* **2012**, *80*, 354–361. [CrossRef]
16. Zhang, D. Synergetic effects of Cu_2O photocatalyst with titania and enhanced photoactivity under visible irradiation. *Acta Chim. Slovaca* **2013**, *6*, 141–149. [CrossRef]
17. Tran, P.D.; Batabyal, S.K.; Pramana, S.S.; Barber, J.; Wong, L.H.; Loo, S.C. A cuprous oxide-reduced graphene oxide (Cu_2O-rGO) composite photocatalyst for hydrogen generation: Employing rGO as an electron acceptor to enhance the photocatalytic ac-tivity and stability of Cu_2O. *Nanoscale* **2012**, *4*, 3875–3878. [CrossRef] [PubMed]
18. Zhang, Z.; Dua, R.; Zhang, L.; Zhu, H.; Zhang, H.; Wang, P. Carbon-layer-protected cuprous oxide nanowire arrays for efficient water reduction. *ACS Nano* **2013**, *7*, 1709–1717. [CrossRef] [PubMed]
19. Zhao, Y.; Wang, W.; Li, Y.; Zhang, Y.; Yan, Z.; Huo, Z. Hierarchical branched Cu_2O nanochains with enhanced photocatalytic activity and stability for H_2 production. *Nanoscale* **2014**, *6*, 195–198. [CrossRef]
20. Zhu, H.; Zhang, C.; Liu, S.; Tang, Y.; Yin, Y. Effects of nanoparticle clustering and alignment on thermal conductivities of Fe_3O_4 aqueous nanofluids. *Appl. Phys. Lett.* **2006**, *89*, 023123. [CrossRef]
21. Zeiny, A.; Jin, H.; Bai, L.; Lin, G.; Wen, D. A comparative study of direct absorption nanofluids for solar thermal applications. *Sol. Energy* **2018**, *161*, 74–82. [CrossRef]
22. Chen, W.; Zou, C.; Li, X. An investigation into the thermophysical and optical properties of SiC/ionic liquid nanofluid for direct absorption solar collector. *Sol. Energy Mater. Sol. Cells* **2017**, *163*, 157–163. [CrossRef]
23. Wang, M.; Shan, X.; Yang, N. Understanding length dependences of effective thermal conductivity of nanochains. *Phys. Lett. A* **2012**, *376*, 3514–3517. [CrossRef]
24. Zhu, D.H.; Yu, W.; Du, H.X.; Chen, L.F.; Li, Y.; Xie, H.Q. Thermal conductivity of composite materials containing copper nanochains. *J. Nanomater.* **2016**, *2016*, 3089716. [CrossRef]
25. Hochbaum, A.; Yang, P. Semiconductor nanowires for energy conversion. *Chem. Rev.* **2010**, *110*, 527–546. [CrossRef] [PubMed]
26. Elisa, S.; Vallejo, J.P.; David, C.; Luis, L. Functionalized graphene nanoplatelet-nanofluids for solar thermal collectors. *Sol. Energy Mater. Sol. Cells* **2018**, *185*, 205–209.
27. Xu, B.; Zhou, J.; Ni, Z.; Zhang, C.; Lu, C. Synthesis of novel microencapsulated phase change materials with copper and copper oxide for solar energy storage and photo-thermal conversion. *Sol. Energy Mater. Sol. Cells* **2018**, *179*, 87–94. [CrossRef]
28. Xia, H.; Zinke-Allmang, M. Rate equation approach to the late stages of cluster ripening. *Phys. A Stat. Mech. Appl.* **1998**, *261*, 176–187. [CrossRef]
29. LaMer, V.K.; Dinegar, R.H. Theory, Production and mechanism of formation of monodispersed hydrosols. *J. Am. Chem. Soc.* **1950**, *72*, 4847–4854. [CrossRef]
30. Yabuki, A.; Tanaka, S. Electrically conductive copper film prepared at low temperature by thermal decomposition of copper amine complexes with various amines. *Mater. Res. Bull.* **2012**, *47*, 4107–4111. [CrossRef]
31. Chen, Z.-Z.; Shi, E.-W.; Zheng, Y.-Q.; Li, W.-J.; Xiao, B.; Zhuang, J.-Y. Growth of hex-pod-like Cu_2O whisker under hydrothermal conditions. *J. Cryst. Growth* **2003**, *249*, 294–300. [CrossRef]
32. Galwey, A.K.; Jamieson, D.; Brown, M.E. Thermal decomposition of three crystalline modifications of anhydrous copper(II) formate. *J. Phys. Chem.* **1974**, *78*, 2664–2670. [CrossRef]
33. Behura, A.K.; Gupta, H.K. Efficient direct absorption solar collector using nanomaterial suspended heat transfer fluid. *Mater. Today Proc.* **2020**, *22*, 1664–1668. [CrossRef]
34. Zeng, J.; Xuan, Y. Tunable full-spectrum photo-thermal conversion features of magnetic-plasmonic Fe_3O_4/TiN nanofluid. *Nano Energy* **2018**, *51*, 754–763. [CrossRef]
35. Fang, X.P. *Research on Heat and Mass Transfer Mechanisms and Optical Character of Nanofluids*; Nanjing University of Science and Technology: Nanjing, China, 2013.
36. Tajik Jamal-Abad, M. Thermal conductivity of Cu and Al-Water nanofluids. *Int. J. Eng.* **2013**, *26*, 821–828.
37. Sadri, R.; Ahmadi, G.; Togun, H.; Dahari, M.; Kazi, S.N.; Sadeghinezhad, E.; Zubir, N. An experimental study on thermal con-ductivity and viscosity of nanofluids containing carbon nanotubes. *Nanoscale Res. Lett.* **2014**, *9*, 151. [CrossRef]
38. Nabil, M.; Azmi, W.; Hamid, K.A.; Zawawi, N.; Priyandoko, G.; Mamat, R. Thermo-physical properties of hybrid nanofluids and hybrid nanolubricants: A comprehensive review on performance. *Int. Commun. Heat Mass Transf.* **2017**, *83*, 30–39. [CrossRef]
39. Daviran, S.; Kasaeian, A.; Tahmooressi, H. Evaluation of clustering role versus brownian motion effect on the heat conduc-tion in nanofluids: A novel approach. *Int. J. Heat Mass Transf.* **2017**, *108*, 822–829. [CrossRef]
40. Song, D.; Hatami, M.; Wang, Y.; Jing, D.; Yang, Y. Prediction of hydrodynamic and optical properties of TiO_2/water suspen-sion considering particle size distribution. *Int. J. Heat Mass Transf.* **2016**, *92*, 864–876. [CrossRef]
41. Lin, J.; Cai, X.; Liu, Z.; Liu, N.; Xie, M.; Zhou, B.; Wang, H.; Guo, Z. Anti-liquid-interfering and bacterially antiadhesive strategy for highly stretchable and ultrasensitive strain sensors based on cassie-baxter wetting state. *Adv. Funct. Mater.* **2020**, *30*, 2000398. [CrossRef]

coatings

MDPI

Article

Corrosion Behavior of FeB-30 wt.% $Al_{0.25}$FeNiCoCr Cermet Coating in Liquid Zinc

Xiaolong Xie [1,2], Bingbing Yin [1,2,*], Fucheng Yin [1,2] and Xuemei Ouyang [1,2]

1 School of Materials Science and Engineering, Xiangtan University, Xiangtan 411105, China; 201831550103@smail.xtu.edu.cn (X.X.); fuchengyin@xtu.edu.cn (F.Y.); ouyangxuemei@xtu.edu.cn (X.O.)
2 Key Laboratory of Materials Design and Preparation Technology of Hunan Province, Xiangtan University, Xiangtan 411105, China
* Correspondence: yinbingbing1990@xtu.edu.cn; Tel.: +86-0731-58298119

Abstract: The corrosion of galvanizing equipment parts by liquid zinc is an urgent problem that needs solving. In this work, FeB-30 wt.% $Al_{0.25}$FeNiCoCr cermet coating was deposited on the surface of 316L stainless steel by AC-HVAF to protect galvanizing equipment parts from corrosion by liquid zinc. The microstructures and phase compositions of powders and the coating were determined by SEM, EDS, and XRD in detail. Additionally, the microhardness, fracture toughness, abrasion wear resistance, and corrosion resistance of the coating to liquid zinc were also studied. The results indicate that the abrasion wear resistance and corrosion resistance of the coating are much better than that of the 316L stainless steel substrate. The failure of the coating in liquid zinc is mainly due to the penetration of liquid zinc into macro-cracks, which causes the coating to peel off.

Keywords: AC-HVAF; FeB; HEAs; coating; corrosion; liquid zinc

Citation: Xie, X.; Yin, B.; Yin, F.; Ouyang, X. Corrosion Behavior of FeB-30 wt.% $Al_{0.25}$FeNiCoCr Cermet Coating in Liquid Zinc. *Coatings* **2021**, *11*, 622. https://doi.org/10.3390/coatings11060622

Academic Editor: Albano Cavaleiro

Received: 25 April 2021
Accepted: 19 May 2021
Published: 22 May 2021

1. Introduction

Hot dip galvanizing is an effective method to prevent the atmospheric corrosion of steel [1,2]. However, liquid zinc is corrosive to almost all single metals and most alloys. Moreover, this corrosion is usually accompanied by wear. The corrosion and wear caused by liquid zinc are serious problems in the hot dip galvanizing industry [3]. During continuous hot dip galvanizing, important components such as stable rolls and sink rolls are corroded and worn by liquid zinc for a long time [4]. One of the widely used materials to produce these important parts is 316L stainless steel [5]. However, it was not particularly designed for galvanizing; thus, its corrosion and wear resistance need to be strengthened. To prolong its lifetime, MoB/CoCr [6,7] and WC/Co [8–10] cermets coatings are deposited on the steel using high-velocity oxygen fuel (HVOF) method [11–14]. There are still some issues to be resolved such as depositing even MoB/CoCr or WC/Co coating on 316L stainless steel. Except for the low-carbon WC/Co coating using Co_3W_3C and Co_6W_6C as bonding phases, most cermet coatings are composed of hard phases and metal bonding phases. On one hand, bonding phases, cobalt, nickel, and other metallic elements or alloys are more easily eroded by liquid zinc than hard phases, which is the main factor leading to failure in liquid zinc. On the other hand, the significant diversity in the thermal expansion coefficient (TEC) between a coating and a 316L stainless steel substrate is the major reason for the failure [15]. Therefore, the key to extending the usage life is to reduce or avoid the corrosion of a bonding phase by liquid zinc and to select the ceramic whose thermal expansion coefficient is close to that of 316L stainless steel as a hard phase.

The research results of Tsipas et al. showed that the corrosion resistance of FeB to liquid zinc is suitable [15]. Additionally, the TECs of FeB ($\alpha_{FeB} = 23 \times 10^{-6}$/K) [16] and 316L stainless steel ($\alpha_{316L} = 19.3 \times 10^{-6}$/K) [17] are close. However, brittleness and cracks form easily and expand, which lead to the failure of FeB in liquid zinc. Results showed that corrosion resistance of FeB alloy can be enhanced by alloying with Mo or W; however, it was still highly brittle [18,19]. The brittleness of FeB can be improved by using Co as the bonding phase, but it is easily eroded by liquid zinc, which is the primary reason for failure [20].

In recent years, high-entropy alloys have been fabricated to replace the traditional metal bonding phases of Ti(CN)-based [21], WC-based [22], and TiB_2-based [23] cermets, and they are considered to be beneficial for the sintering of cermets [23–25]. In addition, high-entropy alloys [26–29] have better fracture toughness, corrosion resistance, and wear resistance than cobalt, nickel, and other metals or alloys. The majority of high-entropy alloys consists of Al, Fe, Ni, Co, Cr, and other common elements, and the preparation process of high-entropy alloys, including ball mill and vacuum sintering, is common. Therefore, adopting high entropy alloy instead of cobalt and nickel as a bonding phase is not only low-cost, but the fracture toughness, wear resistance, and corrosion resistance are also improved.

In this study, FeB-30 wt.% $Al_{0.25}$FeNiCoCr cermet coating was prepared on 316L stainless steel substrate by activated combustion high-velocity air fuel (AC-HVAF). AC-HVAF is a technology, which uses compressed air and fuel to generate high-speed airflow to heat powders to a molten or semi-molten state, and at the same time accelerate the powders to more than 700 m/s, and then hit the treated substrate [30]. The powder particles impacted on the substrate surface are deformed by pressing to form a laminated sheet, which adheres to the substrate surface, and then cools and accumulates continuously, finally forming a layered coating. Compared with coatings prepared by HVOF, coatings deposited by AC-HVAF [30–32] have high density and low oxide content. In addition, the replacement of pure oxygen with compressed air can decrease the cost of preparation [33]. The cermet coating uses FeB as a hard phase, which has good corrosion resistance to liquid zinc, and $Al_{0.25}$FeNiCoCr high-entropy alloy as a bonding phase to strengthen the fracture toughness. In general, the content of the bonding phase accounting for the content of cermets is 8–30 wt.% [34]. In order to improve the brittleness of FeB as much as possible, the content of the bonding phase was designed to be 30 wt.%. In this work, the fracture toughness, abrasion wear property, and corrosion behavior of FeB-30 wt.% $Al_{0.25}$FeNiCoCr cermet coating in liquid zinc were studied.

2. Experiment

2.1. Preparation of FeB-30 wt.% $Al_{0.25}$FeNiCoCr Powders

We added 2.9 wt.% aluminum, 24.06 wt.% iron, 25.35 wt.% nickel, 25.35 wt.% cobalt, 22.34 wt.% chromium powders with an average particle size of 2 μm, and 99.9 wt.% purity to a ball mill (YXQM-4L) (MITR, Changsha, China). Prior to ball milling, polyethylene glycol was added to mixed powders with a mass ratio to the powders of 1:100. The ball mill parameters are shown in Table 1. After ball milling, the powders were dried for 20 h in a drying chamber (DZ-1BCIV) (Taiste, Tianjin, China) at 100 °C. The powders were then sintered with a vacuum furnace (ZM-30-16) (Chenhua, Shanghai, China), ground with a mortar, and sieved with a 60-mesh stainless steel sieve. Afterward, 30 wt.% $Al_{0.25}$FeNiCoCr powders and 70 wt.% FeB powders with a particle size ranging from 1 to 5 μm with 99.9 wt.% purity were ball-milled. FeB-30 wt.% $Al_{0.25}$FeNiCoCr powders were prepared through the same approaches used for $Al_{0.25}$FeNiCoCr powders, apart from the ball milling time of 3 h, a ball ratio of 3:1, and a sieve of 400 mesh. The vacuum sintering process of $Al_{0.25}$FeNiCoCr powders and FeB-30 wt.% $Al_{0.25}$FeNiCoCr powders is shown in Figure 1.

Table 1. Ball mill process parameters of $Al_{0.25}$FeNiCoCr powders.

Parameter	Value
Method	wet ball milling
Media	anhydrous ethanol
Time (h)	50
Rotation speed (rpm)	300
Ball-to-powder weight ratio	10:1

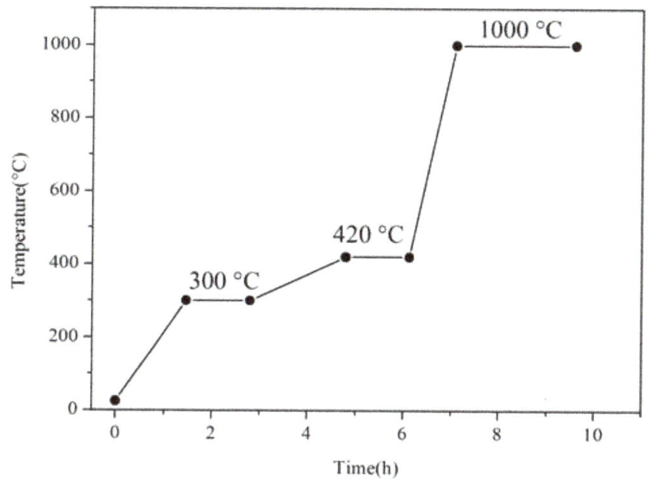

Figure 1. Vacuum sintering process of $Al_{0.25}$FeNiCoCr powders and FeB-30 wt.% $Al_{0.25}$FeNiCoCr powders.

2.2. Preparation of FeB-30 wt.% $Al_{0.25}$FeNiCoCr Coating

The 316L stainless steel was used as the substrate and its dimensions were 100 × 100 × 3 mm^3. In order to enhance the bond strength between the substrate and the coating, the substrate was sandblasted with alumina sand to dislodge impurities on the substrate surface and improve its coarseness before spraying. The cermet powders were sprayed onto a cermet coating with AC-HVAF equipment (AK06) (Kermetico Inc., Benicia, CA, USA). The spraying process parameters are shown in Table 2. Usually, a thermal spray coating thickness ranges from 100 to 1000 μm [34]. If the coating is too thick, it will have complex stress state and cracks will form easily. If a coating is too thin, liquid zinc penetrates the coating more directly and quickly through pores, thus reducing the corrosion resistance to liquid zinc. Therefore, the thickness of the coating was designed to be 260 μm in this work.

Table 2. Spraying process parameters of FeB-30 wt.% $Al_{0.25}$FeNiCoCr coating.

Parameter	Value
Air flow (L/min)	89
Propane flow (L/min)	82.5
Nitrogen flow (L/min)	23
Hydrogen flow (L/min)	30
Combustion chamber pressure (psi)	76
Rotating speed of powder feeding (rpm)	8
Spray angle (°)	90
Spray distance (mm)	180
Lance transversal speed (mm/s)	900
Step away from the spray gun (mm)	3

2.3. Microhardness and Fracture Toughness Tests of the Coating

A Vickers microhardness tester (MH-5L) was used to measure the microhardness of the coating before and after immersion in liquid zinc. The load was 1.96 N and the loading time was 15 s. In order to ensure the accuracy of test results, the measurements were recorded at five different locations and the average microhardness values were obtained. The indentation method was used to measure the fracture toughness of the coating. After testing the Vickers microhardness of the coating, indentation and two cracks appeared on the surface of the coating, and then the length of two diagonal lines of the indentation and the length of the cracks were measured under a scanning electron microscope (SEM), and 10 sets of data were measured and averaged to obtain the fracture toughness value. The fracture toughness of the coating is calculated using the following equation [35]:

$$K_{IC} = 0.203 \times \sqrt{\frac{a_1 + a_2}{2}} \times \left(\frac{c_1 + c_2}{a_1 + a_2}\right)^{\frac{-3}{2}} \times \mathrm{HV} \tag{1}$$

where K_{IC} is the fracture toughness (MPa·m$^{1/2}$), a_1 and a_2 are the half length of the indentation diagonal (mm), c_1 and c_2 are the half length of the cracks (mm), and HV is the Vickers hardness of the cermet (GPa).

2.4. Abrasion Wear Experiment of the Coating and the Substrate

The abrasion wear performance of FeB-30 wt.% $Al_{0.25}$FeNiCoCr coating and 316L stainless steel substrate was detected by a wet rubber wheel abrasion test machine (MLS-225) (YIHUA, Jinan, China). The experiment parameters are shown in Table 3. Both the coating and the substrate were worn 6 times with each abrasion wear of 200 revolutions. After each abrasion wear experiment, the mass was recorded using an electronic scale (FA1104) (Liangping, Shanghai, China) with a precision of 0.1 mg. The mass loss after each abrasion wear test was obtained by weighing the mass before and after each experiment.

Table 3. Abrasive wear test parameters of FeB-30 wt.% $Al_{0.25}$FeNiCoCr coating and the substrate.

Parameter	Value
Load (N)	24.5
Abrasive (feature)	polygonal river sand
Contents (weight)	2000 g river sand and 2000 g water
Rubber wheel diameter (mm)	250
Rubber wheel speed (rpm)	220

2.5. Immersion Experiment in Liquid Zinc

FeB-30 wt % $Al_{0.25}$FeNiCoCr coating samples were processed into cuboids with a size of 16 × 16 × 3.2 mm^3. The heat was supplied by the furnace (SG2-7.5-10) (Rongfeng, Shanghai, China), and the static immersion experiments in liquid zinc were conducted in graphite crucibles placed in the furnace. In order to study corrosion behavior of the coating in liquid zinc, samples were immersed in liquid zinc at 450 °C for 1, 3, 5, and 10 days. The average thickness of specimens before and after the immersion test were gauged with eight measurements. The corrosion rate was calculated using the following equation [36]:

$$R = (a - b)/t \tag{2}$$

where R is the corrosion rate, a is the original thickness of the coating, b is the coating thickness after immersion in liquid zinc, and t is the immersion time.

2.6. Material Characterizations

The phase compositions of specimens were determined using X-ray powder diffraction (XRD) analysis (Rigaku Ultima IV, Tokyo, Japan; 40 kV and 40 mA with Cu Kα radiation. Among them, The samples were scanned in the 2θ ranging from 20° to 90° with the step space of 0.02° and scanning speed of 2 °/min. The microstructures of the specimens were observed using a scanning electron microscope (SEM, JSM-6360LV) (JEOL Ltd., Tokyo, Japan). The extra high tension (EHT) was 20 kV, I probe was 400 PA, and spot size was 450. Energy dispersive spectroscopy (EDS, OXFORD INCA) (High Wycombe, U.K.) was also used to identify the chemical compositions of specimens.

3. Results and Discussion

3.1. Morphologies and Phase Compositions of the Powders

Figure 2a, b shows the morphologies of $Al_{0.25}$FeNiCoCr powders after ball milling and vacuum sintering, respectively. We found that the particle size of $Al_{0.25}$FeNiCoCr powders became larger and denser after vacuum sintering. Figure 2c shows the phase compositions of the $Al_{0.25}$FeNiCoCr powders as determined by XRD. After ball milling, the powders consisted of FCC solid solution and BCC solid solution. However, BCC solid solution transformed into FCC solid solution after vacuum sintering.

Figure 2. Morphologies and phase compositions of $Al_{0.25}$FeNiCoCr powders after ball milling 50 h and vacuum sintering: (**a**) ball milling, (**b**) vacuum sintering, and (**c**) phase compositions.

Figure 3a,b shows the morphologies of FeB-30 wt.% $Al_{0.25}$FeNiCoCr powders after ball milling and vacuum sintering, respectively. After ball milling, the powders had a multi-angular shape; after sintering, the powders became more compact and more spherical, which were beneficial to thermal spraying. The phase compositions of FeB-30 wt.% $Al_{0.25}$FeNiCoCr powders after ball milling and vacuum sintering are shown in Figure 3c. No new phase formed during the ball milling. After vacuum sintering, the powders comprised (Fe, Al, Ni, Co, Cr)B and (Fe, Al, Ni, Co, $Cr)_2B$ solid solutions, as shown in Figure 3c. A part of FeB was transformed to Fe_2B during the vacuum sintering. Moreover, the elements in this high-entropy alloy were completely dissolved into FeB and Fe_2B. The bonding phase did not exist in the powders, which had two influences on the coating's performance: cracks were more likely to occur and grow, and the problem of the binder being easily corroded by liquid zinc was solved.

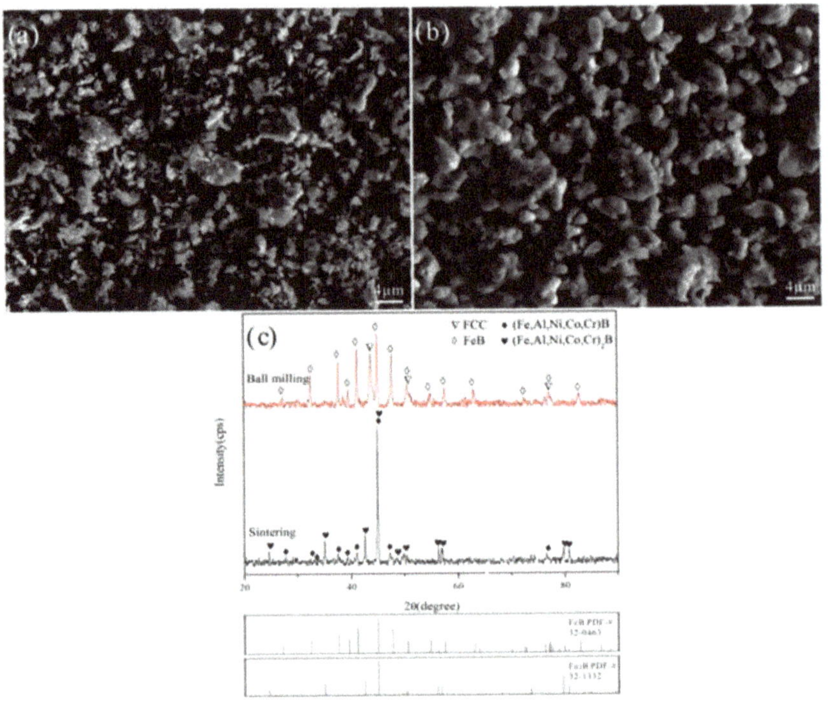

Figure 3. Morphologies and phase compositions of FeB-30 wt.% $Al_{0.25}$FeNiCoCr powders after ball milling and vacuum sintering: (**a**) ball milling, (**b**) vacuum sintering, and (**c**) phase compositions.

3.2. Microstructure Characterizations and Phase Compositions of the Coating

As shown in Figure 4a, there are some sandwiched alumina between the coating and the substrate. This might have been caused by sandblasting. The sprayed powders presented a molten or semi-melted state at high temperatures. When it quickly struck the substrate, it rapidly solidified and plastically deformed. It also led to some pores on the coating, as shown in Figure 4b. In Figure 4c, XRD shows that the coating was still composed of $(FeAlNiCoCr)_2B$ and (FeAlNiCoCr)B phases, indicating that no phase transitions had occurred during thermal spraying.

3.3. Microhardness and Fracture Toughness of the Coating

The microhardness values of the coating ranged from 827 to 859 HV, which is much lower than that of FeB (approximately 2100 HV) and Fe_2B (approximately 1800 HV) [37], which was caused by the addition of high-entropy alloy elements and the existence of pores formed during the thermal spraying process. According to Equation (1), the fracture toughness values of the coating ranged from 5.27 to 5.89 $MPa{\cdot}m^{1/2}$. Since the elements in this high-entropy alloy were completely dissolved in FeB and Fe_2B, the fracture toughness of the coating was significantly higher than that of FeB (1.79 ± 0.70 $MPa{\cdot}m^{1/2}$) and Fe_2B (2.42 ± 0.66 $MPa{\cdot}m^{1/2}$) [38].

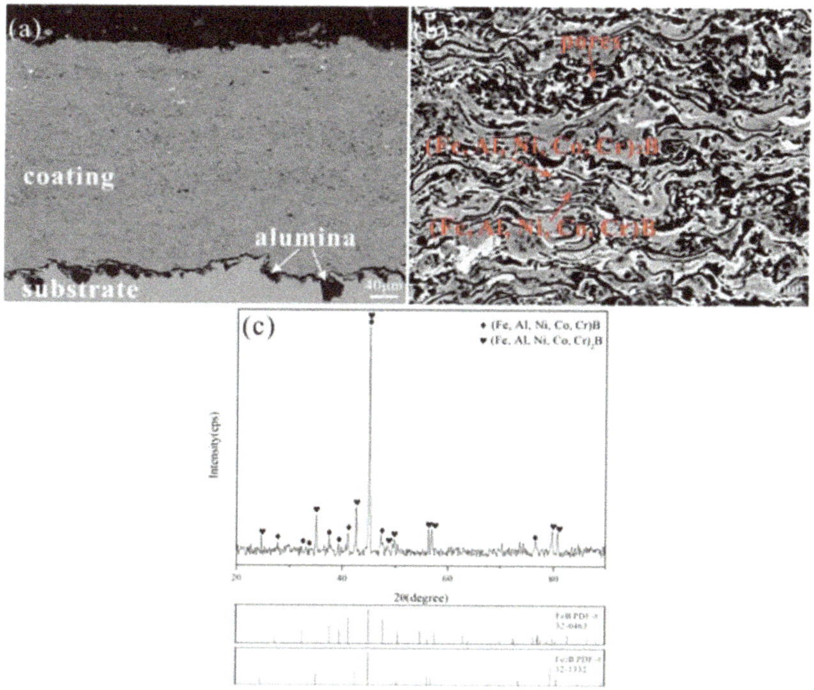

Figure 4. Morphologies and phase compositions of FeB-30 wt.% $Al_{0.25}FeNiCoCr$ coating after ball milling and vacuum sintering: (**a**) ball milling, (**b**) vacuum sintering, and (**c**) phase compositions.

3.4. Abrasion Wear of the Coating and the Substrate

As shown in Figure 5a, furrow-like wear marks appeared on the coating after the abrasion wear test because a small amount of particles in the coating peeled off due to abrasive particles, and some holes formed on the surface of the coating. These particles further wore the coating and formed more furrow-like wear marks. Figure 5b shows the surface morphology of the 316L stainless steel substrate after the abrasion wear test. It can be seen that there are more furrow-like wear marks on the substrate than on the coating. Figure 6 is a graph showing the relationship between their mass loss after each wear and the revolutions. Since the coating and substrate initially had high surface roughness, their mass loss was large after the first wear, and then gradually reduced. Compared with the substrate, the abrasion wear resistance of the coating was significantly improved, being three times as much as that of the substrate.

3.5. Corrosion Behavior of the Coating in Liquid Zinc

According to Equation (2), the corrosion rate of the coating was calculated by measuring the thickness change of the coating immersed in 450 °C liquid zinc for 1, 3, 5, and 10 days. Figure 7 is a diagram of the relationship between thickness loss, corrosion rate, and corrosion time. The corrosion rate of this coating in liquid zinc bath at 450 °C was 24.19 μm/day, which is lower than that of 316L stainless steel in liquid zinc bath at 460 °C with a value of 115.44 μm/day [39]. Figure 7 shows that the coating thickness did not change in the initial period, indicating that the coating still maintained good resistance to liquid zinc. However, with the immersion time extended to 3 days, the thickness loss and corrosion rate of the coating began to increase. In the period of 5 to 10 days, the thickness loss of the coating increased significantly, and the corrosion rate increased gradually. At this time, the corrosion resistance to liquid zinc was considerably reduced.

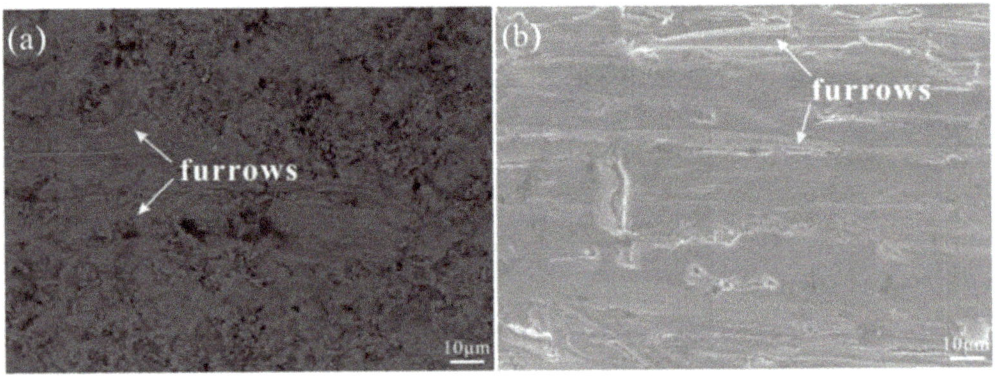

Figure 5. Surface morphologies of FeB-30 wt.% $Al_{0.25}FeNiCoCr$ coating and 316L stainless steel substrate after abrasive wear for 1200 revolutions: (**a**) coating, (**b**) 316L stainless steel.

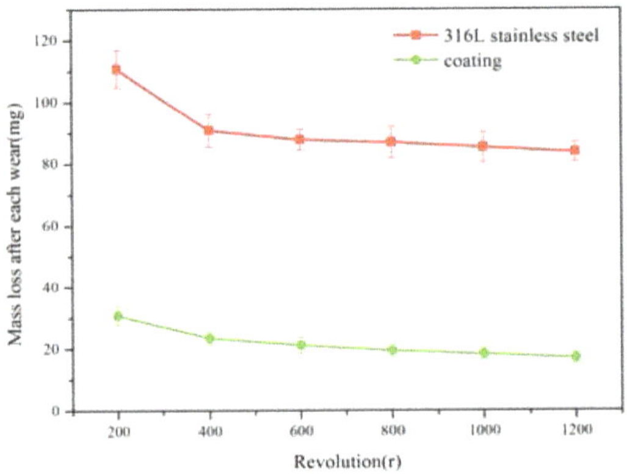

Figure 6. Comparison of abrasive wear properties between the coating and 316L stainless steel.

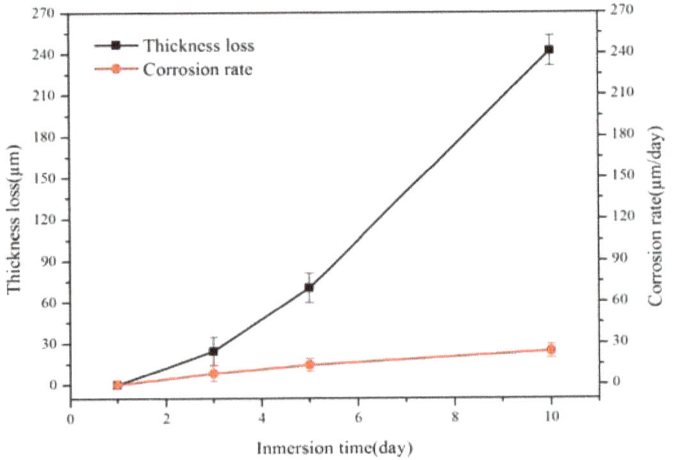

Figure 7. Corrosion resistance of FeB-30 wt.% $Al_{0.25}FeNiCoCr$ coating immersed in liquid zinc at 450 °C.

Figure 8 shows the interface morphology of the coating immersed in liquid zinc for various times. During immersion in liquid zinc, as shown in Figure 9, the relationship between the microhardness of the coating near the liquid zinc and immersion time is closely related to the corrosion process near cracks. After immersion for 1 day, liquid zinc and the coating maintained a good boundary, but micro-cracks were generated under the influence of residual stress and thermal stress, and the microhardness of the coating near the liquid zinc was almost the same as that of the coating before immersion in liquid zinc. Micro-cracks grew into macro-cracks, and liquid zinc more rapidly invaded into the coating along macro-cracks after being immersed for three days. In addition, liquid zinc was filled in these macro-cracks, causing local stress concentration and promoting the propagation of cracks. In turn, this caused more coating to peel off. A small amount of elements such as Fe in the coating were dissolved by liquid zinc to form porous corrosion products like δ phase (91.3 wt.% Zn, 6.62 wt.% Fe, 0.21 wt.% Al, 0.68 wt.% Ni, 0.61 wt.% Co, and 0.58 wt.% Cr), which can fix peeled coatings to resist the corrosion by liquid zinc. The porosity of the coating near the liquid zinc increased and the microhardness value of the coating decreased. After soaking for 5 days, the speed at which the coating peeled off was accelerated, and more coating fell off and flowed into liquid zinc. The pores of the coating near the liquid zinc continued to increase, and the microhardness value further decreased. After immersing for 10 days, the coating was almost completely eroded by liquid zinc, and the remaining coating was too thin to test its microhardness value, but the substrate was not affected by liquid zinc and remained intact.

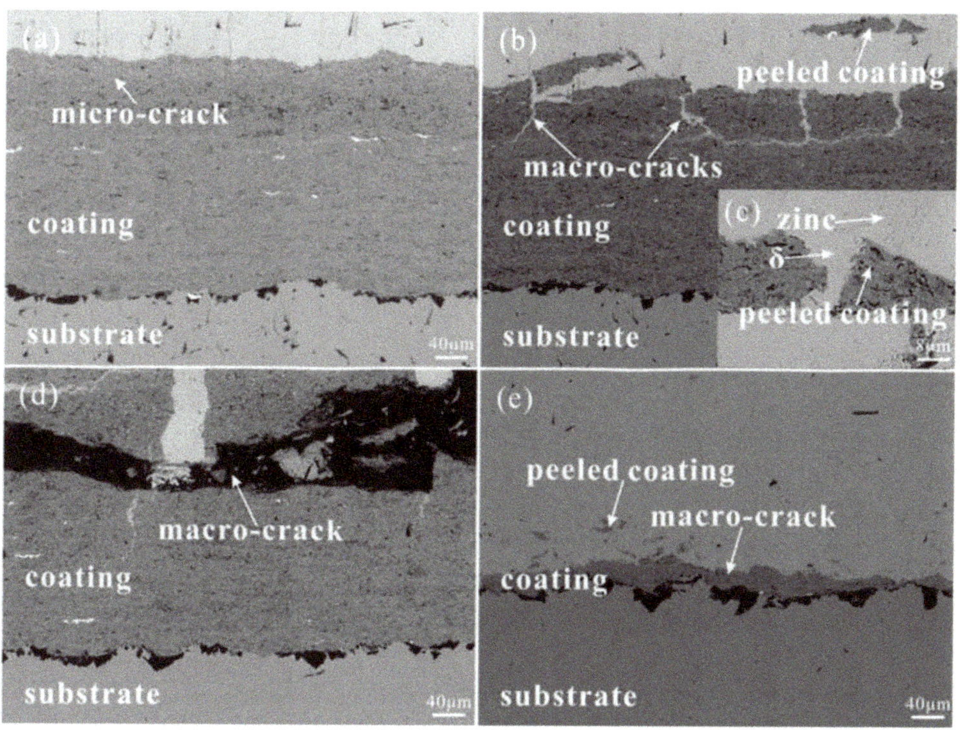

Figure 8. Cross-section morphologies of the coating after the immersion experiment in liquid zinc at 450 °C for different times (**a**) 1, (**b**) 3, (**c**) 3, (**d**) 5, and (**e**) 10 days.

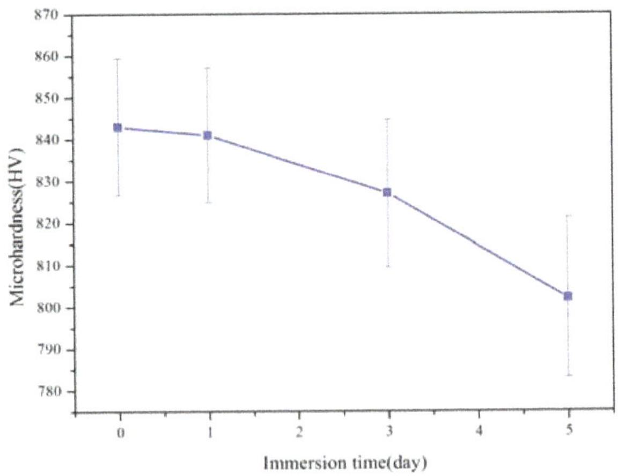

Figure 9. The relationship between the microhardness of the coating near the liquid zinc and immersion time.

In order to further study the corrosion behavior of the coating in liquid zinc, area scanning analysis was performed of Fe and Zn elements at the corrosion interface between the coating and liquid zinc after 3 days of corrosion, as shown in Figure 10. It can be seen that Fe was mainly distributed in the substrate and the coating, including spalled coatings; however, Zn was mainly distributed in liquid zinc, corrosion products, and macro-cracks. Liquid zinc penetrated into the coating through macro-cracks, and the distribution of Zn in the coating was minimal, indicating that (Fe, Al, Ni, Co, Cr)B and (Fe, Al, Ni, Co, Cr)$_2$B solid solutions were not easily corroded by liquid zinc. Therefore, the failure of the coating in liquid zinc was dominated by the penetration of liquid zinc into macro-cracks, and it was accompanied by the dissolution of elements in the coating by liquid zinc.

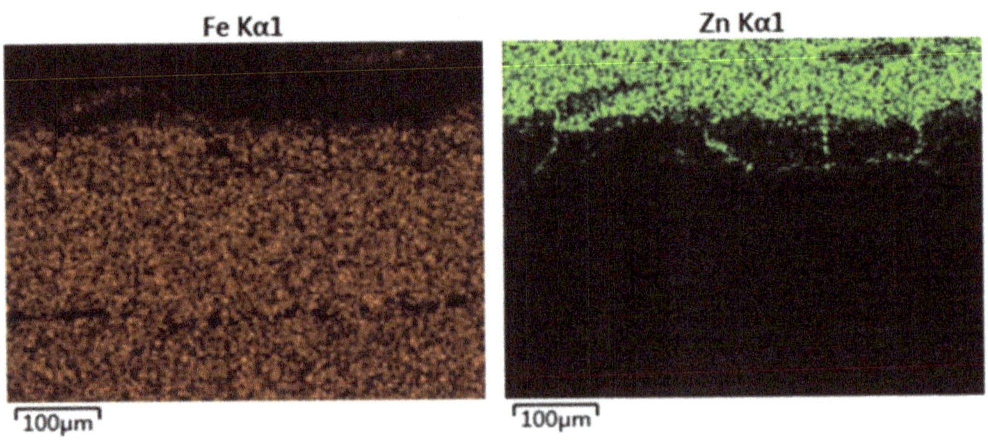

Figure 10. Area scanning analysis of Fe and Zn of the coating immersed in liquid zinc for 3 days.

4. Conclusions

FeB-30 wt.% $Al_{0.25}$FeNiCoCr coating was prepared by AC-HVAF, which consists of (Fe, Al, Ni, Co, Cr)B and (Fe, Al, Ni, Co, Cr)$_2$B solid solutions. The microhardness values of the coating range from 827 to 859 HV, and the fracture toughness values of the coating range from 5.27 to 5.89 MPa·m$^{1/2}$. The abrasion wear resistance of the coating is three times higher than that of 316L stainless steel. Compared with the 316L stainless steel substrate, the corrosion resistance of the coating to liquid zinc was significantly enhanced by about five times. The durability of the coating with a thickness of 260 μm in liquid zinc at 450 °C exceeds 10 days. Under the effect of residual stress and thermal stress, micro-cracks form in the coating when the coating is immersed in liquid zinc. With the increase in immersion time, micro-cracks propagate to macro-cracks. A small amount of elements such as Fe in the coating are dissolved by liquid zinc to form porous corrosion products such as δ phase, which increases the porosity of the coating near the liquid zinc and decreases the microhardness value of the coating. Liquid zinc penetrates the coating along cracks, leading to local stress concentration and further propagation of cracks, causing the coating to peel off and fail.

Author Contributions: Conceptualization, B.Y. and X.X.; methodology, B.Y. and X.X.; formal analysis, X.X. and B.Y.; data curation, X.O.; writing—Original draft preparation, X.X.; writing—Review and editing, F.Y. and B.Y.; supervision, B.Y. and F.Y.; project administration, B.Y. and F.Y. All authors have read and agreed to the published version of the manuscript.

Funding: This work was funded by the National Science Foundation of the China (Nos. 52001268, 51771160, and 11904307), the Scientific Research Fund of Hunan Provincial Educational Department (Nos. 2018JJ4057 and 18C0111), the Natural Science Foundation of Hunan province (Grant No. 2019JJ50576), and the Hunan Postgraduate Research and Innovation Project (No. CX20190478).

Institutional Review Board Statement: Not applicable.

Informed Consent Statement: Not applicable.

Data Availability Statement: Data sharing not applicable.

Conflicts of Interest: The authors declare no conflict of interest.

References

1. Cai, X.; Huang, Y.; Li, Y.; Zhao, L. Production process and technology development of hot-dip galvanizing. *Appl. Mech. Mater.* **2014**, *488*, 61–65. [CrossRef]
2. Cook, T. Hot-dip galvanizing technology. *Met. Finish.* **2000**, *98*, 19–28. [CrossRef]
3. Shibli, S.; Meena, B.; Remya, R. A review on recent approaches in the field of hot dip zinc galvanizing process. *Surf. Coat. Technol.* **2015**, *262*, 210–215. [CrossRef]
4. Ren, X.; Mei, X.; She, J.; Ma, J. Materials resistance to liquid zinc corrosion on surface of sink roll. *J. Iron Steel Res. Int.* **2007**, *14*, 130–136. [CrossRef]
5. Bobde, S.; Kshirsagar, S. Improving the sink roll life in galvalume using material AT101 & the various thermal-spray coating on SS3L6L roll surface. *Int. J. Soft Comput. Eng.* **2013**, *3*, 282–286.
6. Lv, H.; Nie, P.; Yan, Y.; Wang, J.; Sun, B. Microstructure and interfacial adhesion of high velocity oxy-fuel-sprayed mob-cocr alloy coating on 316L stainless steel. *Surf. Interface Anal.* **2009**, *41*, 725–729. [CrossRef]
7. Lv, H.; Wang, J.; Yan, Y.; An, Q.; Nie, P.; Sun, B. Characterisation of detonation sprayed Mo–Co–Cr–B alloy coatings. *Mater. Sci. Technol.* **2010**, *26*, 950–955. [CrossRef]
8. López, A.; Rams, J. Protection of carbon steel against molten aluminum attack and high temperature corrosion using high velocity oxygen-fuel WC–Co coatings. *Surf. Coat. Technol.* **2015**, *262*, 123–133. [CrossRef]
9. Mi, P.; Zhao, H.; Wang, T.; Ye, F. Sliding wear behavior of hvof sprayed WC-(Nano-WC–Co) coating at elevated temperatures. *Mater. Chem. Phys.* **2018**, *206*, 1–6. [CrossRef]
10. Tomita, T.; Takatani, Y.; Kobayashi, Y.; Harada, Y.; Nakahira, H. Durability of WC/Co sprayed coatings in molten pure zinc. *ISIJ Int.* **1993**, *33*, 982–988. [CrossRef]
11. Karaoglanli, A.; Oge, M.; Doleker, K.; Hotamis, M. Comparison of tribological properties of HVOF sprayed coatings with different composition. *Surf. Coat. Technol.* **2017**, *318*, 299–308. [CrossRef]
12. Aw, P.; Tan, A.; Tan, T.; Qiu, J. Corrosion resistance of tungsten carbide based cermet coatings deposited by high velocity oxy-fuel spray process. *Thin Solid Films* **2008**, *516*, 5710–5715. [CrossRef]
13. Peat, T.; Galloway, A.; Toumpis, A.; Harvey, D. Evaluation of the synergistic erosion-corrosion behaviour of HVOF thermal spray coatings. *Surf. Coat. Technol.* **2016**, *299*, 37–48. [CrossRef]
14. Wielage, B.; Wank, A.; Pokhmurska, H.; Grund, T.; Rupprecht, C.; Reisel, G.; Friesen, E. Development and trends in HVOF spraying technology. *Surf. Coat. Technol.* **2006**, *201*, 2032–2037. [CrossRef]
15. Tsipas, D.; Triantafyllidis, G.; Kiplagat, J.; Psillaki, P. Degradation behaviour of boronized carbon and high alloy steels in molten aluminium and zinc. *Mater. Lett.* **1998**, *37*, 128–131. [CrossRef]

16. Ozdemir, O.; Usta, M.; Bindal, C.; Ucisik, A. Hard iron boride (Fe_2B) on 99.97 wt.% pure iron. *Vacuum* **2006**, *80*, 1391–1395. [CrossRef]
17. Mizuno, H.; Junya, K. MoB/CoCr cermet coatings by HVOF spraying against erosion by molten Al-Zn alloy. *J. Ther. Spray Technol.* **2007**, *16*, 404–413. [CrossRef]
18. Ouyang, X.; Chen, G.; Yin, F.; Liu, Y.; Zhao, M. Effect of molybdenum on the microstructures of as-cast Fe-B alloys and their corrosion resistance in liquid zinc. *Corrosion* **2017**, *73*, 942–952. [CrossRef]
19. Liu, X.; Wang, M.; Yin, F.; Ouyang, X.; Li, Z. Effects of tungsten addition on the microstructure and corrosion resistance of Fe-3.5B alloy in liquid zinc. *Materials* **2017**, *10*, 399. [CrossRef]
20. Ye, P.; Yin, F.; Liu, Y.; Ouyang, X.; Xie, X. Corrosion resistance of liquid zinc FeB/Co cermet coating deposited by AC-HVAF. *Chin. J. Nonferrous Met.* **2018**, *28*, 782–791.
21. Fang, Y.; Chen, N.; Du, G.; Zhang, M.; Zhao, X.; Cheng, H.; Wu, J. High-temperature oxidation resistance, mechanical and wear resistance properties of Ti(C,N)-based cermets with $Al_{0.3}$CoCrFeNi high-entropy alloy as a metal binder. *J. Alloy. Comp.* **2020**, *815*, 152486. [CrossRef]
22. Velo, I.; Gotor, F.; Alcalá, M.; Real, C.; Córdoba, J. Fabrication and characterization of WC-HEA cemented carbide based on the CoCrFeNiMn high entropy alloy. *J. Alloy. Comp.* **2018**, *746*, 1–8. [CrossRef]
23. Fu, Z.; Koc, R. Processing and characterization of TiB_2–TiNiFeCrCoAl high-entropy alloy composite. *J. Am. Ceram. Soc.* **2017**, *100*, 2803–2813. [CrossRef]
24. Cheng, H.; Liu, X.; Tang, Q.; Wang, W.; Yan, X.; Dai, P. Microstructure and mechanical properties of FeCoCrNiMnAl_x high-entropy alloys prepared by mechanical alloying and hot-pressed sintering. *J. Alloy. Comp.* **2019**, *775*, 742–751. [CrossRef]
25. Ye, Q.; Yang, G.; Yang, B. Effect of aging on microstructure and property of AlCoCrFe$Mo_{0.05}Ni_2$ high entropy alloy. *Mater. Sci. Eng. A* **2019**, *760*, 1–6. [CrossRef]
26. Tsai, M.; Yeh, J.W. High-entropy alloys: A critical review. *Mater. Res. Lett.* **2014**, *2*, 107–123. [CrossRef]
27. Zhang, Y.; Zuo, T.; Tang, Z.; Gao, M.; Dahmen, K.; Liaw, P.; Lu, Z. Microstructures and properties of high-entropy alloys. *Prog. Mater. Sci.* **2014**, *61*, 1–93. [CrossRef]
28. Ye, Y.; Wang, Q.; Lu, J.; Liu, C.T.; Yang, Y. High-entropy alloy: Challenges and prospects. *Mater. Today* **2016**, *19*, 349–362. [CrossRef]
29. Miracle, D.; Senkov, O. A critical review of high entropy alloys (HEAs) and related concepts. *Acta Mater.* **2017**, *122*, 448–511. [CrossRef]
30. Sadeghimeresht, E.; Markocsan, N.; Nylén, P. Microstructural characteristics and corrosion behavior of HVAF- and HVOF-sprayed Fe-based coatings. *Surf. Coat. Technol.* **2017**, *318*, 365–373. [CrossRef]
31. Guo, R.; Zhang, C.; Chen, Q.; Yang, Y.; Li, N.; Liu, L. Study of structure and corrosion resistance of Fe-based amorphous coatings prepared by HVAF and HVOF. *Corros. Sci.* **2011**, *253*, 2351–2356. [CrossRef]
32. Bolelli, G.; Berger, L.; Börner, T.; Koivuluoto, H.; Lusvarghi, L.; Lyphout, C.; Markocsan, N.; Matikainen, V.; Nylén, P.; Sassatelli, P.; et al. Tribology of HVOF- and HVAF-sprayed WC–10Co4Cr hardmetal coatings: A comparative assessment. *Surf. Coat. Technol.* **2015**, *265*, 125–144. [CrossRef]
33. Jacobs, L.; Hyland, M.; Bonte, M. Study of the influence of microstructural properties on the sliding-wear behavior of HVOF and HVAF sprayed WC-cermet Coatings. *J. Therm. Spray Technol.* **1999**, *8*, 125–132. [CrossRef]
34. Ikeda, H.; Yanagimoto, K. Surface coating material for liquid zinc bath member, production method thereof, and molten zinc bath member. U.S. Patent 8927111, 6 January 2015.
35. Xie, J. Study on TiB_2-TaC Ceramic Tool Material and Its Oxidation Resistance. Master's Thesis, Taiyuan University of Technology, Taiyuan, China, 2019. (In Chinese).
36. Ma, S.; Xing, J.; Fu, H.; He, Y.; Yu, B.; Li, Y.; Bai, Y. Interface characteristics and corrosion behaviour of oriented bulk Fe_2B alloy in liquid zinc. *Corros. Sci.* **2014**, *78*, 71–80. [CrossRef]
37. Allaoui, O.; Bouaouadja, N.; Saindernan, G. Characterization of boronized layers on a XC38 steel. *Surf. Coat. Technol.* **2006**, *201*, 3475–3482. [CrossRef]
38. Kulka, M.; Makuch, N.; Piasecki, A. Nanomechanical characterization and fracture toughness of FeB and Fe_2B iron borides produced by gas boriding of Armco iron. *Surf. Coat. Technol.* **2017**, *325*, 515–532. [CrossRef]
39. Liu, X.; Barbero, E.; Xu, J.; Burris, M.; Chang, K.; Sikka, V. Liquid metal corrosion of 316L, Fe_3Al, and FeCrSi in molten Zn–Al baths. *Metall. Mater. Trans. A* **2005**, *36*, 2049–2058. [CrossRef]

MDPI
St. Alban-Anlage 66
4052 Basel
Switzerland
Tel. +41 61 683 77 34
Fax +41 61 302 89 18
www.mdpi.com

Coatings Editorial Office
E-mail: coatings@mdpi.com
www.mdpi.com/journal/coatings

www.ingramcontent.com/pod-product-compliance
Ingram Content Group UK Ltd.
Pitfield, Milton Keynes, MK11 3LW, UK
UKHW060121010826
14089UKWH00006B/98